Dear Leanne,

thanks for all your help and friendship over the years!

Ute 26/8/16

AF678312

The Dwarf and Mouse Lemurs of Madagascar

The dwarf and mouse lemurs of Madagascar are two very species-rich lemur genera, yet there is a relative paucity of information on this primate family in published literature. In this first ever treatment of the Cheirogaleidae, international experts are brought together to review and integrate our current knowledge of the behavior, physiology, ecology, genetics, and biogeography of these species.

A wide range of direct and indirect research methods that are currently used to study these cryptic nocturnal solitary foragers is described. By uniting often disparate research on captive and free-ranging taxa and synthesizing recent methodological advances, this book provides new insights that will encourage further studies of this fascinating primate family.

This synthesis will provide an incentive for more integrative studies of the Cheirogaleidae in captivity and in the wild, enabling the impacts of deforestation and other factors to be identified and directions for future conservation efforts to be established.

Shawn M. Lehman is an associate professor in the Department of Anthropology at the University of Toronto. Specifically focusing on lemurs in Madagascar, his primary research objective is to integrate evolutionary ecology and conservation biogeography to model primate responses to anthropogenic disturbances.

Ute Radespiel is a senior lecturer at the Institute of Zoology, University of Veterinary Medicine Hannover, Germany. Her research centers upon determining the factors that drive and influence the evolution of social structures and lemur population dynamics at various spatial scales.

Elke Zimmermann is Professor and Chair of the Institute of Zoology at the University of Veterinary Medicine Hannover, Germany. Her research focuses on behavioral ecology with an aim to discovering the processes behind lemur adaptation and evolution.

Cambridge Studies in Biological and Evolutionary Anthropology

Consulting editors

C. G. Nicholas Mascie-Taylor, *University of Cambridge*

Robert A. Foley, *University of Cambridge*

Series editors

Agustín Fuentes, *University of Notre Dame*

Sir Peter Gluckman, *The Liggins Institute, The University of Auckland*

Nina G. Jablonski, *Pennsylvania State University*

Clark Spencer Larsen, *The Ohio State University*

Michael P. Muehlenbein, *Indiana University, Bloomington*

Dennis H. O'Rourke, *The University of Utah*

Karen B. Strier, *University of Wisconsin*

David P. Watts, *Yale University*

Also available in the series

53. *Technique and Application in Dental Anthropology* Joel D. Irish & Greg C. Nelson (eds.) 978 0 521 87061 0
54. *Western Diseases: An Evolutionary Perspective* Tessa M. Pollard 978 0 521 61737 6
55. *Spider Monkeys: The Biology, Behavior and Ecology of the Genus Ateles* Christina J. Campbell 978 0 521 86750 4
56. *Between Biology and Culture* Holger Schutkowski (ed.) 978 0 521 85936 3
57. *Primate Parasite Ecology: The Dynamics and Study of Host–Parasite Relationships* Michael A. Huffman & Colin A. Chapman (eds.) 978 0 521 87246 1
58. *The Evolutionary Biology of Human Body Fatness: Thrift and Control* Jonathan C. K. Wells 978 0 521 88420 4
59. *Reproduction and Adaptation: Topics in Human Reproductive Ecology* C. G. Nicholas Mascie-Taylor & Lyliane Rosetta (eds.) 978 0 521 50963 3
60. *Monkeys on the Edge: Ecology and Management of Long-Tailed Macaques and their Interface with Humans* Michael D. Gumert, Agustín Fuentes & Lisa Jones-Engel (eds.) 978 0 521 76433 9
61. *The Monkeys of Stormy Mountain: 60 Years of Primatological Research on the Japanese Macaques of Arashiyama* Jean-Baptiste Leca, Michael A. Huffman & Paul L. Vasey (eds.) 978 0 521 76185 7
62. *African Genesis: Perspectives on Hominin Evolution* Sally C. Reynolds & Andrew Gallagher (eds.) 978 1 107 01995 9
63. *Consanguinity in Context* Alan H. Bittles 978 0 521 78186 2
64. *Evolving Human Nutrition: Implications for Public Health* Stanley Ulijaszek, Neil Mann & Sarah Elton (eds.) 978 0 521 86916 4
65. *Evolutionary Biology and Conservation of Titis, Sakis and Uacaris* Liza M. Veiga, Adrian A. Barnett, Stephen F. Ferrari & Marilyn A. Norconk (eds.) 978 0 521 88158 6
66. *Anthropological Perspectives on Tooth Morphology: Genetics, Evolution, Variation* G. Richard Scott & Joel D. Irish (eds.) 978 1 107 01145 8

67. *Bioarchaeological and Forensic Perspectives on Violence: How Violent Death is Interpreted from Skeletal Remains* Debra L. Martin & Cheryl P. Anderson (eds.) 978 1 107 04544 6
68. *The Foragers of Point Hope: The Biology and Archaeology of Humans on the Edge of the Alaskan Arctic* Charles E. Hilton, Benjamin M. Auerbach & Libby W. Cowgill (eds.) 978 1 107 02250 8
69. *Bioarchaeology: Interpreting Behavior from the Human Skeleton, 2nd Ed.* Clark Spencer Larsen 978 0 521 83869 6 & 978 0 521 54748 2
70. *Fossil Primates* Susan Cachel 978 1 107 00530 3
71. *Skeletal Biology of the Ancient Rapanui (Easter Islanders)* Vincent H. Stefan & George W. Gill (eds.) 978 1 107 02366 6
72. *Demography and Evolutionary Ecology of Hadza Hunter-Gatherers* Nicholas Blurton-Jones 978 1 107 06982 4

The Dwarf and Mouse Lemurs of Madagascar

Biology, Behavior and Conservation Biogeography of the Cheirogaleidae

Edited by

SHAWN M. LEHMAN
University of Toronto, Canada

UTE RADESPIEL
University of Veterinary Medicine Hannover, Foundation, Germany

ELKE ZIMMERMANN
University of Veterinary Medicine Hannover, Foundation, Germany

CAMBRIDGE
UNIVERSITY PRESS

University Printing House, Cambridge CB2 8BS, United Kingdom

Cambridge University Press is part of the University of Cambridge.

It furthers the University's mission by disseminating knowledge in the pursuit of education, learning and research at the highest international levels of excellence.

www.cambridge.org
Information on this title: www.cambridge.org/9781107075597

© Cambridge University Press 2016

This publication is in copyright. Subject to statutory exception and to the provisions of relevant collective licensing agreements, no reproduction of any part may take place without the written permission of Cambridge University Press.

First published 2016

Printed in the United Kingdom by TJ International Ltd. Padstow Cornwall

A catalogue record for this publication is available from the British Library

Library of Congress Cataloguing in Publication Data
Names: Lehman, Shawn M., editor.
Title: The dwarf and mouse lemurs of Madagascar: biology, behavior, and conservation biogeography of the Cheirogaleidae / edited by Shawn M. Lehman, University of Toronto, Canada, Ute Radespiel, University of Veterinary Medicine Hannover, Foundation, Germany, Elke Zimmermann, University of Veterinary Medicine Hannover, Foundation, Germany.
Description: New York : Cambridge University Press, 2016. |
Series: Cambridge studies in biological and evolutionary anthropology |
Includes bibliographical references and index.
Identifiers: LCCN 2015043107 | ISBN 9781107075597 (hardback)
Subjects: LCSH: Cheirogaleidae.
Classification: LCC QL737.P933 D83 2016 | DDC 599.8/3–dc23
LC record available at http://lccn.loc.gov/2015043107

ISBN 978-1-107-07559-7 Hardback

Cambridge University Press has no responsibility for the persistence or accuracy of URLs for external or third-party internet websites referred to in this publication, and does not guarantee that any content on such websites is, or will remain, accurate or appropriate.

Contents

Color plates are to be found between pp. 460 and 461

Contributors

Curswan A. Andrews
Department of Zoology & Entomology, University of Fort Hare, South Africa

Andrea Baden
Centre ValBio, Madagascar

Marina B. Blanco
Duke Lemur Center, USA

Kate F. Carstens
Percy FitzPatrick Institute of African Ornithology, University of Cape Town, South Africa

Brooke Crowley
Department of Geology, Department of Anthropology, University of Cincinnati, USA

Kathrin H. Dausmann
Department of Animal Ecology & Conservation, University of Hamburg, Germany

Anja M. Deppe
Institute for the Conservation of Tropical Environments, Stony Brook University, USA

Lance A. Durden
Department of Biology, Georgia Southern University, USA

Claudia Fichtel
Behavioral Ecology & Sociobiology Unit, German Primate Center, Germany

Fabien Génin
Department of Zoology & Entomology, University of Fort Hare, South Africa

Laurie R. Godfrey
Department of Anthropology, University of Massachusetts, USA

Colin Groves
School of Archaeology & Anthropology, Australian National University, Australia

Laura Heck
Paläontologisches Institut und Museum, Universität Zürich, Switzerland

James P. Herrera
Department of Mammalogy, American Museum of Natural History, USA

Jason M. Kamilar
Department of Anthropology, University of Massachusetts, USA

Peter M. Kappeler
Behavioural Ecology & Sociobiology Unit, German Primate Center (DPZ), Germany

Sabine B.R. Kästner
Small Animal Clinic, University of Veterinary Medicine Hannover, Germany

Sharon E. Kessler
Department of Anthropology & McGill School of Environment, McGill University, Canada

Shawn M. Lehman
Department of Anthropology, University of Toronto, Canada

Runhua Lei
Grewcock Center for Conservation and Research, Omaha's Henry Doorly Zoo & Aquarium, USA

Edward E. Louis, Jr.
Grewcock Center for Conservation and Research, Omaha's Henry Doorly Zoo & Aquarium, USA

Tongasoa Lydia
Department of Paleontology and Biological Anthropology, University of Antananarivo, Madagascar

Judith C. Masters
Department of Zoology & Entomology, University of Fort Hare, South Africa

Nadine Mestre-Francés
Université de Montpellier, Paris, France

Kathleen M. Muldoon
Department of Anatomy, Arizona College of Osteopathic Medicine, Midwestern University, USA

Leanne T. Nash
School of Human Evolution and Social Change, Arizona State University, USA

K. Anne-Isola Nekaris
Nocturnal Primate Research Group, School of Social Sciences and Law, Oxford Brookes University, UK

Jean-Luc Picq
UFR de Psychologie, Université Paris 8, France

Ute Radespiel
Institute of Zoology, University of Veterinary Medicine Hannover, Germany

Hajarimanitra Rambeloarivony
Department of Zoology & Entomology, University of Fort Hare, South Africa

Emilienne Rasoazanabary
Department of Anthropology, Stony Brook University, USA

Rodin M. Rasoloarison
Département de Biologie Animale, Université d'Antananarivo, Madagascar

E. Johanna Rode-Margono
Nocturnal Primate Research Group, School of Social Sciences and Law, Oxford Brookes University, UK

Kathrin S. Schaber
University of Veterinary Medicine, Institute of Zoology, Hannover, Germany

Alexandra F. Schütter
Small Animal Clinic, University of Veterinary Medicine Hannover, Germany

Christoph Schwitzer
Bristol Zoological Society, UK

Doreen Schwochow
Department of Animal Breeding and Genetics, Swedish University of Agricultural Sciences, Sweden

Travis S. Steffens
Department of Anthropology, University of Toronto, Canada

Christina Strube
Institute of Parasitology, University of Veterinary Medicine Hannover, Germany

Sandra Thorén
Institute of Zoology, University of Veterinary Medicine Hannover, Germany

Julia Tünsmeyer
Small Animal Clinic, University of Veterinary Medicine Hannover, Germany

Kim Valenta
Department of Anthropology & McGill School of Environment, McGill University, Canada

Jean-Michel Verdier
Université de Montpellier, Paris, France

David W. Weisrock
Department of Biology, University of Kentucky, USA

Patricia C. Wright
Department of Anthropology, Stony Brook University, USA

Anne Yoder
Department of Biology, Duke University, USA

Elke Zimmermann
Institute of Zoology, University of Veterinary Medicine Hannover, Germany

Sarah Zohdy
College of Veterinary Medicine, Auburn University, USA

Foreword

The group of primates of the subfamily Cheirogaleidae (Malagasy primates; family Lemuridae) are among the most interesting animals alive today. They originated approximately 25–30 million years ago; they are all Malagasy and are nocturnal in activity; they include the smallest of the living primates; they include the only known obligate hibernators within the primates; and, especially like the other nocturnal primate species, their taxonomy has been greatly expanded within the past two decades (Yoder *et al.*, Chapter 1; Groves, Chapter 2).

This book, *The Dwarf and Mouse Lemurs of Madagascar: Biology, Behavior and Conservation Biogeography of the Cheirogaleidae*, is edited by Shawn M. Lehman, Ute Radespiel, and Elke Zimmermann. It examines their evolution, taxonomy, and genetics; the methods used for studying captive and wild cheirogaleids; their behavior and ecology; the sensory ecology, communication system, and cognition of these animals; and their conservation biology. The book contains 27 chapters with an array of over 40 exceptional international authors, specializing in anthropology, biology, psychology, veterinary medicine, and primatology.

When research on the ecology and behavior of the Malagasy primates was just beginning, over 50 years ago, Jean-Jacques Petter and Arlette Petter-Rousseaux knew of only six species in three genera of cheirogaleids (Petter, 1962; Petter-Rousseaux, 1967). Now, over 30 species in 5 genera are recognized today. Changes have occurred because of "traditional" (morphological) taxonomic research. This has been related, since the mid-1980s, mainly to specific-mate recognition systems of the vocalizations of different nocturnal primate species. However, since around 2000, DNA sequencing has played a great role in recognizing different species (Yoder *et al.*, Chapter 1; Groves, Chapter 2).

Biologists have found that the molecular and morphological diversity related to biogeographic patterns is more complex in Cheirogaleidae than was previously thought. Related to a number of factors during their evolution, the family had initial genus-level divergence in the Oligocene–Miocene epochs, followed by a widespread Pliocene–Pleistocene species-level radiation. Louis, Jr. and Lei (Chapter 3) argue that there was a basal split within Cheirogaleidae at approximately 29 million years ago: with *Phaner* splitting at 28.17 million years ago, *Cheirogaleus* at 23.77 million years ago, *Allocebus* at 17.94 million years ago, and lastly *Mirza* from *Microcebus* at 14.81 million years ago. Andrews *et al.* (Chapter 4) argue that frugivory evolved convergently in Eocene stem primitive primates (strepsirrhines) and monkeys (haplorhines) from faunivorous ancestors. They assume that "the angiosperm evolutions model of primate evolution" (Sussman, 1991; Sussman *et al.*, 2013) provides a good context for the extensive euprimate radiations that occurred after 55 Ma. They also show that mouse lemurs have evolved small size only relatively recently (< 8 Ma), making this species an unsuitable model for a common primate ancestor.

Recent research by Herrera *et al.* (Chapter 5) shows that size differences in sympatric species of cheirogaleids enable them to avoid competition and exploit different resources. Discrete differences in female genital morphology indicate that sexual isolation also has occurred between sympatric species.

In the second part of the book, the authors show that these small primates can be used for the study of human disease. Here, Kästner *et al.* (Chapter 6) indicate that the smallest primates, *Microcebus* (the mouse lemurs), are an excellent model for research on human disease and modern imaging methods. The mainstay for short-term field procedures is anesthesia with the inhalation agents of isoflurane and sevoflurane. This allows induction and maintenance of anesthesia with stable physiologic conditions over several hours. *Microcebus murinus*, for example, is an exceptional model for Alzheimer's disease and bovine spongiform encephalopathy. This species displays most of the pathognomonic lesions observed in Alzheimer's disease (AD). Furthermore, its small size allows the maintenance of large breeding colonies and greater numbers for cognitive or behavioral tests (Verdier and Mestre-Francés, Chapter 7). Zimmermann *et al.* (Chapter 8) measure several life-history traits of *Microcebus* to assess the extent of intra- and interspecific variation and potential constraints. They find that a shortened annual cycle, of 8 months vs. 12 months, does not affect litter size, but leads to accelerated aging. In this genus, life-history patterns linked to a female's energy allocation to reproduction are species-specific and phylogenetically variable, whereas those related to aging are interspecifically invariant and phylogenetically constrained. In relationship to trapping, Deppe *et al.* (Chapter 9) assessed the effects of weather and lunar cycle on the activity pattern of brown mouse lemurs (*Microcebus rufus*). Trapping was highest during dryer, full moon nights, and trapping success increased with decreasing temperatures. Infectious diseases in wild primates are now receiving a great amount of research attention and mouse lemur parasitology is well positioned to become a new model in this research (Kessler *et al.*, Chapter 10). Zohdy and Durden (Chapter 11) provide a comprehensive overview of the known ectoparasites of the Cheirogaleidae. Understanding parasite–host associations has the potential to reveal novel ecological information about these lemurs, and their potential role as vectors of disease agents.

The third part of the book provides new data on the ecology and behavior of the Cheirogaleidae. There are new data on the morphology, behavior, ranging patterns, and habitat of the newly named and little known small species *Mirza zaza* (Margono *et al.*, Chapter 12). There is also a study of the mouse lemur, *Microcebus griseorufus*, that lives in the western reserve of Beza Mahafaly (Rasoazanabary and Godfrey, Chapter 13). This species lives in many habitats from xerophytic environments to more mesic forests, with little overlap in tree species composition. The mouse lemurs living in different forests are different in adult body mass, aspects of their cheiridial (foot and hand) morphology, and sleeping and feeding behavior. Although typically living in xerophytic forest habitats, they appear to do "well" in more mesic forests. The mouse lemurs have distinct and parallel seasonal shifts in both feeding and sleeping behavior. During the dry season, mouse lemurs feed more on gums and prefer tree holes for sleeping; during the rainy season, they rely more on insects and fruits and prefer open nests.

There is also a study of intra- and interspecific isotopic variation of the mouse lemurs (*Microcebus murinus* and *M. ravelobensis*) of northwestern Madagascar (Heck *et al.*, Chapter 14), showing that seasonal environmental changes influence the feeding ecology of these animals. Stable isotope ratios of carbon and nitrogen (δ13C and δ15N) offer a promising complementary methodological tool for diet. Isotope values for plants were influenced by the canopy effect, plant species, and study site. These factors influenced body mass, study site choice, and, to some extent, sex and seasonality. Seasonal changes can be related to dietary shift from a more omnivorous diet in the wet season to a more herbivorous diet in the dry season. Some degree of dietary plasticity may enable these species to avoid interspecific competition in times of food scarcity.

Finally, there is also a study of patterns of resource use in two sympatric mouse lemur species (the gray mouse lemur, *Microcebus murinus*, and the golden-brown mouse lemur, *M. ravelobensis*; Thorén *et al.*, Chapter 15). The lemurs live in the forest reserve of Ankarafantsika National Park, a dry deciduous forest located in northwestern Madagascar. The authors studied the food resources used by the *Microcebus* and found that: (1) food resources were typically shared by multiple individuals; (2) at least 5 of 14 foods were used by both species; (3) foods were used in succession (i.e., only one individual at a resource at a time); and (4) the majority of social encounters resulted in conflicts with decided outcomes, but aggressive behaviors were only observed in the minority of encounters. Basically, as in other primates (Sussman and Garber, 2011), the two species avoid competing over shared food resources by partitioning resources on a temporal scale. Studies like this help us understand the adaptive flexibility and possible environmental constraints of these animals.

Recent studies indicate that lepilemurids and cheirogaleids shared an ancestor in the late Eocene. It has been shown that the small size in cheirogaleids involved at least three episodes of parallel phyletic dwarfing, with the *Microcebus* lineage diverging in the late Miocene. These studies carry major implications for the evolutionary interpretations of these animals. Génin and Masters (Chapter 16) review the information and reinterpret it within the new phylogenetic framework for *M. murinus*. The suite of adaptations to environmental hypervariability and habitat fragmentation in *Microcebus*, such as the pleiotropic body size reduction by means of truncated ontogeny (progenesis), are among these changes.

The different species of *Cheirogaleus* are very flexible in the *modus operandi* and timing of hibernation. Even within species differences are considerable, depending on habitat and climatic parameters (Dausmann and Blanco, Chapter 17). In the dry forest, species hibernate in tree hollows with various insulation properties. In the rainforest, species retreat underground for hibernation. There are extensive consequences on hibernation parameters and patterns. Underground hibernacula and hibernation in well-insulated hollows in large trees cause energy expenditure and body temperature (T_b) bouts that are almost constant and spontaneous arousals can occur regularly. In thinner trees, insulation capacities are decreased, and energy expenditure and T_b fluctuate with the ambient temperatures (T_a) to various degrees. The colder the habitat, the greater the saving of energy through hibernation. Hibernation was originally adapted to, and runs best at, mostly constant ambient temperatures. It seems conceivable that hibernation in tree hollows could have been the original *modus operandi* in ancestral *Cheirogaleus*. This implies that hypometabolism in some species is truly independent of ambient temperatures, and fluctuating temperatures might be advantageous, allowing for extended, continuous bouts of hypometabolism. This would avoid interruptions by active arousals.

Part IV of this book focuses on the sensory ecology, communication, and cognition of the Cheirogaleidae. Valenta and Lehman (Chapter 18) show that the diminutive size, activity patterns, and distinct sensory ecology of the mouse lemur represent a unique frugivorous guild and make this animal an important seed disperser for numerous species of plants. Mouse lemurs rely heavily on olfaction and choose fruits based on the presence of certain micronutrients. New studies suggest that solitary foraging and crypsis are effective predator avoidance strategies for many nocturnal primates (Fichtel, Chapter 19). However, they also evolved specialized antipredator behaviors that are well adapted to the different hunting styles of their predators. Just as with other primates (Hart, 2000), the Cheirogaleidae employ specific tactics to reduce the risk of encounters with predators, such as predator-sensitive foraging, choosing safe sleeping sites, and vigilance. Tactics employed during or after encountering predators, such as alarm calling, fleeing, and confronting or mobbing predators, have also been described in some cheirogaleids. Picq (Chapter 20) summarizes the numerous behavioral tasks that have been specifically designed for mouse lemurs (*M. murinus*) in order to assess cognitive aging. These animals are good at numerous behavioral tasks: they easily master classical laboratory tests on cognition, possessing cognitive capacities. In fact, they far exceed common expectations for the species. They often age in a way that parallels aging patterns in humans and, thus, the neurobiological mechanisms underlying cognitive decline could be similar. For example, procedural memory is spared, executive function is especially vulnerable to the aging process, and declarative-like memory is only affected in a small set of the aged population. In time, these animals could provide a unique opportunity for understanding the mechanisms that drive cognitive impairment in humans.

Studying the acoustic divergence in communication of mouse lemurs, Zimmermann (Chapter 21) finds that calls used in mating and group reunion diverge most within and between populations and species, whereas calls used in the alarm and disturbance contexts show no species-specific variation. This study shows that for speciation and evolution a mixture of neutral and adaptive responses drives acoustic divergence in these species. Kessler *et al.* (Chapter 22) studied kin selection in mouse lemurs as a driving factor in the evolution of primate social complexity. They find that recent field studies on sleeping group associations, infant rearing behavior, and population genetics show that the key to the success of the mouse lemur radiation is likely to have been their flexibility. It appears that close maternal kin networks have been the foundation from which more complex social systems could evolve.

In the final section of the book, the subject of the conservation biology of the cheirogaleids is explored. Kamilar *et al.* (Chapter 23) studied the species diversity of six mouse lemur species. They found that climate, especially rainfall and temperature, plays a significant role in *Microcebus* distribution and in understanding their current biogeography. They also point out the possible future shifts and reduction in distribution related to current climate change. Lehman (Chapter 24) examined the edge effect on tree dendrometrics data (tree height, diameter at breast height, and canopy size), abiotic data (ambient temperature, percent humidity, wind speed, and light intensity), and mouse lemur densities in western dry forests. Edge effects are the changes in community structure that can occur between different populations living at the boundaries of two distinct habitats. He found that, contrary to previous studies, *M. murinus* had higher density estimates in edge habitats and *M. ravelobensis* had higher densities in interior habitats. He notes that the edge effect on mouse lemurs may differ markedly in eastern humid forests and western dry forests in Madagascar.

Steffens and Lehman (Chapter 25) studied the effects of plant dendrometrics, habitat diversity, and anthropogenic disturbance on two sympatric species of *Microcebus*. They found a significant, positive correlation between density and abundance for both primate species. Variations in the density of *M. murinus* were due to plant basal area, stem density, and the distance from continuous forest. Variations in *M. ravelobensis* density resulted from basal area, mean tree height, total disturbance, and the distance from continuous forest. The results indicate that the link between plant dendrometrics and *M. murinus* abundance and density may reflect sleeping site requirements and dietary preferences, whereas edge effects may explain abundance and density in *M. ravelobensis*. An understanding of animal density and abundance, in a heterogeneous landscape, is critical to understanding the dynamics of population biogeography. Out of 25 species of mouse lemur, one species, *Microcebus murinus*, occurs throughout large parts of southern to northwestern Madagascar. Radespiel (Chapter 26) shows that this species has a higher seasonal plasticity in feeding habits and, in northwestern Madagascar, they have a relatively higher ability to move about freely and migrate in areas of habitat disturbance or discontinuity. This would enable *M. murinus* to expand into new and unpredictable environments such as temporarily available and highly seasonal highland habitats.

Lehman *et al.* (Chapter 27) show that recent conservation assessments indicate high extinction risks for many cheirogaleids. Of the 32 species of cheirogaleids whose conservation status was assessed by the authors, 6 were assumed to be among the least threatened species, 6 were vulnerable, 16 were endangered, and 4 were assessed as critically endangered. Anthropogenic factors (for example, various forms of farming) were related to the problem of conservation of species with relatively small geographic ranges. Such things as additional spatial data on habitat breadth, abundance, and responses to varying kinds and intensities of anthropogenic disturbance are needed on all species. There were no major underlying biotic or abiotic correlates related to species living in or out of the currently protected animal area system in Madagascar. The authors recommend that further studies must be related to determining overall species richness, standardization of research methods, and how populations of each species respond to landscape patterns of forest fragmentation, forest loss, and edge effects.

This book brings the biology, behavior, and conservation biogeography of the Cheirogaleidae up to the standards of other research in primatology. Research on the evolution, taxonomy, and genetics; the methods used for studying captive and wild primates and their relationship to human biology and medicine; the natural behavior and ecology of the animals; their sensory ecology, communication system and cognition; and their conservation biology are all active areas of research in primate ecology and evolution. *The Dwarf and Mouse Lemurs of Madagascar: Biology, Behavior and Conservation Biogeography of the Cheirogaleidae* has added a great deal to our knowledge of these animals and of their primate relatives. The editors, Shawn M. Lehman, Ute Radespiel, and Elke Zimmermann, have done an exceptional job of collecting these papers.

Robert W. Sussman
St. Louis

References

Hart D. 2000. Primates as Prey: Ecological, Morphological, and Behavioral Relationships Between Primate Species and their Predators. PhD Dissertation, Washington University, St. Louis, Missouri.

Petter J-J. 1962. Recherches sur l'éthologie des Lémuriens Malgaches. *Memoires Museum Natural History, Paris* 27:1–146.

Petter-Rousseaux A. 1967. Recherches sur la Biologie de la Réproduction des Primate Inférieurs. *Mammalia* 26(Suppl 1):1–88.

Sussman RW. 1991. Primate origins and the evolution of the Angiosperms. *American Journal of Primatology* 23:209–223.

Sussman RW, Garber PA. 2011. Cooperation, collective action, and competition in primate social interactions. In CJ Campbell, A Fuentes, KC Mackinnon, SK Bearder, RSM Stumpf (eds.), *Primates in Perspective*, 2nd edition (pp. 587–599). Oxford University Press, New York.

Sussman RW, Tab Rasmussen D, Raven PH. 2013. Rethinking primate origins again. *American Journal of Primatology* 75:95–106.

Part I

Cheirogaleidae: evolution, taxonomy, and genetics

1 Cheirogaleid diversity and evolution: big questions about small primates

Anne D. Yoder, David W. Weisrock, Rodin M. Rasoloarison, and Peter M. Kappeler

Introduction

The family Cheirogaleidae is arguably the most interesting group of primates alive today. Within this single clade, hypothesized to have originated approximately 25–30 Mya, we find the world's smallest living primate (genus *Microcebus*), one species that went "missing" for more than three decades (genus *Allocebus*), the only known obligate hibernator within the primates (genus *Cheirogaleus*), the only primate species that preys upon other members of its phylogenetic family (genus *Mirza*), and also, a taxonomic system that has exploded within the past two decades. This taxonomic explosion has been decidedly lopsided, however. Whereas the genus *Allocebus* has remained monotypic, containing the single species *A. trichotis* since its original description in 1875 (Günther 1875), the genus *Microcebus* (mouse lemurs) has gone from a two species system as recently as 1993 to one that that now contains more than 20 recognized species. This apparent skew in species-level diversity cries out for further exploration. Is it an artifact of organismal and geographic sampling bias, with certain species and ecosystems preferentially sampled, or is it based in biology, with some branches of the cheirogaleid tree (namely, the mouse lemurs) intrinsically more prone to evolutionary divergence? An exploration of these themes and questions is our goal in this chapter.

The first genus-level phylogeny of the cheirogaleid lemurs was published by Rumpler *et al.* (1994) and has remained virtually unchanged in the subsequent decades. Using karyotype data and restriction fragment analysis, the authors found strong support for the phylogeny illustrated in Figure 1.1. Notably, Rumpler and Albignac (1972) had long before discovered that the karyotype of *Phaner* (2n = 46) is quite distinct from that of the other four genera (2n = 66), leading those authors to propose a two-subfamily taxonomy of the Cheirogaleidae, the monotypic Phanerinae (including only the genus *Phaner*) and the Cheirogaleinae (comprising the four remaining genera). More recent molecular phylogenetic analyses have sampled more densely at the species level and have yielded fresh insights into interspecific relationships within the various genera, while leaving the "skeleton" of the

The Dwarf and Mouse Lemurs of Madagascar: Biology, Behavior and Conservation Biogeography of the Cheirogaleidae, eds. S.M. Lehman, U. Radespiel, E. Zimmermann. Published by Cambridge University Press. © Cambridge University Press 2016.

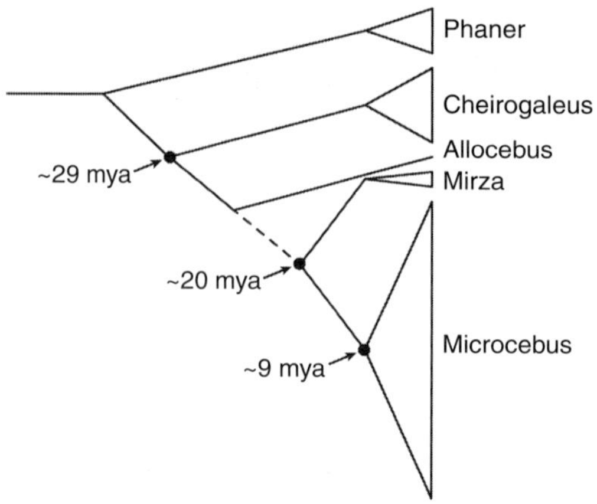

Figure 1.1 A generalized phylogeny for the Cheirogaleidae. Relationships among genera represent a consensus across multiple phylogenetic studies. Branches within multispecies genera are collapsed for simplicity, with clade size proportional to taxonomic diversity. Branch lengths are not proportional to time due to a lack of divergence time estimates for nodes involving *Allocebus* and *Phaner*. Mean estimated divergence times are presented for some nodes (Yang and Yoder, 2003; Yoder and Yang, 2004; Thiele *et al.*, 2013). The dashed line highlights the fact that, in the multilocus study of Weisrock *et al.* (2012), the number of gene trees that support the *Mirza–Microcebus* clade is roughly similar to the number of gene trees supporting an alternative placement of *Allocebus* and *Microcebus* in a clade. In contrast, all remaining branches received support from the majority of sampled loci.

phylogeny unchanged. The stability of the cheirogaleid clade has made divergence time estimations feasible. Divergence time analysis has consistently shown the basal radiation of mouse lemurs to be on the order of 10–9 Mya (Yang and Yoder, 2003; Thiele *et al.*, 2013). Interestingly, the Thiele *et al.* (2013) study found the dwarf lemur radiation to be of approximately the same age, with a basal divergence date of 9.6 Mya (depending upon the loci examined). These authors concluded that the genus *Cheirogaleus* contains deeply divergent lineages "which are considerably older than several species of mouse lemur" (p. 602). Similar conclusions were reached by Springer *et al.* (2012), although with slightly younger age estimates with ~7 Mya for mouse lemurs and ~9 Mya for dwarf lemurs. Unfortunately, none of the divergence time studies to date has included the basal lineage, genus *Phaner*, and thus all age estimates of the ancestral cheirogaleid radiation will be underestimates. That said, all studies agree in finding diversification within the Cheirogaleidae to have originated by the late Oligocene, at least.

Although it is true that several molecular phylogenetic studies have shown a weak relationship between *Phaner* and *Lepilemur* (e.g., Roos *et al.*, 2004; Springer *et al.*, 2012; Masters *et al.*, 2013), this result is likely to be an artifact of the rapid rate of mitochondrial evolution perhaps exacerbating the effects of long-branch attraction (Felsenstein, 1978; Hillis, 1996; Huelsenbeck, 1997; Wiens and Hollingsworth, 2000). Rather, a synthetic view of recent molecular phylogenetic studies indicates that it is the entire cheirogaleid clade that is the sister to the genus *Lepilemur* (e.g., see Masters *et al.*, 2013; Yoder, 2013 for recent reviews). Looking more closely at the cheirogaleid clade, the dwarf lemurs (Groeneveld *et al.*, 2009, 2010; Thiele *et al.*, 2013) and the mouse lemurs (Yoder *et al.*, 2000; Heckman *et al.*, 2006, 2007; Weisrock *et al.*, 2010) have shown complicated taxonomic expansions and rearrangements (Tables 1.1 and 1.2). In the majority of studies that have examined both mitochondrial (mtDNA) and nuclear (nDNA) DNA, it is typical to find poorly supported internal nodes and a high level of gene tree discordance across the loci under investigation (Heckman *et al.*, 2007; Weisrock *et al.*, 2010). Such results are typical of species radiations that are both recent and explosive (Moore, 1995; Knowles and Carstens, 2007; Shaffer and Thomson, 2007). Within the mouse lemur clade, several studies have shown strong support for three deep lineages, one that contains *M. murinus* plus *M. griseorufus*; another deeply diverged lineage represented by *M. ravelobensis*, *M. danfossi*, and *M. bongolavensis*; and a third lineage that is composed of all other mouse lemur species including strong support for a distal subclade composed of *M. berthae*, *M. rufus*, and *M. myoxinus* (Figure 1.2; Heckman *et al.*, 2006, 2007; Weisrock *et al.*, 2010, 2012). This latter subclade is especially intriguing given that *M. rufus* (an eastern, rainforest-adapted animal) is markedly divergent both ecologically and geographically from *M. berthae* and *M. myoxinus*, both of which occur in the dry deciduous forests of western Madagascar.

The chapter by Groves (Chapter 2) gives a comprehensive summary of the finest details of cheirogaleid taxonomy, leaving little need for us to cover the same ground. Rather, we examine here the evidence that has driven the dazzling proliferation of species designations in one genus (*Microcebus*) while leaving others essentially unchanged since their original descriptions (i.e., *Allocebus*, *Mirza*, and *Phaner*). Not terribly surprisingly, the proliferation of mouse lemur species coincides closely with the advent of readily available DNA sequences via the polymerase chain reaction (PCR) and with the development of user-friendly software for phylogenetic analysis of these sequences (e.g., Swofford, 1990). An empirical investigation of the mouse lemur radiation has verified the theoretical predictions of recent and rapid species diversification (Weisrock *et al.*, 2012). These authors, in an effort to apply a multilocus approach to reconstructing a species-level phylogeny for the mouse lemurs, found that the lingering effects of incomplete lineage sorting within the mouse lemur radiation severely compromise our ability to conduct standard phylogenetic analysis. That study, despite its inability to resolve the mouse lemur species tree, nonetheless came to a fundamental conclusion: depending upon which alleles were selected for concatenation in the multilocus analysis, phylogenetic resolutions could differ, often dramatically, and with convincing statistical support. Thus, mouse lemurs show the classic hallmarks of a rapid species radiation wherein phylogenetic relationships will be difficult to reconstruct.

Table 1.1 Evidence for and history of taxonomic revisions to the genus *Microcebus* (mouse lemurs).

Latin binomial	Common name	Original evidence	Synonyms/Revisions	Publication
M. murinus	Gray mouse lemur	Morphology		Miller (1777)
M. rufus	Brown mouse lemur	Morphology	*M. smithii* (Gray, 1842)	Geoffroy (1834)
M. myoxinus	Pygmy mouse lemur	Morphology	(Peters, 1852)	Schmid and Kappeler (1994)
M. ravelobensis	Golden-brown mouse lemur	Morphology		Zimmermann *et al.* (1998)
M. tavaratra	Northern rufous mouse lemur	Morphology		Rasoloarison *et al.* (2000)
M. sambiranensis	Sambirano mouse lemur	Morphology		Rasoloarison *et al.* (2000)
M. griseorufus	Reddish-gray mouse lemur	Morphology	Kollman (1910)	Rasoloarison *et al.* (2000)
M. berthae	Madame Berthe's mouse lemur	Morphology	*M. myoxinus* (Schmid and Kappeler, 1994)	Rasoloarison *et al.* (2000)
M. griseorufus		Morphology		Rasoloarison *et al.* (2000)
M. lehilahytsara	Goodman's mouse lemur	Morphology; mtDNA		Kappeler *et al.* (2005)
M. mittermeieri	Mittermeier's mouse lemur	Morphology; mtDNA		Louis *et al.* (2006)
M. jollyae	Jolly's mouse lemur	Morphology; mtDNA		Louis *et al.* (2006)
M. simmonsi	Simmons' mouse lemur	Morphology; mtDNA		Louis *et al.* (2006)
M. mamiratra	Claire's mouse lemur	Morphology; mtDNA	*M. lokobensis* (Olivieri *et al.*, 2007)	Andriantompohavana *et al.* (2006)
M. margotmarshae	Margot Marsh's mouse lemur	Morphology; mtDNA		Andriantompohavana *et al.* (2006)
M. bongolavensis	Bongolava mouse lemur	Morphology; mtDNA		Olivieri *et al.* (2007)
M. danfossi	Danfoss' mouse lemur	Morphology; mtDNA		Olivieri *et al.* (2007)
M. lokobensis	Lokobe mouse lemur	Morphology; mtDNA	*M. mamiratra* (Andriantompohavana *et al.*, 2006)	Olivieri *et al.* (2007)
M. arnholdi	Arnhold's mouse lemur	Morphology; mtDNA		Louis *et al.* (2008)
Microcebus spp.		Morphology; mtDNA		Radespiel *et al.* (2008)
M. macarthurii	MacArthur's mouse lemur	Morphology; mtDNA		Radespiel *et al.* (2008)
M. gerpi	Gerp's mouse lemur	Morphology; mtDNA		Radespiel *et al.* (2012)
M. marohita	Marohitra mouse lemur	Morphology; mtDNA; nDNA		Rasoloarison *et al.* (2013)
M. tanosi	Anosy mouse lemur	Morphology; mtDNA; nDNA		Rasoloarison *et al.* (2013)

Note: All species descriptions take geographic distributions into account as evidence for species designation.

Table 1.2 Evidence for and history of taxonomic revisions to the genus *Cheirogaleus* (dwarf lemurs).

Latin binomial	Common name	Original evidence	Publication
C. medius	Fat-tailed dwarf lemur	Morphology	Saint-Hilaire (1812)
C. adipicaudatus	Southern fat-tailed dwarf lemur	Morphology	Grandidier (1868)
C. major	Greater dwarf lemur	Morphology	Saint-Hilaire (1812)
C. crossleyi	Furry-eared dwarf lemur	Morphology	Grandidier (1870)
C. sibreei	Sibree's dwarf lemur	Morphology	Forsyth Major (1896)
C. ravus	Greater iron gray dwarf lemur	Morphology	Groves (2000)
C. minusculus	Lesser iron gray dwarf lemur	Morphology	Groves (2000)
C. lavasoensis	Lavasoa dwarf lemur	Morphology; mtDNA; nDNA	Thiele *et al.* (2013)

Note: All species descriptions take geographic distributions into account as evidence for species designation.

Lumping and splitting

Prior to the ready acquisition of DNA sequences for phylogenetic analysis, primatologists had favored a two-species taxonomy of mouse lemurs that included *M. murinus*, a long-eared gray animal from the southern and western regions of Madagascar, and *M. rufus*, a short-eared reddish animal from the east. Martin (1972), in particular, made note of the differing habitats and ecological constraints defining the two species, with *M. murinus* inhabiting dry deciduous and xerophytic forest and specializing on insectivory, and *M. rufus* inhabiting humid rainforest and showing dietary tendencies toward omnivory. This taxonomic stability first quavered with the description of a third, measurably smaller, species from the dry deciduous forests of western Madagascar (Schmid and Kappeler, 1994). Initially, this third species was referred to as *M. myoxinus*, although this designation was to be shortly overturned by Rasoloarison *et al.* (2000). In the interim, a fourth and much larger species was identified in northwestern Madagascar and designated as *M. ravelobensis* (Zimmermann *et al.*, 1998). Thus, even though new mouse lemur species were being recognized and described at an increasing rate throughout the 1990s, the pace was rather moderate when species designations relied upon combined assessments of ecology and morphology.

The pace accelerated abruptly with the introduction of genetic data into the species discovery process. With the combined efforts of a morphological team lead by Rasoloarison *et al.* (2000) and a genetics team led by Yoder *et al.* (2000), the rate of taxonomic revision accelerated rapidly with the description of five new species. Along with the revised taxonomy of *M. myoxinus*, now recognized as *M. berthae*, so began the era of seemingly outlandish taxonomic proliferation. In two decades, *Microcebus* expanded from a genus containing 2 species to one containing at least 20

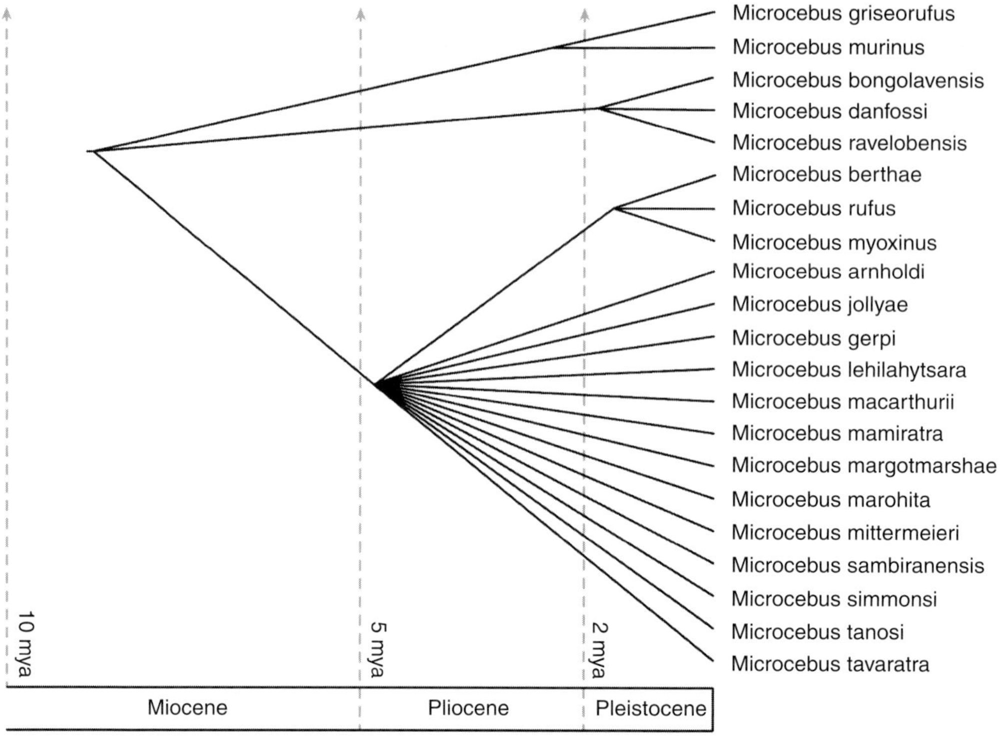

Figure 1.2 A phylogeny depicting our current best understanding of phylogenetic relationships within *Microcebus*. As the figure indicates, considerable uncertainty remains regarding the interrelationships among mouse lemur species. Branches are scaled to be proportional to time.

species (Kappeler *et al.*, 2005; Andriantompohavana *et al.*, 2006; Louis *et al.*, 2006, 2008; Olivieri *et al.*, 2007; Radespiel *et al.*, 2008, 2012; Rasoloarison *et al.*, 2013), and with many of these based primarily if not entirely on small DNA data sets.

Not surprisingly, there has been resistence from the primatological community, with assertions made of "a remarkable lack of introspection" (Tattersall, 2007) as well as concerns that "species are based solely on evidence of genetic distance and diagnostic characters of mitochondrial DNA sequences sampled from a few individuals per location" (Markolf *et al.*, 2011). We take these concerns seriously, and indeed, this is the impetus for our contribution to this very special volume. Here, we wish to address the issue of species recognition both specifically – asking how many species of mouse lemurs are there, and are they "real"? – and more generally, examining the criteria and analytical framework for recognizing species. Although it is something of a tautology to say that species are the product of speciation, it is useful to remind ourselves that "speciation is a multi-level process unfolding through time and space" (Abbott *et al.*, 2013, p. 231). In this vein, de Queiroz (2007) has elegantly made the point that by attempting to pin a name to an organismal unit that is the product of an ongoing process, we are setting ourselves up for disagreement and controversy. Here, we couch our discussion in the light of Simpson's Evolutionary Species Concept as expressed in Weisrock *et al.* (2010). That is, a "lineage (an ancestral-descendant sequence of populations) evolving separately from others and with its own unitary evolutionary role and tendencies" (Simpson, 1961). As for how to identify these lineages as species, de Queiroz (2007, p. 879) again describes matters succinctly by pointing out that "all of the properties formerly treated as secondary species criteria are relevant to species delimitation to the extent that they provide evidence of lineage separation ... and thus more lines of evidence are associated with a higher degree of corroboration."

In other words, an integrative approach that combines genetic distance, morphometric distinction, behavioral variation, and biogeographic separation establishes the lines of evidence supporting species recognition hypotheses (Markolf *et al.*, 2013). Moreover, the careful assessment of this evidence can justify the recognition of species identities, even when there is clear-cut indication of limited gene flow among and between hypothesized species (Yoder, 2014), a phenomenon that is clearly of significance for the mouse lemur radiation (Hapke *et al.*, 2011; Rakotondranary *et al.*, 2011a).

What is it about mouse lemurs?

Why is it that mouse lemurs show such clear patterns of genetic divergence and evolutionary isolation even though they are morphologically and ecologically so similar? One putative explanation relates to their nocturnal habits. Primatologists have sporadically discussed the possible relationship between nocturnality and cryptic speciation for several decades, although it is an essay on bats that offers the most detailed analysis. In a perceptive essay on the issue, Jones (1997) describes the biological complexities of cryptic species, particularly as they relate to nocturnality. In Jones' view, the lack of visual information concomitant with a noctunal lifestyle will be compensated by other means of interindividual signaling and communication such as acoustic and olfactory cues. Moreover, Jones predicted that biologists are significantly underestimating evolutionary diversity when we rely solely on visual information (i.e., morphological characters) as our guide for species identification, asserting that untold amounts of genetic biodiversity will be underappreciated using such methods. Jones anticipated that "advances in molecular biology may allow insights into speciation in cryptic species" (p. 345), a scenario that has been unfolding over the past two decades as we have investigated the mouse lemur radiation. Despite the appeal of Jones' cryptic speciation hypothesis, we must note that all cheirogaleid primates are nocturnal, so the apparent imbalance in species diversity within the clade cannot be reduced to this single explanation. Moreover, and even more complicatedly, biologists have found it difficult to explain how primate species that occupy the same ecological niche – as mouse lemurs appear to do – and are difficult if not impossible to distinguish morphologically can coexist sympatrically (Radespiel *et al.*, 2006; Dammhahn and Kappeler, 2008b; Rakotondranary and

Ganzhorn 2011; Rakotondranary *et al.*, 2011a, 2011b; Thoren *et al.*, 2011b). As we have gained increased knowledge of the precise details of mouse lemur behavioral ecology, however, these mysteries are resolving.

Mouse lemurs as a model system

Rather than view these conundrums as obstacles to biodiversity discovery, we believe that mouse lemurs can more usefully be presented as a model system for investigating the often subtle and overlooked biological forces that can drive species radiations and maintain species boundaries (although these forces may act differentially, with some as "drivers" and others as "maintainers"). Beyond the intriguing idea of looking towards alternative signaling mechanisms of olfactory and acoustic communication for driving and maintaining species boundaries (e.g., Braune *et al.*, 2005; Scheumann *et al.*, 2007; Braune *et al.*, 2008; Leliveld *et al.*, 2011; Hohenbrink *et al.*, 2012; Yoder *et al.*, 2014), it is becoming increasingly clear that painstaking investigations of behavior and ecology are necessary for disentangling the relevant parameters (e.g., see Radespiel, Chapter 26).

Among mouse lemurs, *Microcebus murinus* is unique in having a relatively vast geographic distribution. Whereas all other *Microcebus* species have geographically limited ranges – with some restricted to isolated forest fragments – *M. murinus* is distributed along most of the western half and along the south of the island. It is presently not known what are the genotypic and phenotypic traits possessed by *M. murinus* that apparently confer an adaptive flexibility that has promoted its spread across vast areas of Madagascar (but see Chapter 26). Like other species within the genus, *M. murinus* is known to exhibit episodic bouts of torpor, presumably in the face of resource limitations. Unlike other species, however, there appears to be a sex bias to this behavior that is distinctive. For example, *M. murinus* populations have been shown to exhibit different torporing patterns both by age and by sex (Schmid and Kappeler, 1998; Schmid, 1999; Terrien *et al.*, 2010a, 2010b), with one study finding that whereas 73% of females undergo prolonged torpor bouts, only 19% of males show the same behavior (Schmid, 1999). Over the extensive range of *M. murinus*, the species has been confirmed or hypothesized to overlap with multiple congeners including *M. griseorufus* (its sister species), *M. berthae*, *M. myoxinus*, *M. bongolavensis*, and *M. ravelobensis*. And although it remains to be confirmed, there is the potential for overlap with *M. tavaratra*, *M. sambiranensis*, and *M. danfossi* given the latitudinal distribution of these species. Irrespective of their sympatric status, all species in the genus compared to *M. murinus* show much smaller species ranges, with some showing alarming patterns of geographic restriction and microendemism.

Here, we review the literature on the behavioral ecology of *Microcebus*, focusing on three regions of sympatric overlap between *M. murinus* and *M. griseorufus*, *M. berthae*, and *M. ravelobensis*, respectively. Notably, all of these investigations of ecological and behavioral distinction in sympatry involve *M. murinus*, by far the most geographically widespread of the recognized species of mouse lemur. Potentially, this introduces an unwanted bias to the generalities and conclusions that can be drawn from a synthetic view of these studies, but at present, they are the best available. Moreover, despite their relatively limited taxonomic scope, these studies provide insight into the subtlety and power of environmental forces that can create almost invisible niche heterogeneity wherein each species can occupy a unique ecological domain.

Mouse lemur niche heterogeneity: from southern to northern Madagascar

Lessons from *Microcebus griseorufus*

One example of this pattern of sympatry couples *M. murinus* with *M. griseorufus*, with multiple lines of evidence for distribution overlap and potential hybridization between the two species in southern Madagascar. Within their respective ranges, *M. murinus* and *M. griseorufus* exhibit marked differences in abundance. *M. murinus* is typically common throughout its range and has been studied extensively since its description (Kappeler and Rasoloarison, 2003). Conversely, *M. griseorufus* has been recorded from relatively few locations within a narrow range in the southwest of the island in the recent past and remains poorly studied (Genin, 2008; Bohr *et al.*, 2011; Kobbe *et al.*, 2011). A recent study (Blair *et al.*, 2014) used a multilocus coalescent framework to test the hypothesis that the relatively limited range of *M. griseorufus* relative to *M. murinus* is the consequence of peripatric speciation wherein *M. griseorufus* was initially formed via the isolation of a relatively small ancestral population that was on the geographic limits of a more widely distributed ancestral *M. murinus* population. The study rejected this hypothesis soundly, finding instead that historical demographic data much better fit a model of allopatric divergence from a range-restricted common ancestor in southwestern Madagascar, with subsequent range expansions for *M. murinus*. Whether due to ecological constraint (Bohr *et al.*, 2011) or interspecific competition, *M. griseorufus* is presently restricted to the arid spiny forest in the south, whereas *M. murinus* has successfully expanded northward throughout much of western Madagascar and to limited areas in the southeast.

Microcebus griseorufus is special within the mouse lemur radiation for apparently thriving in one of the most challenging environments inhabited by any primate: the dry spiny forests of the southwest. Rainfall and accordingly resources are unpredictable both in amount and distribution, and this species has developed a number of physiological and behavioral strategies for coping with this uncertainty. A four-year study of one population showed that individuals used daily torpor and expressed opportunistic seasonal fattening when food availability was high, reverting to dependence on plant exudates when experiencing periods of regional drought (Genin, 2008). Another long-term study showed that *M. griseorufus* shows high levels of seasonal as well as individual flexibility in thermoregulation (Kobbe *et al.*, 2011). Whereas all individuals remained normothermic during the rainy season when resources are abundant, heterothermy ranging from irregular short torpor bouts, to regular daily torpor, to prolonged torpor of a few days, and even hibernation over several weeks were observed during the dry season. When compared with

M. murinus, *M. griseorufus* shows an interesting combination of habitat fidelity, strongly preferring spiny forest to gallery or transitional forest types, as well as greater flexibility in thermoregulation.

Early genetic work found that *M. griseorufus* and *M. murinus* show species-specific mtDNA haplotypes, which at the time was regarded as evidence for reproductive isolation in sympatry (Yoder *et al.*, 2002). More recent and sophisticated genetic analyses that examined both mitochondrial and nuclear autosomal loci have revealed a much more complicated picture of genetic exchange between the two species (Gligor *et al.*, 2009; Hapke *et al.*, 2011). In the 2009 study, the authors found that whereas mitochondrial haplotypes displayed a sharply delimited boundary at the eastern edge of spiny forest, nuclear loci yielded evidence for a mixed ancestry of mouse lemurs in the ecotone between the spiny forest habitat of *M. griseorufus* and the littoral forest habitat of *M. murinus*. In this case, the authors concluded that asymmetric nuclear introgression was due to male-biased dispersal, divergent environmental selection, and an expansion of dry spiny forest in the course of aridification (Gligor *et al.*, 2009). In the 2011 study, in another field setting, the authors found that whereas *M. griseorufus* is restricted to the spiny forest habitat, *M. murinus* occurs in the gallery forest but locally invades the dryer habitat of *M. griseorufus*. In this setting, the authors found evidence for bidirectional introgressive hybridization, in contrast to the asymmetric patterns previously observed. In comparing the two settings and the coincident patterns of introgressive gene flow, the authors found that hybridization could enhance the adaptability of mouse lemurs without necessarily leading to a loss of distinctiveness (Hapke *et al.*, 2011), a finding very much in keeping with emerging views of the genomics of speciation (Seehausen *et al.*, 2014; Yoder, 2014). Finally, and in keeping with the "sine qua non" of speciation theory, a recent study by Sommer *et al.* (2014) has found that hybrids formed by *M. murinus* and *M. griseorufus* show reduced fitness as measured by higher parasite loads.

To summarize, it appears that interactions between *M. murinus* and *M. griseorufus* can differ depending upon ecological and environmental context. This has been confirmed empirically: when the two species exist in allopatry, both behave as ecological generalists, yet when they occur in sympatry, both species show distinct microhabitat preferences, allowing them to partition the environment (Rakotondranary and Ganzhorn, 2011). Moreover, stable isotope analysis revealed that their dietary niches diverge most markedly in the dry "lean" season (Rakotondranary *et al.*, 2011b). The results pertaining to *M. griseorufus* and *M. murinus* have been extended to a three-species comparison in southeastern Madagascar where they co-occur with *M. rufus*. This study further confirmed the preference for gallery forest in *M. murinus* and spiny forest for *M. griseorufus*, adding the observation that *M. rufus* shows a strong fidelity to evergreen humid forest (Rakotondranary *et al.*, 2011a).

Finally, with regard to heterothermy, *M. griseorufus* was found to exhibit even greater flexibility than *M. murinus*, with individuals exhibiting variability in energy-saving strategies ranging from irregular short torpor bouts, regular daily torpor, prolonged torpor of a few days, and hibernation over several weeks as a function of individual variation in fat reserves (Kobbe *et al.*, 2011). Moreover, *M. griseorufus* also exhibits similar dietary flexibility between habitats and seasons (Crowley *et al.*, 2014), so that the relative ecological success of *M. murinus* is even more enigmatic. Taken together, these patterns of range expansion, ecological flexibility, and potential competitive advantage make this pair of species a particularly interesting system for future studies within the mouse lemur radiation.

Lessons from *Microcebus berthae*

As *M. griseorufus* overlaps with *M. murinus* in the south, so does *M. berthae* in the west, although with a considerably smaller area of geographic overlap (Schäffler and Kappeler, 2014). Detailed field studies are emerging to illustrate that just as with *M. griseorufus*, *M. berthae* and *M. murinus* partition their environment in subtle though biologically meaningful ways. Both species are omnivorous and experience similar seasonal fluctuations in food supply (Schwab and Ganzhorn, 2004); nonetheless, they exhibit fine-grained differences in annidation. Feeding niche overlap is high in terms of food categories, but stable isotope analyses revealed niche differentiation in fruit and animal matter ($\delta^{15}N$), but not in basal resources ($\delta^{13}C$; Dammhahn and Kappeler, 2010, 2014). *M. berthae* is more insectivorous and has a narrower feeding niche than *M. murinus*, which includes more fruits and gum in its diet (Dammhahn and Kappeler, 2008a), resulting in different intraspecific competitive regimes and spatial requirements (Dammhahn and Kappeler, 2010). In addition, *M. berthae* does not hibernate and is less able to use spontaneous torpor than *M. murinus* (Ortmann *et al.*, 1997; Schmid *et al.*, 2000). These studies lend further support to the notion of a strong propensity for *M. berthae* and *M. murinus* to partition the environment spatially by microhabitat, with the two species showing a non-overlapping "checkerboard" pattern of spatial segregation in sympatry (Schwab and Ganzhorn, 2004), and with *M. berthae* occupying home ranges nearly twice as large as those of *M. murinus* (Dammhahn and Kappeler, 2005). More recent analyses indicate that the distribution and abundance of the two mouse lemur species may also depend on that of other cheirogaleid members of their local community, with *Cheirogaleus medius* locally displacing *M. murinus* and *Mirza coquereli* acting as an opportunistic predator of *M. murinus* (Schäffler *et al.*, 2015). Thus, coexistence of mouse lemurs here and elsewhere may also be stabilized by additional factors and species.

Lessons from *Microcebus ravelobensis*

Perhaps the most intensively studied of the areas of species overlap is that in the far northwest of Madagascar where *M. murinus* and *M. ravelobensis* occur sympatrically. A range of behavioral differences diagnoses the two species. Spatially, densities of *M. murinus* have been shown to increase with altitude while those of *M. ravelobensis* decrease (Rakotondravony and Radespiel, 2009). As with the other two species comparisons, detailed ecological studies have shown distinct differences in microhabitat choice and species-specific patterns of spatial distribution (Rendigs *et al.*, 2003), including choice of nest site type and composition of nesting groups. Whereas *M. murinus* tends towards female-only sleeping groups, *M. ravelobensis* is more likely to show mixed-sex sleeping groups (Radespiel *et al.*, 2003). Dietary distinctions are also detectable, although they are subtle (Radespiel *et al.*, 2006; Sehen *et al.*, 2010; Thoren *et al.*, 2011b). In direct interactions, *M. murinus* is dominant when paired with *M. ravelobensis*, despite a lack of a significant size difference (Thoren *et al.*, 2011a).

Surprisingly, *M. ravelobensis* also appears to violate at least two of the "rules" that govern mouse lemur behavior. Whereas studies of other species (primarily, of *M. murinus*) consistently show patterns of a strong bias towards male dispersal and female philopatry (Schliehe-Diecks *et al.*, 2012), male dispersal in *M. ravelobensis* is significantly delayed (Radespiel *et al.*, 2009). The greatest surprise, however, relates to female dominance. Whereas behavioral studies of all other mouse lemur species show strong patterns of female dominance, *M. ravelobensis* does not (Eichmueller *et al.*, 2013). Whether and how these behavioral differences contribute to species coexistence remains to be explored in more detail.

Now what?

The section above will, we hope, convince the reader that the remarkable taxonomic proliferation of mouse lemur species has biological merit. It appears that these animals are using an array of behavioral, dietary, physiological, and signaling mechanisms to divide up their world and maintain their separate evolutionary trajectories, thus destined for unique evolutionary endpoints. However, many challenges remain towards fully comprehending this fascinating species radiation. Their diminutive size, nocturnal habits, and remote geographic distribution require determined and sometimes extraordinary effort on the part of field biologists who wish to study these petite primates. We predict that the coming years will continue to provide rich insight into the unique ecological and behavioral traits of these animals that in turn will further inform our understanding of the mechanisms that have driven and continue to maintain species boundaries.

But what can be done immediately, and over the coming few years, to deepen our understanding of species diversity, not just of mouse lemurs, but of the entire cheirogaleid clade? As was predicted by Jones (1997), genetic data have opened up a previously obscured world of biodiversity in the cryptic mouse and dwarf lemur radiations. Presently, these studies provide a very lopsided view of rampant speciation on one branch in the clade (i.e., the mouse lemurs) and virtual evolutionary stasis in others (e.g., *Allocebus*). This begs for further exploration. Is it a function of intrinsic biology? Do mouse lemurs have some innate genomic or phenotypic properties that predispose them to divergent speciation? Or is it rather a function of sampling bias and experimental approach?

Recent work focused on the dwarf lemurs (genus *Cheirogaleus*) appears to suggest the latter, at least to some appreciable extent. Thiele *et al.* (2013) forcefully make the case that taxonomic partitioning can strongly influence interpretations of species range limits and genetic diversity among any number of other important measures of biodiversity. By directly comparing homologous data within the same analytical framework for mouse and dwarf lemurs (e.g., see their figure 4), these authors are able to make a strong case for claiming either that mouse lemur diversity is overestimated, or dwarf lemur diversity is underestimated. We suspect that the most likely explanation is "a bit of both." Accordingly, the conclusion must be drawn that much work lies ahead before we can claim that the full diversity of the cheirogaleid clade is known.

So how do we proceed? From the earliest phases of "mouse lemur mania" there have been repeated calls for an integrative approach to species discovery, one that takes into account morphology, behavior, ecology, sensory signaling, geography, and autosomal as well as sex-linked genetic loci (e.g., Yoder *et al.*, 2000, 2005; Heckman *et al.*, 2006; Weisrock *et al.*, 2010; Markolf *et al.*, 2011; Zimmermann and Radespiel, 2014). The collection of behavioral and ecological data is arduous and time-consuming, however, with results slow to emerge. We therefore make the urgent call that the community of scientists interested in species discovery within this fascinating clade of primates adopt a uniform genetic "toolkit" such that each individual, population, and putative species can be examined within a general and universal framework, much as has been recently illustrated in the study of Thiele *et al.* (2013). Moreover, when depositing genetic data, investigators should be assiduous in also depositing their aligned matrices. It is well understood that the process of sequence alignment is rife with assumptions. When investigators are required to download individual gene sequences, concatenate and align them, assumptions made (either computationally or by eye) can impact downstream results, especially with regard to the estimation of branch lengths in phylogenetic analysis – the very evidence upon which so many species hypotheses have been founded (e.g., Yoder *et al.*, 2000; Weisrock *et al.*, 2010; Thiele *et al.*, 2013). If such standards can be developed and implemented, we can enter a new era of analysis wherein investigators work as a community, united in the goals of biodiversity discovery and conservation of the Cheirogaleidae as a whole. This approach will be both collegial and scientifically rigorous.

Acknowledgments

We would like to thank the editors for their invitation to contribute to this special volume on the cheirogaleid primates and for their patience during the preparation of our manuscript. Our contribution was significantly improved thanks to their editorial oversight and by the constructive critique of two anonymous reviewers. We also thank and recognize the Malagasy institutions and collaborators who have made our work possible over the years, namely the Département Biologie Animale, CAFF-CORE and the Ministry of Forest and the Environment. This is Duke Lemur Center publication #1284.

References

Abbott R, Albach D, Arntzen JW, *et al.* 2013. Hybridization and speciation. *Journal of Evolutionary Biology* 26:229–246.

Andriantompohavana R, Zaonarivelo JR, Engberg SE, *et al.* 2006. Mouse lemurs of north-western Madagascar with a description of a new species at Lokobe Special Reserve. *Occasional Papers Museum of Texas Tech University* 259.

Blair C, Heckman KL, Russell AL, Yoder AD. 2014. Multilocus coalescent analyses reveal the demographic history and speciation patterns of mouse lemur sister species. *BMC Evolutionary Biology* 14:57.

Bohr YEMB, Giertz P, Ratovonamana YR, Ganzhorn JU. 2011. Gray-brown mouse lemurs (*Microcebus griseorufus*) as an example of distributional constraints through increasing desertification. *International Journal of Primatology* 32:901–913.

Braune P, Schmidt S, Zimmermann E. 2005. Spacing and group coordination in a nocturnal primate, the golden brown mouse lemur (*Microcebus ravelobensis*): the role of olfactory and acoustic signals. *Behavioral Ecology and Sociobiology* 58:587–596.

Braune P, Schmidt S, Zimmermann E. 2008. Acoustic divergence in the communication of cryptic species of nocturnal primates (*Microcebus* ssp.). *BMC Biology* 6:1–10.

Crowley BE, Rasoazanabary E, Godfrey LR. 2014. Stable isotopes complement focal individual observations and confirm dietary variability in reddish-gray mouse lemurs (*Microcebus griseorufus*) from southwestern Madagascar. *American Journal of Physical Anthropology* 155:77–90.

Dammhahn M, Kappeler PM. 2005. Social system of *Microcebus berthae*, the world's smallest primate. *International Journal of Primatology* 26:407–435.

Dammhahn M, Kappeler P. 2008a. Comparative feeding ecology of sympatric *Microcebus berthae* and *M. murinus*. *International Journal of Primatology* 29:1567–1589.

Dammhahn M, Kappeler PM. 2008b. Small-scale coexistence of two mouse lemur species (*Microcebus berthae* and *M. murinus*) within a homogeneous competitive environment. *Oecologia* 157:473–483.

Dammhahn M, Kappeler PM. 2010. Scramble or contest competition over food in solitarily foraging mouse lemurs (*Microcebus* spp.): new insights from stable isotopes. *American Journal of Physical Anthropology* 141:181–189.

Dammhahn M, Kappeler PM. 2014. Stable isotope analyses reveal dense trophic species packing and clear niche differentiation in a Malagasy primate community. *American Journal of Physical Anthropology* 153:249–259.

de Queiroz K. 2007. Species concepts and species delimitation. *Systematic Biology* 56:879–886.

Eichmueller P, Thoren S, Radespiel U. 2013. The lack of female dominance in golden-brown mouse lemurs suggests alternative routes in lemur social evolution. *American Journal of Physical Anthropology* 150:158–164.

Felsenstein J. 1978. Cases in which parsimony or compatibility methods will be positively misleading. *Systematic Zoology* 27:401–410.

Genin F. 2008. Life in unpredictable environments: first investigation of the natural history of *Microcebus griseorufus*. *International Journal of Primatology* 29:303–321.

Gligor M, Ganzhorn JU, Rakotondravony D, *et al.* 2009. Hybridization between mouse lemurs in an ecological transition zone in southern Madagascar. *Molecular Ecology* 18:520–533.

Groeneveld LF, Weisrock DW, Rasoloarison RM, Yoder AD, Kappeler PM. 2009. Species delimitation in lemurs: multiple genetic loci reveal low levels of species diversity in the genus *Cheirogaleus*. *BMC Evolutionary Biology* 9:30.

Groeneveld LF, Blanco MB, Raharison JL, *et al.* 2010. MtDNA and nDNA corroborate existence of sympatric dwarf lemur species at Tsinjoarivo, eastern Madagascar. *Molecular Phylogenetics and Evolution* 55:833–845.

Groves CP. 2000. The genus *Cheirogaleus*: unrecognized biodiversity in dwarf lemurs. *International Journal of Primatology* 21:943–962.

Günther, A. 1875. Notes on some mammals from Madagascar. *Proceedings of the Zoological Society of London 1875*, 78–80.

Hapke A, Gligor M, Rakotondranary SJ, Rosenkranz D, Zupke O. 2011. Hybridization of mouse lemurs: different patterns under different ecological conditions. *BMC Evolutionary Biology* 11:297.

Heckman KL, Rasoazanabary E, Machlin E, Godfrey LR, Yoder AD. 2006. Incongruence between genetic and morphological diversity in *Microcebus griseorufus* of Beza Mahafaly. *BMC Evolutionary Biology* 6:98.

Heckman KL, Mariani CL, Rasoloarison R, Yoder AD. 2007. Multiple nuclear loci reveal patterns of incomplete lineage sorting and complex species history within western mouse lemurs (*Microcebus*). *Molecular Phylogenetics and Evolution* 43:353–367.

Hillis DM. 1996. Inferring complex phylogenies. *Nature* 383:130–131.

Hohenbrink P, Radespiel U, Mundy NI. 2012. Pervasive and ongoing positive selection in the vomeronasal-1 receptor (V1R) repertoire of mouse lemurs. *Molecular Biology and Evolution* 29:3807–3816.

Huelsenbeck JP. 1997. Is the Felsenstein zone a fly trap? *Systematic Biology* 46:69–74.

Jones G. 1997. Acoustic signals and speciation: the roles of natural and sexual selection in the evolution of cryptic species. *Advances in the Study of Behavior* 26:317–354.

Kappeler PM, Rasoloarison RM. 2003. *Microcebus*, mouse lemurs, Tsidy. In SM Goodman, JP Benstead (eds.), *The Natural History of Madagascar* (pp. 1310–1315). University of Chicago Press, Chicago.

Kappeler PM, Rasoloarison R, Razafimanantsoa L, Walter L, Roos C. 2005. Morphology, behaviour and molecular evolution of giant mouse lemurs (Mirza spp.) Gray, 1870, with description of a new species. *Primate Report* 71:3–26.

Knowles LL, Carstens BC. 2007. Delimiting species without monophyletic gene trees. *Systematic Biology* 56:887–895.

Kobbe S, Ganzhorn JU, Dausmann KH. 2011. Extreme individual flexibility of heterothermy in free-ranging Malagasy mouse lemurs (*Microcebus griseorufus*). *Journal of Comparative Physiology B Biochemical Systemic and Environmental Physiology* 181:165–173.

Leliveld LMC, Scheumann M, Zimmermann E. 2011. Acoustic correlates of individuality in the vocal repertoire of a nocturnal primate (*Microcebus murinus*). *Journal of the Acoustical Society of America* 129:2278–2288.

Louis EE, Coles MS, Andriantompohavana R, *et al.* 2006. Revision of the mouse lemurs (*Microcebus*) of eastern Madagascar. *International Journal of Primatology* 27:347–389.

Louis EE, Engberg SE, McGuire SM, *et al.* 2008. Rivision of the mouse lemurs, *Microcebus* (Primates, Lemuriformes), of northern and northwestern Madagascar with descriptions of two new species at Montagne d'Ambre National Park and Antafondro Classified Forest. *Primate Conservation* 23:19–38.

Markolf M, Brameier M, Kappeler PM. 2011. On species delimitation: yet another lemur species or just genetic variation? *BMC Evolutionary Biology* 11:216.

Markolf M, Rakotonirina H, Fichtel C, *et al.* 2013. True lemurs ... true species – species delimitation using multiple data sources in the brown lemur complex. *BMC Evolutionary Biology* 13:233.

Martin RD. 1972. A preliminary field-study of the lesser mouse lemur (*Microcebus murinus* J.F. Miller, 1777). *Zeitschrift für Tierpychologie* (Suppl 9):43–89.

Masters JC, Silvestro D, Génin F, DelPero M. 2013. Seeing the wood through the trees: the current state of higher systematics in the Strepsirrhini. *Folia Primatologica* 84:201–219.

Moore WS. 1995. Inferring phylogenies from mtDNA variation: mitochondrial-gene trees versus nuclear-gene trees. *Evolution* 49:718–726.

Olivieri G, Zimmermann E, Randrianambinina B, *et al.* 2007. The ever-increasing diversity in mouse lemurs: three new species in north and northwestern Madagascar. *Molecular Phylogenetics and Evolution* 43:309–327.

Ortmann S, Heldmaier G, Schmid J, Ganzhorn JU. 1997. Spontaneous daily torpor in Malagasy mouse lemurs. *Naturwissenschaften* 84:28–32.

Radespiel U, Ehresmann P, Zimmermann E. 2003. Species-specific usage of sleeping sites in two sympatric mouse lemur species (*Microcebus murinus* and *M. ravelobensis*) in Northwestern Madagascar. *American Journal of Primatology* 59:139–151.

Radespiel U, Reimann W, Rahelinirina M, Zimmermann E. 2006. Feeding ecology of sympatric mouse lemur species in northwestern Madagascar. *International Journal of Primatology* 27:311–321.

Radespiel U, Olivieri G, Rasolofoson DW, *et al.*, 2008. Exceptional diversity of mouse lemurs (*Microcebus* spp.) in the Makira region with the description of one new species. *American Journal of Primatology* 70:1033–1046.

Radespiel U, Juric M, Zimmermann E. 2009. Sociogenetic structures, dispersal and the risk of inbreeding in a small nocturnal lemur, the golden-brown mouse lemur (*Microcebus ravelobensis*). *Behaviour* 146:607–628.

Radespiel U, Ratsimbazafy JH, Rasoloharijaona S, *et al.* 2012. First indications of a highland specialist among mouse lemurs (*Microcebus* spp.) and evidence for a new mouse lemur species from eastern Madagascar. *Primates* 53:157–170.

Rakotondranary SJ, Ganzhorn JU. 2011. Habitat separation of sympatric *Microcebus* spp. in the dry spiny forest of south-eastern Madagascar. *Folia Primatologica* 82:212–223.

Rakotondranary SJ, Hapke A, Ganzhorn JU. 2011a. Distribution and morphological variation of *Microcebus* spp. along an environmental gradient in southeastern Madagascar. *International Journal of Primatology* 32:1037–1057.

Rakotondranary SJ, Struck U, Knoblauch C, Ganzhorn JU. 2011b. Regional, seasonal and interspecific variation in (15)N and (13)C in sympatric mouse lemurs. *Naturwissenschaften* 98:909–917.

Rakotondravony R, Radespiel U. 2009. Varying patterns of coexistence of two mouse lemur species (*Microcebus ravelobensis* and *M. murinus*) in a heterogeneous landscape. *American Journal of Primatology* 71:928–938.

Rasoloarison RM, Goodman SM, Ganzhorn JU. 2000. Taxonomic revision of mouse lemurs (*Microcebus*) in the western portions of Madagascar. *International Journal of Primatology* 21:963–1019.

Rasoloarison RM, Weisrock DW, Yoder AD, Rakotondravony D, Kappeler PM. 2013. Two new species of mouse lemurs (Cheirogaleidae: *Microcebus*) from Eastern Madagascar. *International Journal of Primatology* 34:455–469.

Rendigs A, Radespiel U, Wrogemann D, Zimmermann E. 2003. Relationship between microhabitat structure and distribution of mouse lemurs (*Microcebus* spp.) in northwestern Madagascar. *International Journal of Primatology* 24:47–64.

Roos C, Schmitz J, Zischler H. 2004. Primate jumping genes elucidate strepsirrhine phylogeny. *Proceedings of the National Academy of Sciences of the United States of America* 101:10650–10654.

Rumpler Y, Albignac R. 1972. Cytogenetic study of the endemic Malagasy lemurs subfamily Cheirogaleinae Gregory 1915. *American Journal of Physical Anthropology* 38:261–264.

Rumpler Y, Crovella S, Montagnon D. 1994. Systematic relationships among Cheirogaleidae (Primates, Strepsirhini) determined from analysis of highly repeated DNA. *Folia Primatologica* 63:149–155.

Schäffler L, Kappeler P. 2014. Distribution and abundance of the world's smallest primate, *Microcebus berthae*, in Central Western Madagascar. *International Journal of Primatology* 35:557–572.

Schäffler L, Saborowski J, Kappeler PM, 2015. Agent-mediated spatial storage effect in heterogeneous habitat stabilizes competitive mouse lemur coexistence in Menabe Central, Western Madagascar. *BMC Ecology* 15:7. DOI: 10.1186/s12898-015-0040-1.

Scheumann M, Zimmermann E, Deichsel G. 2007. Context-specific calls signal infants' needs in a strepsirrhine primate, the gray mouse lemur (*Microcebus murinus*). *Developmental Psychobiology* 49:708–718.

Schliehe-Diecks S, Eberle M, Kappeler PM. 2012. Walk the line – dispersal movements of gray mouse lemurs (*Microcebus murinus*). *Behavioral Ecology and Sociobiology* 66:1175–1185.

Schmid J. 1999. Sex-specific differences in activity patterns and fattening in the gray mouse lemur (*Microcebus murinus*) in Madagascar. *Journal of Mammalogy* 80:749–757.

Schmid J, Kappeler PM. 1994. Sympatric mouse lemurs (*Microcebus* spp.) in western Madagascar. *Folia Primatologica* 63:162–170.

Schmid J, Kappeler PM. 1998. Fluctuating sexual dimorphism and differential hibernation by sex in a primate, the gray mouse lemur (*Microcebus murinus*). *Behavioral Ecology and Sociobiology* 43:125–132.

Schmid J, Ruf T, Heldmaier G. 2000. Metabolism and temperature regulation during daily torpor in the smallest primate, the pygmy mouse lemur (*Microcebus myoxinus*) in Madagascar. *Journal of Comparative Physiology B Biochemical Systemic and Environmental Physiology* 170:59–68.

Schwab D, Ganzhorn JU. 2004. Distribution, population structure and habitat use of *Microcebus berthae* compared to those of other sympatric cheirogaleids. *International Journal of Primatology* 25:307–330.

Seehausen O, Butlin RK, Keller I, *et al.* 2014. Genomics and the origin of species. *Nature Reviews Genetics* 15:176–192.

Sehen L, Goetze D, Rajeriarison C, *et al.* 2010. Structural and floristic traits of habitats with differing relative abundance of the lemurs *Microcebus murinus* and *M. ravelobensis* in Northwestern Madagascar. *Ecotropica* 16:15–30.

Shaffer HB, Thomson RC. 2007. Delimiting species in recent radiations. *Systematic Biology* 56:896–906.

Simpson GG. 1961. *Principles of Animal Taxonomy*. Columbia University Press, New York.

Sommer S, Rakotondranary SJ, Ganzhorn JU. 2014. Maintaining microendemic primate species along an environmental gradient – parasites as drivers for species differentiation. *Ecology and Evolution* 4(24):4751–4765.

Springer MS, Meredith RW, Gatesy J, *et al.* 2012. Macroevolutionary dynamics and historical biogeogrphy of primate diversification inferred from a species supermatrix. *PLoS ONE* 7:e49521.

Swofford DL. 1990. *PAUP: Phylogenetic Analysis Using Parsimony*. Illinois Natural History Survey, Champaign, IL.

Tattersall I. 2007. Madagascar's lemurs: cryptic diversity or taxonomic inflation? *Evolutionary Anthropology* 16:12–23.

Terrien J, Ambid L, Nibbelink M, Saint-Charles A, Aujard F. 2010a. Non-shivering thermogenesis activation and maintenance in the aging gray mouse lemur (*Microcebus murinus*). *Experimental Gerontology* 45:442–448.

Terrien J, Perret M, Aujard F. 2010b. Gender markedly modulates behavioral thermoregulation in a non-human primate species, the mouse lemur (*Microcebus murinus*). *Physiology and Behavior* 101:469–473.

Thiele D, Razafimahatratra E, Hapke A. 2013. Discrepant partitioning of genetic diversity in mouse lemurs and dwarf lemurs – biological reality or taxonomic bias? *Molecular Phylogenetics and Evolution* 69:593–609.

Thoren S, Linnenbrink M, Radespiel U. 2011a. Different competitive potential in two coexisting mouse lemur species in Northwestern Madagascar. *American Journal of Physical Anthropology* 145:156–162.

Thoren S, Quietzsch F, Schwochow D, *et al.* 2011b. Seasonal changes in feeding ecology and activity patterns of two sympatric mouse lemur species, the Gray Mouse Lemur (*Microcebus murinus*) and the Golden-brown Mouse Lemur (*M. ravelobensis*), in Northwestern Madagascar. *International Journal of Primatology* 32:566–586.

Weisrock DW, Rasoloarison RM, Fiorentino I, *et al.* 2010. Delimiting species without nuclear monophyly in Madagascar's mouse lemurs. *PLoS ONE* 5:e9883.

Weisrock DW, Smith SD, Chan LM, *et al.* 2012. Concatenation and concordance in the reconstruction of mouse lemur phylogeny: an empirical demonstration of the effect of allele sampling in phylogenetics. *Molecular Biology and Evolution* 29:1615–1630.

Wiens JJ, Hollingsworth BD 2000. War of the iguanas: conflicting molecular and morphological phylogenies and long-branch attraction in iguanid lizards. *Systematic Biology* 49:143–159.

Yang ZH, Yoder AD. 2003. Comparison of likelihood and Bayesian methods for estimating divergence times using multiple gene loci and calibration points, with application to a radiation of cute-looking mouse lemur species. *Systematic Biology* 52:705–716.

Yoder AD. 2013. The lemur revolution starts now: the genomic coming of age for a nonmodel organism. *Molecular Phylogenetics and Evolution* 66:442–452.

Yoder AD. 2014. Gene flow happens. *Evolutionary Anthropology* 23:15–17.

Yoder AD, Yang Z. 2004. Divergence dates for Malagasy lemurs estimated from multiple gene loci: geological and evolutionary context. *Molecular Ecology* 13:757–773.

Yoder AD, Rasoloarison RM, Goodman SM, *et al.* 2000. Remarkable species diversity in Malagasy mouse lemurs (primates, *Microcebus*). *Proceedings of the National Academy of Sciences of the United States of America* 97:11325–11330.

Yoder AD, Burns MM, Genin F. 2002. Molecular evidence of reproductive isolation in sympatric sibling species of mouse lemurs. *International Journal of Primatology* 23:1335–1343.

Yoder AD, Olson LE, Hanley C, *et al.* 2005. A multidimensional approach for detecting species patterns in Malagasy vertebrates. *Proceedings of the National Academy of Sciences of the United States of America* 102:6587–6594.

Yoder AD, Chan LM, dos Reis M, *et al.*, 2014. Molecular evolutionary characterization of a V1R subfamily unique to strepsirrhine primates. *Genome Biology and Evolution* 6:213–227.

Zimmermann E, Radespiel U. 2014. Species concepts, diversity, and evolution in primates: lessons to be learned from mouse lemurs. *Evolutionary Anthropology* 23:11–14.

Zimmermann E, Cepok S, Rakotoarison N, Zietemann V, Radespiel U. 1998. Sympatric mouse lemurs in north-west Madagascar: a new rufous mouse lemur species (*Microcebus ravelobensis*). *Folia Primatologica* 69, 106–114.

2 The taxonomy of Cheirogaleidae: an ever-expanding species list

Colin Groves

Introduction

So what is taxonomy?

Taxonomy is the way we order and classify the natural world. It is more of a specialised field than many people realise. Concepts like "how different does it have to be to be a distinct species/genus/family?" are not relevant; what we want to know is information about evolution and phylogeny.

The basic unit of taxonomy is the species. Species have a real existence: they are evolutionary lineages (de Queiroz, 2007). What we seek is evidence for these lineages, and the way to recognise them is what has been called the Phylogenetic Species Concept (Groves, 2001; Cotterill *et al.*, 2014) which, following de Queiroz's (2007) analysis, would be better dubbed the Diagnosability Criterion. If one population (or group of populations) is consistently (diagnosably) different from others in heritable characters, this is absolute evidence that it forms a separate evolutionary lineage. The proposal that a given population, sample, cluster, or whatever, constitutes a species is therefore a testable hypothesis.

The so-called Higher Categories do not, on the other hand, have a real existence. Since Hennig (1950/1966) there has arisen a general consensus that a genus or a family must be monophyletic, but beyond that there is subjectivity. There has been a proposal to link these categories to time-depths (see, for example, Groves, 2001); thus, a genus should have been separated around the Miocene–Pliocene boundary, and a family should have been separated around the Oligocene–Miocene boundary. There has been surprisingly little discussion about this, which is distressing, because we are supposed to be doing science, and science is about testability.

The concept of Cheirogaleidae

Early zoologists took a while to recognise that dwarf and mouse lemurs are related to each other, rather than, for example, mouse lemurs being Madagascan galagos. More modern arrangements were beginning to be expressed by the mid-1800s. In Gray (1870), the genus *Cheirogaleus* was included in the Lemuridae, and two years

The Dwarf and Mouse Lemurs of Madagascar: Biology, Behavior and Conservation Biogeography of the Cheirogaleidae, eds. S.M. Lehman, U. Radespiel, E. Zimmermann. Published by Cambridge University
Press. © Cambridge University Press 2016.

later Gray (1872) recognised a tribe Cheirogaleina with more or less its present composition.

This general scheme was followed for nearly a century. It was Petter (1965) who first expressed dissatisfaction with placing the Cheirogaleinae in the Lemuridae: "to the ecologist and ethologist their being grouped in the same family ... appears scarcely justified" (p. 314). Rumpler (1974), meanwhile, definitively separated the cheirogaleids from the Lemuridae, splitting the Cheirogaleidae into subfamilies Cheirogaleinae Gregory, 1915 and a new subfamily Phanerinae, basing himself in the latter case on a marked difference in karyotypes, dermatoglyphics, the presence of a throat gland in *Phaner*, and the latter's "peculiar behaviour". Petter *et al.* (1977) followed this up, recognising the Cheirogaleidae with the two subfamilies Cheirogaleinae and Phanerinae, and this seems to have been the point from which primate taxonomy has never looked back.

However, are the Cheirogaleidae a homogeneous (monophyletic) group, and what are the limits of the concept? Chatterjee *et al.* (2009) found that *Phaner* actually forms a clade with *Lepilemur* rather than with other cheirogaleids. Perelman *et al.* (2011) found that there is little evidence that it links with one in preference to the other, but noted a high probability that *Lepilemur* and the Cheirogaleidae do form a clade with respect to other Malagasy lemurs. Masters *et al.* (2013) reviewed the evidence, and proposed to include *Lepilemur* in the family Cheirogaleidae. This is certainly an important topic for discussion in the future.

The relationships of the Cheirogaleidae

In the early 1960s, a discovery was made that would throw the whole question of the relationships of the Cheirogaleidae once again into doubt. What Saban (1963) found was that in the Cheirogaleinae the course of the carotid system through the temporal bone resembles that of the Lorisiformes, and is quite different from that in the other Malagasy lemurs. In most Malagasy lemurs, the common carotid divides into external and internal branches, the latter of which enters the bulla where it divides into a large stapedial branch and a small promontory branch. In the Cheirogaleidae and the Lorisiformes, however, the stapedial is very small (in *Phaner* it is absent altogether) and the promontory very large, and the common carotid gives off in addition an anterior carotid branch which does not enter the bulla but, along with the promontory artery, enters the braincase through a special foramen, the promontory foramen (or, as Szalay and Katz (1973) preferred to regard it, the anterior carotid foramen). Saban did not himself make much of this difference, but it was picked up later, independently, by Szalay and Katz (1973) and by Groves (1974), who considered that this is good evidence for the monophyly of cheirogaleids and lorisiforms. Groves (1974) argued that the Cheirogaleidae should be placed along with the lorises and galagos in a superfamily Lorisoidea. Szalay and Katz (1973) in addition noted that there are some similarities also with the pattern in *Lepilemur*, although the latter does not have an anterior carotid.

This difference was taken up again by Tattersall and Schwartz (1974), who classed the Cheirogaleidae in the infraorder Lorisiformes (the other two infraorders of what they called Prosimii (reverting to an earlier era in primate classification) being Lemuriformes and Indriiformes), and proposed a morphotype and character evolution. The same authors (Schwartz and Tattersall, 1985) later elaborated this scheme, pointing out a few other similarities between cheirogaleids and the lorisid–galagid clade; they also gave detailed descriptions of the cranial and dental characters of the genera and species recognised by them.

The evidence was re-examined in the light of ontogenetic data by Yoder (1992), who deemed that it strongly supported cheirogaleid/lorisiform monophyly; but later the same author (Yoder, 1994) re-examined the evidence in the light of further morphological and molecular data, and this time concluded that the Cheirogaleidae do after all form a clade with other Malagasy lemurs, not with the Lorisiformes.

The genera of the Cheirogaleidae

In Gray (1870), the genus *Cheirogaleus* was awarded three species: *C. milii*, *C. typicus* and *C. smithii*; the grey mouse lemur, on the other hand, was included with a query in the genus *Lepilemur* as *Lepilemur murinus*; *Lepilemur* included only one other species, *L. furcifer*, and was put in the Galagonina. In an appendix to the same work, he separated *C. smithii* into a new genus *Azema*, and he had now become aware that the real *Lepilemur* was a rather different animal, so he separated the former *L. murinus* into a new genus *Murilemur* and *L. furcifer* into another new genus *Phaner*; while he also described a new genus *Mirza* for *M. coquerelii* [*sic*], which he had not mentioned before.

Dunkel *et al.* (2012) gave explanations of Gray's new generic names. Particularly entertaining are the origins of the names *Mirza*, *Phaner* and *Azema*: J.E. Gray seems to have attended a production of a theatrical comedy by W.S. Gilbert in which characters with these names appeared.

By 1872, Gray had got it more or less right: *Opolemur* (for what he had previously called *Cheirogaleus milii*), *Phaner*, *Cheirogaleus*, *Mirza*, *Azema* and *Murilemur* were now all regarded as constituting a tribe Cheirogaleina.

Mivart (1873) outright rejected Gray's new genera, and the next commentator, Forsyth Major (1894), mentioned them only in passing and placed the species *smithi*, *coquereli* and *furcifer* in *Microcebus*, while continuing to separate the *Cheirogaleus medius*-group as a separate genus from the *C. major*-group. In this he was followed by Forbes (1896) and Elliot (1912), while Kollmann (1910) actually placed what is in effect the *C. medius*-group in *Microcebus*. It seems astonishing the way most of these authors relentlessly sank the strikingly distinctive *Phaner* into *Microcebus* while splitting *Cheirogaleus* (or *Chirogale* as they called it) into two. It was Schwarz (1931) who finally recognised *Phaner* as a full genus.

It was likewise Schwarz (1931) who definitively abolished the genus *Opolemur*, stating that "The differences are only gradual, and the essential characters which distinguish the two species *major* and *medius* from *Microcebus* about the same."

Today, we are busily trying to rectify Schwarz's over-lumping, but we must recognise that, in his 1931 paper, he brought order to the group. He recognised only two species in *Microcebus*, three in *Cheirogaleus* and one in *Phaner*.

The first DNA phylogeny of the family (although *Phaner* was not available) was by Pastorini *et al.* (2001), who found the following:

(*Cheirogaleus* (*Allocebus* (*Mirza*, *Microcebus*)))

As an interesting addendum, as late as Elliot (1912), it was *Microcebus* which were called dwarf lemurs and *Cheirogaleus* mouse lemurs; Osman Hill (1953) may have been the first to switch them round.

Microcebus

Early views of species in *Microcebus*

The first mouse lemurs were figured by J.F. Miller (1777, plate 13) and P. Brown (1776, plate 44), the latter a very beautifully executed illustration. Both are easily recognisable as grey mouse lemurs. A description and figure by Buffon (1776), the "rat de Madagascar", is less easily recognisable, but might well also be a grey mouse lemur, and Geoffroy Saint-Hilaire (1796) based the name *Lemur pusillus* on it. This name was not mentioned in his seminal publication which established the basis for modern primate taxonomy (Geoffroy Saint-Hilaire, 1812), where, instead, he began a tradition, which lasted all too long, of regarding the mouse lemur as a kind of bushbaby, giving the name *Galago madagascariensis* to it; he did discuss Miller's description and plate, not repeating Miller's name but calling it *Lemur bicolor* and rejecting it as a primate! Later (Geoffroy Saint-Hilaire, 1828), he returned to his *Lemur pusillus*, but this time made it the type species of a new genus, *Microcebus*. Under that heading, he listed (but did not give a scientific name to) "le Microcèbe roux", describing it as having fleecy red–gold pelage, grey–red on the belly, the tail long and somewhat tufted at the extremity.

As we shall see, the "Microcèbe roux" later achieved a spurious significance. The specimen is still in the Paris Museum. Tattersall (1982) noted that it is in very poor condition, and suggested that it is probably a grey mouse lemur. It is virtually impossible to tell; it is a very unprepossessing object, which Étienne Geoffroy probably would no longer recognise (Figure 2.1).

During the nineteenth century, just one other species of *Microcebus* was described: *Microcebus myoxinus*, by Peters (1852). Forsyth Major (1894), who used the name *Microcebus minor* rather than *M. murinus* (which he regarded as poorly founded) was one of the few late nineteenth/early twentieth-century authors who retained *M. myoxinus* as a valid species; he also recognised *M. smithii* (for this, see below), as well as *M. coquereli* and *M. furcifer*.

Shortly after the turn of the century, Kollman (1910) added a subspecies, *M. minor griseorufus*. It would be the best part of a century before the full biodiversity within the genus began to be realised.

Figure 2.1 "Le Microcèbe roux". (A black and white version of this figure will appear in some formats. For the colour version, please refer to the plate section.)

Schwarz (1931) placed all mouse lemurs (except for *M. coquereli*) in a single species, *M. murinus*, with two subspecies, *M. m. murinus* from "The whole of S.E., E., and W. Madagascar, as far north as the Betsiboka River", and *M. m. rufus* from "The whole of eastern, northern, and east-central Madagascar, including the plateau, as far south as Ft. Dauphin. Also the north-west, down to the Bay of Bombétoka".

Petter (1962) noted the prevailing classification of *Microcebus murinus* into two subspecies (*M. m. murinus* and *M. m. smithii*), but added that he could distinguish a third, coexisting with the grey mouse lemur in the west. The two western forms together would differ from the dark brown–red, short-eared, dark-brown-eyed eastern form in their slightly larger size with a longer tail, lighter colour, larger ears and lighter-coloured eyes. Of the two western forms, the more widely distributed one is grey above generally with some reddish tints, and white below, while the rarer one is more reddish above and yellowish below and in captivity is more active and aggressive. Later (Petter, 1965) he reiterated this difference, specifying that the two western types occur together in the same forest at Ankarafantsika. Later still, with his colleagues, this classification was further elaborated (Petter *et al.*, 1977): provisionally four different forms were distinguished, as follows:

- A reddish form from the east, *Microcebus rufus* E. Geoffroy. There is a population on Nosy Mangabe which has much larger ears.
- Another reddish form from the west (Ankarafantsika, Morondava), much smaller than *M. murinus*, with which it is sympatric in certain regions, and with a relatively longer tail and short ears; although the ears are larger than in *M. rufus*, and it is lighter red.

- A grey form from the west and the south, *Microcebus murinus* (J.F. Miller).
- Finally, a grey and red form from the cliffs of the Angavo (Centre–East). This resembles *M. rufus* but is grey without reddish tones, but the crown of the head and rings around the eyes are contrastingly reddish.

The problem of *Microcebus rufus*

The nomenclature of the eastern species – *smithii* or *rufus* – was not discussed by Petter *et al.* (1977), but was considered in some detail by Tattersall (1982), who concurred that red mouse lemurs do occur sympatrically in places with western grey ones. He traced most names given to mouse lemurs prior to 1840 either to J.F. Miller's (1777) somewhat unsatisfactory figure of *Lemur murinus* or to the beautiful plate 44 of Brown (1776); both of these are clearly grey mouse lemurs.

The problems arise when the correct scientific name for the red mouse lemur is considered. Schwarz (1931) had based his use of the name *rufus* for his eastern subspecies of *M. murinus* on Wagner (1840), on the grounds that Wagner's name was based on the "Microcèbe roux" of Geoffroy Saint-Hilaire (see above), but, as noted by Tattersall (1982), it was not: the name *Microcebus rufus* occurs in Wagner's text only once, in a footnote (1840: 291) when he states that Geoffroy's and Desmarest's *Galago madagascariensis* is "our *Microcebus rufus*". Tattersall, however, continued to use *Microcebus rufus* for the red (eastern) mouse lemur, arguing ("by the exercise of what it would be uncharitable to label sophistry") that the name could instead be dated from Lesson (1840), who described a species *Gliscebus rufus*.

However, all is not right with Lesson's name either. Under the heading "Makiloir roux, *Gliscebus rufus*" he gave a French description, the locality "L'ile de Madagascar", and then a reference to Vigors and Horsfield's (1828) *Cheirogaleus commersonnii* [*sic*], quoting their Latin description, of which his French description was a direct translation. Because *Gliscebus rufus* was the second species in his new genus, the first being the "Makiloir gris, *Gliscebus murinus*", Tattersall (1982) argued that "it is clear that his intention was to distinguish between gray and rufous species of 'Makiloir' ". There is actually no evidence that he even knew of a rufous species of mouse lemur, only that he *thought* he knew of *a* rufous species.

As a matter of fact, Lesson had commented on *Cheirogaleus commersonii* on a previous occasion (Lesson, 1829) when, in a brief review of Vigors and Horsfield (1828), he waxed critical over their comments on gibbons, then continued (in translation): "Nor is their *Cheirogaleus commersonii* at all new. It is the animal which by error M. F. Cuvier described and figured under the name of Maki nain, in October 1821, in the second volume of his *Histoire des Mammifères*." This did not in any way meet with the approval of Vigors and Horsfield, who wrote a sarcastic response, in which they revealed that the animal which they had taken for a *Cheirogaleus* was actually a night monkey (today called *Aotus*), very likely from Brazil (Vigors and Horsfield, 1829). There are a couple of odd things about all this. First, Vigors and Horsfield, while castigating Lesson for pontificating about a specimen he had not actually seen, had to admit that they had got it wrong themselves, but did not admit

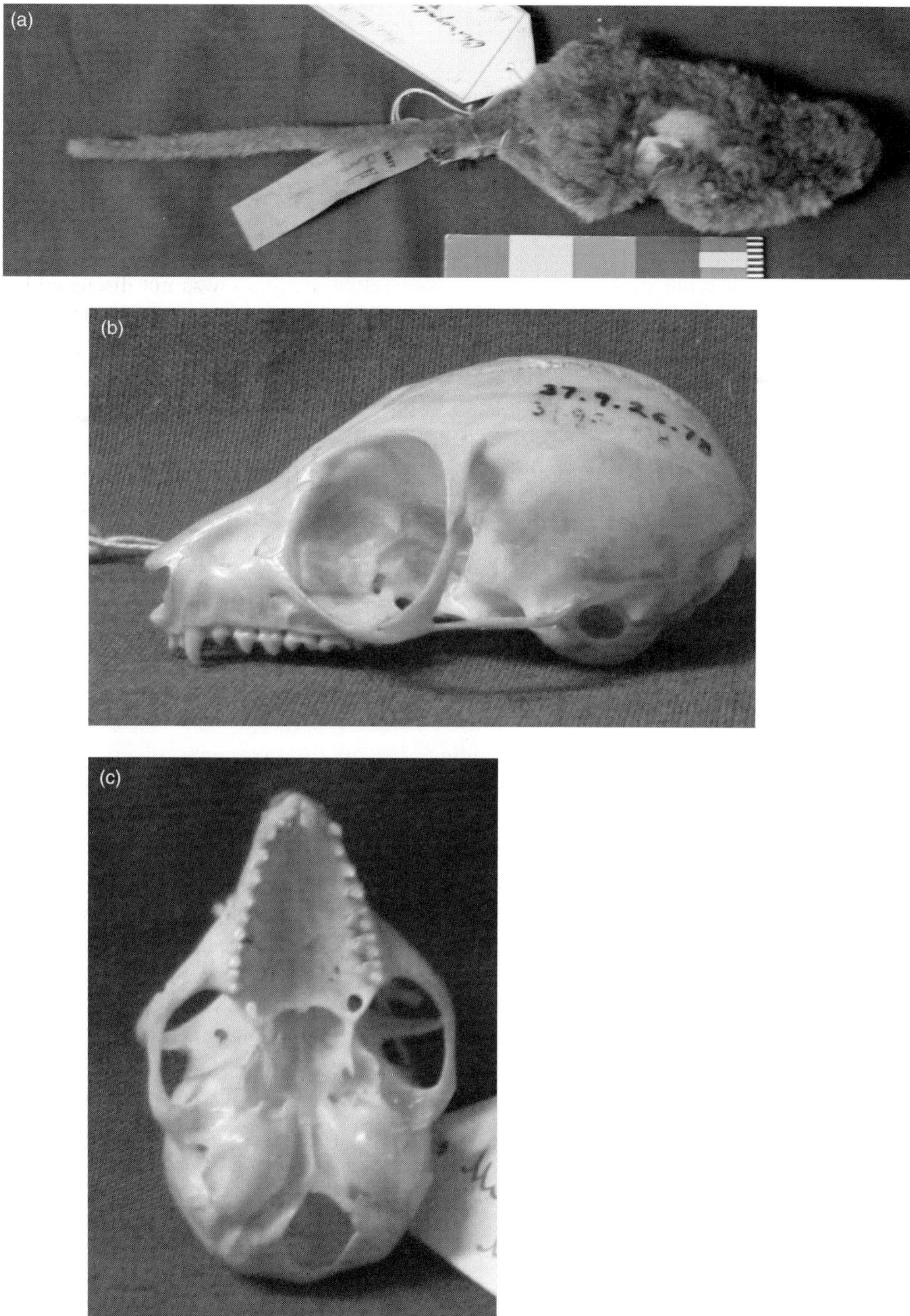

Figure 2.2 British Museum (Natural History), type of *Microcebus smithii* (Gray, 1842); (a) skin; (b) skull; (c) maxillary dentition. (A black and white version of this figure will appear in some formats. For the colour version, please refer to the plate section.)

that they had got it so wrong that they had mistaken a South American monkey for a dwarf lemur, and had even described the supposed dwarf lemur as a new species. Second, Lesson had either not seen their response (or not been informed about it by his colleagues) or had forgotten about it (or had deliberately repressed all memory of it), such that 11 years later he not only mentioned it once again as a dwarf lemur, but even gave it a new name – in the process suppressing Vigors and Horsfield's name!

The only other name, prior to the twenty-first century, supposed to designate an eastern rainforest mouse lemur is *Cheirogaleus smithii* Gray, 1842. This was described as: "Pale brown; streak up the nose and forehead, the chin and beneath paler; tail redder. Hab. Madagascar." The specimen, still in the Natural History Museum in London, is in good condition (Figure 2.2), but there is no indication as to what part of Madagascar it is from; presumably it was thought to be from the east because, until comparatively recently, mouse lemurs of this colour were known only from there. Harper (1940) noted the problems with the name *rufus*, and in order to make the name *smithii* available for the "eastern subspecies" he restricted the type locality to "a few miles north of Fianarantsoa, central Betsileo". In this he was followed by Hill (1953), who mapped the two subspecies *murinus* and *smithii* over more or less the whole of the west and east of Madagascar, respectively.

Modern trends: the extraordinary species diversity within *Microcebus*

It was in the 1990s that the rumblings began: the taxonomic diversity within *Microcebus* had been underestimated. First, Schmid and Kappeler (1994) became aware that there were two species living sympatrically in Kirindy: the common grey *M. murinus* and a much smaller reddish species, which they thought was *M. myoxinus*. In the Ampijoroa Forest Reserve, Zimmermann *et al.* (1998) studied a mouse lemur which they could not identify with any other, and described it as a new species, *Microcebus ravelobensis*. Then came the revision of western mouse lemurs by Rasoloarison *et al.* (2000).

The Rasoloarison *et al.* (2000) revision was a bit of a shock to the system: there was not just one species in western Madagascar, nor even the three that resulted from the Schmid and Kappeler (1994) and the Zimmermann *et al.* (1998) papers, but seven.

- *M. tavaratra* new species, from Ankarana, where it is sympatric with another species, provisionally identified as *M. murinus*.
- *M. sambiranensis* new species, from the Manongarivo Reserve in the Sambirano region, where it had been previously misidentified by authors such as Tattersall (1982) as *M. rufus*.
- *M. ravelobensis* Zimmermann *et al.*, 1998 from Ankarafantsika, where it is sympatric with another species, provisionally identified as *M. murinus*.
- *M. murinus* Miller, 1777, apparently widespread in west and south Madagascar. For obvious reasons, a neotype was designated (from Andranomena, near Morondava).

- *M. myoxinus* Peters, 1852, from the Tsiribihina River north to the Soalala Peninsula. The authors' examination of the type specimen, in the Berlin Museum, showed that it is not in fact the small red species found by Schmid and Kappeler (1994) in Kirindy.
- *M. berthae* new species, from Kirindy – this is the species that had been mistakenly identified by Schmid and Kappeler (1994) as *M. myoxinus.*
- *M. griseorufus* Kollmann, 1910, from Beza Mahafaly Reserve (the locality of a neotype designated by the authors) north to at least Lamboharana, and possibly further south and southeast. At the type locality, it appeared to be sympatric with *M. murinus*, but this was shown to be incorrect by Heckman *et al.* (2006).

Here we have six species of restricted distribution, and one (*M. murinus*) apparently much more widespread; and sympatry was found between four of these new species and *M. murinus* (or the species provisionally identified as that). Yoder *et al.* (2002) later extended these results, finding sympatry between *M. murinus* and *M. griseorufus* also at Berenty, in the spiny forest zone of southernmost Madagascar.

Yoder *et al.* (2000) were able to take mtDNA samples (using the control region) from all seven of these species, as well as samples from two different regions on the eastern side of Madagascar, which they referred to as "*M. rufus* 1" and "*M. rufus* 2". Unexpectedly, an east versus west division was not found; instead, the two southernmost species, *M. murinus* and *M. griseorufus*, formed a clade opposed to a clade of all the others; moreover, the two "*M. rufus*" samples were intermixed among the five more northerly species from the west. This was the first suggestion that the east/west contrast which had been taken for granted by many early authors ought to be re-examined. The finding that *M. murinus* and *M. griseorufus* are sister species is especially interesting, given that they are partially sympatric (see above). The pattern was confirmed with an enlarged data set by Yoder and Heckman (2006).

The first sign that the mouse lemurs of eastern Madagascar were also diverse was in the description of *Microcebus lehilahytsara* from Andasibe by Roos and Kappeler (in Kappeler *et al.*, 2005), which rivals *M. berthae* for the title of the smallest living primate. Radespiel *et al.* (2012) argued that it is the only known highland specialist among mouse lemurs.

Almost as great a shock to the system as the Rasoloarison *et al.* (2000) paper was the one by Louis *et al.* (2006) describing three new species from eastern Madagascar – it was not the new species, which after all were by now more or less expected, but the fact that the main criteria for separating them were DNA sequences, morphological features being essentially secondary although nonetheless significant. This was the first time this had been done in primates. The three species – *M. mittermeieri, M. jollyae* and *M. simmonsi* – all belong to the northern clade of Yoder *et al.* (2000).

The Louis *et al.* (2006) paper also contains the following sentence: "The authors collected the holotype for *Microcebus rufus*, which Lesson (1840) described, in south-central Madagascar just north of Fianarantsoa, west of Ranomafana National Park, which is the location of the type 1 *M. rufus* that Yoder *et al.* (2000; Fig.1 A) discussed" (Louis *et al.*, 2006: 351). Evidently, "neotype" is meant, rather than "holotype", but as we have seen (above), the name *rufus* is unavailable for any mouse lemur: *Microcebus rufus* Wagner, 1840 is a synonym of *M. murinus*, while *Glischropus murinus* Lesson, 1840 does not refer to a mouse lemur at all. The species called *Microcebus rufus* by Louis *et al.* (2006), and *Microcebus rufus* 1 by Yoder *et al.* (2000), will have to be given a new name.

Following hard on the heels of Louis *et al.* (2006), Andriantompohavana *et al.* (2006) examined large series from northwestern Madagascar, confirming the validity of previously described species and describing *Microcebus mamiratra* as new, from Lokobe Reserve on Nosy Be. Olivieri *et al.* (2007) continued the process, also in the northwest, describing *M. bongolavensis*, *M. danfossi* and *M. lokobensis*; but the latter name was given in ignorance of Andriantompohavana *et al.*'s (2006) *M. mamiratra* from the same place, and therefore falls as a synonym of it. The Olivieri *et al.* paper is interesting in its biogeographic approach, showing that the described species fall into previously defined biogeographic subzones and their own defined Inter-River Systems in the northwest. In the following year, Radespiel *et al.* (2008) examined the mouse lemurs of the Makira region, northeastern Madagascar, finding three species in the region, much the commonest being *M. mittermeieri*; they described a second species, *M. macarthurii*, but left the third species undescribed as only a single specimen was available. Louis *et al.* (2008) described *Microcebus margotmarshae* from Antafondro Classified Forest just inland of the range of *M. sambiranensis*, and *M. arnholdi* from the Mt. d'Ambre. Radespiel *et al.* (2012) described the large-sized *Microcebus gerpi* from the Sahafina Forest, on the eastern coast, just below the mountains where *M. lehilahytsara* occurs, supporting the previously expressed idea that the latter is a highland specialist. Most recently, Rasoloarison *et al.* (2013) described the still larger *M. marohita* (overall, the largest known species) from the forest of the same name in the Marolambo district in Toamasina Province, and *M. tanosi* from the Tolagnaro (Fort Dauphin) district, close to the ranges of *M. murinus* and *M. griseorufus*.

There is limited gene flow between some species that are sympatric (Hapke *et al.*, 2011), as turns out to be not uncommon between sympatric congeners, but all species described from 2000 on have turned out to be perfectly distinct (except in the case of the species from Nosy Be, which was inadvertently described twice).

Mouse lemur taxonomy has recently played a significant part in the debates over how much variation occurs within a single taxonomic unit. Simpson *et al.* (1960) maintained that heterogeneity of some sort should be suspected if Coefficients of Variation go above 10 (the Coefficient of Variation (CV) is the standard deviation as a percentage of the mean), and this was reinforced by Yablokov (1966). Groves (1989: 207–212) discussed the matter, concluding that even a CV above 5 should be regarded with suspicion. Recently, a very large sample (181 specimens), collected at Amboasary in 1931, was studied by Cuozzo *et al.* (2013), who showed that the whole sample is referable to *M. griseorufus*. The CV of every craniodental character (n mostly > 120) was well below 10; in fact in every case except for widths of M3 in both jaws, < 5.0 (Cuozzo *et al.*, 2013: table 5a). This enormous sample fully corroborates the idea that metrical variability within a given population or taxonomic unit is strictly limited. By contrast, considerable variation was found by Cuozzo *et al.* (2013) in *M. griseorufus* in both pelage and in nonmetric dental traits.

Mirza

Grandidier (1867) described *Cheirogalus* [*sic*] *coquereli* on the basis of seven specimens from Morondava. Despite Gray's (1870) referral of it to a new genus *Mirza*, giant mouse lemurs were invariably placed in *Microcebus* for a century. Groves (1974) noted that the name *Mirza* "is available ... if subgeneric division is required, although according to Petter (personal communication) it is really quite close to the smaller species". The species was placed in *Cheirogaleus* by Tattersall and Schwartz (1974), but Tattersall (1982) revived the genus *Mirza*.

Until 2005, *Mirza coquereli* was the only known species in the genus, but in that year, in what must be one of the most complete character surveys of a lemur genus (incorporating behaviour and mtDNA sequences), a new species *Mirza zaza* was described by Kappeler and Roos (Kappeler *et al.*, 2005). *M. coquereli* is found in the Morondava district and possibly further south along the east coast, and the smaller, short-tailed, short-eared *M. zaza* is found in the Sambirano region.

Allocebus

Gunther (1875) described and figured *Cheirogaleus trichotis*, noting its long ear-tufts, short tail, caniniform "first false molar" and "pointed, claw-like" nails. He considered it related to *C. smithii*.

Forsyth Major (1894) discussed the species and thought it might be closely related to Grandidier's *Chirogalus crossleyi* and that, if new material became available, the two of them might prove to represent together a new genus. It was unfortunate that no one took up this prescient comment until Petter-Rousseau and Petter's description of *Allocebus*, although Schwarz (1931) did note that it is "less closely allied to the two other species than they are amongst themselves".

The new material for which Forsyth Major (1894) had hoped became available in 1965, and, as anticipated, provided evidence for a new genus. Petter-Rousseaux and Petter (1967) distinguished their new genus *Allocebus* from *Microcebus* and *Cheirogaleus* by its long, caniniform first and second premolars, its small molars, the parallel, not anteriorly convergent, toothrows, and the flattened, not inflated, auditory bullae. The pointed cusps and presence of a distolingual tubercle (small hypocone) on the first and second molars relate it to *Microcebus* rather than to *Cheirogaleus* to which it had previously always been referred. The form of the anterior premolar and of the molars, and the parallel toothrows, were pointed out as similarities (if not close) to *Phaner*.

A. trichotis soon proved to be more widespread than had been feared (Meier and Albignac, 1991), and to extend into highland rainforest (Rakotoarison *et al.*, 1996).

Cheirogaleus

Early views of species in *Cheirogaleus*

A brief taxonomic history of *Cheirogaleus* was given by Groves (2000), and will not be repeated here. It is interesting to contemplate once again the long-lasting belief that two genera were involved: the large species were placed in *Cheirogaleus*, the small ones in *Otolemur* or *Altililemur*.

During the nineteenth century, the following species were described: *major, medius, minor, commersonii* (Wolf, Vigors and Horsfield), *milii, typicus, samati, gliroides, adipicaudatus, crossleyi, thomasi, melanotis* and *sibreei*. Schwarz (1931) simplified this: he recognised three species in *Cheirogaleus*: *C. medius*, *C. major* and *C. trichotis* (this now being referred to a separate genus, *Allocebus*). *C. medius* was divided into two subspecies: *C. medius medius*, known only from Ft. Dauphin, and *C. m. samati*, known only from Morondava. *C. major* was provisionally divided into *C. major major* (regarded as an eastern "race") and *C. m. crossleyi*. For him, Grandidier's *adipicaudatus* was a synonym of *major*, and Forsyth Major's *sibreei* of *crossleyi*.

Petter *et al.* (1977) recognised the usual two species, but recognised subspecies only in *C. major*, which they listed: *C. m. major*, browner in tone, from the forest of the east and the remnants of the forests of the plateau; *C. m. crossleyi*, distinguished by its golden tones and smaller skull, found from north of Maroantsetra on the high crests of the limits of the plateaux as far south as Lake Alaotra; and a third form, seen on the Bongolava, seeming to be intermediate between *C. m. crossleyi* and *C. medius*. Some specimens of the latter had a white tail tip. Although they distinguished these as subspecies, they went on to say that the first two forms "seem to cohabit in certain places", quoting observations made by Peyrieras. *C. medius* was said to be the species of almost all of the forests of the west and south.

However, Tattersall (1982) disbelieved this, and recognised just the two species, with no subspecies. Like Petter *et al.* (1977), he mapped *C. medius* in the south and most of the west, and *C. major* from the whole of the east and north, and the Sambirano where the two species were said to be sympatric. Groves (1989), however, was more impressed by the information given in Petter *et al.* (1977), noting that the evidence of sympatry between *C. m. major* and what they called *C. m. crossleyi* seems to require that the latter be at least provisionally regarded as specifically distinct; "study of museum specimens is needed".

Groves (2000) recounted the history of classification, and proposed a new arrangement with seven species as follows.

- *C. medius* E. Geoffroy Saint-Hilaire, 1812, from the dry forests of the west. The holotype of *C. samati* Grandidier, 1867 was designated as the neotype for this species.
- *C. adipicaudatus* Grandidier, 1868, from Tabiky and Ft. Dauphin, i.e. the southern arid zone. Very like *C. medius*, distinguished by colour details, longer tail and ear, and smaller auditory bulla. Synonym *Opolemur thomasi* Forsyth Major, 1894.
- *C. major* E. Geoffroy Saint-Hilaire, 1812, apparently from the rainforest zone of the east, south of Antongil Bay. The holotype of *Cheirogaleus milii* Geoffroy

Saint-Hilaire, 1828 was designated as neotype of this species (Figure 2.3); S.M. Goodman (personal communication) thinks this likely to have originated from the Tamatave (= Toamasina) district, or between there and Tananarive. Synonym *Cheirogaleus typicus* Smith, 1833.

- *C. crossleyi* Grandidier, 1870, apparently between Vohima and Lake Alaotra, inland of the distribution of *C. major*. Synonym *Chirogale melanotis* Forsyth Major, 1894.
- *C. sibreei* Forsyth Major, 1896, from Ankeramadinika (London), Imerina (Berlin) and Pasandava (Leiden).
- *C. ravus* new species, from Tamatave and Mahambo. Distinguished from *C. major* by colour, short tail, and details of the dentition; skull somewhat resembling

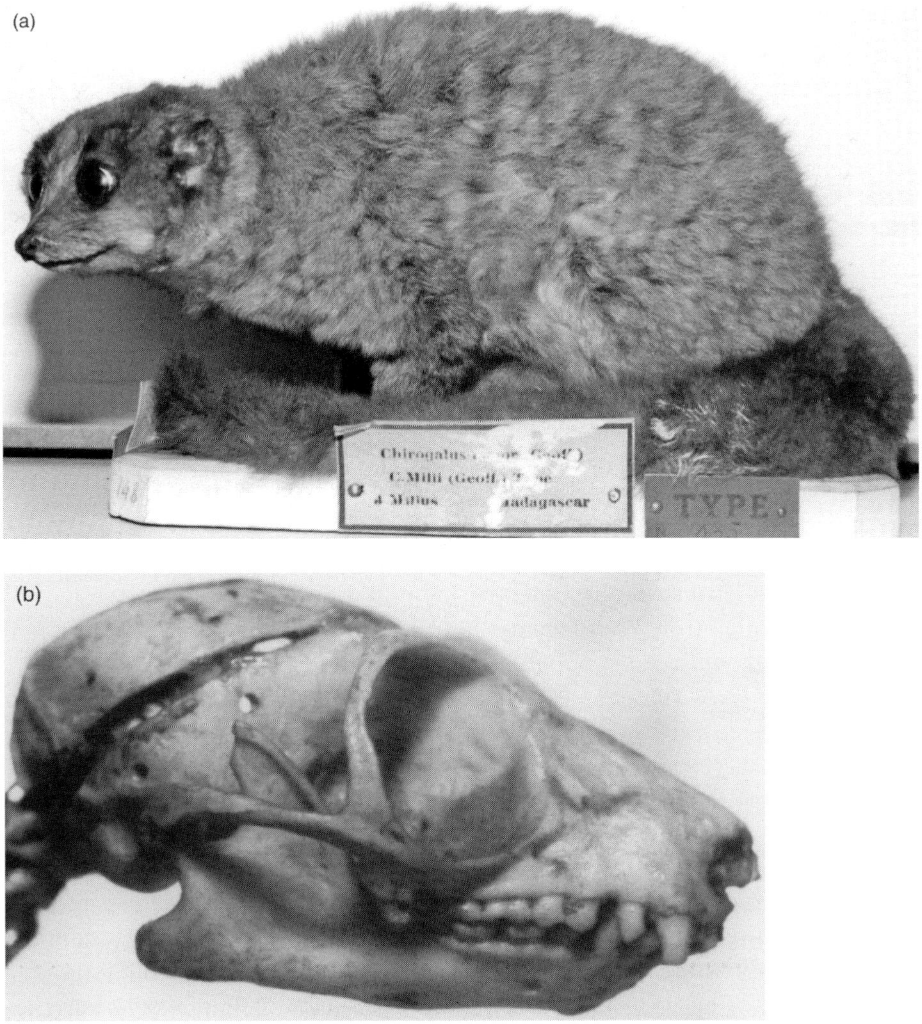

Figure 2.3 Holotype of *Cheirogaleus milii*, neotype of *Cheirogaleus major* MNHNP no.148: (a) skin; (b) skull. (A black and white version of this figure will appear in some formats. For the colour version, please refer to the plate section.)

C. crossleyi. Based on two specimens in the London collection, three from Paris, four from Berlin and two from Leiden.

- *C. minusculus* new species, from Ambositra (known from only a single specimen, London). Groves (2000) thought that this "is certainly the species ... in the Bongolava massif", which Petter *et al.* (1977) had classed as a small form of *C. major*. (This was illustrated by Thalmann (2000), and is not *C. minusculus* after all, but looks rather like *C. crossleyi*).

I should add that I have recently been able to study the US collections. In the Museum of Comparative Zoology, Harvard, I found, unexpectedly, quite a number of specimens of lemurs from the Grandidier collection, among them a specimen, MCZ 44946 skin and skull of a male from Antsianaka, which is probably the type of *C. crossleyi* Grandidier, 1870.

Twenty-first-century studies of *Cheirogaleus*: from diversity to deflation and back again

The first test of the multiple-species model was a DNA study by Hapke *et al.* (2005), who found three well-differentiated species in the far southeastern region, near Ft. Dauphin: *C. major*, *C. minor* and *C. crossleyi*. Not only was the latter therefore confirmed genetically, but its distribution had proved to extend much further south than indicated by specimens studied by Groves (2000).

The second test was less supportive. Groeneveld *et al.* (2009) sequenced DNA from numerous field samples, and also from 17 museum specimens, including three of *C. adipicaudatus*, one of the Paris specimens attributed to *C. ravus* and one of the two Leiden specimens attributed to *C. sibreei*. (They also included 25 GenBank samples.) As far as mtDNA is concerned, they found three clades, which they identified as follows: A, *medius*; B, *major*; C, *crossleyi*. Interestingly, BM 1939.1289, a juvenile *crossleyi*, clustered with A, suggesting hybridisation; and from the field data, too, they found potential hybridisation between *crossleyi* and *major*, as an individual from their locality 25 (Ambalavero, Andrambovato, 21.5°S, 47.4°E) that is morphometrically *major* is genetically *crossleyi*.

The three mitochondrial clades of Groeneveld *et al.* (2009) proved about equally related to each other; according to the authors, they "largely correspond to the three species recognised prior to the taxonomic revisions of Groves" (p. 3), a nonsensical statement considering that they had previously stated (correctly) that the genus had "consisted of two species from the 1930s until the turn of the last century" (p. 2).

Their three clades were likewise separate for the nuclear gene *fiba* (Figure 2.3); but only Clade C was separate for their other nuclear genes, *adora* 3 (Figure 2.3) and *vWF* (Figure 2.4). The population structure based on these three nuclear loci (Figure 2.5) showed much variation between populations within each of them, thus "our results do not rule out the possibility of additional species of dwarf lemurs in Madagascar" (p. 7). There was a particularly good correlation with geography, as they pointed out, within the Clade A lineage, where some western and northwestern populations clustered separately from populations from the northeast (see p. 10). At Sambava (14.4°S, 50.2°E) in the northeast, *medius* and *crossleyi* seem to be sympatric, and genetically both were largely separated from others of the same species (in the case of *medius*, also including nearby Beroboka, locality 7).

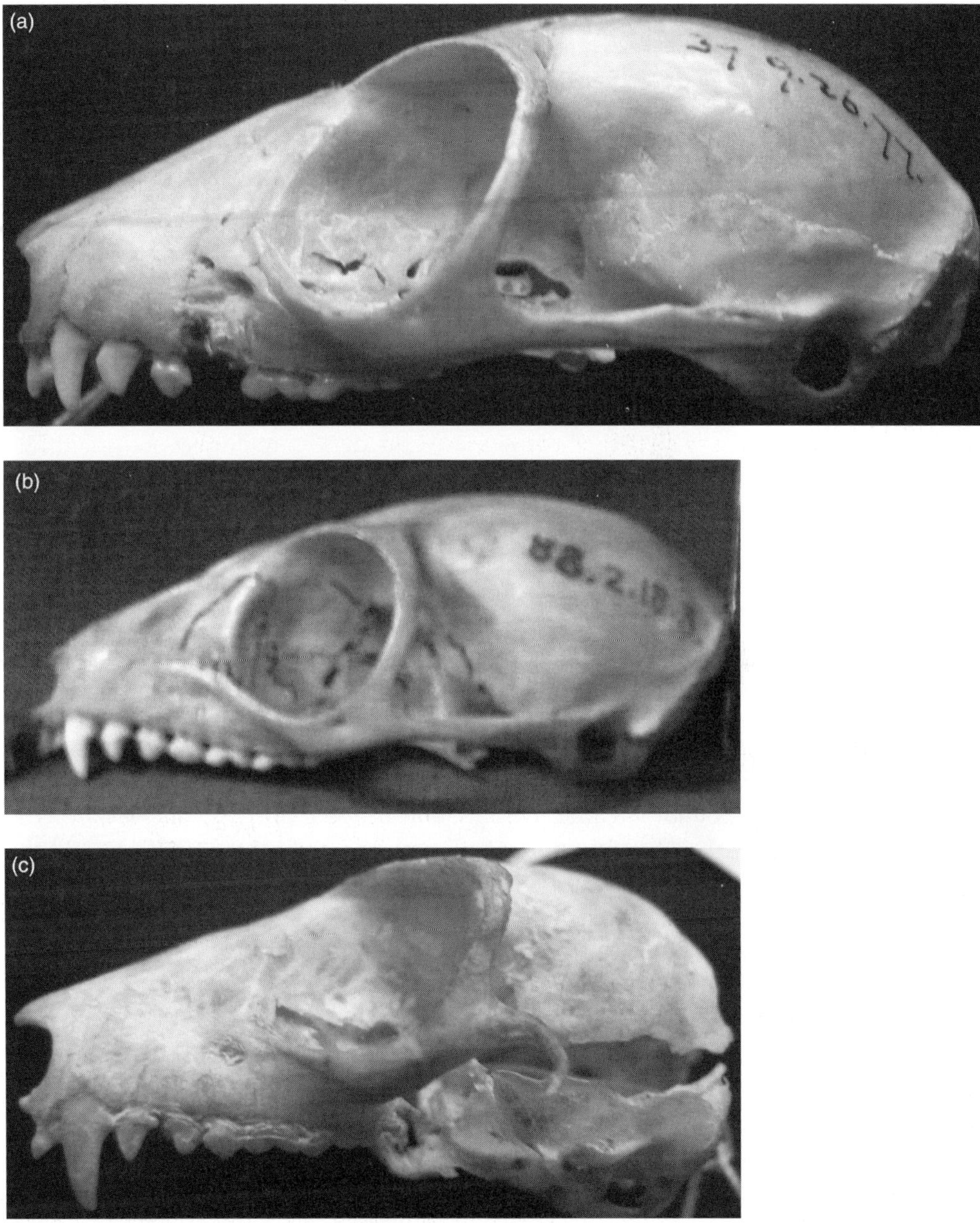

Figure 2.4 Skulls of type specimens of *Cheirogaleus* spp.: (a) *C.* cf. *major* (type of *C. typicus* Smith, 1833); (b) *C. ravus* Groves, 2000 (BM(NH) 88.2.18.3); (c) *C. minusculus* Groves, 2000 (BM(NH) 11.6.21.1);

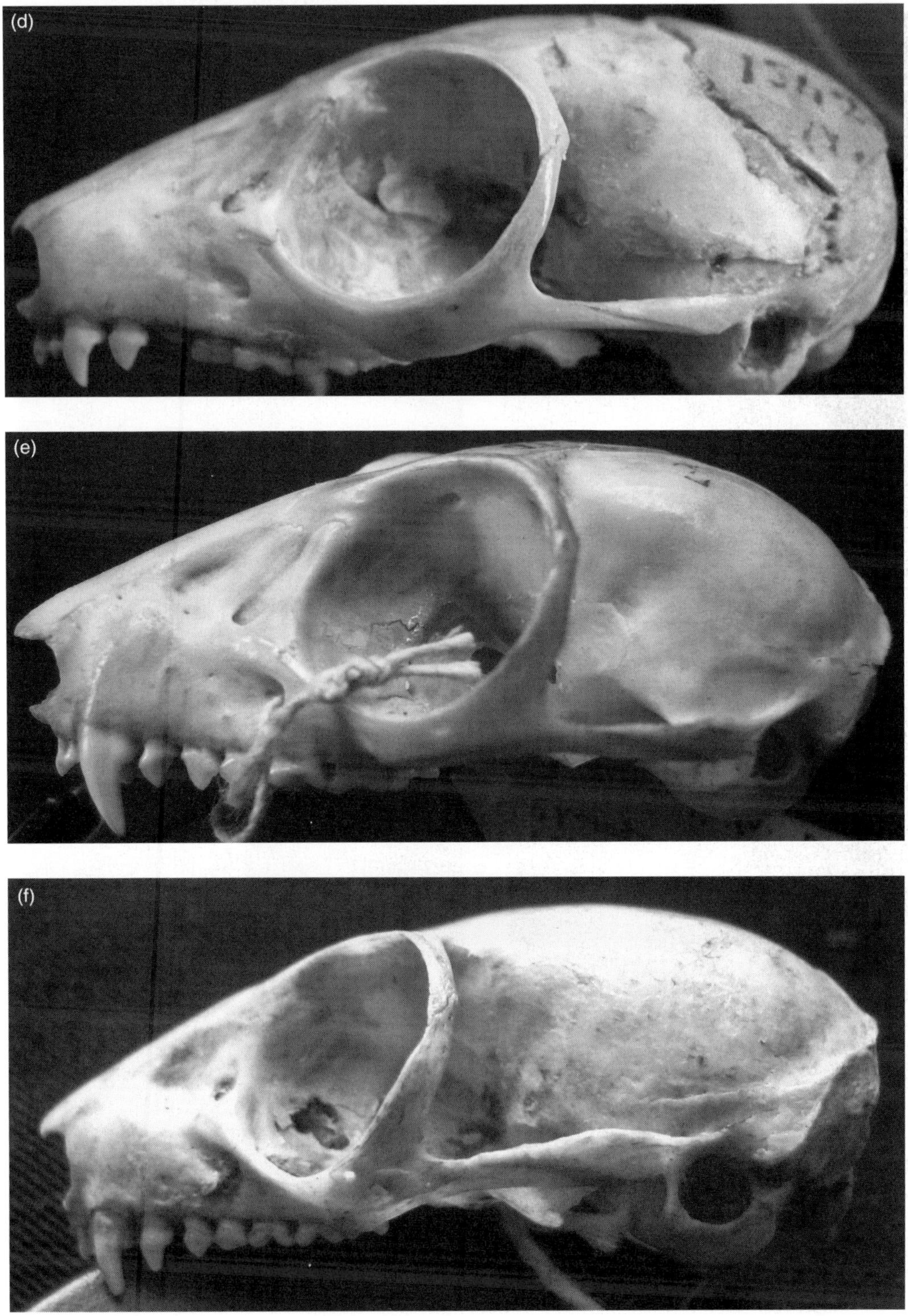

Figure 2.4 (*cont.*) (d) *C. crossleyi* (type of *C. melanotis* Forsyth Major, 1894, BM(NH) 70.5.5.24); (e) *C. sibreei* Forsyth Major, 1896 (97.9.1.160); (f) *C.* cf. *medius* (not a type).

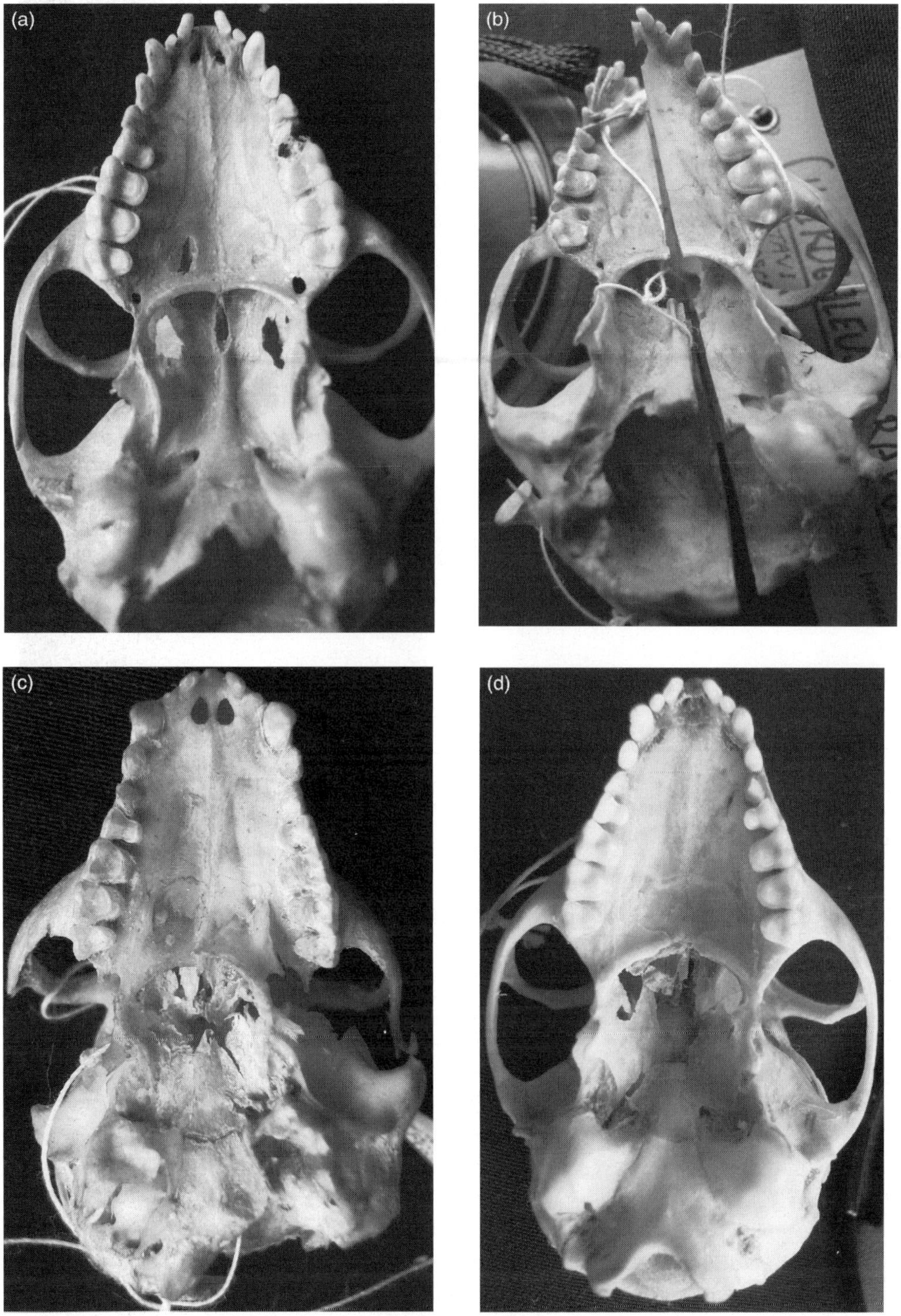

Figure 2.5 Maxillary dentitions of type specimens of *Cheirogaleus* spp.: (a) *C.* cf. *major* (type of *C. typicus* Smith, 1833); (b) *C. ravus* Groves, 2000 (BM(NH) 88.2.18.3); (c) *C. minusculus* Groves, 2000 (BM(NH) 11.6.21.1); (d) *C. crossleyi* (type of *C. melanotis* Forsyth Major, 1894, BM(NH) 70.5.5.24);

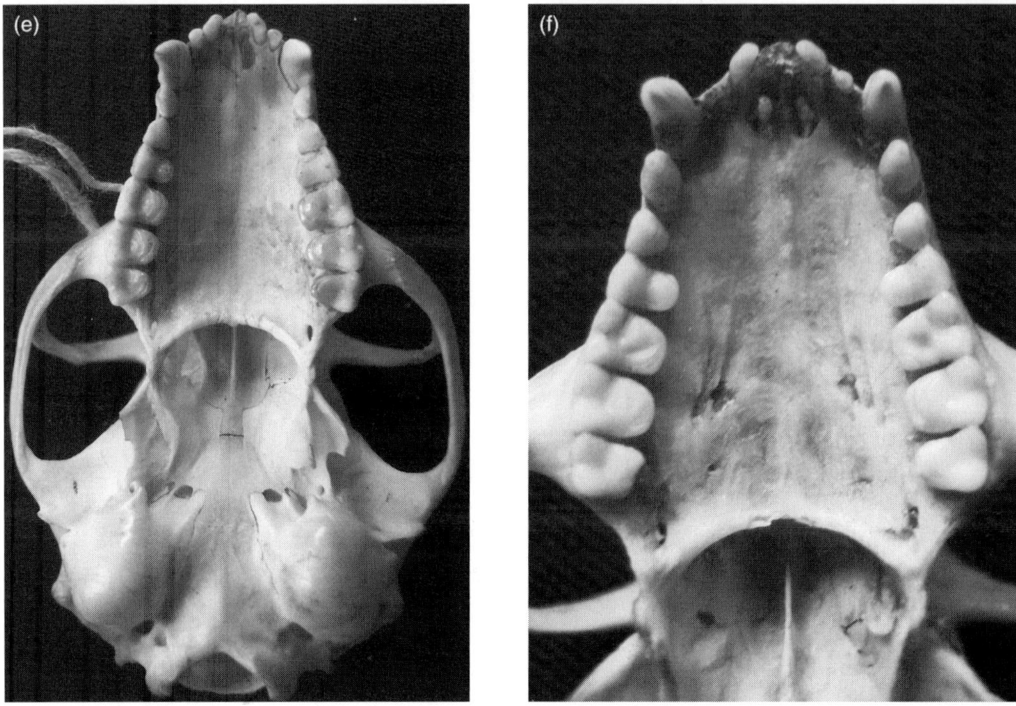

Figure 2.5 (*cont.*) (e) *C. sibreei* Forsyth Major, 1896 (97.9.1.160); (f) *C.* cf. *medius* (BM(NH) 48.166, not a type).

Despite the authors' acknowledgement that there is potential species-level differentiation within their three clades, their paper has been generally understood to have whittled the number of species of *Cheirogaleus* from seven down to three; indeed, the Abstract states clearly – and quite contrary to their statement quoted above – that "there are no clear subdivisions that would warrant additional specific status" and they even seemed to suggest, in a discussion at the end, that there is no need to worry about conservation if there are only three species. Inspection of their figure 2 shows that four of the six terminal clades of Clade A are fully resolved or at 99% posterior probability, the fourth at 95%, and only one (containing a single specimen) is equivocal; all three terminal clades of Clade B and all five of Clade C are similarly fully resolved or at 99%. Moreover, each one of these clades is geographically clustered. This was noticed by Mittermeier *et al.* (2010), who opined that it appears likely that more species will be recognised in future.

My recent study of the collections in the US museums has given me a greater appreciation of variability, and restudy of the Leiden collection in this light suggests that the specimens in Leiden attributed to *C. ravus* are probably in fact *C. major*, and those attributed to *C. sibreei* are not of that species but are similar (but not identical) to *C. medius* and *major*. Thus, Groeneveld *et al.* (2009) were absolutely correct in identifying one of the Leiden "*sibreei*" as *C. major*.

A later study by Groeneveld *et al.* (2010a) reported morphometrics of *Cheirogaleus* external features, skulls and teeth. Craniodental characters were entered into two principal component analyses, and external characters into a third, although only one of these analyses was described in detail and illustrated (Figure 2.3). In this, the specimens formed three major scatters, more or less corresponding to what Groeneveld *et al.* (2009) called *C. medius*, *C. major* and *C. crossleyi*, with two specimens identified as *C. sibreei* (one of them being the type), as well as some unidentified specimens, falling outside of any cluster. Noticeably, most of the *medius* specimens can be seen to form a restricted cluster just in one part of their scatter, leaving just a few, plus a number of unidentified specimens (as well as, interestingly, the type of *C. minusculus*), plotted well away from this main cluster. As in the genetic analysis (Groeneveld *et al.*, 2009), specimens of *C. crossleyi* were found that resembled *C. major*, perhaps suggesting some hybridisation. This is true also in my own data set (Figure 2.1). Unfortunately, the type of *C. ravus* was not included in their analyses, in either craniodental or external measurements; one of the distinguishing features of this species is that the tail is equal to, not longer than, the head and body, but curiously the body measurements cells for this species in table S2a are empty.

This paper also gave, in the Supporting Information, two tables (tables S2 a and b) of statistics for the species as recognised by Groves (2000) and by the authors of the paper. A very large number of the CVs in both tables approach or are above 10 in all three sizeable species hypodigms, implying very strongly that there are indeed multiple species within each of them.

The first of the species sunk by Groeneveld *et al.* (2009, 2010a) to be restored was *C. sibreei*. Blanco *et al.* (2009) discovered it living in sympatry with *C. crossleyi* in a highland forest at Tsinjoarivo, finding differences in genitalia and other characters between them, and corroborating their provisional identification by comparing the skull of a specimen which died in captivity with the type skull. Groeneveld *et al.* (2010b) found that the mtDNA and two nDNA sequences of *C. sibreei* are at least as different from the other three species as these are from each other. *C. sibreei*, thus, is confirmed from only two localities, Andasivodihazo (Tsinjoarivo, 1660 m, 19°41′S, 47°40.6′E) and Ankeramadinika (1408 m), the type locality; these are some 100 km apart. *C. crossleyi* occurs at Vatateza 10 km southeast of Andasivodihazo, and also in the Tsinjoarivo district, at 1396 m (Groeneveld *et al.*, 2010b). The third potential locality, Imerina, remains to be investigated.

The search for the true species diversity within *Cheirogaleus* continues. Thiele *et al.* (2013) obtained plentiful new samples and used the same mitochondrial sequence and the same three nuclear sequences as did Groeneveld *et al.* (2009), finding deep divisions within all three major clades (very deep within *medius* and *crossleyi*, less so within *major*). They also found that *C. sibreei* is probably sister to all others, and that *medius* is sister to *major* + *crossleyi*. The most significant finding, however, was that there are indeed further species to be recognised, at least within *medius* and *crossleyi*. Given that their locality sample CmeD would represent *C. medius*, they proposed designating samples CmeA + CmeB as *Cheirogaleus* sp.

Bekaraoka Sambava, and CmeC as *Cheirogaleus* sp. Ambanja. Again, given that their locality samples CcrA and CcrB are likely to represent *Cheirogaleus crossleyi sensu stricto*, they proposed provisionally designating CcrC as *Cheirogaleus* sp. Ranomafana Andrambovato, while CcrD they felt able to describe at once, designating it *Cheirogaleus lavasoensis*, with type locality Petit Lavasoa; the entire range appears restricted to the southern slopes of the Lavasoa Mountains, northwest of Ft. Dauphin.

For interest and potential future reference, I here figure skulls and maxillary dentitions of (mainly) type specimens in *Cheirogaleus* (Figures 2.4 and 2.5), and give tables of external, cranial and dental measurements (Tables 2.1, 2.2 and 2.3).

Table 2.1 Univariate and bivariate statistics in the *C. major* group. a, external measurements; b, relative tail length; c, cranial measurements; d, maxillary molar lengths.

a

		hb	tail	hf	ear
Antongil	Mean	248.00	295.00	51.333	23.000
	N	3	3	3	3
	SD	17.088	21.794	8.1445	.0000
Tamatave	Mean	240.00	240.00		
	N	1	1		
Ft. Dauphin	Mean	257.00	284.50	58.500	25.000
	N	2	2	2	2

b

	Mean	*N*	SD
Antongil	118.97	3	4.451
Tamatave	100.00	1	.
Ft. Dauphin	110.77	2	2.519

c

		gtl	basl	cbl	teeth	biorb
Antongil	Mean	57.150	50.200	54.300	20.850	37.025
	N	4	3	3	4	4
	SD	2.7331	2.8583	3.1607	1.4663	1.0532
Tamatave	Mean	54.500	44.325	49.567	19.900	33.800
	N	4	4	3	5	5
	SD	4.0158	4.5705	5.3003	1.5460	3.3106
Ft. Dauphin	Mean	54.400	48.100	52.500	16.700	35.600
	N	1	1	1	1	1

Table 2.1 (*cont.*)

d

		m1	mt2	mt3
Antongil	Mean	4.100	4.225	3.600
	N	4	4	4
	SD	.1633	.2500	.0816
Tamatave	Mean	4.120	4.160	3.600
	N	5	5	5
	SD	.4207	.4722	.4062
Ft. Dauphin	Mean	3.700	3.400	3.300
	N	1	1	1
	SD	.	.	.

Notes: hb, head+body length; tail, tail length; hf, hindfoot length; ear, ear length; gtl, greatest skull length; basl, basal skull length; cbl, condylobasal length; teeth, maxillary toothrow length; biorb, biorbital width.

Table 2.2. Univariate and bivariate statistics in the *C. medius* group. a, external measurements and relative tail length; b, cranial measurements; c, maxillary molar lengths.

a

		hb	tail	reltail	hf	ear
Ampijoroa	Mean	187.33	190.67	103.35	54.667	20.333
	N	3	3	3	3	3
	SD	24.826	9.452	17.837	28.8675	2.8868
Morondava	Mean	180.15	177.08	98.44	32.308	17.192
	N	13	13	13	13	13
	SD	8.009	10.484	6.918	6.8482	1.3156
Ft. Dauphin	Mean	197.50	202.50	102.53	42.500	21.500
	N	2	2	2	2	2

b

		gtl	basl	cbl	biorb	teeth	bullal	bullabr	naspal
Ampijoroa	Mean	43.300	37.400	40.850	28.933	14.867	7.850	6.300	9.300
	N	2	2	2	3	3	2	2	2
	SD				2.5482	1.9757			
Morondava	Mean	40.200	34.539	37.745	25.948	14.157	8.769	5.697	8.726
	N	29	28	29	29	21	29	29	19
	SD	1.4018	1.2288	1.4176	.9808	.3723	.3771	.3550	.5801
Ft. Dauphin	Mean	41.500	35.750	36.500	26.750	13.100	7.100	6.250	9.700
	N	2	2	2	2	2	1	2	2

(*continued*)

Table 2.2 (*cont.*)

c

		m 1	m 2	m 3
Ampijoroa	Mean	2.950	3.125	2.650
	N	4	4	4
	SD	.3109	.2754	.2380
Morondava	Mean	2.722	2.844	2.419
	N	27	27	27
	SD	.1368	.1649	.1415
Ft. Dauphin	Mean	3.200	2.750	2.300
	N	2	2	1

Notes: hb, head+body length; tail, tail length; hf, hindfoot length; ear, ear length; gtl, greatest skull length; basl, basal skull length; cbl, condylobasal length; teeth, maxillary toothrow length; biorb, biorbital width; reltail, tail length relative to head+body; bullal, auditory bulla length; bullabr, auditory bullae breadth; naspal, perpendicular height from palate to nasion.

Table 2.3 Univariate and bivariate statistics in the *C. crossleyi* group. a, cranial measurement; b. maxillary molar lengths.

a

		gtl	basl	cbl	teeth	biorb
Mt. d'Ambre	Mean	47.600		43.700	16.000	28.900
	N	1		1	1	1
Vohima	Mean	50.400	43.250	47.750	17.650	31.275
	N	4	4	4	4	4
	SD	1.5727	2.7012	2.2752	1.4295	1.5174
	Minimum	49.0	39.7	45.5	15.6	29.3
	Maximum	52.0	46.2	50.3	18.9	33.0
Alaotra	Mean	53.575	46.400	50.767	19.350	33.750
	N	4	3	3	4	4
	SD	2.0304	1.8248	2.2811	1.6921	1.7861
	Minimum	52.0	45.2	49.4	17.3	31.9
	Maximum	56.4	48.5	53.4	21.1	36.1
Perinet	Mean	53.075	45.275	49.550	16.400	33.325
	N	4	4	4	5	4
	SD	5.3873	5.1610	5.2246	2.3969	4.1612
	Minimum	45.2	37.8	41.9	12.5	27.3
	Maximum	56.7	49.0	53.0	18.8	36.8
Ft. Dauphin	Mean	55.900	49.500	51.900	17.800	36.000
	N	1	1	1	1	1

Table 2.3 *(cont.)*

		bullal	bullabr	brainht	naspal
Mt. d'Ambre	Mean	7.000	7.000	16.800	9.700
	N	1	1	1	1
Vohima	Mean	8.450	6.825	17.375	10.625
	N	4	4	4	4
	SD	.3317	.3500	1.6091	1.4245
	Minimum	8.0	6.4	15.8	9.1
	Maximum	8.7	7.2	19.0	12.2
Alaotra	Mean	8.800	7.675	17.325	12.375
	N	4	4	4	4
	SD	.4243	.2363	.5679	1.1615
	Minimum	8.4	7.5	16.5	11.4
	Maximum	9.3	8.0	17.8	14.0
Perinet	Mean	8.675	7.550	17.300	11.600
	N	4	4	4	4
	SD	.4992	.9110	.7165	1.6186
	Minimum	8.0	6.2	16.3	9.2
	Maximum	9.2	8.1	18.0	12.7
Ft. Dauphin	Mean	9.000	6.200	17.300	12.000
	N	1	1	1	1
	SD	.	.	.	.
	Minimum	9.0	6.2	17.3	12.0
	Maximum	9.0	6.2	17.3	12.0

b

		m1	m2	m3
Mt. d'Ambre	Mean	3.700	3.700	3.200
	N	1	1	1
Vohima	Mean	3.525	3.700	3.225
	N	4	4	4
	SD	.1258	.0816	.1258
	Minimum	3.4	3.6	3.1
	Maximum	3.7	3.8	3.4
Alaotra	Mean	3.950	4.000	3.675
	N	4	4	4
	SD	.3873	.3916	.1893
	Minimum	3.5	3.6	3.4
	Maximum	4.4	4.5	3.8
Perinet	Mean	3.360	3.480	3.200
	N	5	5	5
	SD	.4336	.3271	.3082
	Minimum	2.8	3.0	3.0
	Maximum	3.8	3.8	3.7

(continued)

Table 2.3 (*cont.*)

b

		m1	m2	m3
Ft. Dauphin	Mean	3.500	3.800	3.400
	N	1	1	1

Notes: gtl, greatest skull length; basl, basal skull length; cbl, condylobasal length; teeth, maxillary toothrow length; biorb, biorbital width; bullal, auditory bulla length; bullabr, auditory bullae breadth; naspal, perpendicular height from palate to nasion; brainht, braincase height (basion to vertex).

A preliminary look at metrical diversity in *Cheirogaleus*

When divided into species-groups and then into locality groups, the number of skulls and skins available for morphological assessment in *Cheirogaleus* is still very small: too small for discriminant analysis, so here I present three principal components analyses. Principal components analysis (PCA) is an exploratory method, so the analyses presented here should be regarded as giving some guidance only. As all such methods use intercorrelations (or covariances) between the raw measurements to create new, uncorrelated (orthogonal) variates, it is necessary to separate variables based on growth (here, craniometric variables) from those not based on growth (here, tooth measurements). In the present case, only cranial measurements are used in the PCA.

As discovered for DNA by Thiele *et al.* (2013), divisions within the *C. major* group are less than within the other two groups (Figure 2.6a). As far as the *C. medius* group goes, there is slightly more evidence for diversity: specimens from the Ampijoroa and Toliara regions are well separated, largely by size (bulla length being negatively correlated with size), from the large cluster from the Morondava region, and Toliara has a lower-crowned skull. Here, the Morondava group corresponds to group CmeD of Hapke *et al.* (2011); in the available skulls sampled, there are none corresponding to the other DNA groups, while Hapke *et al.* (2011) had no specimens from the Ampijoroa or Toliara regions. There are also considerable separations within the *C. crossleyi* group; there is a gradient of increasing size from Mt. d'Ambre (Hapke *et al.*'s CcrA) to Vohima to the others, and within the others Ft. Dauphin (presumably group CcrD, *C. lavasoensis*) is distinguished from the Alaotra and Perinet groups (CcrB) by its higher braincase.

Table 2.1 gives the statistics for the raw measurements. In the *C. major* group, the cranial narrowness and the short tail of the Tamatave sample (*C. ravus*) stand out, as do the small teeth of the Ft. Dauphin sample. In the *C. medius* group, there are some odd differences in hindfoot length, which could be real or could be due to different measurement methods; and Ft. Dauphin has a short toothrow. In the *C. crossleyi* group, we see the size difference pointed out before, and the large posterior molars of the Alaotra group.

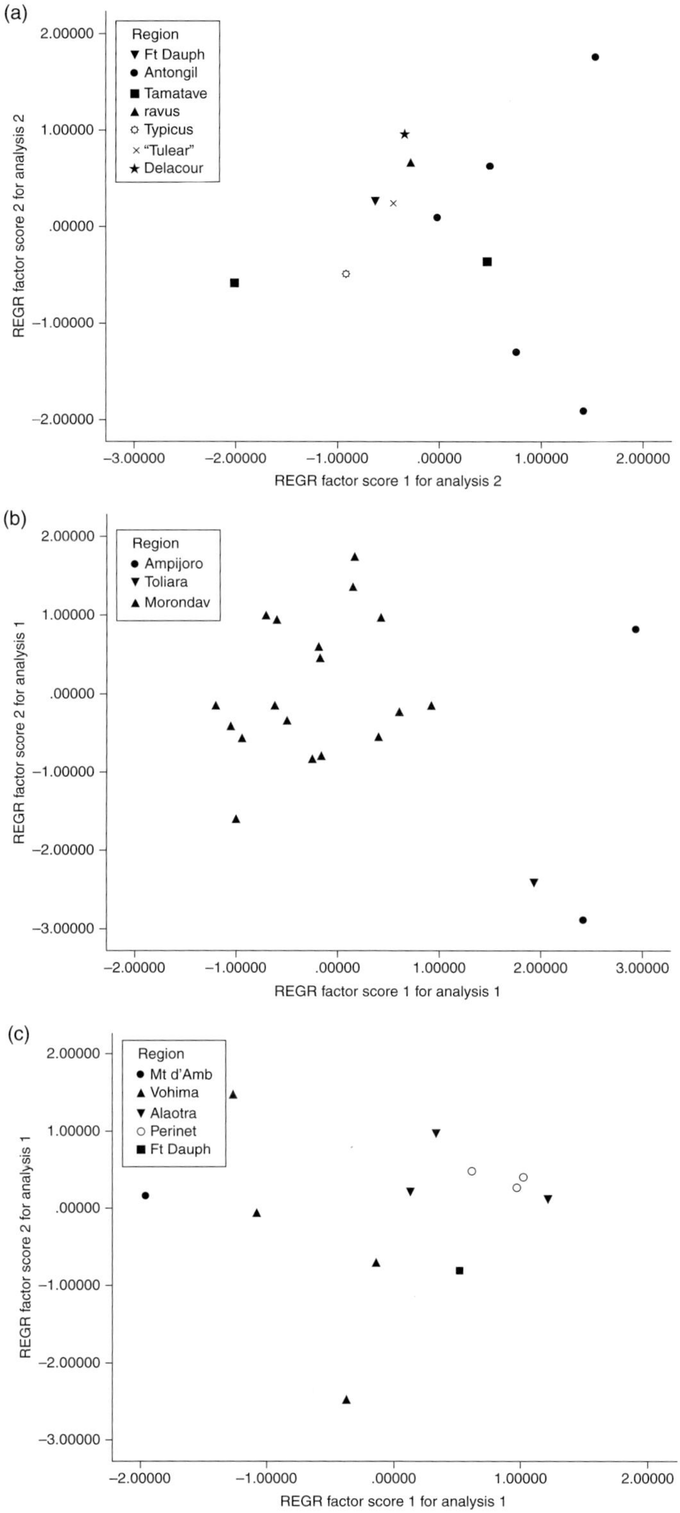

Figure 2.6 Principal components analyses: (a) *C. major* group; (b) *C. medius* group; (c) *C. crossleyi* group.

Table 2.4 *Cheirogaleus* principal component analyses. a, *C. major* group; b, *C. medius* group; c, *C. crossleyi* group.

a

Total variance explained

Component	Initial eigenvalues			Extraction sums of squared loadings		
	Total	% of Variance	Cumulative %	Total	% of Variance	Cumulative %
1	3.432	57.192	57.192	3.432	57.192	57.192
2	1.264	21.072	78.264	1.264	21.072	78.264

Component matrix[a]

	Component	
	1	2
gtl	.859	−.061
biorb	.885	.096
bullal	.265	.904
bullabr	.808	.331
brainht	.735	−.543
naspal	.805	−.174

[a] Two components extracted.

b

Total variance explained

Component	Initial eigenvalues			Extraction sums of squared loadings		
	Total	% of Variance	Cumulative %	Total	% of Variance	Cumulative %
1	4.900	61.251	61.251	4.900	61.251	61.251
2	1.375	17.188	78.439	1.375	17.188	78.439

Component matrix[a]

	Component	
	1	2
gtl	.942	.202
biorb	.892	.260
bullal	−.264	.719
bullabr	.725	−.090
brainht	.589	−.559
naspal	.778	−.496
basl	.890	.321
cbl	.936	.281

[a] Two components extracted.

Table 2.4 (*cont.*)

c

Total variance explained

Component	Initial eigenvalues			Extraction sums of squared loadings		
	Total	% of Variance	Cumulative %	Total	% of Variance	Cumulative %
1	4.421	63.164	63.164	4.421	63.164	63.164
2	1.237	17.666	80.830	1.237	17.666	80.830

Component matrix[a]

	Component	
	1	2
gtl	.974	-.012
biorb	.937	-.120
bullal	.838	-.100
bullabr	.572	.548
brainht	.481	-.668
naspal	.627	.652
cbl	.969	-.199

[a] Two components extracted.

Larger samples may be available in the Tananarive collection (and just possibly could be obtained from owl pellets?), and if so then the powerful method of discriminant analysis may become possible.

Phaner

The first fork-crowned lemur was described by de Blainville (1841) as *Lemur furcifer*, and it was thereafter transferred to *Cheirogaleus* and then to *Lepilemur* before being awarded a genus of its own by Gray (1870), but Mivart (1873) referred it back to *Cheirogaleus* and Forsyth Major (1894) transferred it to *Microcebus*. It was Schwarz (1931) who first restored its generic separation. Rumpler (1974) proposed that it should be placed in a separate, monotypic subfamily Phanerinae, citing features of the karyotype, dermatoglyphics, glands and behaviour. Groves (1974) endorsed this, and noted in addition the absence of an os planum in the orbit, the absence of the stapedial artery, and differences in the foot structure.

Potential geographic variation was briefly noted by Petter *et al.* (1977), but not seriously studied until Groves and Tattersall (1991), who recognised four subspecies. These are strikingly and consistently distinct and were raised to species rank by Groves (2001).

P. furcifer, from the Masoala Peninsula, is darker and browner dorsally, and more yellowish ventrally, than others, with extremely well-marked head forks; the dorsal stripe does not reach the base of the tail; the tail is darkened only on the terminal third, at most; the extremities are dark. To judge from incomplete material, this is evidently large in size, although smaller than *P. electromontis*.

P. pallescens, from the southwest, approximately between the Fiherenana and Tsiribihina Rivers, is pale grey–fawn with relatively poorly defined forks and dorsal stripe, although the latter reaches the rump; the tail is dark for at least its distal half; the extremities are only slightly darker. The smallest species, the tail is much longer than the head and body.

P. parienti, from the Sambirano, is medium brown dorsally and strongly buffy ventrally, with well-marked head forks and dorsal stripe, which extends to the base of the tail; only the distal third of tail is dark, and there may be a white tail tip. Slightly larger than *P. pallescens*. The tail is apparently longer than the head and body. The teeth are large compared to the skull.

P. electromontis, from Ambohitra (Mt. d'Ambre), is relatively light with well-defined forks and dorsal stripe, the latter extending onto the rump; the extremities are only slightly darker; the tail is dark only on the distal third. Size is larger than the others, and the tail is considerably longer than the head and body.

Groves and Tattersall (1991) suggested that other taxa would probably prove to be recognisable. For example, a specimen from Namoroka was provisionally included within *P. pallescens* but seems to be larger in size and the tail is not longer than the head and body; a specimen from Vohémar is extremely large in size; and animals from the south of Madagascar remain to be described.

Conclusions

Our understanding of cheirogaleid taxonomy is still far from complete; so far, only *Microcebus* has been to any degree well served by both morphological and molecular analysis, although it seems not unlikely that more remains to be discovered even about this now multispecies complex. Outstanding problems include the following.

Are there further species of *Microcebus* yet to be discovered? At least one of the known species turns out to be a montane specialist – are there other habitat-specific taxa, so far undetected?

The species at present known as *Microcebus rufus* needs to be called something else. It may prove to be the same as John Edward Gray's *Azema smithii* – or it may not. At any rate, it is curious that one of the (genetically) best-known species of mouse lemur may be without a name!

Should *Microcebus* be divided into two or more genera? The identity of *Azema smithii* is crucial to a satisfactory resolution of this question.

Have we exhausted the diversity within *Mirza*?

Might there be taxonomic diversity within *Allocebus*?

In the light of Thalmann (2000) and Thiele *et al.* (2013), how many species remain to be identified in *Cheirogaleus*?

What is the taxonomic status of the "mystery" populations of *Phaner*?

The documentation of taxonomic diversity is of the highest importance for conservation. Small-bodied lemurs like dwarf and mouse lemurs might be expected to show more diversity than larger, more mobile forms like lemurids and indriids, and so to demarcate smaller, more specific areas. Without accurate taxonomic knowledge, such special areas could be lost, and with them whole species. If a small area contains a special species of *Microcebus*, should we look for special species of tenrec, or chameleon, or gecko in that particular area as well? At the same time, the identification of these small geographic units will add to our understanding of the biogeography of Madagascar as a whole. As McNeely (2006) argued, the systems and species approaches to conservation are tightly integrated, and feed off each other, and we must not neglect either.

Note added in proof

After the above was written, the outline of a new classification of *Cheirogaleus* has been published (Lei *et al.*, 2014, 2015). The four major species-groups were amply confirmed, and characterised morphologically. The description of several new species was anticipated, and one of these, *Cheirogaleus andysabini*, a species of the *C. crossleyi* group from the Mt. d'Ambre, has more recently been described.

References

Andriantompohavana R, Zaonarivelo JR, Engberg SE, *et al.* 2006. Mouse lemurs of north-western Madagascar with a description of a new species at Lokobe Special Reserve. *Occasional Papers Museum of Texas Tech University*, 259:1–24.

Blanco MB, Godfrey LR, Rakotondratsima M, *et al.* 2009. Discovery of sympatric dwarf lemur species in the high altitude forest of Tsinjoarivo, Eastern Madagascar: implications for biogeography and conservation. *Folia Primatologica* 80:1–17.

Brown P. 1776. *New Illustrations of Zoology, containing Fifty coloured Plates of new, curious and non-descript Birds, with a few Quadrupeds, Reptiles and Insects. Together with a short and Scientific Description of the same.* B. White, London.

Buffon, GL LeClerc, Compte de. 1776. *Histoire Naturelle, générale et particulière, avec la description du Cabinet du Roy.* 13. Imprimerie Royale, Paris.

Chatterjee HJ, Ho SYW, Barnes I, Groves C. 2009. Estimating the phylogeny and divergence times of primates using a supermatrix approach. *BMC Evolutionary Biology* 9:259.

Cotterill FPD, Taylor PJ, Gippoliti S, Bishop JM, Groves CP. 2014. Why one century of phenetics is enough: response to 'are there really twice as many bovid species as we thought?' *Systematic Biology* 63(5):819–832.

Cuozzo FP, Rasoazanabary E, Godfrey LR, *et al.* 2013. Biological variation in a large sample of mouse lemurs from Amboasary, Madagascar: implications for interpreting variation in primate biology and paleobiology. *Journal of Human Evolution* 64:1–20.

de Blainville HMD. 1841. Ostéographie des Primatès. Makis (*Lemur*). In *Ostéographie des Mammifères.* Baillière, Paris.

de Queiroz K. 2007. Species concepts and species delimitation. *Systematic Biology* 56:879–886.

Dunkel AR, Zijlstra JS, Groves C. 2012. Giant rabbits, marmosets, and British comedies: etymology of lemur names, part I. *Lemur News* 16:64–70.

Elliot DG. 1912. *Review of the Primates.* 3 volumes. American Museum of Natural History, New York.

Forbes HO. 1896. *A Hand-book to the Primates.* 2 volumes. Edward Lloyd Ltd, London.

Forsyth Major CI. 1894. Über die Malagassischen Lemuriden-Gattungen *Microcebus, Opolemur,* und *Chirogale. Novitates Zoologicae* 1:2–38.

Geoffroy Saint-Hilaire E. 1796. Mémoire sur les rapports naturels des Makis *Lemur,* L. et description d'une espèce nouvelle de Mammifère. *Magasin Encyclopétique* 1:20–50.

Geoffroy Saint-Hilaire E. 1812. Suite au Tableau des Quadrumanes. Second Famille. Lemuriens. Strepsirrhini. *Annales du Muséum d'Histoire Naturelle, Paris* 19:156–170.

Geoffroy Saint-Hilaire E. 1828. *Cours de l'Histoire Naturelle des Mammifères.* Muséum National d'Histoire Naturelle, Paris.

Grandidier A. 1867. Mammifères et oiseaux nouveaux découverts à Madagascar et décrits par M. Alfred Grandidier. *Revue et Magasin de Zoologie* 19:84–88.

Gray, J. E. 1842. Descriptions of some new genera and fifty unrecorded species of Mammalia. *Annals & Magazine of Natural History* 10:255–267.

Gray JE. 1870. *Catalogue of Monkeys, Lemurs and Fruit-eating Bats in the Collection of the British Museum.* British Museum Trustees, London.

Gray JE. 1872. Notes on *Propithecus, Indris,* and other lemurs (*Lemurina*) in the British Museum. *Proceedings of the Zoological Society of London* 1872:846–859.

Groeneveld LF, Weisrock DW, Rasoloarison RM, Yoder AD, Kappeler PM. 2009. Species delimitation in lemurs: multiple genetic loci reveal low levels of species diversity in the genus *Cheirogaleus. BMC Evolutionary Biology* 9:30.

Groeneveld LF, Rasoloarison RM, Kappeler PM. 2010a. Morphometrics confirm taxonomic deflation in dwarf lemurs (Primates: Cheirogaleidae), as suggested by genetics. *Zoological Journal of the Linnaean Society* 161:229–244.

Groeneveld LF, Blanco MB, Raharison J-L, *et al.* 2010b. MtDNA and nDNA corroborate existence of sympatric dwarf lemur species at Tsinjoarivo, eastern Madagascar. *Molecular Phylogenetics and Evolution* 55:833–845.

Groves CP. 1974. Taxonomy and phylogeny of Prosimians. In RD Martin, GA Doyle, AC Walker (eds.), *Prosimian Biology* (pp 449–473). Duckworth, London.

Groves CP. 1989. *A Theory of Human and Primate Evolution.* Oxford University Press, Oxford.

Groves CP. 2000. The genus *Cheirogaleus*: unrecognised biodiversity in dwarf lemurs. *International Journal of Primatology* 21:943–962.

Groves CP. 2001. *Primate Taxonomy.* Smithsonian Institution Press, Washington.

Groves CP, Tattersall I. 1991. Geographical variation in the Fork-marked Lemur, *Phaner furcifer* (Primates, Cheirogaleidae). *Folia Primatologica* 56:39–49.

Gunther A. 1875. Notes on some mammals from Madagascar. *Proceedings of the Zoological Society of London* 1875:78–80.

Hapke A, Fietz J, Nash SD, *et al.* 2005. Biogeography of dwarf lemurs: genetic evidence for unexpected patterns in southeastern Madagascar. *International Journal of Primatology* 26:873–901.

Hapke A, Gligor M, Rakotondranary SJ, Rosencrantz D, Zupke O. 2011. Hybridisation of mouse lemurs: different patterns under different ecological conditions. *BMC Evolutionary Biology* 11:297.

Harper F. 1940. The nomenclature and type localities of certain Old World mammals. Part 1. *Journal of Mammalogy* 41:191–203.

Heckman KL, Rasoazanabary E, Machlin E, Godfrey LR, Yoder AD. 2006. Incongruence between genetic and morphological diversity in *Microcebus griseorufus* of Beza Mahafaly. *BMC Evolutionary Biology* 6:98.

Hennig W. 1950. *Phylogenetic Systematics.* (English translation 1966.)

Hill WCO. 1953. *Primates, Comparative Anatomy and Taxonomy, I – Strepsirrhini, a Monograph.* Edinburgh University Press, Edinburgh.

Kappeler PM, Rasoloarison RM, Razafimanantsoa L, Walter L, Roos C. 2005. Morphology, behaviour and molecular evolution of giant mouse lemurs (*Mirza* spp.) Gray, 1870, with description of a new species. *Primate Report* 71:3–26.

Kollmann M. 1910. Note sur les genres *Chirogale* et *Microcebus*. *Bulletin du Museum National d'Histoire Naturelle* 16:301–304.

Lei R, Frasier CL, McLain AT, *et al.* 2014. Revision of Madagascar's dwarf lemurs (Cheirogaleidae: *Cheirogaleus*): designation of species, candidate species status and geographic boundaries based on molecular and morphological data. *Primate Conservation* 28:1–8.

Lei R, McLain AT, Frasier CL, *et al.* 2015. A new species in the genus *Cheirogaleus* (Cheirogaleidae). *Primate Conservation* 29:1–12.

Lesson R-P. 1829. Observations sur quelques Mammifères conserves dans le Cabinet de la Société de Zoologie, par N.A.Vigors and Thomas Horsfield (*Zoological Journal*; n° XIII, avril–juin 1828, p. 105). *Bulletin des Sciences Naturelles et de Géologie* 16:453–455.

Lesson R-P. 1840. *Species des Mammifères Bimanes et Quadrumanes; suivi d'un Mémoire sur les Oryctéropes*. J.-B. Baillière, Paris.

Louis EE, Coles, MS, Andriantompohavana R, *et al.* 2006. Revision of the mouse lemurs (*Microcebus*) of eastern Madagascar. *International Journal of Primatology* 27:347–389.

Louis, EE, Engberg SE, McGuire SM, *et al.* 2008. Revision of the mouse lemurs, *Microcebus* (Primates, Lemuriformes), of northern and northwestern Madagascar with descriptions of two new species at Montagne d'Ambre National Park and Antafondro Classified Forest. *Primate Conservation* 23:19–38.

Masters JC, Silvestro D, Génin F, DelPero M. 2013. Seeing the wood through the trees: the current state of higher systematics in the Strepsirhini. *Folia Primatologica* 84:201–219.

McNeely JA. 2006. Systems or species? Approaches to conservation for the 21st century. *Integrative Zoology* 2:86–95.

Meier B, Albignac R. 1991. Rediscovery of *Allocebus trichotis* Günther 1875 (primates) in Northeast Madagascar. *Folia Primatologica* 56:57–63.

Miller JF. 1777. *Various Subjects of Natural History wherein are delineated Birds, Animals, and many curious Plants: with the Parts of Fructification of each plant, all of which are drawn and coloured from Nature*. Letterpress, London.

Mittermeier RA, Louis EE, Richardson M, *et al.* 2010. *Lemurs of Madagascar*. Third edition. Conservation International, Washington, DC.

Mivart StG. 1873. On *Lepilemur* and *Cheirogaleus*, and on the zoological rank of the *Lemuroidea*. *Proceedings of the Zoological Society of London* 1873:484–510.

Olivieri G, Zimmermann E, Randrianambinina B, *et al.* 2007. The ever-increasing diversity in mouse lemurs: three new species in north and northwestern Madagascar. *Molecular Phylogenetics and Evolution* 43:309–327.

Pastorini J, Martin RD, Ehresmann P, Zimmermann E, Forstner MRJ. 2001. Molecular phylogeny of the lemur family (Primates) based on mitochondrial DNA sequences. *Molecular Phylogenetics and Evolution* 19:45–56.

Perelman P, Johnson WE, Roos C, *et al.* 2011. A molecular phylogeny of living primates. *PLoS Genetics* 7(3):e1001342.

Peters WCH. 1852. *Naturwissenschaftliche Reise nach Mozambique auf befehl Seiner Majestät des Königs Friedrich Wilhelm IV in den Jahren 1842 bis 1848 ausgeführt. Zoologie. I. Säugethiere*. G. Reimer, Berlin.

Petter J-J. 1962. Recherches sur l'écologie et l'éthologie des lémuriens malgaches. *Mémoires du Muséum National d'Histoire Naturelle N.S., ser. A, Zoologie* 27:1–146.

Petter J-J. 1965. The lemurs of Madagascar. In I DeVore (ed.), *Primate Behaviour: Field Studies of Monkeys and Apes* (pp. 292–319). Holt, Rinehart and Winston, New York.

Petter J-J, Albignac R, Rumpler Y. 1977. *Faune de Madagascar, 44: Mammifères Lémuriens (Primates Prosimiens)*. ORSTOM/CNRS, Paris.

Petter-Rousseaux A, Petter J-J. 1967. Contribution à la systematique des *Cheirogaleinae* (lemuriens malgaches), *Allocebus*, gen.nov., pour *Cheirogaleus trichotis* Günther 1875. *Mammalia* 31:574–582.

Radespiel U, Olivieri G, Rasolofoson DW, *et al.* 2008. Exceptional diversity of mouse lemurs (*Microcebus* spp.) in the Makira region with the description of one new species. *American Journal of Primatology* 70:1033–1046.

Radespiel U, Ratsimbazafy JH, Rasoloharijaona S, *et al.* 2012. First indications of a highland specialist among mouse lemurs (*Microcebus* spp.) and evidence for a new mouse lemur species from eastern Madagascar. *Primates* 53:157–170.

Rakotoarison N, Zimmermann H, Zimmermann E. 1996. First discovery of the hairy-eared dwarf lemur (*Allocebus trichotis*) in a highland rain forest of eastern Madagascar. *Folia Primatologica* 68:86–94.

Rasoloarison RM, Goodman SM, Ganzhorn JU. 2000. Taxonomic revision of mouse lemurs (*Microcebus*) in the western portions of Madagascar. *International Journal of Primatology* 21:963–1019.

Rasoloarison RM, Weisrock DW, Yoder AD, Rakotondravony D, Kappeler PM. 2013. Two new species of mouse lemurs (Cheirogaleidae: *Microcebus*) from Eastern Madagascar. *International Journal of Primatology* 34:455–469.

Rumpler Y. 1974. Cytogenetic contributions to a new classification of lemurs. In RD Martin, GA Doyle, AC Walker (eds.), *Prosimian Biology* (pp. 865–869). Duckworth, London.

Saban R. 1963. Contribution à l'étude de l'os temporal des Primates. *Mémoires du Muséum National d'Histoire Naturelle N.S., ser. A, Zoologie*, 29:1–377.

Schmid J, Kappeler PM. 1994. Sympatric mouse lemurs (*Microcebus* spp.) in western Madagascar. *Folia Primatologica* 63:162–170.

Schwartz JH, Tattersall I. 1985. Evolutionary relationships of living lemurs and lorises (Mammalia, Primates) and their potential affinities with European Eocene Adapidae. *Anthropological Papers of the American Museum of Natural History New York*, 60, part 1.

Schwarz E. 1931. A revision of the genera and species of Madagascar Lemuridae. *Proceedings of the Zoological Society of London* 1931:399–426.

Simpson GG, Roe A, Lewontin RC. 1960. *Quantitative Zoology*. Harcourt, Brace & Co., Inc., New York.

Szalay FS, Katz CC. 1973. Phylogeny of lemurs, galagos and lorises. *Folia Primatologica* 19:88–103.

Tattersall I. 1982. *The Primates of Madagascar*. Columbia University Press, New York.

Tattersall I, Schwartz JH. 1974. Craniodental morphology and the systematics of the Malagasy lemurs (Primates, Prosimii). *Anthropological Papers of the American Museum of Natural History New York* 52, part 3.

Thalmann U. 2000. Greater dwarf lemurs from the Bongolava (central western Madagascar). *Lemur News* 5:33–35.

Thiele D, Razafimahatratra E, Hapke A. 2013. Discrepant partitioning of genetic diversity in mouse lemurs and dwarf lemurs – biological reality or taxonomic bias? *Molecular Phylogenetics and Evolution* 69:593–609.

Vigors NA, Horsfield T. 1828. Observations on some of the Mammalia contained in the Museum of the Zoological Society. *Zoological Journal* 4:105–113.

Vigors NA, Horsfield T. 1829. Notes respecting some species of Mammalia referred to by Mr. Vigors and Dr. Horsfield in the XIIIth No. of this Journal. *Zoological Journal* 5:134–141.

Wagner JA. 1840. Die Säugthiere in Abbildungen nach der Natur, mit Beschreibungen von D. Johann Christian Daniel von Schreber, Fortgesetzt von Dr. Johann Andreas Wagner. Supplementband 1. Erlangen: Expedition des Schreber'schen Säugthier- und des Esper'schen Schmetterlingswerkes.

Yablokov AV. 1966. *Variability of Mammals*. Amerind, New Delhi.

Yoder AD. 1992. The applications and limitations of ontogenetic comparisons for phylogeny reconstruction: the case of the strepsirrhine internal carotid artery. *Journal of Human Evolution* 23:183–195.

Yoder AD. 1994. Relative position of the Cheirogaleidae in strepsirrhine phylogeny: a comparison of morphological and molecular methods and results. *American Journal of Physical Anthropology* 94:25–46.

Yoder AD, Heckman KL. 2006. Mouse lemur phylogeography revises a model of ecogeographic constraint in Madagascar. In SM Lehman and JG Fleagle (eds.), *Primate Biogeography* (pp. 255–268). Springer, New York.

Yoder AD, Rasoloarison RM, Goodman SM, *et al.* 2000. Remarkable species diversity in Malagasy mouse lemurs (Primates, *Microcebus*). *Proceedings of the National Academy of Sciences, USA* 97:11325–11330.

Yoder AD, Burns MM, Génin F. 2002. Molecular evidence of reproductive isolation in sympatric sibling species of mouse lemurs. *International Journal of Primatology* 23:1335–1343.

Zimmermann E, Cepok S, Rakotoarison N, Zietemann V, Radespiel U. 1998. Sympatric mouse lemurs in north-west Madagascar: a new rufous mouse lemur species (*Microcebus ravelobensis*). *Folia Primatologica* 69:106–114.

3 Mitogenomics of the family Cheirogaleidae and relationships to taxonomy and biogeography in Madagascar

Edward E. Louis, Jr. and Runhua Lei

Introduction

Due to its remarkable percentage of endemic species and the continuing pressure from human encroachment and poverty, Madagascar has been listed as one of the world's major biodiversity hotspots (Myers, 2000). The Malagasy strepsirrhine primates, the lemurs, are particularly susceptible to extinction due to anthropogenic activities (Jernvall and Wright, 1998). Taxonomic revisions and ensuing shifts in geographic distributions warrant the need to consistently re-evaluate the conservation status of lemur species as new information becomes available (Martin, 2000). In less than three decades, the number of described lemur species has grown dramatically, from 36 species in 1982 (Tattersall, 1982) to more than 100 species now (Tattersall, 2013; Thiele *et al.*, 2013). This taxonomic increase, especially in the family Cheirogaleidae, is due to advancements in molecular genetic technology, as well as improvements in analytical tools and intensive field investigation (Tattersall, 2007; Thalmann, 2007; Mittermeier *et al.*, 2008; Lei *et al.*, 2012). Conservation decisions and priorities need to be regularly revised due to drastic habitat loss and changes in species distribution and abundances. This is especially true in the family Cheirogaleidae, which has been shown to harbor a large amount of cryptic species diversity.

Historical summary of Cheirogaleidae taxonomy and phylogeny

Cheirogaleidae (Gray, 1873) includes five genera: *Microcebus*, *Mirza*, *Allocebus*, *Cheirogaleus*, and *Phaner*, ranging in size from about 30 g to 600 g (Mittermeier *et al.*, 2010). All cheirogaleid lemurs are nocturnal, move quadrupedally, and have elongated bodies and short legs. This family has experienced a dramatic increase in the number of recognized species, primarily within the genus *Microcebus* (Mittermeier *et al.*, 2008). Initially classified as a single species, *M. murinus* (Miller 1777) was divided into two regionally defined species, *M. murinus* in the west

The Dwarf and Mouse Lemurs of Madagascar: Biology, Behavior and Conservation Biogeography of the Cheirogaleidae, eds. S.M. Lehman, U. Radespiel, E. Zimmermann. Published by Cambridge University Press. © Cambridge University Press 2016.

and *M. rufus* in the east (Tattersall, 1982). In the last two decades, the genus *Microcebus* has expanded from 2 to 21 species based primarily on mitochondrial DNA (mtDNA) fragment sequences and morphological data (Schmid and Kappeler, 1994; Zimmermann *et al.*, 1998; Rasoloarison *et al.*, 2000, 2013; Kappeler *et al.*, 2005; Andriantompohavana *et al.*, 2006; Louis *et al.*, 2006a, 2008; Olivieri *et al.*, 2007; Radespiel *et al.*, 2008, 2012).

The family Cheirogaleidae includes two regionally distributed genera, *Mirza* and *Allocebus*, in western and eastern Madagascar, respectively. Kappeler *et al.* (2005) described one new species in the genus *Mirza*, *M. zaza*, in northwestern Madagascar. The genus *Allocebus* contains a single species, *A. trichotis*, distributed along the eastern coast of Madagascar from the Mangoro River to Marojejy (Markolf *et al.*, 2011). In *Phaner*, Groves and Tattersall (1991) initially recognized four subspecies based on morphological data. Groves (2001) subsequently elevated these four subspecies to full species status.

Although closely related and ecologically similar to *Microcebus*, no new species of *Cheirogaleus* had been described for more than a century until Groves' (2000) revision of the genus in which seven species were accepted based on morphological analyses of museum specimens. Utilizing genetic and morphological data, Groeneveld *et al.* (2009, 2011) afterwards rejected two of these species, *C. adipicaudatus* and *C. ravus*, as being synonymous with *C. medius* and *C. major*, respectively. Although *C. medius* and *C. crossleyi* are represented by extremely divergent mitochondrial haplotypes, Groeneveld *et al.* (2009, 2011) characterized this diversity as being intraspecific. Furthermore, Groeneveld *et al.* (2010) utilized mtDNA and nuclear DNA sequence data to verify the species *C. sibreei*, discovered in Tsinjoarivo, 80 km south-southeast of Antananarivo (Blanco *et al.*, 2009). Recently, Thiele *et al.* (2013) described a new species, *C. lavasoensis*, based on genetic and morphological data. However, several authors have questioned whether this dramatic increase in lemur diversity is a case of taxonomic inflation (Tattersall, 2007, 2013; Thalmann, 2007; Groeneveld *et al.*, 2009; Markolf *et al.*, 2011). Ongoing field and laboratory investigations in Cheirogaleidae indicate that there will be further significant changes in the taxonomy of this family over the next few years (Thiele *et al.*, 2013).

Although the family Cheirogaleidae within infraorder Lemuriformes has been confirmed as monophyletic and closely related to Lepilemuridae by behavioral, chromosomal, molecular, and morphological data (Dene *et al.*, 1976; Rumpler *et al.*, 1989; Stanger-Hall, 1997; Yoder, 1997; Groves, 2001), the evolutionary relationships within Cheirogaleidae have been disputed (e.g., Kappeler *et al.*, 2005; Louis *et al.*, 2006a; Olivieri *et al.*, 2007; Groeneveld *et al.*, 2010; Radespiel *et al.*, 2012; Thiele *et al.*, 2013). Only two different karyotypes (*Phaner* 2N: 46; all other genera 2N: 66) were found in Cheirogaleidae (Rumpler and Albignac, 1973; Rumpler *et al.*, 1995, 1998). Tattersall and Schwartz (1974) found that *Phaner* was distinct from *Microcebus* and *Cheirogaleus* based on morphological data. Yoder *et al.* (1996) grouped *Cheirogaleus* as a sister clade to *Microcebus* and *Mirza* according to morphological and molecular data. However, Stanger-Hall (1997) argued that *Microcebus* should be sister to a clade of *Mirza* and *Phaner* with *Cheirogaleus* as the basal lineage based on morphological and behavioral data. Immunological data grouped *Phaner* and *Cheirogaleus* together with *Microcebus* as the basal lineage (Sarich and Cronin, 1976). However, *Allocebus* was not included in any of these studies. Vocalization data showed closer affinities of *Allocebus* to a *Cheirogaleus–Microcebus* clade than to *Phaner* (Rakotoarison *et al.*, 1996). Based on morphological and behavioral data, Stanger (1993) found that *Cheirogaleus* grouped with an *Allocebus–Microcebus–Mirza* clade, with the last two genera forming a subclade, and *Phaner* as the basal lineage. Crovella *et al.* (1995) utilized highly repetitive DNA data to confirm Stanger's (1993) study, but questioned the position of *Allocebus*. Utilizing about 2000 base pairs of mtDNA sequence data, Pastorini *et al.* (2001) found that *Mirza* and *Microcebus* formed a clade representing the sister group of *Allocebus*, with *Cheirogaleus* as the basal lineage, but they did not include *Phaner* in their study. Based on transposable element data, Roos *et al.* (2004) found a similar phylogenetic relationship of the genera in Cheirogaleidae. Chatterjee *et al.* (2009), Fabre *et al.* (2009) and Springer *et al.* (2012) confirmed the phylogenetic relationships of four genera in Cheirogaleidae, (*Cheirogaleus* (*Allocebus* (*Microcebus*, *Mirza*))), but found that *Phaner* and *Lepilemur* clustered together as a clade, thus challenging the phylogenetic position of *Phaner* and the monophyly of the family Cheirogaleidae. In the study of the *OPN1SW* gene in nocturnal lemurs, phylogenetic analyses indicated similar phylogenetic relationships between different genera, but *Phaner* was grouped in Cheirogaleidae instead of Lepilemuridae (Veilleux *et al.*, 2013). Incomplete data and/or insufficient sampling may be contributing to ongoing controversy in the phylogenetic relationships of Cheirogaleidae.

Lemur biogeographic hypotheses

In the study of biogeographic patterns of Madagascar's biodiversity, several hypotheses have been proposed to explain the current pattern of lemur speciation. Martin (1972) first presented the ecological biogeographic hypothesis based on phytogeographic criteria and similarities between lemur communities that divided Madagascar into six regions. Comparable or sister lemur species were known to occupy forests on the eastern and western halves of Madagascar related to the prevailing forest type; thus, ecological adaptions would arise for the eastern humid rainforest and the western dry deciduous forest. The large central plateau composed of savanna grasslands and limited forest cover was thought to act as a formidable barrier. Additionally, a superimposed climatic gradient overlays Madagascar's topography as rainfall decreases from north to south on both eastern and western halves, culminating in the southern dry spiny forest. Furthermore, Martin (1972) apportioned domains within the eastern and western forests based on variations in lemur communities, with rivers acting as barriers to dispersal and gene flow. Barrier rivers originally proposed, such as the Tsiribihina, Betsiboka, Maevarano, and Mahavavy du Nord Rivers in the west and the Mangoro River in the east, were later corroborated by molecular data (Pastorini *et al.*, 2003; Louis *et al.*, 2006a; Ramaromilanto *et al.*, 2009). Yoder and Heckman (2006), utilizing mtDNA in their study of the genus *Microcebus*, revised the Ecogeographic Constraint model (hereafter known as the "EC" hypothesis) by showing that the eastern and western mouse lemurs were not exclusively restricted to interpopulation gene flow while advocating for comprehensive sampling to test this result further.

In their study of the biogeographic patterns of leaf chameleons (genus *Brookesia*), conducted primarily in northern Madagascar, Raxworthy and Nussbaum (1995) proposed five geographic regions defined by higher altitudinal topographies separated by lower lying intervening dry forest and characterized by high endemism caused by Pleistocene climatic fluctuations. They reasoned that glacial fluctuations caused fragmentation of habitat resulting in multiple allopatric sister-taxon pairs based on similar morphological characters (hereafter known as the mountain refuge or "MR" hypothesis). Townsend *et al.* (2009) followed this study with a comprehensive study of leaf chameleons using mtDNA and nuclear sequence data, testing biogeographic hypotheses, but sister-taxon divergence date estimates showed a firm Miocene age, thus rejecting the temporal component of the MR hypothesis.

Subsequently, Wilmé *et al.* (2006) analyzed Madagascar's biogeography and unique microendemism from a context of watersheds. Low-elevation sources tended to define Centers of Endemism (COEs), whereas watersheds connected to Madagascar's three highest peaks generally included more widespread species. Wilmé *et al.* proposed that during periods of the late Tertiary and Pleistocene glacial peaks, aridification caused contraction of moist environments which would impact lower elevations more severely than higher environments, leading to retraction and fragmentation of these lower elevation watersheds. Conversely, the watersheds connected to high-elevation water sources were shielded and served as "retreat-dispersion watersheds" (RDWs), providing a refuge and source of biodiversity expansion as climates cycled to more favorable conditions (hereafter referred to as the watershed contraction "WC" hypothesis).

Lastly, the asymmetrical topography of Madagascar with the mountainous high elevations of the eastern rainforest to the gently sloping gradient of the western dry deciduous forest has been proposed as a factor in influencing speciation and microendemism (hereafter referred to as the riverine barrier or "RB" hypothesis; Martin, 1972; Ganzhorn *et al.*, 2006). Martin's (1972) biogeographic description of the EC hypothesis was primarily based on the more conspicuous diurnal lemurs. The family Cheirogaleidae represents five genera of small nocturnal lemurs that have limited home ranges or territories, so riverine barriers could potentially influence species formation as both boundaries to movement and conveniently delineated conduits along which populations could move into new environments.

Divergence time estimates

Although evidence of subfossil lemurs has been recovered in Madagascar dating between 26,000 and a few hundred years old, no lemur fossils have been found earlier than the Quaternary (Godfrey and Jungers, 2003; Wells, 2003; Godinot, 2006; Godfrey *et al.*, 2010). Due to this lack of fossils, evidence regarding the biogeographical distribution history of the extant lemurs of Madagascar cannot be corroborated precisely (Godinot, 2006; Masters *et al.*, 2006; Stevens and Heesy, 2006; Tattersall, 2006a, 2006b; Yoder and Nowak, 2006). Various hypotheses have been presented for how lemurs colonized Madagascar. These include Gondwanan vicariance (Arnason *et al.* 1998), temporary land bridges between mainland Africa and Madagascar (McCall, 1997), and the use of vegetation rafts to cross the Mozambique Channel (Kappeler, 2000). Molecular clock estimates have also been utilized to determine the evolutionary origin of lemurs (Yoder, 1997; Poux and Douzery, 2004; Roos *et al.*, 2004; Chatterjee *et al.*, 2009; Fabre *et al.*, 2009; Perelman *et al.*, 2011; Springer *et al.*, 2012).

Based on cytochrome *b* sequence data, Roos *et al.* (2004) estimated the divergence date at 38 million years ago (mya, 31–50) for Cheirogaleidae, and in the same study built a phylogeny using short interspersed elements (SINEs) as markers. Yoder and Yang (2004) utilized multiple gene sequence data including mitochondrial and nuclear genes and estimated the divergence date at 29 mya (26.7–31.8). According to a molecular supermatrix from previously published GenBank sequences, Springer *et al.* (2012) estimated the divergence date for Cheirogaleidae at 21.84 mya (17.39–26.62). Incongruence among these estimates demonstrates the difficulty with which they are calculated, which has been postulated to be mainly due to the fact that different genes have different evolutionary rates (Glazko and Nei, 2003). With a lack of fossils from the late Cretaceous to the late Quaternary, molecular genetic data are particularly important for establishing the evolutionary patterns and relationships of fauna in Madagascar (Poux *et al.*, 2005; Samonds *et al.*, 2012).

The reliability of pattern and molecular genetic data

Complete mitochondrial genomes, or mitogenomes, contain both non-coding and coding regions. The coding regions include 13 genes encoding oxidative phosphorylation-related polypeptides, 2 rRNAs (12S rRNA and 16S rRNA), and 22 tRNAs (Anderson *et al.*, 1981). Because of their rapid time to coalescence, relatively high substitution rates, lack of recombination, haploidy state, and high copy number, the analysis of mitochondrial genomes with respect to size, genome architecture, gene content/arrangement, exon/intron structure, genetic code, and secondary structure of rRNA and tRNA genes has been an attractive source for molecular phylogenetic studies of animal taxa at different taxonomic levels (Avise, 1998; Boore, 2006). Furthermore, valuable phylogenetic signals harbored in mitochondrial protein-coding sequence data can be utilized to further clarify uncertain relationships within and between different taxa (Zardoya and Meyer, 1996; Bourlat *et al.*, 2008).

In lemurs, mtDNA sequences have been used extensively to confirm phylogenetic relationships as well as define and delineate species (e.g. Pastorini *et al.*, 2001; Louis *et al.*, 2006a; Olivieri *et al.*, 2007; Radespiel *et al.*, 2012). These mtDNA studies only utilized two or three gene fragments or subunit genes (partial or complete; Pastorini *et al.*, 2003; Andriaholinirina *et al.*, 2006; Rabarivola *et al.*, 2006; Craul *et al.*, 2007). Furthermore, researchers have utilized different combinations of mtDNA sequence fragments. The non-recombining and maternally inherited characteristics of mtDNA have led to some criticism of its utility in phylogenetic studies, especially involving problems of discordance between mtDNA and nuclear DNA with geographic patterns (Toews and Brelsford, 2012). However, Meiklejohn *et al.* (2014) found well-supported topologies using mitogenomes. Analyses involving individual sections resulted in discordant topologies despite efforts to reduce the incongruence among regions by partitioning by codon position, using mixture and codon-based models, RY coding, and excluding potentially misleading sites and the application of multiple analytic strategies.

Moreover, direct comparisons between the various lemur data sets cannot be formulated without eliminating significant available portions of sequence and/or cytological data. An unfortunate secondary byproduct is the generation of taxonomic confusion between taxa from overlapping regions (Andriantompohavana *et al.*, 2006; Louis *et al.*, 2006b; Craul *et al.*, 2007; Olivieri *et al.*, 2007). With the development of sequencing technology, especially next-generation sequencing, an increasing number of phylogenetic studies on primates have utilized mitogenomes (e.g., Matsui *et al.*, 2009; Schrago *et al.*, 2012; Finstermeier *et al.*, 2013; Zinner *et al.*, 2013; Pozzi *et al.*, 2014). As the advancement of next-generation sequencing technology has dramatically reduced the cost and difficulty of sequencing large fragments, Powell *et al.* (2013) suggested using complete mitogenomes instead of just 1–2 kb of mtDNA sequence in phylogenetic studies (Louis and Lei, 2014).

Mitogenomic sequences have proven to be useful in resolving the phylogenies of closely related species and facilitating the molecular dating of divergence events within a phylogeny (Raaum *et al.*, 2005; Matsui *et al.*, 2009; Chan *et al.*, 2010; Matsudaira and Ishida, 2010; Bjork *et al.*, 2011; Finstermeier *et al.*, 2013). In this study, we generated 56 complete cheirogaleid mitogenomes, utilizing next-generation sequencing to construct a robust phylogeny of the family Cheirogaleidae, and to estimate the divergence times between lineages using relaxed Bayesian methods and a global molecular clock. We also used these data to review taxonomic relationships within Cheirogaleidae, as well as the strength of RB, WC, and MR biogeographic speciation hypotheses.

Methods

Sample collection

All lemurs investigated in this study were free-ranging animals that were live-trapped, hand-caught or immobilized with a CO_2 projection rifle or blowgun and subsequently sedated with 10 mg/kg of Telazol® (Fort Dodge Animal Health, Overland Park, Kansas; Table 3.1; Figure 3.1). Four 2.0-mm biopsies and 1.0 cm^3/kg of whole blood were collected from each animal and stored in room temperature storage buffer (0.1 M Tris, 0.1 M Na_2EDTA, 2% (vol/vol) SDS; Seutin *et al.*, 1991) until transferred to the laboratory for storage at −80°C. We recorded the GPS coordinates to accurately document the capture locality of each animal, so that individuals could be released where they were initially captured (Figure 3.1; Table 3.1). Samples were selected for next-generation sequencing and mitogenomic data generation to represent all nominal species of lemurs in the family Cheirogaleidae, except *Phaner furcifer, Microcebus marohita, M. gerpi, Cheirogaleus lavasoensis*, and *C. minusculus*, because samples could not be obtained for these species.

All collection and export permits were obtained from Madagascar National Parks, formerly Association Nationale pour la Gestion des Aires Protégées, and the Ministère de l'Environnement des Eaux et Forêts of Madagascar. All rules and regulations were followed according to Malagasy law. Samples were imported to the USA under Convention on International Trade in Endangered Species (CITES) permits from the US Fish and Wildlife Service. The Institutional Animal Care and Use Committee of Omaha's Henry Doorly Zoo & Aquarium approved capture and sampling procedures (#12–101).

PCR amplification, DNA sequencing and assembly

A complete list of individuals and source DNA are presented in Table 3.1. The localities of all Cheirogaleidae individuals are displayed in Figure 3.1. Genomic DNA was isolated from the samples using a standard phenol-chloroform-isoamyl extraction protocol (Sambrook *et al.*, 1989). Two methodologies were followed to generate the mitogenomes based on either Illumina MiSeq (Illumina, Inc., San Diego, CA, USA) or Roche 454 sequencing technologies (Roche Life Sciences, Inc., Basel, Switzerland). For the Roche 454 sequencing protocol, two overlapping PCR fragments of approximately 8.5 and 9.5 kb that comprise the entire mitogenome were amplified using the AccuPrime™ Taq DNA Polymerase High Fidelity kit (Life Technologies, Grand Island, NY, USA). The primers used to amplify the two PCR fragments for all Cheirogaleidae species are shown in Table 3.2.

PCR was run with an initial step of 94°C for 2 min; with cycles of 10 s at 94°C; 15 s of 3 cycles each at 62°C, 60°C, 58°C, 56°C and 30 cycles at 50°C; and 8.5 min extension at 65°C. We electrophoresed the amplified samples on a 1.2% agarose gel to verify the PCR products and purified them using QIAquick PCR purification kit (QIAGEN, Valencia, CA). PCR products were quantified using a NanoDrop instrument (Thermo Scientific, Wilmington, MA), and the two PCR fragments from each individual sample were pooled in equimolar ratios. PCR products were then converted into bar-coded 454 sequencing libraries based on the Parallel-Tagged-Sequencing protocol (Meyer *et al.*, 2007, 2008) or into bar-coded Illumina sequencing libraries with Illumina's Nextera sample preparation kit (Meyer and Kircher, 2010). Pooled DNA libraries were sequenced on the 454 Flx Sequencing platform or Illumina MiSeq sequencing system. Sequencing reads for each sample were sorted according to their molecular bar code (Meyer *et al.*, 2008; Meyer and Kircher, 2010). A *de novo* assembly of the 454 data for each sample was produced utilizing the Newbler v2.7 assembly program of the Roche 454 software tools and ABYSS v1.32 assembler (Simpson *et al.*, 2009). A *de novo* assembly of the Illumina data for each sample was produced utilizing the ABYSS assembler (Simpson *et al.*, 2009). The consensus

Table 3.1 Cheirogaleid field samples included in this study. Assignment of clades for *Microcebus* accompanies text in the Results.

ID	Taxon	Locality	Clade	Latitude	Longitude
AMB5.41	*Microcebus arnholdi*	Montagne d'Ambre	Clade 7	−12.47728	49.22244
AMB5.43	*Microcebus arnholdi*	Montagne d'Ambre	Clade 7	−12.51239	49.18931
AND6.4	*Microcebus tanosi*	Andohahela (Manangotry)	Clade 4	−24.76494	46.86233
ANK25	*Microcebus ravelobensis*	Ankarafantsika	Clade 3	−16.36206	46.76661
BEMA7.11	*Microcebus myoxinus*	Tsingy de Bemaraha	Clade 6	−19.05731	44.78067
BEMA7.17	*Microcebus myoxinus*	Tsingy de Bemaraha	Clade 6	−19.04656	44.78053
BEZ7.11	*Microcebus griseorufus*	Beza-Mahafaly (Amptanambo)	Clade 1	−23.69503	44.60144
BIBO7.4	*Microcebus bongolavensis*	Ambodimahabibo	Clade 3	−15.50342	47.48811
CAR5.3	*Microcebus tavaratra*	Ankarana (Mahamasina)	Clade 5	−12.96508	49.13358
DOG8.1	*Microcebus tanosi*	Midongy du Sud (Beharena Sagnira)	Clade 4	−23.52192	47.08831
GAR6	*Microcebus sambiranensis*	Manongarivo	Clade 7	−14.02369	48.27233
HABO6.12	*Microcebus species nova #1*[a]	Mahabo	Clade 4	−23.18486	47.73800
HAZO5.9	*Microcebus murinus*	Ihazofotsy	Clade 1	−24.89056	46.54969
JAR1	*Microcebus mittermeieri*	Anjanaharibe-Sud	Clade 6	−14.79586	49.46475
KIAN3.2	*Microcebus jollyae*	Kianjavato	Clade 2	−21.37808	47.86983
KIAN3.3	*Microcebus jollyae*	Kianjavato	Clade 2	−21.37808	47.86983
KIBO26	*Microcebus murinus*	Tsiombikibo	Clade 1	−16.05439	45.81369
KIR6.30	*Microcebus berthae*	Kirindy	Clade 6	−20.06506	44.65117
KIR6.32	*Microcebus murinus*	Kirindy	Clade 1	−20.06475	44.56650
KOA7.2	*Microcebus danfossi*	Andalirano	Clade 3	−15.21342	47.91375
LAVA7	*Microcebus murinus*	Analalava	Clade 1	−22.58872	45.12922
LOKO4.37	*Microcebus mamiratra*	Lokobe	Clade 7	−13.40469	48.30311
MAS6.11	*Microcebus species nova #2*	Masoala (Masiaposa)	Clade 6	−15.66739	49.96006
NARA8.1	*Microcebus species nova #3*[b]	Mananara-Nord (Ambavala)	Clade 3	−16.55825	49.73322
NOSY82	*Microcebus macarthurii*	Nosy Mangabe	Clade 3	−15.49836	49.76256
POLO5.19	*Microcebus simmonsi*	Tampolo	Clade 4	−17.28747	49.40856
RANO247	*Microcebus rufus*	Ranomafana (Talatakely)	Clade 6	−21.24833	47.42406
RANO250	*Microcebus rufus*	Ranomafana (Talatakely)	Clade 6	−21.24833	47.42406
TAD4.32	*Microcebus lehilahytsara*	Mantadia	Clade 6	−18.83867	48.43106

(continued)

Table 3.1 *(cont.)*

ID	Taxon	Locality	Clade	Latitude	Longitude
TAF06.7	*Microcebus margotmarshae*	Antafondro	Clade 7	-14.04686	48.21936
TRA8.85	*Microcebus rufus*	Andringitra (Camp 4)	Clade 6	-22.20678	46.97197
VEV7.13	*Microcebus species nova* #3[a]	Vevembe	Clade 4	-22.79467	47.18758
AMB5.22	*Cheirogaleus crossleyi*	Montagne d'Ambre		-12.52731	49.17331
ANK5.12	*Cheirogaleus medius*	Ankarana		-16.36206	46.76661
BEMA7.19	*Cheirogaleus medius*	Tsingy de Bemaraha		-19.04525	44.77772
DOG8.2	*Cheirogaleus major*	Midongy du Sud (Beharena Sagnira)		-23.52464	47.09236
LAKI5.18	*Cheirogaleus major*	Lakia		-21.51558	47.91147
MAS6.10	*Cheirogaleus major*	Masoala (Masiaposa)		-15.67189	49.96617
POLO5.20	*Cheirogaleus major*	Tampolo		-17.28747	49.40858
GAR8	*Cheirogaleus crossleyi*	Manongarivo		-14.02369	48.27233
MIZA7.1	*Cheirogaleus crossleyi*	Maromizaha		-18.95694	48.49236
HIH7.3	*Cheirogaleus medius*	Anjiamangirana		-15.21642	47.75189
KIBO7.9	*Cheirogaleus medius*	Tsiombikibo		-16.04886	45.81067
LAVA1	*Cheirogaleus medius*	Analalava		-22.59242	45.13336
MAR30	*Cheirogaleus medius*	Mariarano		-15.47992	46.69333
TRA8.81	*Cheirogaleus crossleyi*	Andringitra Ambarongy		-22.22269	47.01889
ZOM6.2	*Cheirogaleus medius*	Zombitse		-22.88631	44.69375
JOZO4.17	*Cheirogaleus sibreei*	Anjozorobe		-18.46789	47.94131
TVY7.23	*Allocebus trichotis*	Ambatovy		-18.85458	48.29886
TVY7.95	*Allocebus trichotis*	Ambatovy		-18.84922	48.29794
TAF06.14	*Mirza zaza*	Antafondro		-14.03836	48.22744
ZOMB53	*Mirza coquereli*	Zombitse		-22.88189	44.69528
AMB5.37	*Phaner electromontis*	Montagne d'Ambre		-12.52736	49.17422
ANK5.11	*Phaner electromontis*	Ankarana		-16.37011	46.77769
KIR6.22	*Phaner pallescens*	Kirindy		-20.06603	44.65489
TAF06.11	*Phaner parienti*	Antafondro		-14.04658	48.22483

Note: [a]Louis *et al.*, 2006a; [b]Radespiel *et al.*, 2008.

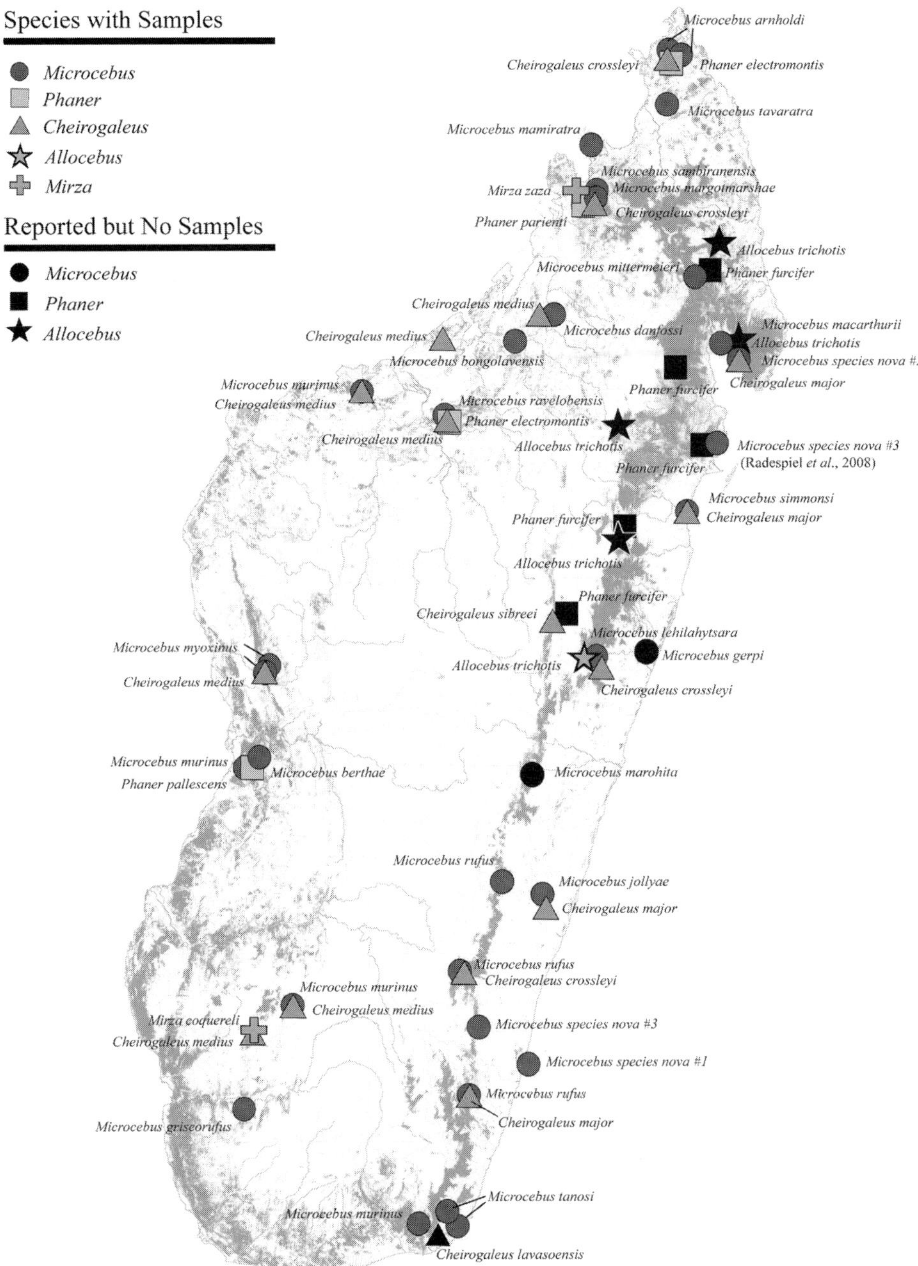

Figure 3.1 Map of sampling localities of the cheirogaleid lemurs of Madagascar. Circles, square, triangle, star, and cross represent the genera *Microcebus*, *Phaner*, *Cheirogaleus*, *Allocebus*, and *Mirza*, respectively. Solid black symbols denote published localities for different species with no molecular data available for this study. Detailed information for locality sites is presented in Table 3.1 or from Meier and Albignac (1989, 1991), Rakotoarison *et al.* (1996), Rakotoarison (1998), Schütz and Goodman (1998), Garbutt (2000), Goodman and Raselimanana (2002), Mittermeier *et al.* (2010), Radespiel *et al.* (2012), and Rasoloarison *et al.* (2013). (A black and white version of this figure will appear in some formats. For the colour version, please refer to the plate section.)

Table 3.2 Primers for PCR.

Genus	Fragment	Forward primer	Reverse Primer
Microcebus	1	CACYATAGACGCTCAAGAAGTRGAAAC	GYTGGTTTCACGGAGGATGGTAG
	2	ATCGTACATACATYTCTTGYCCACAT	GGTTGCYTGATTYAGGCGTCC
	2	ATWGTACATACATTTCTTGYCCACAT	GGTTGCTTGRTTTAGGCGYCC
Cheirogaleus	1	CACGATAGATGCCCAAGAAGTAGAAAC	GYTGGTTTCACGGAGGATGGTAG
	1	CTYATTGCCCTWCCATCACTACG	GYTGGTTTCACGGAGGATGGTAG
	1	CCTTATTGCCCTTCCATCACTACG	GYTGGTTTCACGGAGGATGGTAG
	1	CTTATTGCCCTCCCATCATTACG	GYTGGTTTCACGGAGGATGGTAG
	2	AACRGGCAGTCCTTGATCTTACAG	CTGGGCGGGAGGTTATTAGG
	2	CCGTGCATTGTATGTTCYCCC	GCGTCCTGGGATRGCGTCTG
	2	GCATACGGCCCCAGACCAC	GGGTTGTTTGATTTAGGCGTC
	2	GTCTTCCCACATCGCCATACATC	CTTGGTTTAGTCGTCCTGGAATGG
Allocebus	1	AGATGCCGCTTCCCCTATTATAGAG	GYTGGTTTCACGGAGGATGGTAG
	2	CACCCACAACAACACAGCAGC	CCACAGATTTCCGAACATTGACC
Mirza	1	ACTGAACTYACCCACACAAGCAC	ATTATGTGGGCGGGTTGCTG
	2	GCCTTTCCCCATACATTAATCACC	GGTTGCTTGGTTTAGGCGTCC
Phaner	1	CACAGAACTCACCCATACAAGCAC	TGAAGTAAGAACCAGATGCCAGG
	2	CTACCATCCTCCGTGAAACCAG	TGGGATAGCGTCAGTCTTTACCG

sequence from each assembly was reassembled with the 454 or Illumina data by BWA v0.59 (Li and Durbin, 2009). All sequences in this study have been submitted to GenBank with accession numbers KM112286–KM112341.

Phylogenetic analysis

All 56 mitogenomes were annotated utilizing MITOS (Bernt *et al.*, 2013). All mtDNA-encoded genes, rRNA and tRNA, were aligned against the homologous regions of *Lepilemur hubbardorum* (Lei *et al.*, 2010). The tRNA genes were identified with ARWEN v1.2 (Laslett and Canback, 2008) and verified by comparing GenBank tRNA gene sequences from *L. hubbardorum* (Lei *et al.*, 2010).

The complete mitogenomes of all extant lemur genera (56 cheirogaleids and 9 other species), along with 32 other outgroup primate species (Table 3.3), were aligned utilizing MAFFT v6.0 set to the default parameters (Katoh *et al.*, 2002). Initial sequence comparisons and measures of variability were performed using MEGA v4.0 (Tamura *et al.*, 2007). All of the coding genes were aligned individually before concatenation. Only the heavy chain protein-coding sequences (ND1, ND2, COI, COII, ATPase8, ATPase6, COIII, ND3, ND4L, ND4, ND5, and CytB) were utilized for subsequent analyses, which hereafter are referred to as HP. Overlapping or ambiguous start/stop codon regions in the alignment were removed (Matsui *et al.*, 2009). We applied the strategy of Matsui *et al.* (2009) to deal with the overlapping sequence regions (e.g., due to the overlap region between the *ATP8* and *ATP6* genes, the shared region from *ATP6* was removed).

Table 3.3 Information on the studied species including mitogenome length and base composition of individual mitogenomes.

ID	Taxon	Length (bp)	A%	T%	C%	G%
AMB5.41	*Microcebus arnholdi*	17008	0.33496	0.30709	0.22954	0.12841
AMB5.43	*Microcebus arnholdi*	17006	0.33435	0.30624	0.22986	0.12954
AND6.4	*Microcebus tanosi*	16821	0.33732	0.30325	0.23263	0.12681
ANK25	*Microcebus ravelobensis*	16814	0.33686	0.29850	0.23641	0.12823
BEMA7.11	*Microcebus myoxinus*	17014	0.33437	0.30158	0.23381	0.13025
BEMA7.17	*Microcebus myoxinus*	16964	0.33465	0.30193	0.23361	0.12980
BEZ7.11	*Microcebus griseorufus*	16895	0.33690	0.29991	0.23534	0.12785
BIBO7.4	*Microcebus bongolavensis*	16817	0.33793	0.29785	0.23702	0.12719
KOA7.2	*Microcebus danfossi*	16809	0.33839	0.29645	0.23874	0.12642
CAR5.3	*Microcebus tavaratra*	16824	0.33583	0.30153	0.23431	0.12833
DOG8.1	*Microcebus tanosi*	16990	0.33726	0.30277	0.23267	0.12731
GAR6	*Microcebus sambiranensis*	16778	0.33472	0.30731	0.22929	0.12868
HABO6.12	*Microcebus species nova #1*[a]	16991	0.33730	0.30646	0.22865	0.12760
HAZO5.9	*Microcebus murinus*	17083	0.33607	0.30118	0.23327	0.12949
JAR1	*Microcebus mittermeieri*	16857	0.33482	0.30421	0.23177	0.12920
KIAN3.2	*Microcebus jollyae*	16969	0.33691	0.29790	0.23590	0.12929
KIAN3.3	*Microcebus jollyae*	16969	0.33691	0.29790	0.23590	0.12929
KIBO26	*Microcebus murinus*	17018	0.33482	0.30192	0.23293	0.13033
KIR6.30	*Microcebus berthae*	17078	0.33283	0.30466	0.23094	0.13157
KIR6.32	*Microcebus murinus*	16994	0.33700	0.30217	0.23202	0.12881
LAVA7	*Microcebus murinus*	16975	0.33555	0.30197	0.23328	0.12919
LOKO4.37	*Microcebus mamiratra*	16910	0.33601	0.30615	0.22986	0.12797
MAS6.11	*Microcebus species nova #2*	16859	0.33555	0.30328	0.23240	0.12877
NARA8.1	*Microcebus species nova #3*[b]	16993	0.33296	0.30730	0.22845	0.13129
NOSY82	*Microcebus macarthurii*	17012	0.33282	0.30696	0.22831	0.13185
POLO5.19	*Microcebus simmonsi*	16796	0.33758	0.30263	0.23279	0.12699
RANO247	*Microcebus rufus*	17442	0.32926	0.30398	0.23048	0.13628
RANO250	*Microcebus rufus*	17320	0.33054	0.30266	0.23199	0.13482
TAD4.32	*Microcebus lehilahytsara*	16996	0.33502	0.30437	0.23123	0.12938

(continued)

Table 3.3 *(cont.)*

ID	Taxon	Length (bp)	A%	T%	C%	G%
TAF06.7	*Microcebus margotmarshae*	16821	0.33625	0.30700	0.22924	0.12752
TRA8.85	*Microcebus rufus*	17466	0.32967	0.30362	0.23085	0.13586
VEV7.13	*Microcebus species nova* #3[a]	16996	0.33743	0.30660	0.22858	0.12738
AMB5.22	*Cheirogaleus crossleyi*	16707	0.33531	0.27898	0.26240	0.12330
ANK5.12	*Cheirogaleus medius*	16725	0.32945	0.29291	0.24777	0.12987
BEMA7.19	*Cheirogaleus medius*	16718	0.33096	0.28819	0.25380	0.12705
DOG8.2	*Cheirogaleus major*	16577	0.33359	0.28702	0.25560	0.12379
LAKI5.18	*Cheirogaleus major*	16591	0.33343	0.28588	0.25604	0.12465
MAS6.10	*Cheirogaleus major*	16596	0.33309	0.28501	0.25723	0.12467
POLO5.20	*Cheirogaleus major*	16594	0.33331	0.28510	0.25708	0.12450
GAR8	*Cheirogaleus crossleyi*	16706	0.33431	0.27936	0.26176	0.12457
MIZA7.1	*Cheirogaleus crossleyi*	16702	0.33397	0.27733	0.26308	0.12561
HIH7.3	*Cheirogaleus medius*	16731	0.33058	0.28922	0.25282	0.12737
KIBO7.9	*Cheirogaleus medius*	16718	0.33138	0.28795	0.25350	0.12717
LAVA1	*Cheirogaleus medius*	16721	0.33078	0.28796	0.25399	0.12727
MAR30	*Cheirogaleus medius*	16741	0.33134	0.28815	0.25387	0.12664
TRA8.81	*Cheirogaleus crossleyi*	16729	0.33469	0.27898	0.26230	0.12404
ZOM6.2	*Cheirogaleus medius*	16721	0.33084	0.28766	0.25417	0.12732
JOZO4.17	*Cheirogaleus sibreei*	16439	0.33293	0.28901	0.25160	0.12647
TVY7.23	*Allocebus trichotis*	17046	0.33069	0.28347	0.25678	0.12906
TVY7.95	*Allocebus trichotis*	16878	0.33150	0.28256	0.25809	0.12786
TAF06.14	*Mirza zaza*	16897	0.33337	0.29668	0.24164	0.12831
ZOMB53	*Mirza coquereli*	17066	0.33113	0.29995	0.23802	0.13090
AMB5.37	*Phaner electromontis*	17078	0.32820	0.28827	0.25290	0.13064
ANK5.11	*Phaner electromontis*	16818	0.32792	0.29082	0.25086	0.13040
KIR6.22	*Phaner pallescens*	16738	0.32973	0.28707	0.25338	0.12982
TAF06.11	*Phaner parienti*	16789	0.32843	0.28918	0.25177	0.13062
DASI5.23	***Avahi laniger***	**16973**	**0.31450**	**0.26277**	**0.27809**	**0.14464**

ANOSIB13	*Indri indri*	17121	0.31844	0.27685	0.26844	0.13627
ANJZ5	*Propithecus diadema*	17083	0.32963	0.26687	0.27361	0.12990
MERY9	*Eulemur rubriventer*	17274	0.32517	0.28193	0.26126	0.13164
TSINJ44	*Hapalemur griseus griseus*	16897	0.32680	0.29479	0.24886	0.12955
KIAN8.4	*Prolemur simus*	17192	0.33143	0.29438	0.24762	0.12657
RANO5.3	*Varecia variegata variegata*	17051	0.33341	0.28509	0.25940	0.12210
ISA7	*Lemur catta*	17056	0.33191	0.28817	0.25399	0.12594
ZAH21	*Lepilemur mustelinus*	16957	0.32311	0.26101	0.28378	0.13210
EAST1	*Daubentonia madagascariensis*	16526	0.32573	0.25747	0.28749	0.12931
Otolemur crassicaudatus	*Otolemur crassicaudatus*	16828	0.31531	0.25333	0.29481	0.13656
Galago senegalensis	*Galago senegalensis*	16796	0.32252	0.26727	0.28084	0.12938
Galago moholi	*Galago moholi*	16651	0.32238	0.26815	0.28016	0.12930
Nycticebus coucang	*Nycticebus coucang*	16765	0.32186	0.27080	0.26961	0.13773
Nycticebus bengalensis	*Nycticebus bengalensis*	16736	0.32134	0.27109	0.26948	0.13809
Loris tardigradus	*Loris tardigradus*	16777	0.34112	0.26673	0.26638	0.12577
Loris lydekkerianus	*Loris lydekkerianus*	16792	0.34022	0.26816	0.26536	0.12625
Perodicticus potto	*Perodicticus potto*	16638	0.33147	0.25147	0.28651	0.13054
Perodicticus edwardsi	*Perodicticus edwardsi*	16619	0.33221	0.24989	0.28768	0.13021
Cephalopachus bancanus	*Cephalopachus bancanus*	16928	0.33040	0.27948	0.26636	0.12376
Carlito syrichta	*Carlito syrichta*	17006	0.32289	0.28578	0.25979	0.13154
Callicebus cupreus	*Callicebus cupreus*	16560	0.31570	0.29185	0.26159	0.13086
Callithrix geoffroyi	*Callithrix geoffroyi*	16506	0.32703	0.26833	0.26984	0.13480
Aotus lemurinus	*Aotus lemurinus*	16578	0.32634	0.28423	0.26125	0.12818
Ateles belzebuth	*Ateles belzebuth*	16557	0.33279	0.27656	0.26436	0.12629
Cebus albifrons	*Cebus albifrons*	16551	0.32874	0.28059	0.26319	0.12748
Saimiri sciureus	*Saimiri sciureus*	16472	0.32176	0.26979	0.27532	0.13314
Theropithecus gelada	*Theropithecus gelada*	16542	0.31731	0.25063	0.30474	0.12731
Papio hamadryas	*Papio hamadryas*	16521	0.31445	0.24877	0.30549	0.13129
Mandrillus sphinx	*Mandrillus sphinx*	16542	0.32505	0.25414	0.29803	0.12278
Macaca fascicularis	*Macaca fascicularis*	16575	0.31807	0.25080	0.30353	0.12760
Macaca mulatta	*Macaca mulatta*	16564	0.31719	0.25054	0.30409	0.12817

(*continued*)

Table 3.3 (*cont.*)

ID	Taxon	Length (bp)	A%	T%	C%	G%
Chlorocebus pygerythrus	*Chlorocebus pygerythrus*	16337	0.31976	0.25415	0.30085	0.12524
Cercopithecus albogularis	*Cercopithecus albogularis*	16543	0.31941	0.25624	0.29940	0.12495
Colobus guereza	*Colobus guereza*	16648	0.32052	0.27198	0.27829	0.12920
Procolobus verus	*Procolobus verus*	16605	0.31876	0.28407	0.26697	0.13020
Trachypithecus obscurus	*Trachypithecus obscurus*	16560	0.31981	0.29306	0.25870	0.12844
Semnopithecus entellus	*Semnopithecus entellus*	16533	0.31737	0.29232	0.25839	0.13192
Pygathrix nemaeus	*Pygathrix nemaeus*	16559	0.32315	0.28824	0.26016	0.12845
Presbytis melalophos	*Presbytis melalophos*	16543	0.32727	0.29269	0.25473	0.12531
Nasalis larvatus	*Nasalis larvatus*	16570	0.32511	0.28365	0.26554	0.12571
Rhinopithecus roxellana	*Rhinopithecus roxellana*	16549	0.32389	0.29089	0.25669	0.12853
Symphalangus syndactylus	*Symphalangus syndactylus*	16512	0.31323	0.25012	0.30699	0.12966
Hylobates lar	*Hylobates lar*	16472	0.30591	0.23956	0.31757	0.13696
Hylobates agilis	*Hylobates agilis*	16500	0.30394	0.23921	0.31848	0.13836
Nomascus leucogenys	*Nomascus leucogenys*	16478	0.30926	0.23977	0.31551	0.13545
Pongo pygmaeus	*Pongo pygmaeus*	16389	0.30551	0.23778	0.32442	0.13228
Gorilla gorilla	*Gorilla gorilla*	16412	0.30947	0.25256	0.30703	0.13094
Pan paniscus	*Pan paniscus*	16563	0.31329	0.25273	0.30695	0.12703
Pan troglodytes	*Pan troglodytes*	16554	0.31134	0.25178	0.30802	0.12885
Homo sapiens	*Homo sapiens*	16569	0.30925	0.24715	0.31269	0.13091

Note: [a]Louis *et al.*, 2006a; [b]Radespiel *et al.*, 2008. The species in bold were outgroups.

Phylogenetic trees were constructed from these data sets with maximum likelihood (ML) and Bayesian inference methods, utilizing MrBayes v3.2.2 (Huelsenbeck and Ronquist, 2001; Ronquist and Huelsenbeck, 2003) and PhyML v3.0 (Guindon and Gascuel, 2003; Guindon *et al.*, 2005) program packages. The optimal nucleotide substitution model was chosen using the Akaike information criterion (AIC) as implemented in Modeltest v3.7 (Posada and Crandall, 1998). The ML analyses were performed under the GTR+I+G model selected by Modeltest v3.7 with PAUP Version 4.0b10 software (Swofford, 2001). We performed 500 bootstrap replicates with TBR branch swapping to test the support for nodes in the topology using PhyML v3.0 (Guindon and Gascuel, 2003; Guindon *et al.*, 2005).

For the Bayesian inference analyses, two simultaneous Markov chain Monte Carlo (MCMC) runs with four chains and 5,000,000 generations were performed under the GTR+I+G model selected by MrModeltest v2.2 (Nylander, 2004). For every 100 generations, the tree with the best likelihood score was saved, resulting in 50,000 trees. Topologies prior to –ln likelihood of equilibrium were discarded as burn-in and clade posterior probabilities (PPs) were computed from the remaining trees. These trees were condensed into a majority rule consensus tree using PAUP* 4.0b10 (Swofford, 2001). Branch supports were assigned as PPs on the consensus tree.

Divergence date estimation

Dating of the divergence events within Cheirogaleidae was estimated using a molecular clock approach and several external fossil calibration points. We used the fossil-based divergence dates between *Homo* and *Pan*, which was estimated at 5–8 mya (Senut *et al.*, 2001; Brunet *et al.*, 2002); *Theropithecus* and *Papio*, dated at 3.5–7 mya with a mean of 4 (Tosi *et al.*, 2005; Ting *et al.*, 2008); *Pongo* and the *Homo*/*Pan* clade, dated at approximately 12.5–18 mya with a mean of 14 (Lebatard *et al.*, 2008); hominoids and cercopithecoids, dated at approximately 25.2–34 mya (Benefit and McCrossin, 2002; Kelley, 2002); and Lorisiformes, dated at approximately 36.9 mya (Seiffert *et al.*, 2003). The above-mentioned calibration points were used as uniform, lognormal, lognormal, uniform, and uniform priors, respectively, to obtain the posterior distribution of the estimated divergence times. Divergence times were estimated using the Bayesian approach implemented in the program BEAST v1.8.0 (Drummond and Rambaut, 2007). The unrelated lognormal relaxed clock was implemented in all divergence time estimations. Under this model, rates were allowed to vary among branches without *a priori* assumption of autocorrelation between adjacent branches (Drummond *et al.*, 2006). This model allows sampling of the coefficient for variation of rates, which reflects the degree of departure from a global clock. For the sake of comparison, the analysis was repeated with the assumption of a global molecular clock. The model GTR+I+G was used in BEAST with the Yule speciation priors for branching rates. Two simultaneous replicates were run for 100 million generations with tree and parameter sampling occurring every 10,000 generations.

Two independent analyses of each data set were compared to assess convergence in Tracer v 1.5 (Rambaut and Drummond, 2007) after excluding the first five million samples as burn-in, and were combined for subsequent estimations of the parameters. The effective sample size (ESS) for each parameter was greater than 200; therefore, an appropriate number of steps were discarded as burn-in. The results of the two analyses were combined with acceptable mixing and convergence to the stationary distribution checked using Tracer. Subsequently, the sampling distributions were combined (25% burn-in) using the software LogCombiner 1.5 in the BEAST package (Drummond and Rambaut, 2007). Using TreeAnnotator in the BEAST software package, the maximum-clade-credibility tree topology was identified and given mean node heights calculated from the posterior distribution of the tree and visualized in FigTree v1.3.1 (Rambaut, 2008).

Hypothesis testing

We use the nominal members of the family Cheirogaleidae to test three biogeographic hypotheses (WC, RB, and MR) for Madagascar, utilizing the temporal and qualitative spatial information generated in this study. The temporal predictions were statistically tested based on the divergence dates of the various genera and species in the family Cheirogaleidae. Spatial modeling as in Townsend *et al.* (2009) was not quantitatively investigated in this study, because the sample set for each genus was not comprehensive for each species' distributions for this study, but only consisted of representative taxa from which mitogenomes were generated. Furthermore, species richness and riverine barriers were examined within the context of these three biogeographic hypotheses.

The process to test these hypotheses begins with a phylogeny of the family Cheirogaleidae to corroborate the sister-species pairs using mitogenomic data. Subsequently, we estimate dates of divergence of the family Cheirogaleidae to predict whether chronological events are represented by recent climatic cycles (Pliocene–Pleistocene) as expected by the MR and WC hypotheses. The RB hypothesis could not be evaluated in this context because there is no temporal component for river systems. Furthermore, as Townsend *et al.* (2009) indicated, identifying the major rivers is straightforward, but detecting which of the smaller-sized rivers are also barriers requires assembling multiple parameters which are not available, and thus is beyond the scope of this study. However, patterns can be established from haplotype relationships (mitogenomes) that demonstrate reciprocal monophyly with respect to some of the potential barriers between two localities (Yoder and Heckman, 2006).

We use the spatial data of the family Cheirogaleidae to identify which COE and/or RDW each species occupies. The WC model of Wilmé *et al.* (2006) predicts that sister species should essentially occupy adjacent COEs or a COE to an adjacent RDW with the species in the RDW having a wider geographic range. The MR model of Raxworthy and Nussbaum (1995) would predict similar sister-taxa relationships, except that sister taxa tend to occupy mountain forest areas separated by lower

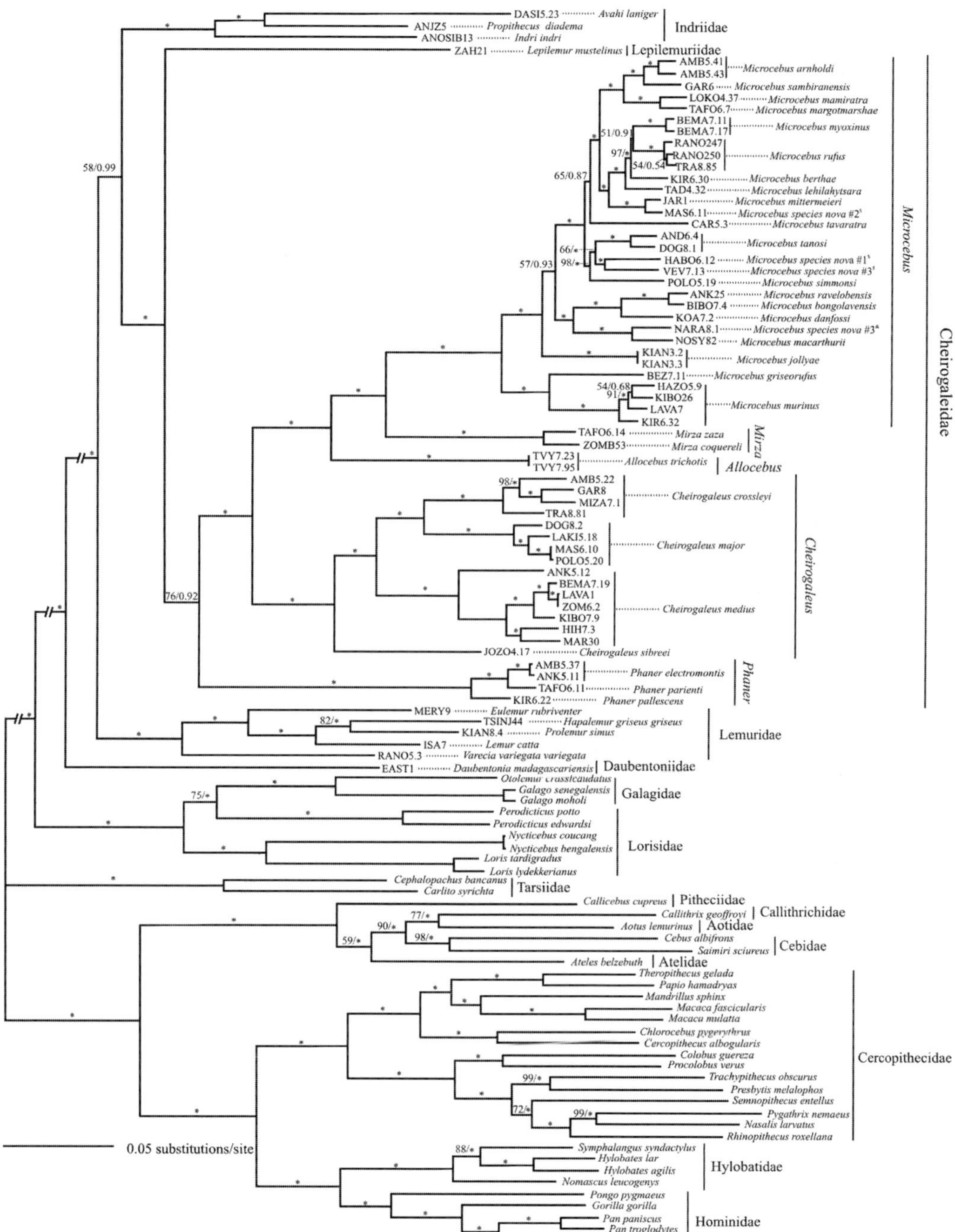

Figure 3.2 Phylogenetic relationships between cheirogaleid species inferred from the maximum likelihood and Bayesian approaches of the mtDNA genome heavy-stranded coding region sequences of the 56 Cheirogaleidae individuals, the 9 outgroup lemur species, and the 32 primate outgroup taxa. Numbers above the dash represent posterior probability support; numbers below the dash represent bootstrap support. Asterisks indicate 100% support; nodes with 100% support in both Bayesian and ML frameworks are labeled with single asterisks.

elevation forest. Although the MR hypothesis of Raxworthy and Nussbaum (1995) was restricted to northern Madagascar, we use the entire Cheirogaleidae sample set in this study to evaluate its fit.

This study did not comprise samples representing the entire distribution for each species for both the molecular and spatial analyses. Furthermore, some genera from the family Cheirogaleidae were not evaluated for each model as follows: (1) samples representing a specific species or its entire distribution were unavailable (*Microcebus marohita, M. gerpi*; *Phaner furcifer*, and *Cheirogaleus lavasoensis* and *C. minusculus*); (2) the distribution of a genus is restricted to western Madagascar (genus *Mirza*) or eastern Madagascar north of the Mangoro River (genus *Allocebus*); and (3) the genus *Phaner* because only western species were available for this study (the eastern species, *Phaner furcifer*, was not available).

Results

Cheirogaleid mitogenome sequence characteristics

We generated complete mitogenome sequences from 56 cheirogaleid individuals (Figure 3.1; Table 3.3). From each individual of the 40 cheirogaleid individuals sequenced by 454 Flx sequencing system technologies, we obtained an average of 35,677 tagged reads with an average length of 776 bp, yielding approximately 27,646 kb of sequence data corresponding to 1657-fold coverage. From each individual of 16 cheirogaleids sequenced by Illumina MiSeq sequencing technologies, we obtained an average of 524,260 paired-end reads with a length of 150 bp, yielding approximately 157,845 kb of sequence data corresponding to 9463-fold coverage. The length of all newly sequenced cheirogaleid mitogenomes varied from 16,439 (*C. sibreei*) to 17,466 bp (*M. rufus*), which are typical for primates, but the GC-content varied slightly among taxa (35.60%–38.87%, Table 3.3). The length differences among cheirogaleid mitogenomes mostly derived from the control region or D-Loop, which displays a highly variable number of repeats. All newly generated mitogenomes consisted of 22 tRNA genes, 2 rRNA genes, 13 protein-coding genes and the control region in the order typical for primates. By combining the 56 new mitogenomes with 51 additional primate mitogenomes, the data set represents all 5 living lemur families; all 15 recognized living lemur genera; and 29 of the 34 currently nominal cheirogaleid species except *Microcebus gerpi* and *M. marohita*, *Cheirogaleus lavasoensis* and *C. minusculus*, and *Phaner furcifer*.

Phylogenetic analyses

The same topology of the phylogenetic tree with high support values was obtained in both ML and Bayesian inference (BI) methods, except in a small number of clades (Figure 3.2). Lemurs appeared as a monophyletic clade with Daubentoniidae as the basal lineage. The sister-group relationship between

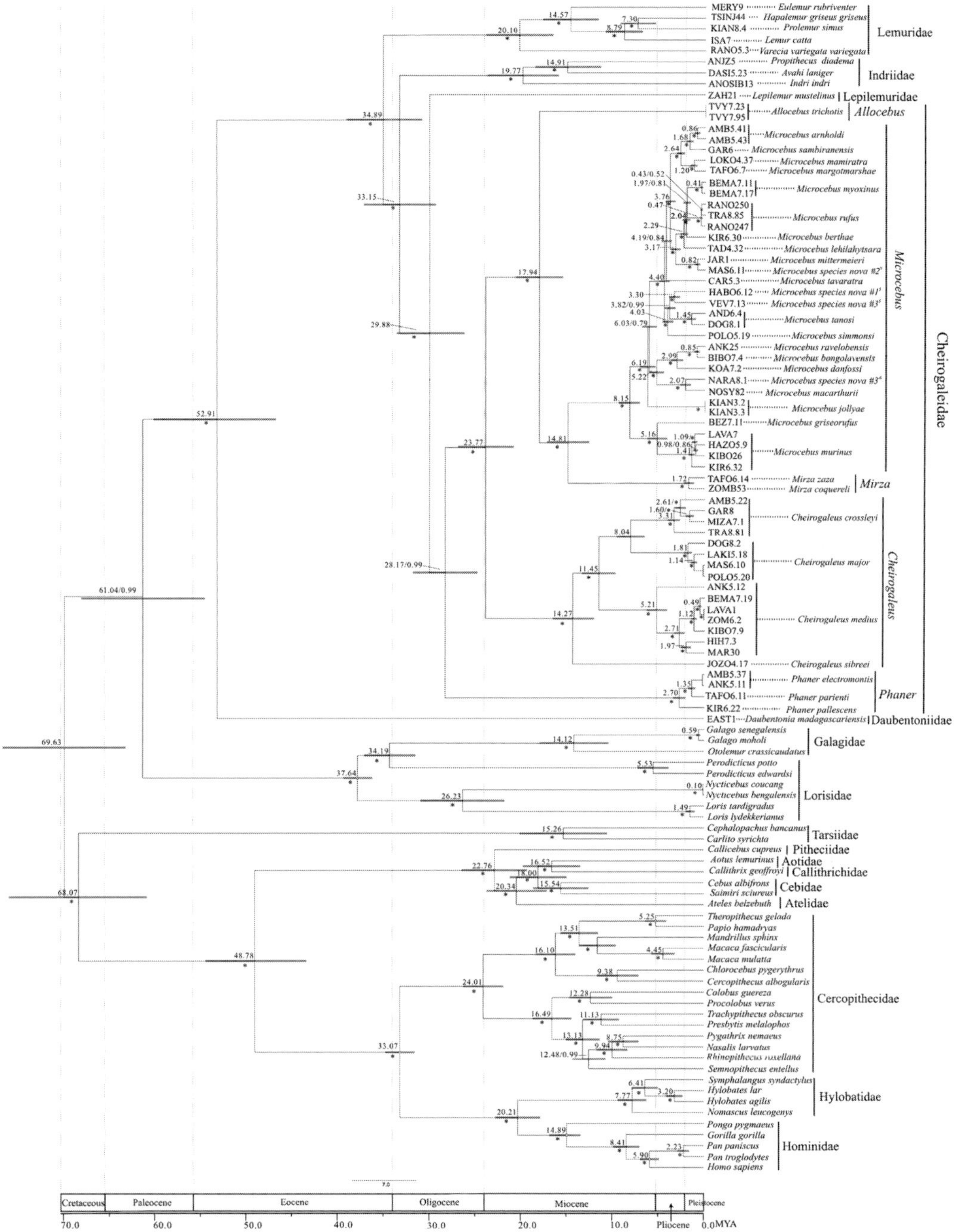

Figure 3.3 Estimated divergence dates for primates based on the mtDNA genome heavy-stranded coding region sequences of the 56 Cheirogaleidae individuals, the 9 outgroup lemur species, and the 32 primate outgroup taxa utilizing a Bayesian relaxed molecular clock analysis with BEAST (Drummond and Rambaut, 2007). The gray bars display the 95% HPD (highest probability density) interval of node ages. Open circles indicate nodes where fossil evidence was used to calibrate the chronogram. Numbers below the dash represent posterior probabilities and asterisks indicate posterior probabilities of 1.00.

Cheirogaleidae and Lepilemuridae are supported by both ML and BI analyses with 100% BS and 1.0 PP, respectively. Within the five families, Cheirogaleidae is clustered with Lepilemuridae, then Indriidae, then Lemuridae, and lastly Daubentoniidae. Among the five genera of Cheirogaleidae, all genera appeared as monophyletic clades and were shown to diverge in the following order: *Microcebus*, *Mirza*, *Allocebus*, *Cheirogaleus*, and lastly *Phaner*. *Mirza zaza* and *M. coquereli* form a single clade. Among the genus *Phaner*, *P. electromontis*, *P. parienti*, and *P. pallescens* diverged sequentially. Within the genus *Cheirogaleus*, each of the four species represented form a well-supported monophyletic clade, with *C. sibreei* as the basal taxon. Following the divergence of *C. sibreei*, there was a separation between *C. medius* and the *C. major*/*C. crossleyi* clade. *Cheirogaleus major* and *C. crossleyi* are each monophyletic and diverged from a common ancestor during the Miocene. The analysis of mitogenomes from 19 of 21 species of *Microcebus*, along with 4 proposed new species, produces a well-supported phylogeny in which each species is represented as a terminal node. Following the divergence of *M. murinus*/*M. griseorufus* (Clade 1), *M. jollyae* (Clade 2) diverged from the rest of the *Microcebus* species (Table 3.1). *Microcebus ravelobensis*/*M. bongolavensis* clustered with *M. danfossi* and then formed a clade (Clade 3) with *M. macarthurii* and *Microcebus species nova* #3. *Microcebus tanosi*, *Microcebus species nova* #1, *Microcebus species nova* #3 and *M. simmonsi* formed a distinct clade (Clade 4). *Microcebus tavaratra* formed its own distinct clade (Clade 5). *Microcebus myoxinus* formed a sister relationship with *M. rufus*, then clustered with *M. berthae*, *M. lehilahytsara*, and *M. mittermeieri*/*Microcebus species nova* #2 (Clade 6). *Microcebus arnholdi*, *M. sambiranensis*, and *M. mamiratra*/*M. margotmarshae* formed a distinct clade (Clade 7). Unlike the other genera, *Allocebus* only occurred as a single taxon. However, the distribution of the genus *Allocebus* extends from Marojejy to Mantadia/Analamazaotra/Mitsinjo Reserve and only a southern population was represented in this study (Figure 3.1).

The average p-distance of the HP sequence among cheirogaleid genera was from 0.144, between *Microcebus* and *Mirza*, to 0.196 between *Phaner* and *Allocebus* (Table 3.4). The average p-distance of the HP sequence among the genus *Cheirogaleus* was from 0.093, between *C. crossleyi* and *C. major*, to 0.123 between *C. sibreei* and *C. crossleyi* or *C. medius* (Table 3.5). The average p-distance of the HP sequence within each *Cheirogaleus* species was from 0.022, in *C. major*, to 0.042 in *C. crossleyi* and *C. medius*. The average p-distance of the HP sequence among the genus *Microcebus* was from 0.016, between *M. bongolavensis* and *M. ravelobensis*, to 0.103 between *M. tavaratra* and *M. griseorufus* (Table 3.6). The p-distance of the HP sequence within each *Microcebus* species was less than 0.024 (e.g. in *M. tanosi*). The average p-distance of the HP sequence among three *Phaner* species was 0.025, between *P. electromontis* and *P. parienti*; 0.044, between *P. electromontis* and *P. pallescens*; 0.045, between *P. pallescens* and *P. parienti*. The average p-distance of the HP sequence between *Mirza coquereli* and *M. zaza* was 0.03.

Table 3.4 Average p-distance of the HP sequence in heavy strand of mtDNA among and within the five genera of Cheirogaleidae.

	Microcebus	*Cheirogaleus*	*Allocebus*	*Mirza*	*Phaner*
Microcebus	**0.072**				
Cheirogaleus	0.167	**0.092**			
Allocebus	0.158	0.167	**0**		
Mirza	0.144	0.167	0.153	**0.03**	
Phaner	0.195	0.188	0.196	0.192	**0.031**

Table 3.5 Average p-distance of the HP sequence in heavy strand of mtDNA among and within species of the genus *Cheirogaleus*. There is no genetic distance among species (n/c), because it is only one sample.

	C. crossleyi	*C. medius*	*C. major*	*C. sibreei*
C. crossleyi	**0.042**			
C. medius	0.116	**0.042**		
C. major	0.093	0.114	**0.022**	
C. sibreei	0.123	0.123	0.121	**n/c**

Table 3.6 Average p-distance of the HP sequence in heavy strand of mtDNA among and within species of the genus *Microcebus*. 1, *M. arnholdi*; 2, *M. tanosi*; 3, *M. ravelobensis*; 4, *M. myoxinus*; 5, *M. griseorufus*; 6, *M. bongolavensis;* 7, *M. danfossi*; 8, *M. tavaratra*; 9, *M. sambiranensis*; 10, *M. species nova* #1 (Louis *et al.*, 2006a); 11, *M. murinus*; 12, *M. mittermeieri*; 13, *M. jollyae*; 14, *M. berthae*; 15, *M. mamiratra*; 16, *M. species nova* #2; 17, *M. species nova* #3 (Radespiel *et al.*, 2008); 18, *M. macarthurii*; 19, *M. simmonsi*; 20, *M. rufus*; 21, *M. lehilahytsara*; 22, *M. margot-marshae*; 23, *M. species nova* #3 (Louis *et al.*, 2006a).

	1	2	3	4	5	6	7	8	9	10	11
1	**0.015**										
2	0.058	**0.024**									
3	0.082	0.078	**n/c**								
4	0.056	0.057	0.078	**0.008**							
5	0.097	0.096	0.100	0.097	**n/c**						
6	0.081	0.077	0.016	0.078	0.1	**n/c**					
7	0.079	0.074	0.050	0.078	0.099	0.050	**n/c**				
8	0.069	0.066	0.084	0.066	0.103	0.084	0.085	**n/c**			
9	0.030	0.061	0.083	0.058	0.096	0.082	0.080	0.070	**n/c**		
10	0.058	0.051	0.078	0.057	0.095	0.077	0.077	0.067	0.059	**n/c**	
11	0.094	0.093	0.099	0.095	0.076	0.098	0.099	0.100	0.094	0.093	**0.021**
12	0.053	0.053	0.077	0.044	0.094	0.077	0.074	0.065	0.054	0.054	0.093
13	0.077	0.073	0.084	0.076	0.102	0.084	0.082	0.084	0.078	0.075	0.098
14	0.055	0.056	0.079	0.035	0.096	0.078	0.076	0.066	0.056	0.055	0.095
15	0.046	0.062	0.085	0.058	0.098	0.084	0.080	0.073	0.048	0.063	0.095
16	0.054	0.054	0.077	0.046	0.094	0.077	0.075	0.067	0.055	0.056	0.092
17	0.078	0.074	0.077	0.077	0.101	0.078	0.076	0.084	0.078	0.074	0.098

(continued)

Table 3.6 (*cont.*)

	1	2	3	4	5	6	7	8	9	10	11
18	0.079	0.075	0.079	0.078	0.100	0.079	0.076	0.084	0.079	0.075	0.098
19	0.060	0.055	0.082	0.059	0.097	0.081	0.078	0.070	0.061	0.056	0.094
20	0.055	0.055	0.080	0.034	0.094	0.079	0.078	0.066	0.058	0.054	0.095
21	0.053	0.054	0.078	0.035	0.094	0.078	0.075	0.064	0.055	0.054	0.092
22	0.045	0.063	0.085	0.059	0.098	0.083	0.081	0.072	0.048	0.063	0.095
23	0.057	0.049	0.078	0.057	0.094	0.077	0.076	0.066	0.059	0.047	0.092

	12	13	14	15	16	17	18	19	20	21	22	23
1												
2												
3												
4												
5												
6												
7												
8												
9												
10												
11												
12	n/c											
13	0.076	0.000										
14	0.045	0.075	n/c									
15	0.057	0.081	0.058	n/c								
16	0.015	0.075	0.047	0.058	n/c							
17	0.076	0.082	0.076	0.081	0.076	n/c						
18	0.076	0.084	0.076	0.082	0.077	0.036	n/c					
19	0.058	0.077	0.059	0.065	0.059	0.079	0.079	n/c				
20	0.046	0.076	0.033	0.059	0.047	0.078	0.078	0.060	0.009			
21	0.043	0.074	0.033	0.056	0.045	0.074	0.074	0.057	0.035	n/c		
22	0.057	0.080	0.059	0.024	0.057	0.082	0.082	0.065	0.059	0.057	n/c	
23	0.054	0.074	0.053	0.061	0.055	0.075	0.076	0.054	0.055	0.053	0.061	n/c

Dates of divergence estimations

Utilizing five fossil calibration points, we estimated mtDNA divergence dates within the family Cheirogaleidae (Figure 3.3 and Table 3.7). The divergence time between Lemuriformes and Lorisiformes was estimated as 61.04 mya. Daubentoniidae was estimated to have diverged from other lemurs at 52.91 mya, followed by a split between Lemuridae and Indriidae/Lepilemuridae/Cheirogaleidae at 34.89 mya, then between Indriidae and Lepilemuridae/Cheirogaleidae at 33.15 mya, and lastly between Lepilemuridae and Cheirogaleidae at 29.88 mya. Within Cheirogaleidae, the estimated divergence date between *Phaner* and *Cheirogaleus*/*Allocebus*/*Mirza*/*Microcebus* was

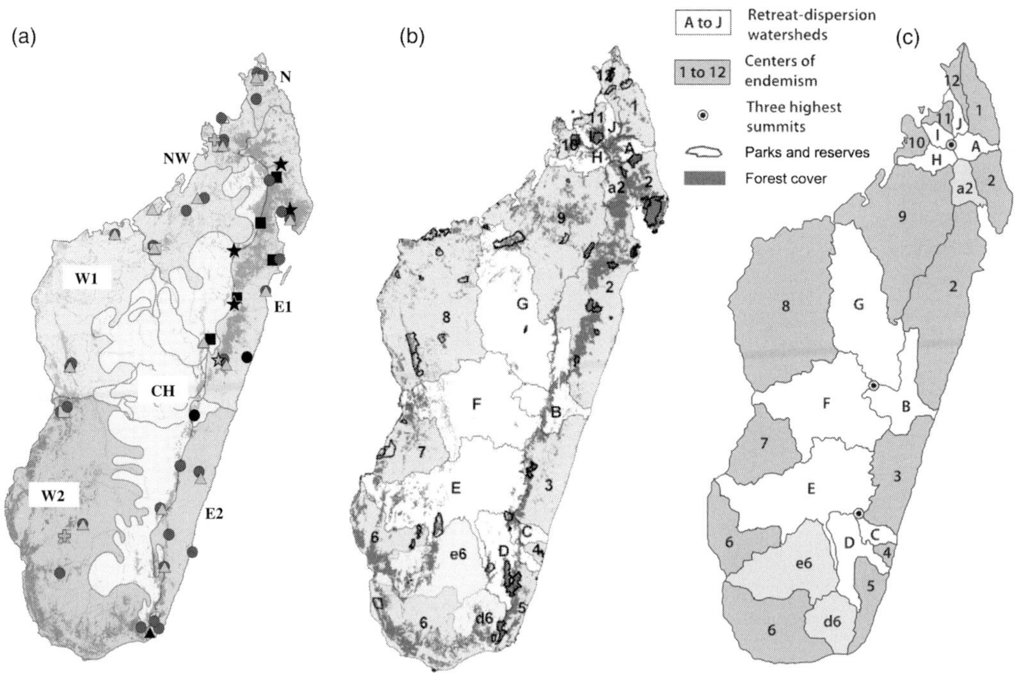

Figure 3.4 Biogeographic regions based on hypotheses proposed to explain biodiversity patterns in Madagascar. (a) Modified map presenting the seven biogeographic regions of lemur distributions based on phytogeographic criteria and similarities of lemur communities as proposed by Martin (1972, 1995), Thalmann (2000) and Ganzhorn *et al.* (2006). (b) Biogeographic map presenting "retreat-dispersion watersheds" (RWD) and Centers of Endemism (COE) modified from Wilmé *et al.* (2006). (c) Modified map presented in Townsend *et al.* (2009) showing RWD and COE for leaf chameleons (*Brookesia*). (A black and white version of this figure will appear in some formats. For the colour version, please refer to the plate section.)

28.17 mya, between *Cheirogaleus* and *Allocebus/Mirza/Microcebus* 23.77 mya, between *Allocebus* and *Mirza/Microcebus* 17.94 mya, and between *Mirza* and *Microcebus* 14.81 mya. The inferred timescale placed the time to most recent common ancestor (TMRCA) of extant Cheirogaleidae from the late Oligocene to the early Miocene (Figure 3.2). Without comparing samples from its entire range, the divergence date of the genus *Allocebus* is set firmly in the Miocene. The radiation of *Phaner* began with an initial divergence at 2.70 mya, followed by a split between *P. parienti* and *P. electromontis* at 1.35 mya. The split between *M. zaza* and *M. coquereli* was estimated at 1.72 mya. Within *Cheirogaleus*, the basal split of *C. sibreei* from *C. medius/C. major/C. crossleyi* was at 14.27 mya, followed by *C. medius* from *C. major/C. crossleyi* at 11.45 mya, lastly by *C. major* and *C. crossleyi* at 8.04 mya. Within *Microcebus*, Clade 1 split from Clades 2–6 at 8.15 mya, followed by Clade 2 from Clades 3–6 at 6.19 mya, then by Clade 3 from Clades 4–6 at 6.03 mya, then by Clade 4 from Clades 5–6 at 4.40 mya, and finally by Clade 5 from Clade 6 at 3.76 mya. Within the Family Cheirogaleidae the genera *Phaner*, *Mirza*, and *Microcebus* exhibited a rapid radiation during the Pliocene and Pleistocene climatic cycles. The genera *Allocebus* and *Cheirogaleus* experienced periods of speciation earlier, during the Miocene.

Table 3.7 Divergence ages with credibility intervals (in parentheses) based on highest posterior density.

Lineage divergence	CHA	PER	SPR	This study
Cheirogaleidae/ Lepilemuridae		39.76 (29.81–50.73)	27.63 (22.91–30.99)	29.88 (26.31–33.63)
Microcebus/Mirza		14.32 (7.80–21.88)	13.19 (9.66–17.29)	14.81 (12.70–17.26)
Allocebus/ Microcebus–Mirza			16.12 (11.16–20.87)	17.94 (15.54–20.68)
Microcebus			6.92 (4.97–8.81)	8.15 (7.06–9.31)
Cheirogaleus			9.34 (4.86–17.77)	14.27(12.27–16.73)
Cheirogaleus/Microcebus–Mirza–Allocebus		24.99 (15.07–34.52)	21.84 (17.39–26.62)	23.77(20.87–26.94)
Lemuriformes/ Chiromyiformes		63.97 (47.41–79.92)	49.96 (45.16–53.78)	52.91 (46.60–59.86)
Lemuriformes	32.4	32.92 (22.39–43.81)	31.77 (26.73–36.64)	34.89 (30.96–30.01)
Cheirogaleidae	23.6	24.99 (15.07–34.52)	21.84 (17.39–26.62)	28.17 (24.90–31.91)
Indriidae	20.6	17.43 (9.83–25.51)	18.45(11.71–25.51)	19.77 (15.89–23.59)
Lemuridae	21.3	26.19 (15.95–35.55)	20.73 (15.28–25.97)	20.10 (16.48–23.72)
Strepsirrhini	51.6	68.74 (58.77–76.61)	54.23 (48.75–57.22)	61.04 (54.27–67.79)
Lorisiformes	37.5	40.34 (35.17–45.61)	34.74 (27.90–38.24)	37.64 (36.90–39.99)
Lorisidae		36.98 (31.06–42.66)	32.28 (25.58–36.31)	26.23 (21.65–30.80)
Galagidae		19.90 (12.82–27.49)	22.67 (16.2–32.22)	34.19 (31.36–36.94)

Three sets of dates were calculated in the relaxed molecular clock analysis using the estimates or the origin of the primate clade taken from Chatterjee *et al.* (2009; CHA; origin = 77.5 million years), Perelman *et al.* (2011; PER; origin = 87.2 million years) and Springer *et al.* (2012; SPR; origin = 87.18 million years) as calibration settings.

Fit of biogeographic hypotheses

This study demonstrated species-level divergence times compatible for a Pliocene/ Pleistocene rapid radiation for the genera *Microcebus*, *Phaner*, and *Mirza* and *C. medius*, which satisfies the temporal component of both the WC (Wilmé *et al.*, 2006) and MR hypotheses (Raxworthy and Nussbaum, 1995). However, this study also demonstrated species-level divergence times consistent within the Miocene for the genus *Allocebus* and *C. major, C. crossleyi, and C. sibreei*, which potentially rejects the temporal component of both WC and MR hypotheses.

We compared the localities of the family Cheirogaleidae samples utilized in this study to the modified maps of Martin (1972), Wilmé *et al.* (2006) and Townsend *et al.* (2009) as shown in Figure 3.4 and summarized in Table 3.8. We accept the current nominal species as valid taxa in order to test the assumptions of the WC by determining and evaluating which region, COE, or RDW each species occupies (Figures 3.4 and 3.5; Table 3.8). Based on the available sampling locality data, we suggest the following status for the species examined in this study. Of the 19 recognized *Microcebus* species and 4 proposed species, 16 are limited to single COEs, 2 are found in 2 COEs, 3 are limited to a single RDW, and 2 are found in 2 RDWs. Of

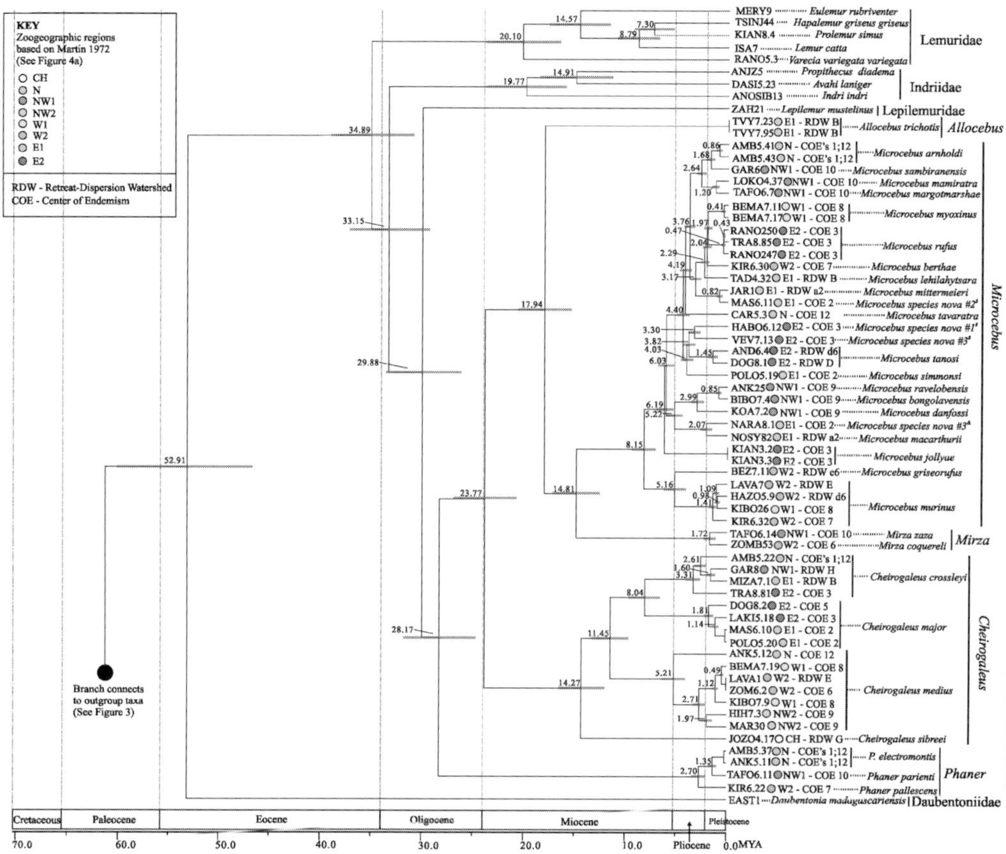

Figure 3.5 Truncated view of Figure 3.3 of estimated divergence dates for the family Cheirogaleidae in Lemuriformes based on the mtDNA genome heavy-stranded coding region sequences of the 56 Cheirogaleidae individuals and the 9 outgroup lemur species utilizing a Bayesian relaxed molecular clock analysis with BEAST (Drummond and Rambaut, 2007). The gray bars display the 95% HPD (highest probability density) interval of node ages. Colored circles following taxa designation correspond to biogeographic regions in Figure 3.4a. Abbreviations designate specific regions in Figure 3.4a and COE and RDW for each lemur taxon located in Figures 3.4b and 3.4c. (A black and white version of this figure will appear in some formats. For the colour version, please refer to the plate section.)

the four *Cheirogaleus* species, all are found in multiple COEs and two are found in two RDWs. Both *Mirza* and *Phaner* species are limited to a single COE except for *Phaner electromontis*, which straddles two COEs. The monotypic genus *Allocebus* is limited to a single RDW. Although this study presents a significant number of novel mitogenomes for the family Cheirogaleidae, it does not quantitatively represent the distributions of all species to evaluate the biogeographic hypothesis models robustly, especially for the genera *Cheirogaleus*, *Phaner*, *Mirza*, and *Allocebus*. However, there are sufficient data for the genus *Microcebus* for selected sister-taxon relationships that can be evaluated for the biogeographic hypotheses (Table 3.9).

Table 3.8 Summary of biogeography hypothesis models for the family Cheirogaleidae. Biogeographic regions (Bio-R) and Centers of Endemism (COE) or "retreat-dispersion watersheds" (RDW) are designated for the genera *Microcebus* (*M.*), *Cheirogaleus* (*C.*), *Mirza*, and *Phaner* (*P.*) correlated to Figures 3.4 and 3.5. River boundaries are listed if known or designated as Not Assigned (N/A) if the species' distribution is not verified or not strongly influenced by river barriers. A Pliocene/Pleistocene (Pli/Pleist) radiation which supports the biogeography hypothesis or Miocene radiation which could potentially reject a biogeography hypothesis is also listed for each species.

Species information			Distribution		Biogeography hypotheses models			
ID	Species	Locality	Bio-R	COE or RDW	Riverine boundary	Riverine boundary	Watershed contraction	Mountain refuge
AMB5.41	*M. arnholdi*	Montagne d'Ambre	N	COEs 1;12	Irodo	Mananjeba	Pli/Pleist	Pli/Pleist
AMB5.43	*M. arnholdi*	Montagne d'Ambre	N	COEs 1;12	Irodo	Mananjeba	Pli/Pleist	Pli/Pleist
CAR5.3	*M. tavaratra*	Ankarana	N	COEs 1;12	Irodo	Manambato	Pli/Pleist	Pli/Pleist
LOKO4.37	*M. mamiratra*	Lokobe	NW1	COE 11	Mahavavy-Nord[c]	Sambirano	Pli/Pleist	Pli/Pleist
GAR6	*M. sambiranensis*	Manongarivo	NW1	COE 10	Sambirano	Andranomalaza	Pli/Pleist	Pli/Pleist
TAFO6.7	*M. margotmarshae*	Antafondro	NW1	COE 10	Andranomalaza	Maevarano	Pli/Pleist	Pli/Pleist
KOA7.2	*M. danfossi*	Andalirano	NW1	COE 9	Maevarano	Sofia	Pli/Pleist	Pli/Pleist
BIBO7.4	*M. bongolavensis*	Ambodimahabibo	NW1	COE 9	Sofia	Mahajamba	Pli/Pleist	Pli/Pleist
ANK25	*M. ravelobensis*	Ankarafantsika	NW1	COE 9	Mahajamba	Betsiboka	Pli/Pleist	Pli/Pleist
BEMA7.11	*M. myoxinus*	Tsingy de Bemaraha	W1	COE 8	N/A	N/A	Pli/Pleist	Pli/Pleist
BEMA7.17	*M. myoxinus*	Tsingy de Bemaraha	W1	COE 8	N/A	N/A	Pli/Pleist	Pli/Pleist
KIBO26	*M. murinus*	Tsiombikibo	W1	COE 8	N/A	N/A	Pli/Pleist	Pli/Pleist
KIR6.30	*M. berthae*	Kirindy	W2	COE 7	Tsiribihina	Morondava	Pli/Pleist	Pli/Pleist
KIR6.32	*M. murinus*	Kirindy	W2	COE 7	N/A	N/A	Pli/Pleist	Pli/Pleist
LAVA7	*M. murinus*	Analalava	W2	RDW E	N/A	N/A	Pli/Pleist	Pli/Pleist
HAZO5.9	*M. murinus*	Ihazofotsy	W2	RDW d6	N/A	N/A	Pli/Pleist	Pli/Pleist
BEZ7.11	*M. griseorufus*	Beza-Mahafaly	W2	RDW e6	Onilahy	Mandrare	Pli/Pleist	Pli/Pleist

JAR1	*M. mittermeieri*	Anjanaharibe-Sud	E1	RDW a2	Bemarivo	Voloina	Pli/Pleist	Pli/Pleist
MAS6.11	*M. species nova #2*	Masoala	E1	COE 2	Masoala[d]	Ankavana	Pli/Pleist	Pli/Pleist
NOSY82	*M. macarthurii*	Nosy Mangabe	E1	RDW a2	Antainambalana	Voloina	Pli/Pleist	Pli/Pleist
NARA8.1	*M. species nova #3*[b]	Mananara-Nord	E1	COE 2	N/A	N/A	Pli/Pleist	Pli/Pleist
POLO5.19	*M. simmonsi*	Tampolo	E1	COE 2	Maningory	Rongaronga	Pli/Pleist	Pli/Pleist
TAD4.32	*M. lehilahytsara*	Mantadia	E1	RDW B	Rongaronga	Mangoro	Pli/Pleist	Pli/Pleist
RANO247	*M. rufus*	Ranomafana	E2	COE 3	Mananjary	Sandrananta	Pli/Pleist	Pli/Pleist
RANO250	*M. rufus*	Ranomafana	E2	COE 3	Mananjary	Sandrananta	Pli/Pleist	Pli/Pleist
TRA8.85	*M. rufus*	Andringitra	E2	COE 3	Mananjary	Sandrananta	Pli/Pleist	Pli/Pleist
KIAN3.2	*M. jollyae*	Kianjavato	E2	COE 3	Mananjary	Sahambavy	Pli/Pleist	Pli/Pleist
KIAN3.3	*M. jollyae*	Kianjavato	E2	COE 3	Mananjary	Sahambavy	Pli/Pleist	Pli/Pleist
HABO6.12	*M. species nova #1*[a]	Mahabo	E2	COE 3	Manampatrana	Mananara	Pli/Pleist	Pli/Pleist
VEV7.13	*M. species nova #3*[a]	Vevembe	E2	COE 3	Manampatrana	Mananara	Pli/Pleist	Pli/Pleist
DOG8.1	*M. tanosi*	Midongy du Sud	E2	RDW D	Mananara	Mandrare	Pli/Pleist	Pli/Pleist
AND6.4	*M. tanosi*	Andohahela	E2	RDW d6	Mananara	Mandrare	Pli/Pleist	Pli/Pleist
AMB5.22	*C. crossleyi*	Montagne d'Ambre	N	COEs 1;12	N/A	N/A	Miocene	Miocene
ANK5.12	*C. medius*	Ankarana	N	COE 12	N/A	N/A	Pli/Pleist	Pli/Pleist
BEMA7.19	*C. medius*	Tsingy de Bemaraha	W1	COE 8	N/A	N/A	Pli/Pleist	Pli/Pleist
DOG8.2	*C. major*	Midongy du Sud	E2	COE 5	N/A	N/A	Miocene	Miocene
LAKI5.18	*C. major*	Lakia	E2	COE 3	N/A	N/A	Miocene	Miocene
MAS6.10	*C. major*	Masoala	E1	COE 2	N/A	N/A	Miocene	Miocene
POLO5.20	*C. major*	Tampolo	E1	COE 2	N/A	N/A	Miocene	Miocene
GAR8	*C. crossleyi*	Manongarivo	NW1	RDW H	N/A	N/A	Miocene	Miocene
MIZA7.1	*C. crossleyi*	Maromizaha	E1	RDW B	N/A	N/A	Miocene	Miocene
HIH7.3	*C. medius*	Anjiamangirana	NW2	COE 9	N/A	N/A	Pli/Pleist	Pli/Pleist
KIBO7.9	*C. medius*	Tsiombikibo	W1	COE 8	N/A	N/A	Pli/Pleist	Pli/Pleist

(*continued*)

Table 3.8 *(cont.)*

Species information			Distribution		Biogeography hypotheses models			
ID	Species	Locality	Bio-R	COE or RDW	Riverine boundary	Riverine boundary	Watershed contraction	Mountain refuge
LAVA1	*C. medius*	Analalava	W2	RDW E	N/A	N/A	Pli/Pleist	Pli/Pleist
MAR30	*C. medius*	Mariarano	NW2	COE 9	N/A	N/A	Pli/Pleist	Pli/Pleist
TRA8.81	*C. crossleyi*	Andringitra	E2	COE 3	N/A	N/A	Miocene	Miocene
JOZO4.17	*C. sibreei*	Anjozorobe	CH	RDW G	N/A	N/A	Miocene	Miocene
ZOM6.2	*C. medius*	Zombitse	W2	COE 6	N/A	N/A	Pli/Pleist	Pli/Pleist
TVY7.23	*A. trichotis*	Ambatovy	E1	RDW B	N/A	N/A	Miocene	Miocene
TVY7.95	*A. trichotis*	Ambatovy	E1	RDW B	N/A	N/A	Miocene	Miocene
TAF06.14	*Mirza zaza*	Antafondro	NW1	COE 10	N/A	N/A	Pli/Pleist	Pli/Pleist
ZOMB53	*Mirza coquereli*	Zombitse	W2	COE 6	N/A	N/A	Pli/Pleist	Pli/Pleist
AMB5.37	*P. electromontis*	Montagne d'Ambre	N	COEs 1;12	Irodo	Mananjeba	Pli/Pleist	Pli/Pleist
ANK5.11	*P. electromontis*	Ankarana	N	COEs 1;12	Irodo	Mananjeba	Pli/Pleist	Pli/Pleist
TAF06.11	*P. parienti*	Antafondro	NW1	COE 10	N/A	N/A	Pli/Pleist	Pli/Pleist
KIR6.22	*P. pallescens*	Kirindy	W2	COE 7	N/A	N/A	Pli/Pleist	Pli/Pleist

Note: [a]Louis *et al.*, 2006a; [b]Radespiel *et al.*, 2008; [c]Nosy Be is the primary locality for *M. mamiratra*; [d]Masoala represents the Masoala Peninsula as a potential boundary. Shaded rows represent selected regional groups of *Microcebus* that have more thorough data and are discussed in the context of biogeographic hypothesis models.

Table 3.9 Summary of analyses of biogeographic hypotheses for selected *Microcebus* sister taxa. Sister taxa A–C represent evaluations of geographically adjacent sister taxa. Sister taxa D–F represent evaluations of phylogenetic sister taxa. "Yes" indicates agreement with the respective biogeographic hypothesis. Shaded blocks are shared boundary rivers between phylogenetic and/or geographic sister taxa.

Sister taxa	Species	Ecogeographic constraint (EC)		Riverine barrier (RB)		Watershed contraction (WC)		Mountain refuge (MR)	
		Biogeographic region		Boundary river A	Boundary river B	Pliocene/ Pleistocene divergence	COE/RDW	Pliocene/ Pleistocene divergence	Mountain refugia
A	*M. arnholdi*	N	Yes	Irodo	Mananjeba	Yes	COE 1 & 12	Yes	Yes
	M. tavaratra	N	Yes	Irodo	Manambato	Yes	COE 1 & 12	Yes	No
B	*M. sambiranensis*	NW1	Yes	Sambirano	Andranomalaza	Yes	COE 10	Yes	No
	M. margotmarshae	NW1	Yes	Andranomalaza	Maevarano	Yes	COE 10	Yes	No
C	*M. danfossi*	NW1	Yes	Maevarano	Sofia	Yes	COE 9	Yes	No
	M. bongolavensis	NW1	Yes	Sofia	Mahajamba	Yes	COE 9	Yes	No
D	*M. arnholdi*	N	Yes	Irodo	Mananjeba	Yes	COE 1 & 12	Yes	Yes
	M. sambiranensis	NW1	Yes	Sambirano	Andranomalaza	Yes	COE 10	Yes	No
E	*M. mamiratra*	NW1	Yes	Mahavavy-Nord	Sambirano	Yes	COE 11	Yes	No
	M. margotmarshae	NW1	Yes	Andranomalaza	Maevarano	Yes	COE 10	Yes	No
F	*M. bongolavensis*	NW1	Yes	Sofia	Mahajamba	Yes	COE 9	Yes	No
	M. ravelobensis	NW1	Yes	Mahajamba	Betsiboka	Yes	COE 9	Yes	No

Discussion

Using mitogenomic sequences to analyze the phylogenetic relationships of 56 extant cheirogaleid species belonging to five genera, along with nine genera from the other four lemur families, the mitochondrial phylogeny of Lemuriformes was robustly described with strong support values for a majority of the nodes. Our results were largely supported by previous publications based on supermatrices, nuclear sequence and transposable element data (Roos *et al.*, 2004; Chatterjee *et al.*, 2009; Fabre *et al.*, 2009; Perelman *et al.*, 2011; Springer *et al.*, 2012; Veilleux *et al.*, 2013). According to supermatrix data presented by Chatterjee *et al.* (2009), Fabre *et al.* (2009) and Springer *et al.* (2012), a sister-group relationship was found between Lepilemuridae and the genus *Phaner*, thereby leading to questions about the monophyly of Cheirogaleidae. With approximately 60% of species data missing within these analyses, this taxonomic grouping can be regarded as weakly supported. In our study, *Phaner* clustered within Cheirogaleidae as the basal lineage. Although the sister-group relationship between Indriidae and Lemuridae was proposed by Chatterjee *et al.* (2009), Perelman *et al.* (2011) argued that a trichotomy of Lemuridae, Indriidae and the Lepilemuridae/Cheirogaleidae lineage was unresolved because of relatively weak nodal support. Finstermeier *et al.* (2013) utilized mitogenomic data to investigate the phylogeny of living primates and found that the relationships of these four lemur families were unresolved, but limited sampling may have impacted these analyses. The phylogenetic analyses based on mitogenome sequences revealed sister-group relationships between Lepilemuridae and Cheirogaleidae, followed by Indriidae, then Lemuridae, but with moderate support value for the relationship between this final pairing (58/0.99; Figure 3.2). A data set of mitogenomes in conjunction with nuclear DNA data, along with morphological data, would provide the clearest possible resolution of this issue. The phylogenetic relationships of the five genera of *Cheirogaleus* were in agreement with Chatterjee *et al.* (2009), Fabre *et al.* (2009) and Springer *et al.* (2012), except that *Phaner* was grouped in Cheirogaleidae, instead of Lepilemuridae, which was congruent with phylogenetic analyses of Veilleux *et al.* (2013) utilizing the *OPN1SW* gene in nocturnal lemurs and Roos *et al.* (2004) utilizing the transposable element data set in Strepsirrhini.

The phylogenetic relationships of four surveyed *Cheirogaleus* species, based on mitogenomic sequences, were congruent with Groeneveld *et al.* (2009, 2010) and Thiele *et al.* (2013), but with higher statistical support. Mitogenomic data indicated strong mitochondrial phylogeographic structure, with extremely divergent haplotypes, within *C. medius* and *C. crossleyi*, which agreed with Hapke *et al.* (2005), Groeneveld *et al.* (2009, 2010) and Thiele *et al.* (2013). Although Hapke *et al.* (2005) and Groeneveld *et al.* (2009) attributed this diversity to the intraspecific level, we found a 4.2% sequence variation among both *C. medius* and *C. crossleyi*, which is larger than the genetic distance between *M. coquereli* and *M. zaza* (3%), and *P. electromontis* and *P. pallescens* (2.5%), but similar to the genetic distance between *P. electromontis* and *P. parienti* (4.4%) and *P. pallescens* and *P. parienti* (4.5%). We agree with Thiele *et al.* (2013) that there is more species diversity harboring in

Cheirogaleus, especially in populations from Ankarana National Park, Montagne d'Ambre Special Reserve, and Ranomafana National Park.

Previous studies based on the partial mitogenome sequence data and/or nuclear DNA sequences failed to resolve phylogenetic relationships among *Microcebus* (Louis *et al.*, 2006a; Olivieri *et al.*, 2007; Louis *et al.*, 2008; Radespiel *et al.*, 2008, 2012; Weisrock *et al.*, 2010). In this study, a highly resolved mitochondrial phylogeny was obtained for the nominal 19 *Microcebus* species and 4 distinct lineages by analyzing divergence between species using the whole mitogenome. This provides additional structural support for the species-status of the 23 previously described lineages in this genus (Louis *et al.*, 2006a; Olivieri *et al.*, 2007; Louis *et al.*, 2008; Radespiel *et al.*, 2008, 2012; Rasoloarison *et al.*, 2013).

Based on mtDNA partial genome sequence data, Yoder and Heckman (2006) suggested a primary north–south split between phylogenetic lineages among *Microcebus* species, instead of an east–west disjunction. However, our complete mitogenome sequence data revealed a more complicated evolutionary pattern in genus *Microcebus* (Figure 3.2). Multiple longitudinal exceptions to the EC hypothesis which was predicated on a strict eastern and western split in the genus *Microcebus* were identified as follows: *M. ravelobensis*, *M. bongolavensis*, and *M. danfossi* from the northwest and *M. species nova #*3 Radespiel *et al.* (2012) and *M. macarthurii* from the northeast; *M. myoxinus*, and *M. berthae* from the central west and *M. lehilahytsara* and *M. rufus* from the central east and southeast, respectively. With the majority of rivers in Madagascar coursing along an east–west axis with their sources at higher elevation on the central plateau, it is logical that the evolutionary pattern of genus *Microcebus* follows an east–west disjunction across latitudinal regions, instead of a north–south split. Each distinct lineage or clade in *Microcebus* is separated by riverine boundaries, but river corridors have obviously allowed longitudinal dispersal during favorable conditions followed by contraction and isolation of populations.

Different hypotheses have been utilized to explain the speciation of taxa in Madagascar, based upon watersheds, riverine barriers, montane refugia, western rainforest refugia, and ecogeographical constraints (Martin, 1972; Raxworthy and Nussbaum, 1995; Goodman and Ganzhorn, 2004; Wilmé *et al.*, 2006; Townsend *et al.*, 2009; Vences *et al.*, 2009). The temporal component of the WC and MR hypotheses was confidently accepted for the genera *Microcebus*, *Phaner*, and *Mirza* and one species of *Cheirogaleus*, *C. medius*. However, the species-level divergence times for the genus *Allocebus* and *C. major*, *C. crossleyi*, *and C. sibreei* were consistent with a Miocene radiation which potentially rejects the temporal component of both WC and MR hypotheses. With the prospective expansion of species diversity within the genus *Cheirogaleus*, species-level divergence times could subsequently shift for the *C. major* and *C. crossleyi* groups to a Pliocene/Pleistocene radiation (Thiele *et al.*, 2013). Thus, we cannot fully reject the temporal component of the MR or WC hypotheses for the genus *Cheirogaleus*. Even so, to our knowledge, this finding in lemurs represents the first study in Madagascar to report species-level radiations compatible with a widespread Pleistocene–Pliocene speciation (Townsend *et al.*, 2009 and references therein).

Comparing the localities of the nominal species to modified maps of Martin (1972), the findings of Wilmé *et al.* (2006) and Townsend *et al.* (2009) are extremely difficult to interpret for multiple reasons. First, the entire distributions of the member taxa of the five genera in the family Cheirogaleidae are not comprehensively represented in the molecular data or spatial mapping. Additionally, as mentioned previously, the taxonomy of the genera *Microcebus* and *Cheirogaleus* are certain to undergo further revision, so the current distribution maps are not accurate. Thus, allocating species in COEs and RDWs is currently difficult. Therefore, the spatial data presented are not comprehensively useful in significantly evaluating the three hypotheses. However, the evaluation of biogeographic hypotheses for selected sister taxa in the genus *Microcebus* is possible (Table 3.9), assessing them utilizing geographically adjacent taxa versus phylogenetically supported comparisons. As the Wilmé *et al.* (2006) WR hypothesis was based on extant species distributions structured in a non-phylogenetic context with climatic, hydrologic and topographic data, we compared a small subset of the closest geographically aligned *Microcebus* species as presented in Table 3.9A–C. For instance, the evaluation of *M. arnholdi*/*M. tavaratra* revealed that the geographically adjacent taxa belong to the same northwestern COE, live in the northern biogeographic region (N), and are separated by a relatively small river (Irodo). Evidence also indicates speciation on mountain refugia for only *M. arnholdi*. The evaluation of the geographically adjacent sister pair *M. sambiranensis*/*M. margotmarshae* showed that these species belong to the same COE, are separated by a relatively small river (Andranomalaza), and live in the same biogeographic region (NW1). There is no support for speciation on mountain refugia for either species. The evaluation of the geographically adjacent sister pair *M. danfossi*/*M. bongolavensis* showed that these species belong to the same COE, are separated by a large river (Sofia), and live in the same biogeographic region (NW1). There is no support for speciation on mountain refugia for either species. Although the historical relationships between the geographically adjacent *Microcebus* species are not incorporated in this manner, a single river barrier could be easily demonstrated.

However, phylogenetic relationships are important to identify how factors influence divergence and species formation, along with the high degree of microendemism found in Madagascar biodiversity. The evaluation of phylogenetic sister taxa *M. arnholdi*/*M. sambiranensis* showed that both species belong to different but not adjacent COEs, are separated by a large distance, live in different biogeographic regions, and have evidence supporting a speciation on a mountain refugia in the case of *M. arnholdi*. With an assumption of a stable geographic position of populations and the expectation that sister taxa would occupy adjacent COEs, these phylogenetic sister taxa are located quite a distance from each other. The evaluation of the phylogenetic sister taxa *M. mamiratra*/*M. margotmarshae* showed that both species belong to adjacent COEs, are separated by a large river (Sambirano), occupy the same biogeographic region (NW1) in the northwest, and have no evidence indicating speciation on different mountain refugia. Additionally, *M. mamiratra* is located on the island of Nosy Be. The evaluation of phylogenetic sister taxa *M. ravelobensis*/*M. bongolavensis* showed that both species belong to the same western COE, are separated by a large river (Mahajamba), and have no evidence indicating speciation on different mountain refugia. These limited numbers of evaluations based on phylogenetic context uncover different combinations of accepted and rejected criteria as only some of the assumptions of each biogeographic hypothesis are met for these *Microcebus* sister taxa. Riverine barriers are obviously important to speciation in this genus. Additionally, rivers such as Sofia, Betsiboka, Tsiribihina, and Mangoky act as corridors that provide dispersal and gene flow pathways between opposite coasts (Pastorini *et al.*, 2003; Louis *et al.*, 2006a). Additional samples will be needed to address true distributions across Madagascar and to evaluate the biogeographic hypotheses for the entire family Cheirogaleidae.

In conclusion, our study is the first to utilize complete mitogenomic analyses to elucidate the phylogeny of the family Cheirogaleidae. The phylogenetic history revealed by high-resolution mitogenomic analyses provides important information about the evolutionary history of Cheirogaleidae, and will contribute to the conservation of these species by providing opportunities to apply for species-specific conservation grant funding and increasing awareness of their plight. Future studies should concentrate on increased comprehensive sampling of all taxa and ecosystems, along with mitogenomic and nuclear DNA loci data, to continue to explore the evolutionary processes and diversity of Madagascar's Cheirogaleidae.

Acknowledgments

We thank the Madagascar National Parks, formerly Association Nationale pour la Gestion des Aires Protégées, and Ministère de l'Environnement des Eaux et Forêts for sampling permission. We are grateful to the Ahmanson Foundation, the Theodore F. and Claire M. Hubbard Family Foundation, Conservation International, the Primate Action Fund, the Margot Marsh Biodiversity Foundation, and the National Geographic Society for financial assistance. We thank Dr. Logan Kistler, George H. Perry and the Penn State Huck Institutes DNA Core Laboratory, and MOgene LC for assistance with sequence data collection and analyses. We would like to thank Adam T. McLain, Cynthia Frasier, Justin Taylor and Carolyn Bailey for providing helpful comments on an earlier version of the drafts and Azure Ginter for her graphics expertise in producing the figures. Finally, we would be completely amiss without acknowledging the significant contributions of the reviewers and editors who contributed valuable insights, comments and suggestions which vastly improved our chapter contribution.

References

Anderson S, Bankier AT, Barrell BG, *et al.* 1981. Sequence and organization of the human mitochondrial genome. *Nature* 290:457–465.

Andriaholinirina N, Fausser J-L, Roos C, *et al.* 2006. Molecular phylogeny and taxonomic revision of the sportive lemurs (*Lepilemur*, Primates). *BMC Evolutionary Biology* 6:17.

Andriantompohavana R, Zaonarivelo JR, Engberg SE, *et al.* 2006. Mouse lemurs of northwestern Madagascar with a description of a new species at Lokobe Special Reserve. *Occasional Papers of the Museum of Texas Tech University* 259:1–24.

Arnason U, Gullberg A, Janke A. 1998. Molecular timing of primate divergences as estimated by two nonprimate calibration points. *Journal of Molecular Evolution* 47:718–727.

Avise JC. 1998. The history and purview of phylogeography: a personal reflection. *Molecular Ecology* 7:371–379.

Benefit BR, McCrossin ML. 2002. The Victoriapithecidae, Cercopithecoidea. In WC Hartwig (ed.), *The Primate Fossil Record* (pp. 241–253). Cambridge University Press, Cambridge.

Bernt M, Donath A, Jühling F, *et al.* 2013. MITOS: improved *de novo* metazoan mitochondrial genome annotation. *Molecular Phylogenetics and Evolution* 69:313–319.

Bjork A, Liu W, Wertheim JO, Hahn BH, Worobey M. 2011. Evolutionary history of chimpanzees inferred from complete mitochondrial genomes. *Molecular Biology and Evolution* 28:615–623.

Blanco MB, Godfrey LR, Rakotondratsima M, *et al.* 2009. Discovery of sympatric dwarf lemur species in the high altitude forest of Tsinjoarivo, Eastern Madagascar: implications for biogeography and conservation. *Folia Primatologica* 80:1–17.

Boore JL. 2006. Requirements and standards for organelle genome databases. *OMICS* 10:119–126.

Bourlat SJ, Nielsen C, Economou AD, Telford MJ. 2008. Testing the new animal phylogeny: a phylum level molecular analysis of the animal kingdom. *Molecular Phylogenetics and Evolution* 49:23–31.

Brunet M, Guy F, Pilbeam D, *et al.* 2002. A new hominid from the Upper Miocene of Chad, Central Africa. *Nature* 418:145–151.

Chan Y-C, Roos C, Inoue-Murayama M, *et al.* 2010. Mitochondrial genome sequences effectively reveal the phylogeny of *Hylobates* gibbons. *PLoS ONE* 5:e14419.

Chatterjee HJ, Ho SYW, Barnes I, Groves C. 2009. Estimating the phylogeny and divergence times of primates using a supermatrix approach. *BMC Evolutionary Biology* 9:259.

Craul M, Zimmermann E, Rasoloharijaona S, Randrianambinina B, Radespiel U. 2007. Unexpected species diversity of Malagasy primates (*Lepilemur* spp.) in the same biogeographical zone: a morphological and molecular approach with the description of two new species. *BMC Evolutionary Biology* 7:83.

Crovella S, Montagnon D, Rumpler Y. 1995. Highly repeated DNA sequences and systematics of Malagasy primates. *Human Evolution* 10:35–44.

Dene H, Goodman M, Prychodko W, Moore GW. 1976. Immunodiffusion systematics of the Primates. III. The Strepsirrhini. *Folia Primatologica* 25:35–61.

Drummond AJ, Rambaut A. 2007. BEAST: Bayesian evolutionary analysis by sampling trees. *BMC Evolutionary Biology* 7:214.

Drummond AJ, Ho SYW, Phillips MJ, Rambaut A. 2006. Relaxed phylogenetics and dating with confidence. *PLoS Biology* 4:e88.

Fabre PH, Rodrigues A, Douzery EJP. 2009. Patterns of macroevolution among Primates inferred from a supermatrix of mitochondrial and nuclear DNA. *Molecular Phylogenetics and Evolution* 53:808–825.

Finstermeier K, Zinner D, Brameier M, *et al.* 2013. A mitogenomic phylogeny of living primates. *PLoS ONE* 8: e69504.

Ganzhorn JU, Goodman SM, Nash S, Thalmann, U. 2006. Lemur biogeography. In J Fleagle, SM Lehman (eds.), *Primate Biogeography: Progress and Prospects* (pp. 229–254). Kluwer Academic Press, New York.

Garbutt N. 2000. Brief observations of hairy-eared dwarf lemur (*Allocebus trichotis*) in Analamazaotra Special Reserve, eastern Madagascar. *Lemur News* 6:37.

Glazko GV, Nei M. 2003. Estimation of divergence times for major lineages of primate species. *Molecular Biology and Evolution* 20:424–434.

Godfrey LR, Jungers WL. 2003. The extinct sloth lemurs of Madagascar. *Evolutionary Anthropology* 12:252–263.

Godfrey LR, Jungers WL, Burney DA. 2010. Subfossil lemurs of Madagascar. In L Werdelin, WJ Sanders (eds.), *Cenozoic Mammals of Africa* (pp. 351–368). University of California Press, Berkeley.

Godinot M. 2006. Lemuriform origins as viewed from the fossil record. *Folia Primatologica* 77:446–464.

Goodman SM, Ganzhorn JU. 2004. Biogeography of lemurs in the humid forests of Madagascar: the role of elevational distribution and rivers. *Journal of Biogeography* 31:47–55.

Goodman SM, Raselimanana AP. 2002. The occurrence of *Allocebus trichotis* in the Parc National de Marojejy. *Lemur News* 7:21–22.

Groeneveld LF, Weisrock DW, Rasoloarison RM, Yoder AD, Kappeler PM. 2009. Species delimitations in lemurs: multiple genetic loci reveal low levels of species diversity in the genus *Cheirogaleus*. *BMC Evolutionary Biology* 9:30.

Groeneveld LF, Blanco MB, Raharison J-L. 2010. MtDNA and nDNA corroborate existence of sympatric dwarf lemur species at Tsinjoarivo, eastern Madagascar. *Molecular Phylogenetics and Evolution* 55:833–845.

Groeneveld LF, Rasoloarison RM, Kappeler PM. 2011. Morphometrics confirm taxonomic deflation in dwarf lemurs (Primates: Cheirogaleidae), as suggested by genetics. *Zoological Journal of the Linnean Society* 161:229–244.

Groves CP. 2000. The genus *Cheirogaleus*: unrecognized biodiversity in dwarf lemurs. *International Journal of Primatology* 21:943–962.

Groves CP. 2001. *Primate Taxonomy*. Smithsonian Institution Press, Washington, DC.

Groves CP, Tattersall I. 1991. Geographical variation in the fork-marked lemur, *Phaner furcifer* (Primates, Cheirogaleidae). *Folia Primatologica* 56:39–49.

Guindon S, Gascuel O. 2003. A simple, fast, and accurate algorithm to estimate large phylogenies by maximum likelihood. *Systematic Biology* 52:696–704.

Guindon S, Lethiec F, Duroux P, Gascuel O. 2005. PHYML Online – a web server for fast maximum likelihood-based phylogenetic inference. *Nucleic Acids Research* 33:W557–W559.

Hapke A, Fietz J, Nash SD, *et al.* 2005. Biogeography of dwarf lemurs: genetic evidence for unexpected patterns in southeastern Madagascar. *International Journal of Primatology* 26: 873–901.

Huelsenbeck JP, Ronquist F. 2001. MRBAYES: Bayesian inference of phylogeny. *Bioinformatics* 17:754–755.

Jernvall J, Wright PC. 1998. Diversity components of impending primate extinctions. *Proceedings of thc National Academy of Sciences of the United States of America* 95:11279–11283.

Kappeler PM. 2000. Lemur origins: rafting by groups of hibernators? *Folia Primatologica* 71:422–425.

Kappeler PM, Rasoloarison RM, Razafimanantsoa L, Walter L, Roos C. 2005. Morphology, behaviour and molecular evolution of giant mouse lemurs (*Mirza* spp.) Gray, 1970, with description of a new species. *Primate Report* 71:3–26.

Katoh K, Misawa K, Kuma K, Miyata T. 2002. MAFFT: a novel method for rapid multiple sequence alignment based on fast Fourier transform. *Nucleic Acids Research* 30:3059–3066.

Kelley J. 2002. The hominoid radiation in Asia. In WC Hartwig (ed.), *The Primate Fossil Record* (pp. 369–384). Cambridge University Press, Cambridge.

Laslett D, Canback B. 2008. ARWEN: a program to detect tRNA genes in metazoan mitochondrial nucleotide sequences. *Bioinformatics* 24:172–175.

Lebatard AE, Bourles DL, Duringer P, *et al.* 2008. Cosmogenic nuclide dating of *Sahelanthropus tchadensis* and *Australopithecus bahrelghazali*: Mio-Pliocene hominids from Chad. *Proceedings of the National Academy of Sciences of the United States of America* 105:3226–3231.

Lei R, Shore GD, Brenneman RA, *et al.* 2010. Complete sequence and gene organization of the mitochondrial genome for Hubbard's sportive lemur (*Lepilemur hubbardorum*). *Gene* 464:44–49.

Lei R, Rowley TW, Zhu L, *et al.* 2012. PhyloMarker – a tool for mining phylogenetic markers through genome comparison: application of the mouse lemur (genus *Microcebus*) phylogeny. *Evolutionary Bioinformatics* 8:423–435.

Li H, Durbin R. 2009. Fast and accurate short read alignment with Burrows-Wheeler transform. *Bioinformatics* 25:1754–1760.

Louis Jr EE, Lei R. 2014. Defining species in an advanced technological landscape. *Evolutionary Anthropology* 23:18–20.

Louis Jr EE, Coles MS, Andriantompohavana R, *et al.* 2006a. Revision of the mouse lemurs (*Microcebus*) of eastern Madagascar. *International Journal of Primatology* 27:347–389.

Louis Jr EE, Engberg SE, Lei, R, *et al.* 2006b. Molecular and morphological analyses of the sportive lemurs (Family Megaladapidae: Genus *Lepilemur*) reveals 11 previously unrecognized species. *Texas Technical University Museum Special Publications* SP49:1–47.

Louis Jr EE, Engberg SE, McGuire SM, *et al.* 2008. Revision of the mouse lemurs, *Microcebus* (Primates, Lemuriformes), of northern and northwestern Madagascar with descriptions of two new species at Montagne d'Ambre National Park and Antafondro Classified Forest. *Primate Conservation* 23:19–38.

Markolf M, Brameier M, Kappeler PM. 2011. On species delimitation: yet another lemur species or just genetic variation? *BMC Evolutionary Biology* 11:216.

Martin RD. 1972. Adaptive radiation and behaviour of the Malagasy lemurs. *Philosophical Transactions of the Royal Society London B* 264:295–352.

Martin RD. 1995. Prosimians: from obscurity to extinction? In L Alterman, K Izard, GA Doyle (eds.), *Creatures of the Dark* (pp. 535–563). Plenum Press, New York.

Martin RD. 2000. Origins, diversity, and relationships of lemurs. *International Journal of Primatology* 21:1021–1049.

Masters JC, de Wit MJ, Asher RJ. 2006. Reconciling the origins of Africa, India and Madagascar with vertebrate dispersal scenarios. *Folia Primatologica* 77:399–418.

Matsudaira K, Ishida T. 2010. Phylogenetic relationships and divergence dates of the whole mitochondrial genome sequences among three gibbon genera. *Molecular Phylogenetics and Evolution* 55:454–459.

Matsui A, Rakotondraparany F, Munechika I, Hasegawa M, Horai S. 2009. Molecular phylogeny and evolution of prosimians based on complete sequences of mitochondrial DNAs. *Gene* 441:53–66.

McCall RA. 1997. Implications of recent geological investigations of the Mozambique Channel for the mammalian colonization of Madagascar. *Proceedings of the Royal Society London B* 264:663–665.

Meier B, Albignac R. 1989. Hairy-eared dwarf lemur rediscovered (*Allocebus trichotis*). *Primate Conservation* 10:30.

Meier B, Albignac R. 1991. Rediscovery of *Allocebus trichotis* Günther 1875 (Primates) in northeast Madagascar. *Folia Primatologica* 56:57–63.

Meiklejohn, KA, Danielson, MJ, Faircloth, BC, *et al.* 2014. Incongruence among different mitochondrial regions: a case study using complete mitogenomes. *Molecular Phylogenetics and Evolution* 78:314–323.

Meyer M, Kircher M. 2010. Illumina sequencing library preparation for highly multiplexed target capture and sequencing. *Cold Spring Harbor Protocols* DOI: 10.1101/pdb.prot5448.

Meyer M, Stenzel U, Myles S, Prüfer K, Hofreiter M. 2007. Targeted high-throughput sequencing of tagged nucleic acid samples. *Nucleic Acids Research* 35:e97.

Meyer M, Stenzel U, Hofreiter M. 2008. Parallel tagged sequencing on the 454 platform. *Nature Protocols* 3:267–278.

Mittermeier RA, Ganzhorn JU, Konstant WR, et al. 2008. Lemur diversity in Madagascar. *International Journal of Primatology* 29:1607–1656.

Mittermeier RA, Louis Jr EE, Richardson M, *et al.* 2010. *Lemurs of Madagascar* (3rd ed.). Conservation International, Arlington.

Myers N. 2000. Biodiversity hotspots for conservation priorities. *Nature* 403:853–858.

Nylander JAA. 2004. MrModeltest v2.2. Program distributed by the author.

Olivieri G, Zimmermann E, Randrianambinina B, *et al.* 2007. The ever-increasing diversity in mouse lemurs: three new species in north and northwestern Madagascar. *Molecular Phylogenetics and Evolution* 43:309–327.

Pastorini J, Martin RD, Ehresmann P, Zimmermann E, Forstner MRJ. 2001. Molecular phylogeny of the lemur family Cheirogaleidae (Primates) based on mitochondrial DNA sequences. *Molecular Phylogenetics and Evolution* 19:45–56.

Pastorini J, Thalmann U, Martin RD. 2003. A molecular approach to comparative phylogeography of extant Malagasy lemurs. *Proceedings of the National Academy of Sciences of the United States of America* 100:5879–5884.

Perelman P, Johnson WE, Roos C, *et al.* 2011. A molecular phylogeny of living primates. *PLoS Genetics* 7:e1001342.

Posada D, Crandall KA, 1998. MODELTEST: testing the model of DNA substitution. *Bioinformatics* 14:817–818.

Poux C, Douzery EJP. 2004. Primate phylogeny, evolutionary rate variations, and divergence times: a contribution from the nuclear gene IRBP. *American Journal of Physical Anthropology* 124:1–16.

Poux C, Madsen O, Marquard E, *et al.* 2005. Asynchronous colonization of Madagascar by the four endemic clades of primates, tenrecs, carnivores, and rodents as inferred from nuclear genes. *Systematic Biology* 54:719–730.

Powell AFLA, Barker FK, Lanyon SM. 2013. Empirical evaluation of partitioning schemes for phylogenetic analyses of mitogenomic data: an avian case study. *Molecular Phylogenetics and Evolution* 66:69–79.

Pozzi L, Hodgson JA, Burrell AS, *et al.* 2014. Primate phylogenetic relationships and divergence dates inferred from complete mitochondrial genomes. *Molecular Phylogenetics and Evolution* 75:165–183.

Raaum RL, Sterner KN, Noviello CM, Stewart CB, Disotell TR. 2005. Catarrhine primate divergence dates estimated from complete mitochondrial genomes: concordance with fossil and nuclear DNA evidence. *Journal of Human Evolution* 48:237–257.

Rabarivola C, Zaramody A, Fausser J-L, *et al.* 2006. Cytogenetic and molecular characteristics of a new species of sportive lemur from northern Madagascar. *Lemur News* 11:45–49.

Radespiel U, Olivieri G, Rasolofoson DW, *et al.* 2008. Exceptional diversity of mouse lemurs (*Microcebus* spp.) in the Makira Region with the description of one new species. *American Journal of Primatology* 70:1–14.

Radespiel U, Ratsimbazafy JH, Rasoloharijaona S, *et al.* 2012. First indications of a highland specialist among mouse lemurs (*Microcebus* spp.) and evidence for a new mouse lemur species from eastern Madagascar. *Primates* 53:157–170.

Rakotoarison N. 1998. Recent discoveries of the hairy-eared dwarf lemur (*Allocebus trichotis*). *Lemur News* 3:21.

Rakotoarison N, Zimmermann H, Zimmermann E. 1996. First discovery of the hairy-eared dwarf lemur (*Allocebus trichotis*) in a highland rain forest of Eastern Madagascar. *Folia Primatologica* 68:86–94.

Ramaromilanto B, Lei R, Engberg SE, *et al.* 2009. Sportive lemur diversity at Mananara-Nord Biosphere Reserve, Madagascar. *Texas Tech Occasional Papers* OP286:1–22.

Rambaut A. 2008. FigTree v1.3.1. http://tree.bio.ed.ac.uk/software/figtree/.

Rambaut A, Drummond AJ. 2007. Tracer v1.5. http://tree.bio.ed.ac.uk/software/tracer/.

Rasoloarison RM, Goodman SM, Ganzhorn JU. 2000. Taxonomic revision of mouse lemurs (*Microcebus*) in the western portions of Madagascar. *International Journal of Primatology* 21:963–1019.

Rasoloarison RM, Weisrock DW, Yoder AD, Rakotondravony D, Kappeler PM. 2013. Two new species of mouse lemurs (Cheirogaleidae: *Microcebus*) from eastern Madagascar. *International Journal of Primatology* 34:455–469.

Raxworthy CJ, Nussbaum RA. 1995. Systematics, speciation and biogeography of dwarf chameleons (*Brookesia*, Reptilia, Squamata, Chamaeleontidae) of northern Madagascar. *Journal of Zoology* 235:525–558.

Ronquist F, Huelsenbeck JP. 2003. MRBAYES 3: Bayesian phylogenetic inference under mixed models. *Bioinformatics* 19:1572–1574.

Roos C, Schmitz J, Zischler H. 2004. Primate jumping genes elucidate strepsirrhine phylogeny. *Proceedings of the National Academy of Sciences of the United States of America* 101:10650–10654.

Rumpler Y, Albignac R. 1973. Cytogenetic study of the endemic Malagasy lemurs subfamily Cheirogaleinae Gregory 1915. *American Journal of Physical Anthropology* 38:261–264.

Rumpler Y, Warter S, Ishak B, Dutrillaux B. 1989. Chromosomal evolution in primates. *Human Evolution* 4:157–170.

Rumpler Y, Warter S, Hauwy M, *et al.* 1995. Cytogenetic study of *Allocebus trichotis*, a Malagasy prosimian. *American Journal of Primatology* 36:239–244.

Rumpler Y, Ganzhorn JU, Tomiuk J, Leipoldt M, Warter S. 1998. A cytogenetic study of *Microcebus myoxinus*. *Folia Primatologica* 69:307–311.

Sambrook J, Fritch EF, Maniatus T. 1989. *Molecular Cloning: A Laboratory Manual* (2nd ed.). Cold Spring Harbor Press, New York.

Samonds KA, Godfrey LR, Ali JR, *et al.* 2012. Spatial and temporal arrival patterns of Madagascar's vertebrate fauna explained by distance, ocean currents, and ancestor type. *Proceedings of the National Academy of Sciences of the United States of America* 109:5352–5357.

Sarich VM, Cronin JE. 1976. Molecular systematics of the primates. In M Goodman, RE Tashian (eds.), *Molecular Anthropology* (pp. 141–170). Plenum, New York.

Schmid J, Kappeler PM. 1994. Sympatric mouse lemurs (*Microcebus* spp.) in western Madagascar. *Folia Primatologica* 63:162–170.

Schrago CG, Menezes AN, Moreira MAM, Pissinatti A, Seuánez HN. 2012. Chronology of deep nodes in the neotropical primate phylogeny: insights from mitochondrial genomes. *PLoS ONE* 7:e51699.

Schütz H, Goodman SM. 1998. Photographic evidence of *Allocebus trichotis* in the Reserve Spéciale d'Anjanaharibe-Sud. *Lemur News* 3:21–22.

Seiffert ER, Simons EL, Attia Y. 2003. Fossil evidence for an ancient divergence of lorises and galagos. *Nature* 422:421–424.

Senut B, Pickford M, Gommery D, *et al.* 2001. First hominid from the Miocene (Lukeino Formation, Kenya). *Comptes Rendus de l'Académie des Sciences* 332:137–144.

Seutin G, White BN, Boag PT. 1991. Preservation of avian blood and tissue samples for DNA analyses. *Canadian Journal of Zoology* 69:82–90.

Simpson JT, Wong K, Jackman SD, *et al.* 2009. ABySS: a parallel assembler for short read sequence data. *Genome Research* 19:1117–1123.

Springer MS, Meredith RW, Gatesy J, *et al.* 2012. Macroevolutionary dynamics and historical biogeography of primate diversification inferred from a species supermatrix. *PLoS ONE* 7:e49521.

Stanger KF. 1993. Vocalizations of some cheirogaleid prosimians evaluated in a phylogenetic context. In L Alterman, GA Doyle, MK Izard (eds.), *Creatures of the Dark: The Nocturnal Prosimians* (pp. 353–376). Plenum, New York.

Stanger-Hall KF. 1997. Phylogenetic affinities among the extant Malagasy lemurs (Lemuriformes) based on morphology and behavior. *Journal of Mammal Evolution* 4:163–194.

Stevens NJ, Heesy CP. 2006. Malagasy primate origins: phylogenies, fossils, and biogeographic reconstructions. *Folia Primatologica* 77:419–433.

Swofford DL. 2001. *PAUP*, Phylogenetic Analysis Using Parsimony (*and Other Methods)*, Version 4.0b5. Sinauer Associates, Sunderland, MA.

Tamura K, Dudley J, Nei M, Kumar S. 2007. MEGA4: molecular evolutionary genetics analysis (MEGA) software version 4.0. *Molecular Biology and Evolution* 24:1596–1599.

Tattersall I. 1982. *The Primates of Madagascar*. Columbia University Press, New York.

Tattersall I. 2006a. Historical biogeography of the strepsirrhine primates of Madagascar. *Folia Primatologica* 77:477–487.

Tattersall I. 2006b. Origin of the Malagasy strepsirrhine primates. In L Gould, ML Sauther (eds.), *Lemurs: Ecology and Adaptation* (pp. 3–17). Springer, New York.

Tattersall I. 2007. Madagascar's lemurs: cryptic diversity or taxonomic inflation? *Evolutionary Anthropology* 16:1–40.

Tattersall I. 2013. Species-level diversity among Malagasy lemurs. In J Masters, M Gamba, F Genin (eds.), *Leaping Ahead: Advances in Prosimian Biology* (pp. 11–20). Springer, New York.

Tattersall I, Schwartz JH. 1974. Craniodental morphology and the systematics of the Malagasy lemurs (Primates, Prosimii). *Anthropological Papers of the American Museum of Natural History* 52:137–192.

Thalmann U. 2000. Lemur diversity and distribution in western Madagascar – inferences and predictions using a cladistic approach. In WR Louren, SM Goodman (eds.), *Diversité et Endemisme à Madagascar* (pp. 191–202). Mémoires de la Société de Biogéographie, Paris.

Thalmann U. 2007. Biodiversity, phylogeography, biogeography and conservation: lemurs as an example. *Folia Primatologica* 78:420–443.

Thiele D, Razafimahatrata E, Hapke A. 2013. Discrepant partitioning of genetic diversity in mouse lemurs and dwarf lemurs – biological reality or taxonomic bias? *Molecular Phylogenetics and Evolution* 69:593–609.

Ting N, Tosi AJ, Li Y, Zhang YP, Disotell TR. 2008. Phylogenetic incongruence between nuclear and mitochondrial markers in the Asian colobines and the evolution of the langurs and leaf monkeys. *Molecular Phylogenetics and Evolution* 46:466–474.

Toews DPL, Brelsford A. 2012. The biogeography of mitochondrial and nuclear discordance in animals. *Molecular Ecology* 21:3907–3930.

Tosi AJ, Detwiler KM, Disotell TR. 2005. X-chromosomal window into the evolutionary history of the guenons (Primates: Cercopithecini). *Molecular Phylogenetics and Evolution* 36:58–66.

Townsend TM, Vieites DR, Glaw F, Vences M. 2009. Testing species-level diversification hypotheses in Madagascar: the case of microendemic *Brookesia* leaf chameleons. *Systemic Biology* 58:641–656.

Veilleux CC, Louis EE, Bolnick DA. 2013. Nocturnal light environments influence differential selection for color vision in nocturnal lemurs. *Molecular Biology and Evolution* 30:1420–1437.

Vences M, Wollenberg KC, Vieites DR, Lees DC. 2009. Madagascar as a model region of species diversification. *Trends in Ecology & Evolution* 24:456–465.

Weisrock DW, Rasoloarison RM, Fiorentino I, *et al.* 2010. Delimiting species without nuclear monophyly in Madagascar's mouse lemurs. *PLoS ONE* 5:e9883.

Wells NA. 2003. Some hypotheses on the Mesozoic and Cenozoic paleoenvironmental history of Madagascar. In SM Goodman, JP Bestead (eds.), *The Natural History of Madagascar* (pp. 397–410). The University of Chicago Press, Chicago.

Wilmé L, Goodman ST, Ganzhorn JU. 2006. Biogeographic evolution of Madagascar's microendemic biota. *Science* 312:1063–1065.

Yoder AD. 1997. Back to the future: a synthesis of strepsirrhine systematics. *Evolutionary Anthropology* 6:11–22.

Yoder AD, Heckman K. 2006. Mouse lemur phylogeography revises a model of ecogeographic constraint in Madagascar. In J Fleagle, SM Lehman (eds.), *Primate Biogeography: Progress and Prospects* (pp. 255–268). Kluwer Academic Press, New York.

Yoder AD, Nowak M. 2006. Has vicariance or dispersal been the predominant biogeographic force in Madagascar? Only time will tell. *Annual Review of Ecology, Evolution, and Systematics* 37:405–431.

Yoder AD, Yang Z. 2004. Divergence dates for Malagasy lemurs estimated from multiple gene loci: geological and evolutionary context. *Molecular Ecology* 13:757–773.

Yoder AD, Cartmill M, Ruvolo M, Smith K, Vilgalys R. 1996. Ancient single origin for Malagasy primates. *Proceedings of the National Academy of Sciences of the United States of America* 93:5122–5126.

Zardoya R, Meyer A. 1996. Phylogenetic performance of mitochondrial protein-coding genes in resolving relationships among vertebrates. *Molecular Biology and Evolution* 13:933–942.

Zimmermann E, Cepok S, Rakotoarison N, Zietemann V, Radespiel U. 1998. Sympatric mouse lemurs in north-western Madagascar: a new rufous mouse lemur species (*Microcebus ravelobensis*). *Folia Primatologica* 69:106–114.

Zinner D, Wertheimer J, Liedigk R, Groeneveld LF, Roos C. 2013. Baboon phylogeny as inferred from complete mitochondrial genomes. *American Journal of Physical Anthropology* 150:133–140.

4 Why cheirogaleids are bad models for primate ancestors: a phylogenetic reconstruction

Curswan A. Andrews, Hajarimanitra Rambeloarivony, Fabien Génin, and Judith C. Masters

Introduction

Primate origins and dietary evolution

The provenance, timing, and environmental circumstances of the origin of primates and their subsequent dispersal are among the most heavily contested subjects in primate evolution. Several authors have contributed to the development of adaptive theories of primate origins (Jones, 1916; Szalay, 1968, 1972; Cartmill, 1974, 1992; Szalay and Dagosto, 1980, 1988; Sussman, 1991), proposing different models to explain the evolution of the unique combination of characteristics associated with the "adaptive shift" that marked the divergence of the primate lineage. In most recent models, diet plays the central role – not too surprisingly, as dietary evolution is one of the cornerstones for explaining the emergence of most mammalian lineages. The only recent model that does not prioritize dietary adaption is that of Szalay and Dagosto (1980, 1988), who proposed that grasping extremities and nails on the digits evolved together with leaping adaptations to facilitate grasp-leaping locomotion. All other models construe the defining primate characteristics as adaptations for food acquisition, usually for a single "ancestral diet," despite the diversity and versatility of modern primate dietary adaptations. Two opposing models of primate dietary evolution enjoy majority support today, both of which owe more to hypothetical scenarios of evolution than they do to fossil specimens. Diet is not simple to read from the fossil record, but the fact that teeth are preserved more frequently than any other body parts, and reflect dietary composition at least in part, provides some insight into ancient primate diets. Additionally, diet coevolves with body size and locomotion, and these additional characteristics can inform our interpretations of fossil diets. Fossil data largely confirm that primate dietary diversity evolved early. Eocene strepsirrhines were clearly insectivorous, frugivorous,

The Dwarf and Mouse Lemurs of Madagascar: Biology, Behavior and Conservation Biogeography of the Cheirogaleidae, eds. S.M. Lehman, U. Radespiel, E. Zimmermann. Published by Cambridge University Press. © Cambridge University Press 2016.

both insectivorous and frugivorous, and folivorous, all of which required specialist adaptations (Kirk and Simons, 2001).

Unfortunately, the fossil record for primates – and particularly for strepsirrhines – is largely incomplete. According to Soligo and Martin (2007), about 25 Ma are missing from the record, and the gaps include some key periods in primate evolution. We know very little of the evolution of primates prior to the Eocene (Silcox *et al.*, 2007), and less about the transition between the first "primates of modern aspect" (Euprimates), represented by the extinct Eocene adapiforms and omomyoids, to crown lineages that seem to appear suddenly in the Neogene. These lacunae complicate reconstructions of the origins and early evolution of the major primate clades.

Hypotheses of primate origins and early evolution

Phylogenetic systematics and molecular phylogenetics have transformed the field of evolutionary biology over the past four decades, but before these methods became widespread, models of ancestral states were derived from hypothetical scenarios of the role of natural selection and adaptation in driving the divergence of the primate clade from its sister taxa, and the emergence of the characters used to define the crown group and at least some fossils. These scenarios still form the context of most comparative studies of early primate evolution, and cheirogaleids play a major role in some reconstructions. We discuss currently popular models of primate dietary evolution below.

(a) Cartmill's visual predation hypothesis

Charles-Dominique and Martin (1970) proposed that mouse lemurs and dwarf galagos share a suite of social, ecological and behavioral traits likely to have been present in the common ancestor of strepsirrhines and, possibly, in ancestral primates. Their proposal inspired the most popular scenario of primate origins, elaborated by Cartmill (1972, 1974, 1992) as what is generally termed the "visual predation hypothesis": i.e., primate characteristics, like grasping hands and feet, claws transformed into nails, forward-pointing orbits and stereoscopic vision, and the enlargement of the brain are viewed as adaptations not only to an arboreal lifestyle, but to a primarily insectivorous diet involving a peculiar form of prey capture characteristic of extant small-bodied strepsirrhines. Cartmill posited that ancestral primates hunted prey by gripping tree branches with their feet, and launching their bodies forward to capture flying insects with their hands, in much the same way that living cheirogaleids, lorises and bushbabies do.

Critics of Cartmill's scenario (Soligo, 2006; Masters *et al.*, 2007; Soligo and Martin, 2007) based their reasoning on phylogenetic reconstructions of ancestral body size which contradicted the hypothesis of a small-bodied ancestor. In particular, Masters *et al.* (2013, 2014) have demonstrated that Malagasy cheirogaleids, often used as living analogues of ancestral primates, evolved relatively recently from a larger-sized ancestor, and reduced their body sizes over time (dwarfing).

(b) Sussman's angiosperm–primate diffuse coevolution hypothesis

Cartmill's model was advanced as a counter to Szalay's (1968, 1972) earlier proposal that primates diverged from their insectivorous ancestors by developing molars capable of crushing plant material, particularly fruit. This idea was developed further by Sussman (1991), who proposed that the ancestral primates had coevolved diffusely with the early lineages of flowering plants (angiosperms); their grasping extremities and the presence of nails on their digits, therefore, evolved for the gathering and consumption of fruit and plant parts. Sussman supported his model with the observation that modern primates evolved contemporaneously with frugivorous bats, plant-feeding birds and herbivorous mammals.

The concept of coevolution is based on the fact that food is not only a resource, but at least part of another living organism which is also a product of evolution. Interactions between animals and their food may be strictly exploitative and antagonistic with no reciprocal benefit for the food species, in which case plants and insects will evolve to defend themselves against predators by developing chemical (toxicity and deterrent taste) or mechanical (spines and other armaments) defenses. Over time, exploitative relationships can evolve towards more reciprocal (coevolved) relationships. Coevolution was first defined as strict one-to-one associations between species, but "diffuse coevolution" involves multiple partners and can affect entire lineages (Janzen, 1980). Sussman (1991) proposed that many primate characteristics, including their social systems, evolved as products of such coevolution, based on the temporal coincidence between the evolution of euprimate characteristics and the first modern angiosperms bearing large, fleshy fruits. Fruit pulp is likely to have evolved as a reward for potential seed dispersers, while the low-energy yield (Hladik, 1977) and toxicity of leaves and flowers evolved as defenses against folivores. The latter proposal is difficult to test, as the putative defenses have variable effects on different herbivorous "predator" species (Cornell and Hawkins, 2003).

Sussman's model was based on a literal reading of the fossil record with its relatively recent date for euprimate origins (Palaeocene/Eocene boundary), compared with estimations derived from molecular dating techniques. The divergence of the primate clade is highly likely to have antedated the Eocene angiosperm radiation, although the diverse euprimate radiation first recorded at 55 Ma may indeed fit this model. Rasmussen (1990) synthesized the theories of Cartmill and Sussman, theorizing that grasping evolved primarily for the location and harvest of fruits, while orbital convergence evolved secondarily to aid in the visual predation of insects (see also Sargis, 2002).

Gummivory/exudativory and dietary evolution in Cheirogaleidae

Nash's exudativorous precursor to folivory hypothesis

Nash (1986) devised a scenario based on Cartmill's visual predation model, in which ancestral faunivory led either to faunivorous–frugivorous or to faunivorous–gummivorous descendants, and adaptations to gummivory (i.e., the consumption of gums) preadapted descendant lineages to folivory. The reverse hypothesis was recently proposed by Génin *et al.* (2010) and Masters *et al.* (2013, 2014): that gummivorous strepsirrhines – and mouse lemurs in particular – are likely to be descended from folivorous ancestors, so that adaptations for folivory may have served as essential precursors to the adoption of gummivory.

Plant exudates and strepsirrhine exudativory

Botanists distinguish three main types of exudates: latexes (caustic, often milky fluids exuded when small branches or leaves are cut), gums (soluble and usually transparent or amber in color), and resins (non-soluble hydrocarbon compounds secreted mostly by coniferous trees, of variable coloration) (Nussinovitch, 2009). Latexes are difficult to define (Power, 2010), but are sometimes consumed as by-products: for instance, Verreaux's sifakas (*Propithecus verreauxi*) consume *Euphorbia* cladodes (Euphorbiaceae) (F. Génin, pers. obs.) and sportive lemurs (*Lepilemur leucopus* and *L. petteri*) consume *Cynanchum* cladodes and *Marsdenia* and *Pentopetia* leaves (Asclepiadaceae), all of which produce latex (Nash, 1998; Dröscher and Kappeler, 2014). Despite their apparently negative characteristics as a food source, gums are the only exudates consumed by primates as primary resources. They are comprised mainly of β-linked complex polysaccharides, which are difficult for mammals to digest without the help of commensal microbes (Power, 2010). They tend to be poor in protein, so that most gummivores also consume insects and small prey (Hladik *et al.*, 1980). Finally, they are scentless and tasteless, both to humans and to *Otolemur crassicaudatus*, an opportunistic gummivore (J. Ward, pers. comm.). They are exuded by trees and lianas of certain families, particularly Fabaceae and Combretaceae, as a result of internal infestations by wood-boring larvae. By contrast, resins contain toxic alkaloids, and are used by the trees to protect wounds from external attack. The distinctive smell of most resins comes from terpenes, which are highly toxic in concentrated doses. Primates may consume the resinous gums of trees belonging to certain families like the Burseraceae and the Anacardiaceae, but these compounds differ from most resins in that they have both a soluble and an insoluble fraction (Nussinovitch, 2009).

The evolutionary origins of gummivory are obscure. Gum-feeding is restricted to two mammalian groups: the gliding and striped possums of the Petauridae family of Australia, which are primarily insectivorous but occasionally consume the gum of *Acacia* and *Eucalyptus* spp. (Ashwell, 2010), and primates. Among the latter, gum-feeding is practised by members of seven strepsirrhine genera (*Allocebus, Galago, Euoticus, Microcebus, Nycticebus, Otolemur*, and *Phaner*; Nash, 1986), as well as three New World monkey genera (*Cebuella* (the pygmy marmosets) and *Callithrix* spp., both of which are more specialized gummivores than *Saguinus* spp., which consume gum facultatively). In addition, a few Old World monkeys (species of *Papio, Erythrocebus*, and *Chlorocebus*) have been observed consuming gum on occasion (Nash, 1986).

Very little is known about the interactions between gummivores and their plant hosts, and no studies have investigated the potential for their coevolution. Gouging bark and making a tree vulnerable to invasion by disease vectors is unlikely to be beneficial to the tree, but gum-scraping could carry associated benefits: gum-scrapers could assist by feeding on xylophagous larvae and actively hunting the adult insects that lay their eggs through the bark. Patterns of dental adaptation for gummivory are not clear-cut. Living obligate gummivores (*Euoticus*, *Phaner*) show striking convergences in their elongate, narrow tooth-combs and caniniform upper anterior premolars, as well as the shapes of their mandibles (Schwartz and Tattersall, 1985; Burrows and Nash, 2010), that are clearly adaptive. Other, less-specialized gummivores (*Galago*, *Microcebus*, *Nycticebus*, *Otolemur*), however, seem to manage very well without such modifications. The lineages of *Euoticus* and *Phaner* are very ancient: according to molecular dating methods, the needle-clawed galagos diverged in the late Eocene (Pozzi *et al.*, 2014), while *Phaner* dates to the Oligocene (Masters *et al.*, 2013). The ages of the lineages of the more opportunistic gummivores are apparently more recent (Masters *et al.*, 2013; Pozzi *et al.*, 2014). Additionally, the fact that a lineage or genus had diverged from its sister taxon does not imply that the species had already adopted gummivory.

In this chapter we explore the relative likelihoods of Cartmill's visual predation hypothesis, Sussman's angiosperm–primate diffuse coevolution hypothesis, Nash's exudativorous precursor to folivory hypothesis, and its antithesis, the folivorous precursor to exudativory hypothesis recently proposed by Masters *et al.* (2014), within a well-supported phylogenetic context. We tested these hypotheses of dietary transition, and reconstructed ancestral dietary states and body sizes for divergences between major lineages.

Methods

Scenarios of dietary evolution

We devised four possible scenarios for primate dietary evolution (Figure 4.1) based on: (a) the visual predation hypothesis (Cartmill); (b) the angiosperm coevolution hypothesis (Sussman); (c) the faunivorous–gummivorous precursor to folivory (Nash); and (d) the reverse postulate of a folivorous precursor to gummivory in cheirogaleids (Masters *et al.*, 2014). In accordance with the precepts of coevolution, we considered that the evolutionary relationship between primates and angiosperms involved at least three steps: (1) the antagonistic interaction between leaf-eaters and leaves, as well as seed-predators and seeds; (2) the coevolution of flowers and pollinators; and (3) the coevolution of fleshy fruits and seed dispersers. Most evolutionary transitions from one dietary syndrome to another were treated as reversible, because of the numerous species we encountered with mixed or intermediate diets. Irreversible transitions were only considered in the reconstruction based on our own hypothesis (Masters *et al.*, 2014) with respect to the faunivory–folivory and folivory–gummivory transitions, because their reversals would involve considerable changes in body size, as well as digestive and dental innovations. The results of these reconstructions were compared with a null hypothesis that all transitions were equally probable (i.e., the control).

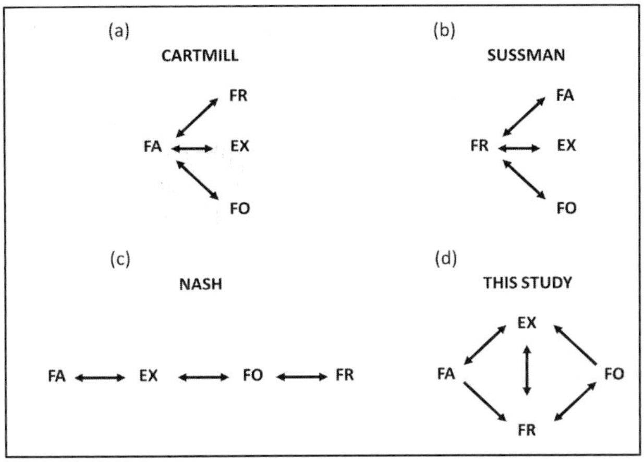

Figure 4.1 Possible scenarios of dietary evolution in primates: (a) Cartmill's small-bodied insectivorous ancestor; (b) Sussman's angiosperm–primate diffuse coevolution theory (larger-bodied frugivorous ancestor); (c) Nash's faunivory–exudativory–folivory transformation series; and (d) model reconstructed for this study based on the hypothesis of Masters *et al.* (2014). (FA, faunivory; EX, exudativory; FO, folivory; FR, frugivory.)

Dietary categorization

We used anatomical adaptations, specifically locomotor, craniodental, and alimentary adaptations, to define primate dietary categories. Data were gathered from our own observations, as well as from Mittermeier *et al.* (2010), Butynski *et al.* (2013) and Rowe and Myers (2014), and are available from the authors. We considered four fundamental "dietary syndromes" (rather than strict diets because of multiple overlaps, see below): faunivory, folivory, frugivory, and exudativory. Despite this precaution, several species could not be characterized by a single dietary syndrome and were considered to have mixed diets. The four dietary syndromes are defined below.

(a) *Faunivory including insectivory*
Living faunivorous primates exhibit a variety of adaptations that are unlikely to be symplesiomorphic because they are associated with at least three kinds of specialized diets. The first is associated with very small size and sharp teeth adapted to the killing and consumption of live insects and small vertebrates, and is observed in *Tarsius*, *Galagoides*, and *Microcebus*. All three taxa have been observed catching insects in flight with their hands, although *Tarsius* is also known as a formidable hunter of vertebrates, including toxic reptiles (Niemitz, 1984). Charles-Dominique and Martin (1970) and Cartmill (1972, 1974, 1992) proposed that this kind of faunivory was the plesiomorphic primate state. The two other kinds of primate faunivory involve specializations for the ambush capture and consumption of slow, potentially toxic insects (*Perodicticus*, *Arctocebus*, and *Loris*), or for the location and extraction of cryptic wood-borers (*Daubentonia*).

(b) *Folivory*
Etymologically, folivory refers to the consumption of leaves; however, leaves and flowers have the same embryological origin, and folivores often consume flowers. This may be an ancient dietary pattern: the first angiosperms had either very small flowers, or large edible flowers and small fruits (Friis *et al.*, 2010, 2011). Ancient *Magnolia* trees, for instance, may use their fleshy petals to attract pollinators in the same way that more recently evolved angiosperms use edible pulp to attract seed dispersers. Most living folivorous primates also consume unripe (often toxic) fruits (e.g., proboscis monkeys (*Nasalis larvatus*) and golden langurs (*Trachypithecus geei*); Das *et al.*, 2008; Meijaard *et al.*, 2008). Our working definition of folivory, therefore, was based on a suite of adaptations to a diet rich in leaves, including large cheek teeth with sharp shearing crests, and long guts.

(c) *Frugivory including granivory*
Fruits and seeds are often consumed simultaneously, and many frugivorous primates also consume the seeds of non-fleshy fruits (e.g. *Trachypithecus vetulus*; Hladik, 1977). Because they are full of energy reserves, seeds are likely to have been attractive dietary options for mammals prior to the evolution of fleshy fruits and the probable evolution of seed toxicity in the Eocene. Most frugivorous primates also consume buds and young leaves. Therefore, our working definition of frugivory was based on a suite of adaptations to a diet rich in fruits and/or seeds, including relatively small cheek teeth with bunodont cusps, and relatively short guts.

(d) *Exudativory including gummivory*
Exudates include gum, but also a variety of other liquids, such as nectar, honey, and the sap-based secretions of moth-bugs (Flatidae, Homoptera). Exudativory, therefore, is also more a dietary syndrome than a restrictive diet. Gum exudation is limited to a few angiosperm families such as Anacardiaceae, Burseraceae, Combretaceae, and Fabaceae, all of which belong to the subclass Rosidae (Génin *et al.*, 2010). Gummivorous primates either gouge bark to induce gum flow (callitrichines and slow lorises) or scrape gum already exuded after prior damage (bushbabies and cheirogaleids; Richard, 1985; Génin *et al.*, 2010). Our working definition of gummivory, based on the ecological convergences observed between gougers and scrapers, included dental specializations for gum-gathering and digestive adaptations for gum digestion (e.g., capacious caeca).

Phylogenetic methods of ancestral state reconstruction

Ancestral state reconstruction has long been a goal of evolutionary biologists, but advances in phylogenetic philosophy and methodology in recent decades – particularly since the development of molecular systematics – have revolutionized this practice. The comparative method, with the assistance of molecular phylogenetics, has provided a versatile new tool for deciphering the past. We used a recent, well-supported phylogeny (Springer *et al.*, 2012) to assess the likelihood of the scenarios of primate dietary evolution that currently enjoy support among primate

Table 4.1 Log Bayes factors and the significance they confer on the models compared.

Log Bayes factor	Significance
<2	Weak difference
>2	Positive difference
5–10	Strong difference
>10	Very strong difference

Log BF = 2(log (harmonic mean (complex model)) – log (harmonic mean (simple model))).

evolutionary biologists, coupled with a Bayesian method of ancestral state reconstruction to estimate the diets of the ancestors subtending various major primate clades: i.e., the ancestor to the entire primate clade, the strepsirrhine ancestor, and the ancestors to the most gummivorous extant species, particularly the cheirogaleids.

To accommodate for uncertainties associated with divergence times, we based our study on four trees published by Springer *et al.* (2012). We estimated Pagel's lambda using the Phytools package (Revell, 2012) in R (R Core Team, 2013) to assess the phylogenetic signal inherent in the data. A lambda value of 0 indicates that the data contain no phylogenetic signal, while a lambda close to 1 indicates a strong signal. We reconstructed ancestral states according to the Multistate-Markov chain Monte Carlo (MCMC) method using the BayesTraits software (available from www.evolution.rdg.ac.uk; Pagel *et al.*, 2004). Like any other Bayesian analysis, the estimation of priors is crucial. To estimate the priors for the reconstruction we used a subjective and wide pre-prior with a uniform distribution varying between 0 and 100. The chain was run 1,010,000 times with the first 10,000 iterations discarded as burn-in, and sampled every 1000th iteration. The transition rates between dietary syndromes were fixed for one run but allowed to change from one iteration to the next. The 10,000 trees sampled from the 1,000,000 iterations were then used to evaluate the distribution of the priors for use in the reconstruction. Priors were estimated separately for each model of evolution, and the priors used for the ancestral reconstruction had a uniform distribution, the boundaries of which were defined by the whole number immediately below the minimum estimated prior, and the whole number immediately above the maximum estimated prior.

We re-ran the MCMC as before, but this time using the estimated priors. We used the Most Recent Common Ancestor approach (MRCA) to reconstruct the diet at internal nodes and to track dietary evolution within the primate clade (Pagel *et al.*, 2004). Bayes factors (BFs) are calculated as the harmonic means of all the likelihood values generated during the 1,000,000 iterations of the model minus the likelihood values generated during the 1,000,000 iterations of the control (null) model. The significance and meaning of the BFs are reported in Table 4.1. The model that had the best fit (i.e., had the highest likelihood value) was analyzed further, and the reconstruction resulting from this model was represented graphically using the Ape package (Paradis *et al.*, 2004) in R (R Core Team, 2013). Divergence times were estimated from a consensus tree built by BayesTrees (Meade and Pagel, 2011) from the four trees used in the reconstruction.

Results

The dietary data contained a strong phylogenetic signal (lambda = 0.99), giving validity to the reconstructions. Tables 4.2 and 4.3 and Figure 4.2 summarize the results of the reconstructions and show that the control (null) model, allowing unrestricted transitions between dietary syndromes, fitted the phylogeny better than the two classical models (i.e., Cartmill's visual predation hypothesis and Sussman's angiosperm–primate diffuse coevolution hypothesis; Table 4.2). While our results confirm Cartmill's hypothesis of a faunivorous primate ancestor, they do not indicate the small body size required to hunt insects in the "fine branch" microhabitat at the tips of tree branches. Further, our reconstruction suggests that all four dietary syndromes constrain future evolutionary trajectories; i.e., only certain transitions from one syndrome to another are allowed, while others are prohibited. Because exudativory is associated with relatively small size and dental adaptations rather different from those required for eating leaves, it is unlikely to have acted as a precursor to folivory, contradicting Nash's hypothesized gummivory–folivory transition. Likewise, folivory and frugivory are often associated with relatively large size, relatively long guts and specialized dentition that are unlikely precursors for faunivory. Finally, faunivory did not evolve directly into folivory, for similar reasons.

The proposition that the primate ancestor was very small in size is contradicted by the reconstructed ancestral body weight between 1 and 2 kg (Table 4.3; see also Soligo and Martin, 2007; Masters *et al.*, 2007, 2014). This argues against the scenario of an agile hunter moving among fine branches, favoring instead a hypothetical ancestor with a slow metabolism and the ability to digest food items that require considerable processing (e.g., the exoskeletons of insects that consist of complex polysaccharides), and perhaps seeds, as observed in living aye-ayes (*Daubentonia*). These characters would explain the slow life history and the low basal metabolism of living primates (Génin and Masters, Chapter 16).

In our reconstructed tree, the unresolved nodes (i.e., the groups not marked with 'x' in Table 4.3) are those for which no single dietary syndrome makes up ≥ 75%. Their distribution in Figure 4.2 implies that the mixed diets that characterize many extant primate genera and families have been evident throughout the phylogenetic history of the primates. Mixed diets may reveal evolutionary transitions/divergences. For instance, a possible ancestral diet for latest Cretaceous primates may include large insects and seeds, as is observed in living aye-ayes (*Daubentonia*). The high level of specialization of *Daubentonia*'s adaptations to foraging on well-protected foods is likely to be extremely derived, but such specializations may not have been crucial for following a diet of insectivory and seed predation at the time of the *Daubentonia*-lemur divergence (Palaeocene). Plesiadapiforms are a diverse paraphyletic group of extinct, Palaeocene animals, at least some of which comprised the sister taxon to the Euprimates (Silcox, 2007). Many plesiadapiforms had peculiar dentitions with enlarged incisors, and cannot be associated readily with any extant dietary syndrome.

Table 4.2 Results of the phylogenetic reconstruction of primate diets indicating the rates of transition simulated for the three tested models, using five different models, and indicating the statistical support of the model (Lh).

Model	Lh	Lh (harmonic mean)	Sig (log BF)	Prior transition rate	Posterior transition rate*											
					FR–EX	FR–FA	FR–FO	EX–FR	EX–FA	EX–FO	FA–FR	FA–EX	FA–FO	FO–FR	FO–EX	FO–FA
No restriction	−68.5	−69.2	Control	Uniform 0–1	0.46	0.46	0.46	0.46	0.46	0.46	0.46	0.46	0.46	0.46	0.46	0.46
Model	−63.4	−64.6	9.1	Uniform 0–2	0.69	0	0.69	0.69	0.69	0	0.69	0.69	0	0.69	0.69	0
Cartmill	−82.6	−85.4	32.4	Uniform 0–4	0	1.52	0	0	1.52	0	1.52	1.52	1.52	0	0	1.52
Sussman	−70.6	−71.4	4.4	Uniform 0–2	0.63	0.63	0.63	0.63	0	0	0.63	0	0	0.63	0	0
Nash	−74.8	−75.8	13.2	Uniform 0–3	0	0	1.55	0	1.55	1.55	0	1.55	0	1.55	1.55	0

* FR, frugivory; EX, exudativory; FA, faunivory; FO, folivory.

Table 4.3 Results of the phylogenetic reconstruction of ancestral states indicating the following data for major internal nodes: the probability of each dietary syndrome, the mean body mass, and the divergence date estimated from Chatterjee *et al.* (2009) and Springer *et al.* (2012).

Node ancestral state	Frugivory	Exudativory	Faunivory	Folivory	Resolution[1]	Body mass (g)[2]	Dates (My)
Primates	0.07	0.11	0.80	0	x	1974[3]	>67.8
Strepsirrhines	0.15	0.15	0.64	0.06		894	53.1–54.2
Lemuriformes	0.54	0	0	0.46		967	49.4–50.0
Cheirogaleidae–Lepilemuridae–Indriidae	0.13	0	0	0.87	x	[4]	30.6
Cheirogaleidae–Lepilemuridae	0.56	0.07	0	0.36		953	30.6–38.9
Cheirogaleidae without *Phaner*	0.63	0.27	0.09	0		398	27.6–28.7
Mirza–Microcebus	0.31	0.18	0.50	0		203	13.2–19.0
Microcebus first divergence	0.34	0.15	0.50	0		92	6.9–9.8
Lepilemur–Phaner	0.08	0	0	0.88	x	779	22.3–32.1
Lorisidae–Galagidae	0	0.20	0.80	0	x	493	34.7–38.5
Euoticus divergence	0	0.46	0.53	0		338	22.7–26.1
Otolemur divergence	0.96	0	0	0	x	866	6.9–8.8

[1]$P > 0.75$ (see Figure 4.2 for color codes indicated when $P > 0.25$); [2]from Masters *et al.* (2014) (except [3] from Masters *et al.*, 2007); [4]not a clade in Chatterjee *et al.* (2009).

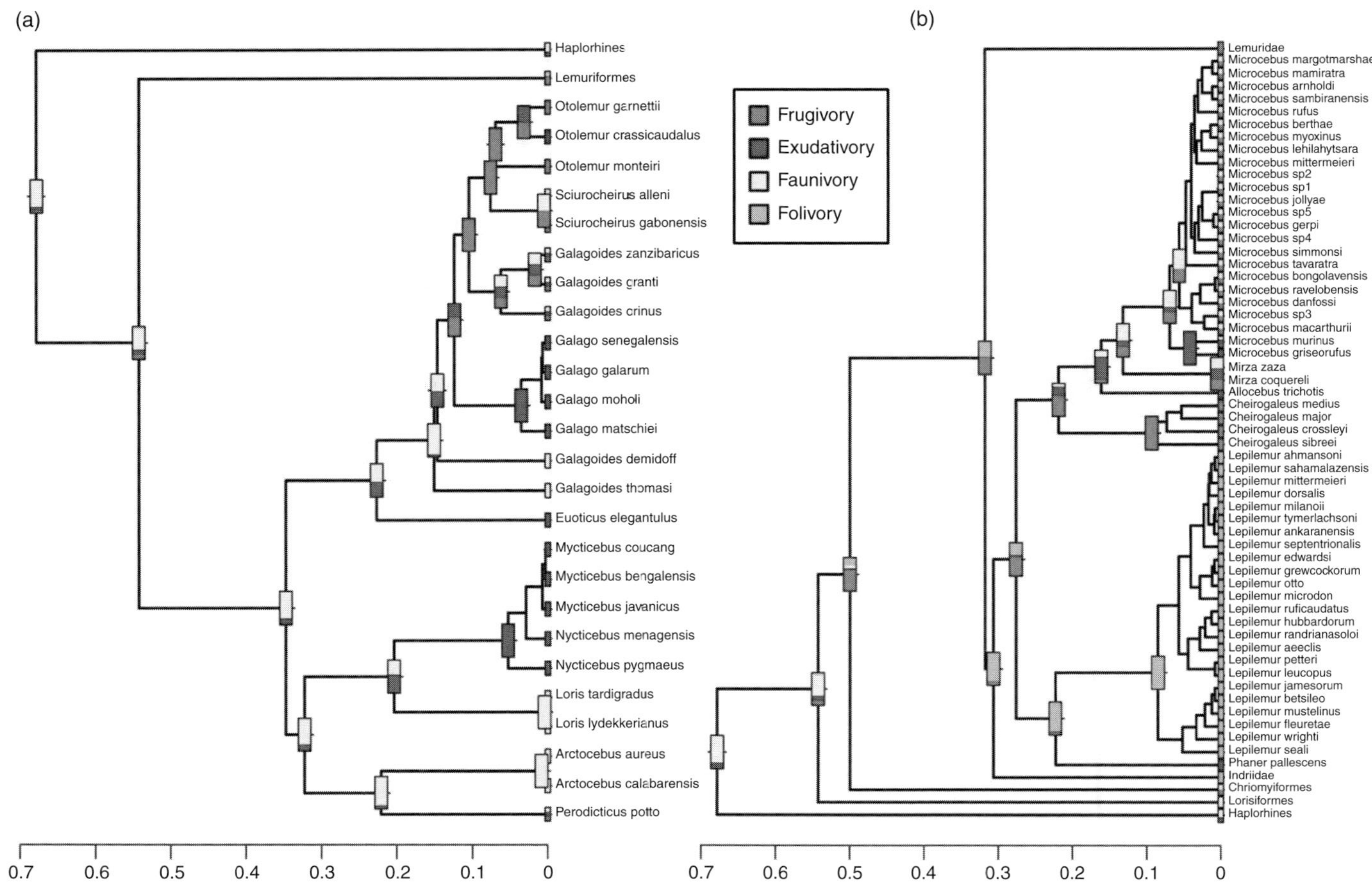

Figure 4.2 Phylogenetic reconstruction of the diets of (a) Lorisoidea, and (b) Cheirogaleidae. The time scales are expressed in 100 Ma. The color codes indicate the probabilities of the reconstructed diets. (A black and white version of this figure will appear in some formats. For the colour version, please refer to the plate section.)

Our reconstructions suggest that the Lemuriformes ancestor had a mixed frugivorous/folivorous diet, and an estimated body mass close to 1 kg. The ancestors of the *Lepilemur*–Cheirogaleidae clade and the Indriidae were folivorous. If this is valid, then exudativory must have evolved from folivory in the case of *Phaner*, although it may have evolved from at least partial frugivory in other cheirogaleids. The Lorisidae–Galagidae ancestor was reconstructed as a faunivore weighing approximately 500 g, although some gummivorous galagids are largely frugivorous (e.g. *Otolemur*, Table 4.3); hence, exudativory in this group may have evolved from faunivory–frugivory. Some degree of faunivory is implicated in the ancestry of most galagid genera, including the gouging, obligate gummivores, *Euoticus*, and the gouging slow lorises, *Nycticebus*. The gum-scraping behavior observed in both cheirogaleids and galagids is likely to have evolved convergently several times from different ancestral states, even within families.

Phylogenetically, gummivory is likely to have evolved in two main phases among the strepsirrhine primates. The lineages that gave rise to the obligate gougers, *Phaner* and *Euoticus*, are ancient (late Eocene–early Oligocene). Facultative gum-scraping evolved relatively more recently, during the dry periods of the Neogene. Both events were synchronous with the diversification of gum trees: e.g., like Mimosoideae in the Oligocene, and Combretaceae in the Miocene–Pliocene in Africa and Madagascar (Maurin, 2009; Bouchenak-Khelladi *et al.*, 2010). Facultative gummivory was accompanied by body-size reduction (dwarfing) in several lineages (Génin *et al.*, 2010; Masters *et al.*, 2014).

Discussion

Body size and dietary evolution

Factors influencing the evolution of body size have long been of interest to biologists, and their study has given rise to generalized rules. For example, Cope's rule holds that body size within animal clades tends to get larger over evolutionary time (Alroy, 1998). Some rules link dietary evolution to body size: Kay's threshold (Kay, 1975) of 500 g is an approximation of the upper body weight limit for insectivorous primates and the lower body weight limit for folivorous primates, because of energetic constraints (Ankel-Simons, 2007). *Daubentonia* does not conform to Kay's rule, and is the largest insectivorous primate, weighing approximately 2 kg; its diet, however, is difficult to categorize, because it also includes significant amounts of hard seeds and some fruits (Mittermeier *et al.*, 2010). Additionally, the evolution of body size is unlikely to have been unidirectional in primates, and evidence for repeated dwarfing exists, with the Cheirogaleidae as a pertinent example. These dwarfing events appear mainly to have been associated with forest fragmentation and drought (Masters *et al.*, 2014). Changes in dietary preferences are a necessary concomitant of dwarfing episodes because of the relationship between body size and metabolism; hence, as the cheirogaleids grew smaller, their diet probably shifted from folivory to a mixture of faunivory and exudativory. This dietary shift was physiologically possible as the gastrointestinal specializations required for digesting insects and gums could have been derived from those adapted to the difficult digestion of leaves and other plant material, which also require fermentation.

Coevolution of consumers and food sources

Another important factor to consider is the origin of the dietary syndromes we have identified; in particular, the consumption of plant exudates. Is exudativory a fall-back diet, as it is often viewed, or is it the result of phytochemical coevolution involving an active role for the consumers – e.g., in plant defense against insect invasion – in much the same way as fruit flesh provides a reward to the consumer in return for seed dispersal? The concept of a reward for services rendered is supported by the fact that all gummivores consume non-toxic gum, even though toxins are present in some exudates, and in some cases where gum is exuded, insects are present. Hence, gum could be a reward for the removal of insects that infest a particular tree. In a similar vein, the early stages of the primate–angiosperm interaction may have involved an association between trees infested with insects and the ancestral primates that consumed them, as well as various plant parts (i.e., exudates, fruits). The simultaneous expansion of gum trees and gummivorous primates in various parts of the world during the Neogene supports this view. Indeed, all gummivores consume insects and catch and consume the adults and the larvae of the wood-borers responsible for gum exudation. This relationship would conform to Sussman's primate–angiosperm coevolution model.

Ancestral state reconstructions: a caveat

Ancestral reconstructions should always be interpreted with caution, as the characters and dietary syndromes used in the models are based on extant taxa, and explore the success of coevolutionary processes in the past on the basis of entirely different, recently derived faunas and floras. The animals and plants of the Palaeocene–Eocene faced very different climatic and habitat conditions from those experienced today, as the plesiadapiforms so eloquently illustrate. We cannot reconstruct history from such analogues; we can only derive broad, generalized models that allow us to devise hypotheses that must be tested using fossil data. It is in this light that we assess that relative likelihood of the models of primate ancestry that enjoy popular support today.

Assessing the validity of currently popular models of primate ancestry

(a) Cartmill's visual predation hypothesis and the evolution of faunivory

While Cartmill's hypothesis of faunivory is supported for the end-Cretaceous ancestor of the primate clade, the reconstructed ancestor is not small-bodied, but medium-sized (1–2 kg) and possibly partially herbivorous. It is likely to have been rather slower-moving than a cheirogaleid, and specialized in the consumption of toxic or social insects, as is seen among the living lorises, or partially granivorous, like the living aye-aye. Tarsiers represent the other group of primates in which faunivory is probably ancient, and likely to have evolved from larger ancestors. In contrast, the classical model ancestral primates, the Malagasy mouse and dwarf lemurs, apparently evolved insectivory and exudativory secondarily from folivory, possibly as recently as the late Miocene or the Pliocene, when the climatic conditions that witnessed their emergence were established.

(b) Sussman's angiosperm–primate diffuse coevolution hypothesis and the evolution of folivory/frugivory

Frugivory is the most widespread dietary preference among primates; the only primates that rarely (if ever) consume fruits are the tarsiers. Our reconstruction supports Sussman's hypothesis of the convergent radiation of folivory–frugivory in several euprimate lineages known to have been present in the Eocene. These include strepsirrhines (certainly among the extinct adapiforms, and probably also in some crown lineages, like the indriids and lemurids) as well as haplorhines. However, Sussman's hypothesis does not apply to the earliest primates, which are likely to have been faunivorous and partly granivorous; our reconstruction implies that faunivory never evolved from frugivory. The fact that *Daubentonia* consumes seeds is also consistent with Sussman's proposal of coevolution, and suggests that granivory may have served as the precursor to frugivory. Living angiosperms defend their seeds against predators, using either chemical (toxicity) and/or mechanical (thick coat) strategies. Because they are full of energy reserves, seeds are likely to have been consumed before the Eocene evolution of fleshy fruits, and most likely evolved their defenses secondarily. Finally, many callitrichines feed on fruit exudates, suggesting that gummivory has evolved from frugivory in this group.

(c) Nash's exudativorous precursor to folivory hypothesis

While our reconstruction supports a dietary transition from either faunivory or frugivory/granivory to exudativory, Nash's postulated transition from gummivory to folivory was not permitted. Of the four groups of extant gummivorous primates (cheirogaleids, galagids, slow lorises, and callitrichines), only the two lorisoid lineages are likely to have derived their exudate feeding from a faunivorous ancestor. The callitrichines appear to have evolved gummivory from a frugivore/granivore, possibly a consumer of hard seeds. The cheirogaleids alone appear to have evolved their exudate feeding from a folivorous precursor that subtended the lemurid–indriid–cheirogaleid (including *Lepilemur*) clade. The convergent evolution of gummivory/exudativory among distinct primate lineages is mirrored by the convergent evolution of reduced body size (dwarfism) in several of these lineages (callitrichines, mouse and dwarf lemurs, and pygmy lorises; Génin *et al.*, 2010; Masters *et al.*, 2014).

A new model of dietary evolution in primates

We propose a new synthetic model of primate dietary evolution that includes elements of the previous models. The clade ancestor was predominantly faunivorous (Cartmill), but of relatively large body size (1–2 kg). The ancestor is hence unlikely to have hunted like a cheirogaleid, but more like a lorisid, possibly including slow-moving, toxic insects in its diet, as well as fruits, seeds or plant parts. The extensive radiation of frugivory/granivory and exudativory after the start of the Eocene (55 Ma) is likely to have occurred convergently in several, already independent primate lineages, in accordance with the precepts of Sussman's diffuse coevolution model between angiosperms and primates.

Our reconstruction indicates that exudativory evolved several times and at various time depths during the last ±75 Ma, under the influence of drying climatic trends and their concomitant spatial (forest fragmentation) and temporal (hypervariability) heterogeneity. The divergence of the euprimates and the plesiadapiforms in the late Cretaceous–early Palaeocene coincided with a warm, dry period during which many xerophytic plants (e.g. Euphorbiaceae) emerged. It is tempting to consider the enlarged, horizontal incisors of many plesiadapiforms as adaptations for gouging, either for cryptic insects or for plant exudates. With respect to living primates, gummivory apparently evolved among the obligate gummivores (*Euoticus*, *Phaner*) in the Eocene/early Oligocene, and among the facultative gummivores (*Allocebus*, *Galago*, *Galagoides*, *Microcebus*, *Nycticebus*, and the callitrichines) in the late Miocene to Pliocene. The late Eocene divergence of *Euoticus* proposed by Pozzi *et al.* (2014) coincides with a period of progressive cooling in Africa indicated by palynological evidence collected in Cameroon (Bobe, 2006); this climatic trend has been related to the progressive isolation of Antarctica following the opening of the Drake Passage (Scher and Martin, 2006), finally leading to the onset of the southern hemisphere glaciation in the early Oligocene (Zachos *et al.*, 1992). Fossil strepsirrhines showing possible adaptations for gummivory have been found in Oligocene deposits in Morocco, Egypt, and Pakistan (Marivaux *et al.*, 2001). A second global cooling and drying event occurred in the late Miocene, and was associated with the expansion of biotas adapted to cope with increased seasonality. The Miocene–Pliocene boundary (5 Ma) was a warm period, followed by the onset of the northern hemisphere glaciation (2–3 Ma), which resulted in spectacular forest regressions at lower latitudes (Bobe, 2006). This period is also likely to have witnessed the onset of modern climatic unpredictability, associated with tropical cyclones and the establishment of the El Niño Southern Oscillation (Fedorov *et al.*, 2010; Sriver, 2010). Gum trees diversified during both of these episodes of aridification, often in the same hypervariable regions (Génin *et al.*, 2010); e.g., the mimosoids (Oligocene) and the Combretaceae (Pliocene) (Maurin, 2009; Bouchenak-Khelladi *et al.*, 2010).

Very little is known about possible interactions between gum-producers and gum-feeders. Gum exudation is limited to a few angiosperm families and gums may have evolved from toxic resins. Older plant taxa, including trees that yield resinous gums (e.g., Anacardiaceae and Burseraceae), seem to use the gums to protect wounds in the bark against external attack, while more recent true gums (e.g., those of Combretaceae and Fabaceae) also expel xylophagous larvae that infest the trees. Only in the second case could gum-feeders offer a service to the plant, by feeding on the larvae and actively hunting the adults when they lay their eggs. These associations may reflect animal–plant associations that go beyond pollination and seed dispersal, and generalize Sussman's (1991) hypothesis of coevolution between primates and angiosperms to include more diverse and older angiosperm partners.

Acknowledgments

CAA and HR received postgraduate bursaries from the South African National Research Foundation, who also funded the research of JCM and FG. Without their support our research would not be possible. We thank Prof. G. de Wet and the indefatigable staff of the Govan Mbeki Research and Development Centre for keeping our research group on the rails. This is Publication No. 4 of the African Primate Initiative for Ecology and Speciation (APIES).

References

Alroy J. 1998. Cope's Rule and the dynamics of body mass evolution in North American fossil mammals. *Science* 280:731–734.

Ankel-Simons F. 2007. *Primate Anatomy: An Introduction*, 3rd edition. Academic Press, New York.

Ashwell KWS. 2010. *The Neurobiology of Australian Marsupials: Brain Evolution in the Other Mammalian Radiation.* Cambridge University Press, New York.

Bobe R. 2006. The evolution of arid ecosystems in eastern Africa. *Journal of Arid Environments* 66:564–584.

Bouchenak-Khelladi Y, Maurin O, Hurter J, van der Bank M. 2010. The evolutionary history and biogeography of Mimosoideae (Leguminosae): an emphasis on African acacias. *Molecular Phylogenetics and Evolution* 57:495–508.

Burrows AM, Nash LT. 2010. Searching for dental signals of exudativory in galagos. In LT Nash, AM Burrows (eds.), *The Evolution of Exudativory in Primates. Developments in Primatology: Progress and Prospects* (pp. 211–233). Springer, New York.

Butynski TM, Kingdon J, Kalina J, editors. 2013. *Mammals of Africa.* Volume 2. Bloomsbury, London.

Cartmill M. 1972. Arboreal adaptations and the origin of the Order Primates. In RH Tuttle (ed.), *The Functional and Evolutionary Biology of Primates* (pp. 97–122). Aldine-Atherton Press, Chicago.

Cartmill M. 1974. Rethinking primate origins. *Science* 184:436–443.

Cartmill M. 1992. New views on primate origins. *Evolutionary Anthropolology* 1:105–111.

Charles-Dominique P, Martin RD. 1970. Evolution of lemurs and lorises. *Nature* 227:257–260.

Chatterjee HJ, Ho SYW, Barnes I, Groves C. 2009. Estimating the phylogeny and divergence times of primates using a supermatrix approach. *BMC Evolutionary Biology* 9:259. doi:10.1186/1471-2148-9-259.

Cornell HV, Hawkins BA. 2003. Herbivore responses to plant secondary compounds: a test of phytochemical coevolution theory. *The American Naturalist* 161(4):507–522.

Das J, Medhi R, Molur S. 2008. *Trachypithecus geei.* In IUCN 2013. IUCN Red List of Threatened Species. Version 2013.2. www.iucnredlist.org. Downloaded on 30 January 2014.

Dröscher I, Kappeler PM. 2014. Competition for food in a solitary foraging folivorous primate (*Lepilemur leucopus*). *American Journal of Primatology* 76:842–854.

Fedorov AV, Brierley CM, Emanuel K. 2010. Tropical cyclones and permanent El Niño in the early Pliocene epoch. *Nature* 463:1066–1070.

Friis EM, Pederson KR, Crane PR. 2010. Diversity in obscurity: fossil flowers and the early history of angiosperms. *Philosophical Transactions of the Royal Society B: Biological Sciences* 365:369–382.

Friis EM, Crane PR, Pederson KR. 2011. *Early Flowers and Angiosperm Evolution.* Cambridge University Press, New York.

Génin F, Masters JC, Ganzhorn JU. 2010. Gummivory in cheirogaleids: primitive retention or adaptation to hypervariable environments? In AM Burrows, LT Nash (eds.), *The Evolution of Exudativory in Primates. Developments in Primatology: Progress and Prospects* (pp. 123–140). Springer Media, New York.

Hladik CM. 1977. A comparative study of the feeding strategies of two sympatric species of leaf-eating monkeys: *Presbytis senex* and *Presbytis entellus.* In TH Clutton-Brook (ed.), *Primate Feeding Ecology: Studies of Feeding and Ranging Behaviour in Lemurs, Monkeys, and Apes* (pp. 324–353). Academic Press, London,

Hladik CM, Charles-Dominique P, Petter J-J. 1980. Feeding strategies of five nocturnal prosimians in the dry forest of the west coast of Madagascar. In P Charles-Dominique, HM Cooper, A Hladik, *et al.* (eds.), *Nocturnal Malagasy Primates: Ecology, Physiology, and Behavior* (pp. 41–73). Academic Press, New York.

Janzen DH. 1980. When is it coevolution? *Evolution* 34:611–612.

Jones FW. 1916. *Arboreal Man.* Edward Arnold, London.

Kay RF. 1975. The functional adaptations of primate molar teeth. *American Journal of Physical Anthropology* 43:195–216.

Kirk EC, Simons EL. 2001. Diets of fossil primates from the Fayum Depression of Egypt: a quantitative analysis of molar shearing. *Journal of Human Evolution* 40(3):203–229. doi.org/10.1006/jhev.2000.0450

Marivaux L, Welcomme J, Antoine P, *et al.* 2001. A fossil lemur from the Oligocene of Pakistan. *Science* 294:587–591.

Masters JC, Lovegrove BG, de Wit MJ. 2007. Eyes wide shut: can hypometabolism really explain the primate colonization of Madagascar? *Journal of Biogeography* 34:21–37.

Masters JC, Silvestro D, Génin F, DelPero M. 2013. Seeing the wood through the trees: the current state of higher systematics in the Strepsirhini. *Folia Primatologica* 84:201–219.

Masters JC, Génin F, Silvestro D, Lister A, DelPero M. 2014. The red island and the seven dwarfs: body size reduction in Cheirogaleidae. *Journal of Biogeography* 41:1833–1847.

Maurin O. 2009. A phylogenetic study of the family Combretaceae with emphasis on the genus *Combretum* in Africa. Dissertation, University of Johannesburg, Johannesburg.

Meade A, Pagel M. 2011. BayesTrees. www.evolution.rdg.ac.uk/BayesTrees.html.

Meijaard E, Nijman V, Supriatna J. 2008. *Nasalis larvatus.* In IUCN 2013. IUCN Red List of Threatened Species. Version 2013.2. www.iucnredlist.org. Downloaded on 30 January 2014.

Mittermeier RA, Louis Jr EE, Richardson M, *et al.* 2010. *Lemurs of Madagascar,* 3rd edition. Conservation International, Washington, DC.

Nash LT. 1986. Dietary, behavioral, and morphological aspects of gummivory in primates. *Yearbook of Physical Anthropology* 29:113–137.

Nash LT. 1998. Vertical clingers and sleepers: seasonal influences on the activities and substrate use of *Lepilemur leucopus* at Beza Mahafaly Special Reserve, Madagascar. *Folia Primatologica* 69(Suppl. 1):204–217.

Niemitz C, ed. 1984. *Biology of Tarsiers.* G. Fischer, Stuttgart.

Nussinovitch A. 2009. *Plant Gum Exudates of the World.* CRC Press, Boca Raton.

Pagel M, Meade A, Barker D. 2004. Bayesian estimation of ancestral character states on phylogenies. *Systematic Biology* 53:673–684.

Paradis E, Claude J, Strimmer K. 2004. APE: Analyses of Phylogenetics and Evolution in R language. *Bioinformatics* 20:289–290.

Power ML. 2010. Nutritional and digestive challenges to being a gum-feeding primate. In LT Nash, AM Burrows (eds.), *The Evolution of Exudativory in Primates. Developments in Primatology: Progress and Prospects* (pp. 25–44). Springer, New York.

Pozzi L, Disotell TR, Masters JC. 2014. A multilocus phylogeny reveals deep lineages within African galagids (Primates: Galagidae). *BMC Evolutionary Biology* 14:72.

R Core Team. 2013. *R: A Language and Environment for Statistical Computing*. R Foundation for Statistical Computing, Vienna, Austria. URL: www.R-project.org/

Rasmussen DT. 1990. Primate origins: lessons from a Neotropical marsupial. *American Journal of Primatology* 22:263–277.

Richard A. 1985. *Primates in Nature*. W.H. Freeman and Company, New York.

Revell LJ. 2012. Phytools: an R package for phylogenetic comparative biology (and other things). *Methods in Ecology and Evolution* 3:217–223.

Rowe N, Myers M. 2014. All the World's Primates. Primate Conservation Inc., Charlestown RI. www.alltheworldsprimates.org. Downloaded on 3 February 2014.

Sargis EJ. 2002. Primate origins nailed. *Science* 298(5598):1564–1565.

Scher HD, Martin EE. 2006. Timing and climatic consequences of the opening of Drake Passage. *Science* 312(5772):428–430.

Schwartz JH, Tattersall I. 1985. Evolutionary relationships of living lemurs and lorises (Mammalia, Primates) and their potential affinities with European Eocene Adapidae. *Anthropological Papers of the American Museum of Natural History* 60(Part 1):100.

Silcox, MT. 2007. Primate taxonomy, plesiadapiforms, and approaches to primate origins. In MJ Ravosa, M Dagosto (eds.), *Primate Origins: Adaptation and Evolution. Developments in Primatology: Progress and Prospects* (pp. 143–178). Springer, New York.

Silcox MT, Sargis EJ, Bloch JI, Boyer DM. 2007. Primate origins and supraordinal relationships: morphological evidence. In W Henke, I Tattersall (eds.), *Handbook of Paleoanthropology*, Volume 2 (pp. 831–860). Springer, Berlin.

Soligo C. 2006. Correlates of body mass evolution in primates. *American Journal of Physical Anthropology* 130:283–293.

Soligo C, Martin RD. 2007. Adaptive origins of primates revisited. *Journal of Human Evolution* 50:414–430.

Springer MS, Meredith RW, Gatesy J, *et al.* 2012. Macroevolutionary dynamics and historical biogeography of primate diversification inferred from a species supermatrix. *PLoS ONE* 7:e49521.

Sriver RI. 2010. Tropical cyclones in the mix. *Nature* 463:1032–1033.

Sussman RW. 1991. Primate origins and the evolution of angiosperms. *American Journal of Primatology* 23:209–223.

Szalay FS. 1968. The beginnings of primates. *Evolution* 22:19–36.

Szalay FS. 1972. Paleobiology of the earliest primates. In RH Tuttle (ed.), *The Functional and Evolutionary Biology of Primates* (pp. 3–35). Aldine-Atherton, Chicago.

Szalay FS, Dagosto M. 1980. Locomotor adaptations as reflected on the humerus of Paleogene primates. *Folia Primatologica* 34:1–45.

Szalay FS, Dagosto M. 1988. Evolution of hallucial grasping in the primates. *Journal of Human Evolution* 17:1–33.

Zachos JC, Breza JR, Wise SW. 1992. Early Oligocene ice-sheet expansion on Antarctica: stable isotope and sedimentological evidence from Kerguelen Plateau, southern Indian Ocean. *Geology* 20(6):569–573.

5 Contact zones and species sympatry in dwarf lemurs (genus *Cheirogaleus*): the roles of ecological adaptation and sexual selection

James P. Herrera, Tongasoa Lydia, and Patricia C. Wright

Introduction

Species divergence is driven by multiple factors, obfuscating the speciation mechanisms responsible for the diversity of life. The speciation mechanisms that dominate the literature are biogeographic models emphasizing allopatric speciation, or adaptive models invoking ecological adaptation and sexual selection (Coyne and Orr, 2004; Nosil, 2012). In the classic biogeographic model, vicariance is the dominant mechanism of speciation, in which previously interbreeding populations are sundered by a dispersal barrier (Ronquist, 1997; Ree and Smith, 2008; Ree and Sanmartín, 2009). The subdivided populations accumulate genetic differences due to drift, leading to reproductive isolation via genetic incompatibilities (Dobzhansky, 1937; Mayr, 1947; Orr, 1996). In the strict vicariant model, no divergent adaptation is necessary and the daughter species retain the niche of the ancestor, because their divergence was due to some physical dispersal barrier only (phylogenetic niche conservatism; Wiens *et al.*, 2010). If the dispersal barriers disappear, the newly formed species may meet at contact zones, but species boundaries are maintained and hybridization is limited by genetic incompatibilities (Coyne and Orr, 1998).

In contrast, adaptive models, especially ecological speciation, posit that species diverge via local adaptation to different habitats (Nosil, 2012). Contact zones exist at parapatric species boundaries, such as in ecotones of the different habitats (Schluter, 2001). Hybridization may be limited by lower fitness of immigrants into suboptimal habitats (prezygotic), or by lower fitness of hybrids because they are maladapted to the preferred habitat of either parent species (postzygotic; Nosil *et al.*, 2005). Immigrants into ecotones and suboptimal habitats may have lower fitness due to competition with neighboring, closely related species that are well adapted to that habitat (Bridle and Vines, 2007). Alternatively, immigrants into the neighbor species' habitats may exhibit ecological character displacement, altering

The Dwarf and Mouse Lemurs of Madagascar: Biology, Behavior and Conservation Biogeography of the Cheirogaleidae, eds. S.M. Lehman, U. Radespiel, E. Zimmermann. Published by Cambridge University Press. © Cambridge University Press 2016.

their phenotypes and ecology to avoid competition and persist (Schluter, 2000; Case *et al.*, 2005; Grant and Grant, 2006). The dynamics in contact zones offer an excellent opportunity to tease apart the subtle differences in geographic speciation mechanisms.

Speciation mechanisms related to behavior, especially sexual selection, are less frequently cited (Panhuis *et al.*, 2001). Reinforcement is the positive selection for trait divergence driven by sexual selection and accentuated when closely related species are found in sympatry compared to in allopatry (Noor, 1999). Sexual selection can reinforce ecological mechanisms of speciation when mate choice is related to habitat preferences (Boughman, 2001); e.g., individuals may select mates that are adapted to their own habitat (Lande, 1982). Mate recognition based on sexual traits may evolve as a pre-mating reproductive barrier when hybridization results in offspring with low fitness (Noor, 1999; Panhuis *et al.*, 2001). Postzygotic sexual selection may also occur when fitness is reduced in hybrids because, for example, hybrid sexual traits or courtship behaviors are not recognized as desirable signals or are maladapted to the environment (Servedio, 2004). Thus, mate choice and sexual selection can also lead to reproductive isolation in addition to environmental adaptation.

Madagascar is a natural living laboratory of evolution. Madagascar contains a disproportionately high species richness given its small size relative to continental biodiversity hotspots (Myers *et al.*, 2000). Further, the species in Madagascar tend to be microendemic, restricted to small geographic ranges isolated by rivers, mountains, or anthropogenic landscape (Vences, 2009). Many biogeographic hypotheses have been proposed for the evolution of microendemism in Madagascar. The unique and contrasting biomes of Madagascar (e.g., eastern rainforest, western dry forest) are relatively ancient (20–60 million years; Wells, 2003), and early researchers hypothesized that the geographically divided biotic domains have their associated assemblage of endemic fauna (Richard and Dewar, 1991; Martin, 1995). Further, several major rivers and topographic features subdivide the domains and act as barriers to gene flow for some species, but not all (Martin, 1972; Pastorini *et al.*, 2003; Goodman and Ganzhorn, 2004a, 2004b; Craul *et al.*, 2008). The distribution of current climate regimes along gradients of elevation and latitude may shape the distribution of organisms via ecological adaptation to different habitats along the gradient (Pearson and Raxworthy, 2009). Montane habitats have also been referred to as “species pumps” because of the high endemism, species richness and diverse ecological niches available (Wollenberg *et al.*, 2008; Vences, 2009). Reinforcing the niche diversity on mountains, there are distinct floral communities at different elevational ranges, creating unique niche space for animals (Lowry *et al.*, 1997). These biogeographic hypotheses suggest that ecological isolation may be the driving cause of evolution in Madagascar. In contrast, the effect of frequent, rapid habitat change due to Pleistocene climate shifts may have resulted in complex speciation patterns and the current distribution of diversity (Wilmé *et al.*, 2006). During glacial maxima, aridification in the lowlands may have isolated forest refugia in watersheds, promoting vicariance, while high-elevation watersheds were areas of retreat and dispersal (Wilmé *et al.*, 2006). This model posits vicariance events in physically isolated populations leading to genetic isolation due to random drift processes and does not invoke an adaptive hypothesis for speciation.

The biogeographic hypotheses for evolution in Madagascar have frequently been based on research from lemurs, the endemic primates. Some lemurs are geographically widespread occurring in multiple biogeographic domains (e.g., *Eulemur rufifrons*, *E. fulvus*), while other species are geographically restricted (e.g., *E. macaco*; Markolf and Kappeler, 2013; Markolf *et al.*, 2013). The family Cheirogaleidae is an exemplary model system for testing biogeographic hypotheses. Mouse lemurs (*Microcebus*) are among the most diverse and widely distributed lemurs in Madagascar. Most *Microcebus* species are microendemics, restricted to regions of Madagascar with hypothesized barriers between them (Weisrock *et al.*, 2010). *M. murinus* is widespread, and there are several regions where *M. murinus* is sympatric with microendemic species like *M. griseorufus* in the south, and *M. ravelobensis* or *M. berthae* in the northwest and west, respectively. In areas of sympatry, *Microcebus* species tend to partition the habitat along microhabitat, dietary, and life-history niche axes (Schwab and Ganzhorn, 2004; Dammhahn and Kappeler, 2008a, 2008b; Rakotondravony and Radespiel 2009; Thorén *et al.* 2011; Radespiel, Chapter 26). Where *Microcebus* species meet at clear ecotones, *M. murinus* appears to be the most generalized, while *M. rufus* is restricted to moist habitats and *M. griseorufus* to the driest habitats (Gligor *et al.*, 2009). The frequency of microendemism, correspondence of ranges to hypothesized Pleistocene refugia, and phylogeography in *Microcebus* seem to mainly support the vicariance model (Yoder and Heckman, 2006; Weisrock *et al.*, 2010; Rakotondranary *et al.*, 2011). In contrast to mouse lemurs, most dwarf lemur species (*Cheirogaleus*) tend to be widespread from the northern to the southern tips of Madagascar and with several zones of sympatry (Hapke *et al.*, 2005; Groeneveld *et al.*, 2010; Thiele *et al.*, 2013). The contrasting biogeographic patterns between *Microcebus* and *Cheirogaleus* are also reflected in the differences in species richness; 21 species of mouse lemurs are recognized (Radespiel *et al.*, 2012; Rasoloarison *et al.*, 2013), while there are only five named species of dwarf lemur (Groeneveld *et al.*, 2010; Thiele *et al.*, 2013). *C. medius* is found in western and southern dry forests primarily, *C. major* and *crossleyi* in eastern rainforests, *C. lavasoensis* is a newly described species found in a restricted region of extreme southeast Madagascar, and *C. sibreei* is found in a restricted zone of central high plateau forest (Figure 5.1, Blanco *et al.*, 2009; Groeneveld *et al.*, 2010; Thiele *et al.*, 2013). There are six published localities where more than one dwarf lemur species coexist, including sites where the dry forest-adapted *C. medius* is found in rainforest, and where rainforest-adapted *C. crossleyi* is found in dry forest (Hapke *et al.*, 2005; Groeneveld *et al.*, 2010). *Cheirogaleus sibreei* is found in high plateau forest sympatric with *C. crossleyi* (Blanco *et al.*, 2009). But how can dwarf lemur species that are so similar in morphology and overall niche (small-bodied nocturnal arboreal omnivores) share the same habitat without competitive exclusion? In this chapter, we present data from two contact zones in which *C. sibreei* and *C. crossleyi* are sympatric. We compare habitat characteristics of sympatric sites to sites in which only *C. crossleyi* is present. We explore morphological differences between these species and among sites to test for evidence of character displacement and mate recognition. Our aim is to investigate the morphological and ecological factors that may be related to speciation.

Hypotheses and predictions

1. There is the potential for sympatric dwarf lemur species in southeast Madagascar. We observed significant variation in pelage coloration and shape among captured dwarf lemurs in the Ranomafana area (Wright and Herrera, unpublished data). We therefore predict that there are multiple species present in sympatry in this region. To test this hypothesis, we quantified and compared the morphology of dwarf lemurs captured in different habitat types. We also compared our data to data collected on dwarf lemurs from a different site for which the species status has been confirmed with molecular evidence.
2. Under the character displacement hypothesis, species are predicted to accentuate differences in their morphology in sympatry compared to in allopatry to avoid niche overlap and competition (Schluter, 2000). We consequently predict that there is significant between-site variation in shape and size within species. We predict species will have a wider size distribution in allopatry than in sympatry. Overlap in size distributions between the species should be minimized in sympatry. Alternatively, no difference in morphometrics and size in sympatric populations would suggest a non-adaptive explanation for speciation or adaptation acting on behavioral or ecological traits not captured by morphology alone.

Methods

Study species

Cheirogaleus crossleyi is a widely distributed species found from the northern to the southern tips of Madagascar. In contrast, *C. sibreei* is geographically restricted. *C. sibreei* was thought to be extinct until 2009, when a single wild population was found in forest fragments at Tsinjoarivo near Antananarivo (Blanco *et al.*, 2009). The taxonomy and morphological descriptions for these species were done by Groves (2000), supported later by morphometric and molecular evidence (Blanco *et al.*, 2009; Groeneveld *et al.*, 2010, 2011). Here, we describe two localities for *C. sibreei*, a Critically Endangered species, that were previously unknown to science.

Study sites

We present data from four sites in southeast Madagascar (Figure 5.1). Two sites are within a hypothesized sympatric zone where morphotypes that match the descriptions for both *C. sibreei* and *C. crossleyi* were captured (SYM1 and 2). The first sympatric site is north of Ranomafana National Park (RNP) in the Ambositra–Vondrozo

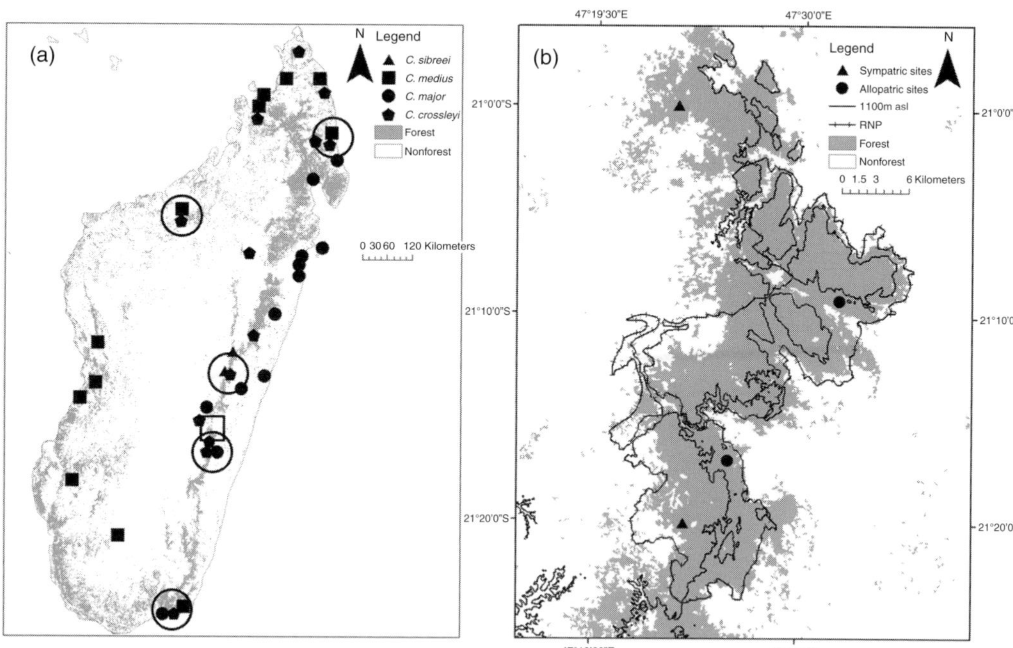

Figure 5.1 Maps displaying: (a) remaining forest cover and localities where the four commonly recognized dwarf lemur species occur. Sites of dwarf lemur species sympatry are outlined in circles. The study sites in this chapter are outlined in one rectangle. (b) Sites surveyed in this study, with sympatric sites (SYM1 north, SYM2 south) marked with triangles and allopatric sites (ALLO1 north, ALLO2 south) marked with circles. The Ranomafana National Park (RNP) boundary is labeled with the dashed lines. A contour line at 1100 m above sea level is drawn to illustrate high plateau forest to the west and eastern slope rainforest to the east of the contour. Landcover data based on Harper *et al.* (2007) classification of Landsat data from 2005. Contour line from digital elevation model from Landsat imagery. Data on distribution of dwarf lemurs from Groeneveld *et al.* (2009, 2010). Maps created with ArcGIS9.2 with the projection WGS 1984 UTM Zone 38S.

corridor, known locally as Ampatsona (SYM1: S21.009, E47.398). The second sympatric site is a high-montane site within RNP, known locally as Maharira (SYM2: S21.334, E47.409). The two sites are approximately 150–200 km south of the only other known *C. sibreei* site, Tsinjoarivo. Both sites are topographically continuous with the central high plateau, west of Madagascar's eastern escarpment and the rainforest slopes. We compare these two sympatric sites to two rainforest sites within RNP on the eastern slopes in which only morphotypes resembling *C. crossleyi* were captured (ALLO1: S21.158, E47.549, known locally as Miaranony, and ALLO2: S21.297, E47.439, Valohoaka).

SYM1 and 2 range in elevation from 1200 to 1400 meters above sea level (asl). The topography is characterized by rolling peaks separated by wide marshes and sharp peaks with rocky outcrops and cliffs. SYM1 consists mostly of secondary and disturbed forest that was used for agriculture and non-commercial logging approximately

60 years ago and some selective logging of hardwoods 40 years ago. In contrast, SYM2 has no history of human disturbance. The trees at SYM1 and 2 are not taxonomically diverse, dominated by only a few genera, especially *Uapaca*, *Polycias*, *Syzigium*, and *Weinnmania*. The forest is thick with saplings and vines, with a low understory and canopy (5–10 m). Rocky outcrops are festooned with *Pachypodia*, *Kalanchoe*, *Polycias*, and *Anthocleista*, similar to western dry forests (Lowry *et al.*, 1997).

In contrast, the eastern slope rainforest sites ALLO1 and ALLO2 range from 600 to 1200 m asl, and are characterized by long slopes and sharp peaks separated by steep, fast-moving rivers. ALLO1 has little human disturbance, while ALLO2 was a site of commercial selective logging in the 1980s and is lightly disturbed. The tree communities at ALLO1 and ALLO2 are diverse, even in disturbed areas. Primary forests are dominated by *Slonea*, *Ocotea*, *Cryptocarya*, *Syzigium*, *Canarium*, *Ficus*, and *Chrysophyllum*, with canopy height between 25 and 40 m. Disturbed forests have similar taxa to primary forests occurring in the understory, while the dominant and biggest trees are *Dombeya*, *Harunga*, and *Croton*, with canopy height ranges between 10 and 25 m.

Live trapping

Data presented here were collected during the rainy seasons (September–March) between September 2012 and December 2013. We established grids of trails for trapping at each site (SYM1 = 10, SYM2 = 5, ALLO1 = 5, ALLO2 = 2). Trap stations were set every 15–50 m (mode = 25 m) along trails. Each trap station had a Sherman brand live-trap (38 cm L × 10 cm W × 11 cm H) placed on branches and vines at heights ranging from 1 to 2 m and a Havahart brand live-trap (81cm L × 25cm W × 30cm H) placed 3–10 m high. At dusk, we opened the traps and baited them with banana and/or pineapple. We checked traps at 22:00 and again at 05:00. Captured dwarf lemurs were not anesthetized; the time period of manipulation was short (5–10 minutes), the animals habituated to the procedure quickly, and when animals were captured in the night, we released them immediately after capture and processing to avoid keeping the animals out of their environment any longer than necessary. Captured animals were weighed, measured, marked (hair clippings on backside) and released at the site of capture. We set traps at each station for 5–10 consecutive nights before moving them to new trails to allow for initial habituation to the trap presence and to allow time to capture a high proportion of individuals in the vicinity of the trap station. Dwarf lemurs are difficult to capture, and capture rates ranged from 0 to 5 individuals/1000 capture attempts. Trapping efforts were not equally distributed among sites (approximate trapping efforts: SYM1 = 3500 capture attempts, SYM2 = 2500, ALLO1 = 2000, ALLO2 = 1000). Despite the uneven distribution of trapping effort and difficulty to capture dwarf lemurs, we are confident that *C. crossleyi* is truly allopatric at sites ALLO1 and 2 for two reasons. First, when *C. sibreei* was present, they were readily captured, often within the first week of trapping, and they more readily entered traps than *C. crossleyi*. Second, trapping was conducted over the course of at least two months and throughout the two-year time period specified above. From our experience, this sampling effort is ample time to determine the presence/absence of cheirogaleid species by trapping. No animals were captured that were clearly pregnant or lactating, and no animals were captured during reproductive months (February–April).

All research complied with the methods permitted by the Madagascar National Parks (permit# 187/13/MEF/SG/DGF/DCB.SAP/SCB) and the International Animal Care and Use Committee at the State University of New York at Stony Brook (project#2012-1959-R1-USDA-Lm). We also practice the Principles for the Ethical Treatment of Non-Human Primates outlined by the American Society of Primatologists.

External morphological descriptions

We recorded the pelage coloration and general description of external morphology as in previous dwarf lemur descriptions (Blanco *et al.*, 2009; Groves, 2000). We inspected the female genital morphology and recorded the characteristics of the vaginal opening and clitoris. *C. sibreei* has an enlarged, pendulous, masculinized clitoris that is hairless and distinctly cleft with the urethral opening occurring within the anterior tip of the cleft. In contrast, in *C. crossleyi*, the clitoris is a small projection that is haired and not masculinized, and the sealed vagina may have some of the vaginal skin protruding at the top of the vaginal opening (Blanco *et al.*, 2009).

Morphometrics

We recorded the following external morphometrics in millimeters on captured individuals following Blanco *et al.* (2009): crown–tail length (from anterior-most point of skull to first caudal vertebrae), tail length (from first to last caudal vertebrae), hind leg length (from hip joint to tarsus) (all with measuring tape), head length (from occipit to tip of nose), head width (maximum axis across parietals), snout length (from distal-most point of eyes to tip of rhinarium), ear length (from inner pinna to tip of ear), and ear width (maximum axis across ear) (all with calipers). These morphometrics are standard in the field (Groves and Harding, 2011), and they are the same as taken by Blanco *et al.* (2009). JPH trained with Blanco during early phases of this research. Included in this data set is a comparative sample of *C. crossleyi* from Tsinjoarivo and RNP and *C. sibreei* from Tsinjoarivo for which the species status has been verified with molecular and morphological data (Blanco, unpublished data; Groeneveld *et al.*, 2010). We use these data to validate which species is represented by the morphotypes in RNP.

Data analyses

Morphotype differentiation and species discrimination

To investigate differences in shape as well as size among morphotypes in our sample, we quantified scale-free shape variables using the ratio of each morphometric variable to the geometric mean of variables (Mosimann correct, Jungers *et al.*, 1995). We checked size-adjusted variables for their distribution and correlation. We found the variables did not deviate from normal distributions, most variables covaried, and we did not identify any extreme bivariate outliers that could be identified as data errors. Four individuals from SYM2 (two *C. sibreei* and two *C. crossleyi*) were excluded from analysis because they were significantly smaller than all other individuals in the sample and were suspected to be juveniles. We compared our morphometric data to those provided by M. Blanco from Tsinjoarivo and a site in RNP not sampled in this study (*C. sibreei*: n = 34 Tsinjoarivo; *C. crossleyi*: n = 43 Tsinjoarivo, n = 14 RNP; data published in Blanco *et al.*, 2009 and unpublished data). The total morphometric data set consists of 83 *C. crossleyi* and 53 *C. sibreei*, a total of 136 individuals.

We conducted a principal components analysis (PCA) on the size-adjusted eight-variable data set, using the variance–covariance matrix. We chose the number of PC axes to retain based on the decreasing variance explained, as visualized in a scree plot. We then entered the PCA scores and size into a discriminant function analysis (DFA) to predict the probability of group membership, assigning group membership (species) *a priori* only to the known *C. crossleyi* and *C. sibreei* samples from Tsinjoarivo and RNP. The new samples from RNP and SYM1 were entered as ungrouped so that the DFA could predict to which group these morphotypes most likely belong. Variables were entered together (rather than, for example, stepwise), prior-probabilities were computed from group sizes, within-group covariance matrices were used, and the leave-one-out classification was used for cross-validation. For the DFA, the same analyses were run on a data set excluding the size metric, as well as on raw morphometrics, and results across analyses were comparable (results not shown). We followed the DFA with multivariate and univariate analyses of variance (MANOVA and ANOVA), comparing variation between species and among sites (two-way analyses) to determine which morphometric variables differ specifically. We conducted the analyses of variance on the full data set as well as on the data set from the southeast only. In these analyses, we entered individuals as either *C. crossleyi* or *C. sibreei* based on the morphological characteristics and the overall results of the DFA. For these statistics, all analyses were done in the R statistical environment (R Core Team, 2014), except the DFA which was run in SPSS® v12 (SPSS Inc., 2003).

Character displacement

We investigated character displacement in our data set from the four sites by comparing the size distributions of the two morphotypes in allopatry and sympatry to a null size distribution generated by bootstrap resampling of the total size data set. In each of 1000 permutations, we drew 26 individuals with replacement, equal to the number of *C. crossleyi* individuals in the data set. Under the character displacement hypothesis, the species that occurs both in sympatry and allopatry is predicted to have a wide size distribution in allopatry that is narrowed in sympatry and that is non-overlapping with the potential competitor in sympatry. We tested if the size distributions differ among species, between allopatric and sympatric populations within species and between populations and the null expectation using Kolmogorov–Smirnov tests of distributions. The bootstrapping and distribution analyses were conducted in R.

Results

Morphological descriptions of captured animals

Capture results are summarized in Table 5.1. In total, we captured 19 individuals that had morphological characteristics typical of *C. sibreei* (n = 7 SYM2, n = 12 SYM1) and 26 individuals matching morphological characteristics of *C. crossleyi* (n = 5 SYM1, n = 5 SYM2, n = 12 ALLO1, n = 4 ALLO2). Photographs of representative individuals for the two species are presented in Figure 5.2. Captured individuals that resembled the description of *C. sibreei* had the following pelage characteristics: the head, dorsum and tail are gray–fawn, with a distinct black dorsal stripe running down the midline. The ventrum is white with darker proximal hairs on the chest and stomach. The neck and chin are white, and the white of the neck extends dorsally forming a collar that does not meet dorsally. The ears are black and haired, and there are distinct, wide, black eye rings. Between the eyes there is a white patch. The eye rings extend down to the muzzle but stop before the snout, which is distinctly more gray than the black eye rings and not heavily haired. The snout is blunt and wide.

In contrast, individuals resembling *C. crossleyi* had the following pelage characteristics: the head, dorsum and tail are brown to red–brown with no dorsal stripe. The tail infrequently has a white tip. The ventrum varies from white to cream in roughly half the captured individuals. In approximately half of the cases (especially SYM2 and ALLO1 cases), the ventrum is distinctly red, variably extending from the breast to the stomach. The ears are black and hairy, and the face is brown, except for the dark black eye rings which extend the length of the long snout and a white patch between the eyes. The chin and cheeks are white to red, depending on the coloration of the ventrum. While the overall appearance of these individuals matches the description for *C. crossleyi*, the red ventrum has not been described before (see Groves, 2000), and was not observed in the Blanco sample from RNP or Tsinjoarivo (M. Blanco, pers. comm.). This color variation will be investigated further in subsequent analyses, especially together with molecular investigation.

The two dwarf lemur morphs also have distinct female genital morphologies. In female individuals that matched the description for *C. sibreei*, the clitoris is enlarged, masculinized, cleft and hairless, as described for *C. sibreei* at Tsinjoarivo (Blanco *et al.*, 2009). In contrast, *C. crossleyi* females had a small, haired clitoris that is not cleft or masculinized, and when the vaginal seal was open, the labia predominate and the clitoris is indistinct (Figure 5.2). These discrete sexual traits may serve as mate recognition signals that allow different species to discriminate potential mates.

Table 5.1 Descriptive statistics of raw morphometric variables. N = number of individuals sampled. All measurements are in millimeters. Tsinjoarivo 1 is the continuous forest on eastern slope (Andasivodihazo) which is an allopatric site where only *C. crossleyi* is found, and Tsinjoarivo 2 is the fragmented forest on the high plateau (Vatateza) which is a sympatric site where *C. crossleyi* and *C. sibreei* occur together. All other localities are from the southeast, either RNP or Ampatsona, north of the park. For the southeast samples collected in this study, SYM1 and SYM2 are the sympatric sites and ALLO1 and ALLO2 are the allopatric sites. Samples from Tsinjoarivo and Talatakely (RNP) are part of the Blanco unpublished data set. The data presented here are: mean, standard deviation (in parentheses) and range. Cells with superscript numbers in common indicate significant differences between sites within species for that variable, including Tsinjoarivo data (within-species ANOVA followed by posthoc Tukey HSD test across sites).

Species and locality	N	Mass (g)	Crown–tail	Tail	Leg	Head length	Head width	Snout length	Ear length	Ear width
C. crossleyi Tsinjoarivo 1	22	395.40 (75.84) 288.00–514.00	245.20^{1} (12.49) 220.00–270.00	258.60^{1} (13.63) 230.00–276.00	102.97^{1} (11.09) 83.56–126.12	57.46 (2.44) 52.20–62.44	35.39 (1.33) 32.64–37.44	16.04 (1.17) 13.44–18.20	22.93 (1.61) 18.46–25.28	11.92^{1} (1.14) 10.00–13.72
C. crossleyi Tsinjoarivo 2	21	339.0 (48.14) 281.60–456.00	247.1^{2} (12.74) 227.00–270.00	253.30^{2} (10.30) 231.00–270.00	106.1^{2} (6.23) 90.30–117.10	56.60 (1.33) 54.00–58.50	31.00 (2.92) 31.00–45.92	16.65^{1} (1.38) 13.54–19.16	22.16^{1} (1.17) 18.90–24.14	11.51^{2} (1.30) 9.60–14.90
C. crossleyi Talatakely (RNP)	14	368.36 (44.44) 311.50–471.50	245.07^{3} (12.82) 228.00–270.00	277.07 (13.04) 255.00–300.00	171.29^{1,2} (11.53) 152.20–200.00	58.73 (1.50) 56.60–61.30	35.90 (1.51) 33.80–39.00	16.75 (1.03) 14.90–19.10	19.06 (4.26) 11.90–26.50	13.79^{1,2} (1.48) 10.98–16.10
C. crossleyi ALLO1	12	368.25 (60.19) 287.00–453.00	231.20^{1,2} (22.68) 205.00–272.00	291.08^{1,2} (17.12) 260.00–325.00	117.67 (21.38) 90.00–165.00	57.44 (4.60) 48.38–66.90	34.55 (2.65) 30.10–40.30	15.25^{1} (2.81) 10.56–19.10	22.71^{1} (3.99) 13.40–29.10	14.66^{2} (3.14) 10.40–22.68
C. crossleyi ALLO2	4	329.00 (13.90) 310.00–342.00	219.25^{1,2} (15.24) 200.00–235.00	277.50 (14.57) 260.00–290.00	104.50 (4.04) 101.00–110.00	57.03 (1.19) 55.87–58.70	35.33 (0.70) 34.90–36.40	13.71^{1} (1.23) 12.30–15.23	23.32^{1} (1.03) 22.24–24.31	15.61^{2} (3.05) 11.52–18.91
C. crossleyi SYM1	5	341.20 (56.90) 274.00–408.00	218.40^{1,2} (9.21) 207.00–231.00	271.40^{1,2} (19.78) 253.00–295.00	113.00 (29.92) 90.00–165.00	57.37 (2.66) 54.68–61.22	34.51 (0.94) 33.30–35.80	16.59 (1.39) 14.68–18.57	24.20 (1.10) 22.77–25.35	14.55^{2} (1.44) 13.22–16.84
C. crossleyi SYM2	5	307.80 (44.97) 270.00–380.00	235.00^{1,2,3} (12.06) 221.00–251.00	297.20 (27.27) 268.00–341.00	116.80 (7.92) 104.00–125.00	56.25 (2.67) 52.92–60.21	34.31 (1.07) 33.50–36.10	15.92 (2.60) 12.32–18.51	24.52 (1.95) 22.77–26.96	14.37^{2} (2.38) 12.05–18.29
C. sibreei Tsinjoarivo 2	12	290.27 (38.91) 249.00–368.30	219.33^{4} (7.09) 210.00–230.00	236.42^{3} (10.13) 224.00–255.00	148.73^{3} (8.92) 135.00–165.00	55.42^{1} (1.13) 53.90–57.18	33.77 (1.51) 30.40–35.60	15.54 (0.76) 14.24–16.76	23.31^{2} (1.14) 21.78–25.18	12.99 (0.87) 11.30–14.62
C. sibreei SYM1	7	212.00 (38.30) 136.00–255.00	191.97^{4} (18.00) 158.00–212.00	219.57^{3} (19.69) 188.00–247.00	97.71^{3} (9.14) 81.00–110.00	50.16^{1} (1.47) 47.55–52.32	32.01 (1.34) 29.30–33.50	11.05 (1.01) 9.97–12.63	22.09^{2} (1.40) 19.55–23.26	14.13 (1.71) 11.74–16.17
C. sibreei SYM2	12	229.55 (27.50) 190.00–275.00	206.58^{4} (8.93) 187.00–215.00	241.67 (13.63) 220.00–265.00	111.33 (15.98) 101.00–160.00	51.43 (2.98) 46.85–57.16	32.13 (1.32) 30.40–35.20	12.89 (1.27) 10.75–14.84	21.64^{2} (1.80) 18.23–23.88	15.07 (1.66) 11.94–17.14

Morphometrics

The raw morphometrics data are summarized in Table 5.1. In the PCA of the total data set including genetically confirmed *C. crossleyi* and *C. sibreei* from central Madagascar, four components were extracted based on decreasing variance observed in a scree plot (Table 5.2, first two components illustrated in Figure 5.3a). Individuals within each species group together, especially *C. sibreei*, but with approximately half the *C. crossleyi* samples overlapping the *C. sibreei* cloud of points. The individuals within species from different sites do not form distinct groups, suggesting the overall species difference is the predominant signal in shape variation.

The DFA using PC scores significantly discriminates variation due to species differences in the data set of genetically confirmed species (discriminant function

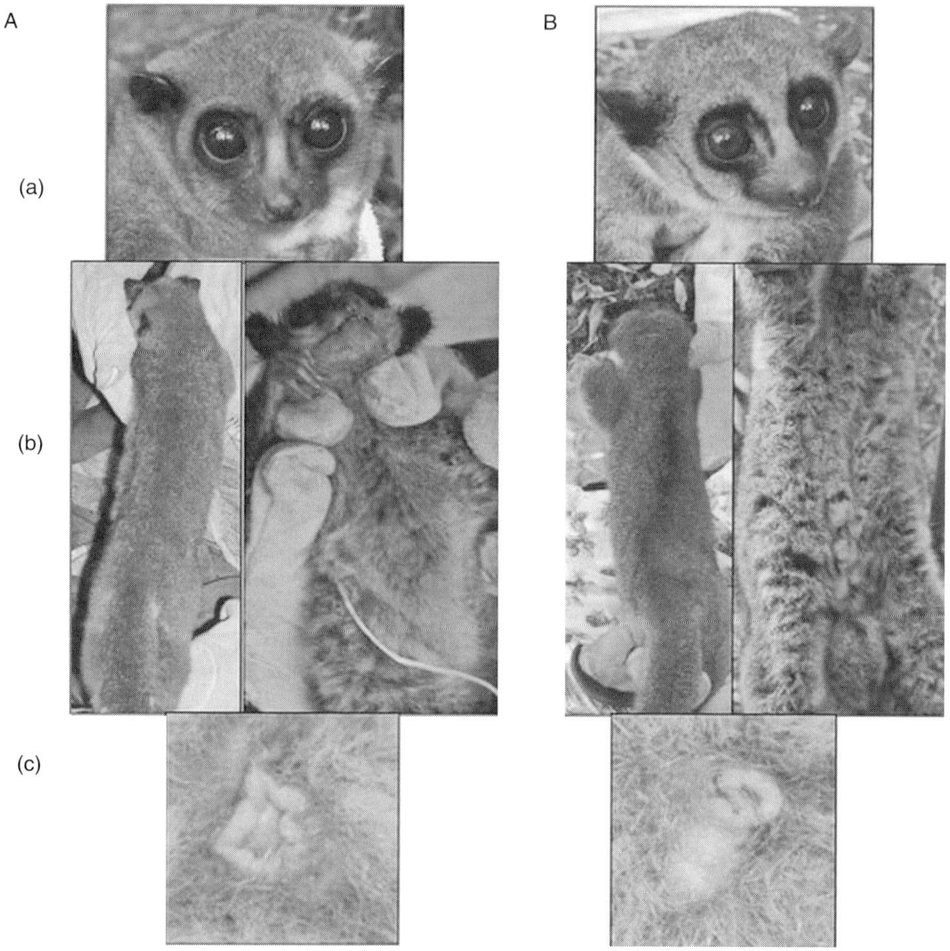

Figure 5.2 Photographs displaying the morphological variation in *Cheirogaleus crossleyi* (A) and *C. sibreei* (B) faces (a), dorsum and ventrum (b), and female genital morphology (c). (A black and white version of this figure will appear in some formats. For the colour version, please refer to the plate section.)

Table 5.2 The loadings of individual morphometrics with each principal component axis.

Original variable	Loading PC1 (31.88%)	Loading PC2 (22.9%)	Loading PC3 (14.11%)	Loading PC4 (10.66%)
Crown–tail length	-0.50	-0.08	0.02	-0.09
Tail length	-0.04	-0.44	0.33	-0.69
Hind leg length	0.25	-0.53	0.08	0.24
Head length	-0.44	0.17	0.31	-0.03
Head width	-0.20	0.48	0.48	0.13
Snout length	-0.40	-0.18	-0.45	0.36
Ear length	0.21	0.45	-0.49	-0.49
Ear width	0.50	0.16	0.32	0.26

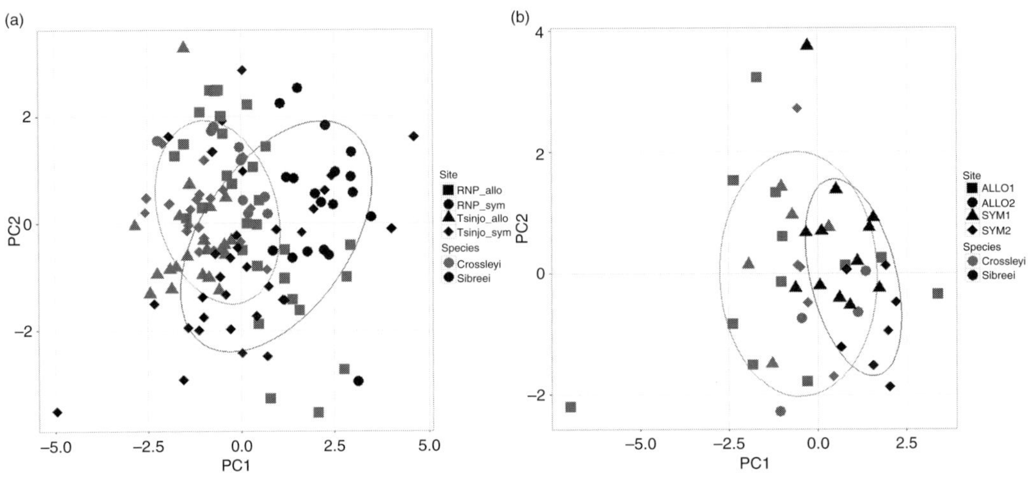

Figure 5.3 Scatterplot of principal component ordination representing variation in the shape of two dwarf lemur species in multivariate space. *Cheirogaleus sibreei* is represented by black data points and *C. crossleyi* by gray data points. (a) Total data set, including samples from Tsinjoarivo (Tsinjo_allo, Tsinjo_sym) and the RNP region (RNP_allo, RNP_sym). (b) The allopatric and sympatric sites sampled by us in the RNP region (SYM1 and 2, ALLO1 and 2). The 95% confidence ellipse for each species encircles the clusters of data points.

eigenvalue = 2.98, Wilks' λ = 0.251, χ^2_9 = 51.794, $p < 0.001$). Cross-validated correct classification of species memberships for the individuals confirmed by genetic analysis to be *C. crossleyi* was 93.8%; correct membership classification for confirmed *C. sibreei* individuals was 100%. These results suggest that the discriminant function can significantly discern known *C. crossleyi* and *C. sibreei* individuals. Results for the unknown individuals captured in this study confirm the general morphological differences; 91.7%–100% of suspected *C. sibreei* were predicted to be *C. sibreei*, and 100% of suspected *C. crossleyi* were predicted to be *C. crossleyi* (Table 5.3).

Table 5.3 Classification matrix of group memberships predicted by discriminant function analysis trained using data from genetically confirmed *C. crossleyi* and *C. sibreei* and predicting the probability of group membership for new samples collected in this study.

Morphotype	% of cases predicted to be *C. crossleyi*	% of cases predicted to be *C. sibreei*
C. crossleyi Tsinjoarivo 1	92.3	7.7
C. crossleyi Tsinjoarivo 2	100	0
C. crossleyi Talatakely	100	0
C. crossleyi ALLO1	100	0
C. crossleyi ALLO2	100	0
C. crossleyi SYM1	100	0
C. crossleyi SYM2	100	0
C. sibreei Tsinjoarivo 2	25	75
C. sibreei SYM1	0	100
C. sibreei SYM2	8.3	91.7

Table 5.4 Results of MANOVA tests to compare the effect of species and site on morphometrics. First, the effects of species and site are tested on the data set including data from Tsinjoarivo and comparing Tsinjoarivo sympatric and allopatric sites to southeast sympatric and allopatric sites described in this chapter. Second, we compare the effects of the species and site differences within only the southeast data set.

Effect	Df	Wilks' lambda	*P*
All data			
Species	1	0.45	<0.001
Site	3	0.11	<0.001
Species × site interaction	1	0.81	0.06
Southeast data set			
Species	1	0.41	<0.001
Site	3	0.40	0.10
Species × site interaction	1	0.88	0.83

The MANOVA results comparing species and sites from Tsinjoarivo and the southeast suggest that there are significant differences in morphometrics due to species and due to site with no significant interaction between species and site (Table 5.4). In the data set from the southeast alone, the MANOVA results suggest a significant difference between species in morphometrics, but no difference among sites and no interaction (Table 5.4). In both data sets, species differences were found in univariate comparisons of crown–tail length, head length, snout length, and ear width. The site difference in the whole data set is due to significant variation in tail, ear and hind limb length, and ear width (Table 5.1).

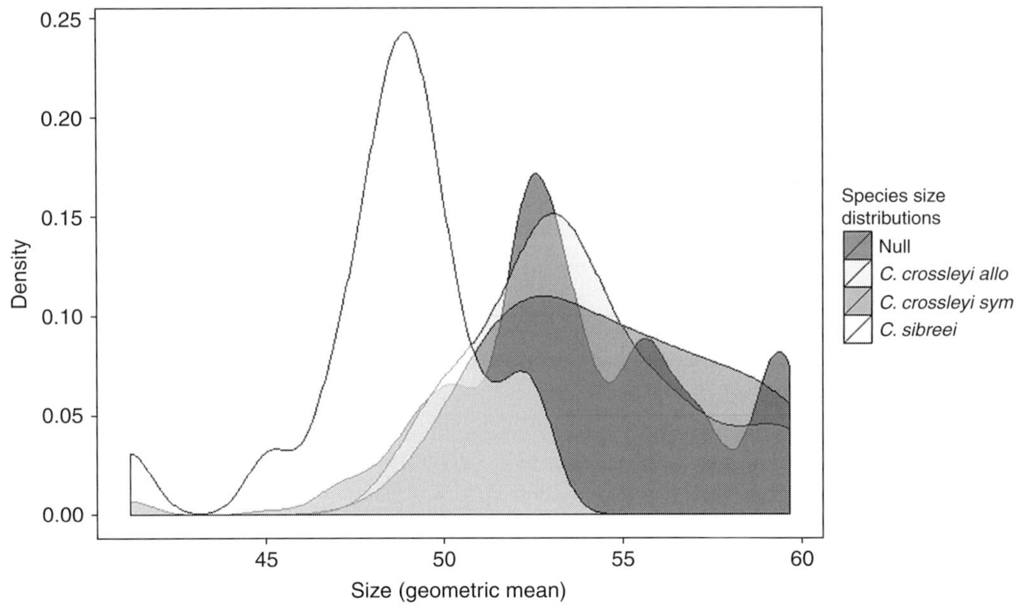

Figure 5.4 Comparison of the densities of size distributions (geometric means) for *C. sibreei* (white), *C. crossleyi* (shades of gray), and a null distribution (black). Under the character displacement hypothesis, we predicted that *C. crossleyi* in sympatry with *C. sibreei* would exhibit a narrower size distribution shifted to larger body size compared to samples from allopatric sites to accentuate the difference with *C. sibreei*.

Character displacement

We compared the size distributions of allopatric and sympatric populations of *C. crossleyi*, sympatric populations of *C. sibreei*, and a null size distribution generated from bootstrapping the combined size data of each species. There is a difference in the size distributions of *C. sibreei* and *C. crossleyi*; *C. sibreei* is significantly smaller than *C. crossleyi* (Kolmogorov–Smirnov test of distributions, $D = 0.765$, $p < 0.001$). The size distributions of *C. crossleyi* from sympatric and allopatric sites do not differ significantly ($D = 0.225$, $p = 0.84$), nor does the distribution of *C. crossleyi* from the sympatric sites differ from the null distribution ($D = 0.189$, $p = 0.87$). The size distribution of *C. crossleyi* in allopatry does not differ from those in sympatry or from the null distribution (Figure 5.4). The results are qualitatively the same whether the size distributions were based on the geometric mean of measurements (Figure 5.4) or mass (not shown). This result suggests that *C. crossleyi* does not shift its phenotype from allopatry to sympatry with *C. sibreei*.

Discussion

Our results suggest ecological adaptation and sexual selection in dwarf lemurs, supported by differences in morphology and ecology between sympatric species.

Here we described two localities where *C. sibreei* is found that were previously unknown to science. Further, at both sites, *C. sibreei* is sympatric with the widely distributed *C. crossleyi*. How these two species coexist in contact zones will give us clues to the evolutionary mechanisms that led to species divergence and diversity. Complementary molecular evidence is needed to test empirically whether the contact zone existed as an ecotone between divergent habitats during ecological speciation or emerged via secondary contact after vicariant speciation. Evidence presented here suggests that the two species meet at ecotones between eastern slope rainforest and western high plateau forest, and that these habitats differ from the eastern slope rainforests where only *C. crossleyi* is found. While rainforest and highland forest are frequently disjunct due to anthropogenic deforestation, at the sites described here they are within a contiguous forest, with no clear barrier to dispersal. The widespread *C. crossleyi* may tolerate high plateau forest, while being more prevalent in eastern slope forest. In contrast, the geographically restricted *C. sibreei* is only found in the western plateau forest, suggesting it is specialized to this habitat.

The two species may not compete in sympatry because we found they are significantly different sizes, which may indicate that they exploit different resources. There was no evidence of character displacement between allopatric and sympatric sites for *C. crossleyi*, suggesting no shift in shape or size to avoid competition when in sympatry. Instead, *C. crossleyi* is consistently bigger than *C. sibreei* at all sites. A similar pattern was observed in sympatric sister species of *Microcebus* in southern Madagascar, with no evidence of morphometric differences within species when in sympatry or allopatry (Rakotondranary *et al.*, 2011); however, those species do not differ in size to the magnitude observed in *C. sibreei* and *C. crossleyi* in this study. The size ratio of *C. crossleyi* to *C. sibreei* is 1.55 in regards to mass and 1.11 in the geometric mean of measurements. The size ratio of 1.3 has been hypothesized to be the minimum difference necessary to minimize competition and allow sympatry (Hutchinson, 1959). The size difference may be enough to minimize competition, as well as serve as a mate recognition trait, explaining why size differences are not accentuated in sympatry compared to in allopatry, as predicted in character displacement (Noor, 1999). Further behavioral and ecological research is needed to identify if and how *C. crossleyi* and *C. sibreei* partition the habitats in sympatry. For example, differences in diet and substrate seem to allow *Microcebus murinus* and *M. berthae* to partition the habitat in sympatry (Schwab and Ganzhorn, 2004; Dammhahn and Kappeler, 2008a), and there are subtle differences in microhabitat preference and local abundance in sympatric populations of *M. murinus* and *M. ravelobensis* (Rakotondravony and Radespiel, 2009).

Mate recognition may reinforce species boundaries in *C. sibreei* and *C. crossleyi*. The differences in pelage, size, and morphology between the species likely serve as signals for species recognition and mate choice. The discrete differences in female genital morphology may suggest a mechanism of mate recognition as well as a trait related to sociosexual systems. The masculinized clitoris seen in *C. sibreei* is an unusual trait among mammals. Masculanized female genitalia are observed in spotted hyenas (*Crocuta crocuta*), spider monkeys (*Ateles* spp.), and several lemurs (e.g., *Lemur catta, Cheirogaleus medius*). In hyenas, the masculinized clitoris is related to high androgen levels during late phases of gestation, and is hypothesized to be related to high female aggression and female dominance (Frank, 1997). Across lemuriforms, the male and female genitalia morphology vary widely and genital structure is related to sexual systems; clitoris length and masculinization are related to female-dominant multi-male social systems (Treatman-Clark, 2006). Dwarf lemurs are reported to be monogamous based on earlier research (Wright and Martin, 1995; Fietz, 2003). Extra-pair copulations are frequent in western dwarf lemurs (Fietz, 2003), however, possibly explaining the long clitoris and penis (Treatman-Clark, 2006). Genital morphology may also be related to scent marking, as well, and scent may be an important factor in mate recognition for these nocturnal primates (Colquhoun, 2011). Further, differences in genital morphology may lead to post-mating isolation due to failure of gametes to meet or zygotes to implant properly (Nosil and Crespi, 2006). Where contact and hybrid zones occur in the genus *Microcebus*, there is typically a signal of male-biased introgression related to male dispersal and female philopatry (Gligor *et al.*, 2009; Weisrock *et al.*, 2010). At the contact zones described here, further molecular testing for gene flow between species is needed and, if present, testing for sex-biased introgression. Other important variables with great consequence on contact zones include sex-biased dispersal, and estimating dispersal distances can be quantified with the help of molecular data (e.g., Kappeler *et al.*, 2002) and lend insights into the causes and consequences of contact zones. Sociosexual systems have a local-level impact on gene-flow that permeates to the ecological boundaries of species (Kappeler, 2008).

The results of this study investigating contact zones in dwarf lemurs suggest an ecology-driven evolutionary scenario for Madagascar. The overall geographic distribution of dwarf lemur species coincides with the current climate model (Pearson and Raxworthy, 2009). Along the eastern escarpment, the current climate model suggests a gradient in the abiotic environment ascending from the eastern coast to the western highlands, and there should be distinct ecological communities along the gradient. *Cheirogaleus crossleyi* covers a wide elevational gradient but most known sites are mid-elevation rainforests (Groeneveld *et al.*, 2009), while *C. sibreei* is found only at higher elevations west of the eastern escarpment (Groeneveld *et al.*, 2010). Evidence from dwarf lemurs also lends support for the mountain–river dispersal barrier hypothesis (Goodman and Ganzhorn, 2004a). The hypothesis predicts species that occur at high elevations should be widespread because they can cross river headwaters; in contrast, low-elevation species are microendemic, restricted by montane barriers at higher elevations and riverine barriers at low elevations. *Cheirogaleus crossleyi* may be widespread because it is able to cross high-elevation river headwaters as well as occupy lower elevations. *Cheirogaleus sibreei* may be geographically restricted because it occurs only at high elevations and is further restricted by the near complete loss of high plateau forest due to anthropogenic activities, especially slash-and-burn agriculture. In the past, however, *C. sibreei* may have been more widespread. With the two new sites described in this chapter, the geographic range of *C. sibreei* is expanded 200 km to the south, and across the Mangoro river, an important hypothesized barrier (Goodman and Ganzhorn, 2004a). If the high plateau was continuously forested in the past, *C. sibreei* may have been distributed throughout the highlands of the Ankaratra and Andringitra mountain ranges. The phylogeographic patterns in dwarf lemur distribution across Madagascar as well as local patterns of gene flow at contact zones will allow us to differentiate between abiotic vicariance models and ecologically adaptive models of speciation in Madagascar.

Acknowledgments

We thank the local people who allowed us to work on their land at Ampatsona (SYM1), and all our local assistants from the villages around RNP and Ampatsona. We thank the Madagascar National Parks for permission and assistance in research performed at RNP, and the Ministry of the Environment and Malagasy government for permission to conduct research in Madagascar. We thank the following institutions for their logistical support: the Malagasy Institut pour la Conservation des Ecosystems Tropicaux, Institute for the Conservation of Tropical Ecosystems, the Centre ValBio research station, and the Stony Brook Research Foundation. We thank M. Blanco for the use of unpublished data, training for JPH early in the research program in Madagascar, comments on this manuscript, and helpful comments throughout the project. We thank the following granting agencies that made this research possible: Primate Conservation, Inc., Primate Action Fund, Margot Marsh Biodiversity Foundation, the Eddie Bauer Explorers Club, the Rufford Small Grants fund, the Mohamed bin Zayed Species Conservation Fund, the American Society of Primatologists, the International Primatological Society, La Conservatoire pour la Protecion des Primates, the National Science Foundation (Graduate Research Fellowship to JPH), the Turner Fellowship and the AGEP scholarship. For scientific advising we thank E. Seiffert, W. Jungers, R. Akçakaya, and L. Dávalos. For helpful comments and discussion of the topics in this chapter we thank L. Taylor. We thank two anonymous reviewers who inspired new insights into the topics discussed in this chapter, especially character displacement, and U. Radespiel for helpful revisions that improved the manuscript. We thank U. Radespiel, E. Zimmermann, and S. Lehman for assembling this volume and inviting our contribution.

References

Blanco MB, Godfrey LR, Rakotondratsima M, *et al.* 2009. Discovery of sympatric dwarf lemur species in the high-altitude rain forest of Tsinjoarivo, eastern Madagascar: implications for biogeography and conservation. *Folia Primatologica* 80(1):1–17.

Boughman JW. 2001. Divergent sexual selection enhances reproductive isolation in sticklebacks. *Nature* 411:944–948.

Bridle JR, Vines TH. 2007. Limits to evolution at range margins: when and why does adaptation fail? *Trends in Ecology & Evolution* 22(3):140–147.

Case TJ, Holt RD, McPeek MA, Keitt TH. 2005. The community context of species' borders: ecological and evolutionary perspectives. *Oikos* 108(1):28–46.

Colquhoun IC. 2011. A review and interspecific comparison of nocturnal and cathemeral strepsirrhine primate olfactory behavioural ecology. *International Journal of Zoology* 2011.

Coyne JA, Orr HA. 1998. The evolutionary genetics of speciation. *Philosophical Transactions of the Royal Society of London Biological Sciences* 353(1366):287–305.

Coyne JA, Orr HA. 2004. *Speciation.* Sinauer Associates, Sunderland.

Craul M, Radespiel U, Rasolofoson D, *et al.* 2008. Large rivers do not always act as species barriers for *Lepilemur* sp. *Primates* 49(3):211–218.

Dammhahn M, Kappeler PM. 2008a. Comparative feeding ecology of sympatric *Microcebus berthae* and *M. murinus. International Journal of Primatology* 29(6):1567–1589.

Dammhahn M, Kappeler PM. 2008b. Small-scale coexistence of two mouse lemur species (*Microcebus berthae* and *M. murinus*) within a homogeneous competitive environment. *Oecologia* 157(3):473–483.

Dobzhansky T. 1937. *Genetics and the Origin of Species.* Columbia University Press, New York.

Fietz J. 2003. Primates: *Cheirogaleus*, dwarf lemurs or fat-tailed lemurs. *The Natural History of Madagascar*:1307–1309.

Frank LG. 1997. Evolution of genital masculinization: why do female hyaenas have such a large 'penis'? *Trends in Ecology & Evolution* 12(2):58–62.

Gligor M, Ganzhorn JU, Rakotondravony D, *et al.* 2009. Hybridization between mouse lemurs in an ecological transition zone in southern Madagascar. *Molecular Ecology* 18:520–533.

Goodman SM, Ganzhorn JU. 2004a. Biogeography of lemurs in the humid forests of Madagascar: the role of elevational distribution and rivers. *Journal of Biogeography* 31(1):47–55.

Goodman SM, Ganzhorn JU. 2004b. Elevational ranges of lemurs in the humid forests of Madagascar. *International Journal of Primatology* 25(2):331–350.

Grant PR, Grant BR. 2006. Evolution of character displacement in Darwin's finches. *Science* 313(5784):224–226.

Groeneveld L, Weisrock D, Rasoloarison R, Yoder A, Kappeler P. 2009. Species delimitation in lemurs: multiple genetic loci reveal low levels of species diversity in the genus *Cheirogaleus. BMC Evolutionary Biology* 9(1):30.

Groeneveld LF, Blanco MB, Raharison JL, *et al.* 2010. MtDNA and nDNA corroborate existence of sympatric dwarf lemur species at Tsinjoarivo, eastern Madagascar. *Molecular Phylogenetics and Evolution* 55(3):833–845.

Groeneveld LF, Rasoloarison R, Kappeler PM. 2011. Morphometrics confirm taxonomic deflation in dwarf lemurs (Primates: Cheirogaleuidae), as suggested by genetics. *Zoological Journal of the Linnean Society* 161:229–244.

Groves CP. 2000. The genus *Cheirogaleus*: unrecognized biodiversity in dwarf lemurs. *International Journal of Primatology* 21:943–962.

Groves C, Harding J. 2011. Morphology, morphometrics and taxonomy. In J Setchell, D Curtis (eds.), *Field and Laboratory Methods in Primatology: A Practical Guide* (pp. 169–188). Cambridge University Press, Cambridge.

Hapke A, Fietz J, Nash SD, *et al.* 2005. Biogeography of dwarf lemurs: genetic evidence for unexpected patterns in southeastern Madagascar. *International Journal of Primatology* 26(4):873–901.

Harper GJ, Steininger MK, Tucker CJ, Juhn D, Hawkins F. 2007. Fifty years of deforestation and forest fragmentation in Madagascar. *Environmental Conservation* 34(04):325–333.

Hutchinson GE. 1959. Homage to Santa Rosalia, or why are there so many kinds of animals? *The American Naturalist* 93:137–159.

Jungers WL, Falsetti AB, Wall CE. 1995. Shape, relative size, and size-adjustments in morphometrics. *Yearbook of Physical Anthropology* 38:137–161.

Kappeler PM. 2008. Genetic and ecological determinants of primate social systems. In J Korb, J Heinze (eds.), *Ecology of Social Evolution* (pp. 225–243). Springer, Berlin.

Kappeler PM, Wimmer B, Zinner D, Tautz D. 2002. The hidden matrilineal structure of a solitary lemur: implications for primate social evolution. *Proceedings of the Royal Society of London Series B: Biological Sciences* 269(1502):1755–1763.

Lande R. 1982. Rapid origin of sexual isolation and character divergence in a cline. *Evolution* 36:213–223.

Lowry PP, Schatz GE, Phillipson PB. 1997. The classification of natural and anthropogenic vegetation in Madagascar. In SM Goodman, BD Patterson (eds.), *Natural Change and Human Impact in Madagascar*. Smithsonian International Press, Washington, DC.

Markolf M, Kappeler PM. 2013. Phylogeographic analysis of the true lemurs (genus *Eulemur*) underlines the role of river catchments for the evolution of micro-endemism in Madagascar. *Frontiers in Zoology* 10(1):70.

Markolf M, Rakotonirina H, Fichtel C, *et al.* 2013. True lemurs ... true species–species delimitation using multiple data sources in the brown lemur complex. *BMC Evolutionary Biology* 13(1):233.

Martin R. 1995. Prosimians: from obscurity to extinction? In L Alterman, G Doyle, K Izard (eds.), *Creatures of the Dark: The Nocturnal Prosimians* (pp. 535–563). Springer, New York.

Martin RD. 1972. Adaptive radiation and behaviour of the Malagasy lemurs. *Philosophical Transactions of the Royal Society of London Biological Sciences* 264:295–352.

Mayr E. 1947. Ecological factors in speciation. *Evolution* 1:263–288.

Myers N, Mittermeier RA, Mittermeier CG, da Fonseca GAB, Kent J. 2000. Biodiversity hotspots for conservation priorities. *Nature* 403(6772):853–858.

Noor MA. 1999. Reinforcement and other consequences of sympatry. *Heredity* 83(5):503–508.

Nosil P. 2012. *Ecological Speciation*. Oxford University Press, Oxford.

Nosil P, Crespi BJ. 2006. Ecological divergence promotes the evolution of cryptic reproductive isolation. *Proceedings of the Royal Society B: Biological Sciences* 273(1589):991–997.

Nosil P, Vines TH, Funk DJ. 2005. Reproductive isolation caused by natural selection against immigrants from divergent habitats. *Evolution* 59(4):705–719.

Orr HA. 1996. Dobzhansky, Bateson, and the genetics of speciation. *Genetics* 144(4):1331.

Panhuis TM, Butlin R, Zuk M, Tregenza T. 2001. Sexual selection and speciation. *Trends in Ecology & Evolution* 16:364–371.

Pastorini J, Thalmann U, Martin RD. 2003. A molecular approach to comparative phylogeography of extant Malagasy lemurs. *Proceedings of the National Academy of Sciences of the United States of America* 100(10):5879–5884.

Pearson RG, Raxworthy CJ. 2009. The evolution of local endemism in Madagascar: watershed versus climatic gradient hypotheses evaluated by null biogeographic models. *Evolution* 63(4):959–967.

R Core Team. 2014 *R: A Language and Environment for Statistical Computing*. R Foundation for Statistical Computing, Vienna, Austria.

Radespiel U, Ratsimbazafy J, Rasoloharijaona S, *et al.* 2012. First indications of a highland specialist among mouse lemurs (*Microcebus* spp.) and evidence for a new mouse lemur species from eastern Madagascar. *Primates* 53:157–170.

Rakotondravony R, Radespiel U. 2009. Varying patterns of coexistence of two mouse lemur species (*Microcebus ravelobensis* and *M. murinus*) in a heterogeneous landscape. *American Journal of Primatology* 71(11):928–938.

Rakotondranary SJ, Hapke A, Ganzhorn JU. 2011. Distribution and morphological variation of *Microcebus* spp. along an environmental gradient in southeastern Madagascar. *International Journal of Primatology* 32(5):1037–1057.

Rasoloarison RM, Weisrock DW, Yoder AD, Rakotondravony D, Kappeler PM. 2013. Two new species of mouse lemurs (Cheirogaleidae: *Microcebus*) from eastern Madagascar. *International Journal of Primatology* 34:455–469.

Ree RH, Sanmartín I. 2009. Prospects and challenges for parametric models in historical biogeographical inference. *Journal of Biogeography* 36(7):1211–1220.

Ree RH, Smith SA. 2008. Maximum likelihood inference of geographic range evolution by dispersal, local extinction, and cladogenesis. *Systematic Biology* 57(1):4–14.

Richard AF, Dewar RE. 1991. Lemur ecology. *Annual Review of Ecology and Systematics* 22:145–175.

Ronquist F. 1997. Dispersal-vicariance analysis: a new approach to the quantification of historical biogeography. *Systematic Biology* 46(1):195–203.

Schluter D. 2000. Ecological character displacement in adaptive radiation. *The American Naturalist* 156(S4):S4–S16.

Schluter D. 2001. Ecology and the origin of species. *Trends in Ecology & Evolution* 16(7):372–380.

Schwab D, Ganzhorn JU. 2004. Distribution, population structure and habitat use of *Microcebus berthae* compared to those of other sympatric cheirogalids. *International Journal of Primatology* 25(2):307–330.

Servedio MR. 2004. The evolution of premating isolation: local adaptation and natural and sexual selection against hybrids. *Evolution* 58:913–924.

SPSS Inc. Released 2003. SPSS for Windows, Version 12.0. Chicago, SPSS Inc.

Thiele D, Razafimahatratra E, Hapke A. 2013. Discrepant partitioning of genetic diversity in mouse lemurs and dwarf lemurs – biological reality or taxonomic bias? *Molecular Phylogenetics and Evolution* 69(3):593–609.

Thorén S, Linnenbrink M, Radespiel U. 2011. Different competitive potential in two coexisting mouse lemur species in northwestern Madagascar. *American Journal of Physical Anthropology* 145(1):156–162.

Treatman-Clark K. 2006. The Evolution of the Reproductive System in Strepsirrhine Primates. PhD dissertation, Stony Brook University, Stony Brook.

Vences M, Wollenberg KC, Vieites DR, Lees DC. 2009. Madagascar as a model region of species diversification. *Trends in Ecology & Evolution* 24(8):456–465.

Weisrock DW, Rasoloarison RM, Fiorentino I, *et al.* 2010. Delimiting species without nuclear monophyly in Madagascar's mouse lemurs. *PLoS ONE* 5(3):e9883.

Wells NA. 2003. Some hypotheses on the Mesozoic and Cenozoic paleoenvironmental history of Madagascar. In SM Goodman, JP Benstead (eds.), *The Natural History of Madagascar* (pp. 16–34). University of Chicago Press, Chicago.

Wiens JJ, Ackerly DD, Allen AP, *et al.* 2010. Niche conservatism as an emerging principle in ecology and conservation biology. *Ecology Letters* 13(10):1310–1324.

Wilmé L, Goodman SM, Ganzhorn JU. 2006. Biogeographic evolution of Madagascar's microendemic biota. *Science* 312(5776):1063–1065.

Wollenberg KC, Vieites DR, Van Der Meijden A, *et al.* 2008. Patterns of endemism and species richness in Malagasy cophyline frogs support a key role of mountainous areas for speciation. *Evolution* 62(8):1890–1907.

Wright PC, Martin LB. 1995. Predation, pollination, and torpor in two nocturnal prosimians: *Cheirogaleus major* and *Microcebus rufus* in the rain forest of Madagascar. In L Alterman, G Doyle, K Izard (eds.), *Creatures of the Dark: The Nocturnal Prosimians* (pp. 45–60). Springer, New York.

Yoder AD, Heckman K. 2006. Mouse lemur phylogeography revises a model of ecogeographic constraint in Madagascar. In J Fleagle, SM Lehman (eds) *Primate Biogeography: Progress and Perspectives* (pp. 255–268). Springer, New York.

Part II

Methods for studying captive and wild cheirogaleids

6 How to anesthetize mouse lemurs

Sabine B.R. Kästner, Julia Tünsmeyer, and Alexandra F. Schütter

Introduction

Mouse and dwarf lemurs are the smallest primate species. They may be used in a variety of scientific projects under field or laboratory conditions. Despite trap capture and the small size of mouse and dwarf lemurs, for successful performance of some procedures animals need to be anesthetized (Glander, 2013). In particular, the very small size of mouse lemurs with a body mass of 20–100 g and a body length of 5–13 cm (Louis *et al.*, 2008) requires special considerations for anesthesia and might limit the use of standard veterinary anesthesia equipment and techniques in the field.

To choose the best anesthetic method for a project, different factors such as drugs and available equipment, type, invasiveness and duration of procedure, possible interaction of drugs with measured parameters, and the experience of the scientist should be considered. In the face of modern imaging methods and the use of mouse lemurs as a research model of human disease, more prolonged anesthetics and sensitive monitoring of vital parameters, ventilation, and oxygenation becomes necessary to maintain physiological conditions.

Therefore, the purpose of the current article was to discuss general considerations for anesthesia in an animal of the size of mouse lemurs relating to equipment, and the properties and effects of anesthetic drugs. In addition, a review of published drug doses and drug combinations used for various procedures in mouse lemurs is included.

Material and methods

Publications concerning anesthesia techniques in mouse and dwarf lemurs were found by searching "Pubmed" of the United States National Library of Medicine, "Veterinary Science" of CAB International and Web of Science. These databases were searched for the following words or terms: "mouse lemur," "dwarf lemur," "Microcebus," "Cheirogaleus", alone or in combination with the word "an*esthesia."

The search covered the period from 1959 to January 2014. Articles were scrutinized for description of immobilization or anesthesia techniques. Relevant articles

The Dwarf and Mouse Lemurs of Madagascar: Biology, Behavior and Conservation Biogeography of the Cheirogaleidae, eds. S.M. Lehman, U. Radespiel, E. Zimmermann. Published by Cambridge University Press. © Cambridge University Press 2016.

were selected manually. Abstracts of conference proceedings were included only when material and methods and results were clearly stated.

Results

Vital parameters

Regardless of which anesthetic technique is used, knowledge of basic physiological variables of the species of interest, such as heart rate, respiratory rate, and body temperature, is useful to successfully monitor the animals' response (respiratory and cardiovascular depression) to anesthesia. Based on the influence of drugs, handling or capture stress and activity state, these values can vary considerably. In addition, hematologic (Der Goukassian, 1983) and biochemical (Marchal *et al.*, 2012) data can be useful to determine the health status of animals before entry into a study. Blood biochemical parameters have recently been reported in different age groups of captive gray mouse lemurs (Marchal *et al.*, 2012), showing that renal function may be impaired in old animals, which can influence the excretion of certain drugs.

Currently, little information on physiologic values of heart rate and respiratory rate of mouse lemurs is available. Heart rates between 180 and 300 beats per minute (bpm), with a mean of 250 bpm, are reported for awake gray mouse lemurs determined by chest wall palpation (Guillemin, 1998); however, under sevoflurane anesthesia, without nociceptive stimulation, pulse rate and heart rate determined by plethysmography and ECG recordings varied between 300 and 600 bpm with a mean of 450 bpm at a mean rectal temperature of 36°C (Schütter *et al.*, 2013), but lower heart rates were observed under isoflurane anesthesia at lower body temperatures. Overall, heart rate seems to follow body temperature closely during inhalation anesthesia (Kästner, personal observation). Mean respiratory rate determined by chest movements of awake gray mouse lemurs is about 140 breaths per minute with a range of 110–180 breaths per minute (Guillemin, 1998). Under isoflurane and sevoflurane in oxygen anesthesia for non-invasive procedures, spontaneous respiratory rate varied between 40 and 80 breaths per minute in gray mouse lemurs (Schütter *et al.*, 2013).

The body temperature is controlled between 37.5°C and 37.8 °C in gray mouse lemurs (*M. murinus*) and 38.1°C in the smaller mouse lemur (*M. myoxinus*) during activity states. At rest (sleep), it is lowered to 35.4–36.5°C (Ortmann *et al.*, 1996). During torpor bouts, body temperature drops below 33°C for varying periods of time followed by spontaneous rewarming without detrimental effects (Ortmann *et al.*, 1996; Schmid and Speakman, 2000, 2009; Genin and Perret, 2003). Anesthesia interferes with central and peripheral mechanisms of temperature regulation and changes in body temperature will alter anesthetic requirements (see below).

Route of drug administration

The intraperitoneal (i.p.), intramuscular (i.m.), and subcutaneous (s.c.) routes of drug administration are described in mouse lemurs. Darting is not necessary because of trapping methods and easy restraint; in addition, darting such a small animal would bear a high risk of injury. Venipuncture for blood sampling is described at the saphenous vein of the hind leg (Marchal *et al.*, 2012). Intravenous drug administration has not been described in mouse lemurs, but might be possible (tail or jugular vein) with training and appropriately sized equipment. However, it has to be kept in mind that vascular access will be an invasive procedure in such small animals.

Intramuscular injection of drugs with a pH outside of the physiological range (i.e., ketamine, barbiturates) can be painful and lead to tissue irritation or necrosis, especially when large volumes are injected into a small muscle. Subcutaneous injection results in longer absorption and anesthetic induction times than intramuscular injection, but is better tolerated by the animal with less tissue irritation. Intraperitoneal injection, a common route of application in small laboratory animals, can result in long induction and/or recovery times (Guillemin, 1998), can be very irritating when using acidic or alkaline drugs, and bears the risk of injecting into the intestines (lack of effect, intestinal injury).

Anesthetic premedicants

The necessity of routine anticholinergic premedication is controversial in human and veterinary anesthesia, because of the risk of inducing tachycardia or tachyarrhythmia and intestinal stasis. Atropine and glycopyrronium can prevent procedure- (vagal reflexes) or drug-induced bradycardia, induce bronchodilation and decrease salivation. Drugs such as ketamine and inhalation anesthetics can induce major salivation and bronchial secretion; therefore, an anticholinergic drug is advisable to avoid airway obstruction in such small animals as mouse lemurs. The use of atropine (0.1 mg/kg s.c. 30 minutes before anesthesia induction; Dhenain *et al.*, 1997; Guillemin, 1998) and glycopyrronium (0.02–0.04 mg/kg s.c. 20 minutes before anesthesia induction; Schütter *et al.*, 2013) have been described as satisfactorily effective in gray mouse lemurs. In various species, the advantages of glycopyrronium over atropine are a longer-lasting anticholinergic action, mainly the antisialogue effect (Mirakhur *et al.*, 1978), less-severe tachycardia and lack of central effects (does not cross the blood–brain barrier). Assuming similar properties of these drugs in mouse lemurs, glycopyrronium seems to be preferable for procedures lasting longer than 1 hour.

In small laboratory animals, a sedative premedication before anesthesia induction resulting in anxiolysis and stress reduction is not commonly used because most of the time drug combinations are injected in a mixed syringe to avoid repeated injections. Diazepam is a benzodiazepine with anxiolytic and dose-dependent sedative effects. Doses of 20–30 mg/kg i.m. resulted in deep sedation with lateral recumbency, but not a surgical anesthesia state, and very long recovery times in mouse lemurs. Doses exceeding 30 mg/kg induced respiratory depression, lateral recumbency and a drop in body temperature below 33°C (Guillemin, 1998). Diazepam is often prepared as a propylene glycol formulation which can cause pain and inflammation at injection sides. Therefore, diazepam in this formulation cannot be recommended for i.m. injection. Diazepam can be replaced by the water-soluble benzodiazepine midazolam at comparable dose rates, which is well tolerated after i.m. injection. Midazolam at 0.5–1 mg/kg s.c. prior to inhalation anesthesia induced only very minor sedation in gray mouse lemurs (Schütter *et al.*, 2013).

In case of long benzodiazepine-induced recovery times or severe respiratory depression, the benzodiazepine effects can be antagonized by flumazenil (Amrein *et al.*, 1988). There are no doses published for mouse lemurs, but increments of 0.1 mg/kg (0.001 mg/10 g) flumazenil (i.m./s.c.) can be used to speed up recovery without detrimental effects (Kästner, personal observation).

Injectable anesthetics

Various sedatives, tranquilizers, and anesthetics alone or in combination are described for the immobilization and anesthesia of mouse lemurs. Under field conditions, injectable drugs with a low potential for respiratory depression are preferable.

Ketamine

Ketamine hydrochloride alone or in combination with drugs promoting muscle relaxation is commonly used in various species. The commercially available solutions of ketamine are slightly acidic (pH 3.5–5.5). Ketamine interacts with various receptors and ion channels, but its activity as an *N*-methyl-d-aspartate (NMDA) receptor antagonist seems to be the dominant mechanism to induce loss of consciousness (Franks, 2008). Ketamine induces a dose-dependent alteration in central nervous system (CNS) activity that leads to a cataleptic, dissociative state, characterized by amnesia (humans) with maintained ocular, laryngeal, pharyngeal, pinnal, and pedal reflexes, and increased muscle tone, which seems to be better tolerated in primates than in other animal species. Ketamine as the sole anesthetic agent for very invasive procedures is questionable. Therefore, concurrent administration of benzodiazepines (diazepam, midazolam), acepromazine or an alpha$_2$ adrenergic-agonist (xylazine, medetomidine, dexmedetomidine) with ketamine is often used to reduce the high muscle tone and to induce a state of anesthesia that allows surgical interventions (Table 6.1). Ketamine has unique cardiovascular effects. Unlike other anesthetics, it stimulates the cardiovascular system, resulting in increases in heart rate, blood pressure, and cardiac output. Central stimulation of the sympathetic nervous system is responsible for the cardiovascular stimulation. Concurrent use of sedatives with reduction of sympathetic outflow will attenuate the stimulatory effects of ketamine. Therefore, combinations of ketamine with alpha$_2$ adrenergic-agonists such as xylazine, medetomidine or its active isomer dexmedetomidine, which reduce sympathetic activity, induce significant cardiovascular depression with reduced heart rate and major reduction in cardiac output. Combinations of ketamine with diazepam, midazolam, or the neuroleptic agent acepromazine affect cardiovascular function less severely. Acepromazine interferes with central thermal regulation, has a long duration of action and cannot be antagonized; therefore, it is not recommended in situations where a quick recovery is required. On the other hand, acepromazine has anti-arrhythmic and anxiolytic effects which can be an advantage in stressful situations when capturing or darting non-domestic animals. Xylazine, medetomidine, and dexmedetomidine can be antagonized with atipamezole, an $alpha_2$ adrenergic-antagonist, when anesthesia gets too deep or to speed up recovery.

Ketamine has minimal effects on central respiratory drive, but can induce a so-called "apneustic" pattern of breathing, characterized by periodic breath-holding on inspiration followed by short periods of hyperventilation. The swallow, sneeze, and cough reflexes remain relatively intact after ketamine administration, but "silent" aspiration of gastric contents can still occur with ketamine anesthesia. Bronchial secretion and salivation are increased with ketamine.

Tiletamine–zolazepam

Tiletamine is chemically related to ketamine (phencyclidine derivative, cyclohexanone) with a longer duration of action. Zolazepam is a benzodiazepine (diazepinone minor tranquillizer) with muscle relaxant and anticonvulsant effects. Telazol® (North America) or Zoletil® (Europe) is a proprietary combination of zolazepam and tiletamine at a ratio of 1:1. The preparation comes as a lyophilized powder, which can be reconstituted with saline, 5% dextrose or sterile water. The clear solution is acidic (pH 2.0–3.5) and should be discarded if precipitation occurs. The prepared solution can be stored at room temperature for 4 days, and for 14 days when refrigerated. Telazol® is a controlled substance in the USA (Schedule III), and import requires special permissions in many countries.

Tiletamine alone produces a cataleptic, dissociative state similar to that produced by ketamine. The benzodiazepine, zolazepam, has anticonvulsant and anxiolytic properties and produces muscle relaxation. The combination of tiletamine and zolazepam induce sedation or general anesthesia with similar characteristics to a ketamine–midazolam combination. Intramuscular injection can be painful due to the low pH of the solution. Doses between 5 and 10 mg/kg i.m. have been used in gray mouse lemurs resulting in sedation and immobilization at the lower dose range and surgical anesthesia with 10 mg/kg. Higher doses or repeated dosing carry a high risk of respiratory depression (Guillemin, 1998) and bad recoveries; therefore, re-dosing is not recommended. In tarsiers (*T. syrichta*) 5–8 mg/kg of Telazol® resulted in about 1 h of anesthesia sufficient for electrophysiological recordings. Supplementation with 15 μg/kg dexmedetomidine or 3 mg/kg Telazol® were used in case the depth of anesthesia was insufficient (Ramsier *et al.*, 2012). As with ketamine, hypersalivation is common and can be reduced with atropine or glycopyrronium, if necessary.

Thiopental

Thiopental is a thiobarbiturate. It is a weak organic acid provided as a sodium salt (yellow, crystalline powder) in sealed vials. Anhydrous sodium carbonate is added to prevent precipitation of the free acid with atmospheric carbon dioxide. After reconstitution with sterile water, the solution is very alkaline (pH in the range of 11–14), which makes it extremely irritant at concentrations greater than 2.5%

(Kästner, 2007). Reduction of alkalinity of the solution results in precipitation of the free acid. Thiopental is primarily a hypnotic agent interacting with specific binding site at the $GABA_A$ receptor without antinociceptive effects. Therefore, high doses are required to achieve immobility in response to painful stimuli (defined as anesthesia). In gray mouse lemurs, i.p. injection of thiopental (1% or 2%) resulted in dose-dependent sedation (≤ 80 mg/kg) or anesthesia (150–250 mg/kg) with long recovery times and severe respiratory depression at the high dose range (Guillemin, 1998), and therefore cannot be recommended.

Pentobarbital

Pentobarbital, another barbiturate in an alkaline solution, has been used for anesthesia maintenance in gray mouse lemurs at low concentrations via the i.m. route for electrophysiological recordings (Table 6.1; Hellekant *et al.*, 1993). Similar to thiopental, pentobarbital interacts with $GABA_A$ receptors (Franks, 2008) and has dose-dependent sedative to hypnotic effects without antinociception. Repeated doses lead to drug accumulation and very long recovery times.

Pentobarbital is no longer licensed as an anesthetic and pentobarbital anesthesia is considered obsolete for clinical anesthesia (recovery). Currently, the veterinary products available are euthanasia solutions prepared in high concentrations (20–40% w/v).

Propofol

Propofol is a highly lipophilic, hypnotic alkyl phenol (2,6 diisopropylphenol). Propofol is currently formulated as a white oil-in-water macroemulsion containing intralipid (1% w/v soya bean oil, 1.2% w/v purified egg phosphatide, 2.25% w/v glycerol). This formulation has a pH of 7 and appears as a slightly viscous, milky white liquid. The veterinary products usually come at a concentration of 1% (w/v) but a 2% (w/v) formulation is available for humans. The emulsion is an ideal culture medium for bacteria and promotes exponential growths; open vials should be used within 24 h. Aseptic handling of multiple-use vials is essential to avoid contamination. Intramuscular injection of propofol does not induce anesthesia at reasonable dose rates, but the injection is non-irritating. Propofol is primarily a hypnotic agent mainly interacting with $GABA_A$ receptors (Franks, 2008) without antinociceptive effects. Intraperitoneal application in mouse lemurs resulted in dose-dependent sedation and muscle relaxation. Only at high doses (200 mg/kg) and in combination with diazepam was a true anesthetic state achieved, which was associated with profound respiratory depression (Guillemin, 1998). Even at lower doses, recovery times were very long (8–16 h) and animals developed dehydration. Therefore, propofol cannot be recommended as an anesthetic agent in mouse lemurs by the i.p. route and without respiratory support and oxygen supplementation.

Local anesthetics

For minor superficial procedures, infiltration of common local anesthetics (lidocaine, mepivacaine, ropivacaine, and bupivacaine) can be used. Care should be

Table 6.1 Injectable drugs and drug combinations with dose ranges and route of application for sedation and anesthesia published in mouse lemurs.

Drug	Species	Dose	Route	Comment	Author
Diazepam	*Microcebus murinus*	20–30 mg/kg	i.m.	Sedation, long recovery	(Guillemin, 1998)
		> 30 mg/kg	i.m.	Sedation and lateral recumbency but not surgical, decreased respiratory rate and decreased body temperature, very long recovery Cannot be recommended	
Ketamine alone	*Microcebus murinus* *Cheirogaleus medius* *Cheirogaleus major*	10 mg/kg = 1 mg/100 g	s.c.	Sedation only	(Lahann, 2007)
	Microcebus murinus	20 mg per animal ~300 mg/kg	s.c.	Skin biopsy, transponder implantation	(Hamalainen *et al.*, 2014)
	Microcebus murinus	3 mg per animal ~ 50 mg/kg	s.c.	Sedation only, for 10–20 min	(Schmid and Speakman, 2009)
	Microcebus murinus	< 80 mg/kg	i.m.	Sedation, muscle rigidity	(Guillemin, 1998)
		100–150 mg/kg =10–15 mg/100 g	i.m.	No response to nociceptive stimulation Surgical anesthesia level?, maintained respiration	
	Microcebus murinus	100 mg/kg 10 mg/100 g	i.m.	Implantation of telemetric device into abdominal cavity	(Perret and Aujard, 2001; Dal-Pan *et al.*, 2010; Pifferi *et al.*, 2013)
	Microcebus murinus	100 mg/kg 10 mg/100 g	i.p.	Implantation of telemetric device into abdominal cavity Implantation of EEG (skull) and EMG electrodes (muscle)	(Pifferi *et al.*, 2011, 2012; Rahman *et al.*, 2013)
Xylazine Ketamine	*Microcebus murinus*	6 mg/kg 65 mg/kg	i.m.	Surgical anesthesia, 40–80 min duration	(Guillemin, 1998)

(*continued*)

Table 6.1 (*cont.*)

Drug	Species	Dose	Route	Comment	Author
Xylazine	*Microcebus murinus*	5 mg/kg	i.m.	Bulbectomy	(Schilling and Perret, 1993)
Ketamine		8 mg/kg	i.m.		
Atropine		0.05 mg/kg premedication	s.c.		
Medetomidine		0.25 mg/kg	i.m.	Pain on injection, quick induction, good analgesia, cardiovascular depression, bronchosecretion	(Guillemin, 1998)
Ketamine		130–140 mg/kg	i.m.	Long recovery (not antagonized)	
Acepromazine	*Microcebus murinus*	0.5 mg/60 g (= 8 mg/kg)	i.m.	Electrophysiological recordings	(Hellekant *et al.*, 1993)
Ketamine		5 mg/60g (= 80 mg/kg)	i.m.		
Pentobarbital for maintenance		1.35 mg (0.1 ml) = 22.5 mg/kg every h	i.m.		
Diazepam	*Microcebus murinus*	5 mg/kg	i.p.	Removal of bulb olfactorius	(Perret *et al.*, 2003)
Ketamine		8 mg/kg	i.p.	Bulbectomy	
Atropine		premed 0.05 mg/kg	s.c.		
Diazepam	*Microcebus murinus*	1 mg/100 g	s.c.	Implantation of telemetric device into abdominal cavity	(Le Tallec *et al.*, 2013)
Ketamine		10 mg/100 g	i.p.		
Meloxicam		0.02 mg/100 g	s.c.	Meloxicam for postoperative analgesia	
Diazepam	*Microcebus murinus*	1 mg/100 g	s.c.	Implantation temperature sensor into abdominal cavity	(Terrien *et al.*, 2008)
Ketamine		10 mg/100 g	i.p.		
Diazepam	*Microcebus murinus*	2 mg/100 g	i.m.	Implantation temperature sensor into abdominal cavity	(Giroud *et al.*, 2008)
Ketamine		10 mg/100 g			
Midazolam	*Cheirogaleus medius*	0.25 mg/kg	?	Subcutanoeus implantation and fixation of EEG electrodes	(Krystal *et al.*, 2013)
Ketamine		10 mg/kg			
Tiletamine-Zolazepam Zoletil® Telazol®	*T. syrichta*	5–8 mg/kg Supplemented with dexmedetomidine 15 μg/kg or 3 mg/kg Telazol® if necessary	?	About 1 h anesthesia for electrophysiological recordings (auditory)	(Ramsier *et al.*, 2012)

Telazol® Zoletil®	*Microcebus murinus*	10 mg/kg	i.m.	Rapid induction, rapid recovery after single dose Variable depths Reinjection not recommended because of respiratory depression and bad recoveries	(Guillemin, 1998)
Thiopental	*Microcebus murinus*	80 mg/kg	i.p.	Deep sedation	(Guillemin, 1998)
		150–250 mg/kg	i.p.	Surgical, dose-dependent respiratory depression	
Propofol		40 mg/kg	i.p.	Mild sedation	(Guillemin, 1998)
		80–500 mg/kg	i.p.	Ataxia, no analgesia, high doses with respiratory depression, very long recovery, dehydration	
Propofol Diazepam	*Microcebus murinus*	160 mg/kg 10 mg/kg	i.p. i.m.	Respiratory depression, very long recovery	(Guillemin, 1998)
Pentobarbital	*Microcebus murinus*	10–18 mg/kg	i.p.	Onset 15–20 min, sedation	(Guillemin, 1998)
		20–50 mg/kg	i.p.	Surgical intervention bulbectomy, removal of nasovomeral organ	
		70 mg/kg	i.p.	Dose-dependent respiratory depression, no analgesia, high pH of solution (irritation!)	

Notes: i.m., intramuscular; i.p., intraperitoneal; s.c., subcutaneous.

taken not to induce local anesthetic toxicity (cardiac arrhythmias, seizures, coma, or respiratory arrest) because these small lemurs are easily overdosed. For common domestic animal species, it is recommended not to exceed 4 mg/kg lidocaine or mepivacaine or 2 mg/kg for bupivacaine as total infiltrated dose. The concentration of commercial solutions can be diluted with physiologic saline to allow better spread of the local anesthetic, but lowering the concentration of the local anesthetic too much can result in loss of activity.

Inhalation anesthesia

In contrast to most field studies, research conducted under laboratory conditions often requires medium- to long-term anesthesia lasting up to several hours (Picq *et al.*, 2012). Maintenance of anesthesia through repeated bolus injections of anesthetic drugs is limited by drug accumulation, difficulties controlling depths of anesthesia and prolonged and uncoordinated recoveries. Modern inhalation anesthesia easily allows rapid changes in delivery of the anesthetic agent and anesthetic depth, and therefore reduces the risk of fatal overdosing of the drugs. Anesthesia can be extended over long periods with current inhalation anesthetics as long as physiologic conditions are maintained (body temperature, oxygenation, eucapnia). Oxygen, air, or air/oxygen mixtures can be used as carrier gas. When using air for longer procedures, oxygenation of the animal should be monitored. For very long anesthetics (> 8 h), pure oxygen must be avoided.

The need of special equipment to deliver inhalant anesthetic could be a disadvantage, but anesthetic systems used for small laboratory rodents are available in many research laboratories and can easily be adapted to mouse lemurs.

Inhalation anesthetics

Today, isoflurane ($C_3H_2ClF_5O$) and sevoflurane ($C_4H_3F_7O$) are most commonly used in veterinary anesthesia and are widely available. Both drugs are fluorinated ethers and have comparable pharmacodynamic characteristics with dose-dependent respiratory and cardiovascular depression (Farber *et al.*, 2010; Pagel *et al.*, 2010). Inhalation anesthetics of the halogenated alkanes and ethers group bind to a variety of membrane receptors and ion channels whereby their interaction with $GABA_A$ receptors and the two-pore-domain-K^+ channel family seem to be predominantly involved in the induction of unconsciousness (Franks, 2008). In clinical use, differences are mainly noticed during anesthesia induction and the recovery period. Sevoflurane (partition coefficient at 37°C = 0.65) has a lower blood/gas solubility than isoflurane (partition coefficient at 37°C = 1.46); therefore, anesthesia induction and the speed of change in anesthetic depths with sevoflurane is faster than with isoflurane (Eger II, 2010). The smell of isoflurane is described as pungent and irritant to mucosal membranes, which has recently been associated with activation of transient receptor potential-A1 in the respiratory epithelium with hyperalgesia and neurogenic bronchoconstriction (Eilers *et al.*, 2010). In contrast, the smell of sevoflurane, which is devoid of activation of transient receptor potential-A1 (Eilers *et al.*, 2010), is described as fruity and non-pungent. Therefore, sevoflurane is the preferred anesthetic for mask or chamber induction with a smoother and more rapid anesthesia induction and less struggling, breath-holding and aversive reactions than with isoflurane. A disadvantage of sevoflurane is that it is currently more expensive than isoflurane. To reduce costs, it is possible to induce anesthesia with sevoflurane and change to isoflurane for maintenance of anesthesia; however, this requires the availability of two separate agent-specific vaporizers. Sevoflurane is less potent than isoflurane: the average minimal alveolar concentration required to suppress movement in response to a defined supramaximal nociceptive stimulus for different species lies between 1.2 vol% and 1.6 vol% for isoflurane and 2.1 vol% and 2.8 vol% for sevoflurane (Clarke *et al.*, 2014).

Anesthesia induction

Injectable anesthesia protocols were discussed earlier in this chapter and can also be used as induction agents followed by maintenance with inhalational agents. Induction of anesthesia by inhalation can be performed by mask or chamber induction. In non-domestic animals, handling and manual restraint induces stress, which will prolong anesthetic induction times, and the release of catecholamines can induce cardiac arrhythmias. Therefore, chamber induction is preferred in mouse and dwarf lemurs. If possible, induction in the animals' familiar sleeping or nest box (Figure 6.1) might further reduce stress responses.

Induction chambers for laboratory rodents are commercially available or can be individually manufactured, by preparing a small plastic box (e.g., acrylic glass, PET-g, polypropylene) at least with an inlet for the anesthetic gas mixture and also an outlet for scavenging waste gas (Figure 6.2). Large, voluminous induction chambers will prolong filling the chamber with the desired gas concentration. At least one side of the induction box should be translucent to be able to observe the animal during anesthesia induction. Induction with an inhalant should be quick to avoid an excitatory phase (or keep this phase as short as possible) with possible adverse reactions like struggling, excitations or nausea, retching and vomiting. To shorten anesthesia induction time, a high gas flow (2–4 l/min; oxygen; oxygen/air mixture) and high vaporizer settings (4% isoflurane, 8% sevoflurane) should be used to quickly increase anesthetic gas concentrations inside the induction chamber. Using this forced induction technique, care must be taken to avoid an overdose of inhalant anesthetic. Therefore, close observation of the lemur is mandatory. As soon as lateral recumbency is achieved and the righting reflex (approximately equal to loss of consciousness in human beings) is lost, animals should be removed from the induction chamber and placed in a face mask for anesthesia maintenance with lower gas concentrations (1–2% isoflurane, 2–4% sevoflurane).

In the past, with slow-acting inhalation anesthetics, induction was commonly achieved by placing small laboratory animals in a box and dripping the volatile anesthetic on gauze or cotton wool which was then also placed in the induction

Figure 6.1 Set-up for anesthesia induction using the sleeping boxes of mouse lemurs. Side openings of the box are connected to the anesthetic system supplying the anesthetic gas mixture. A second opening in the box is connected to an active charcoal canister to scavenge the waste gas. The whole system is placed underneath a venting hood (a). Detail of the connection of a Bain breathing system to the sleeping box. The animal can be observed during induction (b). (A black and white version of this figure will appear in some formats. For the colour version, please refer to the plate section.)

box. Using this method, the concentration of the volatile anesthetic cannot be controlled and can reach very high gas concentrations. With the modern inhalation anesthetics (low blood/gas solubility leading to rapid induction and high vapour pressures), a fatal overdose is reached quickly.

Standard masks and nose cones for laboratory rodents could be used, but care has to be taken not to cause ocular injuries, by pressing the mask against the prominent

Figure 6.2 Custom-made induction chamber with a small port for fresh gas inlet and a large port for connecting to an active, negative pressure scavenging system. The translucent material (polypropylene, SAMLA box, IKEA) allows observation of the animal during induction. (A black and white version of this figure will appear in some formats. For the colour version, please refer to the plate section.)

eyes of mouse lemurs. Individual masks can be easily designed and adapted to the mouse lemur anatomy, so that they provide a good fit. The volume of these masks should be as low as possible to avoid carbon dioxide rebreathing.

Tracheal intubation (Figure 6.3) in mouse lemurs can be challenging, but similar techniques and equipment (24G intravenous catheters or silicone tubing; Figure 6.4) as in small laboratory rodents can be used. Tracheal intubation is required if controlled ventilation becomes necessary to maintain physiological blood gas concentrations, because arterial partial pressure of carbon dioxide significantly influences brain perfusion.

Overall, carbon dioxide production in mouse lemurs will be low in relation to gas flows used for inhalation anesthesia. Simple set-ups for laboratory animals provide the anesthetic gas mixture via a single hose. Exhaled carbon dioxide-containing gas is diluted with fresh gas in the nose cone or face mask and leaks out of a loosely fitted mask into the surroundings. This will lead to workplace contamination with anesthetic gas, so a venting hood or other ways of scavenging the gas should be used.

Coaxial hoses, with the inner hose providing the gas to the animal and the expired or wasted gas being actively or passively scavenged from the outer hose, are commercially available for laboratory rodents. For the proper function of these systems, a secure face mask with a good seal is required, which can be achieved,

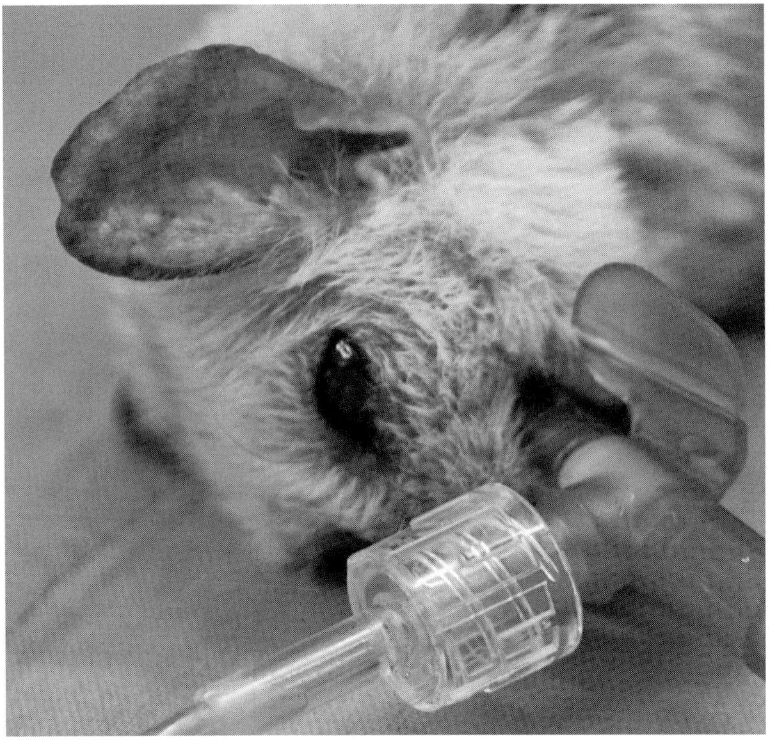

Figure 6.3 Mouse lemur with an endotracheal tube. An intravenous catheter with a side injection port was used. The side port can be used to connect the sampling line of a sidestream capnograph.

for example, by stretching a latex glove over the mask, leaving only a small opening for the nares and the mouth (Figure 6.5). The coaxial Bain breathing system, a non-rebreathing anesthetic system (Clarke *et al.*, 2014), has a similar function, with an inner hose providing fresh gas to the animal and outer hose to collect waste gas (Figure 6.6). The Bain breathing system also contains a reservoir bag, a pressure limiting valve and a connection to a gas-scavenging system (Clarke *et al.*, 2014). A Bain system with 5-m hoses has been used for remote inhalation anesthesia of gray mouse lemurs in an acoustically and electrically isolated chamber (Schopf *et al.*, 2014).

Several hours of isoflurane anesthesia after premedication with atropine (0.1 mg/kg s.c.) is described in gray mouse lemurs (Guillemin, 1998). Anesthesia was induced in an induction chamber with 5 vol% isoflurane in air with a flow rate of 1.5 l/min. A short excitatory phase occurred after 3 min, 7 of 10 lemurs showed retching, and 2 animals vomited. After 6.7 min, the animals were in lateral recumbency and unresponsive to shaking the box and auditory stimuli. Anesthesia was maintained with a 20-ml syringe case as face mask with 2.5% isoflurane for 20 min and then reduced to 1.5% for the remaining time. Respiration was supported with a small animal ventilator at a rate of 50 breaths per minute. Body temperature during

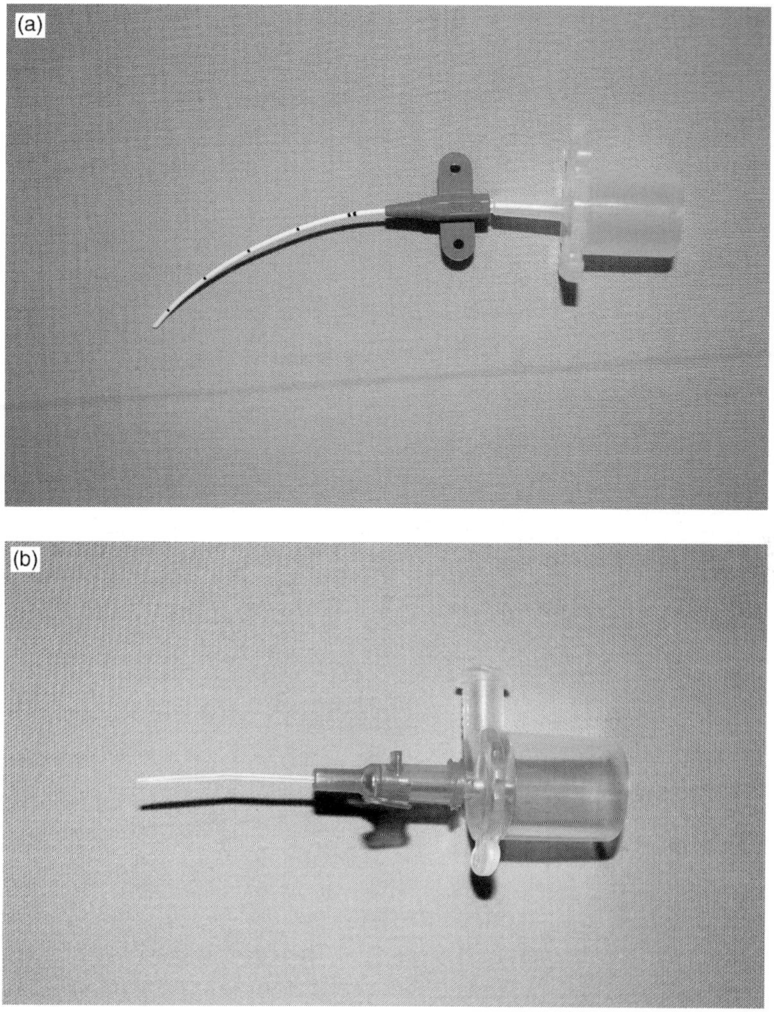

Figure 6.4 Endotracheal tubes made of 24G intravenous catheters. Silicone catheters (a) are very flexible. Teflon or polyethylene catheters are stiff and easier to place. The tube connector incorporates a side port (b) for the capnograph sampling line.

anesthesia was between 33.4 and 34°C. After an anesthetic time of 150 min, mouse lemurs were able to move into sternal recumbency 8.7 min after turning the vaporizer off, and recovery was complete after 19.7 min.

After premedication with glycopyrronium and midazolam, 8% sevoflurane in oxygen at a flow rate of 4 l/min was introduced directly into the sleeping boxes of gray mouse lemurs. Loss of the righting reflex was achieved quicker in geriatric (> 5 years) animals (52.5 ± 11.7 s) than in young (< 5 years) animals (81.5 ± 47.7 s). Induction of anesthesia was smooth, with occasional retching. After 3 h of anesthesia with 3.4% sevoflurane (mask concentration), old and young animals were fully recovered to walking within 10 min. Over the course of anesthesia,

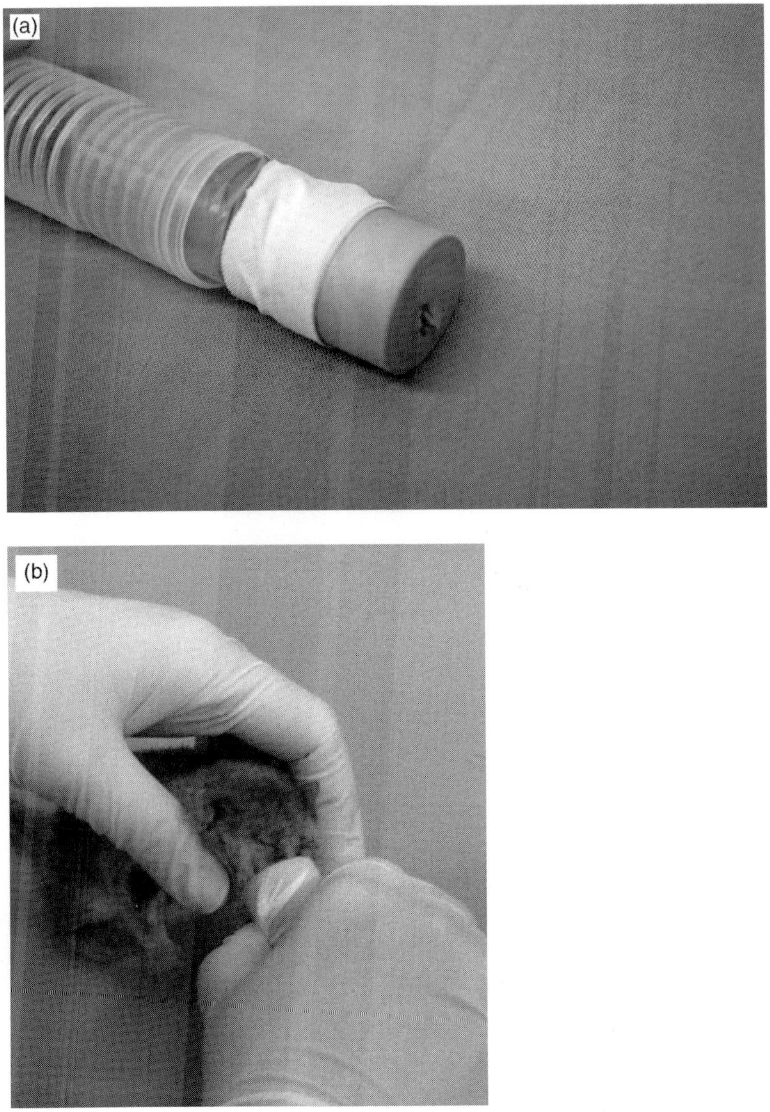

Figure 6.5 Face masks or nose cones can be adapted and sealed by pulling a latex glove over the mask and pinching just a small hole to fit the nares and the mouth (a). Gray mouse lemur breathing from a face mask (b).

animals were spontaneously breathing with an average respiratory rate of 55 breaths per minute (Schütter *et al.*, 2013). Rectal temperatures were maintained between 35 and 36°C.

Mechanical ventilation

To maintain physiological blood carbon dioxide concentrations, intubation and mechanical ventilation becomes necessary. For this purpose, different volume or

Figure 6.6 Coaxial hoses of a Bain non-rebreathing system.

pressure controlled small animal ventilators or piston pumps are available for laboratory rodents of similar size from various companies (Harvard Apparatus, USA; Small Animal Instruments, USA; TSE Systems, Germany; VetEquip, Inc., USA). The ventilator or the pump has to be placed downstream of the vaporizer as modern agent-specific plenum vaporizers pose a high resistance to airflow.

Monitoring and management of anesthesia

Body temperature

During anesthesia, thermoregulation is impaired (reduced metabolism, immobility, breathing cold anesthetic gas mixtures) and a significant drop in body temperature can occur (Guillemin, 1998); therefore, monitoring and regulation of body temperature by external heat sources (heating pads, forced warm air blankets, red light heating lamps) are recommended. Ideally, in small mammals, physiological body temperature should be maintained. However, uncontrolled application of heating equipment is not advisable due to the risk of causing life-threatening hyperthermia. Stress by handling can induce hyperthermia and capture myopathy in wild animals. In addition, the low body mass of mouse lemurs is rapidly heated up by active external heating. Rectal temperature followed changes in heating pad temperature rapidly (Schütter *et al.*, 2013). In isoflurane anesthesia, rectal temperatures of 33–34°C were well tolerated in gray mouse lemurs (Guillemin, 1998). It has to be kept in mind that changes in body temperature will alter anesthetic requirements, heart rate and respiratory rate. With low temperatures, injectable anesthetics are slowly metabolized and recovery will be prolonged. Gas solubility of inhalation anesthetics will be altered with changes in blood and tissue temperature and the required dose of anesthetic is reduced. Therefore, if body temperature drops, the concentration of the inhalation anesthetic has to be reduced to avoid overdose, and vice versa. By downregulating body temperature, anesthetic depth can be increased without changes in the anesthetic dose. As mouse lemurs have the ability to display and tolerate a wide range of body temperatures, including torpor, this can be used to control the anesthetic depth by lowering (down to 32°C) or increasing body temperature.

During anesthesia, body temperature can be measured rectally using a common digital thermometer, but due to the small size of the mouse lemurs, this has to be done with extreme care to avoid damage to the rectum. More convenient are small thermocouple probes for continuous temperature measurements. Access to the rectum is easy, but core temperature might be underestimated at this location, in particular if the probe is placed in feces or is not properly inserted. As in other species, readings from a small, flexible probe placed in the distal half of the esophagus more closely reflect core body temperature.

Monitoring of anesthetic depth

Classical signs to evaluate depth of anesthesia are loss of movement, muscle relaxation, loss of reflex response and attenuation of vegetative responses. Different anesthetics or drug combinations can influence these signs to a varying degree and response to drugs may also depend on species. As with other small mammals, the loss of the righting reflex in mouse lemurs can be used as an indicator of loss of consciousness. Loss of the palpebral reflex (closing the eye in response to touching the eyelids) and loss of jaw tone indicate a deeper state of anesthesia. Loss of the corneal reflex (closing the eye in response to touching the cornea, e.g., with a wet cotton stick) occurs at very deep anesthetic states, being a sign of impending asphyxia. Surgical tolerance can be tested by pinching or clamping a toe with a hemostat and assessing limb withdrawal. Changes in heart rate and respiratory rate are sensitive, but not specific, indicators of anesthetic depth.

Monitoring the cardiovascular system

Clinical evaluation

Monitoring the cardiovascular function by clinical observation is impeded by the small size of mouse lemurs. Auscultation with a stethoscope or simple palpation of the heartbeat at the thoracic wall is possible (Guillemin, 1998), but heart rates above 300 bpm cannot be counted (Longley, 2008). Similarly, palpation of peripheral pulses is usually not practicable. Visual examination of mucous membrane color can give some basic information on oxygenation and perfusion, but might be impeded by pigmentation of the mucous membranes.

Electrocardiogram (ECG)

The ECG displays the electrical activity of the heart (heart rate and rhythm) and the very small amplitudes in mouse lemurs can be amplified easily. Subcutaneous needle electrodes or self-adhesive pediatric surface electrodes on hands and feet can be used in mouse lemurs (Figure 6.7a). Crocodile clamps can squeeze the skin, inducing pain or skin necrosis if left in place over a long period; therefore, their use should be avoided in mouse lemurs. ECG monitors designed for use in humans are often not capable of displaying heart rates higher than 250 bpm; therefore, purpose-designed ECG monitors for laboratory animals should be used (Flecknell, 2009; Figure 6.7b). It has to be kept in mind that an ECG is not a measure of cardiac contractility and a normal ECG trace does not necessarily document adequate tissue perfusion.

Blood pressure

To maintain adequate perfusion of the tissues, a certain driving blood pressure is necessary; therefore, the mean arterial blood pressure is often monitored as a substitute for tissue perfusion pressure. Direct arterial blood pressure measurement

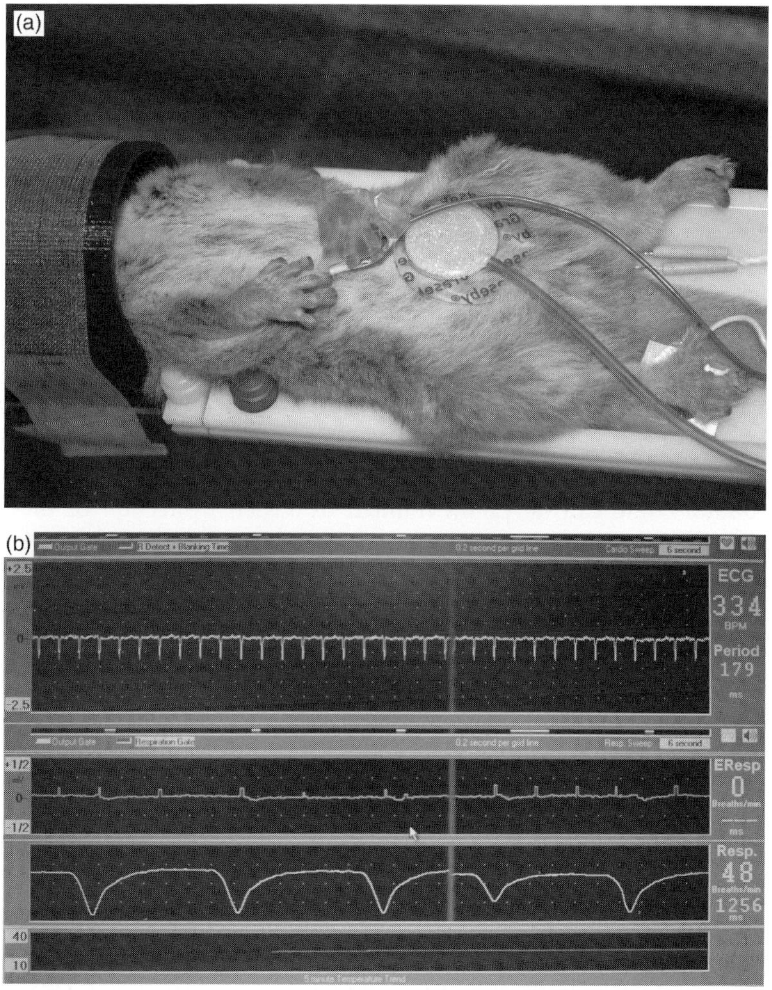

Figure 6.7 (a) Pediatric ECG surface electrodes on one hand and one foot of a gray mouse lemur. An air cushion transmits respiratory excursions to pressure changes. The lemur is equipped for MRI scans. (b) Computer screen with an ECG trace and the respiratory signal. (A black and white version of this figure will appear in some formats. For the colour version, please refer to the plate section.)

requires canulation of an artery, which would need a great deal of skill due to the small size of mouse lemurs, and to our knowledge has not been described in these species yet. Common human pediatric or veterinary oscillometric devices for non-invasive blood pressure measurement are not capable of detecting the high pulse rates of mouse lemurs. For small rodents of comparable size like mice or rats, specifically designed non-invasive blood pressure monitors are available. These monitors use various techniques to detect blood flow, but all of them require an occlusion tail cuff. The accuracy of blood pressure measuring devices using occlusion tail cuffs depends strongly on adequate cuff size (Feng and DiPetrillo, 2009).

Detection of the recurring pulse waves after release of the occlusion of arterial blood flow by the tail cuff can be based on different techniques. Photoplethysmography uses a light source to detect the blood flow and flow detection can be hampered by skin pigmentation. It measures only systolic arterial blood pressure (SAP) and heart rate and is relatively inaccurate. Piezoplethysmography uses piezoelectric ceramic crystals to detect the pulse signal. The technique is very sensitive and allows measurement of blood pressure even at very high pulse rates. In addition, this technique does not interfere with skin pigmentation and might therefore be advantageous for use in mouse lemurs. However, as with photoplethysmography, only SAP and pulse rate are measured. Volume–pressure recording uses a differential pressure transducer to measure the blood volume and blood flow in the tail. This volumetric method measures not only SAP and pulse rate, but also diastolic and mean arterial pressure. It has proven to be reliable and accurate in rodents in different studies with a high correlation with invasive arterial pressure readings (Feng and DiPetrillo, 2009).

Monitoring of respiratory function

The respiratory rate can be easily counted by visual inspection of thoracic wall movement. Also, the respiratory pattern and to a limited degree the depth of respiration can be evaluated by observation. If direct observation is not possible during anesthesia (i.e., during magnetic resonance imaging (MRI)), specifically designed air cushions and pressure sensors for small laboratory animals are available to detect and count thoracic movement (Figure 6.7a). To confirm adequate ventilation or physiologic partial carbon dioxide tensions, arterial or venous blood gas analysis would be the method of choice, but repeated measurements are restricted by the available blood volume of mouse lemurs. The measurement of end-expiratory carbon dioxide partial pressure can be used as a surrogate for blood gas analysis. Capnography allows a relatively non-invasive and continuous monitoring of ventilation and lung perfusion. However, accurate measurement requires endotracheal intubation and care must be taken to avoid adding excessive technical dead space by common airway adapters for the side port of the capnograph. The sampling port should be incorporated in the endotracheal tube connector (Figure 6.4b). Sidestream capnographs continuously sample gas from the airway, and the sampling rate has to be very low in very small animals to avoid dilution of the exhaled gas concentrations with low-minute ventilation. Therefore, common pediatric or veterinary capnographs cannot be used in mouse lemurs. Specifically designed capnography systems with adjustable gas sampling rates are available for small laboratory animals (Small Animal Instruments, USA; Harvard Apparatus, USA). With face masks or nose cones, the diluted carbon dioxide trace can be used to count respiratory rate and give some semi-quantitative information.

Oxygenation

Adequate arterial oxygenation can be assessed by blood gas analysis, but requires arterial puncture or canulation. Pulse oximetry offers the opportunity to detect the saturation of hemoglobin in the arterial blood non-invasively. The measurement principle is based upon differences in the light absorption spectra of oxygenated and deoxygenated hemoglobin. In addition, the working principle depends on the pulsatile nature of the signal to detect the arterial part of it. By measuring the pulsation of the arterial blood, it also provides information about pulse rate and in some monitors also information about pulse signal quality. Similar to the ECG, common pulse oximeters built for humans do not display heart rates > 250 bpm. Therefore, specifically designed monitors for small laboratory animals should be used in mouse lemurs. Alternatively, if a small animal pulse oximeter is not available, a common human or veterinary monitor could be used with the running pulse wave on the screen intermittently being stopped in order to count pulse rates manually. The low pigmentation and minimal hair coat of the hands or feet of mouse lemurs makes them the ideal location to place the sensor clip. During MRI scans, pulse oximeters with fiber-optic or telemetric transmission of the light signal are required and are available for small laboratory animals.

Perianesthetic care

Eye protection

The prominent eyes of mouse lemurs need special attention during anesthesia. Protective reflexes like the palpebral reflex are lost during anesthesia; therefore, care should be taken not to mechanically damage the eyes when handling and positioning the anesthetized lemur. In addition, tear production and blinking is reduced and the cornea can dry out quickly. Lubrication of the eyes with an eye ointment or artificial tears or taping the eyes closed will help to avoid damage to the eyes.

Fasting and fluids

Because induction of anesthesia by inhalation can induce gagging or vomiting, the animals should not be fed close to induction of anesthesia to avoid aspiration of gastric contents, but excessively long fasting periods (> 6 h) are not necessary. Access to water should not be restricted.

In order to prevent dehydration and hypoglycemia in the perianesthetic phase, fluids can be administered. In particular, during procedures of longer duration and when mouse lemurs have not been fed for several hours, fluid and glucose balance should be supported. Common routes of administration in mammals of comparable size are the s.c. and i.p. route, respectively. Depending on the size of the lemur, volumes of 1–3 ml can be easily administered. Commercial maintenance solutions containing 5% dextrose can be mixed with 0.9% NaCl or Hartmann's solution in order to reduce osmolarity and tissue irritation. Other authors recommend mixing 4% dextrose and 0.18% saline (Flecknell, 2009). Intraperitoneal injection bears the risk of organ puncture. However, fluids are more rapidly absorbed from the peritoneum, whereas in compromised animals in a low perfusion state, absorption from s.c. sites may be markedly delayed.

Analgesia

If invasive surgical procedures are planned, postoperative analgesia should be included in the experimental protocol. For very painful surgical procedures, opioids are often used due to their antinociceptive effects, but these drugs markedly enhance respiratory depression caused by general anesthetics. For postoperative pain relief, non-steroidal anti-inflammatory drugs (NSAIDs) are often sufficient and have the advantage of reducing wound swelling and secretions due to their anti-inflammatory effects. A single dose of 0.02 mg/100 g s.c. meloxicam has been used in *Microcebus murinus* after abdominal implantation of a telemetric device (Le Tallec *et al.*, 2013).

As there are no dose rates or dose intervals for analgesic drugs tested or published for mouse lemurs, doses should be deduced from larger, more commonly treated primates or closely related animals. It has to be kept in mind that very small animals like mouse lemurs require higher dose rates on a linear body mass scale than larger species.

Emergency situations

In case of impending or existing respiratory and/or cardiovascular arrest, anesthetic supply should be stopped immediately and the system flushed with oxygen (inhalation anesthesia) or, depending on the anesthetics used, antagonized (Table 6.2). As venous access will most likely not be established, well-perfused tissue, such as the muscles of the front or hind limb, should be used for injection. Injection into the tongue is possible but can lead to swelling and obstruction in such small animals. Intraosseous injection (pelvic bones, long bones of the limbs) can be attempted because intraosseous veins cannot collapse and drugs and fluids are taken up rapidly. Subcutaneous or intraperitoneal injection is not very useful in states of low perfusion because of very slow absorption. Chest compressions between the thumb and the index finger with a frequency of 100–120 might help to maintain some cardiac output and distribution of emergency drugs (Table 6.2).

If available, oxygen should be provided in case of cardiorespiratory compromise. Ventilation with a face mask often is not very successful and large amounts of gas can be pushed into the stomach, which impedes expansion of the lungs. Slowly tilting the animal (head up and down) in a prone position (lying on the back) may provide some "ventilation" because compression and release of the diaphragm by the abdominal organs acts like a piston pump. Doxapram hydrochloride is a respiratory

Table 6.2 Drugs used for emergency situations and to antagonize specific drug effects.

Drug	Dose	Route	Indication	Comment
Adrenaline (epinephrine)	0.0001–0.001 mg/100 g	i.v.; i.o.	Cardiac/circulatory arrest, severe bradycardia	Short effect, repeat injection every 5 min if spontaneous circulation does not reoccur
Atropine	0.01 mg/100 g	i.v.; i.o., i.m.; (s.c.)	Bradycardia, heart rates in mouse and dwarf lemurs below 100–150 bpm can be considered to be too low to maintain adequate cardiac outptut	After i.m. or s.c. injection onset of effect can take several minutes, atropine can induce an initial further decrease in heart rate before the heart rate increases
Doxapram	0.2–0.5 mg/100 g	i.v.; i.o., i.m., (s.c.)	Low respiratory rate, insufficient breathing, apnea Respiratory stimulant with central analeptic effects (lightens anesthesia) when given at high doses	Stimulation of respiration can be short-lived and repeated injection becomes necessary; exceedingly high doses can induce seizures
Atipamezole	0.03–0.1 mg/100 g	i.v.; i.o., i.m., (s.c.)	Antagonizes effects of $alpha_2$ adrenergic drugs such as xylazine, medetomidine, dexmedetomidine, romifidine	Exact doses of the antagonists depend on the dose of the agonist used and the time since injection of the agonist. The recommended doses here are based on average doses used in other small animal species
Flumazenil	0.01–0.1 mg/100 g	i.v.; i.o., i.m., (s.c.)	Antagonizes effects of benzodiazepines like diazepam and midazolam at the $GABA_A$ receptor	
Naloxone	0.05–0.1 mg/100 g	i.v.; i.o., i.m., (s.c.)	Antagonizes effects of opioids at all opioid receptors	

Notes: i.v., intravenous; i.m., intramuscular; i.o., intraosseous; i.p., intraperitoneal; s.c., subcutaneous.

stimulant with central analeptic effects (lightens anesthesia) when given at high doses. Doxapram increases carotid body chemoreceptor discharge by inhibition of K^+ current through pH-sensitive K^+ TASK-1 and TASK-3 channels and leads to an increase in minute ventilation (Cotten *et al.*, 2006). In case the underlying cause of apnea (i.e., too deep anesthesia, high-dose opioid, etc.) is not resolved, the loss of carbon dioxide by hyperventilation can lead to compensatory apnea again and repeated injection can become necessary.

References

Amrein R, Hetzel W, Bonetti EP, *et al.* 1988. Clinical pharmacology of Dormicum (midazolam) and Anexate (flumazenil). *Resuscitation* 16:5–27.

Clarke K, Trim C, Hall L. 2014. Apparatus for administration of anesthetics. In KW Clarke, CM Trim, LW Hall (eds.), *Veterinary Anaesthesia*, 11th edition (pp 209–243). Saunders Elsevier, Edinburgh.

Cotten JF, Keshavaprasad B, Laster MJ, *et al.* 2006. The ventilatory stimulant doxapram inhibits TASK tandem pore (K2P) potassium channel function but does not affect minimum alveolar anesthetic concentration. *Anesthesia and Analgesia* 102:779–785.

Dal-Pan A, Blanc S, Aujard F. 2010. Resveratrol suppresses body mass gain in a seasonal non-human primate model of obesity. *BioMedCentral Physiology* 10:11.

Der Goukassian P. 1983. Hematology of the lesser mouse lemur (*Microcebus murinus*). A preliminary study. *Folia Primatology (Basel)* 41:129–136.

Dhenain M, Michot JL, Volk A, *et al.* 1997. T2-weighted MRI studies of mouse lemurs: a primate model of brain aging. *Neurobiology of Aging* 18:517–521.

Eger II, EI. 2010. Inhaled anesthetics: uptake and distribution. In R Miller, LI Eriksson, LA Fleisher, JP Wiener Kronish, WL Young (eds.), *Miller's Anesthesia.* Churchill Livingstone, Elsevier, Philadelphia.

Eilers H, Cattaruzza F, Nassini R, *et al.* 2010. Pungent general anesthetics activate transient receptor potential-A1 to produce hyperalgesia and neurogenic bronchoconstriction. *Anesthesiology* 112:1452–1463.

Farber, N, Pagel P, Warltier D. 2010. Inhaled anesthetics: pulmonary pharmacology. In R Miller, LI Eriksson, LA Fleisher, JP Wiener Kronish, WL Young (eds.), *Miller's Anesthesia* (pp. 561–594). Churchill Livingstone, Elsevier, Philadelphia.

Feng M, DiPetrillo K. 2009. Non-invasive blood pressure measurement in mice. *Methods in Molecular Biology* 573:45–55.

Flecknell PA. 2009. *Laboratory Animal Anaesthesia* Academic Press, San Diego.

Fleisher, LA, Wiener Kronish, JP, Young, WL (eds.) 2010. *Miller's Anesthesia* (pp. 539–559). Churchill Livingstone, Elsevier, Philadelphia.

Franks NP. 2008. General anaesthesia: from molecular targets to neuronal pathways of sleep and arousal. *Nature Reviews Neuroscience* 9:370–386.

Genin F, Perret M. 2003. Daily hypothermy in captive grey mouse lemurs (*Microcebus murinus*): effects of photoperiod and food restriction. *Journal of Comparative Physiology B* 136:71–81.

Giroud S, Blanc S, Aujard F, *et al.* 2008. Chronic food shortage and seasonal modulations of daily torpor and locomotor activity in the grey mouse lemur (*Microcebus murinus*). *American Journal of Physiology – Regulatory, Integrative and Comparative Physiology* 294:R1958–1967.

Glander KE. 2013. Darting, anesthesia, and handling. In J Eleanor, NB Sterling, ME Blair (eds.), *A Handbook of Techniques in Primate Ecology and Conservation* (pp. 27–39). Oxford University Press, Oxford.

Guillemin J-F. 1998. L'anesthesie d'un lemurien: Le microcebe murin (*Microcebus murinus*). Ecole Nationale Veterinaire D'Alfort. 177 pp.

Hamalainen A, Heistermann M, Fenosoa ZS, *et al.* 2014. Evaluating capture stress in wild gray mouse lemurs via repeated fecal sampling: method validation and the influence of prior experience and handling protocols on stress responses. *General and Comparative Endocrinology* 195:68–79.

Hellekant G, Hladik CM, Dennys V, *et al.* 1993. On the relationship between sweet taste and seasonal body weight changes in a primate (*Microcebus murinus*). *Chemical Senses* 18:27–33.

Kästner S. 2007. Intravenous anaesthetics. In C Seymour, T Duke-Novakovski (eds.), *BSAVA Manual of Canine and Feline Anaesthesia and Analgesia*, 2nd edition (pp. 133–149). British Small Animal Association, Quedgeley.

Krystal AD, Schopler B, Kobbe S, *et al.* 2013. The relationship of sleep with temperature and metabolic rate in a hibernating primate. *PLoS ONE* 8:e69914.

Lahann P. 2007. Feeding ecology and seed dispersal of sympatric cheirogaleid lemurs (*Microcebus murinus, Cheirogaleus medius, Cheirogaleus major*) in the littoral rainforest of south-east Madagascar. *Journal of Zoology* 271:88–98.

Le Tallec T, Perret M, Théry M. 2013. Light pollution modifies the expression of daily rhythms and behavior patterns in a nocturnal primate. *PLoS ONE* 8:e79250.

Longley LA. 2008. Rodent anaesthesia. In LA Longley (ed.), *Anaesthesia of Exotic Pets* (pp. 59–84). Saunders Elsevier, Edinburgh.

Louis EE Jr, Engberg SE, McGuire SM, *et al.* 2008. Revision of the mouse lemurs, *Microcebus* (Primates, Lemuriformes), of northern and northwestern Madagascar with descriptions of two new species at Montagne d'Ambre National Park and Antafondro Classified Forest. *Primate Conservation* 23:19–38.

Marchal J, Dorieux O, Haro L, *et al.* 2012. Characterization of blood biochemical markers during aging in the grey mouse lemur (*Microcebus murinus*): impact of gender and season. *BioMedCentral Veterinary Research* 8:211.

Mirakhur RK, Clarke RS, Elliott J, *et al.* 1978. Atropine and glycopyrronium premedication. A comparison of the effects on cardiac rate and rhythm during induction of anaesthesia. *Anaesthesia* 33:906–912.

Ortmann S, Schmid J, Ganzhorn JU, *et al.* 1996. Body temperature and torpor in a Malagasy small primate, the mouse lemur. In F Geiser, AJ Hulbert, SC Nicol (eds.), *The Tenth International Hibernation Symposium* (pp. 55–61). University of New England Press, Lebanon, NH.

Pagel P, Farber NE, Pracht PF, Warltier DC. 2010. Inhaled anesthetics: cardiovascular pharmacology. In R Miller, LI Eriksson, LA Fleisher, JP Wiener Kronish, WL Young (eds.), *Miller's Anesthesia* (pp. 595–632). Churchill Livingstone, Elsevier, Philadelphia.

Perret M, Aujard F. 2001. Daily hypothermia and torpor in a tropical primate: synchronization by 24-h light-dark cycle. *American Journal of Physiology – Regulatory, Integrative and Comparative Physiology* 281:R1925–1933.

Perret M, Aujard F, Seguy M, *et al.* 2003. Olfactory bulbectomy modifies photic entrainment and circadian rhythms of body temperature and locomotor activity in a nocturnal primate. *Journal of Biological Rhythms* 18:392–401.

Picq JL, Aujard F, Volk A, *et al.* 2012. Age-related cerebral atrophy in nonhuman primates predicts cognitive impairments. *Neurobiology of Aging* 33:1096–1109.

Pifferi F, Dal-Pan A, Languille S, *et al.* 2013. Effects of resveratrol on daily rhythms of locomotor activity and body temperature in young and aged grey mouse lemurs. *Oxidative Medicine and Cell Longevity* 2013:187–301.

Pifferi F, Dal-Pan A, Menaker M, *et al.* 2011. Resveratrol dietary supplementation shortens the free-running circadian period and decreases body temperature in a prosimian primate. *Journal of Biological Rhythms* 26:271–275.

Pifferi F, Rahman A, Languille S, *et al.* 2012. Effects of dietary resveratrol on the sleep–wake cycle in the non-human primate gray mouse lemur (*Microcebus murinus*). *Chronobiology International* 29:261–270.

Rahman A, Languille S, Lamberty Y, *et al.* 2013. Sleep deprivation impairs spatial retrieval but not spatial learning in the non-human primate grey mouse lemur. *PLoS ONE* 8:e64493.

Ramsier MA, Cunningham AJ, Moritz GL, *et al.* 2012. Primate communication in the pure ultrasound. *Biology Letters* 8:508–511.

Schilling A, Perret M. 1993. Removal of the olfactory bulbs modifies the gonadal responses to photoperiod in the lesser mouse lemur (*Microcebus murinus*). *Biology and Reproduction* 49:58–65.

Schmid J, Speakman JR. 2000. Daily energy expenditure of the grey mouse lemur (*Microcebus murinus*): a small primate that uses torpor. *Journal of Comparative Physiology B* 170:633–641.

Schmid J, Speakman JR. 2009. Torpor and energetic consequences in free-ranging grey mouse lemurs (*Microcebus murinus*): a comparison of dry and wet forests. *Naturwissenschaften* 96:609–620.

Schopf C, Zimmermann E, Tünsmeyer J, *et al.* 2014. Hearing and age-related changes in the gray mouse lemur. *Journal of the Association for Research in Otolaryngology* 15:993–1005.

Schütter A, Tünsmeyer J, Schopf C, *et al.* 2013. Cardiorespiratory and anaesthetic characteristics of sevoflurane anaesthesia in young and geriatric grey mouse lemurs (*Microcebus murinus*). Association of Veterinary Anaesthetists Spring Meeting, London: Veterinary Anaesthesia and Analgesia. http://onlinelibrary.wiley.com/doi/10.1111/vaa.12053/pdf. p 17.

Terrien J, Zizzari P, Bluet-Pajot MT, *et al.* 2008. Effects of age on thermoregulatory responses during cold exposure in a nonhuman primate, *Microcebus murinus*. *American Journal of Physiology – Regulatory, Integrative and Comparative Physiology* 295:R696–703.

7 *Microcebus murinus* – a unique primate for modeling human brain disorders, including Alzheimer's disease and bovine spongiform encephalopathy

Jean-Michel Verdier and Nadine Mestre-Francés

Introduction

The continuing increase in life expectancy in the developed world is one of humanity's greatest successes. However, the rising number of older people will have economic and societal consequences, as well as major effects on the occurrence of age-related diseases, especially neurodegenerative disorders such as Alzheimer's disease (AD). Combating AD is one of the most important medical, societal, and economical challenges (Lo *et al.*, 2014). On February 3, 2009, the European Parliament adopted a written declaration on the priorities for fighting AD (No. 80/2008). This action plan was designed to promote research into the causes, prevention, and treatment of this disease.

Unfortunately, despite intensive research, the pathogenetic mechanisms and risk factors of AD are still poorly understood, and no effective treatments are available. One of the reasons underlying the lack of successful AD therapies is the lack of a relevant animal model, despite the large variety of models that have been created (reviewed in Woodruff-Pak, 2008). Transgenic mice are the major animal model for AD studies (Chin, 2011). They have been proven valuable for modeling various aspects of AD neuropathology, such as amyloid-β(Aβ) deposits, neuritic plaques, gliosis, synaptic alterations, and signs of neurodegeneration, as well as associated cognitive changes (McGowan *et al.*, 2006; Kitazawa *et al.*, 2012). Despite these similarities, there are important neuropathological and behavioral differences between these transgenic mouse models and AD in humans. Furthermore, experimental findings and conclusions are highly dependent on model systems and even animal strains, making it difficult to transfer findings to the human condition. Introducing a mutated human gene into a mouse does not necessarily trigger the cascade of events encountered in human disease (Langui *et al.*, 2007). Regarding AD, clinical trials have been stopped because of the death of several patients due to encephalopathies (Robinson *et al.*, 2004). Indeed, due to high social pressures, the results

The Dwarf and Mouse Lemurs of Madagascar: Biology, Behavior and Conservation Biogeography of the Cheirogaleidae, eds. S.M. Lehman, U. Radespiel, E. Zimmermann. Published by Cambridge University Press. © Cambridge University Press 2016.

obtained in transgenic mice have been rapidly and directly translated into human treatments. The subsequent disappointing results demonstrated that observations in these transgenic models cannot necessarily be extrapolated to humans. There are several key criteria to increase the validity of a model, including requiring:

(1) similar symptoms as observed in clinical manifestations;
(2) similar underlying biology; and
(3) a responsiveness to measure clinically effective therapeutic drugs by assessing true readouts.

Based on these needs, new models have emerged (e.g., dogs), which are well suited for longitudinal studies (Cotman and Head, 2008; Head *et al.*, 2008; Barone *et al.*, 2012), but their use is difficult to implement widely. A good "equivalent" animal model for AD (i.e., as close as possible to humans) should exhibit progressive pathology and associated cognitive deficits and should be robust enough to replicate observations by different research teams. Such models are needed for a better understanding of age-associated behavioral and brain abnormalities like those that occur in human AD patients. Indeed, compared to mice, the complex primate brain is better suited to model behavior and cognitive abilities in humans. Indeed, primates can naturally develop Aβ and tau pathology ("AD-like" lesions). Therefore, due to translational applicability, non-human primate (NHP) models, such as macaques (Buccafusco, 2008) or Caribbean vervet (Lemere *et al.*, 2004), are extremely valuable, especially for testing new drugs or treatments (Woodruff-Pak, 2008). Recently, it has been shown that one of the major Aβ deposits in the human AD brain is composed of pyroGlu-3 Aβ (Frost *et al.*, 2013). It is also present in NHPs, but only infrequently and variably in a subset of transgenic mouse models (Frost *et al.*, 2013). However, these models are not time- or cost-effective, and the number of animals available is rather limited.

Compared to monkeys, the gray mouse lemur *Microcebus murinus* provides a unique model. It is relatively inexpensive to maintain and breed, and there are self-sustaining breeding colonies in France and Germany. This small, nocturnal NHP is genetically and behaviorally much closer to humans than rodents, and their lifespan in breeding colonies reaches 8–10 years, with some individuals reaching 12–15 years in captivity (see Chapter 8). This broad lifespan allows for longitudinal studies on aging and associated neurodegenerative diseases or long-term follow-up studies (which are precluded in short-lived rodents). The species has been successfully maintained and bred in captivity since the 1960s, making it one of the few primates for which it is not necessary to capture animals from the wild.

The present article aims to review the utility of mouse lemurs for aging and neurodegenerative disease studies.

The gray mouse lemur is a unique primate model for human brain disorders

As for other NHPs, *M. murinus* is genetically more similar to humans than mice and rats (Janecka *et al.*, 2007). As many human pathogens have coevolved with NHPs, a mutual adaptation of host and pathogens has arisen. For this reason, many pathogenetic studies are most effectively carried out in NHPs because of their greater predictive values as models. Scaling up or preclinical development of therapeutic drugs require an intermediate between rodents, both for toxicity studies and testing drug efficacy. Interestingly, beyond this important fact, the gray mouse lemur has a much longer lifespan than mice. By measuring several parameters, it has been shown that mouse lemur aging mimics that of humans (reviewed in Languille *et al.*, 2012). For example, they develop alterations of sensory systems during aging, including cataract development (Beltran *et al.*, 2007), decreases of major components of the antioxidative system (Rathbun and Holleschau, 1992; Languille *et al.*, 2012), and impaired olfactory information processing (Aujard and Nemoz-Bertholet, 2004). As in humans, mouse lemurs exhibit age-related decreases in motor capacities and coordination (Nemoz-Bertholet and Aujard, 2003). Similarly, alterations of biological rhythms and the endocrine system have been described in both mouse lemurs (Perret, 1997) and humans (Van Someren and Riemersma-Van Der Lek, 2007). Acceleration of seasonal rhythms (i.e., by alternating periods of long and short days over 8 months rather 1 year) affects *M. murinus* survival and longevity (Perret, 1997). This observation means that mouse lemur longevity seems to be correlated to the succession of seasonal cycles rather a determined biological age. This observation is very interesting and has also been proposed in human populations; it has been shown that the accumulation of harsh winters may be responsible for decreased longevity (Robine *et al.*, 2012).

However, one of the most interesting parameters in primate work is the measure of their social interactions, which are essential in mouse lemurs (Radespiel, 2000; Sterling and Radespiel, 2000). Older animals are progressively withdrawn from these interactions (Picq, 1992), and a similar decrease has also been described for rhesus monkeys (Heydecke *et al.*, 1986).

This complex behavior is related to brain organization. Comparisons of the brains of NHPs and humans show similarities not only in terms of anatomy but also in terms of neuronal circuitry, neurochemistry, and functional and cognitive abilities. NHPs provide a more direct and translatable approach to human diseases because they allow the investigation of higher intellectual functions with comparative endpoints (Sutcliffe and Hutcheson, 2012). Several different tests have already been developed with mouse lemurs, such as recognition memory (Picq and Dhenain, 1998), stimulus reward associative learning (i.e., discrimination task: Joly *et al.*, 2006; working memory: Trouche *et al.*, 2010; spatial memory, and set-shifting ability: Picq, 2007; Picq *et al.*, 2012). Very recently, a touchscreen-based procedure has been successfully adapted to mouse lemurs with a more standardized method to evaluate executive functions by pairwise discrimination and reversal learning tasks (Joly *et al.*, 2014).

The results of these tests have led to the conclusion that mouse lemurs mimic the cognitive aging patterns described in humans, especially in the elderly population. The animals develop much more complex social and cognitive skills, modeling the human condition more closely than rodent studies. Also, recent studies in

NHPs have evidenced the primate-specific development of brain circuitry (see for example Sanchez-Gonzalez *et al.*, 2005; Bystron *et al.*, 2006; Dehay and Kennedy, 2007; Garcia-Cabezas *et al.*, 2009). Finally, it is important to mention that not all aged lemurs are affected by behavioral or cognitive dysfunctions (Joly *et al.*, 2006). Some are impaired, but several perform as well as young animals. They clearly behave as humans do with a high degree of intervariability, but they offer the possibility to distinguish between physiological aging and age-associated pathologies as described below.

The gray mouse lemur, a potential "equivalent" model for AD

AD is the most common form of dementia worldwide. It is mainly characterized by the progressive and irreversible loss of neurons located in specific brain areas: the hippocampus and polymodal cortical association areas. AD results in impaired memory, thinking, and behavior and eventually dementia. In terminal AD, most of the cortical brain regions, as well as many subcortical nuclei, are heavily filled with two different types of lesions: Aβ or neuritic plaques and neurofibrillary tangles (NFTs). Aβ plaques are aggregates of the monomeric Aβ peptide that is generated from a larger transmembrane precursor called amyloid precursor protein (APP; Kang *et al.*, 1987). NFTs are essentially made of twisted remnants of tau protein, a microtubule-associated protein essential for maintaining proper cell structure and function (Ballatore *et al.*, 2007). An abnormality in the tau protein disrupts the microtubule network, which is then transformed into abnormal aggregated filaments made up of pathological tau proteins (Grundke-Iqbal *et al.*, 1986).

Distinguishing the molecular and cellular mechanisms of physiological aging and AD as early as possible in the disease process remains one of the most important scientific challenges for developing future treatments. Brain aging mechanisms result from an imbalance between age-related losses and compensatory mechanisms. However, neuronal death is restricted in physiological aging, and impairment in aging is different from that in neurodegenerative diseases (Morrison and Hof, 1997). Changes in synaptic number and structure (e.g., neurotransmission, calcium homeostasis, signaling pathways, inflammation, oxidative stress, etc.) are strongly implicated. However, as a result of adaptive mechanisms, many neuronal properties are preserved. Conversely, in neurodegenerative diseases, there is a brutal collapse of molecular and cellular functions. Ageing remains the main risk factor for sporadic AD, and the evolution from normal aging to AD is associated with other factors, such as hypertension, hypercholesterolemia, diabetes, and vascular alteration. Given this extremely complex pathological situation, future treatment for AD will require elucidation of the causes of the primary events. For example, we need to understand whether neuronal alterations occur independently of Aβ deposition and tau hyperphosphorylation and whether microglial activation and astrocytosis precede neuritic/neuronal changes in AD. Also, what changes occur over time to make the aged brain tissue more vulnerable to AD is a critical question. As a consequence,

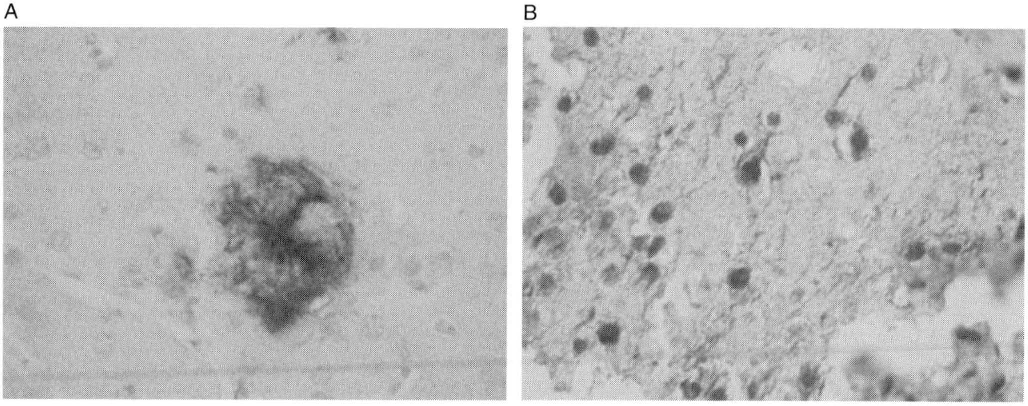

Figure 7.1 Aβ plaques in the occipital cortex (A) and tau-PHF neurons in the piriform cortex (B) of a mouse lemur brain. (A black and white version of this figure will appear in some formats. For the colour version, please refer to the plate section.)

finding a relevant animal model of AD is an important research strategy used in ongoing trials to understand disease pathology and to test new AD treatments.

In view of this, we have pioneered the use of the gray mouse lemur as a reference NHP model to study mechanisms associated with physiological aging and AD etiology for many years. During aging, some of these lemurs (roughly 5–10%) develop signs similar to those of AD patients, including Aβ plaque deposits and tau aggregation (Bons *et al.*, 1995; Delacourte *et al.*, 1995; Mestre-Francés *et al.*, 2000; Figure 7.1).

Therefore, as a natural chronic model, it has progressively gained interest for AD studies that are complementary to mice, which are acute transgenic models. In addition to measuring cognitive deficits (cf. supra), magnetic resonance imaging (MRI) evaluation of cerebral atrophy has also yielded interesting results (Dhenain *et al.*, 2003; Kraska *et al.*, 2011). Measures of cognitive abilities have shown that spatial memory performance is related to hippocampal and entorhinal cortex atrophy (Picq *et al.*, 2012), the two regions primarily affected in AD. This is the first time that we have shown a clear relationship between age-associated cognitive alteration and cerebral atrophy in NHPs. Because aging is the primary risk factor of AD, we analyzed the transcriptome of temporal cortex tissue from young, healthy old, and "AD-like" animals (i.e., Aβ plaques and cortical atrophy, which are pathognomonic signs of AD). We identified 47 genes that discriminated among young, healthy old, and "AD-like" animals. Functional categorization revealed that most of the genes upregulated in healthy old animals and downregulated in "AD-like" animals belonged to metabolic pathways, particularly those underlying protein synthesis. These data suggest the existence of compensatory mechanisms during physiological brain aging that are dysfunctional in "AD-like" animals (Abdel Rassoul *et al.*, 2010). Notably, these transcriptomic approaches allow researchers to distinguish between healthy aged and AD-like animals. Finally, to emphasize the potential of mouse lemurs in AD studies, anti-Aβ vaccination has been launched to improve the safety of Aβ-targeting immunotherapy in humans. Administration of a full-length Aβ self-antigen (Elan/Wyeth AN-1792 trial) resulted in brain and spinal cord inflammation (Orgogozo *et al.*, 2003; Gilman *et al.*, 2005). Consequently, less-toxic Aβ derivatives have been developed to favor a Th2 response promoting antibody production instead of a Th1 response that mediates a cytotoxic T-cell response (Sigurdsson *et al.*, 2001). Using these Aβ derivatives in mouse lemurs (K6Aβ1–30), we observed that they elicit a substantial antibody response, and importantly this effect is reversible, which enhances the safety profile of this approach (Trouche *et al.*, 2009). However, MRI and histology also revealed that microhemorrhages and iron deposits may follow this type of vaccination (Joseph-Mathurin *et al.*, 2013). Interestingly, such findings were not described in the original paper describing a mouse transgenic model (Schenk *et al.*, 1999).

The gray mouse lemur is highly sensitive to prions

Transmissible spongiform encephalopathies are a group of fatal neurodegenerative disorders that affect both humans and animals. The likelihood of spreading bovine spongiform encephalopathy (BSE) to humans through the food chain has been a source of great public health concern. Therefore, the oral route of infection should be studied attentively, especially in NHPs. This risk of transmission is tightly linked to the capacity of the BSE agent to adapt to primates, and the uncertain extent of human exposure to BSE is due to our incomplete knowledge regarding the efficiency of oral infection and the magnitude of any biological barriers to bovine-to-human transmission that may exist. Previous studies have shown that the BSE agent is pathogenic after intracerebral (IC) inoculation in marmosets (Baker *et al.*, 1993, 1998) and cynomolgus macaques (Lasmezas *et al.*, 2001; Yutzy *et al.*, 2007; Ono *et al.*, 2011), but very few studies have examined oral BSE transmission to NHPs (Lasmezas *et al.*, 2005; Yutzy *et al.*, 2007). In this case, the mouse lemur has been shown to be a very useful model.

An atypical form of BSE was recently identified in cattle in Europe (Casalone *et al.*, 2004; Jacobs *et al.*, 2007), North America (Dudas *et al.*, 2010), and Japan (Hagiwara *et al.*, 2007) and was designed as L-type BSE (L-BSE) due to the lower apparent molecular mass of unglycosylated PrP^{res} (PrP resistant) on Western blot as compared with classical BSE. Experimental evidence from studies in primate models and transgenic mice expressing human PrP suggested a higher risk of transmission to humans of the rare and putatively sporadic form of L-BSE than for the classical BSE agent. However, a major unresolved issue concerns the potential transmissibility of the L-BSE agent via the oral route. To address this question, we infected mouse lemurs with L-BSE by an IC or oral route. We demonstrated that the L-BSE agent can be transmitted by the oral route from cattle to both young and adult mouse lemurs (Mestre-Francés *et al.*, 2012). In comparison to IC-inoculated animals, the orally challenged mouse lemurs exhibited longer survival periods, as expected with this route of infection. We also observed less-severe clinical signs with no evidence of motor dysfunction in contrast to IC-challenged animals.

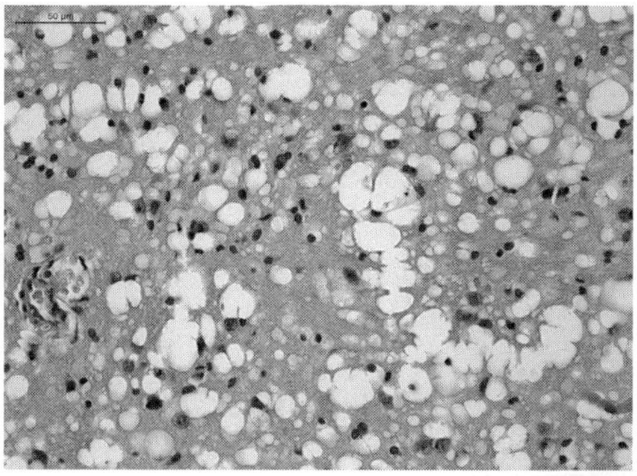

Figure 7.2 Spongiosis in intracerebrally inoculated animals showing large confluent vacuoles. (A black and white version of this figure will appear in some formats. For the colour version, please refer to the plate section.)

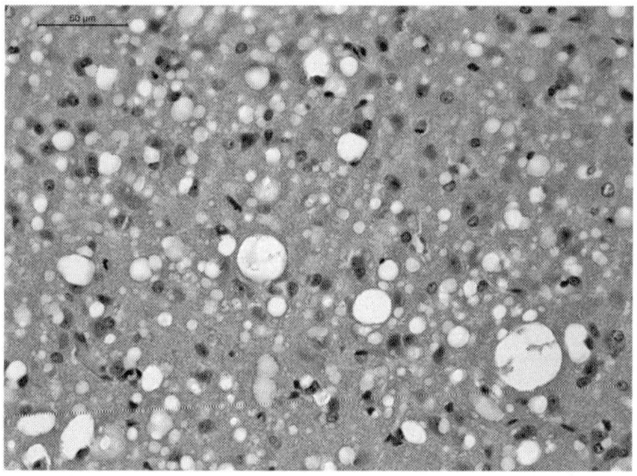

Figure 7.3 Spongiosis in orally infected animals. (A black and white version of this figure will appear in some formats. For the colour version, please refer to the plate section.)

Histopathological analysis revealed severe spongiform changes in the brains of IC-challenged mouse lemurs (Figure 7.2). Specifically, they displayed a pattern of vacuolation characterized by intense spongiosis with many confluent vacuoles. In orally challenged animals, spongiosis was characterized by fewer vacuoles and was restricted to the striatum, thalamus, colliculi (Figure 7.3), and brainstem.

Analysis of PrP^{res} distribution in the brain as assessed by PET-blot showed that in IC-challenged animals, PrP^{res} strongly accumulated in the whole brain (Figure 7.4). In orally challenged animals, strong PrP^{res} accumulation was limited to the striatum and thalamus (Figure 7.5).

PrP^{res} immunodetection with the monoclonal anti-PrP antibody 3F4 showed a dense synaptic pattern associated with plaque-like deposits (Figure 7.6A) and was accompanied by reactive astrocytic gliosis (Figure 7.6B).

Our study has confirmed experimentally the potential risk of interspecies oral transmission of L-BSE. Because a previous study on L-BSE in transgenic mice expressing human PrP suggested an absence of any transmission barrier between

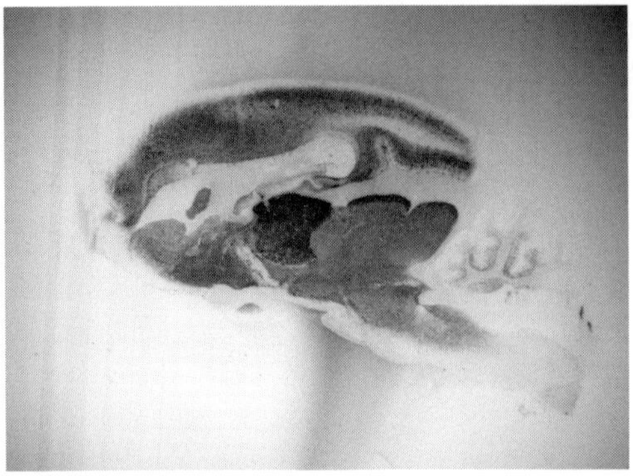

Figure 7.4 PET-blot showing PrP^{res} distribution in the whole brain of IC-inoculated mouse lemurs. (A black and white version of this figure will appear in some formats. For the colour version, please refer to the plate section.)

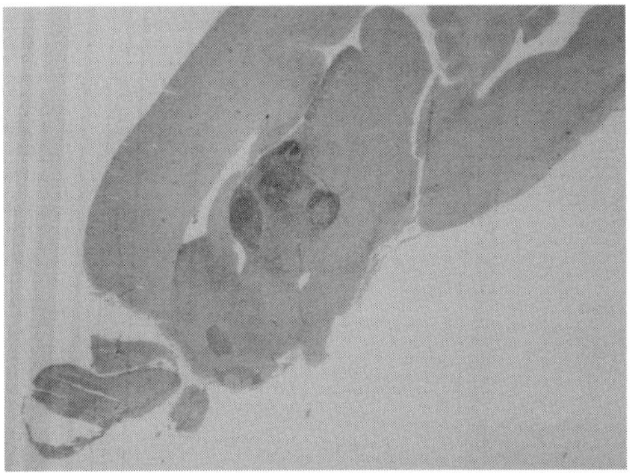

Figure 7.5 PET-blot showing PrP^{res} distribution in the whole brain of orally inoculated mouse lemurs. (A black and white version of this figure will appear in some formats. For the colour version, please refer to the plate section.)

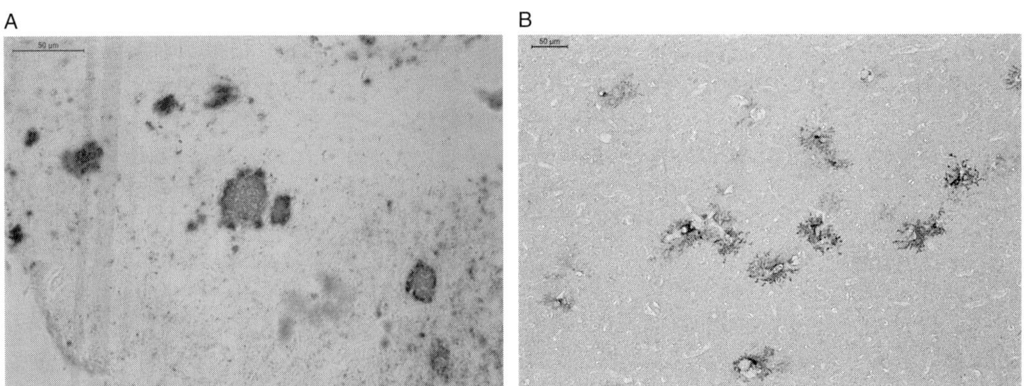

Figure 7.6 Immunodetection of PrP^{res} in plaques with 3F4 antibody (A). GFAP antibody labeling shows reactive astrocytic gliosis (B). (A black and white version of this figure will appear in some formats. For the colour version, please refer to the plate section.)

cattle and humans for this particular BSE strain, it is therefore important to maintain precautionary measures that prevent the entry of potentially infected tissues from cattle into the food chain.

Conclusion

A better understanding of physiological aging processes requires shifting from a purely descriptive scientific endeavor to mechanistic or hypothesis-driven experiments. Such research should be helpful in understanding how to modulate the aging process to prevent age-related diseases, especially neurodegenerative diseases. In this setting, mouse lemurs share common broad behaviors with humans that are helpful for studying aging and neurodegenerative diseases. For instance, monitoring aging parameters over time and identifying the expression of genes involved in brain vulnerability, especially in the very early phases, could provide clues to develop new therapeutic targets. Moreover, the identification of chromosomic regions linked to aging or AD will allow us to identify new candidate markers.

However, there are three main constraints that limit the use of this model. The first is our limited knowledge of the *M. murinus* genome. This problem is likely to be resolved in the next few months as we have already started whole-genome approaches (comparison of wild-type and AD-like genomes, sequencing of genes known to be involved in AD in humans, development of a microsatellite bank, etc.). The second limitation of the model is the number of animals available for experiments. Although the populations are substantially greater than those of other NHPs, to research the mechanisms involved in the occurrence of diseases it is obviously necessary to develop cellular approaches. In addition, it will be important to refine the number of animals used. In view of this, we are now developing cellular models from wild-type and from "AD-like" mouse lemurs to investigate underlying mechanisms responsible for potential dysfunctions. Regarding BSE, we have also developed the use of retina cells and the *in vivo* primate retina to study prion-related neurotoxicity. Indeed, in terms of ocular system, NHPs and humans have unique features, such as macula/fovea, not found in other mammals (Stone and Johnston, 1981). We have demonstrated that the neurotoxic fragment of PrP ($PrP^{106-126}$) triggers rod photoreceptor cell loss by apoptosis. In addition, we observed that retinas injected with this fragment exhibit a significant increase in the number of apoptotic nuclei, mainly in the ganglion cell layer. This study investigated the use of the gray mouse lemur retina as a model system to rapidly test prion-related neurotoxicity and develop therapeutic measures (Torrent *et al.*, 2010). The possibility of conducting fast and competitive studies in an NHP model will certainly provide valuable insight into the biological effects of prions in humans, and thus to the development of improved therapies. In addition, our approach could certainly be transferred to other specific amyloidogenic proteins; therefore, the model is of increasing significance for understanding disease pathogenesis. The third limitation is the relatively low number of "AD-like" animals. As we mentioned earlier, the natural proportion of animals with AD traits (i.e., Aβ plaques and/or cortical atrophy) is

5–10%. This proportion is not sufficient for large numbers of experiments. In order to circumvent this problem, we are crossing AD trait-bearing animals to increase the "AD-like" lineage in our colony.

Given their small size, mouse lemurs can also be used to model other pathologies. In addition, the relatively high number of animals significantly raises the attractiveness of this model. For instance, we are developing approaches in an *M. murinus* model of Parkinson's disease showing motor disabilities. We also feel that its high versatility may be useful for other neurodegenerative diseases, such as spinal cord injury and amyotrophic lateral sclerosis. Although major advances have been made in the last few years, our knowledge of human brain function is still limited, and the use of NHPs is needed to significantly advance neuroscience research.

Finally, the gray mouse lemur may also be relevant for experiments performed currently in much larger primates, including human immunodeficiency virus (Apetrei *et al.*, 2012), type 2 (Wagner *et al.*, 2006) or type 1 (Contreras *et al.*, 2004) diabetes, or cardiovascular research (Shen, 2010). This is obviously not an exhaustive list of possible research areas that could be investigated using gray mouse lemurs.

Acknowledgments

This study is part of a project that has received funding from the European Community's 7th Framework Programme (FP7/2007–2013) under grant agreement no. 278486 (acronym "DEVELAGE"), and the Fonds Unique Interministériel and Région Languedoc-Roussillon under grant agreement no. 110284 (acronym "DiaTrAl"). We thank the animal keepers, Sylvie Rouland, Elisabeth Huetter, and Joël Cuoq, for animal maintenance.

References

Abdel Rassoul R, Alves S, Pantesco V, *et al.* 2010. Distinct transcriptome expression of the temporal cortex of the primate *Microcebus murinus* during brain aging versus Alzheimer's disease-like pathology. *PloS One* 5(9):e12770.

Apetrei C, Pandrea I, Mellors JW. 2012. Nonhuman primate models for HIV cure research. *PLoS Pathogens* 8:e1002892.

Aujard F, Nemoz-Bertholet F. 2004. Response to urinary volatiles and chemosensory function decline with age in a prosimian primate. *Physiology and Behavior* 81:639–644.

Baker HF, Ridley RM, Wells GA. 1993. Experimental transmission of BSE and scrapie to the common marmoset. *Veterinary Record* 132:403–406.

Baker HF, Ridley RM, Wells GA, Ironside JW. 1998. Prion protein immunohistochemical staining in the brains of monkeys with transmissible spongiform encephalopathy. *Neuropathology and Applied Neurobiology* 24:476–486.

Ballatore C, Lee VM, Trojanowski JQ. 2007. Tau-mediated neurodegeneration in Alzheimer's disease and related disorders. *Nature Reviews: Neuroscience* 8:663–672.

Barone E, Mancuso C, Di Domenico F, *et al.* 2012. Biliverdin reductase-A: a novel drug target for atorvastatin in a dog pre-clinical model of Alzheimer disease. *Journal of Neurochemistry* 120:135–146.

Beltran WA, Vanore M, Ollivet F, *et al.* 2007. Ocular findings in two colonies of gray mouse lemurs (*Microcebus murinus*). *Veterinary Ophthalmology* 10:43–49.

Bons N, Jallageas V, Silhol S, *et al.* 1995. Immunocytochemical characterization of Tau proteins during cerebral aging of the lemurian primate *Microcebus murinus. Comptes Rendus de l'Académie des Sciences Série III: Sciences de la Vie* 318:741–747.

Buccafusco JJ. 2008. Estimation of working memory in macaques for studying drugs for the treatment of cognitive disorders. *Journal of Alzheimer's Disease* 15:709–720.

Bystron I, Rakic P, Molnar Z, Blakemore C. 2006. The first neurons of the human cerebral cortex. *Nature Neuroscience* 9:880–886.

Casalone C, Zanusso G, Acutis P, *et al.* 2004. Identification of a second bovine amyloidotic spongiform encephalopathy: molecular similarities with sporadic Creutzfeldt–Jakob disease. *Proceedings of the National Academy of Sciences of the United States of America* 101:3065–3070.

Chin J. 2011. Selecting a mouse model of Alzheimer's disease. *Methods in Molecular Biology* 670:169–189.

Contreras JL, Smyth CA, Curiel DT, Eckhoff DE. 2004. Nonhuman primate models in type 1 diabetes research. *ILAR Journal* 45:334–342.

Cotman CW, Head E. 2008. The canine (dog) model of human aging and disease: dietary, environmental and immunotherapy approaches. *Journal of Alzheimer's Disease* 15:685–707.

Dehay C, Kennedy H. 2007. Cell-cycle control and cortical development. *Nature Reviews: Neuroscience* 8:438–450.

Delacourte A, Sautiere PE, Wattez A, *et al.* 1995. Biochemical characterization of Tau proteins during cerebral aging of the lemurian primate *Microcebus murinus. Comptes Rendus de l'Académie des Sciences Série III: Sciences de la Vie* 318:85–89.

Dhenain M, Chenu E, Hisley CK, Aujard F, Volk A. 2003. Regional atrophy in the brain of lissencephalic mouse lemur primates: measurement by automatic histogram-based segmentation of MR images. *Magnetic Resonance in Medicine* 50:984–992.

Dudas S, Yang J, Graham C, *et al.* 2010. Molecular, biochemical and genetic characteristics of BSE in Canada. *PLoS ONE* 5:e10638.

Frost JL, Le KX, Cynis H, *et al.* 2013. Pyroglutamate-3 amyloid-beta deposition in the brains of humans, non-human primates, canines, and Alzheimer disease-like transgenic mouse models. *American Journal of Pathology* 183:369–381.

Garcia-Cabezas MA, Martinez-Sanchez P, Sanchez-Gonzalez MA, Garzon M, Cavada C. 2009. Dopamine innervation in the thalamus: monkey versus rat. *Cerebral Cortex* 19:424–434.

Gilman S, Koller M, Black RS, *et al.* 2005. Clinical effects of Abeta immunization (AN1792) in patients with AD in an interrupted trial. *Neurology* 64:1553–1562.

Grundke-Iqbal I, Iqbal K, Tung YC, *et al.* 1986. Abnormal phosphorylation of the microtubule-associated protein tau (tau) in Alzheimer cytoskeletal pathology. *Proceedings of the National Academy of Sciences of the United States of America* 83:4913–4917.

Hagiwara K, Yamakawa Y, Sato Y, *et al.* 2007. Accumulation of mono-glycosylated form-rich, plaque-forming PrPSc in the second atypical bovine spongiform encephalopathy case in Japan. *Japanese Journal of Infectious Diseases* 60:305–308.

Head E, Pop V, Vasilevko V, *et al.* 2008. A two-year study with fibrillar beta-amyloid (Abeta) immunization in aged canines: effects on cognitive function and brain Abeta. *Journal of Neuroscience* 28:3555–3566.

Heydecke H, Schwibbe M, Kaumanns W. 1986. Studies on social behaviour of aging rhesus monkeys (*Macaca mulatta*). *Primate Report* 15:41–59.

Jacobs JG, Langeveld JP, Biacabe AG, *et al.* 2007. Molecular discrimination of atypical bovine spongiform encephalopathy strains from a geographical region spanning a wide area in Europe. *Journal of Clinical Microbiology* 45:1821–1829.

Janecka JE, Miller W, Pringle TH, *et al.* 2007. Molecular and genomic data identify the closest living relative of primates. *Science* 318:792–794.

Joly M, Deputte B, Verdier JM. 2006. Age effect on olfactory discrimination in a non-human primate, *Microcebus murinus. Neurobiology of Aging* 27:1045–1049.

Joly M, Ammersdörfer S, Schmidtke D, *et al.* 2014. Touchscreen-based cognitive tasks reveal age-related impairment in a primate aging model, the grey mouse lemur (*Microcebus murinus*). *PLoS One* 9(10):e109393.

Joseph-Mathurin N, Dorieux O, Trouche SG, *et al.* 2013. Amyloid beta immunization worsens iron deposits in the choroid plexus and cerebral microbleeds. *Neurobiology of Aging* 34:2613–2622.

Kang J, Lemaire HG, Unterbeck A, *et al.* 1987. The precursor of Alzheimer's disease amyloid A4 protein resembles a cell-surface receptor. *Nature* 325:733–736.

Kitazawa M, Medeiros R, Laferla FM. 2012. Transgenic mouse models of Alzheimer disease: developing a better model as a tool for therapeutic interventions. *Current Pharmaceutical Design* 18:1131–1147.

Kraska A, Dorieux O, Picq JL, *et al.* 2011. Age-associated cerebral atrophy in mouse lemur primates. *Neurobiology of Aging* 32:894–906.

Langui D, Lachapelle F, Duyckaerts C. 2007. Animal models of neurodegenerative diseases. *Médecine Sciences* 23:180–186.

Languille S, Blanc S, Blin O, *et al.* 2012. The grey mouse lemur: a non-human primate model for ageing studies. *Ageing Research Review* 11:150–162.

Lasmezas CI, Fournier JG, Nouvel V, *et al.* 2001. Adaptation of the bovine spongiform encephalopathy agent to primates and comparison with Creutzfeldt–Jakob disease: implications for human health. *Proceedings of the National Academy of Sciences of the United States of America* 98:4142–4147.

Lasmezas CI, Comoy E, Hawkins S, *et al.* 2005. Risk of oral infection with bovine spongiform encephalopathy agent in primates. *Lancet* 365:781–783.

Lemere CA, Beierschmitt A, Iglesias M, *et al.* 2004. Alzheimer's disease abeta vaccine reduces central nervous system abeta levels in a non-human primate, the Caribbean vervet. *American Journal of Pathology* 165:283–297.

Lo AW, Ho C, Cummings J, Kosik KS. 2014. Parallel discovery of Alzheimer's therapeutics. *Science Translational Medicine* 6:241–245.

McGowan E, Eriksen J, Hutton M. 2006. A decade of modeling Alzheimer's disease in transgenic mice. *Trends in Genetics* 22:281–289.

Mestre-Francés N, Keller E, Calenda A, *et al.* 2000. Immunohistochemical analysis of cerebral cortical and vascular lesions in the primate *Microcebus murinus* reveal distinct amyloid beta1–42 and beta1–40 immunoreactivity profiles. *Neurobiology of Disease* 7:1–8.

Mestre-Francés N, Nicot S, Rouland S, *et al.* 2012. Oral transmission of L-type bovine spongiform encephalopathy in primate model. *Emerging Infectious Diseases* 18:142–145.

Morrison JH, Hof PR. 1997. Life and death of neurons in the aging brain. *Science* 278:412–419.

Nemoz-Bertholet F, Aujard F. 2003. Physical activity and balance performance as a function of age in a prosimian primate (*Microcebus murinus*). *Experimental Gerontology* 38:407–414.

Ono F, Tase N, Kurosawa A, *et al.* 2011. Atypical L-type bovine spongiform encephalopathy (L-BSE) transmission to cynomolgus macaques, a non-human primate. *Japan Journal of Infectious Disease* 64:81–84.

Orgogozo JM, Gilman S, Dartigues JF, *et al.* 2003. Subacute meningoencephalitis in a subset of patients with AD after Abeta42 immunization. *Neurology* 61:46–54.

Perret M. 1997. Change in photoperiodic cycle affects life span in a prosimian primate (*Microcebus murinus*). *Journal of Biological Rhythms* 12:136–145.

Picq JL. 1992. Aging and social behaviour in captivity in *Microcebus murinus. Folia Primatologica* 59:217–220.

Picq JL. 2007. Aging affects executive functions and memory in mouse lemur primates. *Experimental Gerontology* 42:223–232.

Picq JL, Dhenain M. 1998. Reaction to new objects in young and aged grey mouse lemurs (*Microcebus murinus*). *The Quarterly Journal of Experimental Psychology* 51B:337–348.

Picq JL, Aujard F, Volk A, Dhenain M. 2012. Age-related cerebral atrophy in nonhuman primates predicts cognitive impairments. *Neurobiology of Aging* 33:1096–1109.

Radespiel U. 2000. Sociality in the gray mouse lemur (*Microcebus murinus*) in northwestern Madagascar. *American Journal of Primatology* 51:21–40.

Rathbun WB, Holleschau AM. 1992. The effects of age on glutathione synthesis enzymes in lenses of Old World simians and prosimians. *Current Eye Research* 11:601–607.

Robine JM, Herrmann FR, Arai Y, *et al.* 2012. Exploring the impact of climate on human longevity. *Experimental Gerontology* 47:660–671.

Robinson SR, Bishop GM, Lee HG, Munch G. 2004. Lessons from the AN 1792 Alzheimer vaccine: lest we forget. *Neurobiology of Aging* 25:609–615.

Sanchez-Gonzalez MA, Garcia-Cabezas MA, Rico B, Cavada C. 2005. The primate thalamus is a key target for brain dopamine. *Journal of Neuroscience* 25:6076–6083.

Schenk D, Barbour R, Dunn W, *et al.* 1999. Immunization with amyloid-beta attenuates Alzheimer-disease-like pathology in the PDAPP mouse. *Nature* 400:173–177.

Shen YT. 2010. Primate models for cardiovascular drug research and development. *Current Opinion in Investigational Drugs* 11:1025–1029.

Sigurdsson EM, Scholtzova H, Mehta PD, Frangione B, Wisniewski T. 2001. Immunization with a nontoxic/nonfibrillar amyloid-beta homologous peptide reduces Alzheimer's disease-associated pathology in transgenic mice. *American Journal of Pathology* 159:439–447.

Sterling EJ, Radespiel U. 2000. Advances in studies of sociality in nocturnal prosimians. *American Journal of Primatology* 51:1–2.

Stone J, Johnston E. 1981. The topography of primate retina: a study of the human, bushbaby, and new- and old-world monkeys. *Journal of Comparative Neurology* 196:205–223.

Sutcliffe J, Hutcheson D. 2012. Perspectives on the non-human primate touch-screen self ordered spatial search paradigm. In AJ Spink, F Grieco, OE Krips, *et al.* (eds.), Proceedings of Measuring Behavior, August 28–31, 2012. Utrecht, The Netherlands.

Torrent J, Soukkarieh C, Lenaers G, *et al.* 2010. *Microcebus murinus* retina: a new model to assess prion-related neurotoxicity in primates. *Neurobiology of Disease* 39:211–220.

Trouche SG, Asuni A, Rouland S, *et al.* 2009. Antibody response and plasma Abeta1–40 levels in young *Microcebus murinus* primates immunized with Abeta1–42 and its derivatives. *Vaccine* 27:957–964.

Trouche SG, Maurice T, Rouland S, Verdier JM, Mestre-Francés N. 2010. The three-panel runway maze adapted to *Microcebus murinus* reveals age-related differences in memory and perseverance performances. *Neurobiology of Learning and Memory* 94:100–106.

Van Someren EJ, Riemersma-Van Der Lek RF. 2007. Live to the rhythm, slave to the rhythm. *Sleep Medicine Reviews* 11:465–484.

Wagner JE, Kavanagh K, Ward GM, *et al.* 2006. Old world nonhuman primate models of type 2 diabetes mellitus. *ILAR Journal* 47:259–271.

Woodruff-Pak DS. 2008. Animal models of Alzheimer's disease: therapeutic implications. *Journal of Alzheimer's Disease* 15:507–521.

Yutzy B, Holznagel E, Coulibaly C, *et al.* 2007. Time-course studies of 14-3-3 protein isoforms in cerebrospinal fluid and brain of primates after oral or intracerebral infection with bovine spongiform encephalopathy agent. *Journal of General Virology* 88:3469–3478.

8 Life history variation in mouse lemurs (*Microcebus murinus*, *M. lehilahytsara*): the effect of environmental and phylogenetic determinants

Elke Zimmermann, Ute Radespiel, Nadine Mestre-Francés, and Jean-Michel Verdier

Introduction

A central question in life history research is the extent of variation and potential constraints of life history traits within a species and between closely related species. The life history of any species is determined by developmental and reproductive rates as well as reproductive effort spent over its lifetime (Stearns, 1992). These traits are usually assessed by gestation length, number and size of offspring at birth, body mass and age at weaning, patterns of postnatal growth, age at first reproduction, interbirth interval, and lifespan. Intraspecific variation in life history pattern has been found in non-human primate species in the wild, and is often attributed to different socioecological factors such as social status, group size, predation risk, diet, and abiotic environmental conditions (Kappeler and Perreira, 2003; Jones, 2011). Thus, for example, in baboons, it was described that some life history patterns such as age at first reproduction or longevity are affected by rank and the quality of social bonds, matriline, nutritive condition and parasite load (e.g. Altmann and Alberts, 2003; Altmann *et al.*, 2010, 2012; Silk *et al.*, 2010; Gesquière *et al.*, 2011). In a controlled laboratory setting, experimentally induced caloric restriction has been found to affect life history patterns such as longevity in rhesus macaques (Colman *et al.*, 2009; but see Mattison *et al.*, 2012 for contrary results in another colony). In mouse lemurs caloric restriction does not generally seem to affect aging (Dal-Pan *et al.*, 2011), but lifespan was extended under caloric restriction when food was supplemented with resveratrol (Marchal *et al.*, 2013). Interspecific life history variation is common between species of different primate lineages and forms the basis for life history theory (e.g. Stearns, 1992; Kappeler and Perreira, 2003; Zimmermann and Radespiel, 2007, 2014). It is postulated that life history trait variation between closely

The Dwarf and Mouse Lemurs of Madagascar: Biology, Behavior and Conservation Biogeography of the Cheirogaleidae, eds. S.M. Lehman, U. Radespiel, E. Zimmermann. Published by Cambridge University Press. © Cambridge University Press 2016.

related species should be minor, due to a long common history in the past and the same phylogenetic constraints shaping life history (Harvey and Clutton-Brock, 1985; Kamilar and Cooper, 2013). Thus, it is important to understand which determinants modify the variation of life history patterns in a species to get insight into environmental and phylogenetic plasticity.

Mouse lemurs are interesting primate models for studying the variation of life history patterns in primates. They represent the world's smallest primates with a high cryptic species diversity of currently 21 recognized species (Radespiel *et al.*, 2012; Rasoloarison *et al.*, 2013; see also Chapters 10 and 26). Mouse lemurs are nocturnal and arboreal and inhabit dense forest environments in Madagascar. They cope with a harsh and seasonally challenging climate and forest environment, which strongly affects their feeding habits, reproduction, communication, and cognitive skills (see also Chapters 20, 21, and 26). They adapt to these seasonal challenges by living on a seasonably variable mixed diet of fruits, flowers, nectar, gum of various gum trees, honeydew of homopteran larvae, as well as arthropods and small vertebrates (Radespiel *et al.*, 2006; Joly-Radko and Zimmermann, 2010; Thoren *et al.*, 2011), which they locate by employing similar spatial memory skills as found in monkeys and apes (Joly and Zimmermann, 2007, 2011; Joly *et al.*, 2008; Lührs *et al.*, 2009). Moreover, reproduction is photoperiodically triggered so that infants are born at the beginning of the resource-rich rainy season (Schmelting *et al.*, 2000; Wrogemann and Zimmermann, 2001; Wrogemann *et al.*, 2001; Randrianambinina *et al.*, 2003). In the best studied species (see Radespiel, 2006, for review), females and males live rather solitarily and come together only for breeding (*M. berthae*), females form small kin-based sleeping groups, but still range solitarily during the night, as do males (*M. murinus*), or both sexes form mixed-sex sleeping groups (*M. ravelobensis*). Female sleeping group members of *M. murinus* cooperate in infant care, to which males do not contribute (Eberle and Kappeler, 2006). Mouse lemurs face the highest predation risk in lemurs (Scheumann *et al.*, 2007). Predators are nocturnal viverrids, domestic cats and dogs, raptors and owls, as well as snakes. As in other nocturnal primates, acoustic and olfactory signals are of the utmost importance for governing predator avoidance as well as spacing between and cohesion within social groups (Zimmermann *et al.*, 2000; Braune *et al.*, 2005, 2008; Scheumann *et al.*, 2007; Sündermann *et al.*, 2008; Zimmermann 2010, 2013; Kappel *et al.*, 2011; Hohenbrink *et al.*, 2012, 2013; Kessler *et al.*, 2012). Due to their small body size, mouse lemurs can be maintained and bred cost-efficiently under controlled experimental conditions in captivity (Wrogemann and Zimmermann, 2001; Wrogemann *et al.*, 2001; Bons *et al.*, 2006; Lacreuse and Herndon, 2009; Fischer and Austad, 2011; Finch and Austad, 2012; Languille *et al.*, 2012) and currently three captive research colonies of more than 100 mouse lemurs exist worldwide (Brunoy and Montpellier, France, for *M. murinus* and Hannover, Germany, for *M. murinus* and *M. lehilahytsara*). The maximum lifespan of mouse lemurs in the field in Madagascar was reported to be 8 years for the gray (*M. murinus*) and golden-brown mouse lemur (*M. ravelobensis*) in the northwest (Ampijoroa;

own unpubl. data), and for the rufous mouse lemur (*M. rufus*) in the southeast (Ranomafana; Zohdy, 2012), whereas in captivity the maximum longevity was reported to be 18.2 years for the gray mouse lemur (Weigl, 2005). Thus, mouse lemurs represent the fastest aging primates and those with the shortest lifespan, which nevertheless seems to be highly variable. Mouse lemurs are also the focus of long-term field research at several localities in Madagascar, making comparable longitudinal field data available on inter- and intraspecific variation of life history patterns (Ampijoroa: Zimmermann/Radespiel, Lehman; Kirindy: Ganzhorn and Kappeler; Ranomafana: Wright; Beza Mahafaly: Rasoazanabary; Mandena/St. Lucie: Ganzhorn; Berenty: Génin/Masters; see also Chapters 21, 22, and 26).

In this chapter we present new data on life history variation in mouse lemurs, based on two of the world's largest captive colonies. We assume that in captivity, developmental and reproductive rates are near the maximum of the reaction norm for the respective species (see also Lee and Kappeler, 2003) due to zero predation and sufficient food supply, so that we have an excellent opportunity to measure the extent of variation and its potential constraints. By comparing two captive mouse lemur colonies of the same species, the gray mouse lemur (*Microcebus murinus*), experimentally exposed to a photoperiodic regime leading to an 8-month and 12-month annual cycle, respectively, we will explore to what extent an environmental determinant, the photoperiod, affects life history patterns. By comparing the gray and the Goodman's mouse lemur (*M. lehilahytsara*), living under the same photoperiodic regime (12-month annual cycle), we will explore to what extent life history patterns vary phylogenetically. Molecular studies have revealed that the most recent common ancestor of both mouse lemur species may go back as far as 8.9 Ma (Yoder and Yang, 2004).

Housing conditions and photoperiodic regimes of colonies

Because environmental conditions are known to induce variations in life history traits, we describe the housing conditions for both colonies in some detail below.

Montpellier

The Montpellier colony of the gray mouse lemur (*Microcebus murinus*) (hereafter termed colony MMU-M) originates from animals imported from the area of Mandena in southeastern Madagascar in the 1960s. The MMU-M averages about 300 animals born and raised within the breeding colony (École Pratique des Hautes Études, France), license approval no. 34-05-026-FS, according to the guidelines of the French Ministère de l'Agriculture (Decree 87–848), the European Community Directive (2010/63/EU), and the regional ethic committee for animal experimentation (CEEA-LR-1002). Animals are exposed to 24–26°C and 50% relative humidity. All studied mouse lemurs were individually marked by microchips. Animals are kept throughout their life under a reversed dark-night cycle and an artificial accelerated photoperiodic regime (annual cycle: 8 months) with a long-day period

(14 h light/day) for 5 months of the year and a short-day period (10 h light/day) for 3 months. Light is provided during the day by cool fluorescent lamps (360–720 lux) and dimmed red light (2 lux) during the night. The shift from one photoperiod to the other happens on one day.

Food, sufficient for two days, was delivered every other day. Food consists of fresh fruits and a mixture of cereals, milk and egg, enriched with vitamins. Water is available *ad libitum*. Animals are housed in same-sex groups (of fewer than five individuals) or individually depending on the period of the year and the sociability of animals (cage sizes= 200 × 100 × 100 cm^3 or 60 × 40 × 60 cm^3, respectively). Environmental enrichment was provided by providing tree branches and several nest boxes in the cages. Social grouping of mouse lemurs was dependent on breeding purposes and research projects. Usually, breeding pairs of unrelated partners are formed during female estrus until the female is in anestrus. Pairs are separated afterwards. Pregnant females are isolated two weeks before delivery. To control breeding, females are usually not paired before their first cycle and then continue to breed a maximum of seven times throughout their whole life. Body mass is measured every two weeks throughout the whole year. Health status is checked every day.

Hannover

The Hannover colony of the gray mouse lemur (*M. murinus*; hereafter termed MMU-H) averages about 100 mouse lemurs, originating from southern Madagascar about 40 years ago (see Hafen, 1998; Wrogemann and Zimmermann, 2001; Wrogemann *et al.*, 2001). The colony of the Goodman's mouse lemur (*M. lehilahytsara*; hereafter termed colony MLE) averages about 30 lemurs, whose founders originated from the area of Andasibe in eastern Madagascar in 1995 and 1996 (see Wrogemann and Zimmermann, 2001; Wrogemann *et al.*, 2001). All studied mouse lemurs were born in this colony (housing and colony management of the mouse lemur colony is in accordance with the NRC Guide for the Care and Use of Laboratory Animals, the European Directive 2010/63/EU on the protection of animals used for scientific purposes, and the German Animal Welfare Act and corresponding section for animals used for scientific purposes, licensed by the Bezirksregierung Hannover (reference number: AZ 33.9-42502-05-10A080) and by Ordnungsamt, Gewerbe- und Veterinärabteilung, Landeshauptstadt Hannover (AZ 42500/1H)). Each mouse lemur is marked individually by microchips.

Mouse lemurs of both species are maintained under controlled conditions at ambient temperatures of on average 25°C and relative humidity of on average 50–60% in cages equipped with branches and nest boxes (see Wrogemann and Zimmermann, 2001; Wrogemann *et al.*, 2001 for details). Animals are kept throughout their life under a reversed dark-night cycle and an artificial photoperiodic regime (annual cycle: 12-months) with a long-day period (14 h light/day) for 8 months of the year (starting in January) and a short-day period (10 h light/day) for the other 4 months (starting in September). The shift from one photoperiod to the other happens gradually twice a year. In January and September, day length is changed by 30 min twice weekly (see Wrogemann and Zimmermann, 2001; Wrogemann *et al.*, 2001). Light is provided during the day by cool fluorescent lamps (360–780 lux) and dimmed red light (4–19 lux) during the night.

Animals are fed daily with either a mixture of fresh fruits and vegetables or a milk pap protein diet enriched with vitamins (see Wrogemann and Zimmermann, 2001; Wrogemann *et al.*, 2001), supplemented three times a week with insects (mealworms, locusts). Water is provided *ad libitum*. Social grouping of mouse lemurs changed over the years and is dependent on breeding purposes and research projects. Usually, shortly before the onset of the breeding season, unrelated male–female pairs are formed, which are kept together either until shortly before the female gives birth and are then separated, or during the whole breeding season. To control breeding, females are not paired before their first reproductive season. Females of MMU-H are allowed to breed a maximum of eight times throughout their whole life. On a regular basis, same-sex groups of several females or males are also formed which stay together unless there are severe conflicts. Health status is monitored daily, whereas body mass as well as reproductive status is measured once a week throughout the whole year. Infant development of body mass is measured twice per week up to a body mass of 40 g.

Data collection and analyses

For this chapter we investigated data of both colonies and species, respectively. In both colonies, we determined the following four life history traits: body mass from females at the age of first reproduction (according to Wrogemann and Zimmermann, 2001; Wrogemann *et al.*, 2001), age at first reproduction (the age at which a female enters her first estrus during her first year of life, i.e., during her first reproduction), litter size (= number of infants in a litter seen alive for female at the age of first reproduction), and longevity (age at which adult males and females of the respective colonies were found dead for natural reasons or had to be euthanatized because of veterinary reasons, e.g., untreatable diseases such as severe cancer). We furthermore established age- and sex-specific survival curves for MMU for each colony considering the number of dead individuals per age category and excluding neonatal and pre-weaning mortality. For MMU-H, we additionally measured body mass of infants soon after birth (age of 4–7 days) and at the presumed weaning age (45 days; defined as "no more milk in diet" according to Martin, 1972), and gestation length of females at the age of first reproduction (according to Wrogemann and Zimmermann, 2001; Wrogemann *et al.*, 2001). The handling of infants before the age of 4 days would have risked subsequent rejection by their mothers and thus was avoided for ethical reasons.

In both colonies, breeding is controlled so that only a limited sample of data was available for females at the age of first reproduction. Data for infant development and female gestation length were not collected in MMU-M and thus could not be compared within MMU.

For longevity and survival analysis of MMU-M, we included 87 males and 84 females. As the animals were kept on a photoperiodically induced shortened annual cycle for their whole life, the length of the cycle was 8 months instead of 12 months. Longevity is thus given both in cycles and years (e.g., three 8-month cycles are equivalent to 2 years). Only a limited number of females bred in their first year of life (N = 50 females at the age of first reproduction) and could be included in the study.

For MMU-H, we included 96 gray mouse lemurs and 13 Goodman's mouse lemurs in the longevity and survival analysis. For a subsample of 14 MMU-H females and 7 MLE, reproductive data were available during their first year of life, i.e., during their first reproduction. Most females were not paired in their first year of reproduction and thus could not be included in the analysis.

Because not all data for life history traits were normally distributed, we always performed Mann–Whitney U (MWU) tests when exploring differences between species, colonies or sexes. If results were significant, they are in most cases displayed in graphs with medians and quartiles. For comparative reasons with published resources, we decided to present means and standard deviations as well as minimum and maximum values for the respective traits in the text. Pearson's rank correlations were calculated for normally distributed values to assess the relationship between gestation length and female body mass at first reproduction. Statistica 12 (StatSoft, Inc.) was used for the statistical analysis.

Comparison of life history patterns between two colonies of the gray mouse lemur: effect of photoperiod

Body mass at the age of first reproduction

Mean body mass of females at the age of first reproduction in colony MMU-M was 95.2 g (SD 13.9, min = 67, max = 134, N = 50) and thus significantly higher (Figure 8.1) than in MMU-H with 78 g (SD 9.69, min = 61.2, max = 105.1, N = 14).

Age at first reproduction

In MMU-M, the mean age at first reproduction was 6.16 months, the equivalent of 184.8 days (SD 35.78, min = 100 days, max = 285 days, N = 50). One female had her first estrus as early as 3.4 months after birth (born at the end of the 5-month reproductive period, and came into estrus after only a 3-month resilience period of short days) and at a body mass of 90 g. She delivered one healthy singleton. Two others were in estrus 4.6 months after birth, at a body mass of 84 and 88 g, respectively, and gave birth, respectively, to healthy triplets and a singleton. In MMU-H, the mean age at first reproduction was 11.8 months, the equivalent of 349 days (SD 43.88, min = 293 days, max = 426 days, N = 14). Consequently, age of first reproduction differed significantly between the colonies (Figure 8.2).

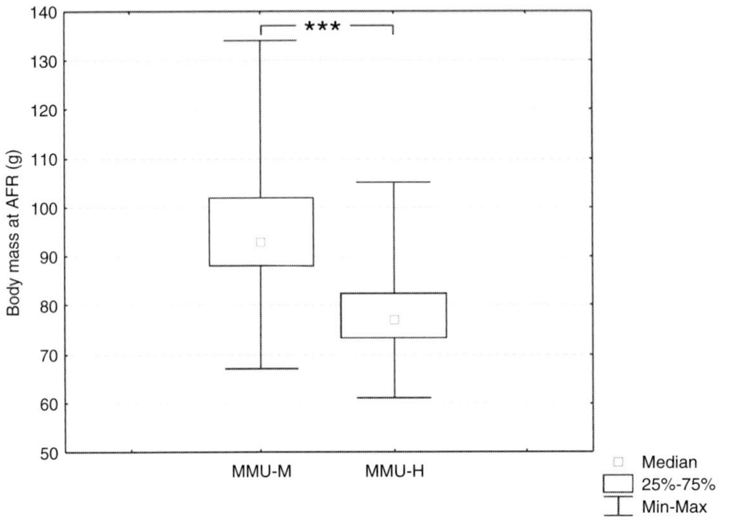

Figure 8.1 Average body mass of females at the age of first reproduction (AFR) is significantly different between MMU-M and MMU-H; MWU, $U = 141.5$, $Z = 4.6618$, $p < 0.00001$. *** indicates a significant difference of $p < 0.001$.

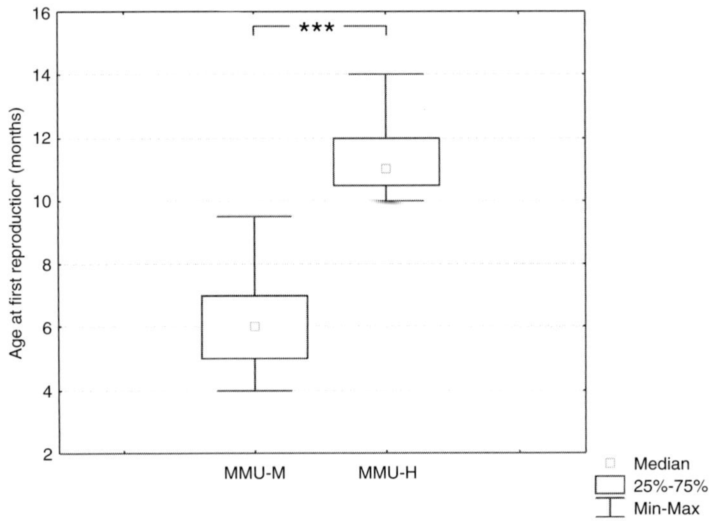

Figure 8.2 Average age of females at first reproduction is significantly different between MMU-M and MMU-H; MWU, $U = 0$, $Z = -5.72356$, *** indicates a significant difference of $p < 0.0000001$.

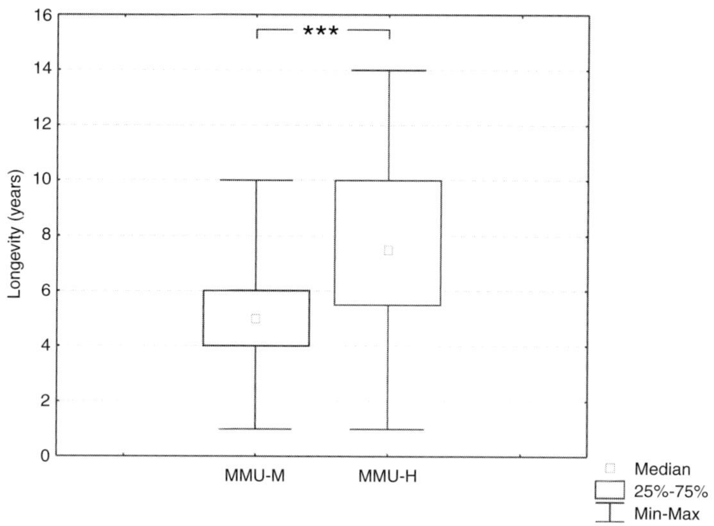

Figure 8.3 Average longevity in years differs significantly between MMU-M and MMU-H; MWU, $U = 3790.5$; $Z = 7.355$, *** indicates a significant difference of $p < 0.0000001$.

Litter size at first reproduction

Mean litter size in MMU-M was 2.2 (SD 0.8, min = 1, max = 4, $N = 50$) infants per litter. Females in MMU-M delivered only once in an 8-month cycle. Mean litter size of females at the age of first reproduction for MMU-H was 2.1 (SD 0.6, min = 1, max = 3, $N = 14$) infants per litter per female. Females delivered one to three times in a 12-month cycle. There was no significant difference in litter size between both colonies (MWU test; n.s.).

Longevity and survival

For MMU-M, the mean lifespan was 5.27 years (SD 1.68, min = 1, max = 10.7, $N = 171$) which was equivalent to 7.56 annual cycles (SD 2.74, min = 1, max = 16, $N = 171$). Thc equivalent numbers were 7.7 years (SD 3.2, min = 1, max = 14, $N = 96$) as well as annual cycles for MMU-H. Thus, longevity differed highly significantly by about 3 years between colonies (Figure 8.3), but no difference was observed when comparing cycles. Within colonies, there was no significant difference between sexes in longevity in MMU-M (MWU test, $Nm = 87$, $Nf = 82$, n.s.), but there was a significant sex difference in MMU-H (MWU test, $Nm = 51$, $Nf = 45$, $U = 823$, $Z = -2.395$, $p = 0.0168$). In the latter colony, females lived on average almost one year longer than males (female: mean = 8.6, SD 2.93, min = 2, max = 14; male: mean = 7.7, SD 3.2, min = 1, max = 13).

The survival curves of the colonies further supported these findings (Figure 8.4a,b). Curve profiles were similar between colonies. When considering years, mouse lemurs exposed to the 8-month annual cycle died earlier compared to the 12-month cycle; however, there was no difference between colonies when considering cycles. Correspondingly, phenotypic signs of aging, such as a whitening of fur, started at about 5 years in MMU-M, and about 7 years in MMU-H. Across both colonies, females showed about one year higher maximum lifespan than males.

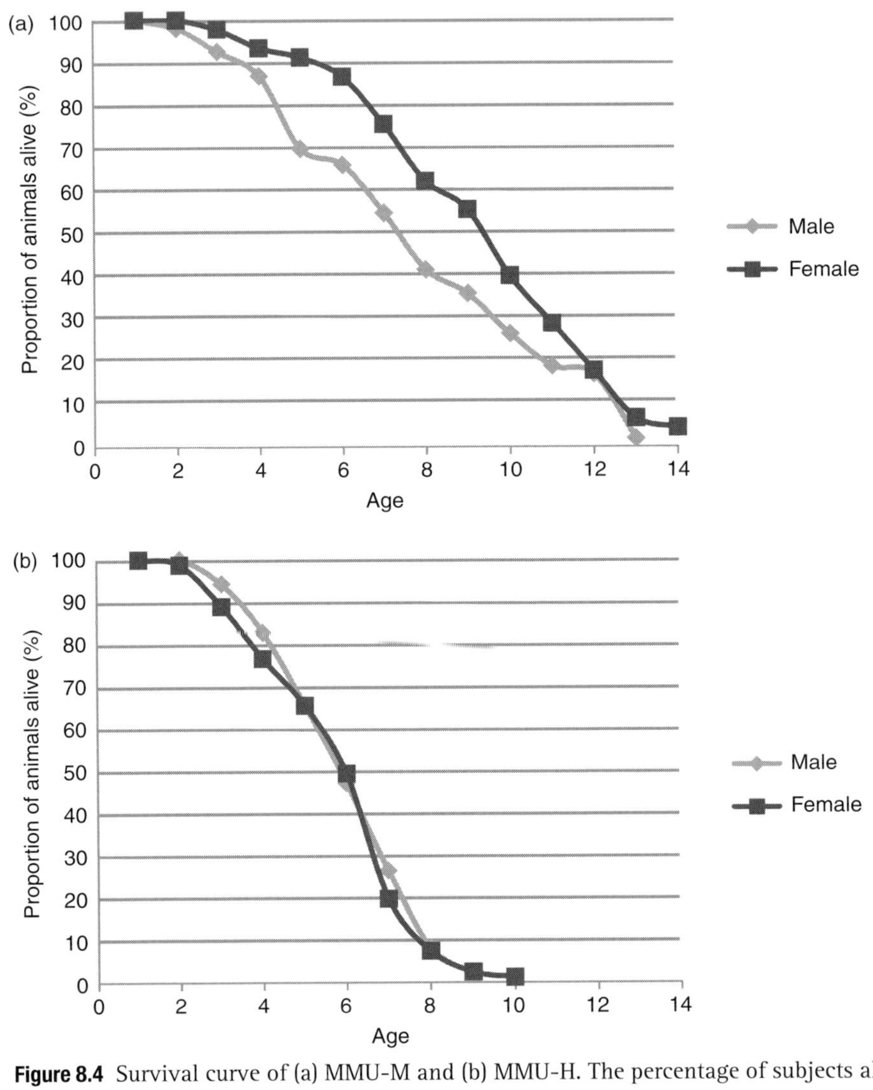

Figure 8.4 Survival curve of (a) MMU-M and (b) MMU-H. The percentage of subjects alive per age category is presented. Age is given in years.

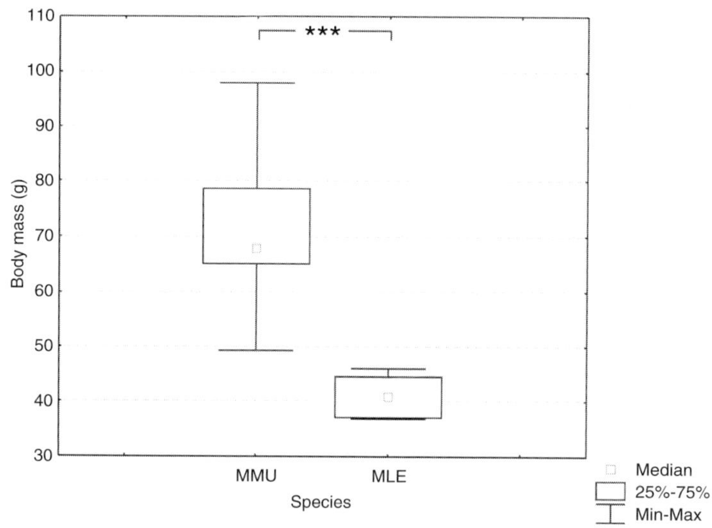

Figure 8.5 Average body mass of females at first reproduction differed significantly between *Microcebus murinus* (MMU = MMU-H) and *M. lehilahytsara* (MLE); MWU, $U = 6$, $Z = 5.087$, *** indicates a significant difference of $p < 0.0000001$.

Comparison of life history patterns between two species of mouse lemurs: effect of phylogeny

Body mass of females at first reproduction

The mean body mass of females at the age of first reproduction was 48.6 g (SD 10.25, min = 35.1, max = 69.8, $N = 18$) for MLE. It was thus significantly lower than in MMU-H (Figure 8.5 and intraspecific paragraph above).

Development of body mass in infants

For MLE, the mean body mass of infants aged 4–7 days was 9 g for males (SD 2.18, min = 5.3, max = 12.1, $N = 18$) and 8.4 g for females (SD 1.26, min = 6.3, max = 10.9, $N = 18$). For MMU-H infants of the same age, the mean body mass was 10.5 g for males (SD 2.25, min = 6, max = 14.9, $N = 20$) and 10.4 g for females (SD 2.17, min = 6.5, max = 14.3, $N = 20$. There was no difference between the sexes within either species (MWU test, n.s. in both cases). Whereas there was no species difference in neonatal male body mass (MWU test, n.s.), gray mouse lemur female neonates were significantly heavier than Goodman's mouse lemur female neonates (MWU test, $U = 73$, $Z = 3.129$, $p < 0.002$).

For MLE, the mean body mass at weaning (at about 35 days after birth) was 24.6 g for males (SD 5.9, min = 15.3, max = 33.4, $N = 18$) and 26.2 g for females (SD 5.39, min = 14.3, max = 36.8, $N = 18$). For MMU-H, it was 37.3 g for males (SD

4.9, min = 21.3, max = 41.8, $N = 20$) and 35.9 g for females (SD 5.4, min = 21.6, max = 45.6, $N = 20$). There was no sex difference within species (MWU test, n.s.). However, median body mass differed significantly between species, both for males and females (MWU test male: $U = 21$, $Z = 4.649$, $p < 0.00001$; female: $U = 34$, $Z = 4.269$, $p < 0.0001$).

Age at first reproduction

The mean age at first reproduction for MLE was 328 days (SD 50.97, min = 274, max = 396, $N = 7$). It was not significantly different from MMU-H at 349 days (see intraspecific paragraph above; MWU test, n.s.).

Gestation length at the age at first reproduction

The mean gestation length, measured as the gestation length of females at first reproduction, was 56 days for MLE (SD 2.27, min = 52, max = 59, $N = 7$) and 62 days for MMU-H (SD 2.99, min = 56, max = 68, $N = 14$). Gestation length showed a significant difference between species (Figure 8.6).

Litter size

The mean litter size of MLE was 2.1 infants (SD 0.378, min = 2, max = 3, $N = 7$). Litter size showed no significant difference between species (see intraspecific paragraph before for MMU-H; MWU test, n.s,)

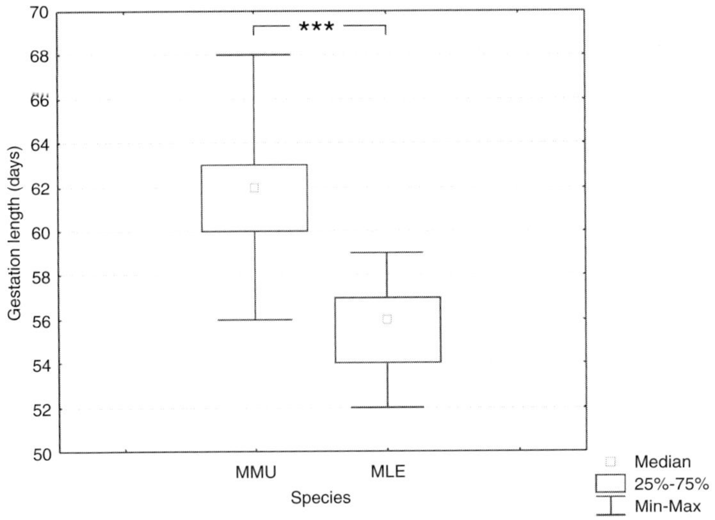

Figure 8.6 Average gestation length of females at first reproduction differs significantly between *Microcebus murinus* (MMU = MU-H) and *M. lehilahytsara* (MLE). MWU, $U = 5$, $Z = 3.6568$, *** indicates a significant difference of $p < 0.001$.

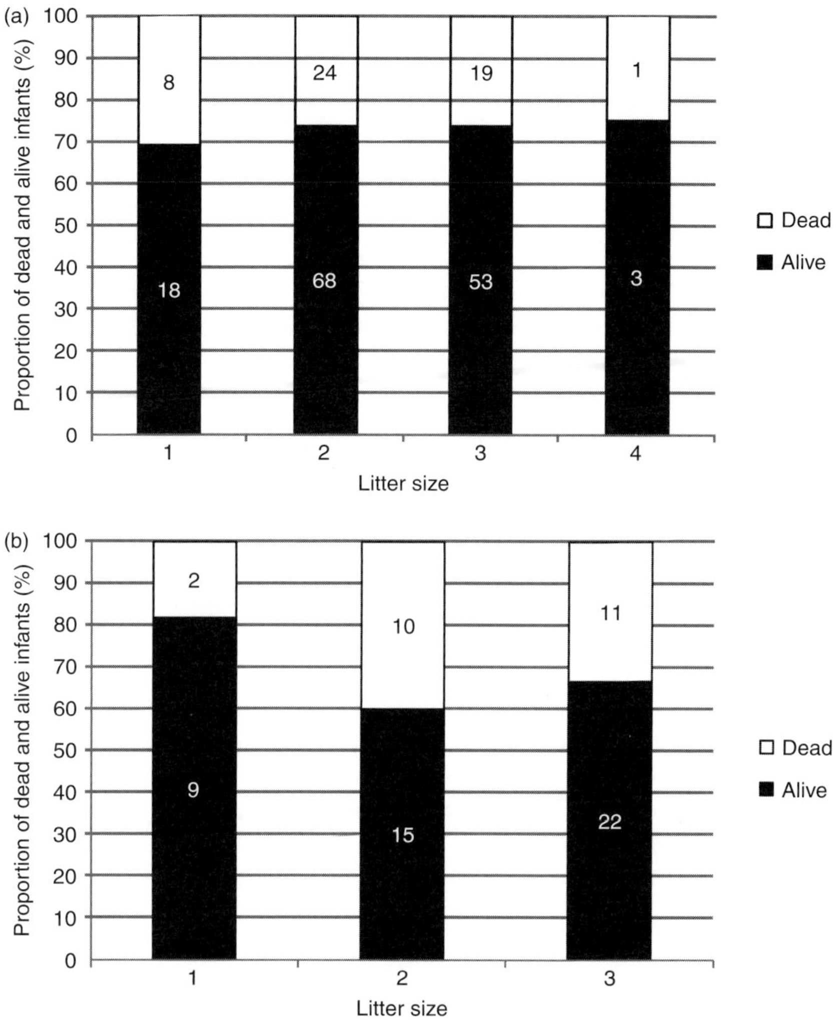

Figure 8.7 Infant survival in (a) MMU-H, and (b) MLE in litters of 1–4 infants. Numbers in bars = absolute number of infants in the respective category.

Infant survival in MMU-H was independent of litter size (Figure 8.7a) and even increased from 69% in litters of one infant to 75% in the one litter of quadruplets. In contrast, and as expected, in MLE infant survival was highest in singletons (82%; Figure 8.7b), and dropped to 60% and 67% in twin and triplet litters, respectively.

Longevity

The mean lifespan for MMU-H was 7.7 years (SD 3.2, min = 1 max = 14, $N = 96$; see above, intraspecific comparisons) and for MLE was 9.1 years (SD 4.1, min = 4, max = 15, $N = 13$). There was no significant difference between species (MWU test, n.s.). The presence of a sex difference in longevity for MMU-H was described above (see intraspecific comparisons). Within MLE, there was no significant difference between sexes in longevity (MWU test, *N*m = 9, *N*f = 4, n.s.), but this finding has to be interpreted very cautiously, because the sample sizes, in particular for females, are still quite low.

Discussion

We found that three of four measured life history traits, namely female body mass and age at first reproduction as well as longevity, varied between colonies of the same mouse lemur species exposed to different environments (photoperiod). Only one trait, litter size, remained stable. A comparison of life history traits of two different species exposed to the same environment (photoperiod) revealed both phylogenetically fixed and open life history traits for the genus *Microcebus*. Three of the six measured traits were phylogenetically invariant, namely age at first reproduction, litter size, and longevity, whereas three traits varied between species, namely female body mass at first reproduction, gestation length and body mass at weaning.

Intraspecific variation of life history patterns

The expression of multiple phenotypic states by a single genome under different environmental conditions is called phenotypic plasticity (Houston and McNamara, 1992) and characterizes the reaction norm of a given species. Phenotypic plasticity is ubiquitous and a widely documented phenomenon in naturally occurring populations (Gotthard and Nylin, 1995), where it is suggested to speed up, slow down, or to have little effect on evolutionary change (e.g. Price *et al.*, 2003). In gray mouse lemurs, our results on the intraspecific variation of life history traits revealed that litter size did not differ between colonies in 12- and 8-month annual cycles, respectively, suggesting that variation of this trait is constrained physiologically and/or phylogenetically, which seems to be typical for primates (Ross, 1998; Zimmermann and Radespiel, 2014). In contrast, female body mass and age at first reproduction, as well as longevity, were significantly affected by a shortened annual cycle and thus can be called phenotypically plastic life history traits. Interestingly, and not previously reported for primates, we show that under photoperiodically triggered shortened annual cycles, the gray mouse lemur is able to adjust its growth, development and reproductive schedules accordingly. Thus, it is able to speed up growth to reach the species-specific body condition for reproduction earlier and to reproduce successfully at a faster rate; however, at the cost of faster chronological aging and a shorter lifespan, suggesting that these three life history traits might be linked to aging. Maturational speed, reflected in the time necessary to reach adult female body mass and age at first reproduction, is also one of the most variable life history traits in humans (Kuzawa and Bragg, 2012), where it was found to be linked to nutritional state. Nutrition was not restricted for either MMU-M or for MMU-H and thus does not seem to help explain our findings in gray mouse lemurs. However, our findings support the hypothesis that in this seasonally active primate, maturation seems to depend strongly on seasonal variation of photoperiod.

Previous research already proposed that longevity in this species depends more on a fixed number of photoperiodically induced seasonal cycles than on a fixed biological age (Perret, 1997). Consequently, if the effect were linear, one would predict that longevity, and probably female age at first reproduction, should be shortened by 33% and mean body mass at first reproduction should have been achieved 33% more rapidly in MMU-M (artificially shortened annual cycle of 8 months) than in MMU-H (normal annual cycle of 12 months). Indeed, this prediction fits quite well for longevity being a bit more than the predicted 2.6 years shorter under a shortened annual cycle. However, no comparable relationship was revealed for female age at first reproduction, which is shortened by much more than the predicted 3.93 months. Likewise, no relationship was obvious for mean female body mass at first reproduction. These findings suggest that the latter two life history traits are linked to different physiological/biochemical processes than those involved in photoperiodically triggered aging. Various physiological/biochemical determinants such as the accumulation of deleterious mutations (e.g. Ackermann *et al.*, 2007), oxidative stress (e.g. Marchal *et al.*, 2013) or mitonuclear protein imbalance (Wolff and Gemmell, 2012; Houtkooper *et al.*, 2013) are thought to be involved as potential mechanisms affecting longevity in animals. To what extent, and how, they also play a role in aging in the gray mouse lemur requires further examination.

All in all, our study revealed that within one cheirogaleid species, the gray mouse lemur, life history traits linked to aging (growth and development indicated by body mass and age at first reproduction as well as longevity) are environmentally plastic, meaning that the respective reaction norms are broad and those traits have a strong potential for adaptive variation under changing environmental conditions. Under natural conditions, however, these patterns are heavily constrained by photoperiod. Because photoperiod varies predictively over the year at any location of the mouse lemur distribution in Madagascar, and is usually associated with predictable annual variations in climate and food abundance (Chapter 26), these environmental constraints selected mouse lemurs for optimizing fitness by adjusting reproduction, growth and development accordingly.

Interspecific variation of life history patterns

Central to mammalian life history theory is the exploration of how energy is allocated towards growth, maintenance, and reproduction throughout the life of individuals belonging to different species (Stearns, 1992; Bogin and Smith, 1996; Roff, 2002; Kappeler and Pereira, 2003; Catlett *et al.*, 2010). If variation across all individuals within a species is smaller than variation between species under the same environmental conditions, the respective pattern is suggested to be genetically fixed within a species, but variable between species and thus phylogenetically open. If variation between different species overlaps, the respective trait can be regarded as phylogenetically fixed.

In our study, the age of females at first reproduction and longevity of both sexes did not show significant differences between the two mouse lemur species. A similar pattern for age of females at first reproduction was reported from captive chimpanzees and bonobos (De Lathouwers and Van Elsacker, 2005). These findings suggest that these aging-related life history traits are phylogenetically fixed. Contrastingly, significant interspecific variation was revealed in patterns linked to the female's energy allocation to reproduction such as female body mass, gestation length at first reproduction, and infant body mass at weaning. One explanation is that gestation length and infant body mass at weaning are allometrically related to body mass of females at first reproduction (see, e.g., Ross, 1998; Dubman *et al.*, 2012; Zimmermann and Radespiel, 2014) and thus can be predicted by the significant difference in body mass between species. Thus, females in the smaller-bodied *M. lehilahytsara* exhibited a significantly lower female body mass at first reproduction coupled with a shorter gestation length and a lower infant body mass at weaning than females in the larger-bodied *M. murinus*. Indeed, body mass at first reproduction predicted gestation length significantly across species (Pearson rank correlation: $N = 21$, $r = 0.60789$, $p < 0.05$). These findings suggest that body-mass related traits are species-specific and thus phylogenetically open. These traits may have evolved as a consequence of selection for specific ecological conditions experienced in the ancestral environment (e.g. Price *et al.*, 2003), and facilitated ecological radiation into new niches for small-bodied primates (e.g. Ford, 1980; Montgomery and Mundy, 2013).

Phenotypic plasticity and aging

Mouse lemurs of MMU-M seem to age at a much more rapid pace than those of MMU-H. If we define mouse lemurs as aged when older than the average lifespan of the colony, then mouse lemurs in MMU-M should be considered as aged when 5 years or older and in MMU-H when 7.7 years and older. Interestingly, phenotypic signs of aging such as whitening of fur (see Figure 8.8) or the presence of eye pathologies such as cataracts (Dubicanac *et al.*, 2013) or hearing deficiencies (Schopf *et al.*, 2014) emerged at about 5 years of age in MMU-M, and at about 7 years of age in MMU-H, further supporting age differentiation. Compared to findings presented in the present study, longevity in another large mouse lemur colony (Brunoy), maintained under similar conditions, was found to be much shorter, about 1.5 years shorter for the 12-month annual cycle and about 1 year shorter for the 8-month cycle (Perret, 1997; Languille *et al.*, 2012). To what extent this variation can be explained by differences in husbandry conditions (e.g., nutrition, social settings) requires further examination. Another surprising result was the higher female than male maximum longevity found in both the MMU-M and MMU-H colony (but see Perret, 1997 and Languille *et al.*, 2012 for no sex-specific differences in maximum longevity in a third MMU colony). A similar sex-related difference in lifespan was reported from industrialized compared to indigenous human populations (e.g. Oeppen and Vaupel, 2002; Borras *et al.*, 2003; Hirokawa *et al.*, 2013). This sex difference in life expectancy of

Figure 8.8 Phenotypic changes during aging in the gray mouse lemur. Displayed is a gray mouse lemur infant of a few days (photo: M. Scheumann), a young adult (2 years) (photo: E. Zimmermann), and an old (10 years) gray mouse lemur (photo: E. Zimmermann). (A black and white version of this figure will appear in some formats. For the colour version, please refer to the plate section.)

human populations was explained by the slower immune system aging in women versus men (Hirokawa *et al.*, 2013). More research on this aspect is necessary to assess to what extent this explanation can be generalized for primates.

Maximum lifespan differs between captive and natural settings. In nature, mouse lemurs of three different species (*Microcebus murinus*/*M. ravelobensis*, *M. rufus*) living in dry deciduous or rain forests, respectively, were reported to have a maximum longevity of 8 years (own unpubl. data; Zohdy and Radespiel, 2012), whereas in captivity, mouse lemurs live much longer, i.e., up to 11 years in MMU-M exposed to an 8-month annual cycle, up to 15 years for both species in MMU-H exposed to a 12-month annual cycle, and up to 18.2 years if taking into account all published data for mouse lemurs (Weigl, 2005). These findings for a shorter life expectancy in nature compared to non-natural conditions coincide with those for rats and mice (Finch and Austard, 2012), and are further supported by survival curves. Survival in these species is high early in adult life and then declines rapidly as senescence progresses (Turturro *et al.*, 1999; Finch and Austard, 2012). We found similar survival curves in *M. murinus* of both colonies, supporting the notion that husbandry is working quite well for this species. In contrast, other non-human primate survival curves from laboratory animals are typically more linear, or even concave, suggesting that husbandry is still at an early stage of development (Finch and Austard, 2012).

Mouse lemurs age more rapidly than any other non-human primate (Finch and Austard, 2012). Furthermore, the aging process can be manipulated and accelerated predictably by annual variations of photoperiod. This makes mouse lemurs a unique primate model for which fascinating new insights into the processes and mechanisms of primate aging and the potential and causation of phenotypic plasticity in primate life history patterns can be expected in the future. In the best of all possible worlds, it would be nice to see a multivariate study that addresses the impacts of environment and genetics on a large number of populations, using information regarding the genetic distance between populations and species and experimentally manipulating populations that are distantly and closely related (Laurie Godfrey, pers. comm.). Our research was intended to provide the first crucial steps for such a comprehensive framework.

Acknowledgments

This study is part of a project which has received funding from the European Community's 7th Framework Programme (FP7/2007–2013) under grant agreement no. 278486 acronym "DEVELAGE". We thank Elisabeth Engelke and Birgit Haßfurther for helping to collect part of the data. Further, we thank the animal keepers, Sylvie Rouland, Elisabeth Huetter, Virginie Sansiquet, and Joel Cuoq, for animal maintenance in Montpellier, and Lisa Früh, Iris Grages, and Johanna Samtlebe for animal maintenance in Hannover, as well as Elisabeth Engelke and Sönke von den Berg for technical support in Hannover.

References

Ackermann M, Schauerte A, Stearns SC, Jenal U. 2007. Experimental evolution of aging in a bacterium. *BMC Evolutionary Biology* 7:126.

Alberts SC, Altmann J. 2012. The Amboseli Baboon Research Project: 40 years of continuity and change. In P Kappeler, DP Watts (eds.), *Long-term Field Studies of Primates* (pp. 261–288). Springer, Berlin.

Altmann J, Alberts SC. 2003. Variability in reproductive success viewed from a life-history perspective in baboons. *American Journal of Human Biology* 15:401–409.

Altmann J, Gesquière L, Galbany J, Onyango PO, Alberts SC. 2010. Life history context of reproductive aging in a wild primate model. *Annals of the New York Academy of Sciences* 1204:127–138.

Bogin B, Smith BH. 1996. Evolution of the human life cycle. *American Journal of Human Biology* 8:703–716.

Bons N, Rieger F, Prudhomme D, Fisher A, Krause K-H. 2006. *Microcebus murinus*: a useful primate model for human cerebral aging and Alzheimer's disease? *Genes, Brain and Behavior* 5:120–130.

Borras C, Sastre J, Garcia-Sala D, *et al.* 2003. Mitochondria from females exhibit higher antioxidant gene expression and lower oxidative damage than males. *Free Radical Biology and Medicine* 34:546–552.

Braune P, Schmidt S, Zimmermann E. 2005. Spacing and group coordination in a nocturnal primate, the golden brown mouse lemur (*Microcebus ravelobensis*): the role of olfactory and acoustic signals. *Behavioral Ecology and Sociobiology* 58:587–596.

Braune P, Schmidt S, Zimmermann E. 2008. Acoustic divergence in the communication of cryptic species of nocturnal primates (*Microcebus* ssp.). *BMC Biology* 6:19.

Catlett KK, Schwartz GT, Godfrey LR, Jungers WL. 2010. "Life History Space": a multivariate analysis of life history variation in extant and extinct Malagasy lemurs. *American Journal of Physical Anthropology* 142:391–404.

Colman RJ, Anderson RM, Johnson SC, *et al.* 2009. Caloric restriction delays disease onset and mortality in rhesus monkeys. *Science* 325:201–204.

Dal-Pan A, Terrien J, Pifferi F, *et al.* 2011. Caloric restriction or resveratrol supplementation and ageing in a non-human primate: first-year outcome of the RESTRIKAL study in *Microcebus murinus*. *Age* 33:15–31.

De Lathouwers M, Van Elsacker L. 2005. Reproductive parameters of female *Pan paniscus* and *P. troglodytes*: quality versus quantity. *International Journal of Primatology* 26:55–71.

Dubicanac M, Joly M, Karck J, Nolte I, Verdier J-M, Zimmermann, E. 2013. Ocular pathologies in aging mouse lemur (*Microcebus murinus*) populations: cataract formation depends on the annual photoperiodic cycle (abstract). 13th Conference of the Gesellschaft für Primatologie, Hamburg, 6.2.-8.2.2013, 52.

Dubman E, Collard M, Mooers AO. 2012. Evidence that gestation duration and lactation duration are coupled traits in primates. *Biology Letters* 8:998–1001.

Eberle M, Kappeler PM. 2006. Family insurance: kin selection and cooperative breeding in a solitary primate (*Microcebus murinus*). *Behavioral Ecology and Sociobiology* 60:582–588.

Finch CE, Austad SN. 2012. Primate aging in the mammalian scheme: the puzzle of extreme variation in brain aging. *Age* 34:1075–1091.

Fischer KE, Austad SN. 2011. The development of small primate models for aging research. *Institute for Laboratory Animal Research Journal* 52:78–88.

Ford SM. 1980. Callitrichids as phyletic dwarfs, and the place of the Callitrichidae in Platyrrhini. *Primates* 21:21–43.

Gesquière LR, Learn NH, Simao MCM, *et al.* 2011. Life at the top: rank and stress in wild male baboons. *Science* 333:357–360.

Gotthard K, Nylin S. 1995. Adaptive plasticity and plasticity as an adaptation – a selective review of plasticity in animal morphology and life-history. *Oikos* 74:3–17.

Hafen TG. 1998. Dialekte bei Lemuren: bioakustische, morphometrische und molekulargenetische Untersuchungen zur intraspezifischen Variabilität beim grauen Mausmaki (*Microcebus murinus*). Dissertation, Universität Konstanz, Konstanz, 153 pp.

Harvey PH, Clutton-Brock TH. 1985. Life history variation in primates. *Evolution* 39:559–581.

Hirokawa K, Utsuyama M, Hayashi Y, *et al.* 2013. Slower immune system aging in women versus men in the Japanese population. *Immunity & Ageing* 10:19.

Hohenbrink P, Radespiel U, Mundy NI. 2012. Pervasive and ongoing positive selection in the vomeronasal-1 receptor (V1R) repertoire of mouse lemurs. *Molecular Biology and Evolution* 29:3807–3816.

Hohenbrink P, Mundy NI, Zimmermann E, Radespiel U. 2013. First evidence for functional vomeronasal 2 receptor genes in primates. *Biology Letters* 9:20121006.

Houston AI, McNamara JM. 1992. Phenotypic plasticity as a state-dependent life-history decision. *Evolutionary Ecology* 6:243–253.

Houtkooper RH, Mouchiroud L, Ryu D, *et al.* 2013. Mitonuclear protein imbalance as a conserved longevity mechanism. *Nature* 497:451–457.

Joly M, Zimmermann E. 2007. First evidence for relocation of stationary food resources during foraging in a strepsirrhine primate (*Microcebus murinus*). *American Journal of Primatology* 69:1045–1052.

Joly M, Zimmermann E. 2011. Do solitary foraging nocturnal mammals plan their routes? *Biology Letters* 7:638–640.

Joly M, Scheumann M, Zimmermann E. 2008. Wild mouse lemurs revisit artificial feeding platforms: implications for field experiments on sensory and cognitive abilities in small primates. *American Journal of Primatology* 70:892–896.

Joly-Radko M, Zimmermann E. 2010. Seasonality in gum and honeydew feeding in gray mouse lemurs. In AM Burrows, LT Nash (eds.), *The Evolution of Exudativory in Primates* (pp. 141–153). Springer, New York.

Jones JH. 2011. Primates and the evolution of long, slow life histories. *Current Biology* 21:R708–R717.

Kamilar JM, Cooper N. 2013. Phylogenetic signal in primate behaviour, ecology and life history. *Philosophical Transactions of the Royal Society of London Series B Biological Sciences* 368:20120341.

Kappel P, Hohenbrink S, Radespiel U. 2011. Experimental evidence for olfactory predator recognition in wild mouse lemurs. *American Journal of Primatology* 73:928–938.

Kappeler PM, Pereira ME. 2003. *Primate Life Histories and Socioecology*. University of Chicago Press, Chicago.

Kessler SE, Scheumann M, Nash LT, Zimmermann E. 2012. Paternal kin recognition in the high frequency/ultrasonic range in a solitary foraging mammal. *BMC Ecology* 12:26.

Kuzawa CW, Bragg JM. 2012. Plasticity in human life history strategy: implications for contemporary human variation and the evolution of genus *Homo*. *Current Anthropology* 53:S369–S382.

Lacreuse A, Herndon JG. 2009. Nonhuman primate models of cognitive aging. In JL Bizon, A Woods (eds.), *Animal Models of Human Cognitive Aging* (pp. 29–58). Springer, New York.

Languille S, Blanc S, Blin O, *et al.* 2012. The grey mouse lemur: a non-human primate model for ageing studies. *Ageing Research Reviews* 11:150–162.

Lee PC, Kappeler PM. 2003. Socioecological correlates of phenotypic plasticity of primate life histories. In PM Kappeler, ME Pereira (eds.), *Primate Life Histories and Socioecology* (pp. 41–65). University of Chicago Press, Chicago.

Lührs M-L, Dammhahn M, Kappeler PM, Fichtel C. 2009. Spatial memory in the grey mouse lemur (*Microcebus murinus*). *Animal Cognition* 12:599–609.

Marchal J, Dal-Pan A, Epelbaum J, *et al.* 2013. Calorie restriction and resveratrol supplementation prevent age-related DNA and RNA oxidative damage in a non-human primate. *Experimental Gerontology* 48:992–1000.

Martin RD. 1972. A laboratory breeding colony of the lesser mouse lemur. In WIB Beveridge (ed.), *Breeding Primates* (pp. 161–171). Karger, Basel.

Mattison JA, Roth GS, Beasley TM, *et al.* 2012. Impact of caloric restriction on health and survival in rhesus monkeys from the NIA study. *Nature* 489:318–321.

Montgomery SH, Mundy NI. 2013. Parallel episodes of phyletic dwarfism in callitrichid and cheirogaleid primates. *Journal of Evolutionary Biology* 26:810–819.

Oeppen J, Vaupel JW. 2002. Broken limits to life expectancy. *Science* 296:1029–1031.

Perret M. 1997. Change in photoperiodic cycle affects life span in a prosimian primate (*Microcebus murinus*). *Journal of Biological Rhythms* 12:136–145.

Price TD, Qvarnström A, Irwin DE. 2003. The role of phenotypic plasticity in driving genetic evolution. *Proceedings of the Royal Society B Biological Sciences* 270:1433–1440.

Radespiel U. 2006. Ecological diversity and seasonal adaptations of mouse lemurs (*Microcebus* spp.). In L Gould, ML Sauther (eds.), *Lemurs: Ecology and Adaptation* (pp. 211–233). Springer, New York.

Radespiel U, Reimann W, Rahelinirina M, Zimmermann E. 2006. Feeding ecology of sympatric mouse lemur species in northwestern Madagascar. *International Journal of Primatology* 27:311–321.

Radespiel U, Ratsimbazafy JH, Rasoloharijaona S, *et al.* 2012. First indications of a highland specialist among mouse lemurs (*Microcebus* spp.) and evidence for a new mouse lemur species from eastern Madagascar. *Primates* 53:157–170.

Randrianambinina B, Rakotondravony D, Radespiel U, Zimmermann E. 2003. Seasonal changes in general activity, body mass and reproduction of two small nocturnal primates: a comparison of the golden brown mouse lemur (*Microcebus ravelobensis*) in northwestern Madagascar and the brown mouse lemur (*Microcebus rufus*) in eastern Madagascar. *Primates* 44:321–331.

Rasoloarison RM, Weisrock DW, Yoder AD, Rakotondravony D, Kappeler PM. 2013. Two new species of mouse lemurs (Cheirogaleidae: Microcebus) from eastern Madagascar. *International Journal of Primatology* 34:455–469.

Roff DA. 2002. *Life History Evolution*. Sinauer Associates, Sunderland, MA.

Ross C. 1998. Primate life histories. *Evolutionary Anthropology* 6:54–63.

Scheumann M, Rabesandratana A, Zimmermann E. 2007. Predation, communication, and cognition in lemurs. In SL Gursky, KAI Nekaris (eds.), *Primate Anti-predator Strategies* (pp. 100–126). Springer, New York.

Schmelting B, Ehresmann P, Lutermann H, Randrianambinina B, Zimmermann E. 2000. Reproduction of two sympatric mouse lemur species (*Microcebus murinus* and *M. ravelobensis*) in north-west Madagascar: first results of a long term study. In WR Lourenço, SM Goodman (eds.), *Diversité et endémisme à Madagascar* (pp. 165–175). Société de Biogéographie, Paris.

Schopf C, Zimmermann E, Tünsmeyer J, *et al.* 2014. Hearing and age-related changes in the gray mouse lemur. *Journal of the Association for Research in Otolaryngology*; DOI 10.1007/s10162-014-0478-4.

Silk JB, Beehner JC, Bergman TJ, *et al.* 2010. Strong and consistent social bonds enhance the longevity of female baboons. *Current Biology* 20:1359–1361.

Stearns SC. 1992. *The Evolution of Life Histories*. Oxford University Press, Oxford.

Sündermann D, Scheumann M, Zimmermann E. 2008. Olfactory predator recognition in predator-naïve gray mouse lemurs (*Microcebus murinus*). *Journal of Comparative Psychology* 122:146–155.

Thoren S, Quietzsch F, Schwochow D, *et al.* 2011. Seasonal changes in feeding ecology and activity patterns of two sympatric mouse lemur species, the gray mouse lemur (*Microcebus murinus*) and the golden-brown mouse lemur (*M. ravelobensis*), in northwestern Madagascar. *International Journal of Primatology* 32:566–586.

Turturro A, Witt WW, Lewis S, *et al.* 1999. Growth curves and survival characteristics of the animals used in the Biomarkers of Aging Program. *Journals of Gerontology. Series A, Biological Sciences and Medical Sciences* 54:B492–B501.

Weigl R. 2005. *Longevity of Mammals in Captivity; From the Living Collections of the World.* Schweizerbart Science Publishers, Stuttgart.

Wolff JN, Gemmell NJ. 2012. Mitochondria, maternal inheritance, and asymmetric fitness: why males die younger. *Bioessays* 35:93–99.

Wrogemann D, Zimmermann E. 2001. Aspects of reproduction in the eastern rufous mouse lemur (*Microcebus rufus*) and their implications for captive management. *Zoo Biology* 20:157–167.

Wrogemann D, Radespiel U, Zimmermann E. 2001. Comparison of reproductive characteristics and changes in body weight between captive populations of rufous and gray mouse lemurs. *International Journal of Primatology* 22:91–108.

Yoder AD, Yang Z. 2004. Divergence dates for Malagasy lemurs estimated from multiple gene loci: geological and evolutionary context. *Molecular Ecology* 13:757–773.

Zimmermann E. 2010 Vocal expression of emotion in a nocturnal prosimian primate group, mouse lemurs. In SM Brudzynski (ed.), *Handbook of Mammalian Vocalization: An Integrative Neuroscience Approach* (pp. 215–225). Academic Press, Oxford.

Zimmermann E. 2013. Primate serenades: call variation, species diversity, and adaptation in nocturnal strepsirrhines. In J Masters, M Gamba, F Génin (eds.), *Leaping Ahead: Advances in Prosimian Biology* (pp. 287–295). Springer, New York.

Zimmermann E, Radespiel U. 2007. Primate life histories. In W Henke, I Tattersall (eds.), *Handbook of Paleoanthropology Vol 2: Primate Evolution and Human Origins* (pp. 1163–1205). Springer, Berlin.

Zimmermann E, Radespiel U. 2014. Primate life histories. In W Henke, I Tattersall (eds.), *Handbook of Paleoanthropology Vol 2: Primate Evolution and Human Origins*, 2nd edition. Springer, Berlin.

Zimmermann E, Vorobieva E, Wrogemann D, Hafen T. 2000. Use of vocal fingerprinting for specific discrimination of grey (*Microcebus murinus*) and rufous mouse lemurs (*Microcebus rufus*). *International Journal of Primatology* 21:837–852.

Zohdy S. 2012. Senescence ecology: Aging in a population of wild brown mouse lemurs (*Microcebus rufus*). Dissertation, Helsinki University Print, Helsinki.

Zohdy S, Radespiel U. 2012. Lifespan and survival in eastern and western mouse lemurs. 24th Congress of the International Primatological Society, 12–17 August 2012, Cancun, Mexico.

9 The effects of the lunar cycle, temperature, and rainfall on the trapping success of wild brown mouse lemurs (*Microcebus rufus*) in Ranomafana National Park, southeastern Madagascar

Anja M. Deppe, Andrea Baden, and Patricia C. Wright

Introduction

Micromammals are notoriously difficult to observe because of their small size, and their cryptic, and often nocturnal lifestyle (Halle and Stenseth, 2001). Therefore, live trapping methods are commonly employed to determine distribution, estimate population densities, and infer animal activity (Gentry *et al.*, 1965; Vickery and Rivest, 1991; Atsalis, 2000; Caro *et al.*, 2001; Radespiel *et al.*, 2001; Schwab and Ganzhorn, 2004; Halle 2006; Prugh and Brashares, 2010). Because trapping is an indirect method of studying animal behavior, results should be interpreted cautiously (Prugh and Brashares, 2010), and should take into account a number of biotic and abiotic factors (Prugh and Golden, 2014). Factors that can affect activity and therefore capture rates of small mammals include predator behavior (Hilton *et al.*, 1999; Michalski and Norris, 2011; Cozzi *et al.*, 2012; Falcy and Danielson, 2013), and the temporal and spatial availability of food and shelter (Kotler *et al.*, 1993; Caro *et al.*, 2001; Davidson and Morris, 2001; Kelt *et al.*, 2004). In addition, possible affects of season, temperature, rainfall, and the lunar cycle must be considered (Kotler *et al.*, 1991; Daly *et al.*, 1992; Sutherland and Predavec, 1999; Stokes *et al.*, 2001; Michalski and Norris, 2011; Maestro and Marinho,2014; Prugh and Golden, 2014).

Small mammals may reduce activity during extreme temperatures (either hot or cold) due to thermoregulatory demands (Gentry and Odum, 1957; Vickery and Bider, 1981; Falcy and Danielson, 2013). Cold is especially dangerous to small mammals because of their susceptibility to hypothermia. This might explain why old-field mice (*Peromyscus polionotus*) were less frequently captured during cold than warm

The Dwarf and Mouse Lemurs of Madagascar: Biology, Behavior and Conservation Biogeography of the Cheirogaleidae, eds. S.M. Lehman, U. Radespiel, E. Zimmermann. Published by Cambridge University Press. © Cambridge University Press 2016.

nights (Gentry and Odum, 1957), and why common voles (*Microtus arvalis*) reduced activity with decreasing temperature (Lehman and Sommersberg, 1980). Some small mammals (Lyman, 1983), including small primates such as mouse lemurs (*Microcebus*) and dwarf lemurs (*Cheirogaleus*), may become completely inactive for periods of time, undergoing hibernation (dwarf lemurs) or periods of torpor (Genin et al., 2005; Blanco and Godfrey, 2014). Mouse lemurs may be in an inactive, torpid state for days, weeks, or even months during the austral winter (Ortmann *et al.*, 1997; Schmid, 2000).

Rain may increase or decrease animal activity. Insectivorous mice and shrews, for example, became more active during rainy nights, possibly because the rain increased insect activity or entered their below-ground shelters (Gentry *et al.*, 1965; Jahoda, 1970; Doucet and Bider, 1974; Vickery and Bider, 1981; Vickery and Rivest, 1991; Stokes *et al.*, 2001; Maestri and Marinho, 2014). Rain can have a negative effect on the activity of small mammals. For example, wet fur exacerbates heat loss, which might explain why common voles ceased activity when rain coincided with low temperatures (Lehman and Sommersberg, 1980). Rain also increases background noise and reduces visibility, making the detection of predators and prey more difficult (Vickery and Rivest, 1991; Hilton *et al.*, 1999; Maestri and Marinho, 2014). Rain further makes it more difficult to observe animals (personal observation), and can thereby affect the human observer's perception of activity.

Lunar illumination is also known to influence animal activity (Gursky, 2003). Moon phases and associated variations in lunar brightness can affect the hunting success of visually oriented predators such as owls, carnivores, and snakes (Dice, 1945; Vermeij, 1982; Clarke, 1983; Bearder *et al.*, 2002; Cozzi *et al.*, 2012; Lillywhite and Brischoux, 2012). A reduction in travel and foraging during moonlit nights has been observed in rodents (Kotler *et al.*, 1991; Bowers *et al.*, 1993; Hughes *et al.*, 1994; Plesner *et al.*, 1995), and particularly in species that live in habitats with no or little cover (Prugh and Golden, 2014) such as deserts (Daly *et al.*, 1992; Price *et al.*, 1994; Upham and Hafner, 2013), meadows (Bouliska, 1995), or secondary forests (Michalski and Norris, 2011). This might be explained by a higher (perceived) predation risk during lunar illumination. Rodents in desert and meadow habitats may be particularly vulnerable because visual predators can detect them more easily, and aerial predators further benefit from an unobstructed attack path. The avoidance of moonlight (lunar phobia) has further been observed in primates (Erkert, 1989; Bearder *et al.*, 2006; Starr *et al.*, 2012; Rode-Margono and Nekaris, 2014), and bats (Fenton *et al.*, 1977; Morrison, 1978; Lang *et al.*, 2006). In bats, moonlight avoidance may, however, be an adaptation to the lunar phobic behavior of their insect prey rather than an antipredation behavior (Erkert, 1974).

An increase in activity during bright nights (lunar philia) has been recorded in a range of animals including rodents (Sutherland and Predavec, 1999; Kotler, 1984), wallabies (Blumstein *et al.*, 2000), primates (Bearder *et al.*, 2002; Gursky, 2003; Eppeley *et al.*, 2015), and carnivores (Cozzi *et al.*, 2012). Lunar philia has been linked to enhanced prey or predator detection in visually oriented species (Bearder *et al.*, 2002; Gursky, 2003; Prugh and Golden, 2014), and to the reduction of feeding competition with dominant, lunar phobic, species (Upham and Hafner, 2013). Lunar philic behavior appears to be particularly common in nocturnal primates including owl monkeys (*Aotus*), galagos (*Galago*), and tarsiers (*Tarsius*), as well as cathemeral lemurs (Erkert, 1974, 1989; Charles-Dominique, 1977; Erkert and Gröber, 1986; Wright, 1989; Nash, 1986, 2007; Donati *et al.*, 2001; Bearder *et al.*, 2002; Gursky, 2003; Fernandez-Duque and Erkert, 2006; Eppeley *et al.*, 2015). Because primates are visually oriented (Kirk and Kay, 2004; Tetreault *et al.*, 2004), they probably benefit from enhanced predator detection during bright nights (Charles-Dominique, 1977; Gursky, 2003; Bearder *et al.*, 2006). Insectivorous or carnivorous species like tarsiers, lorises (*Nycticebus*), and mouse lemurs may further benefit from enhanced visual prey detection (Gursky, 2003; Starr *et al.*, 2012). However, there is also evidence that some nocturnal primates, including lorises and mouse lemurs, may be lunar neutral or lunar phobic, which could be an effect of food availability, vulnerability to predation, or temperatures during the study period (Erkert, 1989; Bearder *et al.*, 2006; Starr *et al.*, 2012; Rode-Margono and Nekaris, 2014).

The purpose of this study was to test the hypothesis that abiotic factors, specifically temperature, rainfall, and the lunar cycle, affect activity patterns and capture rates of the brown mouse lemur (*Microcebus rufus*) in the Talatakely trail system area in Ranomafana National Park (RNP). The brown mouse lemur is a small-bodied (45 g), nocturnal primate found in the southeastern rainforests of Madagascar (Atsalis, 2000), which forages mostly solitarily for fruit, nectar, and insects (Atsalis, 1999a). Mouse lemurs are endemic to Madagascar and inhabit a variety of habitats including rainforest, dry forest, spiny forest, plantations, and human dwellings (Kappeler and Rasoloarison, 2003). Due to their small body size, mouse lemurs across all habitat types are subject to high predation rates (Goodman *et al.*, 1993; Goodman, 2003a). Predators include diurnal and nocturnal raptors (Goodman *et al.*, 1993; Karpanty, 2006; Karpanty and Wright, 2007), viverrids (Hawkins, 2003; Dollar *et al.*, 2007), and snakes (Cadle, 2003; Raxworthy, 2003).

Little is known about predation rates for the brown mouse lemur at our study site. The diurnal raptors found in RNP include the Henst's goshawk (*Accipiter henstii*), the Madagascar buzzard (*Buteo brachypterus*), and the Madagascar harrier hawk (*Polyboroides radiatus*), and predation rates are low at an estimated 4.5% annually (Karpanty, 2006). Carnivorous mammals in RNP that are known to eat mouse lemurs are the diurnal ring-tailed mongoose (*Galidia elegans*; Goodman, 2003b; Deppe *et al.*, 2008), and the cathemeral fosa (*Cryptoprocta ferox*; Dollar *et al.*, 2007). There is no evidence that the nocturnal broad-striped mongoose (*Galidictis fasciata*; Goodman, 2003c) or the nocturnal Malagasy civet (*Fossa fossana*; Kerridge *et al.*, 2003) prey on mouse lemurs, but little is known about these largely terrestrial predators. Owls are one of the most effective predators (Vermeij, 1982), and while they are highly visual, they are able to locate prey in complete darkness by acoustic cues alone (Payne, 1971; Konishi, 1973). In the dry forest, owls appeared to be by far the most dangerous predator to mouse lemurs and were estimated to take nearly a third of the gray mouse lemur population annually (Goodman *et al.*, 1993). Owl species present in RNP are the small Madagascar scops owl (*Otus rutilus*), the medium-sized Madagascar red owl (*Tyto soumagnei*), and the large Madagascar long-eared owl (*Asio madagascariensis*; Hawkins and Goodman, 2003). We do not know owl predation rates at our site, but long-term trap–recapture data suggest that overall, predation rates for the brown mouse lemur are well below those found in gray mouse lemurs (unpublished data). Snakes include various smaller colubrid species, but the large nocturnal Madagascar tree boa (*Sanzinia madagascariensis*) is probably the most dangerous to brown mouse lemurs (Cadle, 2003; Raxworthy, 2003).

Like many mouse lemur species (Ortmann *et al.*, 1997; Schmid, 2001), the brown mouse lemur appears to undergo periods of inactivity (torpor) during the often cold and lean austral winter (personal observation), where temperatures may reach lows of 4°C (Wright and Andriamihaja, 2002). Anecdotal evidence further suggested that brown mouse lemurs decrease activity with increasing rainfall. Based on our observations and what has been found for other primate species, we predicted that brown mouse lemurs are more active, and more likely to be trapped, during warm nights than cold ones, dry nights than rainy nights, and during bright nights than dark nights. We applied an indirect measure, live-trap rates, assuming that a higher trap rate reflects higher general activity.

Methods

We present data gathered over a 4-year period (2004–2007) during Madagascar's austral spring (September–December), resulting in a total of 228 trap nights during which 6718 baited traps yielded 1795 captures. We focus only on data collected during the austral spring because brown mouse lemurs were generally absent from traps during the austral winter (June–August) and fall (March–May), possibly because of inactivity due to low temperatures and high food availability, respectively (Harcourt, 1987; Atsalis, 1999a). During January and February, frequent storms and cyclones prevented research.

Study site

Fieldwork was conducted in Talatakely, a well-established trail system of ~ 9 ha in Ranomafana National Park. The park, established in 1991, comprises 43,500 ha of continuous rainforest located in southeastern Madagascar at 21°16′S latitude and 47°20′E longitude (Wright, 1992; Wright and Andriamihaja, 2002). Elevations range from 500 to 1500 m, and annual rainfall is 4000 mm (RNP/Centre ValBio, unpublished data), most of which falls during the austral summer months from December to March. Based on RNP/Centre ValBio records, temperatures range from lows in June–September (4–12°C) to highs in December–February (30–32°C). Our data were collected in the spring when temperatures ranged from 10 to 20°C at night and 15 to 25°C during the day.

Data collection

Capture–mark–recapture techniques were employed as part of a long-term collaborative project to monitor the behavioral ecology of brown mouse lemur populations within the Talatakely trail system (Blanco and Meyer, 2009). Animals were captured nightly using Sherman live traps (30 × 8 × 10 cm; H.B. Sherman Traps, Inc., Tallahassee, FL, USA). Following standard protocol (Harcourt, 1987; Atsalis, 1999a; Blanco, 2008), traps were set at 25–50 m intervals in pairs along opposing pre-existing trails. Each trap was situated at least 5 m off-trail and fixed to a sturdy, horizontal substrate at ~1.5 m above the ground. Traps were baited with a piece of fresh banana and opened nightly at 1700 hours during 4–5 consecutive nights per week. Traps were checked and re-collected at 2100 hours, because our experience has shown that brown mouse lemurs are most likely to enter the traps within the first 2 h of nightfall. All captured lemurs were brought to the Centre ValBio research station where, with the exception of a few days where the chip scanner was not available, individuals were sexed and identified, or if unknown, marked with AVID microchips for later re-identification (Avid Identification Systems, Inc., Norco, CA, USA). Animals were then released between 2300 and 0100 hours the same night of capture to minimize disturbance. Individuals were released at the site of their original capture. Because the same lemurs were commonly recaptured every night, approximately 30 traps were rotated among 60 fixed locations along the trail to minimize the dependence on provisioning. All methods and procedures complied with protocols approved by the Stony Brook University Institutional Animal Care Committee and adhered to the legal requirements of Madagascar.

Data analysis

Nightly capture success, or trap rate, was calculated by dividing the number of captured individuals by the number of total traps opened that night. Data were found to be non-normally distributed using a Kolmogorov–Smirnov goodness-of-fit test and, for this reason, non-parametric statistics were used. We used a non-parametric Mann–Whitney *U* test to test for differences in trap rates between males and females.

Moon phases were determined using a lunar calendar and were divided into four main categories: full moon (defined as the night of full moon, plus two days prior to and following full moon, i.e., 5 days), new moon (defined as the night of new moon, plus two days prior to and following, i.e., 5 days), waning (the 10 days between full and new moon), and waxing (the 10 days between new and full moon). To test whether moon phase was related to trap rate, we used a non-parametric Kruskal–Wallis test followed by pairwise Mann–Whitney *U* comparisons with Bonferroni corrections.

Rainfall and temperature data were extracted from the CVB database. We tested for relationships between trap rate and both minimum nightly temperature and total nightly rainfall using Spearman's rho correlations. We excluded maximum

Table 9.1 Details of annual and total trapping effort during the study period 2004–2007 for the Talatakely trail system in Ranomafana National Park. The trap rate (in %) states the likelihood of a brown mouse lemur entering a trap, and is calculated by dividing the total number of individuals captured by the total number of traps set during that period.

Trap period	Trap nights	Traps	Captures	Males	Females	Trap rate (%)
Sep–Dec 2004	55	2051	380	269	111	18.5
Oct–Dec 2005	52	1393	364	194	170	26.1
Sep–Nov 2006	49	1246	416	278	138	33.4
Sep–Dec 2007	72	2028	635	471	164	31.3
Total	228	6718	1795	1212	583	26.7

Table 9.2 Total number of brown mouse lemur captures by month, sex, and male to female capture ratio. The trap rate is the capture likelihood (total number of individuals caught divided by the total number of traps set) in percent for both sexes combined.

Month	All	Female	Male	Male:female	Trap rate (%)
September	384	16	368	23.0	19
October	895	315	580	1.8	31
November	452	215	237	1.1	29
December	64	37	27	0.73	21

temperatures measured for each day because low temperatures were most relevant to night-active brown mouse lemurs and because there was a strong correlation between daily maximum and minimum temperatures. Statistical analysis was conducted using SPSS 15.0. The alpha level was set *a priori* at 0.05 unless otherwise noted.

Results

We captured a total of 1795 brown mouse lemurs (68% males, 32% females) during a total of 228 trap nights (Table 9.1). Capture likelihood was highest in October and November (Table 9.2). Males were captured significantly more often than females (Mann–Whitney U = 13,930, $p < 0.001$; Table 9.2). Overall trap rates differed significantly across lunar phases (H = 11.16, p = 0.011; Figure 9.1). Animals were significantly more likely to be captured during full moon nights than during any other phase (Full × Waxing, p = 0.023; Full × New, p = 0.026; Full × Waning, p = 0.030). Rainfall and minimum temperature varied across months (Table 9.3). We found a significant negative relationship between trap rate and both rainfall (Spearman's rho = –0.187, $p < 0.001$) and minimum nightly temperature (–0.146, p = 0.029), such that trap rates increased with decreasing rainfall and decreasing temperatures.

Table 9.3 Monthly average (and ranges) of minimum temperatures and rainfall for the Talatakely trail system in Ranomafana National Park during the study period 2004–2007.

Month	Min. temp (°C)	Rainfall (mm)
September	13.6 (10–17)	7.0 (0–42)
October	14.7 (9–19)	4.7 (0–54)
November	16.9 (10–21)	6.1 (0–58)
December	19.3 (11–22)	13.7 (0–60)
Average	15.5 (9–22)	6.3 (0–60)

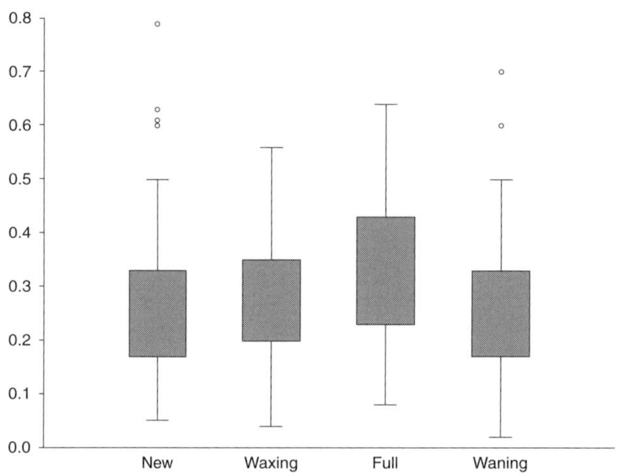

Figure 9.1 Brown mouse lemur trap rate by lunar phase. Box plots represent median ± range. Traps rates were significantly higher during full moon than any other lunar phase.

Discussion

We analyzed trapping data collected over a four-year period in RNP to assess whether rainfall, minimum temperature, or the lunar cycle affected the likelihood of capturing brown mouse lemurs in live traps. We assumed the capture rates were a proxy for general activity, and predicted that activity might vary due to changes in predation risk, foraging efficiency, and thermoregulatory demands. We found that brown mouse lemurs were most likely to be captured during nights around a full moon, and dry nights with lower temperatures. Overall, more than twice as many males than females were captured.

Lunar illumination has been hypothesized to enhance to detection of live prey as well as predators (Gursky, 2003). It has been proposed that visually oriented prey species such as primates may benefit from bright nights because their increased ability to detect predators lowers predation risk, while less-visual species such as rodents may experience an increase in predation risk by visually oriented predators (Gursky, 2003; Bearder *et al.*, 2006; Prugh and Golden, 2014; Rode-Margono and Nekaris,2014; Eppeley *et al.*, 2015). While our prediction was met in that the only moon phase that positively affected capture likelihood was the full moon category, the effect was not very strong (p-values between 0.023 and 0.03). It is thus unclear whether brown mouse lemurs are lunar philic like so many other primates (e.g., Gursky, 2003; Eppeley *et al.*, 2015). We will next discuss how moonlight might affect brown mouse lemur behavior. It is important to note that while our site had been selected logged in the 1980s, the forest has since recovered, with most areas having a closed canopy. In combination with frequent cloud cover, this might make moonlight penetration negligible on many nights. An effect of cover was observed in the paca (*Cuniculus paca*), a Neotropical rodent. Lunar illumination had little effect on pacas inhabiting a dense canopy primary rainforest (Michalski and Norris, 2011), whereas pacas inhabiting a more open canopy secondary forest avoided moonlight, presumably because of a higher (perceived) predation risk (Emmons *et al.*, 1989; Harmsen *et al.*, 2011). Differences in canopy cover may also explain why gray mouse lemurs reduced activity with increasing illumination (Erkert, 1989). Gray mouse lemurs inhabit a more open canopy dry forest habitat, and may thus experience a higher predation risk by visual predators (Goodman *et al.*, 1993). However, the gray mouse lemurs tested were captive animals with no need to avoid predators or to forage, so the behavior observed in the study (Erkert, 1989) may not accurately reflect the behavior of wild individuals.

It is, however, conceivable that brown mouse lemurs benefit from illumination by an enhanced ability to detect food or predators. Brown mouse lemurs commonly hunt flying and crawling insects such as moths and beetles, and even though they rely on hearing to detect insects (Görlitz and Siemers, 2007), their ability to catch them may increase with increasing illumination. Flowers were common during our trapping period, and moonlight may make them easier to detect. The largest benefit from lunar illumination might, however, be derived from an increased ability to detect predators. Brown mouse lemur predation rates by nocturnal predators are not known for our site, but nocturnal predators present include the fosa (*C. ferox*), the Madagascar tree boa (*S. madagascariensis*), and owls (*O. rutilus, T. soumagnei, A madagascariensis*; Goodman, 2003a; Raxworthy, 2003). The large fosa (7–12 kg) probably poses little threat to the small agile brown mouse lemurs during the night, and is extremely rare at our site (Wright, personal observation). While owl hunting success increases with illumination, they can hunt in complete darkness, relying on acoustic signals alone (Konishi, 1973; Clarke, 1983), which enables them to be active across all moon phases. In the rainforest, owls are camouflaged ambush hunters that pounce on their prey from perches, and thus the likelihood to detect an owl may not be affected by levels of illumination. Moreover, predator recognition experiments suggest that owl recognition is not well developed in the brown mouse lemur (Deppe and Wright, 2007), maybe because owls appear to be relatively rare at our site (Wright, personal observation). In contrast, the ability to detect snakes might increase with increasing illumination. For example, while the Madagascar tree boa is a very well-camouflaged predator that is difficult to detect even during the daytime (Deppe, personal observation), experiments suggest that brown mouse lemurs visually recognize snakes (Deppe 2005, 2006). Future studies increasing sample sizes, including more seasons, and expanding experiments may shed more light on how the brown mouse lemur responds to the lunar illumination.

There is increasing evidence that behavioral responses to lunar illumination are variable within species, and may change in response to season, food availability, age, habitat characteristics, and predator presence (e.g. Lockard and Owings, 1974; Kotler, 1984; Emmons *et al.*, 1989; Sutherland and Predavec, 1999; Harmsen *et al.*, 2011; Starr *et al.*, 2012; Upham and Hafner, 2013; Prugh and Golden, 2014; Rode-Margono and Nekaris, 2014). The gray slender loris (*N. lydekkerianus*), for example, was found to be largely lunar neutral, but sometimes vocalized and foraged more for insects during bright nights (Bearder *et al.*, 2006). The Javan slow loris (*N. javanicus*) tended to reduce activity during cold nights, regardless of illumination, suggesting that thermoregulatory considerations affected activity more than moonlight (Rode-Magonos and Nekaris, 2014).

Our prediction that brown mouse lemurs would be more likely to be captured during warmer nights was not met because capture likelihood increased with decreasing temperature. This was surprising because small mammals are susceptible to hypothermia (Falcy and Danielson, 2013), and tend to reduce activity when its cold (e.g., Lehman and Sommersberg, 1980; Rode-Margono and Nekaris, 2014). During the austral winter, June through August, when temperatures may reach lows of 4°C, mouse lemurs, especially females, may enter an inactive state of torpor to conserve energy (Ortmann *et al.*, 1997; Schmid, 2000; Genin *et al.*, 2005). The torpor behavior of the brown mouse lemur is not well understood, but low capture rates suggest prolonged torpor in this species as well (Harcourt, 1987; Atsalis, 1999b; Table 9.2; unpublished data). The seeming preference for cold temperatures found in our study is thus surprising. However, trapping took place in the early evening, where temperatures were well above nightly minimum temperatures. So while it is possible that brown mouse lemurs reduced activity during the coldest parts of the night, this would not be reflected in our study. The austral spring tends to be warm and dry, especially from October on (Table 9.3), but we do not believe it is hot enough to have thermoregulatory consequences, especially after sundown. Moreover, by far the most brown mouse lemurs were captured during October and November. It is important to note that our p-value of 0.029 was relatively high. We do not think that during our study period temperature had an effect on trapping success, but that our result was confounded by other factors, such as insect or predator activity, or sample size.

Rain may increase or decrease animal activity by increasing insect activity, flooding underground shelters, lowering visibility, and increasing background noise (Gentry *et al.*, 1965; Jahoda, 1970; Doucet and Bider, 1974; Vickery and Bider, 1981; Vickery and Rivest, 1991; Hilton *et al.*, 1999; Stokes *et al.*, 2001; Maestri and Marinho, 2014). As expected, rainfall negatively affected capture rates, where captures decreased with increasing rainfall. Brown mouse lemurs may prefer dry nights because detecting prey and predators might be easier. Rainfall, particularly heavy rain, increases background noise and lowers visibility, and may thus negatively affect the ability to see or hear predators and insects. While small mammals have been recorded to increase activity during light rain as a result of increased insect activity (Doucet and Bider, 1974; Vickery and Bider, 1981; Cresswell and Harris, 1988), heavy rain might have negative effects. Flying insects in particular stay out of the rain, and the background noise probably makes it more difficult for brown mouse lemurs to acoustically locate insect prey (Görlitz and Siemers, 2007). Rain may also have thermoregulatory consequences because wet fur exacerbates heat loss (Jahoda, 1970, 1973), and the small brown mouse lemur may avoid rain to conserve energy. One limitation of our finding is that because the rainfall was collected over a 24-h period, we do not know whether it actually rained during trapping hours. However, there is anecdotal evidence that trap rates were low during heavy rain (Deppe, personal observation). In addition, there is often a marked increase in frequency and severity of rainfall in December, which likely contributes to the low trap rates during that time (Deppe, personal observation; Table 9.3).

In summary, our findings suggest that during the months of September–December, brown mouse lemurs are more likely to be captured during full moon nights, dry nights, and lower nightly minimum temperatures. No link has been established between trapping success and activity (Prugh and Brashares, 2010), so it is important to keep in mind that trapping success may not accurately reflect general activity. Nonetheless, the more an animal is moving about, the more likely it is to encounter a trap. However, brown mouse lemurs quickly learn the trap locations and become very "trap happy", where in some cases individuals return to the same trap every night. These individuals might not necessarily be generally active, but may go straight to a known trap for banana. It is also important to note that October and November are the prime mating months at our site. Both males and females are highly active, conspicuous, sometimes travel in groups, and may even be active in broad daylight (Deppe, personal observation). These activity patterns are specific to the mating season, and because half of our data were collected during that time, our findings may not extend to other times of the year. It would be informative to examine whether trapping during other seasons reveals different results with regard to the lunar cycle, rainfall, and temperature. We acknowledge the limitations of this study, particularly its indirect nature of assessing activity. Future studies should aim to include additional seasons and the employment of radiotelemetry techniques or observations to obtain more direct information about brown mouse lemur activity cycles.

Acknowledgments

We acknowledge the government of Madagascar Ministry of Forests, Environment and Tourism and the CAFF/CORE oversight committee for authorization to work in Madagascar. We thank the Institute for the Conservation of Tropical Environments (ICTE/MICET), particularly director Benjamin Andriamihaja, and Madagascar National Parks (MNP) for logistical support. This project would not have been possible without the support of staff and technicians of the Centre ValBio Research Station in Ranomafana National Park, in particular Dr. Anna Feistner, our research technician Rasendrinirina Victor, and Drs. Marina Blanco and Misa Randriamiarisoa. We further thank Hannah Thomas and Paul Carson, Francois Zakamanana, the late Dr. Giselle Randria, and Dr. Jukka Jernvall. No animals were injured as a result of this research. Funding was generously provided by Primate Conservation, Inc., Conservation International, Idea Wild, Stony Brook University, the National Science Foundation's Doctoral Dissertation Improvement Grant, and the University of Helsinki.

References

Atsalis S. 1999a. Seasonal fluctuations in body fat and activity levels in a rain-forest species of mouse lemur, *Microcebus rufus. International Journal of Primatology* 20:883–910.

Atsalis S. 1999b. Diet of the brown mouse lemur (*Microcebus rufus*) in Ranomafana National Park, Madagascar. *International Journal of Primatology* 20:193–229.

Atsalis S. 2000. Spatial distribution and population composition of the brown mouse lemur (*Microcebus rufus*) in Ranomafana National Park, Madagascar, and its implications for social organization. *American Journal of Primatology* 51:61–78.

Bearder S, Nekaris KA, Buzell A. 2002. Dangers in the night: are some nocturnal primates afraid of the dark? In LE Miller (ed.), *Eat or Be Eaten: Predator Sensitive Foraging Among Primates* (pp. 21–43). Cambridge University Press, Cambridge.

Bearder S, Nekaris KA, Curtis DJ. 2006. A re-evaluation of the role of vision in the activity and communication of nocturnal primates. *Folia Primatologica* 77:50–71.

Blanco MB. 2008. Reproductive schedules of female *Microcebus rufus* at Ranomafana National Park, Madagascar. *International Journal of Primatology* 29:323–338.

Blanco MB, Godfrey LR. 2014. Hibernation patterns of dwarf lemurs in the high altitude forest of eastern Madagascar. *Developments in Primatology: Progress and Prospects* pp. 23–42.

Blanco MB, Meyer JS. 2009. Assessing reproductive profiles in female brown mouse lemurs (*Microcebus rufus*) from Ranomafana National Park, southeast Madagascar, using fecal hormone analysis. *American Journal of Primatology* 71:439–446.

Blumstein DT, Daniel JC, Griffin, AS, Evans CS. 2000. Insular tammar wallabies (*Macropus eugenii*) respond to visual but not acoustic cues from predators. *Behavioral Ecology* 11:528–535.

Bouliska A. 1995. Interactions between predation risk and competition: a field study of kangaroo rats and snakes. *Ecology* 76:165–178.

Bowers MA, Jefferson JL, Kuehler, MG. 1993. Variation in giving-up densities of foraging chipmunks (*Tamias striatus*) and squirrels (*Sciurus carolinensis*). *Oikos* 66:229–236.

Cadle JE. 2003. Colubridae, snakes. In SM Goodman, JP Benstead (eds.), *The Natural History of Madagascar* (pp. 997–1004). University of Chicago Press, Chicago.

Caro TM, Brock R, Kelly M. 2001. Diversity of mammals in the Bladen Nature Reserve, Belize, and factors affecting their trapping success. *Mammalian Biology* 66:90–101.

Charles-Dominique P. 1977. *Ecology and Behaviour of Nocturnal Primates.* Columbia University Press, New York.

Clarke JA. 1983. Moonlight's influence on predator/prey interactions between short-eared owls (*Asio flammeus*) and deermice (*Peromyscus maniculatus*). *Behavioral Ecology and Sociobiology* 13:205–209.

Cozzi G, Broekhuis F, McNutt JW, Turnbull LA, MacDonald DW, Schmid B. 2012. Fear of the dark or dinner by moonlight? Reduced temporal partitioning among Africa's large carnivores. *Ecology* 93:2590–2599.

Cresswell WJ, Harris S. 1988. The effects of weather conditions on the movement and activity of badgers (*Meles meles*) in a suburban environment. *Journal of Zoology, London* 216:187–194.

Daly M, Behrends PR, Wilson MI, Jacobs LF. 1992. Behavioural modulation of predation risk: moonlight, avoidance and crepuscular compensation in a nocturnal rodent, *Dipodomys merriami. Animal Behaviour* 44:1–9.

Davidson DL, Morris DW. 2001. Density-dependent foraging effort of deer mice (*Peromyscus maniculatus*). *Functional Ecology* 15:575–583.

Deppe AM. 2005. Visual predator recognition and response in wild brown mouse lemurs (*Microcebus rufus*) in Ranomafana National Park, Madagascar. *American Journal of Primatology* 66(Suppl 1):97–98.

Deppe AM. 2006. Visual snake recognition in wild brown mouse lemurs (*Microcebus rufus*). *American Journal of Primatology* 68(Suppl 1):34.

Deppe AM, Wright PC. 2007. Owl recognition via visual, olfactory, and auditory sensory cues in a small nocturnal primate, *Microcebus rufus*, in Ranomafana National Park, Madagascar. Prosimian Congress, Ithala, p. 43.

Deppe AM, Randriamiarisoa M, Kasprak AH, Wright PC. 2008. Predation on the brown mouse lemur (*Microcebus rufus*) by a diurnal carnvivore, the ring-tailed mongoose (*Galidia elegans*). *Lemur News* 13:17–18.

Dice LR. 1945. Minimum intensities of illumination under which owls can find dead prey by sight. *The American Naturalist* 784:385–415.

Dollar L, Ganzhorn J, Goodman SM. 2007. Primates and other prey in the seasonally variable diet of *Cryptoprocta ferox* in the dry deciduous forest on western Madagascar. In A Gursky, KAI Nekaris (eds.), *Primate Anti-predator Strategies* (pp. 63–76). Springer, New York.

Donati G, Lunardini A, Keppeler PM, Borgognini Tarli SM. 2001. Nocturnal activity in the cathemeral red-fronted lemur (*Eulemur fulvus rufus*), with observations during a lunar eclipse. *American Journal of Primatology* 53:69–78.

Doucet CJ, Bider JR. 1974. The effects of weather on the activity of the masked shrew. *Journal of Mammalogy* 55:348–363.

Emmons LH, Sherman P, Bolster D, *et al.* 1989. Ocelot behaviour in moonlight. In KH Redford, JF Eisenberg (eds.), *Advances in Neotropical Mammalogy* (pp.233–242). Sandhill Crane Press, Gainesville.

Eppely TM, Ganzhorn JU, Donati G. 2015. Cathemerality in a small, folivorous primate: proximate control of diel activity in *Hapalemur meridionalis. Behavioral Ecology and Sociobiology* April: DOI 10.1007/s00265-015-1911-3.

Erkert HG. 1974. Der Einfluss des Mondlichtes auf die Aktivitätsperiodik nachtaktiver Säugetiere. *Oecologia* 14:269–287.

Erkert HG. 1989. Lighting requirements of nocturnal primates in captivity: a chronological approach. *Zoo Biology* 8:179–191.

Erkert HG, Gröber J. 1986. Direct modulation of activity and body temperature of owl monkeys (*Aotus lemurinus griseimembra*) by low light intensities. *Folia Primatologica* 47:171–188.

Falcy MR, Danielson B. 2013. A complex relationship between moonlight and temperature on the foraging behavior of the Alabama beach mouse. *Ecology* 94:2632–2637.

Fenton MB, Boyle NG, Harrison TM, Oxley D J. 1977. Activity patterns, habitat use, and prey selection by some African insectivorous bats. *Biotropica* 9:73–85.

Fernandez-Duque E, Erkert HG. 2006. Cathemerality and lunar periodicity of activity rhythms in owl monkeys of the Argentinian Chaco. *Folia Primatologica* 77:123–138.

Genin F, Schilling A, Perret M. 2005. Social inhibition of seasonal fattening in wild and captive gray mouse lemurs. *Physiology & Behavior* 86:185–194.

Gentry JB, Odum EP. 1957. The effect of weather on the winter activity of old-field rodents. *Journal of Mammalogy* 38:72–78.

Gentry JB, Golley FB, McGinnis JT. 1965. Effect of weather on captures of small mammals. *The American Midland Naturalist* 75:526–530.

Goodman SM. 2003a. Predation on lemurs. In SM Goodman, JP Benstead (eds.), *The Natural History of Madagascar* (pp. 1221–1228). University of Chicago Press, Chicago.
Goodman SM. 2003b. Carnivora: *Galidia elegans*, ring-tailed mongoose, *Vontsira mena*. In SM Goodman, JP Benstead (eds.), *The Natural History of Madagascar* (pp.1351–1354). University of Chicago Press, Chicago.
Goodman SM. 2003c. *Galidictis*, broad striped mongoose, *Vontsira fotsy*. In SM Goodman, JP Benstead (eds.), *The Natural History of Madagascar* (pp. 1355–1357). University of Chicago Press, Chicago.
Goodman SM, O'Connor S, Langrand O. 1993. A review of predation on lemurs: implications for the evolution of social behavior in small, nocturnal primates. In PM Kappeler, JU Ganzhorn (eds.), *Lemur Social Systems and their Ecological Basis* (pp. 51–66). Plenum Press, New York.
Görlitz H, Siemers, BM. 2007. Sensory ecology of prey rustling sounds: acoustical features and their classification by wild grey mouse lemurs. *Functional Ecology* 21:143–153.
Gursky S. 2003. Lunar philia in a nocturnal primate. *International Journal of Primatology* 24:351–367.
Halle S. 2006. Polyphasic activity patterns in small mammals. *Folia Primatologica* 77:15–26.
Halle S, Stenseth NC. 2001. *Activity Patterns in Small Mammals: An Ecological Approach*. Springer, New York.
Harcourt C. 1987. Brief trap/retrap study of the brown mouse lemur (*Microcebus rufus*). *Folia Primatologica* 49:209–211.
Harmsen B, Foster R, Siver SC, Ostro LET, Doncaster CP. 2011. Jaguar and puma activity patterns in relation to their main prey. *Mammalian Biology* 76:320–324.
Hawkins CE. 2003. *Cryptoprocta ferox*, fossa, *Fosa*. In SM Goodman, JP Benstead (eds.), *The Natural History of Madagascar* (pp. 1360–1363). University of Chicago Press, Chicago.
Hawkins CE, Goodman SM. 2003. Introduction to the birds. In SM Goodman, JP Benstead (eds.), *The Natural History of Madagascar* (pp. 1019–1044). University of Chicago Press, Chicago.
Hilton GM, Ruxton GD, Cresswell W. 1999. Choice of foraging area with respect to predation risk in redshanks: the effects of weather and predator activity. *Oikos* 87:295–302.
Hughes JJ, Ward D, Perrin MR. 1994. Predation risk and competition affect habitat selection and activity of Namib desert gerbils. *Ecology* 75:1397–1405.
Jahoda JC. 1970. The effects of rainfall on the activity of *Onychomys leucogaster breviauritus*. *American Zoologist* 10:326.
Jahoda JC. 1973. The effect of the lunar cycle on the activity pattern of *Onchomys leucogaster breviauritus*. *Journal of Mammalogy* 54:544–549.
Kappeler PM, Rasoloarison RM. 2003. *Microcebus*, mouse lemurs. In SM Goodman, JP Benstead (eds.), *The Natural History of Madagascar* (pp. 1310–1315). University of Chicago Press, Chicago.
Karpanty S. 2006. Direct and indirect impacts of raptor predation on lemurs in southeastern Madagascar. *International Journal of Primatology* 27:239–261.
Karpanty S, Wright PC. 2007. Predation on lemurs in the rainforest of Madagascar by multiple predator species: observations and experiments. In A Gursky, KAI Nekaris (eds.), *Primate Anti-predator Strategies* (pp. 77–99). Springer, New York.
Kelt DA, Meserve PL, Nabors LK, Forister ML, Gutierrez JR. 2004. Foraging ecology of small mammals in semiarid Chile: the interplay of biotic and abiotic factors. *Ecology* 85:383–397.
Kerridge FJ, Ralisoamalala RC, Goodman SM, Pasnick SD. 2003. In SM Goodman, JP Benstead (eds.), *The Natural History of Madagascar* (pp. 1363–1365). University of Chicago Press, Chicago.
Kirk EC, Kay RF. 2004. The evolution of high visual acuity in the anthropoidea. In CF Ross, RF Kay (eds.), *Anthropoid Origins: New Visions* (pp. 539–602). Kluwer, New York.
Konishi M. 1973. Locatable and nonlocatable acoustic signals from barn owls. *The American Naturalist* 107:775–785.
Kotler BP. 1984. Risk of predation and the structure of desert rodent communities. *Ecology* 65:689–701.

Kotler BP, Brown, JS, Hasson O. 1991. Factors affecting gerbil foraging behavior and rates of owl predation. *Ecology* 7:22249–22260.

Kotler BP, Brown JS, Mitchell WA. 1993. Environmental factors affecting patch use in two species of gerbilline rodents. *Journal of Mammalogy* 54:256–260.

Lang AB, Kalko EKV, Roemer H, Bockholdt C, Dechmann DKN. 2006. Activity levels of bats and katydids in relation to the lunar cycle. *Behavioral Ecology* 146:659–666.

Lehmann U, Sommersberg CW. 1980. Activity patterns of the common vole, *Microtus arvalis*-automatic recording of behaviour in an enclosure. *Oecologia* 47:61–75.

Lillywhite HB, Brischoux F. 2012. Is it better in the moonlight? Nocturnal activity of insular cottonmouth snakes increases with lunar light levels. *Journal of Zoology* 286:194–199.

Lockard RB, Owings DH. 1974. Seasonal variation in moonlight avoidance by bannertail kangaroo rats. *Journal of Mammalogy* 55:189–193.

Lyman C. 1983. *Hiberation and Torpor in Mammals and Birds.* Academic Press, London.

Maestri R, Marinho JR. 2014. Singing in the rain. Rainfall and moonlight affect daily activity patterns of rodents in a Neotropical forest. *Acta Theriologica* 59:427–433.

Michalski F, Norris D. 2011. Activity pattern of *Cuniculus paca* (Rodentia: Cuniculidae) in relation to lunar illumination and other abiotic variables in the southern Brazilian Amazon. *Zoologia* 28:701–708.

Morrison DW. 1978. Lunar phobia in a neotropical fruit bat, *Artibeus jamaicensis* (Chiroptera: Phyllostomidae). *Animal Behaviour* 26:852–855.

Nash LT. 1986. Influence of moonlight level on traveling and calling patterns in two sympatric species of *Galago* in Kenya. In D Taub, F King (eds.), *Current Perspectives in Primate Social Dynamics* (pp. 357–367). Van Norstrand, New York.

Nash LT. 2007. Moonlight and behavior in nocturnal and cathemeral primates, especially *Lepilemur leucopus*: illuminating possible anti-predator efforts. In S Gursky, KAI Nekaris (eds.), *Primate Anti-predator Strategies* (pp. 173–205). Springer, New York.

Ortmann S, Heldmaier G, Schmid J, Ganzhorn JU. 1997. Spontaneous daily torpor in Malagasy mouse lemurs. *Naturwissenschaften* 84:28–34.

Payne RS. 1971. Acoustic location of prey by barn owls (*Tyto alba*). *Journal of Experimental Biology* 54:535–573.

Plesner Jensen S, Honess P. 1995. The influence of moonlight on vegetation height preference and trappability of small mammals. *Mammalia* 59:35–42.

Price MV, Waser NM, Bass TA. 1984. Effects of moonlight on microhabitat use by desert rodents. *Journal of Mammalogy* 65:353–356.

Prugh LR, Brashares J. 2010. Basking in the moonlight? Effect of illumination on capture success of the endangered giant kangaroo rat. *Journal of Mammalogy* 91:1205–1212.

Prugh LR, Golden CD. 2014. Does moonlight increase predation risk? Meta-analysis reveals divergent responses of nocturnal mammals to lunar cycles. *Journal of Animal Ecology* 83:504–514.

Radespiel U, Sarikaya Z, Zimmermann E. 2001. Sociogenetic structure in a free-living nocturnal primate population: sex-specific differences in the grey mouse lemur (*Microcebus murinus*). *Behavioral Ecology and Sociobiology* 50:493–502.

Raxworthy C. 2003. Boidae, boas. In SM Goodman, JP Benstead (eds.), *The Natural History of Madagascar* (pp. 993–997). University of Chicago Press, Chicago.

Rode-Margano EJ, Nekaris KAI. 2014. Impact of climate and moonlight on a venomous mammal, the Javan slow loris (*Nycticebus javanicus* Geoffroy, 1812). *Contributions to Zoology* 83:217–225.

Schmid J. 2000. Daily torpor in the gray mouse lemur (*Microcebus murinus*) in Madagascar. *Oecologia* 123:175–183.

Schmid J. 2001. Daily torpor in free-ranging gray mouse lemurs (*Microcebus murinus*) in Madagascar. *Folia Primatologica* 22:1021–1031.

Schwab D, Ganzhorn JU. 2004. Distribution, population structure and habitat use of *Microcebus berthae* compared to those of other sympatric cheirogaleids. *Folia Primatologica* 25: 307–330.

Starr C, Mekaris KAI, Leung L. 2012. Hiding from the moonlight: luminosity and temperature affect activity of Asian nocturnal primates in a highly seasonal forest. *PloS ONE* 7:e36396.

Stokes MK, Slade NA, Blair SM. 2001. Influences of weather and moonlight on activity patterns of small mammals: a biogeographical perspective. *Canadian Journal of Zoology* 79:966–972.

Sutherland DR, Predavec M. 1999. The effects of moonlight on microhabitat use by *Antechinus agilis* (Marsupialia: Dasyuridae). *Australian Journal of Zoology* 47:1–17.

Tetreault N, Hakeem A, Allman JM. 2004. The distribution and size of ganglion cells in *Cheirogaleus medius* and *Tarsius syrichta*: implications for the evolution of sensory systems in primates. In CF Ross, RF Kay (eds.), *Anthropoid Origins: New Visions* (pp. 463–475). Kluwer, New York.

Upham NS, Hafner JC. 2013. Do nocturnal rodents in the Great Basin Desert avoid moonlight? *Journal of Mammalogy* 94:59–72.

Vermeij JG. 1982. Unsuccessful predation and evolution. *The American Naturalist* 120:701–720.

Vickery WL, Bider JR. 1981. The influence of weather on rodent activity. *Journal of Mammalogy* 62:140–145.

Vickery WL, Rivest D. 1991. The influence of weather on habitat use by small mammals. *Ecography* 15:205–211.

Wright PC. 1989. The nocturnal primate niche in the New World. *Journal of Human Evolution* 18:635–658.

Wright PC. 1992. Primate ecology, rainforest conservation, and economic development: building a national park in Madagascar. *Evolutionary Anthropology* 1:25–33.

Wright PC, Andriamihaja B. 2002. Making a rain forest park work in Madagascar: Ranomafana National Park and its long-term research commitment. In J Terborgh, C VanSchaik, L Davenport, M Rao (eds.), *Making Parks Work: Strategies for Preserving Tropical Nature* (pp. 112–136). Island Press, Washington, DC.

10 Tiny samples from tiny lemurs: methodological considerations for endoparasite analyses in mouse lemurs

Sharon E. Kessler, Ute Radespiel, Kathrin S. Schaber, and Christina Strube

Introduction

Studies of parasite burden and transmission in wild non-human primates are important for combating emerging infectious diseases in humans, understanding processes of zoonosis, and managing and conserving endangered species (Gillespie *et al.*, 2008). In a biodiversity hotspot facing extreme anthropogenic pressure, lemurs are a high conservation priority (Schwitzer *et al.*, 2014) and the relationship between lemur health and human health is only beginning to be understood (Bublitz *et al.*, 2015; Zohdy *et al.*, 2015). Mouse lemurs (*Microcebus* spp.) are well positioned to become a very useful model for endoparasite studies in lemurs. As a highly specious genus with 21 recognized species, they are widespread across the island of Madagascar, occurring in diverse habitats including rainforests, dry deciduous forests, littoral forests, etc. (Radespiel, 2006; Sommer *et al.*, 2014; Zimmermann and Radespiel, 2014). In addition, species vary (within and between species) in their diets, use of torpor, degree of sociality, and sympatry with congeneric species (Lahann *et al.*, 2006; Radespiel, 2006; Sommer *et al.*, 2014; Chapter 22). This variation, combined with the ease with which they can be trapped, examined, and sampled without anesthesia, thus yielding reasonable sample sizes, makes them attractive for asking a multitude of questions with implications for understanding host–parasite coevolution, transmission dynamics and zoonosis, conservation, and wildlife management.

Mouse lemurs harbor, in particular, vector-borne parasites and direct lifecycle parasites (Radespiel *et al.*, 2015). They may become infected with vector-borne parasites when feeding on arthropods serving as intermediate hosts (Radespiel *et al.*, 2015). In addition, they are vulnerable to parasites with direct lifecycles when feeding in contaminated vegetation or co-sleeping in tree holes/nests with infected individuals who may defecate in the sleeping sites (Sommer *et al.*, 2014; Radespiel *et al.*, 2015). The endoparasite studies conducted to date have included foci on relationships between parasites and seasonality (Raharivololona and Ganzhorn, 2010),

The Dwarf and Mouse Lemurs of Madagascar: Biology, Behavior and Conservation Biogeography of the Cheirogaleidae, eds. S.M. Lehman, U. Radespiel, E. Zimmermann. Published by Cambridge University Press. © Cambridge University Press 2016.

forest fragmentation (Schad *et al.*, 2005), the genetics of parasite resistance (Schad *et al.*, 2005; Schwensow *et al.*, 2010a, 2010b; Sommer *et al.*, 2014), sympatry among host mouse lemur species (Sommer *et al.*, 2014; Radespiel *et al.*, 2015), and parasites with the potential to be zoonotic with human populations (Rasambainarivo *et al.*, 2013; Bublitz *et al.*, 2015; Zohdy *et al.*, 2015). Many of these studies were conducted within the last five years and meeting abstracts suggest that several other studies are underway (e.g., Alldredge *et al.*, 2013, 2014; Rodriguez *et al.*, 2013). As additional studies are conducted, more comparative data will be available and it will become feasible to answer questions dealing with the complexities of host–parasite coevolution, zoonosis, health, and conservation.

The goal of this chapter is to encourage additional research on mouse lemur parasites by presenting data on several methodological considerations that we hope will assist in planning future studies. We encourage readers to consult extensively with detailed, published guidelines for how to collect, preserve, and store field samples and how to prepare samples for analysis and then conduct those analyses so as to maximize the usefulness of the data and the cross-species comparisons that can be made (i.e., Stuart and Strier, 1995; Gillespie, 2006; Gillespie *et al.*, 2008). Exactly what procedures are used will depend on the parasites targeted, questions being asked, and any particular constraints presented by the field site and host species (Stuart and Strier, 1995; Gillespie, 2006; Gillespie *et al.*, 2008). Here we present our data on additional methodological considerations including several that may arise when studying tiny primates which, like mouse lemurs, typically only offer only tiny samples.

In particular, we examine two questions that are highly relevant for small-bodied mammals. (1) Does sample mass predict prevalence and the eggs/gram found in the sample? (2) Given small sample mass, how accurate are fecal egg counts estimated by McMaster methods relative to full counts of all eggs in a given sample? We acknowledge that fecal egg counts are highly problematic measures of infection intensity (Gillespie, 2006); however, as fecal egg counts are (a) often used, and (b) may be potentially less problematic when used to define relative, qualitative categories of infection intensity (i.e., low, moderate, high), we discuss them here. Our focus is not on whether fecal egg counts accurately represent the number of adult individuals infecting a host, but rather on whether a common method for estimating eggs per gram actually reflects the number of eggs in that sample.

We also address two additional questions which are also likely to have widespread applicability. (3) We compare two fixatives: the commercially available PROTO-fix CLR (Alpha-Tec Systems, Vancouver, WA, USA) and 90% ethanol, which is cheap and widely available in pharmacies around the world. Finally, (4) we test whether storage and repeated processing is associated with lower egg counts.

Methods

Sample collection

The samples included in this study were selected from a larger bank of samples at the University of Veterinary Medicine Hannover (Radespiel *et al.*, 2015) in order to evaluate the four methodological issues in question. Samples were collected during the dry season (May–October) of 2010 in the Ankarafantsika National Park, Madagascar in two study sites: JBA, 16°19′S, 46°48′E and JBB, 16°18′S, 46°48′E. *M. murinus* and *M. ravelobensis* are sympatric in JBA and only *M. ravelobensis* is present in JBB. Samples were collected directly from the animals (not from traps) during monthly capture/recapture sessions, weighed, and then preserved either in 90% ethanol, in PROTO-fix CLR (Alpha-Tec Systems, Vancouver, WA, USA), or in formalin (diluted to 10% with bottled water). Ninety percent ethanol was easy to acquire in country and PROTO-fix CLR (Alpha-Tec Systems, Vancouver, WA, USA) is a mixture of ethyl alcohol, methanol, isopropyl, and formaldehyde, which is designed to be more environmentally friendly (no mercury or heavy metals and low formaldehyde content) than the "gold standard" polyvinyl alcohol (Jensen *et al.*, 2000). Samples were stored at ambient temperature for up to 19 months and then transported to the University of Veterinary Medicine Hannover, where they were stored for several months at 4°C before parasite analyses were conducted in the Institute for Parasitology. The following parasites were found in the larger sample bank: four nematode egg types (*Ascaris* sp., *Subulura* sp., strongyle egg type, *Lemuricola* sp.), and two species of tapeworms (*Hymenolepis* spp.) (Radespiel *et al.*, 2015). Because the purpose of this chapter is to discuss methodology, we direct readers interested in the relationship between ecological factors and endoparasites in mouse lemurs to our work addressing these topics (Radespiel *et al.*, 2015).

Sample preparation

Samples were prepared with a NaCl floatation-sedimentation technique (Radespiel *et al.*, 2015). They were centrifuged, weighed, mixed with 2–3 ml of saturated NaCl solution in a mortar, passed through a sieve, and washed with NaCl solution until a defined feces–solution ratio was obtained (7.5 ml per 0.5 g feces).

All samples were analyzed using two methods: estimating eggs per gram using the McMaster method, and conducting a full count in a counting chamber.

Estimated eggs per gram using the McMaster floatation chamber

Aliquots of 0.15 ml were counted in a McMaster counting chamber with a sensitivity of 50 eggs per gram of feces. Afterwards, the sample was transferred to a 50-ml tube and tap water was added until the total volume was 50 ml. Sedimentation occurred overnight and the supernatant was removed.

Samples were divided into two groups based on how they were processed. Group A samples (79 samples) were analyzed using the McMaster method then frozen at −80°C for several months before being thawed and analyzed using the full count method described below. Group B samples were analyzed in direct succession with both methods before being frozen for long-term storage (43 samples).

Full count

The complete and sedimented egg sample was transferred to a 50 mm × 20 mm counting chamber and all eggs were counted and identified using an Axiophote microscope (Carl Zeiss MicroImaging). Identified eggs were photographed using an Olympus CAMEDIA C-5050 Zoom digital camera and cell^B Image Acquisition Software (version 3.1; Olympus Soft Imaging Solutions).

Data analysis

For the purposes of analyzing prevalences, we scored samples as either no eggs present (0) or eggs present (1). When eggs were present, we calculated the number of eggs per gram without separating the eggs by type. All statistical tests were conducted in IBM SPSS Statistics 22 with an alpha of 0.05.

Question 1: sample mass

Using 43 samples which were processed once before being frozen for long-term storage (group B), we conducted a logistic regression following Field (2013) and Spearman's rank correlation. The logistic regression tested whether sample mass can predict the presence of eggs in samples. The Spearman's rank correlation tested for an association between sample mass and egg count using the values from the full count method.

Question 2: McMaster vs. full count

We compared the results of the McMaster method and the full count method using 43 samples from group B. Measurements were paired by sample. We compared the number of eggs/g found using a Wilcoxon matched pairs test.

Question 3: fixative

We compared samples within group A which differed by fixative used (ethanol or PROTO-fix) using a chi-square test for differences in prevalence and a Mann–Whitney *U* test for differences in eggs/g. The two groups were balanced by month, species, study site, and, to the extent possible, sample mass. The ethanol samples and PROTO-fix samples did not have significantly different sample masses (Mann–Whitney *U* test, $U = 497$, $n_{\text{ethanol}} = 32$, $n_{\text{protofix}} = 32$, $p = 0.840$). We did not control for sex because prior work found no sex differences in parasite prevalence and load in these populations (Radespiel *et al.*, 2015). Because formalin was used only in the first few months of collection and prior work did find parasite load differences across months, it is not possible to disentangle the two effects (Radespiel *et al.*, 2015). However, because PROTO-fix is a solution containing ethyl alcohol, methanol, isopropyl, and formaldehyde, we would expect it to perform similarly to both formalin and ethanol.

Question 4: processing and storage

We compared samples from group A and group B using the results of the full count. The groups were balanced by month, species, study site, and, as much as possible, by sample mass. Groups A and B did not have sample masses that were significantly different (Mann–Whitney U test, U = 442.5, n_A = 30, n_B = 30, p = 0.912). As in the tests for differences based on fixative effects, we did not control for sex because prior work found no sex differences in parasite prevalence and load in these populations (Radespiel *et al.*, 2015). Because no fixative effects were found (see results below), we did not control for fixative. We compared prevalence using a chi-squared test and eggs/g using a Mann–Whitney U test.

Results

Question 1: sample mass

Sample mass varied from 0.05 to 1 g and did not significantly predict prevalence (logistic regression, $\chi^2(1)$ = 0.37, p > 0.05, Table 10.1). Sample mass was also not significantly correlated with a higher egg count (Spearman's rank test, rho = 0.105, n = 43, p = 0.503). The smallest infected sample was 0.06 g. However, Figure 10.1 shows that the variance associated with these two variables is highly heteroscedastic. Thus, most small samples contained no eggs, but the variance in the number of eggs increased strongly with increasing sample mass. While various transformations may help correct this (Field, 2013), we present the raw data here to illustrate the underlying pattern that should be considered when planning sample collections.

Question 2: McMaster vs. full count

The McMaster method of estimating eggs per gram of feces did not detect significantly different numbers of eggs than the full count method (Wilcoxon matched pairs test, Z = −1.445, n = 43, p = 0.149). However, within the infected samples, the McMaster values and the full count values were not correlated (Spearman's rank test, rho = 0.171, N = 22, p = 0.446). A visual examination of the data shows that the McMaster method is much more likely to classify an infected sample as infection-free (Figure 10.2). When the full count method was used, 22 samples (51%) were classified as infected. Of these same samples, only 4 (18%) were classified as infected using the McMaster method. No samples classified as infected by the McMaster method were classified as infection-free by the full count method.

Finally, when eggs were detected in the McMaster method, the multiplication factor used to calculate the number of eggs per gram of feces produced much higher estimated egg counts than the full count method did (median eggs/g, McMaster: 500

Table 10.1 Coefficients of the logistic regression model predicting whether a sample was infected or not based upon sample weight.

Variable	B	SE	Wald	Df	Sig.	Odds ratio	95% CI for odds ratio	
							Lower	Upper
Sample weight	−0.873	1.445	0.365	1	0.546	0.418	0.025	7.102
Constant	0.310	0.532	0.339	1	0.560	1.363		

Note: $R^2_{\text{Hosmer and Lemeshow}} = 0.006$, $R^2_{\text{Cox \& Snell}} = 0.009$, $R^2_{\text{Nagelkerke}} = 0.011$.

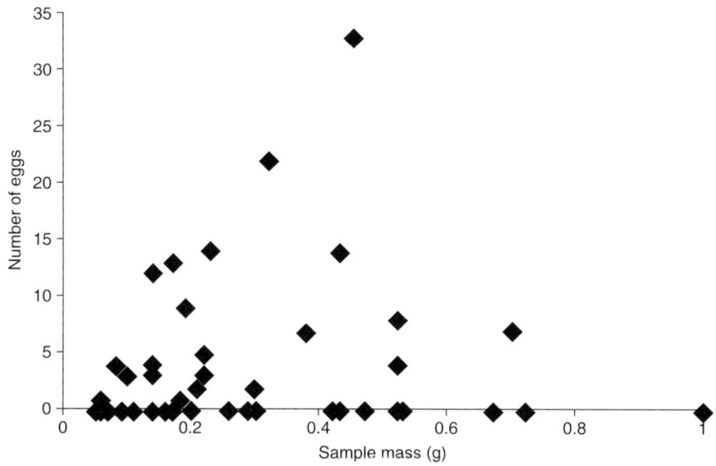

Figure 10.1 A scatterplot showing sample mass on the X-axis and the number of eggs on the Y-axis. The variance in number of eggs found increases with sample mass.

(25% quartile: 200, 75% quartile: 575), full-count: 22.08 (25% quartile: 12.73, 75% quartile: 52.72).

Question 3: fixative

Samples preserved in ethanol and in PROTO-fix did not show significantly different prevalences (chi-square test, $\chi^2 = 1.463$, df = 1, $p = 0.226$) or numbers of excreted eggs per gram (Mann–Whitney U test, $U = 445$, $n_{\text{ethanol}} = 32$, $n_{\text{protofix}} = 32$, $p = 0.214$).

Question 4: storage and re-analysis

Group A samples relative to group B showed a significantly lower prevalence (chi-square test, $\chi^2 = 11.915$, df = 1, $p = 0.001$) and lower numbers of eggs excreted per gram of feces based on the full count data set (Mann–Whitney U test, $U = 215$, $n_A = 30$, $n_B = 30$, $p < 0.001$).

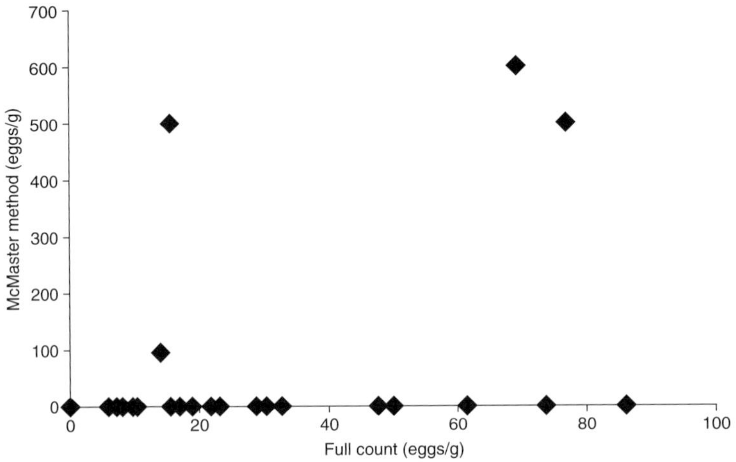

Figure 10.2 A scatterplot showing the number of eggs/g with the full count on the *X*-axis and the McMaster method on the *Y*-axis. The absence of a correlation between the results of the two methods is visible. Only the samples which were classified as infected by at least one of the methods are included.

Discussion

Our results indicate that sample mass, analysis technique (McMaster vs. full count), and storage and re-analysis, but not fixative choice, affected prevalence and egg counts. These results illustrate several methodological issues that are relevant for future endoparasite studies on mouse lemurs and other small mammals.

Although sample mass was not significantly related to infection prevalence or fecal egg count, Figure 10.1 suggests that one of the primary challenges for conducting endoparasite analyses in small animals is obtaining a large enough sample to find eggs that might be present. However, no minimum fecal threshold is noticeable in Figure 10.1, suggesting that there is no easy rule such as "Collect X amount of fecal material, and if eggs are present, you will find them." Therefore, we suggest that researchers simply try to collect samples that are as large as possible. This may require researchers to consider trade-offs between collecting fresh feces exclusively from the animal and also including feces collected from traps or temporarily caging the animal and collecting material from the cages. The drawback of the second two options is that feces collected from traps/cages have an increased chance of having been contaminated by free-living nematodes in the environment (Gillespie, 2006). However, when subjects excrete only small amounts of feces, the risk of obtaining false negatives may be judged more serious than the risk of some environmental contamination. Egg excretion can be variable over time (Gillespie, 2006) and a very small sample, even from an infected animal, carries a greater risk than a larger sample of being egg-free simply by chance. Thus, the researcher must decide what collection protocol would be most suited to the target parasite types and their life-cycles, the body size of the animals, and the questions being asked.

We found that the McMaster method of estimating the number of eggs per gram of feces was an inaccurate estimator of the true numbers of eggs in the samples. Although the counts obtained by the McMaster method and the full count method were not significantly different, they were also not significantly correlated. The lack of a significant difference may be due to the large variance produced by the McMaster method (Figure 10.2). Furthermore, the McMaster method failed to detect infections in 82% of the samples where the full count detected infections. Part of this striking deviation can be explained with the relatively low sensitivity of the McMaster method (50 eggs/g fecal material) compared to the full count method. Based on the full count results of group B, 73% (16 of 22) of our infected samples contained less than the 50 eggs/g fecal material that would necessarily remain undetected by the McMaster method (Figure 10.2). However, this explanation does not explain why, when eggs were found, the McMaster method then estimated far more eggs to be in the sample than were actually present. The 75% quartile of the full count method (52.72 eggs/g) was still nearly four times smaller than the 25% quartile of the McMaster method (200 eggs/g). This difference is large enough that it is likely to cause significant errors, even when egg counts are only being used to estimate qualitative infection intensities (i.e., high, medium, low). Thus, we recommend that full counts be used for mouse lemur samples. While time-consuming, this method is more accurate and is not nearly as time-consuming as it would be for the entirety of a gorilla sample!

We did not find significant differences in prevalence or the number of eggs/g based on the fixative type. Given the low prevalences of the various egg types in our study, it was not possible to test the effects of fixative type by parasite type. Therefore, because formalin is preferred for helminth preservation while polyvinyl alcohol for protozoa and RNAlater for molecular identification of parasites (Gillespie, 2006; Gillespie *et al.*, 2008), we recommend that researchers consider the pros and cons of each the available substances as they pertain to the target parasites, transportation requirements (e.g., plane travel), and local conditions (e.g., availability and disposal requirements of substances).

Repeated processing of samples, storage, and freeze–thaw cycles does appear to be associated with the loss of eggs. This may be due to *Acaris* eggs sticking to the sides of plastic tubes during transfers. In addition, repeated freezing and thawing may damage egg shells with the result that (1) eggs do not float due to the floatation solution flowing into the egg, causing the egg to increase in specific weight, or (2) eggs may break up and no longer be visible. Due to this, we suggest that samples should be analyzed once using the full count method before freezing material that is to be retained for future work (e.g., molecular analyses).

Conclusions

Based on the results of this study we recommend that future studies of mouse lemur endoparasites:

- maximize the amount of sample collected

- conduct full counts of the entirety of the samples collected
- minimize repeated processing of samples and extraneous freeze–thaw cycles.

Although it was not possible here to statistically test the effects of ethanol and PROTO-fix on individual parasite types, no differences were found between the fixatives in total numbers of eggs detected, providing preliminary evidence that both functioned well.

We hope that our results will assist researchers to design future parasitological studies in mouse lemurs including preparing them for some of the issues that may arise when working with tiny primates and their tiny samples. We look forward to when comparisons will be possible within and among mouse lemur species to answer questions pertaining to the effects of their corresponding habitats, social systems, etc., on mouse lemur–parasite evolution. We believe that this will have implications for wildlife conservation and public health in Madagascar.

Acknowledgments

We would like to thank the Malagasy government, Ministère de l'Environnement et des Forêts, and Madagascar National Parks (MNP) for their permission to work in the Ankarafantsika National Park and for their support during our fieldwork. We are very thankful to Solofonirina Rasoloharijaona and Blanchard Randrianambinina from the University of Mahajanga for assistance during all phases of our fieldwork. We thank Alida I.F. Hasiniaina and Lisette Leliveld for assisting with sample collection, and Jhonny Kennedy for guide services. We thank Elke Zimmermann and the Institute of Zoology for long-term support and Sandra Buschbaum for technical support during fecal analyses. The study was funded by an NSF Dissertation Improvement Grant (#0961779, SEK), PEO Scholar Award (SEK), the Animal Behavior Society (SEK), Lewis and Clark Fund of the American Philosophical Society (SEK), American Society of Primatologists (SEK), Sigma Xi (National Chapter, grant #G2009101504, SEK), Sigma Xi (Arizona State University chapter, SEK), Arizona State University Graduate and Professional Student Association (SEK), and the Arizona State University School of Human Evolution and Social Change (SEK) and the Fonds du Recherché du Québec Postdoctoral Fellowship (SEK). Methods were approved by Madagascar National Parks (N101/11/MEF/SG/DGF/DCB.SAP/SCB, N102/11/MEF/SG/DGF/DCB.SA/SCB) and the Arizona State University Institutional Animal Care and Use Committee (Protocol: 10-1077R).

References

Alldredge KL, Irwin MT, Godfrey LR. 2013. Roundworms on the Red Island: gastrointestinal parasite intensity in four lemur species from the Tsinjoarivo region, Madagascar. *American Journal of Physical Anthropology* 150:66–66.

Alldredge KL, Romer ZC, Fitzpatrick MC, Irwin MT. 2014. Strongyloids in the soil: an exploration of gastrointestinal parasite load and geophagy in lemurs at Tsinjoarivo, Madagascar. *American Journal of Physical Anthropology* 153:64–65.

Bublitz DC, Wright PC, Rasambainarivo FT, *et al.* 2015. Pathogenic enterobacteria in lemurs associated with anthropogenic disturbance. *American Journal of Primatology* 77(3):330–337.

Field A. 2013. *Discovering Statistics Using IBM SPSS Statistics*. Sage, London.

Gillespie TR. 2006. Noninvasive assessment of gastrointestinal parasite infections in free-ranging primates. *International Journal of Primatology* 27(4):1129–1143.

Gillespie TR, Nunn CL, Leendertz FH. 2008. Integrative approaches to the study of primate infectious disease: implications for biodiversity conservation and global health. *Yearbook of Physical Anthropology* 51:53–69.

Jensen B, Kepley W, Guarner J, *et al.* 2000. Comparison of polyvinyl alcohol fixative with three less hazardous fixatives for detection and identification of intestinal parasites. *Journal of Clinical Microbiology* 38:1592–1598.

Lahann P, Schmid J, Ganzhorn JU. 2006. Geographic variation in populations of *Microcebus murinus* in Madagascar: resource seasonality or Bergmann's rule? *International Journal of Primatology* 27(4):983–999.

Radespiel U. 2006. Ecological diversity and seasonal adaptations of mouse lemurs (*Microcebus* spp.). In L Gould, ML Sauther (eds.), *Lemurs: Ecology and Adaptation* (pp. 211–234). Springer, New York.

Radespiel U, Schaber K, Kessler SE, Schaarschmidt F, Strube C. 2015. Variations in the excretion patterns of helminth eggs in two sympatric mouse lemur species (*Microcebus murinus* and *M. ravelobensis*) in northwestern Madagascar. *Parasitology Research* 114(3):941–954.

Raharivololona BM, Ganzhorn JU. 2010. Seasonal variations in gastrointestinal parasites excreted by the gray mouse lemur *Microcebus murinus* in Madagascar. *Endangered Species Research* 11(2):113–122.

Rasambainarivo FT, Gillespie TR, Wright PC, *et al.* 2013. Survey of *Giardia* and *Cryptosporidium* in lemurs from the Ranomafana National Park, Madagascar. *Journal of Wildlife Diseases* 49(3):741–743.

Rodriguez IA, Rasoazanabary E, Godfrey LR. 2013. What's eating *Microcebus*? Endo- and ectoparasite ecology of *Microcebus griseorufus* at Beza Mahafaly Special Reserve, Madagascar. *American Journal of Physical Anthropology* 150:234–234.

Schad J, Ganzhorn JU, Sommer S. 2005. Parasite burden and constitution of major histocompatibility complex in the Malagasy mouse lemur, *Microcebus murinus*. *Evolution* 59(2):439–450.

Schwensow N, Dausmann K, Eberle M, Fietz J, Sommer S. 2010a. Functional associations of similar MHC alleles and shared parasite species in two sympatric lemurs. *Infection Genetics and Evolution* 10(5):662–668.

Schwensow N, Eberle M, Sommer S. 2010b. Are there ubiquitous parasite-driven major histocompatibility complex selection mechanisms in gray mouse lemurs? *International Journal of Primatology* 31(4):519–537.

Schwitzer C, Mittermeier RA, Johnson SE, *et al.* 2014. Averting lemur extinctions amid Madagascar's political crisis. *Science* 343(6173):842–843.

Sommer S, Rakotondranary SJ, Ganzhorn JU. 2014. Maintaining microendemic primate species along an environmental gradient – parasites as drivers for species differentiation. *Ecology and Evolution* 4(24):4751–4765.

Stuart MD, Strier KB. 1995. Primates and parasites – a case for a multidisciplinary approach. *International Journal of Primatology* 16(4):577–593.

Zimmermann E, Radespiel U. 2014. Species concepts, diversity, and evolution in primates: lessons to be learned from mouse lemurs. *Evolutionary Anthropology* 23:11–14.

Zohdy S, Grossman MK, Fried IR, *et al.* 2015. Diversity and prevalence of diarrhea-associated viruses in the lemur community and associated human population of Ranomafana National Park, Madagascar. *International Journal of Primatology* 36(1):143–153.

11 A review of ectoparasites in the Cheirogaleidae

Sarah Zohdy and Lance A. Durden

Introduction

Ectoparasites of terrestrial mammals include several groups of insects (lice, fleas, louse flies, and others) and arachnids (mites and ticks) that inhabit the host skin or pelage and have morphological adaptations for this habitat and for host attachment. A diverse assemblage of blood-feeding flies including mosquitoes, sandflies, black flies, biting midges, horse and deer flies and certain muscoid flies have very short-term associations with mammals (only during feeding) and do not have adaptations for living on hosts; these flies will not be discussed in detail in this chapter. Ectoparasites are a diverse component of the Malagasy fauna, and have the potential to impact the health and well-being of their hosts. Because lemurs provide a biotic environment and ecosystem for their ectoparasites, an understanding of lemur ecology can directly include lemur ectoparasite ecology; similarly, an understanding of lemur ectoparasites can provide further information on the health and evolutionary history of their hosts. Unfortunately, there has been no coordinated research effort to investigate these parasites, and most of the research has been directed only at the description of new taxa. Consequently, very little is known about the relationships between cheirogaleid lemurs and their ectoparasites; however, there is great potential for the ecological and systematic relationships between these parasites and their hosts to be revealed.

Ectoparasites include lice (Phthiraptera), fleas (Siphonaptera), louse flies (Diptera), true bugs (Hemiptera), and ticks and mites (Arachnida: Acari). Most mites known to biologists are free-living, predaceous or plant-feeders, but there is a very diverse subgroup that includes parasites of animals. In contrast, all ticks are parasites, feeding primarily on the blood of their hosts and acting as vectors for numerous infectious pathogens. Sucking lice are dorsoventrally flattened, feed exclusively on the blood of their hosts and also spend every stage (eggs, three nymphal instars and adults) of their life cycle on the host. Members of the Siphonaptera (fleas) are bilaterally flattened, wingless insects that feed solely on the blood of birds and

The Dwarf and Mouse Lemurs of Madagascar: Biology, Behavior and Conservation Biogeography of the Cheirogaleidae, eds. S.M. Lehman, U. Radespiel, E. Zimmermann. Published by Cambridge University Press. © Cambridge University Press 2016.

mammals. A few members of the Hemiptera (true bugs) are important blood-feeders and, in the neotropical region, some are vectors of a zoonotic protozoan parasite. Among the Diptera, the vast majority of species are free-living; however, there are several taxa whose adults are adapted for ectoparasitism, such as the louse flies (family Hippoboscidae). The larval stages (maggots) of some Diptera are parasitic and feed subdermally or internally on the living tissues of vertebrates to cause a condition known as myiasis; these larval dipterans will not be considered further in this chapter.

The duration of association between the parasite and host can vary considerably, even in those associations that are obligate. There are numerous examples in which the parasites are in continuous contact with the host (e.g., sucking lice). In these cases, all life-cycle stages are dependent on their association with the host and they cannot usually survive for any length of time away from the host (Durden, 2001). In some cases, only specific life-cycle stages are in continuous association with the host, although the parasite has an absolute requirement for the host. In contrast, some blood-feeding arthropods are only intermittently associated with the host and the rest of their time is spent free-living. In these instances, oviposition is generally away from the host and larval stages develop without dependence on the host. Blood-feeding adult flies such as mosquitoes, tabanids (horse and deer flies), black flies and some muscoid flies have very short interactions with the host yet have a serious impact on the host's health through both direct effects of the feeding activity and indirectly, because they can act as vectors of a variety of pathogens.

In this chapter, we discuss the ectoparasite taxa that have been collected from cheirogaleid lemurs to date: the mites, lice, and hard ticks, and other ectoparasites in Madagascar that have not yet been described from cheirogaleid lemurs, but may also parasitize them. We also discuss how ectoparasites can be used as tools in host ecology, and how they can act as vectors of wildlife disease.

Ectoparasites collected from cheirogaleid lemurs

Ticks

Ticks, like other ectoparasites, are typically found on the head of cheirogaleids, particularly on the ears, and near the eyes, nose, and mouth (Figure 11.1, which shows lice), presumably because, as in most other mammals, effective anti-ectoparasite self-grooming is performed by the host's teeth (Freeland, 1981). At least three species of ticks are known to parasitize cheirogaleids: *Ixodes lemuris*, *Haemaphysalis lemuris*, *Haemaphysalis* sp. and probably *H. simplex* (Table 11.1). Both *I. lemuris* (Figure 11.2) and *H. lemuris* (Figure 11.3) appear to be fairly widespread ectoparasites of several species of lemurs (Uilenberg *et al.*, 1979; Klompen, 2003; Durden *et al.*, 2010; Blanco *et al.*, 2013) with all stages (larvae, nymphs, males and females) parasitizing these hosts. However, the immature stages (larvae and/or nymphs) of other Madagascar ticks, particularly some belonging to the genus *Haemaphysalis*, also parasitize lemurs. As a general rule, immature stages of ticks tend to be less

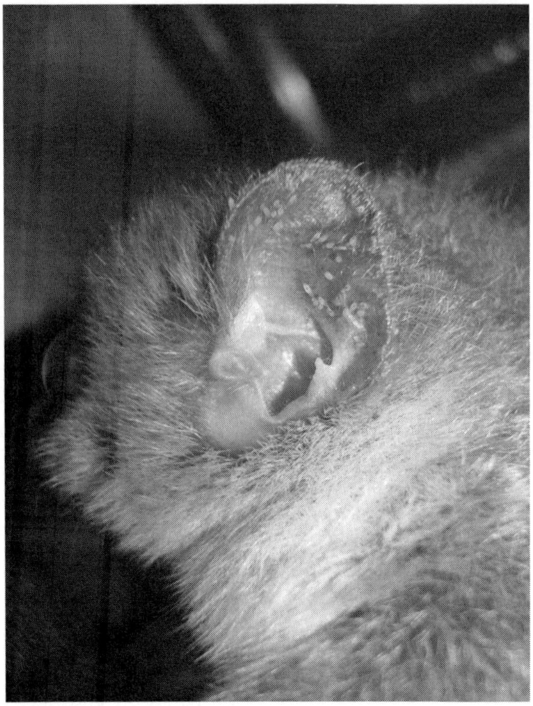

Figure 11.1 Ectoparasites are often found on the head of cheirogaleid lemurs. Lice found on the ears of *Microcebus rufus* at Ranomafana National Park (modified from Durden *et al.*, 2010).

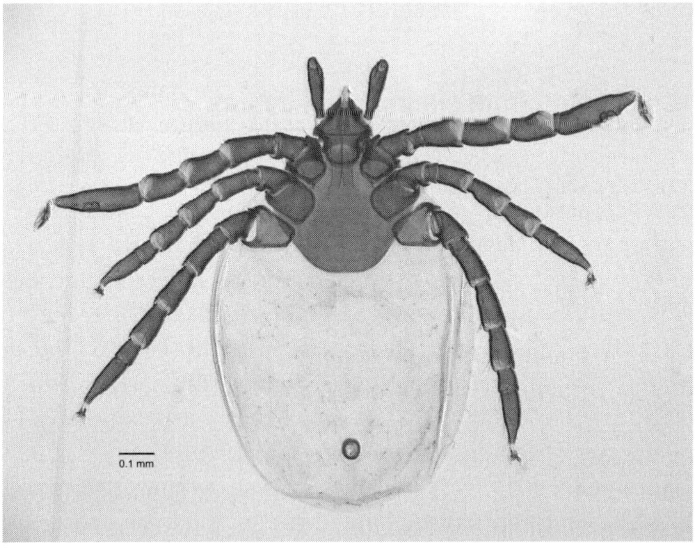

Figure 11.2 Cleared and slide-mounted *Ixodes lemuris* larva as reported in Blanco *et al.* (2013).

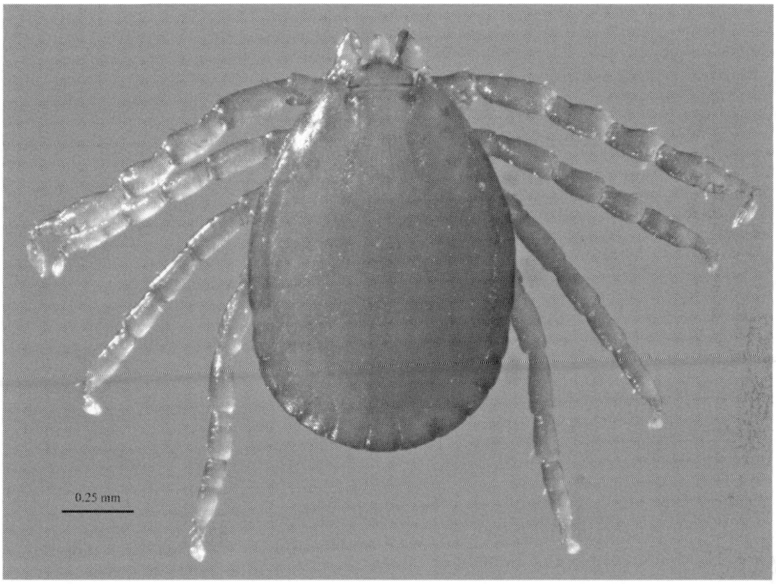

Figure 11.3 Image of a *Haemaphysalis lemuris* nymph collected from *M. rufus* as reported in Durden *et al.* (2010).

host-specific than conspecific adults (Durden, 2006). This certainly holds true for the immature stages of some species of *Haemaphysalis* in Madagascar which typically parasitize other native mammals such as tenrecs as adults, but which parasitize tenrecs, lemurs and other native mammals as immatures (Uilenberg *et al.*, 1979; Klompen, 2003; Durden *et al.*, 2010).

Lice

At least three species of sucking lice are known to parasitize cheirogaleids: *Lemurpediculus verruculosus* from *Microcebus rufus* (Figure 11.4) and two undescribed additional species of *Lemurpediculus*, one each from *Cheirogaleus sibreei* and *C. crossleyi*, respectively (Table 11.1). These lice are typically found on the head of the host, particularly near the ears (Figure 11.1); and every stage of *Lemurpediculus verruculosus* has been found on the testes of male *Microcebus rufus* before and during the breeding season, suggesting this location as an important breeding ground for *Lemurpediculus verruculosus* (Zohdy *et al.*, 2012). The overlap in parasite and host breeding may suggest reproductive synchrony in both parasite and host, which may be induced by an increase in host sex hormones in the imbibed bloodmeal that a louse takes, a phenomenon that occurs in some species of fleas that parasitize rabbits (Rothschild and Ford, 1964). A potential fourth species of *Lemurpediculus* collected from *Microcebus griseorufus* was not identified to species (Rodriguez *et al.*, 2012) and has not been examined by us. Another species, *Lemurpediculus petterorum*, is known to parasitize *Lepilemur* sp. (Megaladapidae) (Durden and Musser, 1994) so the genus *Lemurpediculus* is not exclusively associated with cheirogaleids.

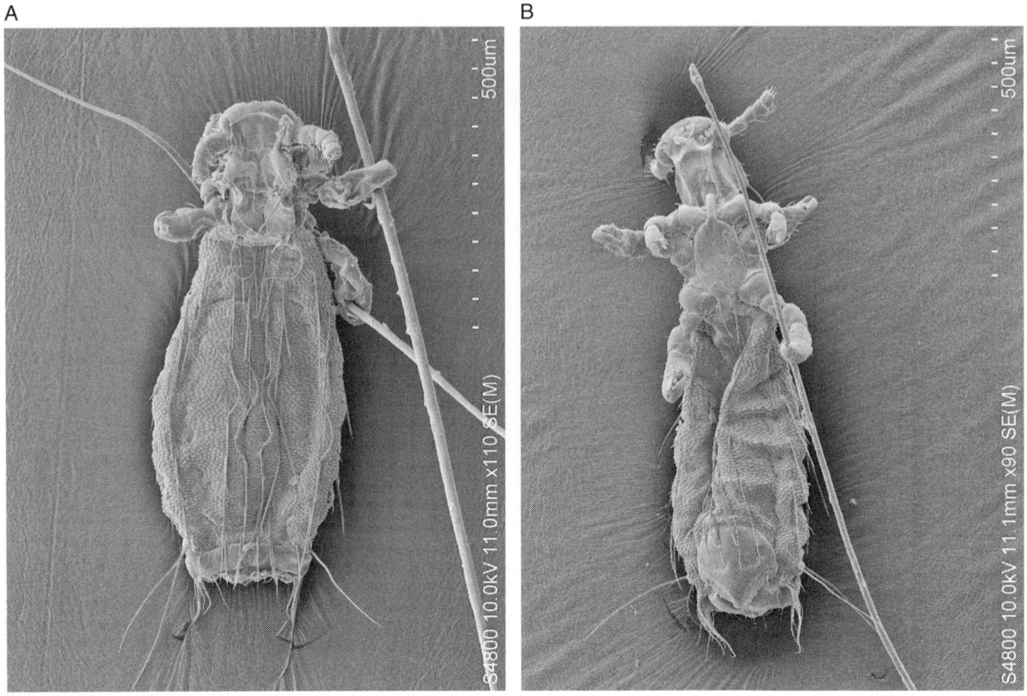

Figure 11.4 Scanning electron micrograph image of *Lemurpediculus verruculosus* female found on the ears of *Microcebus rufus*. (A) Dorsal view. (B) Ventral view.

However, some other (larger) lemurs are parasitized by members of the sucking louse genus *Phthirpediculus*: *Avahi laniger* (Indriidae) by *P. avahidis*, *Eulemur mongoz* (Lemuridae) by *P. brygooi* and *Propithecus edwardsi* (Indriidae) by *P. propitheci* (Durden and Musser, 1994: note, these authors incorrectly listed the host of *P. propitheci* to be *Lepilemur edwardsi*). Sucking lice are permanent, obligately hematophagous ectoparasites of eutherian mammals that are intimately associated with their hosts with many species being host-specific to one species of mammal (or sometimes to a few closely related mammalian species; Durden, 2001). Some sucking lice are not only host-specific to one species, but specific to one region of the body of their specific host species. We predict that new species of sucking lice will be collected from other species of lemurs including cheirogaleids, perhaps with a distinct species of sucking louse on each lemur species. Sucking lice inhabit the skin surface of their hosts where they imbibe blood, but gravid females may move a short distance from the skin to cement their eggs onto selected host hair shafts (Durden, 2001).

To date, just one species of chewing louse, *Trichophilopterus babakotophilus*, has been recorded from lemurs and, contrary to the situation with the apparently host-specific lemur-associated sucking lice, it has been recorded from several different species of lemurs (*Indri indri*, *Propithecus diadema*, *P. verreauxi*, *Eulemur coronatus*, *E. macaco* and *E. mongoz*; Price *et al.*, 2003). These are all relatively large lemurs and, based on current data, it appears that *T. babakotophilus* does not parasitize small lemurs such as cheirogaleids.

Mites

Several species of relatively small mites parasitize cheirogaleids (Table 11.1). As a group, these mites exploit their cheirogaleid hosts in quite different ways. Interestingly, although three species of larger gamasine mites (*Aetholaelaps sylstrai*, *A. trilyssa* and *Liponysella madagascariensis*) are known to parasitize larger lemurs (OConnor, 2003), only smaller mites have thus far been recorded from smaller lemurs such as cheirogaleids.

Five species of atopomelid fur mites belonging to the genus *Listrophoroides* (subgenus *Lemurlistrophoroides*) are known to parasitize cheirogaleids (Table 11.1). Unfortunately, none of the hosts of these mites were identified to species, so one future challenge will be to determine their exact host relationships, because many species of atopomelids are host-specific (Fain, 1976). Atopomelids align themselves along host hair shafts which they grasp with their highly modified anterior legs while they feed on hair surface scales, sebaceous secretions or skin detritus (Krantz, 2009). *Listrophoroides* spp. mites are widely distributed, but 6 of the known 16 subgenera in this genus are endemic to Madagascar where they are intimately associated with endemic mammals including lemurs, tenrecs and nesomyine rodents; conversely, members of the other subgenera that occur outside of Madagascar almost exclusively parasitize murid rodents (OConnor, 2003).

The chortoglyphid mite *Alabidopus microcebus* which parasitizes *Microcebus murinus* is the only known representative of this mainly Australasian mite family known to occur on Madagascar (OConnor, 2003). The deutonymphal (hypopial) stage of *A. microcebus* enters the hair follicles of its lemur host and imbibes host fluids/tissue through its integument (it lacks a mouth at this stage), whereas other stages of this mite are free-living (OConnor, 2003).

Another fur mite recorded from cheirogaleids is the chirodiscid *Lemuroecius cheirogalei*, which parasitizes *Cheirogaleus major* (Table 11.1). This is the only chirodiscid known to parasitize a primate, with most other members of the family occurring on bats (OConnor, 2003). Morphologically, chirodiscids are even more adapted than atopomelids for grasping the hairs of their hosts. In chirodiscids, some distal segments of the anterior legs are fused and expanded into a flattened, curved plate that surrounds the host hair to secure them to their hosts (OConnor, 2003). Like atopomelids, chirodiscids feed on host hair scales, sebaceous secretions or skin particles (Krantz, 2009).

The last mite species known to parasitize cheirogaleids is the psoroptid scab mite, *Cheirogalalges evansi*, which has been recorded from both *Cheirogaleus major* and *C. medius* (Table 11.1). Like other psoroptids, this mite penetrates the skin surface of its hosts with its sharp chelicerae to imbibe fluids. This feeding action induces localized areas of the host skin to transform into weeping lesions which harden into protective scabs under which the mites then feed (Krantz, 2009). Secondary bacterial skin infections can be associated with these lesions.

Table 11.1 Ectoparasites recorded from cheirogaleid lemurs.

Host	Ectoparasite(*)	Reference(s)
"Mouse lemur" (not identified)	*Lemurpediculus verruculosus* (L)	Ward (1951), Durden and Musser (1994)
Cheirogaleus sp.	*Listrophoroides angulatus* (A)	Fain (1970, 1976), OConnor (2003), Bochkov (2010)
(Not identified to species)	*Listrophoroides cheirogaleus* (A)	Fain (1970, 1976), OConnor (2003), Bochkov (2010)
	Listrophoroides gracilis (A)	Fain (1970, 1976), OConnor (2003), Bochkov (2010)
	Listrophoroides mediostrictus (A)	Fain (1970, 1976), OConnor (2003), Bochkov (2010)
	Listrophoroides tenellus (A)	Fain (1970, 1976), OConnor (2003), Bochkov (2010)
Cheirogaleus crossleyi	*Lemurpediculus* n. sp. # 2 (L)	Durden and Blanco (unpublished)
	Ixodes lemuris (T)	Blanco *et al.* (2013)
Cheirogaleus major	*Cheirogalalges evansi* (P)	Fain (1963, 1966), OConnor (2003), Bochkov and OConnor (2006), Bochkov (2010)
	Lemuroecius cheirogalei (C)	Fain (1968, 1976), OConnor (2003), Bochkov (2010)
Cheirogaleus medius	*Cheirogalalges evansi* (P)	Fain (1963, 1966), OConnor (2003), Bochkov and OConnor (2006), Bochkov (2010)
Cheirogaleus sibreei	*Lemurpediculus* n. sp. # 1 (L)	Durden and Blanco (unpublished)
	Ixodes lemuris (T)	Blanco *et al.* (2013)
Microcebus griseorufus	*Lemurpediculus* cf *verruculosus* (L)	Rodriguez *et al.* (2012)
	Haemaphysalis lemuris (T)	Rodriguez *et al.* (2012)
	Haemaphysalis simplex? (T)	Rodriguez *et al.* (2012)
Microcebus murinus	*Alabidopus microcebus* (G)	Fain and Lukoschus (1977), OConnor (2003)
Microcebus rufus	*Lemurpediculus verruculosus* (L)	Durden *et al.* (2010), Zohdy *et al.* (2012)
	Haemaphysalis lemuris (T)	Durden *et al.* (2010), Blanco *et al.* (2013)
	Haemaphysalis sp. (T)	Durden *et al.* (2010)
	Ixodes lemuris (T)	Blanco *et al.* (2013)
Microcebus cf *rufus*	*Ixodes lemuris* (T)	Blanco *et al.* (2013)

* A, atopomelid fur mite (Arachnida, Acari, Atopomelidae); C, chirodiscid fur mite (Arachnida, Acari, Chirodiscídae); G, glycyphagid hypopial mite (Arachnida, Acari, Glycyphagidae); P, psoroptid scab mite (Arachnida, Acari, Psoroptidae); L, sucking louse (Insecta, Phthiraptera, Anoplura, Polyplacidae); T, hard tick (Arachnida, Acari, Ixodidae).

Other ectoparasites

As already noted, additional species of ectoparasites almost certainly await discovery on various species of cheirogaleids (and other lemurs) in Madagascar. No ectoparasites are known from some species of *Cheirogaleus* or *Microcebus* and none are yet known from any species of cheirogaleids belonging to the genera *Allocebus*, *Mirza*, or *Phaner*. Concerted efforts to collect arthropods from these small lemurs is encouraged in order to increase our knowledge of the ectoparasite fauna that depends on these hosts for its continued survival and to determine if any of these ectoparasites transmit pathogens during their feeding activities or debilitate their hosts in other ways.

Some ectoparasites appear to be notably absent from cheirogaleids. For example, two species of endemic hippoboscid flies (*Allobosca crassipes* and *Parabosca alata*) collectively parasitize several species of larger lemurs (*Eulemur macaco*, *E. rubriventer*, *Lepilemur mustelinus*, *Avahi laniger*, *Varecia variegata*, *Propithecus diadema*, *P. edwardsi*, and *P. verreauxi*) in Madagascar (Rahola *et al.*, 2011), but have not been recorded from small lemurs, which could perhaps be more efficient groomers of large flies crawling through their fur. The apparent absence of larger (gamasine) mites on cheirogaleids has already been mentioned. Further, there are no published records of chiggers (family Trombiculidae) from any species of lemurs. This is remarkable and it seems likely that several species of chiggers parasitize lemurs including cheirogaleids. Perhaps the small size of these mites, or the fact that few ectoparasite surveys of lemurs have been undertaken, are factors that contribute to the lack of chigger records from lemurs. Although the chigger fauna of Madagascar is extensive, it has been inadequately studied so far, with only a handful of species reported from the island (Vercammen-Grandjean and Watkins, 1965; OConnor, 2003). Chiggers feed on lysed host tissues through a proteinaceous straw-like feeding tube (a stylostome) that remains embedded in host tissue following chigger detachment, often causing intense pruritus (itching) and reddened host lesions (Krantz, 2009; Mullen and OConnor, 2009).

Some wide-ranging, non-native, introduced ectoparasite species may be recorded from cheirogaleids in the future. For example, fleas are rarely associated with primates, but one introduced species that shows little host specificity, the sticktight flea *Echidnophaga gallinacea*, has already been recorded from the gray bamboo lemur, *Hapalemur griseus* (Duchemin *et al.*, 2003). As its common name implies, this flea attaches to the host for long periods of time with its elongate mouthparts and it could be a threat to several species of lemurs if it continues to spread in Madagascar. Females of another introduced flea, the jigger or chigoe, *Tunga penetrans*, embed subdermally in host tissues to feed and induce suppurating lesions that are prone to secondary bacterial infections. This flea has so far been recorded from humans, pigs, and dogs in Madagascar (Duchemin *et al.*, 2003), which seem to be its preferred hosts, but it is possible that it could also parasitize lemurs in the future.

Parasite–host coevolution

The island of Madagascar is very rich in biodiversity and home to many unique vertebrate species. These vertebrate organisms, such as the cheirogaleid lemurs, play host to their own unique array of parasite species. In the same way that Madagascar's island biogeography has influenced lemur diversity, lemurs, which act as an ecosystem for their parasites, have a similar effect on parasite diversity. In a sense, each lemur is its own island hotspot for a unique array of hitchhiking parasites. Wherever the host goes the parasite communities follow; therefore, an understanding of parasite–host coevolution is necessary in order to understand the role that hosts play as parasite ecosystems and that parasites play in host health and life history.

As obligate parasites, blood-feeding lice require constant contact with their hosts, and every stage of the life cycle occurs on the host. Because of this close relationship, cospeciation between lice and their hosts has occurred over millennia, to the extent that most lice are specialized to one host species, and cannot survive off of the host (Marshall, 1981; Durden, 2001). Because of this host specificity and association, lice can be used as a tool to infer the evolutionary history of their hosts, in this case cheirogaleid lemurs. Because parasites evolve faster than their hosts, they can not only be used as a tool to better understand host speciation events, but they can also be used to investigate host evolutionary history. This information together with information on host evolutionary history can provide a more pristine insight into cospeciation events, and the past, present, and future of parasite–host evolutionary histories.

Parasite–host coevolution can be exploited in other ways to gain further information about host biology. In a previous study (Zohdy *et al.*, 2012), we aimed to better understand the movement in both parasites and hosts in a brown mouse lemur population in central southeastern Madagascar at Ranomafana National Park. To do so, *Microcebus rufus* was first identified as the host of the sucking louse species *Lemurpediculus verruculosus* (Durden *et al.*, 2010). Because sucking lice are obligate ectoparasites that cannot survive off of their hosts, in order to move throughout host populations, lice require host individuals to come into direct contact. Also, louse flies (Hippoboscidae) have not been recorded from mouse lemurs and because phoresy (hitch-hiking to a new host on a larger, more mobile ectoparasite) between sucking lice and louse flies has never been recorded, we could be confident that louse transfer was via direct host contact. As mouse lemurs are nocturnal and arboreal and hence difficult to follow in the rainforests, we exploited this sucking louse trait not only to gain a better understanding of the movement of lice throughout a population of known lemurs, but to better understand the unseen social contacts that brown mouse lemurs have. We did this by marking individual lice on lemurs with a nail-polish dot code to identify the specific mouse lemur that the lice were first found on. By monitoring the movement of the lice between individual mouse lemurs, we were able to create a mouse–lemur social network as revealed by parasite transfers. Further, by tracking the lice and their lemurs, we were able to identify individual lemurs in the population as louse "superspreaders," individuals that had a high turnover of lice and were continually donating their lice to other mouse lemurs, and "supercollectors," individuals that seemed to attract lice the most from other individuals. If lice are vectors of pathogens between mouse lemurs as they are in other hosts, identifying "superspreaders" allows us to find which individual lemurs are the most highly connected to others, and hence could be responsible for a population-wide louse-borne epidemic.

Ectoparasites and host health

The direct effects of ectoparasite infestation are typically associated with parasite-feeding activities and therefore include blood loss, inflammation, immunosuppression, and secondary infection of wounds caused by ectoparasite attachment and feeding. In these cases, ectoparasites can cause external injury to their wildlife hosts, and hence pose a threat to the species. The indirect effects of ectoparasites (as vectors of disease agents) can be fatal to their hosts and can cause population-wide and zoonotic and anthropozoonotic outbreaks. Due to the dietary adaptations that many ectoparasites have, feeding on blood or other tissues/fluids, ectoparasites have the potential to come into contact with pathogens that the hosts are carrying. For example, individual parasites that imbibe host blood (such as ticks, lice, and fleas) come into contact with blood-borne pathogens circulating within the host with every feeding bout. If those parasites then attach to new hosts they have the potential to spread blood-borne disease agents within the host population. It is because of this vector potential that arguably all blood-feeding ectoparasites have the ability to act as vectors. Therefore, identifying and understanding parasite ecology has potential importance with respect to disease ecology and parasite–host interactions. Many ectoparasites (an exception is the causative agent of plague, which is transmitted orally) spread disease by defecating fecal material contaminated with pathogens from the bloodmeal near the point on the skin where the ectoparasite mouthparts were inserted which creates an open wound. If the fecal material is then rubbed into the insertion site the pathogen or parasite can enter the new host's system and this infection can lead to virulence in the host species that can then be passed on by a new ectoparasite.

As all lemurs are threatened due to habitat destruction, it is important to understand the health of cheirogaleid lemurs to better understand how habitat destruction influences lemur health, and how potential disease outbreaks in lemur populations have the potential to cause local extinction events. Because many of the ectoparasites mentioned in this chapter have the potential to spread disease agents in lemur populations (and potentially cause spillover events into human and other animal populations) it is important to first identify what ectoparasite species exist and their prevalence in these lemur populations, and then try to identify whether or not those species can act as vectors. This information, combined with individual lemur morphological and physiological profiles, can help us to better understand which individuals in the host populations are most susceptible to ectoparasites, and what causes that susceptibility. With the population-wide data that are being collected island-wide for cheirogaleid species in Madagascar, there is great potential to understand what characteristics make hosts susceptible to being parasitized. Combining these data with social network information provides the basic framework to pinpoint key individual lemurs in a population and to model potential population-wide transmission routes.

Hard ticks (Ixodidae) are exclusively hematophagous ectoparasites and are second in importance only to mosquitoes globally as vectors of animal and human disease agents (Nicholson *et al.*, 2009). That these ticks have already been found to parasitize cheirogaleid lemurs suggests that they may act as vectors of pathogens in these lemurs as they do in other animals. Hard ticks feed on their hosts at every developmental stage where they are potentially exposed to host pathogens in the bloodmeal. As they molt to the next lifestage and (typically) parasitize a larger host, they have the ability to continue to carry the pathogens through a molt (i.e., transstadial transmission). The tick *I. lemuris* found on mouse and dwarf lemurs in the eastern rainforests is potentially an important vector. Only immature ticks have been found on mouse lemurs, whereas all active stages (larvae, nymphs, and adults) have been found on dwarf lemurs (Blanco *et al.*, 2013), suggesting that mouse lemurs may not be able to support adult ticks (or that they can effectively groom them from their fur) and may play a significant role in tick-borne disease transmission dynamics as both recipients and donors of pathogens transmitted in tick bloodmeals.

Conclusion

Despite the growing information now known about cheirogaleid species, relatively little is known about their ectoparasites or about cheirogaleid–ectoparasite dynamics. Some ectoparasites of cheirogaleids such as mites and sucking lice exhibit host specificity (OConnor, 2003; Durden *et al.*, 2010), whereas others such as ticks are less host-specific (Blanco *et al.*, 2013). Due to the nature of cheirogaleid field studies, these lemur species are often captured, providing an ideal opportunity to examine ectoparasite diversity and ectoparasite–host interactions, especially because mark–recapture data provide the opportunity to collect a rich physiological data set from host species and host individuals, providing invaluable opportunities to study parasite–host interactions. Further research will promote additional understanding of the still-unknown ectoparasite taxa that use cheirogaleid species as an ecosystem at descriptive, evolutionary, and disease ecology levels.

Acknowledgments

We thank Hans Klompen (Ohio State University) and Marina Blanco (Duke University) for sharing specimens and information about ectoparasites of various species of lemurs. We would like to thank the University of Helsinki, Department of Chemistry for use and assistance using the Field Emission Scanning Electron Microscope (FESEM), and Emory University's NIH T32 (FIRST) award for funding.

References

Blanco MB, Elfawal MA, Durden LA, *et al.* 2013. Genetic diversity of ixodid ticks parasitizing eastern mouse and dwarf lemurs in Madagascar, with descriptions of the larva, nymph and male of *Ixodes lemuris* (Acari: Ixodidae). *Journal of Parasitology* 99:11–18.

Bochkov AV. 2010. A review of mammal-associated Psoroptidia (Acariformes: Astigmata). *Acarina* 18:99–260.

Bochkov AV, OConnor BM. 2006. Revision of the genus *Gaudalges* (Acari: Psoroptidae), parasites of Malagasy lemurs. *Acarina* 14:3–20.

Duchemin J-B, Ratovonjato J, Duplantier J-M. 2003. Siphonaptera, fleas, *Parasy Gasy*. In SM Goodman, JP Benstead (eds.), *The Natural History of Madagascar* (pp. 687–692). University of Chicago Press, Chicago.

Durden LA. 2001. Lice (Phthiraptera). In WM Samuel, MJ Pybus, AA Kocan (eds.), *Parasitic Diseases of Wild Mammals*, 2nd edition (pp. 3–17). Iowa State University Press, Ames.

Durden LA. 2006. Taxonomy, host associations, life cycles and vectorial importance of ticks parasitizing small mammals. In S Morand, BR Krasnov, R Poulin (eds.), *Micromammals and Macroparasites: From Evolutionary Ecology to Management* (pp. 91–102). Springer, Tokyo.

Durden LA, Musser GG. 1994. The sucking lice (Insecta, Anoplura) of the world: a taxonomic checklist with records of mammalian hosts and geographical distributions. *Bulletin of the American Museum of Natural History* 218:1–90.

Durden LA, Zohdy S, Laakkonen J. 2010. Lice and ticks of the eastern rufous mouse lemur, *Microcebus rufus*, with descriptions of the male and third instar nymph of *Lemurpediculus verruculosus* (Phthiraptera: Anoplura). *Journal of Parasitology* 96:874–878.

Fain, A. 1963. Les acariens producteurs de gale chez les lèmuriens et singes avec une étude des Psoroptidae (Sarcoptiformes). *Bulletin de l'Instutut Royal des Sciences Naturelles de Belgique* 39:1–125.

Fain A. 1966. Les acariens producteurs de gale chez les lèmuriens et singes II. Nouvelles observations avec description d'une espèce nouvelle. *Acarologia* 8:94–114.

Fain A. 1968. Notes sur trois acariens remarquables (Sarcoptiformes). *Acarologia* 10:86–110.

Fain A. 1970. Diagnoses de nouveaux Listrophorides de la famille Atopomelidae (Acarina: Sarcoptiformes). *Bulletin et Annales de la Société Royale d'Entomologie de Belgique* 106:275–306.

Fain A. 1976. *Faune de Madagascar 42. Arachnides, Acariens, Astigmata, Listrophoroidea.* ORSTOM, CNRS, Paris. 131 pp.

Fain A, Lukoschus FS. 1977. New endofollicular or subcutaneous hypopi from mammals (Acarina: Astigmata). *Acarologia* 19:484–493.

Freeland WJ. 1981. Functional aspects of primate grooming. *Ohio Journal of Science* 81:173–177.

Klompen H. 2003. Ixodida, ticks, *Kongona*. In SM Goodman, JP Benstead (eds.), *The Natural History of Madagascar* (pp. 588–593). University of Chicago Press, Chicago.

Krantz GW. 2009. Habits and habitats. In GW Krantz, DE Walter (eds.), *A Manual of Acarology*, 3rd edition (pp. 64–82). Texas Tech University Press, Lubbock.

Marshall AG. 1981. *The Ecology of Ectoparasitic Insects.* Academic Press, London.

Mullen GR, OConnor BM. 2009. Mites (Acari). In GR Mullen, LA Durden (eds.), *Medical and Veterinary Entomology*, 2nd edition (pp. 433–492). Elsevier, Amsterdam.

Nicholson WL, Sonenshine DE, Lane RS, Uilenberg G. 2009. Ticks (Ixodoidea). In GR Mullen, LA Durden (eds.), *Medical and Veterinary Entomology*, 2nd edition (pp. 493–542). Elsevier, Amsterdam.

OConnor BM. 2003. Acariformes: parasitic and commensal mites of vertebrates. In SM Goodman, JP Benstead (eds.), *The Natural History of Madagascar* (pp. 593–602). University of Chicago Press, Chicago.

Price RD, Hellenthal RA, Palma RL, Johnson KP, Clayton DH (eds.) 2003. The chewing lice: world checklist and biological overview. *Illinois Natural History Survey Special Publication* No. 24. 501 pp.

Rahola N, Goodman SM, Robert V. 2011. The Hippoboscidae (Insecta: Diptera) from Madagascar, with new records from the "Parc National de Midongy Befotaka." *Parasite* 18:127–140.

Randimby F, Duplantier JM, Rantovonjato J, *et al.* 2001. Analyse de la spécificité parasitaire des tiques. Intérêt de la situation de Madagascar et des échantillonnages approfondis. *Archives de l'Institut Pasteur de Madagascar* 67:49–52.

Rodriguez IA, Rasoazanabary E, Godfrey LR. 2012. Multiple ectoparasites infest *Microcebus griseorufus* at Beza Mahalafy Special Reserve, Madagascar. *Madagascar Conservation and Development* 7:45–48.

Rothschild M, Ford B. 1964. Breeding of the rabbit flea (*Spilopsyllus cuniculi*) (Dale) controlled by the reproductive hormones of the host. *Nature* 201:103–104.

Uilenberg G, Hoogstraal H, Klein JM. 1979. Les tiques (Ixodoidea) de Madagascar et leur rôle vecteur. *Archives de L'Institut Pasteur de Madagascar*, Numéro Spécial, Antanarivo, Madagascar, 153 pp.

Vercammen-Grandjean PH, Watkins SG. 1965. *Schoutedenichia gigantea* and *S. tiptoni*, two new and extraordinary species from Madagascar (Acarina: Trombiculidae). *Acarologia* 7:683–694.

Ward RA. 1951. Description of a new species of Anoplura (*Lemurphthirus verruculosus*) from a Madagascar lemur. *Entomological News* 62:190–192.

Zohdy S, Kemp AD, Durden LA, Wright PC, Jernvall J. 2012. Mapping the social network: tracking lice (*Lemurpediculus verruculosus*) in a wild primate population (*Microcebus rufus*) to infer social contacts and vector potential. *BioMed Central Ecology* 12:4.

Part III

Cheirogaleidae: behavior and ecology

12 Morphology, behaviour, ranging patterns and habitat use of the northern giant mouse lemur *Mirza zaza* in Sahamalaza, northwestern Madagascar

E. Johanna Rode-Margono, Christoph Schwitzer, and K. Anne-Isola Nekaris

Introduction

Madagascar is one of the most important biodiversity hotspots in the world, underpinned by its high proportion of endemic species and high rates of deforestation of about 1500 km^2 per year (Myers *et al.*, 2000; Moat and Smith, 2007). Since 1997 until today, the number of described lemur taxa has risen from 50 to 107 species and subspecies (Schwitzer *et al.*, 2013). In 2012, a new IUCN Red List assessment revealed that 94% of the then 103 described lemur taxa had been placed in one of the Threatened categories (Schwitzer *et al.*, 2013). Apart from geographic distribution and population densities, thorough knowledge of the biology and ecology, such as habitat needs or life-history data, are crucial to design conservation measures (Schwab and Ganzhorn, 2004; Merker *et al.*, 2005; Schwitzer *et al.*, 2013). In particular, the behaviour and ecology of nocturnal lemur species are poorly studied (Chapter 19); up until the reassessment in 2012, 37 of the 40 lemur taxa assessed as Data Deficient on the IUCN Red List were nocturnal (Schwitzer *et al.*, 2013).

One of these poorly studied species is the nocturnal northern giant mouse lemur *Mirza zaza* Kappeler & Roos, family Cheirogaleidae, which was distinguished from *M. coquereli* Grandidier as a unique species in 2005 (Kappeler *et al.*, 2005). *M. zaza* is a nocturnal lemur of approximately 300 g that lives in the dry deciduous forests of northwestern Madagascar up to the transition zone to the Sambirano evergreen rainforest domain in the north (Kappeler *et al.*, 2005; Markolf *et al.*, 2008). The species is IUCN Red-Listed as Endangered (IUCN, 2014). The only population estimates so far yielded the highest encounter rates in degraded forests with a high number of mango trees (*Mangifera indica*, family Anacardiaceae), the fruits of which may be an important food source for *M. zaza* (Markolf *et al.*, 2008;

The Dwarf and Mouse Lemurs of Madagascar: Biology, Behavior and Conservation Biogeography of the Cheirogaleidae, eds. S.M. Lehman, U. Radespiel, E. Zimmermann. Published by Cambridge University Press. © Cambridge University Press 2016.

Mittermeier, pers. comm.). Western dry forest is one of the forest types declining the most quickly in Madagascar, with ongoing threats to remaining and already heavily fragmented forests where *M. zaza* occurs (Schwitzer and Lork, 2004; Schwitzer *et al.*, 2007).

In order to determine species distribution boundaries, taxonomy and biogeography must be properly understood. Several morphological variables differ between *M. zaza* and *M. coquereli*, *M. zaza* generally being smaller but having slightly larger testes (Kappeler *et al.*, 2005). The geographical distribution of *M. zaza* seems to be limited by two large rivers, the Maevarano River in the south and the Mahavavy in the north (Markolf *et al.*, 2008), and could be further separated by the Sambirano River crossing the currently assumed distribution of *M. zaza*, as large rivers are suggested to serve as species boundaries (Craul *et al.*, 2007).

Information on the behaviour and ecology of *M. zaza* is still scarce. In contrast to *M. coquereli* that has never been observed to share nests, *M. zaza* sleeps in group-exclusive nests with 2–8 individuals of all sex and age classes, including several closely related and unrelated males (Kappeler *et al.*, 2005; Rode *et al.*, 2013a). Home ranges of *M. coquereli* at Kirindy are on average 4 ha and overlap to a various extent with neighbouring animals (Kappeler, 1997a), but ranging patterns are unknown for *M. zaza*. Nest trees of *M. zaza* are characterised by large and tall trees with a high number of lianas (Rode *et al.*, 2013a).

Information about home range use and habitat requirements is crucial for conservation planning (Merker *et al.*, 2005; Neilson *et al.*, 2006; Schwitzer *et al.*, 2007). A home range is generally defined as the area traversed by the individual in its normal activities (Burt, 1943). Home range sizes give an indication about the species' space requirements that should be incorporated in habitat management plans (Crompton and Andau, 1986). Home range size may increase in less-suitable or lower-quality habitat, due to an animal's need to find sufficient resources (VanderWerf, 1993; Wiens and Zitzmann, 2003; Merker *et al.*, 2005). Home range patterns can also be used to define the rather elusive social organisation of nocturnal primate species (Harcourt and Nash, 1986; Müller, 1998; Nekaris and Bearder, 2011). Information about habitat use can be used to model species distribution and abundance (Rendigs *et al.*, 2003; Chouteau, 2004; Rode *et al.*, 2013b). Knowledge about the importance of certain microhabitat variables can help to more effectively manage endangered populations (Crompton and Andau, 1986; Rendigs *et al.*, 2003; Chouteau, 2004).

In this study, we provide preliminary information on intraspecific variations in the morphology, behaviour and ecology of *M. zaza* by (1) describing and comparing the morphology of *M. zaza* from Sahamalaza Peninsula to *M. zaza* from Ambato, two areas geographically separated by the large river Sambirano; (2) providing a description of behaviour including an activity budget, influence of lunar luminosity on activity, and feeding behaviour; (3) determining group and individual home range size and overlap; and (4) investigating habitat use in terms of height of the animal, position in trees and microhabitat preferences.

Methods

Ethics statement

This study was conducted under permit from the Madagascan Ministère de l'Environnement et des Forêts (Autorisation de recherche #231/11/MEF/SG/DGF/DCB.SAP/SLRSE du 27/04/10) and was ethically approved by the Welfare & Research Advisory Board of the Bristol, Clifton and West of England Zoological Society.

Study site

The study took place at Ankarafa Forest (14°22'64.2"S, 47°45'31.5"E), Sahamalaza, northwestern Madagascar, during the dry season for 10 weeks in 2010 from the beginning of May until the middle of July (Figure 12.1). Ankarafa Forest is situated on the Sahamalaza Peninsula, province Mahajanga, within the boundaries of the Sahamalaza–Iles Radama National Park (Aire Protégée Terrestre, Marine et Côtière; 26,000 ha), which is part of the country's protected area network managed through Madagascar National Parks (MNP). The Sahamalaza Peninsula is also a UNESCO Biosphere Reserve (1533 km^2), and is biogeographically located in a transition zone between the Sambirano evergreen rainforest domain in the north and the western dry deciduous forest region in the south (Schwitzer, 2005). The strict seasonal climate of this zone is represented by a dry, cool season from May to October and a rainy, hot season from November to April (Schwitzer *et al.*, 2007; Seiler *et al.*, 2013). Mean annual rainfall is 1600 mm, and yearly average temperature fluctuates around 28.0°C, with a maximum average temperature of 32.0°C in November and a minimum average of 20.6°C in August (Schwitzer *et al.*, 2007). The vegetation of the semi-humid forest shows characteristics of the two very different domains, with tree heights of up to 25 m (Schwitzer *et al.*, 2006). In contrast to the western dry deciduous forests, most trees in Sahamalaza do not shed their leaves. Due to anthropogenic disturbance (traditional slash and burn agriculture and clearance for cattle herds), only fragments of primary and secondary forest remain, isolated by savannah areas (Schwitzer and Lork, 2004; Schwitzer *et al.*, 2007). Other lemur species living in Sahamalaza are the blue-eyed black lemur *Eulemur flavifrons*, the aye-aye *Daubentonia madagascariensis*, the Sambirano lesser bamboo lemur *Hapalemur occidentalis*, the fat-tailed dwarf lemur *Cheirogaleus medius* and the Sahamalaza sportive lemur *Lepilemur sahamalazensis*. The species are all threatened by hunting and habitat loss (Schwitzer *et al.*, 2006). The Sahamalaza region has been listed as a priority site for conservation in the Lemur Conservation Strategy 2013 (Schwitzer *et al.*, 2013).

A primary and a secondary forest fragment were selected as study sites as the animals were known to be present there. The primary forest fragment (10 ha) had been degraded by a forest fire, but is in the process of regeneration (Seiler *et al.*, 2013). The secondary forest fragment (13 ha) mainly represents "mature" secondary forest with a small part of primary forest (Seiler *et al.*, 2013).

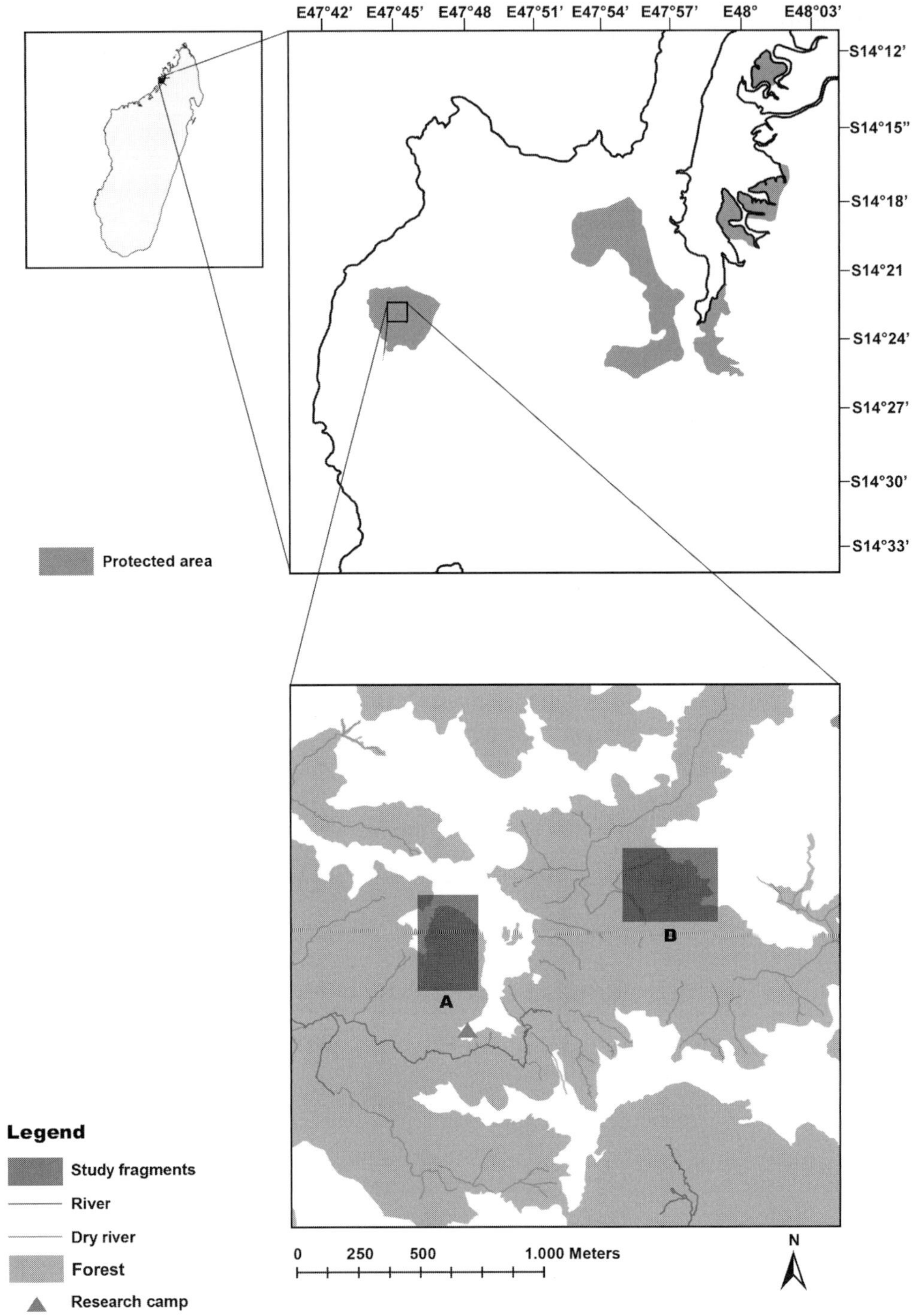

Figure 12.1 Map of Ankaraka Forest, Sahamalaza Peninsula, northwestern Madagascar, indicating the degraded primary forest fragment (A) and mature secondary forest fragment (B) where the study took place.

Data collection

We captured *M. zaza* using 24–30 live traps (Tomahawk Live Traps size 12, Tomahawk, USA) in both forests. Following the advice of experienced guides, the traps were placed systematically in the two forest fragments. We placed them firmly on branches or in vegetation tangles at approximately 1.5 m height. We baited the traps with banana in the evening and checked throughout the night and the latest at the next early morning (Kappeler *et al.*, 2005).

We anaesthetized the animals briefly with ketamine (0.01 ml/100 g body mass, 100 mg/ml; Lahann, 2008). We took standard morphometric measurements of adult *M. zaza* including body mass, body length, tail length, head length and width, canine height, ear length and width, hind foot, hand, upper and lower hind- and forelimb, and testes length and width (following Schmid and Kappeler, 1994). We used a standard scale (precise to 1 g), tape measure and an electronic calliper (Maplin, UK). Testis volume (V) was calculated with the formula for a regular ellipsoid

$$V = 1/6 * (\pi W^2 L)$$

where W was the average width of both testes and L was the length of the longest testis (Kappeler, 1997b).

Relative testis volume was calculated as the mean testis volume divided by mean body mass (Schwab, 2000). Age was classified according to body weight with 150–199 g juvenile, 200–249 g subadult and above 250 g adult (Kappeler *et al.*, 2005). Means of the measurements of first captures and recaptures to remove radio-collars were compared to means from Kappeler *et al.* (2005) for *M. zaza* animals from Ambato Peninsula north of Sahamalaza (13°25'S, 48°29'E) and *M. coquereli* animals from Kirindy in the south (20°40'S, 44°39'E). If not indicated otherwise, all measurements are in mm.

We cut an individual ring into the fur of the tail for easier identification during radio-tracking (Kappeler, 1997a). Animals were fitted with TW3 rubber-coated cable tie radio-collars (Biotrack Ltd, UK, 3–4 g, max. 2% of body weight). All collars were removed at the end of the study. We used a TR-4 receiver (Telonics Inc., USA; frequency range 150.545–150.969 MHz) and flexible Yagi antenna (Biotrack Ltd, UK) to locate the animals during the night. Home ranges were determined by nightly follows of focal animals (Pimley *et al.*, 2005).

Two teams followed the animals six days per week during two night shifts (18:00–00:00 hours and 00:00–05:00 hours). Before 18:00 and after 05:00 nests were observed. We used instantaneous behaviour sampling with 5-min intervals (Altmann, 1974), and the behaviour categories "moving" (continuous, directed movement from one location to another), "resting or sleeping" (remain stationary with eyes open or closed), "feeding and foraging" (movement associated with searching for food or actual consumption of food items), "grooming" (lick or scratch own fur) and "social behaviour" (including playing, chasing, allogrooming). We also determined the height above ground of the animal. As direct observations were very difficult due to dense canopy, we noted behaviour whenever the animal was seen after the interval had passed, not waiting until the next 5-min point. We changed the focal animal after approximately 2 hours of observation in order to receive a representative sample of each animal over the whole field period. As luminosity we recorded the exact percentage of the moon illuminated when above the horizon, using the programme MOONDV version 1 (Thomas, 1998). When below the horizon an illumination of 0 was recorded. To have a clear division between dark and bright nights we grouped luminosity into the categories "dark" (no moon) and "bright" (more than 50%). Details about feeding behaviour and food items were collected *ad libitum*.

We aimed for a 15-min interval between GPS locations of the focal animal to determine home ranges (Harris *et al.*, 1990), but due to erratic encounters and regular loss of the animal we also collected data after slightly shorter intervals. We took only GPS locations of direct sightings, but applied close triangulation in some cases. We used a Garmin GPS 60 (Garmin, UK). We did not use locations with a lower precision than 15 m and mean precision was 8.3 ± 2.1 m (n = 116). Home ranges were calculated with the programme RANGES 8 (Anatrack Ltd, UK) as 100% minimum convex polygon (MCP) and 95% fixed kernel (KM) with smoothing multiplier set to 1 and matrix set to the number of cells (n = 40). MCP draws a line around the outmost locations of an area to build a polygon (Kenward, 2001). Even though this method is known to include areas that might have never been visited by the animal and thus might overestimate the home range (Kenward, 2001; Pimley *et al.*, 2005), it was chosen for better comparability as it is used in many ecological studies (Harris *et al.*, 1990). Kernel estimators depend on the use intensity of different areas. Different centres of activity are possible, excursions of the animal are excluded and, thus, home range estimates are more accurate (Kenward, 2001; Pimley *et al.*, 2005). Each nest site used by the animals was included only once as we controlled the nest sites every day for nest analysis (cf. Rode *et al.*, 2013a) and wanted to avoid an overrepresentation of these areas. Group home ranges refer to home ranges of one nest group. A nest group was defined as all animals sleeping in one nest in the majority of observations (Rode *et al.*, 2013a). Home range size and the percentage overlap of neighbouring home ranges were analysed with RANGES 8. The calculated home range size depends on the number of locations sampled (Harris *et al.*, 1990; Seaman *et al.*, 1999). Enough locations are sampled when home range size reaches an asymptote and does not increase when further locations are added. RANGES 8 offers an automatic incremental area analysis that shows how area increases with increasing sample size. In this study, the home range size for several animals did not approach an asymptote. This can be the case if animals shift their home ranges or simply when not enough locations were sampled (Harris *et al.*, 1990). However, we included the results as minimum home range sizes.

We collected data on trees used by the animals during the night when they were active, and on generally available (random) trees. We identified used trees every 15 min when we took a GPS. We determined random trees by calculating random coordinate pairs lying between the minimum and maximum of the group home range's easting and northing coordinates and choosing the nearest tree to this point. We omitted random points falling outside the group home ranges or in unused habitat (adjacent savannah). We recorded the following variables on each tree: species, height, diameter at breast height (DBH), crown diameter (horizontally at the broadest point), number of lianas and connectivity (number of directly connected trees). Additionally, we sampled tree density with the point centred quarter method (Ganzhorn, 2003; Rode *et al.*, 2013b) by measuring the distance from the centre tree (used and random, respectively). We measured DBH and distance between trees to the nearest centimetre, while tree height and crown diameter were estimated to the nearest metre and averaged between the estimates of two observers. We distinguished small trees (stems between 5 and 10 cm DBH) and large trees (stems larger than 10 cm DBH). Although usually only trees above 10 cm DBH are considered in habitat description, we included smaller-sized trees as *M. zaza* has been observed using smaller vegetation as well (Ganzhorn, 2003). Central tendencies are reported as medians due to a skewed frequency distribution.

Data analyses

All body measurements were normally distributed. Because no difference in variables occurred between sexes, animals were combined and one-sample *t*-tests computed to compare variables to means of animals from Ambato and Kirindy (Dytham, 2011). We used Pearson's chi-squared tests to investigate whether the part of the night (before and after midnight) affected the activity budget and Fisher's exact test to determine whether luminosity affected the activity of animals. For both tests, we excluded the observations before 19:00 hours and after 5:00 hours to account for effects of twilight. For testing the effect of the part of the night, we additionally excluded observations between 23:00 hours and 1:00 hours. For testing the effect of moonlight, we grouped behaviour categories into "inactive" (resting and sleeping) and "active" (all other categories).

Preferences are assumed if use intensity of certain microhabitat variables differs from that of generally available microhabitat (Johnsen, 1980; Adrianasolo *et al.*, 2006; Neilson *et al.*, 2006; Thomas and Taylor, 2006; Rode *et al.*, 2013a), thus we compared the use of used and random trees. As none of the habitat variables were normally distributed and transformations failed to achieve normality, differences between used and random trees were calculated with non-parametric Mann–Whitney *U* tests (two-tailed).

All statistical tests were performed with SPSS 17.0 (IBM, US) and followed Dytham (2011). We set the significance level to $p < 0.05$.

Results

Morphology

In total we captured 12 animals at Ankarafa Forest (Table 12.1). During the first capture period we caught and radio-collared eight animals. During recapture, more

Table 12.1 *Mirza zaza* observed between May and July 2010. Identification code, stable nest group the animal was belonging to and total number of GPS locations taken during nightly follows are given. Total numbers of GPS fixes include nest waypoints while nest locations were listed separately.

Nest group	Focal individual	Sex	Reproductive age	Radio-tracked		Total waypoints	Nest waypoints
				From	To		
1	M1	Male	Adult	20/05/2010	08/07/2010	19*	1
1	M2	Male	Adult	26/05/2010	09/07/2010	12*	1
1	F(S)1	Female	Subadult	25/05/2010	09/07/2010	80	1
1	M3	Male	Adult	27/05/2010	10/07/2010	19*	1
2	F2	Female	Adult	26/05/2010	11/07/2010	14*	3
2	F(J)3	Female	Juvenile	(Captured 10/07/2010)		–	–
3	M4	Male	Adult	27/05/2010	13/07/2010	83	3
3	M5	Male	Adult	28/05/2010	12/07/2010	49	3
3	F(S)4	Female	Subadult	28/05/2010	13/07/2010	90	3
4	M6	Male	Adult	(Captured 10/07/2010)		–	–
4	M(S)7	Male	Subadult	(Captured 13/07/2010)		–	–
4	F5	Female	Adult	(Captured 17/07/2010)		–	–

* Home range sizes did not reach an asymptote.

towards the dry season where food is scarcer, we recaptured all radio-collared animals plus four new animals that were not radio-tracked in one week. Animals caught first entered the traps again up to five times. The animals belonged to four different nest groups. Morphometric measurements of *M. zaza* from Sahamalaza are listed in Table 12.2. Comparing Sahamalaza animals to measurements of *M. zaza* from Ambato and *M. coquereli* from Kirindy (Kappeler *et al.*, 2005), we found several significant differences in morphometric measurements. Differences between *M. zaza* from Sahamalaza and *M. coquereli* from Kirindy were confirmed as in Kappeler *et al.* (2005), with *M. coquereli* showing higher measurements for body length, tail length, ear length, humerus and radius but shorter testis length. Comparing *M. zaza* from Sahamalaza to Ambato, the body length of Sahamalaza individuals was significantly smaller, while measurements of the tail length, head width and length, hind foot, tibia, humerus, and testis length were all significantly larger (8 of 13 variables).

Behaviour

We followed the animals for 192 h and were able to locate them for 44 h. The animals were moving in 65.3% of the direct observations, feeding or foraging 14.6%, resting or sleeping 13.4%, grooming 5.7% and engaged in social activities 1.0% (Figure 12.2). Frequencies of behaviour categories differed between the first and the second part of the night (χ^2 = 11.380, df = 3, p = 0.010, n = 207) with less feeding and foraging in the second part of the night. We could not detect a significant effect of moonlight on activity (Fisher's exact test, p = 0.150, n = 182).

Table 12.2 Descriptive statistics (median, mean and SD) for 18 morphometric variables of adult male and female *Mirza zaza* from Sahamalaza National Park. Asterisks mark significant differences to means of animals from Ambato (Kappeler *et al.*, 2005). L, length; W, width; V, volume; SD, standard deviation of the mean.

	Combined sex				Males				Females			
	Median	Mean	SD	N	Median	Mean	SD	N	Median	Mean	SD	N
Weight (g)	284	283.13	17.65	8	290.3	283.92	20.81	6	280.8	280.75	0.35	2
Head body (cm)	22	22.31*	0.75	8	22.3	22.5	0.77	6	21.8	21.75	0.35	2
Tail (cm)	30	30.28*	0.54	8	30.1	30.17	0.54	6	30.6	30.63	0.53	2
Head L	56.8	56.23*	1.34	7	56.2	55.92	1.49	5	57.0	57.03	0.28	2
Head W	32.8	33.10*	2.17	7	32.9	33.44	2.14	5	32.3	32.26	2.8	2
Canine L	4.3	4.01	0.67	7	3.9	3.86	0.76	5	4.4	4.38	0.11	2
Canine W	2.5	2.34[a]	0.69	4	2.6	2.34	0.84	3	2.3	2.34		1
Ear L	27.1	27.08	1.38	8	27.1	26.98	1.46	6	27.4	27.37	1.57	2
Ear W	16.0	16.08[a]	0.77	8	15.9	15.91	0.82	6	16.6	16.56	0.48	2
Foot	57.6	57.11*	2	8	57.2	56.69	2.1	6	58.4	58.37	1.35	2
Hand	33.6	33.09[a]	2.87	8	33.6	32.61	3.17	6	34.5	34.53	1.38	2
Femur	60.9	60.71	4.49	8	60.9	61.24	4.57	6	59.1	59.11	5.44	2
Tibia	70.9	71.27*	3.35	8	71.6	71.84	3.62	6	69.6	69.55	2.26	2
Humerus	47.5	47.37*	2.43	8	48.0	47.82	2.38	6	46.0	46.04	2.88	2
Radius	44.1	44.85	2.99	8	44.1	44.38	2.46	6	46.3	46.27	5.2	2
Testis L					29.0	27.31*	4.16	6				
Testis W					15.9	15.17[a]	3.03	6				
Testis V (mm^3)					3906.9	3640.70[a]	1511.91	6				

[a] Not tested against Kappeler *et al.* (2005).

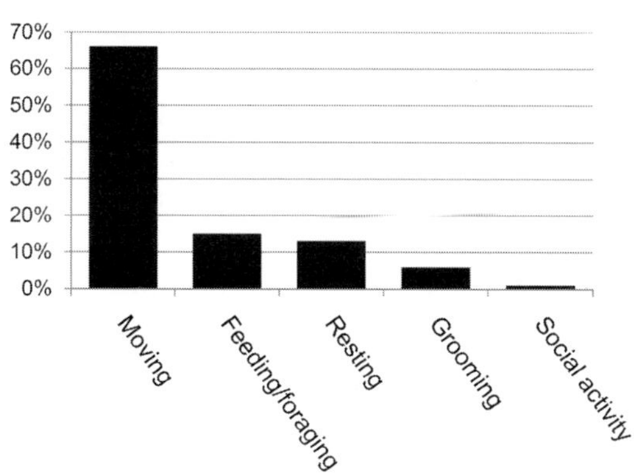

Figure 12.2 Activity budget of eight adult and subadult individuals of *Mirza zaza* (n = 314).

We were able to observe details of 15 feeding bouts of 5 different animals. Three times they fed on the fruits of *Gardenia rutenbergiana* (family Rubiaceae) (two animals); they opened the fruits and let the shell fall to the ground. Twice one animal fed on the gum of an unidentified tree with the local name "Talandoa".

Figure 12.3 Animals of *Mirza zaza* were observed licking homopteran Flatidae larvae (left) and their sugary excretions at Ankarafa, Sahamalaza. (A black and white version of this figure will appear in some formats. For the colour version, please refer to the plate section.)

The gum leaked from a wound on the tree bark. Once, an animal licked at a crack in a bamboo (family Poaceae). We could observe one individual licking the sugary excretions of homopteran Flatidae larvae and cochineals that colonised the base of a small unidentified tree for about 30 min (Figure 12.3). We saw the sympatric blue-eyed lemur feeding on these during the day. Finally, we have five observations where animals seemed to feed on leaves covered with a white substance (three animals); and four observations where they licked branches and lianas (two animals). Due to the white coating of the leaves, we believe that these nine observations actually presented or involved licking larvae secretions.

Home range sizes

The MCP method yielded group home range sizes of 0.97 ha and 2.34 ha for groups 1 and 3, respectively, whereas the KM yielded 0.52 ha and 1.59 ha, respectively. Because these groups lived in different forests there was no overlap. The home range of F2 (the single collared individual of group 2 and thus regarded as a separate nest group here), as calculated using the MCP method, overlapped with 14% of nest group 1's home range, while the home range of nest group 1 overlapped with 7% of F2's home range. Using the KM there was no overlap between nest group 1 and F2.

Individual home range sizes were 1.28 ± 0.71 ha using MCP and1.02 ± 0.75 ha using KM (n = 8; Table 12.3). Mean individual home ranges increased to 1.44 ± 0.55 ha (MCP, n = 3) and 1.32 ± 0.85 ha (KM, n = 4) when home range sizes that did not reach an asymptote were excluded. There were differences between individuals but no consistent differences between sexes. Individual home ranges calculated with MCP overlapped to a high extent in both nest groups (Figure 12.4), but less when calculated with KM (Figure 12.5, Table 12.4).

Table 12.3 Individual home ranges for eight animals of *Mirza zaza*, as calculated using MCP and KM.

ID	Home range (ha)	
	MCP	KM
M1	0.58*	0.44*
M2	0.51*	0.28*
F1	0.84	0.31
M3	0.63*	0.57*
F2	1.99*	1.58*
M4	1.92	1.82
M5	2.21*	2.20
F3	1.55	0.95
Mcan ± SD	1.28 ±0.71	1.02 ± 0.75
Mean ± SD (excluding*)	1.44 ± 0.55	1.32 ± 0.85
Male mean ± SD	1.17 ± 0,82	0.95 ± 0.63
Female mean ± SD	1.46 ± 0,58	1.06 ± 0.88

* Home range size of animal did not reach an asymptote and present minimum home ranges. To account for this, a mean without these animals was calculated as well.

Table 12.4 Home range overlap between individuals of groups 1 and 3, respectively, as calculated with MCP and KM. As overlap can be calculated from the perspective of each of the two pair partners, the table shows two different values for each pair above and below the diagonal. Overlap always refers to the perspective of individuals listed in the columns.

Group 1					Group 3			
Percentage overlap between MCP home ranges								
	M1	M2	F1	M3		M4	M5	F3
M1		58	90	75	M4		95	72
M2	65		98	74	M5	82		65
F1	62	60		67	F3	89	92	
M3	69	60	90					
Percentage overlap between KM home ranges								
	M1	M2	F1	M3		M4	M5	F3
M1		37	49	44	M4		91	51
M2	59		46	69	M5	75		43
F1	70	41		42	F3	100	99	
M3	35	34	23					

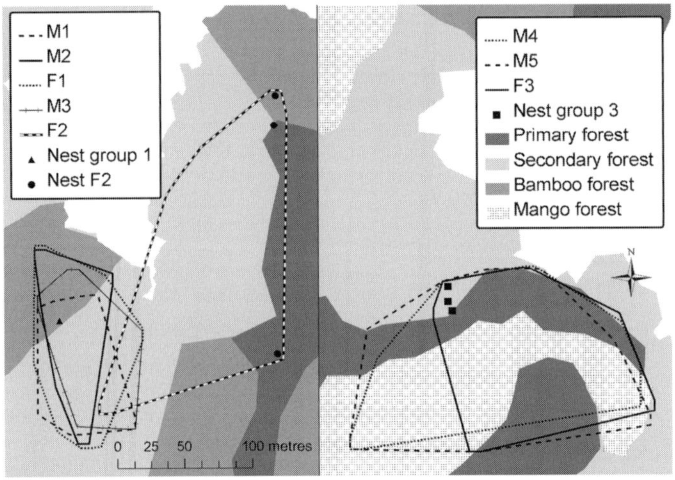

Figure 12.4 Individual home ranges for groups 1, 2 and 3 in two forest fragments, a primary and a secondary forest, and overlap, using the MCP method. Known nest sites are indicated. We defined mango forest as containing more than 10% mango trees. White areas depict non-forest matrix.

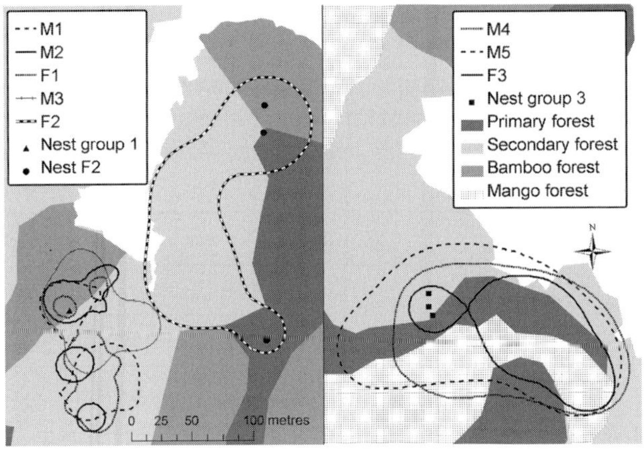

Figure 12.5 Individual home ranges for groups 1, 2 and 3 in two forest fragments, a primary and a secondary forest, and overlap, using the KM. Known nest sites are indicated. We defined mango forest as containing more than 10% mango trees. White areas depict non-forest matrix.

Vertical strata and position in trees

The animals were found at a median height of 7 m (range 0–25 m, n = 326). They were observed at heights of 1 m or lower on several occasions. On most occasions we saw *M. zaza* on branches (45.3%) or tree trunks (28.4%). They were seen less often in the periphery of the canopy (12.7%). The rest of the time we observed them on lianas and in tree forks.

Microhabitat preferences

Used large trees in the secondary forest had significantly higher DBH, larger crown diameter and higher connectivity than random trees (Table 12.5). Large trees in the primary forest and small trees in both forests showed no difference in any variable between used and random trees. In the primary forest, both large trees and small trees showed a smaller distance to the centre tree (as an indication of tree density) in used microhabitat compared to random microhabitat.

Table 12.5 Medians for five variables of used and random trees in two forest fragments with primary and secondary forest. Empty cells indicate where no data were available. Used trees were compared to random trees in each forest. Significance according to Mann–Whitney *U* tests. $^{*}p < 0.05$, $^{**}p < 0.01$, $^{***}p < 0.001$.

	Median			
	Used 1° forest	Random 1° forest	Used 2° forest	Random 2° forest
Height of large trees (m)	12.00	12.00	12.00	13.00
Height of small trees (m)	6.00	7.25	9.00	8.00
DBH of large trees (cm)	17.63	16.62	18.78*	15.28
DBH of small trees (cm)	7.32	6.89	7.00	6.90
Crown diameter of large trees (m)	6.00	6.00	6.00**	5.00
Crown diameter of small trees (m)	3.00	3.00	2.50	2.50
Lianas of large trees (no.)	2		2	2
Lianas of small trees (no.)	1		2	1
Connectivity of large trees (m)	3		4***	2
Connectivity of small trees (m)	2		2	1
N large trees	55	DBH: n = 164 Height and crown diameter: n = 65	45	228
N small trees	15	DBH: n = 150 Height and crown diameter: n = 63	11	232
Distance of large trees (m)	3.98**	4.69	3.90	3.85
Distance of small trees (m)	2.98***	4.07	2.96	3.25
N of large trees	204	DBH: n = 196 Height and crown diameter n = 60	212	208
N of small trees	204	DBH: n = 182 Height and crown diameter n = 60	212	208

Discussion

Morphology

In this study, the morphology of *M. zaza* from Sahamalaza was compared to *M. coquereli* in Kirindy (Kappeler *et al.*, 2005), and all significant results suggest smaller body measurements but bigger testes, confirming the previous differences found between the two species (Kappeler *et al.*, 2005). The animals captured in Sahamalaza for this study also showed significant differences in 8 of 13 variables in comparison to the population of *M. zaza* in Ambato. Animals from Sahamalaza had a significantly shorter body length, while tail length, head width and length, hind foot, tibia, humerus, and testis volume were all significantly larger than in Ambato. Although differences can stem from inter-observer differences between the studies or small sample size, differences found in several variables lead us to conclude that the Sahamalaza population differs morphologically from the Ambato population. According to Craul *et al.* (2007), the phylogenetic relationships within *Lepilemur* follows an "eight Inter-River System" with the River Sambirano separating two different species. Although this geographic barrier may also cause a certain difference between populations of *M. zaza* to the south and the north of the river, morphological differences do not warrant the description of separate species without additional information about genetic, biological and ecological differences (Yoder *et al.*, 2000).

Behaviour

The behaviour of *M. zaza* was very difficult to observe due to the dense canopy and fast movement of animals. Despite almost 200 h of observation, we could see the behaviour for about 26 h (314 observations). Thus, the activity budget must be seen as preliminary behaviour data. In agreement with what Pages (1978) found for *M. coquereli*, *M. zaza* showed less feeding and foraging in the second part of the night. Pages (1978) also observed more social activities during the second half of the night. This could not be confirmed for this study, probably due to its preliminary nature and low sample size. Although most nocturnal primates exhibit a higher activity level in bright nights (lunar philia), "hiding" in bright nights (lunar phobia) is typical for most other small mammals as a predator avoidance strategy (reviewed in Prugh and Golden, 2013; Saldaña-Vázquez and Munguía-Rosas, 2013). Because *M. coquereli* is heavily preyed upon by the Madagascar long-eared owl (*Asio madagascariensis*) and the fossa (*Cryptoprocta ferox*) (Rasoloarison *et al.*, 1995), *Mirza* spp. seem to be highly vulnerable to those nocturnal predators. Thus, we expected *M. zaza* to adopt a lunar phobic strategy to avoid detection by fossas and owls. However, we could not detect an effect of moonlight on activity level in our study, and observation should be intensified to detect a possible influence of lunar illumination on this species.

Due to the preliminary nature of our study and the relatively difficult observation of the animals, we had only a limited number of feeding events where we could identify the food item. These included fruits, insect secretions and plant exudates. *M. coquereli* is known to feed on arthropods, small birds, mammals, amphibians and reptiles, eggs, fruit, flowers, buds, gums and insect secretions. They were reported to spend 60% of their feeding time and 50% of the feeding observations licking at branches of the vines *Hippocratea* or *Elachyptera minimiflora* colonised by aggregations of homopteran Flatidae larvae and cochineals (Petter, 1978; Hladik *et al.*, 1980; Pages, 1980). This diet was suggested to allow the relatively small-bodied *M. coquereli* to stay active during the dry season (Hladik *et al.*, 1980), which would be especially advantageous in areas with a pronounced dry season such as in the dry western forests where *M. coquereli* occurs. *M. zaza* in this study lived in a transition zone between the wet Sambirano area and the dry western forests with possibly less severe dry seasons. Nevertheless, we observed different animals of *M. zaza* feeding on this larva and its secretions one time directly and nine times by animals licking branches and lianas, indicating that this food source is still important. It would be interesting to compare the use of Flatidae between both species and habitats.

Spatial ranging patterns

Compared to *M. coquereli* with a mean individual home range size of 4 ha (Kappeler, 1997a), individual and group home ranges of *M. zaza* were relatively small. Because the MCP method usually overestimates home range size especially with a low number of GPS locations, the home range sizes reported here might be even smaller. On the other hand, some of our home range sizes did not reach an asymptote, and it was difficult to follow animals as no systematic paths were established in order not to destroy remaining forest, and many areas were impenetrable due to thick understorey and erosion holes (Rode-Margono, pers. obs.). Thus, true home ranges might be larger than reported.

Especially when lacking direct behaviour observation as is often the case for nocturnal mammals, the spatial distribution of individuals is used to infer social organisation (Mueller and Thalman, 2000; Nekaris, 2003; Nekaris and Bearder, 2011). Group home range overlap of *M. zaza* using the MCP method revealed only slight overlap of the two neighbouring groups, whereas using KM, there was no overlap at all. The relative exclusiveness of group home ranges was supported by the fact that we saw only one uncollared animal (together with M5 at the periphery of its individual home range) while following focal animals during the whole study period (Rode-Margono, pers. obs.). Individual home ranges overlapped extensively from 60% to 98% (MCP) for group members. Similar overlaps were reported for lemurs living in social monogamous pairs like in *Phaner furcifer* (average overlap 82%; Schülke and Kappeler, 2003), *Cheirogaleus major* and *C. medius* (61–96% overlap between group members; Lahann, 2008) or different species of *Lepilemur* (between 81% and 89% from the female's perspective, between 43% and 61% from the male's perspective; Zinner *et al.*, 2003; Hilgartner *et al.*, 2012; Droescher and Kappeler, 2013). Nest groups of *M. zaza* can be regarded as social groups including nest-sharing male dyads consisting of either related or unrelated individuals, which suggests that *M. zaza* lives in dispersed multi-male/multi-female groups (Rode *et al.*, 2013a). The high spatial overlap of group home ranges of individuals in one nest group supports these results.

Grouping patterns such as social nesting groups, high overlap of individual home ranges, and small home ranges may also be a result of limitation of crucial ecological resources, either through natural variations such as food availability or anthropogenic effects like habitat degradation or fragmentation (Bowers and Matter, 1997; Hanya *et al.*, 2006; Cristóbal-Azkarate and Arroyo-Rodríguez, 2007). Regarding habitat degradation and fragmentation, for instance, the home range of group 1 was limited on three sides by another animal's range, savannah and relatively homogenous bamboo forest (Figures 12.4 and 12.5). However, as we observed affiliative social behaviour within groups, especially during returns to the nest (Rode *et al.*, 2013a), we do not think that the social organisation is unusual in this species. The individual F3, for instance, spent 20% of the observations in close proximity (< 5 m) with one of the males and quiet "Hn" calls were noticed very frequently (Rode-Margono, pers. obs.). *Mirza* and *Cheirogaleus* emit these calls during locomotion and meetings with familiar individuals suggesting that animals stay in close contact during the night (Pages, 1980; Stanger, 1993). Finally, onset of the mating season may also affect the home range sizes, especially of males (Kappeler *et al.*, 1997a; Schwab, 2000). The mating season of *M. zaza* is believed to fall between July and August (Kappeler *et al.*, 2005), and thus would begin at the end of our study. During the mating season, males of species with promiscuous mating systems and scramble competition such as *Mirza* spp. or *Microcebus* spp. roam wider, have larger home ranges and do excursions in search of receptive females (Kappeler, 1997; Eberle and Kappeler, 2004).

Habitat use

In this study, *M. zaza* ranged at heights between 0 and 25 m, with a median height of 7 m. *M. coquereli* were detected at heights of 4.5 m (range 1–8 m; Pages, 1978) and occasionally foraged on the ground (Pages, 1980). The apparent flexibility in height use seems to be an advantage because vertical stratification is suggested to mitigate interspecific competition in lemurs (Lahann, 2008).

When comparing trees used during the animals' general nocturnal activity and random trees, in the mature secondary forest, animals preferred trees with a larger DBH, larger crown diameter and higher connectivity between trees. Furthermore, animals preferred microhabitats with a higher tree density as indicated by the distance between trees in the degraded primary forest. Although no preference for trees with many lianas was apparent in this study, *M. zaza* preferred larger and higher nesting trees with a higher amount of lianas (Rode *et al.*, 2013a). This preference for more covered and well-connected habitat may be related with the high vulnerability of animals to predation (Price *et al.* 1984; Prugh and Golden, 2013;

Saldaña-Vázquez and Munguía-Rosas, 2013). Due to the difficulty of observing the behaviour of animals, we used general activity to determine "used" trees and thus the preference for certain tree characteristics. In future studies it would be interesting to investigate microhabitat preferences especially for feeding, inactivity or social behaviour.

Conclusions and conservation recommendations

We found several morphological differences between *M. zaza* in Sahamalaza and *M. zaza* in Ambato, north of the Sambirano River, and recommend a thorough genetic comparison based on more samples. *M. zaza* proved to be difficult to observe and our preliminary result that animals are lunar neutral should be reviewed. Home range sizes are relatively small and ranging patterns indicate that animals form dispersed groups with two of three groups including several males. Based on our results and the fact that much of the habitat in Sahamalaza is already highly fragmented, we recommend protecting the remaining habitat, especially intact dense forest areas with high vegetation cover. We encourage more intense studies on behaviour, ranging patterns, the influence of degraded and fragmented habitat on social organisation and ecological requirements in order to assist effective conservation planning.

Acknowledgments

Many thanks to the Association Européenne pour l'Etude et la Conservation des Lémuriens (AEECL), Bristol Zoological Society and the staff of the Madagascar Institute for the Conservation of Tropical Environments for their support. Special thanks go to Rode-Margono's Malagasy research counterpart, the local guides at the Ankarafa Research Station and PhD student Melanie Seiler who helped with the organisation of the field work. This study was supported by the MSc Primate Conservation at Oxford Brookes University, Conservation International's Primate Action Fund and Primate Conservation, Inc.

References

Altmann J. 1974. Observational study of behavior: sampling methods. *Behaviour* 49(3/4):227–267.

Andrianasolo TH, Andrianjazalahatra TL, Rakotondranary SJ, *et al.* 2006. Habitat utilisation of nocturnal lemurs in evergreen littoral forests of different degrees of degradation. In C Schwitzer, S Brandt, O Ramilijaona, *et al.* (eds.), *Proceedings of the German-Malagasy Research Cooperation in Life and Earth Sciences* (pp. 151–159). Concept Verlag, Berlin.

Bowers MA, Matter SF. 1997. Landscape ecology of mammals: relationships between density and patch size. *Journal of Mammalogy* 78:999–1013.

Burt WH. 1943. Territoriality and home range concepts as applied to mammals. *Journal of Mammalogy* 24(3):346–352.

Chouteau P. 2004. The impacts of logging on the microhabitats used by two species of couas in the western forest of Madagascar. *Comptes Rendus Biologies* 327(12):1157–1170.

Craul M, Zimmermann E, Rasoloharijaona S, Randrianambinina B, Radespiel U. 2007. Unexpected species diversity of Malagasy primates (*Lepilemur* spp.) in the same biogeographical zone: a morphological and molecular approach with the description of two new species. *BMC Evolutionary Biology* 7(1):83.

Cristobal-Azkarate J, Arroyo-Rodriguez V. 2007. Diet and activity pattern of howler monkeys (*Alouatta palliata*) in Los Tuxtlas, Mexico: effects of habitat fragmentation and implications for conservation. *American Journal of Primatology* 69(9):1013.

Crompton RH, Andau PM. 1986. Locomotion and habitat utilization in free-ranging *Tarsius bancanus*: a preliminary report. *Primates* 27(3):337–355.

Droescher I, Kappeler PM. 2013. Defining the low end of primate social complexity: the social organization of the nocturnal white-footed sportive lemur (*Lepilemur leucopus*). *International Journal of Primatology* 34(6):1225–1243.

Dytham C. 2011. *Choosing and Using Statistics: A Biologist's Guide.* Wiley & Blackwell, Oxford.

Eberle M, Kappeler PM. 2004. Sex in the dark: determinants and consequences of mixed male mating tactics in *Microcebus murinus*, a small solitary nocturnal primate. *Behavioral Ecology and Sociobiology* 57(1):77–90.

Ganzhorn JU. 2003. Habitat description and phenology. In JM Setchell, DJ Curtis (eds.), *Field and Laboratory Methods in Primatology* (pp. 40–56). Cambridge University Press, Cambridge.

Hanya G, Kiyono M, Yamada A, *et al.* 2006. Not only annual food abundance but also fallback food quality determines the Japanese macaque density: evidence from seasonal variations in home range size. *Primates* 47(3):275–278.

Harcourt CS, Nash LT. 1986. Social organization of Galagos in Kenyan coastal forests: I. *Galago zanzibaricus. American Journal of Primatology* 10(4):339–355.

Harris S, Cresswell WJ, Forde PG, *et al.* 1990. Home-range analysis using radio-tracking data – a review of problems and techniques particularly as applied to the study of mammals. *Mammal Review* 20(2–3):97–123.

Hilgartner R, Fichtel C, Kappeler PM, Zinner D. 2012. Determinants of pair-living in red-tailed sportive lemurs (*Lepilemur ruficaudatus*). *Ethology* 118(5):466–479.

Hladik CM, Charles-Dominique P, Petter JJ. 1980. Feeding strategies of five nocturnal prosimians in the dry forest of the west coast of Madagascar. In P Charles-Dominique, HM Cooper, A Hladik, *et al.* (eds.), *Nocturnal Malagasy Primates: Ecology, Physiology, and Behaviour* (pp. 41–73). Academic Press, New York.

IUCN. 2014. IUCN Red List of Threatened Species. Online: www.iucnredlist.org. Accessed 31 August 2014.

Johnson DH. 1980. The comparison of usage and availability measurements for evaluating resource preference. *Ecology* 61(1):65–71.

Kappeler PM. 1997a. Intrasexual selection in *Mirza coquereli*: evidence for scramble competition polygyny in a solitary primate. *Behavioral Ecology and Sociobiology* 41(2):115–127.

Kappeler PM. 1997b. Intrasexual selection and testis size in strepsirrhine primates. *Behavioral Ecology* 8(1):10.

Kappeler PM, Rasoloarison RM, Razafimanantsoa L, Walter L, Roos C. 2005. Morphology, behaviour and molecular evolution of giant mouse lemurs (*Mirza* spp.) GRAY, 1870, with description of a new species. *Primate Report* 71:3–26.

Kenward R. 2001. *A Manual for Wildlife Radio Tagging.* Academic Press, London.

Lahann P. 2008. Habitat utilization of three sympatric cheirogaleid lemur species in a littoral rain forest of southeastern Madagascar. *International Journal of Primatology* 29(1):117–134.

Markolf M, Kappeler PM, Rasoloarison R. 2008. Distribution and conservation status of *Mirza zaza*. *Lemur News* 13:37–40.

Merker S, Yustian I, Mühlenberg M. 2005. Responding to forest degradation: altered habitat use by Dian's tarsier *Tarsius dianae* in Sulawesi, Indonesia. *Oryx* 39(2):189–195.

Moat S, Smith PP. 2007. *Atlas of the Vegetation of Madagascar.* Kew Publishing, Royal Botanic Gardens, Kew, UK.

Mueller AE, Thalmann URS. 2000. Origin and evolution of primate social organisation: a reconstruction. *Biological Reviews* 75(3):405–435.

Müller AE. 1998. A preliminary report on the social organisation of *Cheirogaleus medius* (Cheirogaleidae; Primates) in north-west Madagascar. *Folia Primatologica* 69(3):160–166.

Myers N, Mittermeier RA, Mittermeier CG, da Fonseca GAB, Kent J. 2000. Biodiversity hotspots for conservation priorities. *Nature* 403(6772):853–858.

Neilson K, Curran JM, Towns DR, Jamieson H. 2006. Habitat use by chevron skinks (*Oligosoma homalonotum*) (Sauria: Scincidae) on Great Barrier Island, New Zealand. *New Zealand Journal of Ecology* 30(3):345–356.

Nekaris KAI. 2003. Spacing system of the Mysore slender loris (*Loris lydekkerianus lydekkerianus*). *American Journal of Physical Anthropology* 121(1):86–96.

Nekaris KAI, Bearder SK. 2011. The lorisiform primates of Asia and mainland Africa: diversity shrouded in darkness. In CJ Campbell, A Fuentes, KC MacKinnon, SK Bearder, RM Stumpf (eds.), *Primates in Perspective*, 2nd edition (pp. 34–54). Oxford University Press, Oxford.

Pages E. 1978. Home range, behaviour and tactile communication in a nocturnal Malagasy lemur *Microcebus coquereli*. In DA Chivers, KA Joysey (eds.), *Recent Advances in Primatology* (pp. 171–177). Academic Press, London.

Pages E. 1980. Ethoecology of *Microcebus coquereli* during the dry season. In P Charles-Dominique, HM Cooper, A Hladik, *et al.* (eds.), *Nocturnal Malagasy Primates: Ecology, Physiology, and Behaviour* (pp. 97–116). Academic Press, New York.

Petter JJ. 1978. Ecological and physiological adaptations of five sympatric nocturnal lemurs to seasonal variation in food production. In DA Chivers, KA Joysey (eds.), *Recent Advances in Primatology* (pp. 211–223). Academic Press, London.

Pimley ER, Bearder SK, Dixson AF. 2005. Home range analysis of *Perodicticus potto edwardsi* and *Sciurocheirus cameronensis*. *International Journal of Primatology* 26(1):191–206.

Price MV, Waser NM, Bass TA. 1984. Effects of moonlight on microhabitat use by desert rodents. *Journal of Mammalogy* 65(2):353–356.

Prugh LR, Golden CD. 2013. Does moonlight increase predation risk? Meta-analysis reveals divergent responses of nocturnal mammals to lunar cycles. *Journal of Animal Ecology* 83(2):504–514.

Rasoloarison RM, Rasolonandrasana BPN, Ganzhorn JU, Goodman SM. 1995. Predation on vertebrates in the Kirindy Forest, western Madagascar. *Ecotropica* 1:59–65.

Rendigs A, Radespiel U, Wrogemann D, Zimmermann E. 2003. Relationship between microhabitat structure and distribution of mouse lemurs (*Microcebus* spp.) in northwestern Madagascar. *International Journal of Primatology* 24(1):47–64.

Rode EJ, Nekaris KAI, Markolf M, *et al.* 2013a. Social organisation of the northern giant mouse lemur *Mirza zaza* in Sahamalaza, north western Madagascar, inferred from nest group composition and genetic relatedness. *Contributions to Zoology* 82:71–83.

Rode EJ, Stengel CJ, Nekaris KAI. 2013b. Habitat assessment and species niche modeling. In E Sterling, N Bynum, M Blair (eds.), *Primate Ecology and Conservation* (pp. 79–102). Oxford University Press, Oxford.

Saldaña-Vázquez RA, Munguía-Rosas MA. 2013. Lunar phobia in bats and its ecological correlates: a meta-analysis. *Mammalian Biology – Zeitschrift für Säugetierkunde* 78(3):216–219.

Schmid J, Kappeler PM. 1994. Sympatric mouse lemurs (*Microcebus* spp.) in western Madagascar. *Folia Primatologica* 63(3):162–170.

Schülke O, Kappeler PM. 2003. So near and yet so far: territorial pairs but low cohesion between pair partners in a nocturnal lemur, *Phaner furcifer*. *Animal Behaviour* 65(2):331–343.

Schwab D. 2000. A preliminary study of spatial distribution and mating system of pygmy mouse lemurs (*Microcebus cf myoxinus*). *American Journal of Primatology* 51(1):41–60.

Schwab D, Ganzhorn JU. 2004. Distribution, population structure and habitat use of *Microcebus berthae* compared to those of other sympatric cheirogalids. *International Journal of Primatology* 25(2):307–330.

Schwitzer C. 2005. Die Avifauna des Naturschutzgebiets "Sahamalaza–Iles Radama", Nordwest-Madagaskar. *Zeitschrift des Kölner Zoo* 48(3):133–44.

Schwitzer C, Lork A. 2004. "Projet Sahamalaza-iles Radama": Ein internationales Schutzprojekt für den Sclater's Maki (*Eulemur macaco flavifrons* GRAY, 1867). *Zeitschrift des Kölner Zoo* 2:75–84.

Schwitzer C, Mittermeier RA, Davies N, *et al.* 2013. *Lemurs of Madagascar: a strategy for their conservation 2013–2016*. IUCN SSC Primate Specialist Group, Bristol Conservation and Science Foundation, and Conservation International, Bristol.

Schwitzer C, Schwitzer N, Randriatahina GH, Rabarivola C, Kaumanns W. 2006. "Programme Sahamalaza": new perspectives for the in situ and ex situ study and conservation of the blue-eyed black lemur (*Eulemur macaco flavifrons*) in a fragmented habitat. In C Schwitzer, S Brandt, O Ramilijaona, *et al.* (eds.), *Proceedings of the German-Malagasy Research Cooperation in Life and Earth Sciences* (pp. 135–149). Concept Verlag, Berlin.

Schwitzer N, Randriatahina GH, Kaumanns W, Hoffmeister D, Schwitzer C. 2007. Habitat utilization of blue-eyed black lemurs, *Eulemur macaco flavifrons* (Gray, 1867), in primary and altered forest fragments. *Primate Conservation* 22(1):79–87.

Seaman DE, Millspaugh JJ, Kernohan BJ, *et al.* 1999. Effects of sample size on kernel home range estimates. *Journal of Wildlife Management* 63:739–747.

Seiler M, Holderied M, Schwitzer C. 2013. Effects of habitat degradation on sleeping site choice and use in Sahamalaza sportive lemurs (*Lepilemur sahamalazensis*). *International Journal of Primatology* 34:260–280.

Stanger KF. 1993. Structure and function of the vocalizations of nocturnal prosimians (Cheirogaleidae). Tübingen University, Tübingen, Germany.

Thomas RM. 1998. MoonDV software. Moon version 1.

Thomas DL, Taylor EJ. 2006. Study designs and tests for comparing resource use and availability II. *Journal of Wildlife Management* 70(2):324–336.

VanderWerf EA. 1993. Scales of habitat selection by foraging ‘elepaio in undisturbed and human-altered forests in Hawaii. *Condor* 95(4):980–989.

Wiens F, Zitzmann A. 2003. Social structure of the solitary slow loris *Nycticebus coucang* (Lorisidae). *Journal of Zoology* 261(1):35–46.

Yoder AD, Rasoloarison RM, Goodman SM, *et al.* 2000. Remarkable species diversity in Malagasy mouse lemurs (Primates, *Microcebus*). *Proceedings of the National Academy of Sciences* 97(21):11325–11330.

Zinner D, Hilgartner RD, Kappeler PM, Pietsch T, Ganzhorn JU. 2003. Social organization of *Lepilemur ruficaudatus*. *International Journal of Primatology* 24(4):869–888.

13 Living in riverine and xeric forests: *Microcebus griseorufus* at Beza Mahafaly, southwestern Madagascar

Emilienne Rasoazanabary and Laurie R. Godfrey

Introduction

Twenty-one species of mouse lemurs are now recognized in Madagascar (Radespiel *et al.*, 2012; Rasoloarison *et al.*, 2013). They live in a wide variety of habitats, from the xerophytic forests of the south and southwest (Yoder *et al.*, 2002; Rasoazanabary, 2004; Génin, 2008, Kobbe and Dausmann, 2009) to the dry deciduous forests of the west and north (Ganzhorn and Schmid, 1998; Schmid and Kappeler, 1998; Zimmermann *et al.*, 1998), and to the low-, mid-, and high-altitude rainforests of the east (Atsalis, 1999; Lahann *et al.*, 2006; Radespiel *et al.*, 2008, 2012; Gligor *et al.*, 2009; Blanco, 2011). Mouse lemurs do not live in high-altitude *Erica*-dominated heathland.

The xerophytic (or "spiny") forests of southwestern Madagascar are some of the driest and most seasonal environments in all of Madagascar (Ratsirarson *et al.*, 2001; Fenn, 2003; Génin, 2008; Kamilar and Muldoon, 2010). Dewar and Richard (2007) give it high marks for "unpredictability" – both in terms of month-to-month and year-to-year variation. Succulent, dry-loving plants belonging to the families Didiereaceae and Euphorbiaceae dominate the xerophytic bush of the Spiny Thicket Ecoregion (Fenn, 2003; Burgess *et al.*, 2004). Only one species of mouse lemur thrives in these habitats, *Microcebus griseorufus*.

Among the western mouse lemur species, *Microcebus murinus* is the best studied. It has a broad geographic distribution and has been hailed as wide-niched. It lives in sympatry with other mouse lemur species that appear to have much narrower habitat preferences, including *M. bongolavensis* in the north (Radespiel *et al.*, 2008), *M. ravelobensis* in the northwestern dry deciduous forest of Ampijoroa (Zimmermann *et al.*, 1998), and *M. berthae* in the western dry deciduous forest of Kirindy (Schmid and Kappeler, 1994; Schwab and Ganzhorn, 2004). It also lives in sympatry with *M. griseorufus* in the southern dry forest of Berenty (Yoder *et al.*, 2002) and the southeastern humid forest at Andohahela (Rakotondranary and Ganzhorn, 2011). At

The Dwarf and Mouse Lemurs of Madagascar: Biology, Behavior and Conservation Biogeography of the Cheirogaleidae, eds. S.M. Lehman, U. Radespiel, E. Zimmermann. Published by Cambridge University Press. © Cambridge University Press 2016.

Kirindy, *M. murinus* prefers the moister forest habitats bordering the Kirindy River (Yoder *et al.*, 2002; Rasoazanabary, 2006), and at Berenty Private Reserve, it avoids xerophytic forest (Yoder *et al.*, 2002). In southeastern Madagascar, there is evidence of hybridization between *M. griseorufus* and *M. murinus* in the transitional forest corridor connecting the xerophytic forests (with *M. griseorufus*) located west of Ampihamy to the wetter littoral forests (with *M. murinus*) located east of Sarikady (Gligor *et al.*, 2009). Hybrids of the two are found in the transitional forest between dry and xerophytic forests at Andohahela National Park (Rakotondranary and Ganzhorn, 2011), but there is no *M. griseorufus* in the mesic forest. Mandena and other southeastern coastal sites harbor populations of *M. murinus*.

Microcebus murinus has been reported to tolerate very disturbed forests (Ganzhorn, 1995; Ganzhorn and Schmid, 1998; Radespiel and Raveloson, 2001; Hapke, 2005). However, it does not inhabit the xerophytic forests of the extreme south and southwest (Génin, 2008), and it may be disappearing from patchy mesic forests in the south (Yoder *et al.*, 2002). During the 1990s, *M. murinus* skeletal remains were identified in owl pellets collected in the vicinity of Beza Mahafaly (Goodman *et al.*, 1993a, 1993b); presumably the animal was hunted by owls in the more mesic forests in the region. However, no individuals belonging to this species were present at the Beza Mahafaly Special Reserve (hereafter BMSR) when research on *M. griseorufus* at this site began in 2003 (Heckman *et al.*, 2006). Mouse lemurs captured in 1931 in a xerophytic forest at Amboasary (and previously identified as *M. murinus*) are actually *M. griseorufus* (Cuozzo *et al.*, 2013).

Microcebus griseorufus has been reported in xerophytic forests along the western coast of Madagascar (Tsimanampetsotsa, Lamboharana, Mikea, Itampolo, Vohombe, Behara, and Namoroka; see Wilmé *et al.*, 2006; Kobbe and Dausmann, 2009). In the south, it occupies the xerophytic forests at Berenty (Yoder *et al.*, 2002; Génin, 2008), and Amboasary (Cuozzo *et al.*, 2013). In south-central Madagascar, it lives in isolated forests such as BMSR. It is also found at Mahavelo and Ampihamy in southeastern Madagascar. Whereas *M. griseorufus* normally inhabits xerophytic forests, it has also invaded contiguous mesic forests in some places. BMSR is one such place, where *M. griseorufus* lives in the mesic and xerophytic forests. Another is the transitional forests (Sarikady, Ankilivalo, and Anjà) between the littoral forest of the east (at Mandena) and the easternmost xerophytic forests of Mahavelo and Ampihamy (Gligor *et al.*, 2009). Another is Tsimanampetsotsa, where *M. griseorufus* lives in xerophytic as well as more mesic, though still dry, forests.

At BMSR, the mesic (Parcel 1, also called "gallery") and xerophytic (Parcel 2, also called "spiny") forests are quite distinct (Figure 13.1). Ground cover is thin or absent in many parts of the xerophytic forest, but present, at least in patches, in the mesic forest. The ground is noticeably more shaded in the mesic than in the xerophytic forest due to differences in the diameter of mature tree crowns. The terrain of the mesic forest is flat, whereas the xerophytic forest floor is steep in places and punctuated with rocky outcrops. The soil of the mesic forest tends to be moist, even during the dry season, because of its proximity to the Sakamena River. The water table is sufficiently high to maintain a fairly dense forest throughout the year,

Figure 13.1 The two forest habitats at Beza Mahafaly. Top row: mesic (gallery) forest, from left to right, wet season and dry season (Emilienne Rasoazanabary placing trap). Bottom: xerophytic (spiny) forest, from left to right, wet season and dry season. (Photo credits: top left, Emilienne Rasoazanabary; top right, Brooke Crowley; bottom left, Brooke Crowley; bottom right, Laurie Godfrey.) (A black and white version of this figure will appear in some formats. For the colour version, please refer to the plate section.)

particularly in the eastern half of the mesic forest, nearest the river, despite the fact that the river is dry most of the year. It is also the case that the mesic forest at Beza has some common plants (such as *Euphorbia tirucalli*, *Salvadora* spp., *Quivisianthe papionae*, and *Grewia* spp.) that are characteristic of more xerophytic forests elsewhere, and absent from more mesic forests to the east (including the mesic forest at Berenty).

This chapter focuses on the morphological and behavioral variation of *Microcebus griseorufus* in Parcels 1 and 2 at BMSR. It is based on data collected simultaneously in both parcels between October 2006 and September 2007. Our goals are to: (1) document the physical environments, structure, and phenology (i.e., the timing of periodic events such as fruiting) of the mesic and xerophytic forests at BMSR, focusing particularly on the plant resources used by *Microcebus*; (2) investigate the physical (morphometric) and behavioral differences between adult *M. griseorufus* in the two habitats; and (3) explore the relationships between habitat, morphology, and behavior. Are there morphological differences between mouse lemurs living in the different forests? If so, are these differences related to the differences in forest characteristics, or to the ways in which they use those forests? Are there seasonal shifts in behavior, and if so, are they related to or independent of site characteristics? To what extent are mouse lemurs selective in their use of trees for feeding and sleeping in their different habitats? What are the conservation implications of their plasticity and selectivity?

Methods

Botanical survey

At the start of our research, we established a total of 8 botanical plots (each 10 × 100 m) in the mesic (4 plots) and xerophytic forests (4 plots) at BMSR, following the methods established in Ratsirarson *et al.* (2001). At each forest, two botanical plots were located within areas selected to trap and follow focal individuals (designated "main animal study areas") and two were located outside the main animal study areas but nevertheless within each parcel (designated "random study areas"). In November, 2006, we tagged all plants in these botanical plots with DBH (diameter at breast height, approximately 4.5 feet above ground level) ≥ 5 cm, and recorded their heights and the size (height and diameter) of their crowns. Using monthly surveys we tracked phenological changes of each tagged plant. The status of the leaves (emergent, mature, senescent, shed), flowers (absent, budding, mature), and fruits (absent, present including immature and mature) of each plant was recorded.

Capture–recapture

We used capture–recapture methods to sample *M. griseorufus* at the BMSR during 12 successive months from October 9, 2006 to September 30, 2007. During this time period, we set 33,120 traps in each of the two forests. Sampling was conducted simultaneously in both forests by two teams of three people each. We used Sherman live traps (7.7 × 7.7 × 30.5 cm) to capture mouse lemurs. At dusk (between 5 and 6 p.m.), traps baited with bananas were set both on the ground and in trees at heights of about 1–2 m and they were checked in the morning between 5 and 7 a.m. (at dawn). Traps that did not have mouse lemurs in them had the uneaten bananas removed and they were then closed for the day. All traps containing captured animals were collected for cleaning at the campsite, and were cleaned before reuse. While they were transported to the camp, the animals were left in the traps. At this hour of the morning, during the cold, dry season, most were in a state of torpor, but during the rainy season, they were awake and active. Before handling captured animals, animal handlers used gloves to prevent the transmission of diseases. Before taking measurements, each animal received a safe dosage of Telazol (usually 0.001 mg, following procedures established by Rensing, 1999). Dosages were based on prior tests on mouse lemurs at Kirindy. Microchips (Transponder ID 100, UK) were inserted to identify animals, and they were monitored and fed, if necessary, while in the camp during the daylight hours. At dusk, each was returned to the location where it had been captured.

We used two trapping schedules ("intensive daily sampling" and "reduced sampling"), depending on the month. During months of intensive sampling (October, January, May, and September), 180 traps were set each day for 30 consecutive days at each main sampling area. During each of the remaining 8 months, 60 traps were set per day for 15 days in selected random sampling areas (outside the main sampling area but inside the parcel), and 180 traps were set in the main sampling areas for 3 days at each month's end. The main sampling areas were about 6.20 ha while random sampling areas were about 20 m × 20 m. In the main sampling areas, traps were placed 25 m apart in all directions, while traps in the random sampling areas were only approximately 7 m apart. These shorter distances were used to prevent trap loss, as ground cover in random sampling areas was often quite dense.

Morphometric measurements

Captured individuals were identified by sex and weighed with a spring scale balance (± 1 g). At the time of first capture, a full set of morphometric measurements was taken using digital calipers (Table 13.1). Immature individuals, captured largely in May and recognizable as such due to their immature limb proportions (especially short arms) and low body mass (usually under 28 g), were eliminated from statistical analyses. We checked for differences between left and right sides and found none; therefore, unless there was an injury and one side was abnormal, only the left side was used in statistical analyses.

Radiotracking and focal individual observations

Focal individual follows were conducted simultaneously by teams of three people working in each forest during the rainy season (11/22/06 to 04/04/07, 453 h) and

Table 13.1 List of traits measured for each individual.

Head length	From the tip of the nose to the back of the head
Bizygomatic width	The width of the head, across the zygomatic processes, perpendicular to head length
Body length	From the back of the head to the base of the tail
Tail length	From the base of the tail to the tip of the most distal palpable vertebra, hair tufts excluded
Canine height	From the base of the canine to the tip of the crown
Ear length	From the top to the bottom of the pinna (maximum)
Arm length	From above the greater tuberosity of the humerus to the olecranon process (palpated) of the ulna (shoulder to elbow)
Forearm length	From the inside of the elbow to the proximal end of the wrist
Hand length	From the base of the palm to the distal end of the longest digit (digit 4), nail excluded
Palm length	From the base of the palm to the base of the proximal phalanx of digit 3
Palm width	From the medial to the lateral edge of the palm at the base of digit 1
Manus digit 1 length	From the base to the tip of digit 1, nail excluded (phalanges only)
Manus digit 2 length	From the base to the tip of digit 2, nail excluded (phalanges only)
Manus digit 3 length	From the base to the tip of digit 3, nail excluded (phalanges only)
Manus digit 4 length	From the base to the tip of digit 4, nail excluded (phalanges only)
Manus digit 5 length	From the base to the tip of digit 5, nail excluded (phalanges only)
Thigh length	From above the greater trochanter of the femur, palpated, (hip joint) to the knee
Leg length	From the proximal tibia (posterior face, below the knee) to the calcaneal tuberosity, palpated
Foot length	From the calcaneal tuberosity, palpated, to the tip of the longest toe, nail excluded (digit 4)
Sole length	From the calcaneal tuberosity, palpated, to the base of digit 3
Sole width	Perpendicular to sole length at the metatarsophalangeal joints
Pedis digit 1 length	From the base to the tip of digit 1, nail excluded (phalanges only)
Pedis digit 2 length	From the base to the tip of digit 2, claw excluded (phalanges only)
Pedis digit 3 length	From the base to the tip of digit 3, nail excluded (phalanges only)
Pedis digit 4 length	From the base to the tip of digit 4, nail excluded (phalanges only)
Pedis digit 5 length	From the base to the tip of digit 5, nail excluded (phalanges only)

then during the dry season (05/19/07 to 08/26/07, 727 h). During this period, 39 individuals were fitted with radio collars (8 females and 8 males in the rainy season, and 11 females and 12 males in the dry season). These individuals were equally distributed between sites (20 in the mesic forest and 19 in the xerophytic forest). The radio collars that we used (i.e., TW4-button cell tags, Biotrack, Wareham, UK) were designed to be least obstructive to small arboreal animals; each weighed only 2 g and accommodated the wire within the collar band. All radio-collared individuals were followed using a portable TR-4 receiver (Telonics, Impala, AZ). One of us (ER) directed the simultaneous collection by two teams of observers (including ER) of focal observational data on feeding.

Sleeping sites: beginning on the morning of the first day of focal individual observation, we followed collared individuals' radio signals to the nests where they slept. We recorded the observation date, and identified, flagged, and numbered the individual plants used as nests. On every focal individual sampling day, we recorded the plant species used for sleeping and the location of the nest. We also recorded the height of the nest above the ground and nest type (open tangled vegetation or tree hole).

Feeding behavior: on every focal individual sampling night, at each forest, two radio-collared individuals were followed. Observations began as soon as the individual left its nest (or, if the individual was already active, as soon as the individual could be found). Focal individual sampling continued according to three schedules that were equally employed across habitats: either from dusk until 2 a.m., from midnight until mouse lemurs returned to their sleeping sites, or all night long. Every 2 min (using instantaneous point sampling, following Altmann, 1974), we identified whether the radio-collared focal individual was feeding, and if yes, the foods consumed (leaves, fruits, flowers, insects or insect secretions, or gums). Gums were defined as any exudates, including sap. We observed variation in gum consumption (e.g., gouging, or utilization of plants that were already damaged), but we did not record the frequency of each. Features of the plants on which animals fed were recorded (species, height above the ground, and support diameter and orientation at feeding location). Most focal feeding observations were made between 1830 and 0200 hours during the rainy season and between 1730 and 0200 hours during the dry season, as these were the hours mouse lemurs were generally active.

Data were collected in accordance with the University of Massachusetts, Amherst IACUC Protocol No. 27-17-01 to Emilienne Rasoazanabary, and all procedures involving animal capture and handling adhered to the principles for ethical treatment of non-human primates of major primatological societies and to the legal requirements for animal treatment as established and monitored in Madagascar.

Data analysis

The general characteristics of the plants in the two forests were assessed using standard ecological descriptors: (1) species richness or the total number of plant species sampled at each forest, (2) species richness as a percentage of the total number of plant species sampled at all locations, (3) Simpson's Index, which measures the probability that two individual plants randomly sampled from a forest will belong to the same species, and (4) Simpson's Index of Diversity, which captures the probability that two individual plants sampled from a forest will not belong to the same species and thus gives an indication of species diversity. Higher values of Simpson's Index of Diversity mean higher diversity (Simpson, 1949).

Differences in the frequencies of the various tree species in the botanical plots for the two forests were tested using χ^2; we also used χ^2 to test differences in the two forests between plant species used by mouse lemurs for feeding and sleeping. We did not fit full random effects models (accounting for individual differences) in determining the significance of differences in diet by habitat or season because of sampling imbalance in our focal individual-follow feeding data (in turn caused by collar damage or the early death or disappearance of a few focal individuals). When some individuals have low representation, a full random effects model can exhaust the existing degrees of freedom and the error terms cannot be statistically justified. Instead, we treated all 2-min feeding scan points as independent observations and used χ^2 to test for differences in dietary behavior by site and season. We did check for evidence of distortion via pseudoreplication (that is, we searched for bias introduced by overrepresentation of unusual individuals), and found none (for further information, see the discussion in Crowley et al., 2014).

Discriminant function analysis was used to determine whether the forests could be distinguished by their general phenological (timing of leafing, fruiting, flowering) and structural (crown heights, crown diameters, tree heights, DBH) characteristics and to determine whether mouse lemurs from the two forest habitats differ in morphological traits. These tests were supplemented by a series of t-tests and ANOVAs to explore specific relationships. We also measured feeding and sleeping site selectivity by calculating the difference between the % of focal individual samples involving use of a particular tree species and that same tree species' relative abundance in the forest (as recorded in botanical plots). We designated the dry season as May–October and the rainy season as November–April.

Results

Habitat structure, tree density, tree diversity, and phenology

A total of 971 individual plants with DBH $\geq$ 5 cm belonging to 41 species were recorded in the two forests (Table 13.2). The total number of such plants (which included bushes as well as trees) is considerably higher in the xerophytic forest (n = 609) than the mesic forest (n = 362). Significant differences between the two forests were found in their average height, diameter at breast height, crown height, and crown diameter (Table 13.3). Mesic forest trees are taller, wider, and have larger crowns (in both height and diameter) than xerophytic forest trees.

Only one-quarter (24.4%) of the 41 plant species are found in both forests, but there is virtually no difference in the total number of plant species found in each (26 for the mesic forest and 28 for xerophytic forest). Because on average these plants are bigger and taller in the mesic forest than in the xerophytic forest, there are fewer individuals per plot in the mesic forest. Tree/bush density (number of individuals per hectare) is therefore much higher in the xerophytic (1522 trees/ha) than the mesic forest (905 trees/ha). The values of Simpson's Index of Diversity (0.91 for the mesic forest and 0.87 for the xerophytic forest) show that, despite the smaller number of plants with diameters $\geq$ 5 cm in each plot, the mesic forest is more diverse than the xerophytic forest.

Table 13.2 List of plant species ≥ 5 cm DBH in botanical plots, with number of individuals of each species included in phenological surveys.

Genus and species	Mesic *N*	Spiny *N*	Total
Acacia bellula	26	–	26
Acacia polyphylla	3	–	3
Alantsilodendron humbertii	–	2	2
Alluaudia procera	–	127	127
Bridelia sp.	1	–	1
Cedrelopsis grevei	2	78	80
Commiphora aprevalii	7	23	30
Commiphora brevicalyx	–	8	8
Commiphora marchandii	3	1	4
Commiphora rombe	1	50	51
Commiphora simplicifolia	–	2	2
Diospyros sakalavarum	1	–	1
Euphorbia decosei	–	9	9
Euphorbia stenoclada	–	5	5
Euphorbia tirucallii	54	13	67
Fernandoa madagascariensis	2	–	2
Grewia franciscana	25	4	29
Grewia grevei	1	1	2
Grewia leucophylla	19	–	19
Gyrocarpus americanus	1	140	141
Hymenodactyon decary	–	9	9
Macpersonia gracilis	–	5	5
Mahafanogne	–	4	4
Mimosa delicantuta	–	4	4
Operculicaria decaryi	–	9	9
Pachypodium geayi	–	8	8
Pachypodium rutenbergianum	–	10	10
Phyllanthus decoryanus	–	39	39
Physena sessiliflora	1	–	1
Quivisianthe papionae	19	–	19
Rhigozum madagascariensis	4	6	10
Rhopalocarpus lucidus	36	–	36
Salvadora angustifolia	15	3	18
Salvadora sp.	16	–	16
Strychnos madagascariensis	4	–	4
Suregada chauvetiae	46	23	69
Tallinella grevei	4	–	4
Tamarindus indica	53	–	53
Terminalia fatraea	1	14	15
Terminalia seyrigii	17	6	23
Uncarina grandidieri	–	6	6
Totals	362	609	971

Table 13.3 Physical characteristics of trees and bushes in the botanical plots of two forest habitats (*t*-tests with unpooled variances for samples with unequal variance).

Measurement	Mesic forest mean (± SD)	Xerophytic forest mean (± SD)	*t*, df	*P*
Height (m)	8.47 (± 3.19)	5.94 (± 2.64)	12.7, 649.3	<0.001
DBH (cm)	12.51 (± 9.07)	10.67 (± 5.87)	3.4, 539.7	0.001
Crown height (top to bottom, in m)	3.17 (± 2.15)	2.00 (± 1.00)	8.5, 325.4	<0.001
Crown diameter (m)	2.68 (± 1.69)	1.47 (± 0.64)	10.9, 280.3	<0.001

Discriminant function analysis of phenological data for 709 individual trees and bushes in the mesic (n = 238 plants) and xerophytic (n = 471 plants) forests summarizes the differences between the two forests succinctly. This function, which is highly significant (Wilks' lambda = 0.22, χ^2 = 1045.1, df = 33, $P < 0.001$), distinguishes mesic (with strongly negative scores) from xerophytic forest trees and bushes (with positive scores) with great confidence. Post-hoc classification success was 95.6%; with cross-validation, the success rate drops very little to 95.5%. The variables that correlate most strongly positively with this function are leafing scores for the dry season (June, July, August, and September, with stronger correlations later in that season). Crown diameter, total height, and the height from top to bottom of the crown have the strongest negative correlations with this function. This means that, during the dry season (especially late in the dry season), the xerophytic forest trees and bushes are more likely to have lost their leaves than are those from the mesic forest. Furthermore, xerophytic forest trees and bushes are more likely to be shorter and to have smaller crowns than those from the mesic forest.

The fruiting schedules in both forests show pronounced seasonality (Figure 13.2). Fruit production peaks in November in both forests habitats (as reflected in the mean fruiting scores for each month), and drops during the month of December. It remains low from January to April. No fruiting occurred in the xerophytic forest from May to September. In the mesic forest, fruiting occurs throughout the year, but drops below 2% during the very dry months. The November peak was higher for the mesic than the xerophytic forest; December fruiting was relatively higher in the xerophytic forest.

Mouse lemur capture and recapture data

Overall, between October 2006 and September 2007, we captured a total of 184 individual *M. griseorufus* within the two forests (116 in the mesic forest and 68 in the xerophytic forest). Of the individuals captured, 151 were adult (102 in the mesic forest and 49 in the xerophytic forest), and 33 were immature. By the end of the year, our recapture rates were exceedingly high (over 90% of captured individuals were recaptures), suggesting that we had sampled almost all individuals at each site.

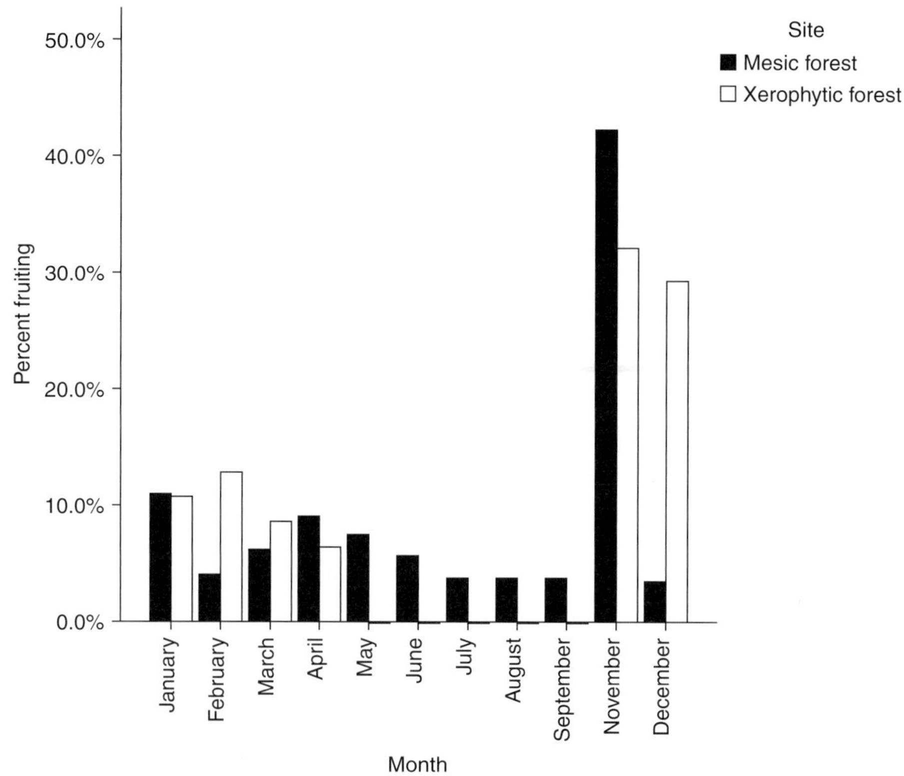

Figure 13.2 For each site, the percentage of trees or bushes that are fruiting, or that display some combination of emerging, unripe and/or ripe fruit. Black bars: mesic forest; open bars: xerophytic forest.

Mouse lemur morphometrics

Analysis of the morphometric data revealed site differences for adults of both sexes. Mean body mass of adult *M. griseorufus* at BMSR was over 10% higher in the mesic forest than in the xerophytic forest. The average adult body mass by site was 49.5 ± 10.9 g at the mesic forest and 44.1 ± 6.0 g in the xerophytic forest (t = 4.0, df = 149.5, $P < 0.001$). Skull length also differed significantly by forest; the average adult skull length was 33.3 ± 1.3 mm at the mesic forest and 31.0 ± 1.1 mm in the xerophytic forest (t = 11.6, df = 153, $P < 0.001$). Adult females were significantly heavier than males (t = 3.1, df = 80.8, $P < 0.01$) and they had significantly longer skulls (t = 2.8, df = 83.2, $P < 0.01$).

Other morphometric traits differed significantly by forest habitat. These are summarized in a discriminant function analysis that successfully distinguishes among individual mouse lemurs from the two forests on the basis of their measurements alone (post-hoc classification success = 99.3%; under cross-validation, this drops to 98.7%). This function is highly significant (χ^2 = 360, df = 26, $P < 0.001$); individuals from the xerophytic forest (with strongly positive scores) are separated from individuals from the mesic forest (with negative scores).

Individuals with positive scores on this function (i.e., those from the xerophytic forest) have relatively short digits 3, 4, and 5 of the foot and hand, and a relatively long pollex and hallux (and thus a more clamp-like hand and foot), while individuals with negative scores on this function have relatively long digits 3, 4, and 5 of the hand and foot and relatively short pollex and hallux (and thus a more hook-like hand and foot). Individuals from the xerophytic forest also have shorter but relatively broader skulls and shorter tails, and those from the mesic forest have longer but relatively narrower skulls and longer tails. The significance of these differences was confirmed by independent samples *t*-tests.

Mouse lemur behavior: resource usage patterns

Mouse lemurs in the region of BMSR have omnivorous diets, mainly comprising insects, fruits, and gums. Individual mouse lemurs were found feeding on plant parts belonging to 39 plant species during the two focal individual sampling periods. Sleeping was observed in 43 tree species. Of these, a total of 21 species were used for both feeding and sleeping (Table 13.4). In all, mouse lemurs used 61 plant species for either feeding or sleeping (or both), of which approximately half (n = 30) were not recorded in our botanical plots. The latter includes plants that have small DBH values (vines, for example) which were present but not sampled in the botanical plots. Plants with small DBH (≤ 5 cm) were sometimes used as resources by mouse lemurs. However, mouse lemurs favor plants with large DBH for feeding and sleeping; thus every plant species that was exploited often (for feeding, sleeping, or both) was actually represented in the botanical plot data.

Feeding behavior

Mouse lemurs from the two forests differed significantly in diet ($\chi^2 = 134.4$, df = 4, $P < 0.001$). Mouse lemurs in the mesic forest fed more on fruits and less on insects; mouse lemurs in the xerophytic forest consumed gums and insects to a greater degree than "expected," and they consumed much less fruit. Gums comprised 47.9% of the diet of *M. griseorufus* at BMSR, but more so in the xerophytic forest (59.2%) than the mesic forest (40.8%). In both forests combined, adult females fed significantly more on fruits than did adult males ($\chi^2 = 7.6$, df = 1, $P < 0.01$).

Mouse lemurs from the two forests also differ significantly in plant species eaten ($\chi^2 = 1830.56$, df = 53, $P < 0.001$). Plant species that were particularly important food resources in the mesic forest included *Rhopalocarpus lucidus* (20.9% of all focal feeding records in this forest), *Acacia bellula* (20.1%), *Grewia leucophylla* (14.2%), *Bridelia* sp. (14.0%), and *Euphorbia tirucalli* (5.8%). Plant species that were particularly important food resources in the xerophytic forest included *Terminalia fatraea* (33.9% of all focal feeding records in this forest), *Alantsilodendron humbertii* (20.5%), *Cedrelopsis grevei* (7.3%), *Mimosa delicantuta* (5.2%), and *Operculicaria decaryi* (4.5%). Note the reliance on *Terminalia fatraea* and *Alantsilodendron humbertii* in the xerophytic forest (54.4% of all feeding observations combined) but not in the mesic forest, where *Terminalia* is very rare and *Alantsilodendron* even more

Table 13.4 Plant species used by mouse lemurs for feeding (F) or sleeping (S) in the two forests.

Plant species	Usage	Feeding		Sleeping	
		Mesic	Spiny	Mesic	Spiny
Acacia bellula	F,S	++	–	++	–
*Acacia minutifolia**	F	+	–	–	–
Acacia polyphylla	F,S	+	–	+	–
Alantsilodendron humbertii	F,S	–	++	+	+
Albizzia tulearensis	S	–	–	–	+
Alluaudia procera	S	–	–	–	++
Bridelia sp.	F	++	+	–	–
Byttneria sp.*	F	–	+	–	–
Calopikis sp.*	F	–	+	–	–
Cedrelopsis grevei	F,S	–	+	–	+
*Clerodendrum emirnense**	F	–	+	–	–
Commiphora aprevalii	F	–	+	–	–
Commiphora brevicalyx	F,S	+	+	+	+
Commiphora rombe	S	–	–	–	+
*Cynanchum arenarium**	F	–	+	–	–
*Dialium madagascariensis**	S	–	–	+	–
*Enterospermum pruinosum**	S	–	–	+	–
Euphorbia decosei	S	–	–	–	+
Euphorbia stenoclada	S	–	–	–	+
Euphorbia tirucalli	F,S	+	+	++	+
Gonocripta grevei	S	–	–	+	–
Grewia franciscana	F,S	–	+	+	–
Grewia grevei	S	–	–	+	+
Grewia leucophylla	F,S	++	–	++	–
Grewia sp. 1*	S	–	–	+	–
Grewia sp. 2*	S	–	–	–	+
*Grewia triflora**	F	+	–	–	–
*Grewia tuleariensis**	F	+	+	–	–
Gyrocarpus americanus	F,S	–	+	–	+
Hippocratea anguistipetal	F	+	–	–	–
Hymenodactyon decaryi	F,S	–	+	–	+
Ipomae majungensis	S	–	–	–	+
Lafikosy*	S	–	–	+	–
Mahafanogne	S	–	–	–	+
Maragnatolaka	F	–	+	–	–
*Metaporana parvifolia**	S	–	–	+	+
*Microsteira diotostigma**	F	+	–	–	–
Mimosa delicatula	F,S	–	+	–	+
Olax sp.*	F	+	–	–	–
Operculicaria decaryi	F	–	+	–	–
*Opuntia dilleri**	F	–	+	–	–
Pentarhopalopilia sp.*	S	–	–	+	–
Pentopetio sp.	S	–	–	+	–

(*continued*)

Table 13.4 (*cont.*)

Plant species	Usage	Feeding		Sleeping	
		Mesic	Spiny	Mesic	Spiny
Phyllanthus decoryanus	F,S	–	+	–	+
Physena sessiliflora	F,S	+	–	+	–
*Quivisianthe papionae**	F,S	+	–	+	–
Rhigozum madagascariensis	F,S	–	+	+	+
Rhopalocarpus lucidus	F,S	+	+	+	–
*Rhoupellina boivini**	S	–	–	+	–
Roipitike	F	+	+	–	–
Salvadora angustifolia	F,S	+	–	++	+
Salvadora sp.	S	–	–	+	–
Scutia myrtina	F,S	+	–	+	–
Seta*	S	–	–	+	–
Suregada decidua	F,S	+	+	+	+
*Tallinella grevei**	S	–	–	+	+
Tamarindus indica	F	++	–	–	–
Terminalia fatraea	F,S	+	++	–	++
*Terminalia tricristata**	F	–	+	–	–
Terminalia seyrigii	F,S	+	+	+	+
*Vanilla madagascariensis**	S	–	–	–	+

++, more than 10% of feeding or sleeping observations in forest; +, observed, but not often; –, not observed.

* Not recorded in botanical plots.

so. *Alantsilodendron humbertii* and *Terminalia fatraea* are favored for gum feeding in the xerophytic forest and *Acacia bellula* and *Rhopalocarpus lucidus* are favored for gum feeding in the mesic forest. Mouse lemurs fed on the fruit of 23 plant species, three of which (*Bridelia* sp., *Grewia leucophylla*, and *Terminalia fatraea*) were frequently visited.

Mouse lemurs were found feeding significantly higher above the ground in the mesic forest than in the xerophytic forest (Table 13.5). Differences in feeding height correlate with differences in plant heights and diameters at breast height (Table 13.5).

Food selection varied significantly by season (χ^2 = 740.77, df = 4, $P < 0.001$). During the rainy season, mouse lemurs ate mostly insects and fruits. *Microcebus griseorufus* fed on gums throughout the year, but dependence on gums increased dramatically during the dry season. Gum consumption comprised 79.0% of dry season feeding observations cf. 24.8% of rainy season feeding observations, whereas fruit and insect consumption (34.4% and 39.6% of rainy season feeding observations, respectively) was relatively high during the rainy season.

Plant species utilized for gum consumption also differed by season (χ^2 = 204.67, df = 34, $P < 0.001$). In the mesic forest, *Acacia bellula* was consumed mostly during

Table 13.5 Characteristics of trees or bushes in which mouse lemurs fed at the two forests.

Measurement	Mesic forest	Xerophytic forest	*t*, df	*P*
Height at feeding	6.7 ± SD 3.3 m	3.6 ± SD 1.6 m	28.0, 1477	<0.001
Mean height of trees selected for feeding	8.8 ± SD 3.1 m	5.5 ± SD 2.8 m	12.7, 511.8	<0.001
DBH of trees used for feeding	12.9 ± SD 9.1cm	7.3 ± SD 2.9 cm	10.3, 402.4	<0.001

the dry season, as it is one of the few tree species in this forest that provides gums when other foods are not available. This species does not occur in the xerophytic forest.

Leaves and flowers were consumed in very small quantities. Eleven plant species were visited for leaf consumption, only one of which (*Salvadora angustifolia*) was visited repeatedly (and then only when leaves were very young, during the rainy season).

Sleeping sites

Microcebus griseorufus slept either in naturally tangled vegetation ("open nests") or in natural holes in living or dead trees (Figure 13.3). There are significant differences in the types of nests chosen by forest ($\chi^2 = 44.84$, df = 1, $P < 0.001$), with mouse lemurs in the xerophytic forest using more tree holes, independently of season. In the xerophytic forest, 61.9% of all sleeping site locations were tree holes, whereas that percentage decreased to 48.0% in the mesic forest.

There are enormous differences in the plant species selected for sleeping by forest ($\chi^2 = 1880.14$, df = 45, $P < 0.001$). In the mesic forest, the most commonly exploited plants were *Euphorbia tirucalli* (26.7% of all focal sleeping records at this forest), *Grewia leucophylla* (17.3%), *Acacia bellula* (13.2%), *Tamarindus indica* (11.4%), and *Salvadora angustifolia* (8.6%). In the xerophytic forest, the most commonly exploited sleeping trees were *Alluaudia procera* (34.4%), *Terminalia fatraea* (16.4%), *Euphorbia stenoclada* (8.6%), *Cedrelopsis grevei* (5.4%), and *Alantsilodendron humbertii* (4.5%). The favored tree at the xerophytic forest, *Alluaudia procera*, does not exist in the mesic forest. Both open nests and tree holes are located significantly higher above the ground in the mesic forest, where the mean height of trees is also greater than in the xerophytic forest (Table 13.6). In both forests, mouse lemurs used open nests that are higher above the ground than are tree holes, likely because they use the canopy for open nests.

There are also significant seasonal differences in plant species used for sleeping ($\chi^2 = 760.0$, df = 45, $P < 0.001$). During the hot, rainy season, mouse lemurs used nests that were higher in the trees (mean height = 3.91 ± SD 1.83 m) than those used during the dry season (mean height = 2.93 ± SD 1.71 m, $t = 13.3$, df = 2292, $P < 0.001$). Types of nests also differ by season ($\chi^2 = 672.09$, df = 1, $P < 0.001$). Open nests are strongly preferred during the rainy season (73.9% of all nests), whereas

Figure 13.3 Comparison of nest types. Left: tree hole in a dead tree (*Alantsilodendron humbertii*). Right: open nest of tangled vegetation in a famata tree (*Euphorbia tirucalli*). Photo credits: Emilienne Rasoazanabary. (A black and white version of this figure will appear in some formats. For the colour version, please refer to the plate section.)

Table 13.6 *t*-Tests of differences between measurements of trees or bushes selected for sleeping at the two forests.

Measurement	Mesic forest mean (SD)	Xerophytic forest mean (SD)	*t*, df	*P*
Mean height of all nests	3.91 (± 2.15) m	2.82 (± 1.16) m	15.3, 1878.0	<0.001
Mean height of open nests	4.12 (± 2.15) m	3.32 (± 1.02) m	08, 964.58	<0.001
Mean height of tree holes	3.67 (± 2.13) m	2.52 (± 1.13) m	11.69, 832.3	<0.001
Mean height of sleeping trees or bushes (based on samples in botanical plots)	8.47 (± 3.19) m	5.94 (± 2.64) m	11.36, 670.9	<0.001

tree holes are preferred during the dry season (79.9% of all nests). The shift occurs both in the mesic and xerophytic forests: for the mesic forest, χ^2 =263.8 (df = 1, $P < 0.001$); for the xerophytic forest, χ^2 = 459.1 (df = 1, $P < 0.001$). This seasonal shift is associated with a decrease (from the wet to the dry season) in the number of individual plants used for sleeping, as not all plants have holes appropriate for such purposes (Table 13.7). In total, 24.1% of the 29 species used in the dry season are abandoned during the wet season. Conversely, 43.6% of the 39 species used as nests during the rainy season are abandoned during the dry season.

Table 13.7 Nest-type preference during the rainy and dry seasons at the two forests.

Trait	Mesic forest, rainy season	Mesic forest, dry season	Xerophytic, rainy season	Xerophytic, dry season
# Nests	174	105	157	45
Mean height	4.45 (±2.26) m	3.46 (±1.95) m	3.36 (±0.98) m	2.30 (±1.07) m
# Plant species	28	13	17	19
# Open nests	147	51	114	7
# Tree holes	32	54	34	36
# Plant species with both nest types	5	3	9	2
Plant species with 40 or more records of use for sleeping	*Euphorbia tirucalli* *Acacia bellula* *Tamarindus indica*	*Grewia leucophylla*	*Alluaudia procera* *Euphorbia stenoclada* *Cedrelopsis grevei* *Phyllanthus decoryanus*	*Terminalia fatraea* *Alluaudia procera* *Alantsilodendron humbertii* *Rhigozum madagascariensis*

Resource selectivity in the mesic and xerophytic forests

Table 13.8 lists the nine plant species that individually comprise more than 10% of focal individual observations for sleeping or feeding at either the mesic or xerophytic forest. The table gives the % of focal individual observations for feeding and sleeping on each of these plant species and their relative abundances within each of those forests (the latter based on % of plants recorded in the botanical plots). All nine of these species were indeed recorded in botanical plots. In both forests, mouse lemurs sleep in plant species that are relatively abundant; thus they are not particularly selective in their choice of trees for sleeping. The correlation between the relative abundance of the top sleeping trees and these plants' relative use during focal individual observation is high for both the mesic forest (Pearson's $r = 0.78$, $n = 9$, $P = 0.01$) and the xerophytic forest (Pearson's $r = 0.93$, $n = 9$, $P < 0.001$).

The same is not true of trees or bushes selected for feeding. Here, the correlations between relative preferred plant abundance in the forest and relative preferred plant use by focal individuals for feeding are low for both mesic ($r = 0.18$) and xerophytic ($r = -0.10$) forests ($n = 9$, not significant for both). Thus, mouse lemurs are selective in their feeding behavior, selecting trees that are not necessarily abundant in the forest.

Habitat and morphology

The mesic forest contains trees and bushes that are, on average, taller and with larger crowns (both in terms of crown diameter and height) than those from the xerophytic forest. To test the hypothesis that mouse lemurs in the mesic forest actually use trees that are taller and larger crowned than do those in the xerophytic forest, we compared the dimensions of those tree species most often selected for

Table 13.8 Plant species most favored for feeding or sleeping by focal individuals.

Species	Relative abundance* in mesic forest (%)	Relative abundance* in xerophytic forest (%)	% of focal animal records in specified forest on specified plant			
			Feed		Sleep	
			Mesic	Spiny	Mesic	Spiny
Alantsilodendron humbertii	0	0.3	0	20.5	1.1	4.5
Alluaudia procera	0	20.9	0	0	0	34.4
Terminalia fatraea	0.3	2.3	0	33.9	0	16.4
Euphorbia tirucalli	14.9	2.1	5.8	0	26.7	4.0
Acacia bellula	7.2	0	20.1	0	13.2	0
Rhopalocarpus lucidus	9.9	0	20.9	0	5.4	0
Tamarindus indica	14.6	0	0	0	11.7	0
Grewia leucophylla	5.2	0	14.2	0	17.1	0
Bredelia sp.	0.3	0	14.0	0	0	0

* Percentages of individuals in the botanical sampling plots.

feeding and sleeping. We also compared the dimensions of branches used as supports when feeding on different food types, specifically testing the hypothesis that mouse lemurs use smaller supports when feeding on fruits and insects (as occurs more often in the mesic forest) than when feeding on gums (as occurs more often in the xerophytic forest).

Indeed, the preferred resource plants in the mesic forest are significantly taller and have larger crowns than the preferred resource plants in the xerophytic forest (Table 13.9). Supports used for exudate feeding are significantly larger than those used for feeding on fruits and insects (Table 13.10, $P < 0.001$). Gum feeding supports were located significantly closer to the ground ($P < 0.001$) than those used for feeding on fruits and insects (which themselves do not differ significantly in height).

Discussion

Our data demonstrate that: (1) *Microcebus griseorufus* at BMSR is far more selective in its choice of feeding trees than in its choice of sleeping trees; (2) there are strong and similar seasonal shifts in preferred sleeping site type and in diet in both the mesic and xerophytic forests; (3) the mesic and xerophytic forest mouse lemurs differ strikingly in morphological (especially cheiridial) characteristics; and (4) the two forest populations of mouse lemurs differ significantly in adult body mass and in adult skull size.

Feeding selectivity and seasonal dietary shifts

In general, mouse lemurs are flexible omnivores. Their primary staples are insects, fruits, and gums (Martin, 1972, 1973; Barre *et al.*, 1988; Corbin and Schmid, 1995;

Table 13.9 *t*-Tests comparing metric characteristics of plant species most often used for either feeding or sleeping by focal individuals.*

Variable	Mesic forest			Xerophytic forest			*t* value, df, and *P***
	Mean	SD	SE	Mean	SD	SE	
Height (m)	9.01	3.36	0.24	6.55	3.93	0.32	6.3, 343, <0.001
Crown height (m)	3.57	2.58	0.21	2.22	1.27	0.11	5.6, 217.04, <0.001
DBH (cm)	14.93	9.82	0.71	12.55	5.93	0.48	2.8, 315.93, 0.006
Crown diameter (m)	3.09	1.91	0.16	1.33	0.53	0.05	10.3, 160.55, <0.001

*Sample sizes vary because not all of the individual trees could be measured for all traits.
**When the Levene's *F* test of equality of variances demonstrated that the variances of the two samples being compared were significantly different, we did not pool the variances in testing the significance of differences of the means. This affected the degrees of freedom but in no case did it affect the significance of the result.

Table 13.10 Dimensions of supports used for feeding on insects, fruits, and gums, with ANOVA.

Variable	Resource			ANOVA *F*, df, and *P*
	Fruits *N*, mean, SD	Insects *N*, mean, SD	Gums *N*, mean, SD	
Height at feeding (m)	447, 5.7, 3.1	708, 5.3, 2.8	1084, 4.6, 3.0	27.33, 2, <0.001
Branch diameter (cm)	447, 1.6, 1.4	670, 2.3, 3.4	1081, 7.3, 4.8	507.42, 2, <0.001

Génin, 2001, 2003; Lahann, 2007; Dammhahn and Kappeler, 2008). At BMSR, females, the dominant sex, consume more fruit than males. Fruit production peaks early in the rainy season at BMSR, just as elsewhere across Madagascar (e.g., in dry deciduous forests like Kirindy or Ankarafantsika: Dammhahn and Kappeler, 2008; Sato, 2012; or eastern rainforests like Ranomafana: Atsalis, 1999). What differs is the availability of fruit at other times. Because the climate at BMSR is drier and more seasonal than at these other forests, fruit production is extremely limited during the dry season, and the relative abundance of insects depends at least in part on the relative abundance of fruit. Mouse lemurs at BMSR must feed on gums in order to compensate for the scarcity of fruits and insects during the dry season. Génin (2008) reports a similar phenomenon for *M. griseorufus* at Berenty.

Thus, it is not surprising that at BMSR, feeding selectivity is greater for gum trees than for trees used for any other purpose (dietary or otherwise). This is true in both the xerophytic and the mesic forest, but selectivity is greatest in the xerophytic forest where only two species, *Terminalia fatraea* and *Alantsilodendron humbertii*, are used for gum extraction, and where the availability of fruit is exceptionally low, especially during the dry season.

Other primates living in dry and highly seasonal habitats rely on gums (Nash, 1986; Isbell, 1998; Heymann and Smith, 1999; Peres, 2000; Radespiel *et al.*, 2006;

Nowack *et al.*, 2013). For other species of mouse lemur, particularly those living in the west (like *M. murinus*), gums become important fallback or keystone foods (Génin, 2003). Competition for access to gum trees between *M. murinus* and gum specialist *Phaner furcifer* has been reported at Kirindy forest (Génin, 2003); here, gums can comprise up to 75% of the diet of mouse lemurs during the dry season. At the same site, *M. berthae* relies to a lesser extent on gums, but nevertheless must depend on them during the dry season (Dammhahn and Kappeler, 2008).

Seasonal shifts in nest type

Mouse lemurs at BMSR are far less selective when seeking sites for sleeping; the plant species selected for sleeping are important only to the extent that some species may be more likely than others to provide appropriate nests. However, in any given season, mouse lemurs at BMSR do show selectivity for nest type. Explanations for mouse lemur nest selectivity in other forests have focused primarily on predator avoidance and thermoregulation. Ganzhorn and Schmid (1998) and Radespiel *et al.* (1998, 2003) evaluated nest "quality" primarily in terms of protection against predators, as predation risk is high for mouse lemurs (Goodman *et al.*, 1993a, 1993b). If predation were primary, tree holes would be preferred to open nests due to their better protection, and holes that are high in trees might be safer than holes that are closer to the ground. Other researchers have embraced the same set of assumptions (Radespiel *et al.*, 1998; Rasoloharijaona *et al.*, 2003, 2008). However, in a study of *Microcebus murinus* at Ampijoroa, Lutermann *et al.* (2010) found that nest choice is seasonal and relates more to thermoregulatory factors than to protection against predators. Thermoregulation was also implicated in a study of nest construction and use by *M. ravelobensis* at Ampijoroa (Thorén *et al.*, 2010).

These latter results match our own: the seasonal shift in nest-type preference can be understood in terms of thermoregulatory factors. Tree holes, which provide greater insulation than open tangles of vegetation, are preferred during the cold dry season, and open tangled nests are preferred during the hot rainy season. In addition, nests closer to the ground are preferred during the cold dry season and nests higher in the trees are preferred during the hot rainy season. Tree holes that are nearer to the ground may be less exposed to chilling morning winds and may provide a thermoregulatory advantage when temperatures are low. Conversely, during the hot wet season, higher tree holes and open nests that are more exposed to winds may provide relief from the very high midday temperatures that mouse lemurs encounter while they sleep. The thermoregulatory hypothesis may explain why at BMSR the largest seasonal shift in nest height occurs in the xerophytic forest. Because of the lack of canopy shading in xerophytic thickets, these forests can experience the largest seasonal shifts in temperature. During the hot rainy season, temperatures inside well-insulated nests can be very high (Schmid, 1998). Nests cannot be assessed in terms of absolute "quality" when their utility varies seasonally with ambient temperature.

Similar seasonal shifts (with preference for open nests during the hot, wet season and tree holes during the cool, dry season) may characterize other lemurs. Tree holes function as "winter residences" for *Lepilemur sahamalazensis* (Seiler *et al.*, 2013). *L. mustelinus* also prefer tree holes during the cool, dry season and into the beginning of the wet season (Rasoloharijaona *et al.*, 2008). *Avahi occidentalis* also show seasonal changes in sleeping site selection (Ramanankirahina *et al.*, 2012). Génin (2010) reports that at Berenty, *M. griseorufus* females with young prefer tree holes, and there is a seasonal pattern in the selection of sleeping sites here too.

Morphometric variation

Perhaps our least anticipated result is that mesic and xerophytic forest *Microcebus griseorufus* adults at BMSR would differ in cheiridial proportions. Variation of this sort among mouse lemur populations has never before been reported and warrants further study. Among primates, digit length and proportions vary tremendously, and that variation is strongly correlated with posture and locomotion. The striking metric differences between the cheiridia of *M. griseorufus* in the mesic and xerophytic forests of BMSR appear to be related to the substrates that they must navigate in their respective forests when traveling to their sleeping and feeding sites (see below). However, unless gene flow is very limited for mouse lemurs living in the mesic and xerophytic forests of BMSR, we would not expect selection to have acted independently on their digits. Without some barrier to gene flow, site-specific differences in digit proportions cannot have resulted from natural selection, and are more likely to reflect developmental differences. There is a growing literature on how environmental stimuli affect bone modeling and remodeling, and can trigger subtle shifts in the timing of gene expression. Furthermore, the first digit differs from all other digits in the genetic control of its expression (see, for example, Wagner and Vargas, 2008). On the basis of mitochondrial DNA, Heckman *et al.* (2006) assert that strong gene flow connects the mouse lemurs in the two forests at BMSR. The study of their nuclear DNA is needed to confirm this, as the mitochondrial haplotypes may be old. During our study period (2003–2007), no individual initially trapped in either the mesic or xerophytic forest was later trapped in the other.

Even if the basis of the morphometric differences remains unclear, the morphometric differences themselves seem to correspond to the observed patterns of habitat use in both forest types. When mouse lemurs are in the high canopy (as they are more often in the mesic forest), they generally use branches that are small in diameter. Having relatively longer digits 3–5 increases prehensility, and is therefore advantageous when navigating on small branches. In the xerophytic forest, mouse lemurs tend to travel on larger branches and they prefer open nests and tree holes that are closer to the ground. Nest location matters; open nests that are closer to the ground provide larger supports and more intertwined lianas than do those that are high above the ground. On larger supports, using the fingers and toes as clamps can be advantageous.

When mouse lemurs feed on fruit (which they also do more often in the mesic forest), they are often using the most peripheral branches, which are also small in diameter. In contrast, in the xerophytic forest where mouse lemurs feed more often at lower heights and more often on gums, they are more frequently using larger branches or tree trunks. Most gums are consumed while individuals are clinging vertically to a relatively large trunk or branch. In contrast, insects and fruits are more likely consumed on small peripheral branches.

An explanation for the observed differences in adult body mass and skull size of *M. griseorufus* in the mesic and xerophytic forests at BMSR is also lacking, and also requires more research. We suspect again these differences are developmental, and that the larger size of *M. griseorufus* in the mesic forest may relate to the greater availability of sugar-rich fruit there. Another possible explanation relates to differences in the timing of growth and development at the two sites. If the two forests differ in the relative number of survivors from different birth seasons (at Beza, December and January vs. April and May), or if they differ in the timing of the birth seasons, the mouse lemurs living in the two forests may also differ in average adult size. Late-born individuals may remain small throughout their lives because their growth was interrupted by short day exposure. In fact, there is some evidence supporting this hypothesis. Vaginal swelling was observed three weeks earlier in the gallery forest than in the spiny forest, and infants were observed three weeks earlier. In addition, while pregnant females were observed at both sites in April, more pregnant females were observed in the spiny forest than in the gallery forest. We do not have good data on infant mortality from the two sites.

Alternatively (or additionally), access to fruits by lactating mothers and weanlings may be the critical factor. Focal individual data confirming differences in fruit consumption at the two forests are supported by stable isotope data (Crowley *et al.*, 2014). In this light, it is interesting that female *M. griseorufus* at BMSR, who consume more fruits than males, are larger in body mass than males. Génin *et al.* (2010) report the same phenomenon for male and female *M. griseorufus* from Berenty. Bohr *et al.* (2011) describe similar differences between populations of *M. griseorufus* living in the more mesic "dry forest" vs. the yet drier portions of the reserve (spiny bush) at Tsimanampetsotsa. Just as at BMSR, there is more available fruit in the more mesic forest; mouse lemurs from these wetter forests are in "better condition."

Conservation implications

In summary, despite the extraordinary plasticity exhibited by *Microcebus griseorufus* in its exploitation of very different habitats, that plasticity operates under specific constraints – a need to find plants that yield nutritional exudates, and a need to find nests that shield appropriately against cold external conditions and maintain comfortable internal temperatures.

Of all mouse lemur species, *M. griseorufus* is most tolerant of dry habitats. If, under climate change, those habitats become even drier, fruiting will likely become even more seasonally constricted and severe droughts more frequent. We know which plant species are most important to *M. griseorufus*, and we know their relative abundance in their respective forests. We need to understand how these plants respond to drought. We also know that *M. griseorufus* can survive in more "mesic" (although still dry) forests; in some ways it appears to do quite well in those forests. However, if this species is forced into particular forests simply because they are protected, it is likely that populations will become increasingly fragmented. Survival will depend on access to fruit and to appropriate gum-producing trees. Access to critical resources will be paramount.

By studying the morphology and behavior of closely related populations of mouse lemurs living in very different forests, we are beginning to understand the breadth of and limits to adaptive flexibility of *Microcebus griseorufus*. If we can understand how well they can adapt to environments that are atypical for the species (such as the mesic forest at BMSR), we should be able to better make conservation policy recommendations.

Acknowledgments

We thank the Madagascar National Parks (MNP, formerly called ANGAP or Association Nationale pour la Gestion des Aires Protégées), the Ministère de l'Environnement de Madagascar, and the Département des Eaux et Forêts at the École Supérieure des Sciences Agronomiques (ESSA), Université d'Antananarivo, for giving us permission to conduct fieldwork at Beza Mahafaly Special Reserve (BMSR). We thank the Chief of the BMSR for logistical support and the Scientific Coordinator for his participatory effort in this research. For help in the field, we thank students from the Universities of Antananarivo and Toliara, Viviane and Tsibara, respectively. We are also very grateful to local communities, particularly local guides (Ralaivao, Elahavelo, Efitiria, Enafa, Edabo, Edward, Etinahy, Filaoke, Fanomezana, Dada, Etala, Ebolo) who collected data day and night. This manuscript was substantially improved by comments on an earlier version from two anonymous reviewers and the editors of this volume. Financial support for this research was provided by the Margot Marsh Biodiversity Foundation, the American Society of Primatologists, Primate Conservation, Inc., Wildlife Conservation Society, the National Geographic Society, and the Wenner-Gren Foundation for Anthropological Research (to Emilienne Rasoazanabary) and a Guggenheim Fellowship (to Laurie Godfrey).

References

Altmann J. 1974. Observational study of behavior: sampling methods. *Behaviour* 49:227–267.

Atsalis S. 1999. Seasonal fluctuation in body fat and activity level in a rain-forest species of mouse lemur, *Microcebus rufus*. *International Journal of Primatology* 20:883–910.

Barre V, Lebec A, Petter J-J, Albignac R. 1988. Étude du microcèbe par radiotracking dans la forêt de l'Ankarafantsika. In L Rakotovao, V Barre, J Sayer (eds.), *L'Équilibre des Écosystèmes Forestiers à Madagascar. Actes d'un Séminaire International* (pp. 61–71). IUCN, Gland.

Blanco MB. 2011. Timely estrus in wild brown mouse lemur females at Ranomafana National Park, southeastern Madagascar. *American Journal of Physical Anthropology* 145:311–317.

Bohr YE-MB, Giertz P, Ratovonamana YR, Ganzhorn JU. 2011. Gray-brown mouse lemurs (*Microcebus griseorufus*) as an example of distributional constraints through increasing desertification. *International Journal of Primatology* 32:901–913.

Burgess ND. 2004. *Terrestrial Ecoregions of Africa and Madagascar: A Conservation Assessment.* Island Press, Washington.

Corbin GD, Schmid J. 1995. Insect secretions determine habitat use patterns by a female lesser mouse lemur (*Microcebus murinus*). *American Journal of Primatology* 37:317–324.

Crowley BE, Rasoazanabary E, Godfrey LR. 2014. Stable isotopes complement focal individual observations and confirm dietary variability in reddish-gray mouse lemurs (*Microcebus griseorufus*) from southwestern Madagascar. *American Journal of Physical Anthropology* 155:77–90.

Cuozzo FP, Rasoazanabary E, Godfrey LR, *et al.* 2013. Biological variation in a large sample of mouse lemurs from Amboasary, Madagascar: implications for interpreting variation in primate biology and paleobiology. *Journal of Human Evolution* 64:1–20.

Dammhahn M, Kappeler PM. 2008. Comparative feeding ecology of sympatric *Microcebus berthae* and *M. murinus. International Journal of Primatology* 29:1567–1589.

Dewar RE, Richard AF. 2007. Evolution in the hypervariable environment of Madagascar. *Proceedings of the National Academy of Sciences, USA* 104:13723–13727.

Fenn MD. 2003. The spiny forest ecoregion. In SM Goodman, JP Benstead (eds.), *The Natural History of Madagascar* (pp. 1525–1529). University of Chicago Press, Chicago.

Ganzhorn JU. 1995. Low-level forest disturbance effects on primary production, leaf chemistry, and lemur populations. *Ecology* 76:2084–2096.

Ganzhorn JU, Schmid J. 1998. Different population dynamics of *Microcebus murinus* in primary and secondary deciduous dry forests of Madagascar. *International Journal of Primatology* 19:785–796.

Génin F. 2001. Gummivory in mouse lemurs during the dry season in Madagascar. *Folia Primatologica* 72:119–120.

Génin F. 2003. Female dominance in competition for gum trees in the grey mouse lemur *Microcebus murinus. La Terre et la Vie* 58:397–410.

Génin F. 2008. Life in unpredictable environments: first investigation of the natural history of *Microcebus griseorufus. International Journal of Primatology* 29:303–321.

Génin F. 2010. Who sleeps with whom? Sleeping association and socio-territoriality in *Microcebus griseorufus. Journal of Mammalogy* 91:942–951.

Génin FGS, Masters JC, Ganzhorn JU. 2010. Gummivory in cheirogaleids: primitive retention or adaptation to hypervariable environments? In AM Burrows, LT Nash (eds.), *The Evolution of Exudativory in Primates* (pp. 123–140). Springer, New York.

Gligor M, Ganzhorn JU, Rakotondravony D, *et al.* 2009. Hybridization between mouse lemurs in an ecological transition zone in southern Madagascar. *Molecular Ecology* 18:520–533.

Goodman SM, Langrand O, Raxworthy CJ. 1993a. Food habits of the Madagascar long-eared owl *Asio madagascariensis* in two habitats in southern Madagascar. *Ostrich* 64:79–85.

Goodman SM, Langrand O, Raxworthy CJ. 1993b. The food habits of the barn owl *Tyto alba* at three sites on Madagascar. *Ostrich* 64:160–171.

Hapke A. 2005. Populationsgenetik und Differenzierung von Arten der Gattungen Mausmaki, *Microcebus* (E Geoffroy St. Hilaire, 1828) und Katzenmaki, *Cheirogaleus* (E Geoffroy St Hilaire, 1812) in Südost-Madagaskar. Dissertation, University of Hamburg, Hamburg, Germany.

Heckman KL, Rasoazanabary E, Machlin E, Godfrey LR, Yoder AD. 2006. Incongruence between genetic and morphological diversity in *Microcebus griseorufus* of Beza Mahafaly. *BMC Evolutionary Biology* 6:98.

Heymann EW, Smith AC. 1999. When to feed on gums: temporal patterns in gummivory in wild tamarins, *Saguinus mystax* and *Saguinus fuscicollis* (Callithricinae). *Zoo Biology* 18:459–471.

Isbell LA. 1998. Diets for a small primate: insectivory and gummivory in the (large) patas monkey (*Erythrocebus patas pyrrhonotus*). *American Journal of Primatology* 45:381–398.

Kamilar JM, Muldoon KM. 2010. The climatic niche diversity of Malagasy primates: a phylogenetic perspective. *PLoS ONE* 5(6):e11073. doi:10.1371/J.pone.0011073

Kobbe S, Dausmann KH. 2009. Hibernation in Malagasy mouse lemurs as a strategy to counter environmental challenge. *Naturwissenschaften* 96:1221–1227.

Lahann P. 2007. Feeding ecology and seed dispersal of sympatric cheirogaleid lemurs (*Microcebus murinus, Cheirogaleus medius, Cheirogaleus major*) in the littoral rainforest of south-east Madagascar. *Journal of Zoology* 271:88–98.

Lahann P, Schmid J, Ganzhorn JU. 2006. Geographical variation in populations of *Microcebus murinus* in Madagascar: resource seasonality or Bergmann's rule? *International Journal of Primatology* 27:983–999.

Lutermann H, Verburgt L, Rendigs A. 2010. Resting and nesting in a small mammal: sleeping sites as a limiting resource for female grey mouse lemurs. *Animal Behaviour* 79:1211–1219.

Martin RD. 1972. A preliminary field study of the lesser mouse lemur (*Microcebus murinus* JF Miller, 1777). *Advances in Ethology* 9:43–89.

Martin RD. 1973. A review of the behaviour and ecology of the lesser mouse lemur (*Microcebus murinus*, JF Miller 1777). In RP Michael, J Crook (eds.), *Comparative Ecology and Behaviour of Primates* (pp. 1–68). Academic Press, London.

Nash LT. 1986. Dietary, behavioral, and morphological aspects of gummivory in primates. *Yearbook of Physical Anthropology* 29:113–137.

Nowack J, Wippich M, Mzilikazi N, Dausmann KH. 2013. Surviving the cold, dry period in Africa: behavioral adjustments as an alternative to heterothermy in the African lesser bush-baby (*Galago moholi*). *International Journal of Primatology* 34:49–64.

Peres CA. 2000. Identifying keystone plant resources in tropical forest: the case of gums from *Parkia* pods. *Journal of Tropical Ecology* 16:287–317.

Radespiel U, Raveloson H. 2001. Preliminary study on the lemur communities at three sites of dry deciduous forest in the Reserve Naturelle d'Ankarafantsika. *Lemur News* 6:22–24.

Radespiel U, Cepok S, Zietemann V, Zimmermann E. 1998. Sex-specific usage patterns of sleeping sites in grey mouse lemurs (*Microcebus murinus*) in northwestern Madagascar. *American Journal of Primatology* 46:77–84.

Radespiel U, Ehresmann P, Zimmermann E. 2003. Species-specific usage of sleeping sites in two sympatric mouse lemur species (*Microcebus murinus* and *M. ravelobensis*) in northwestern Madagascar. *American Journal of Primatology* 59:139–151.

Radespiel U, Reimann W, Rahelinirina M, Zimmermann E. 2006. Feeeding ecology of sympatric mouse lemur species in northwestern Madagascar. *International Journal of Primatology* 27:311–321.

Radespiel U, Olivieri G, Rasolofoson DW, *et al.* 2008. Exceptional diversity of mouse lemurs (*Microcebus* spp.) in the Makira region with the description of one new species. *American Journal of Primatology* 70:1033–1046.

Radespiel U, Ratsimbazafy JH, Rasoloharijaona S, *et al.* 2012. First indications of a highland specialist among mouse lemurs (*Microcebus* spp.) and evidence for a new mouse lemur species from eastern Madagascar. *Primates* 53:157–170.

Rakotondranary SJ, Ganzhorn JU. 2011. Habitat separation of sympatric *Microcebus* spp. in the dry spiny forest of south-eastern Madagascar. *Folia Primatologica* 82:212–213.

Ramanankirahina R, Joly M, Zimmermann E. 2012. Seasonal effects on sleeping site ecology in a nocturnal pair-living lemur (*Avahi occidentalis*). *International Journal of Primatology* 33:428–439.

Rasoazanabary E. 2004. A preliminary study of mouse lemurs in the Beza Mahafaly Special Reserve, southwest Madagascar. *Lemur News* 9:4–7.

Rasoazanabary E. 2006. Male and female activity patterns in *Microcebus murinus* during the dry season at Kirindy Forest, western Madagascar. *International Journal of Primatology* 27:437–64.

Rasoloarison RM, Weisrock DW, Yoder AD, Rakotondravony D, Kappeler PM. 2013. Two new species of mouse lemurs (Cheirogaleidae: *Microcebus*) from eastern Madagascar. *International Journal of Primatology* 34:455–469.

Rasoloharijaona S, Rakotosamimanana B, Randrianambinina B, Zimmermann E. 2003. Pair-specific usage of sleeping sites and their implications for social organization in a nocturnal Malagasy primate, the Milne Edwards' sportive lemur (*Lepilemur edwardsi*). *American Journal of Physical Anthropology* 122:251–258.

Rasoloharijaona S, Randrianambinina B, Zimmermann E. 2008. Sleeping site ecology in a rain-forest dwelling nocturnal lemur (*Lepilemur mustelinus*): implications for sociality and conservation. *American Journal of Primatology* 70:247–253.

Ratsirarson J, Randrianarisoa J, Ellis E, *et al.* 2001. *Beza Mahafaly: Ecologie et réalités socio-économiques. Recherche pour le développement. Série Sciences Biologiques No. 18.* Centre d'Information et de Documentation Scientifique et Technique, Antananarivo.

Rensing S. 1999. Immobilization and anesthesia of nonhuman primates. *Primate Report* 55:33–38.

Sato H. 2012. Seasonal fruiting and seed dispersal by the brown lemur in a tropical dry forest, north-western Madagascar. *Journal of Tropical Ecology* 29:61–69.

Schmid J. 1998. Tree holes used for resting by gray mouse lemurs (*Microcebus murinus*) in Madagascar: insulation capacities and energetic consequences. *International Journal of Primatology* 19:797–809.

Schmid J, Kappeler PM. 1994. Sympatric mouse lemurs (*Microcebus* spp.) in western Madagascar. *Folia Primatologica* 63:162–170.

Schmid J, Kappeler PM. 1998. Fluctuating sexual dimorphism and differential hibernation by sex in a primate, the gray mouse lemur (*Microcebus murinus*). *Behavioral Ecology and Sociobiology* 43:125–132.

Schwab D, Ganzhorn JU. 2004. Distribution, population structure and habitat use of *Microcebus berthae* compared to those of other sympatric cheirogaleids. *International Journal of Primatology* 25:307–330.

Seiler M, Holderied M, Schwitzer C. 2013. Effects of habitat degradation on sleeping site choice and use in Sahamalaza sportive lemurs (*Lepilemur sahamalazensis*). *International Journal of Primatology* 34:260–280.

Simpson EH. 1949. Measurement of diversity. *Nature* 163:688.

Thorén S, Quietzsch F, Radespiel U. 2010. Leaf nest use and construction in the golden-brown mouse lemur (*Microcebus ravelobensis*) in the Ankarafantsika National Park. *American Journal of Primatology* 72:48–55.

Wagner GP, Vargas AO. 2008. On the nature of thumbs. *Genome Biology* 9:213.

Wilmé L, Goodman SM, Ganzhorn JU. 2006. Biogeographic evolution of Madagascar's microendemic biota. *Science* 312:1063–1065.

Yoder AD, Burns MM, Génin F. 2002. Molecular evidence of reproductive isolation in sympatric sibling species of mouse lemurs. *International Journal of Primatology* 23:1335–1343.

Zimmermann E, Cepok S, Rakotoarison N, Zietemann V, Radespiel U. 1998. Sympatric mouse lemurs in north-west Madagascar: a new rufous mouse lemur species (*Microcebus ravelobensis*). *Folia Primatologica* 69:106–114.

14 Determinants of isotopic variation in two sympatric mouse lemur species from northwestern Madagascar

Laura Heck, Brooke Crowley, Sandra Thorén, and Ute Radespiel

Introduction

Malagasy habitats undergo vast seasonal changes in temperature and rainfall regimes, which in turn strongly influence the biology and life history of their inhabitants (Wright, 1999). The western dry deciduous forests show a yearly alternation between a distinct dry season and a hot and humid rainy season (Jury, 2003). Correspondingly, the phenology of dry forest trees – i.e., the availability of leaves, flowers and fruits – changes substantially over the course of a year (Sorg and Rohner, 1996; Thorén *et al.*, 2011). As a result, plant food resources are not available evenly throughout the yearly cycle, and a relatively resource-poor dry season can be distinguished from a relatively resource-rich rainy season (Hladik *et al.*, 1980).

The length of both seasons varies across Madagascar due to a geographical gradient in minimal temperatures and annual rainfall. Adaptive responses of lemurs to this environmental seasonality include seasonal changes in general activity (Schmid and Kappeler, 1998; Wright, 1999; Dausmann *et al.*, 2004), shelter usage pattern (Ramanankirahina *et al.*, 2012), body mass (Hladik *et al.*, 1980; Schmid and Kappeler, 1998; Atsalis, 1999; Randrianambinina *et al.*, 2003), reproduction (Hladik *et al.*, 1980; Wright, 1999), and feeding regime (Hladik *et al.*, 1980; Thorén *et al.*, 2011). It is known, for example, that diademed sifakas (*Propithecus diadema*) shift their diet from fruits and seeds to non-fruit foods, such as leaves and flowers, and decrease food ingestion and general feeding time during the dry season (Irwin *et al.*, 2014). This ability to flexibly change behavior or resource usage patterns in view of certain environmental changes is the result of adaptive plasticity (Nussey *et al.*, 2007). Adaptive plasticity in seasonal feeding regimes can be expected to be highest in omnivorous species that can, in principle, switch between different sources of energy and protein by consuming different food plant species or varying the proportion of animal prey in the diet. Moreover, omnivores can consume

The Dwarf and Mouse Lemurs of Madagascar: Biology, Behavior and Conservation Biogeography of the Cheirogaleidae, eds. S.M. Lehman, U. Radespiel, E. Zimmermann. Published by Cambridge University Press. © Cambridge University Press 2016.

different parts, such as fruits and gums, of the same food plant species according to the availability of these items.

Mouse lemurs (*Microcebus* spp.) are a genus of omnivorous, solitary foragers that can be found in all Malagasy forest types. Twenty-one mouse lemur species have been described so far (Mittermeier *et al.*, 2010; Radespiel *et al.*, 2012; Rasoloarison *et al.*, 2013). Coexistence of two species has been recorded in forests across the island. For example, *M. murinus* and *M. griseorufus* co-occur in southern Madagascar (Gligor *et al.*, 2009; Rakotondranary *et al.*, 2010, 2011), *M. murinus* and *M. berthae* are found together in southwestern Madagascar (Dammhahn and Kappeler, 2005, 2008a, 2008b), and *M. murinus* and *M. ravelobensis* live in partial sympatry in northwestern Madagascar where this study is situated (Zimmermann *et al.*, 1998; Rendigs *et al.*, 2003; Rakotondravony and Radespiel, 2009).

Seasonal changes in the diet of mouse lemurs have already been described for several species (*M. murinus*: Lahann, 2006; Dammhahn and Kappeler, 2008a; Thorén *et al.*, 2011; *M. ravelobensis*: Thorén *et al.*, 2011; *M. griseorufus*: Génin, 2008; Rakotondranary *et al.*, 2011; Crowley *et al.*, 2014). For example, it was shown that in Kirindy, western Madagascar, *M. murinus* switches from an insect-rich diet in the beginning of the wet season to a more fruit-rich diet towards the end of the wet season (Dammhahn and Kappeler, 2008a). In the Ankarafantsika National Park, northwestern Madagascar, Thorén *et al.* (2011) observed that *M. ravelobensis* relied heavily on insect secretions and less on fruits in the late dry season and early rainy season, whereas *M. murinus* shifted from exudates as its main food source to fruits at the beginning of the rainy season.

Although observational data have provided valuable evidence for mouse lemur diets, feeding budgets determined by observational data alone always remain patchy and incomplete. Feeding studies on mouse lemurs notoriously suffer from limited contact times, as a small, 60 g, arboreal, nocturnal animal cannot be continuously observed in a forest. Published contact times vary from 30% to 70% of the actual activity time in different studies and species (Génin, 2003; Dammhahn and Kappeler, 2008a; Thorén *et al.*, 2011). Complementary evidence for seasonal variations of feeding habits is, therefore, urgently needed to complement incomplete records provided by direct behavioral observations.

Stable isotope ratios of carbon and nitrogen ($\delta^{13}C$ and $\delta^{15}N$) offer a promising approach. A growing number of studies show that isotope analyses can be used to identify ecological feeding niches (e.g., bats: Popa-Lisseanu *et al.*, 2007; birds: Hobson, 1995; martens: Ben-David *et al.*, 1997; lemurs: Rakotondranary *et al.*, 2011; Crowley *et al.*, 2013, 2014; Dammhahn and Kappeler, 2010, 2014). This is possible because stable isotope ratios in an animal's fur, for example, contain information about ingested foods (Szepanski *et al.*, 1999; Kelly, 2000). An increase in $\delta^{13}C$ values of ca. +5‰ between plants and herbivore fur (+1–2‰ between higher trophic levels), and an increase in $\delta^{15}N$ values of +2–5‰ between trophic levels can be used to roughly identify an animal's trophic level (DeNiro and Epstein, 1978, 1981; Crowley *et al.*, 2011). Because isotope values increase with increasing trophic level, an omnivorous individual that eats a mix of plant and animal matter is expected to have higher carbon and nitrogen isotope values than one that is completely herbivorous. It has to be noted, however, that the isotopic signature of individuals may also be influenced by body mass and nutritional status (e.g., wild dover sole: Rau *et al.*, 1981; earthworms: Schmidt *et al.*, 2004).

Food plants can also differ in their isotope values depending on several biotic and abiotic factors (reviewed in Evans, 2007; Marshall *et al.*, 2007). First, leaves and fruits in the canopy may have higher $\delta^{13}C$ values than those close to the ground (van der Merwe and Medina, 1989; Crowley *et al.*, 2012). This "canopy effect" is due to differences in sunlight exposure, relative humidity, and the assimilation of CO_2 produced by microbes in the soil. Second, soil composition, temperature and the availability of water influence the isotope composition of plants. In particular, it has been shown that well-nourished plants have lower $\delta^{13}C$ values than plants with nutritional deficits or water stress (O'Leary, 1981). As a result, carbon isotope values for C_3 plants are typically lower in mesic than in xeric environments due to a higher stomatal conductance (Winter *et al.*, 1982; Swap *et al.*, 2004; Crowley *et al.*, 2011). The $\delta^{15}N$ values in plants depend on relative humidity, temperature, soil chemistry, moisture content, mycorrhizal associations, and the ability to use symbiotic bacteria to fix nitrogen directly from the atmosphere (legumes; Ambrose, 1991; Amundson *et al.*, 2003; Schmidt and Stewart, 2003; Evans, 2007; Marshall *et al.*, 2007; Crowley *et al.*, 2011). For example, Muzuka (1999) found low $\delta^{15}N$ values in humid soils due to a high discrimination against ^{15}N by plant cells. He further described high $\delta^{15}N$ values in plants that grew on dry, sandy soils or on overgrazed soils. As the available organic matter (and associated nitrogen compounds) is washed out very quickly from these soils, plants are likely forced to assemble every possible source of nitrogen (i.e., they cannot discriminate against ^{15}N), which leads to increased $\delta^{15}N$ values (Handley *et al.*, 1999). Because many of the factors that affect isotope values in vegetation can change over short distances, isotope values can vary dramatically among different microhabitats. For example, Codron *et al.* (2005) determined that $\delta^{15}N$ values for plants growing in different microhabitats can vary by up to 4‰, and Crowley *et al.* (2012) found that substantial isotopic differences exist between the forest interior and its edges. Additionally, different plant species may also differ in their isotopic content (Zhang and Cregg, 1996; Murray *et al.*, 1998), and small differences in the isotopic composition of fruits and leaves may exist (O'Leary, 1981; Heaton, 1999; Cernusak *et al.*, 2009).

The aim of this study is to analyze the seasonal variations in the isotopic content of fur from two mouse lemur species (*M. murinus* and *M. ravelobensis*) and to identify potential sources of isotopic variation in mouse lemur food plants in the Ankarafantsika National Park, northwestern Madagascar. These species have roughly the same body size (Zimmermann *et al.*, 1998), both are omnivorous solitary foragers (Radespiel *et al.*, 2006; Thorén *et al.*, 2011), and both form stable sleeping groups based on matrilineal relatedness (Radespiel *et al.*, 2001, 2009). On the other hand, both species have been shown to differ in microhabitat preferences (Rendigs *et al.*, 2003; Rakotondravony and Radespiel, 2009) as well as sleeping site preference (Radespiel *et al.*, 2003; Thorén *et al.*, 2010), and these differences have been used to explain differences in their regional distribution (Rakotondravony and Radespiel, 2009). Species-specific differences in their seasonal feeding ecology have been suggested based on direct observational data, although both species also show substantial dietary overlap (Thorén *et al.*, 2011).

Caut *et al.* (2008) discovered that a dietary shift in rats is reflected in the isotope ratios of their fur (basal parts) after approximately 40 days. This moderate turnover rate combined with the easy and non-invasive collection of fur makes it a perfect choice for studies of seasonal changes in feeding habits (e.g. Caut *et al.*, 2008; Dammhahn and Kappeler, 2010; Rakotondranary *et al.*, 2011; Crowley *et al.*, 2014). In this study, we aim to analyze the isotopic content of mouse lemur fur with respect to (1) season, (2) species (*M. murinus* vs. *M. ravelobensis*), (3) sex, (4) body mass and (5) study site. We include isotope values from plants to control for baseline isotopic variability related to abiotic and biotic variables. We evaluate how isotopic variability in plants is related to (1) position in tree (canopy vs. understory), (2) plant part (fruit vs. leaves), (3) study site, and (4) plant species. Seasonal samples of plants were not available. Therefore, this parameter could not be evaluated.

Material and methods

Study sites

This study was conducted in the Ankarafantsika National Park in northwestern Madagascar (Figure 14.1). The dry season (May to October) is generally cooler and lacks precipitation almost completely, whereas the rainy season (November to April) is hot and humid with heavy rains in January and February. The mean annual rainfall is 1100–2255 mm (Rakotondravony and Radespiel, 2009). The mean annual temperature is 27°C with highest temperatures in October and November (maximum monthly mean 37°C) and lowest temperatures in June and July (minimum monthly mean 16°C) (Schmelting, 2000). We collected samples from two study sites, JBA and JBB, within the limits of the park. The first study site, JBA (16°19′S, 46°48′E), has a size of 30.6 ha, and is located on a plateau approximately 190 m above sea level (Sehen *et al.*, 2010; Figure 14.1). The forest at JBA is situated on sandy soil with low water-holding capacity. A grid system of trails with intersections approximately every 50 m allows access. The second site, JBB (16°15′S, 47°02′E), is situated north of Lake Ravelobe at a lower altitude (89 m a.s.l.). With only 5.06 ha, this area is distinctly smaller than JBA. JBB has a grid system of trails with intersections every 25 m. Its forest grows on reddish argilliferous soils and is partly disturbed due to previous anthropogenic influences (e.g., introduced plants) and past forest fires (Sehen *et al.*, 2010). During heavy rains from November to April, it is temporarily inundated by water. JBA and JBB also differ in vegetation structure and floristic composition, details of which are provided elsewhere (see Rendigs *et al.*, 2003; Sehen *et al.*, 2010; Chanu *et al.*, 2012).

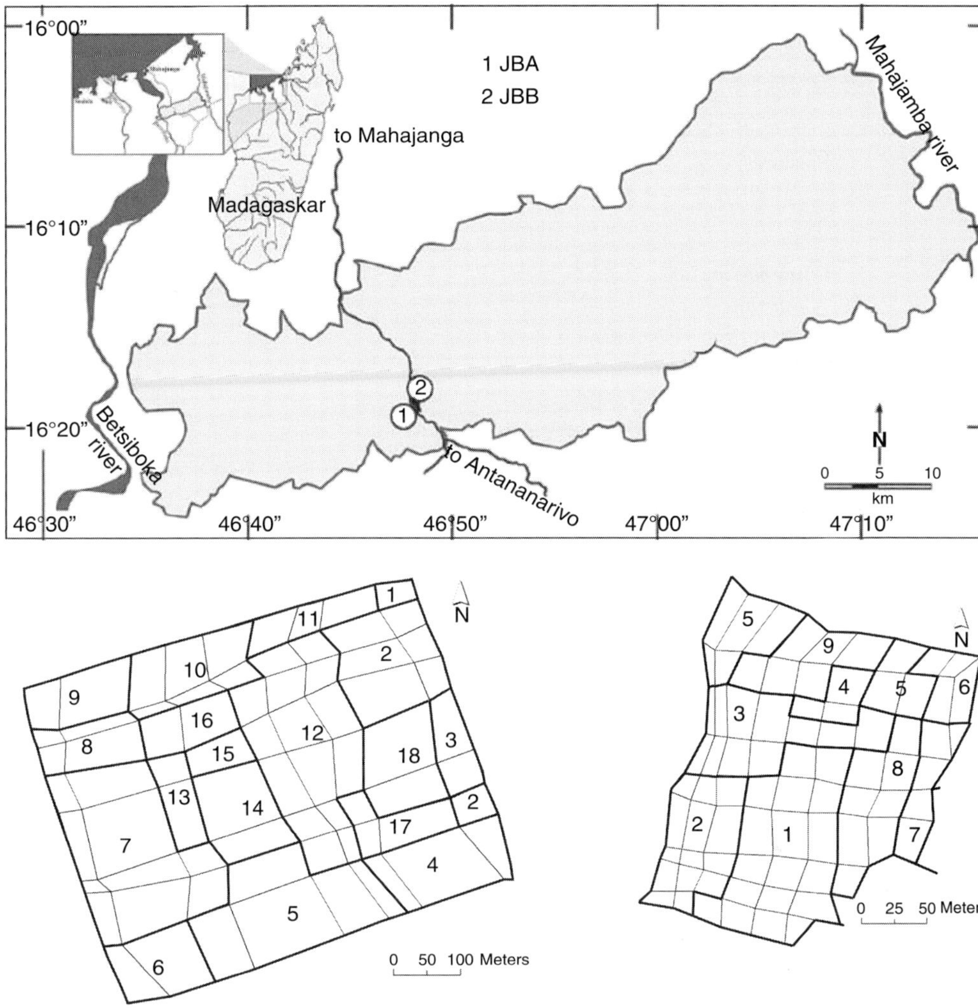

Figure 14.1 Ankarafantsika National Park and the two study sites (1 and lower left: JBA, 2 and lower right: JBB; adapted from Sehen *et al.*, 2010); the lines in the two enlarged maps mark the grid system in each study site; thick lines mark the limits of the vegetation units used for sampling (see Methods for details); the two sites are not shown proportionally (see scale).

Plant collection

A total of 446 plant samples (344 leaf samples, 102 fruit samples) were collected from April to May 2010 by Herisoa V. Razakanirina. This included samples from one abundant plant species (*Securinega perrieri*: Sehen *et al.*, 2010) and 13 mouse lemur food plant species (Thorén *et al.*, 2011), leading to a total of 14 plant species belonging to 11 different plant families (Table 14.1). Only one of these (*Rothmannia reniformis*) was sampled in both study sites, whereas nine species were exclusively collected in JBA and four species were taken from JBB only (Table 14.1). Chanu *et al.* (2012) compared the botanical characteristics of JBA and JBB. They identified

Table 14.1 Collected plant samples sorted by species, study site, and known consumer. *Scientific species name not known; **species determination was not possible; [1]very common plant species, but not known as a food source of mouse lemurs; raw sample size: number of analyzed samples for each species; canopy sample size: number of analyzed samples for each species collected at the canopy.

Family	Species	Site	Food plant of		Raw sample size	Canopy sample size
			M. murinus	*M. ravelobensis*		
Cannelaceae	*Cinnamosma fragrans*	JBA		X	26	13
Celastraceae	*Mystroxylon aethiopicum*	JBA	X		24	12
Chrysobalanaceae	*Grangeria porosa*	JBB		X	18	9
Euphorbiaceae	*Securinega perrieri*[1]	JBA			14	7
Fabaceae	*Baudouinia fluggeiformis*	JBA	X	X	64	37
	Rohimena*	JBA	X		22	11
Meliaceae	*Malleastrum gracile*	JBB		X	18	9
Oleaceae	*Noronhia boinensis*	JBA	X	X	97	57
Rubiaceae	*Canthium* sp.**	JBA		X	12	6
	Rothmannia reniformis	JBB/JBA	X	X	44	20
	Psychotria sp.**	JBB	X	X	17	10
Saliaceae	*Scolopia inappendiculata*	JBA	X	X	36	18
Sapindaceae	*Molinaea retusa*	JBB		X	18	9
Sphaerosepalaceae	*Rhopalocarpus similis*	JBA	X	X	36	18
Total					446	236

multiple "vegetation units" within each site (JBA: $n = 18$; JBB: $n = 9$) based on qualitative and semi-quantitative comparisons of physiognomy and floristic composition (Figure 14.1). Up to five samples from different individuals were collected for each plant species in each vegetation unit in order to cover potential spatial heterogeneity within both sites. To facilitate direct isotopic comparisons between (1) food plant species, and (2) study sites, and to obtain the largest possible sample, leaves were collected from all plant specimens, and ripe fruits were obtained in addition whenever available. From each tree, about 2 g of leaves were collected from the canopy and from about 1.3 m above the ground ($n = 170$ samples at each height). In total, we obtained 102 fruit samples from 4 plant species (*Baudouinia fluggeiformis*: $n = 29$, *Noronhia boinensis*: $n = 62$, *Psychotria* sp.: $n = 9$, and *Rothmannia reniformis*: $n = 2$). Fruit samples from the other plant species were not available during the collection period. Differences in the number of samples per species (see Table 14.1) are due to differences in species abundance in vegetation units or to a lack of fruits during the sampling period. All samples were dried at low temperatures (below 65°C) in an oven and kept dry with silica gel.

Fur collection

Gray mouse lemurs (*M. murinus*) were trapped monthly in JBA and golden-brown mouse lemurs (*M. ravelobensis*) were trapped monthly in JBA and JBB from September to November 2007, and from March to October 2008 (no captures in May 2008). For detailed capture information see Thorén *et al.* (2011). All animals were captured and released close to trail crossings. Fur samples were collected from the tail of each animal and stored dry. A total of 239 fur samples were included in this study: 62 samples belonged to *M. murinus* (29 female, 33 male samples) and 177 to *M. ravelobensis* (n = 86 from JBA: 42 female, 44 male samples; n = 91 from JBB: 43 female, 48 male samples). Samples were obtained from a total of 197 individuals (JBA_{mur} = 52, JBA_{rav} = 66, JBB_{rav} = 79), of which 36 individuals were sampled more than once (median= 2 times, max. = 4 times). Individuals never contributed more than one sample to each season. The majority of these samples were previously analyzed for a different study by Crowley *et al.* (2011); however, 45 samples (JBA_{mur} = 12 (8 males, 4 females), JBA_{rav} = 27 (15 males, 12 females), JBB_{rav} = 6 (4 males, 2 females)) were specifically processed and analyzed for this study.

Sample preparation and analysis

Dried plant material was ground to a powder using an agate mortar and pestle. Fur samples were cleaned with methanol (Crowley *et al.*, 2011). Approximately 5 mg of each plant sample and 700 μg of fur were weighed into tin boats, combusted and analyzed for $\delta^{13}C$ and $\delta^{15}N$ values on a Finnigan Thermo Electron Delta+XP continuous flow system (Bremen, Germany) connected to a Carlo Erba elemental analyzer (Milan, Italy) at the University of California, Santa Cruz Stable Isotope Laboratory, USA. Analytical precision was assessed using replicates of International Atomic Energy Agency (IAEA) acetanilide and a gelatin laboratory standard.

Data analyses

Prior to each statistical analysis, the data were tested for normality with a Shapiro–Wilk test. In case of normality, parametric tests were subsequently used. If this was not the case, non-parametric alternatives were used. Multivariate statistics could not be employed for the analyses of the plant samples due to too many empty cells. Isotopic differences between leaf samples collected in the canopy and the understory and between leaf and fruit samples were compared at the level of individual trees using parametric paired t-tests. Isotope values for plant samples from both study sites and for fruits from the canopy and understory were compared with non-parametric Mann–Whitney U tests. We did not have enough fruit samples from the same tree (canopy and understory) for paired analyses. Isotopic differences among plant species were evaluated with parametric analysis of variance (ANOVA) or non-parametric Kruskal–Wallis ANOVA tests. Each significant ANOVA was followed by a post-hoc Tukey test. In case of a significant result in the Kruskal–Wallis ANOVA, non-parametric Mann–Whitney *U* tests were used for pairwise post-hoc comparisons.

We checked for isotopic differences in fur samples collected during the two sample years (2007 and 2008) using Mann–Whitney *U* tests. Because the isotope values did not differ significantly among years for *M. murinus*, data from both 2007 and 2008 were pooled. However, because $\delta^{13}C$ values for *M. ravelobensis* differed significantly between 2007 and 2008, only data from 2008 were used for interspecific comparisons.

Based on a previous feeding study (Thorén *et al.*, 2011), three different "feeding seasons" (seasons 1, 2, and 3 described below) were initially defined to categorize the samples. Given the cumulative isotopic record of fur for rats (Caut *et al.*, 2008) and the complete regrowth of fur after three months for both *M. murinus* and *M. ravelobensis* (Thorén, pers. obs.), we surmise that samples from season 1 (March and April) incorporate food digested during the rainy season. Samples from season 2 (June and July) represent the transition phase between rainy and dry season including the early dry season, and samples from season 3 (September and October) represent the second half of the dry season. Because the *M. ravelobensis* samples from September and October differed significantly between 2007 and 2008 in their $\delta^{13}C$ values (see above), we defined a fourth season for this particular dataset (season 0; September and October 2007). This further subdivision was not necessary for *M. murinus* or for *M. ravelobensis* $\delta^{15}N$ values.

We used general linear models (GLM) to identify the intraspecific determinants of isotopic variation for mouse lemur fur. Carbon and nitrogen isotope values for *M. murinus* (analyses 1 and 2: $n = 62$) and *M. ravelobensis* (analyses 3 and 4: $n = 170$) were analyzed with respect to the independent variables season, sex, body mass and study site (note that study site could only be included for *M. ravelobensis*).

To evaluate possible interspecific differences in isotope values, we used two additional GLMs to assess the influence of three independent variables on the $\delta^{13}C$ and $\delta^{15}N$ values for lemurs from JBA: species, season, and the continuous covariate body mass. Significant GLM results were complemented with post-hoc Tukey tests. To test for a possible influence of season and species on body mass, we ran a two-way ANOVA followed by a complementary Tukey test. Finally, we used Pearson correlations to evaluate how body mass is related to the carbon and nitrogen isotope ratios of fur. All statistical analyses were performed using STATISTICA (Version 6.1, StatSoft, Inc., Tulsa, USA) and the significance for all tests was set at alpha = 0.05.

Results

Isotopic variation in plant tissues

Leaf samples collected from the canopy had significantly higher $\delta^{13}C$ values than leaves collected at 1.3 m from the ground (paired *t*-test: $n = 170$ for each

Table 14.2 Mean $\delta^{13}C$ and $\delta^{15}N$ values and ±1 SD for each plant species at each study site. *scientific species name not known; **species determination was not possible; [1]very common plant species, but not known as a food source for mouse lemurs.

Family	Species	JBA		JBB	
		$\delta^{13}C$	$\delta^{15}N$	$\delta^{13}C$	$\delta^{15}N$
Cannelaceae	*Cinnamosma fragrans*	−29.9 (±1.0)	9.3 (±0.6)		
Celastraceae	*Mystroxylon aethiopicum*	−31.4 (±0.8)	10.4 (±0.8)		
Chrysobalanaceae	*Grangeria porosa*			−31.5 (±1.1)	1.6 (±0.9)
Euphorbiaceae	*Securinega perrieri*[1]	−28.3 (±0.8)	10.2 (±0.8)		
Fabaceae	*Baudouinia fluggeiformis*	−28.6 (±1.3)	11.8 (±0.9)		
	Rohimena*	−30.7 (±0.8)	10.8 (±0.5)		
Meliaceae	*Malleastrum gracile*			−31.2 (±1.1)	3.1 (±1.9)
Oleaceae	*Noronhia boinensis*	−28.6 (±0.9)	10.8 (±0.9)		
Rubiaceae	*Canthium* sp.**	−29.9 (±0.8)	10.7 (±0.3)		
	Rothmannia reniformis	−29.7 (±0.5)	14.8 (±2.7)	−30.6 (±0.5)	6.4 (±0.5)
	Psychotria sp.**			−29.9 (±1.3)	4.2 (±0.7)
Saliaceae	*Scolopia inappendiculata*	−28.7 (±0.9)	10.9 (±1.0)		
Sapindaceae	*Molinaea retusa*			−30.1 (±0.7)	2.33 (±1.0)
Sphaerosepalaceae	*Rhopalocarpus similis*	−29.7 (±0.9)	9.9 (±1.0)		

group, $t = 4.70$, $p < 0.0001$). No such difference was found in $\delta^{15}N$ values (paired t-test: $n = 170$ for each group, $t = 1.89$, $p > 0.05$). Similar results were obtained for fruit samples (Mann–Whitney U test; $\delta^{13}C$: $n_{high} = 95$, $n_{low} = 6$, $U = 66$, $p = 0.0006$; $\delta^{15}N$: $n_{high} = 95$, $n_{low} = 6$, $U = 175$, $p > 0.05$). In view of these differences, only the data with the larger sample size, i.e., the samples collected from the canopy, were used for further analyses. Paired t-tests indicate that there were no significant differences in $\delta^{13}C$ or $\delta^{15}N$ values between canopy leaves and fruits ($\delta^{13}C$: $n = 33$, $t = 1.33$, $p > 0.05$; $\delta^{15}N$: $n = 33$, $t = -0.75$, $p > 0.05$). Therefore, leaf and fruit samples were merged for all subsequent analyses. For individual plants with more than one sample (e.g., both leaves and fruit from the canopy), mean isotope values were used. This procedure led to a new plant sample size of $n = 236$ for all subsequent analyses (196 JBA, 40 JBB, Table 14.1).

Plant samples collected from JBA had significantly higher $\delta^{13}C$ and $\delta^{15}N$ values than those from JBB (Mann–Whitney U test: $n_{JBA} = 196$, $n_{JBB} = 40$, $U_{\delta 13C} = 1556$, $p_{\delta 13C} < 0.0001$; $U_{\delta 15N} = 1$, $p_{\delta 15N} < 0.0001$). In JBA, mean isotope values for individual plant species (see Table 14.2) varied from −28.3 ± 0.8‰ to −31.4 ± 0.8‰ for $\delta^{13}C$, and 9.3 ± 0.6‰ to 14.8 ± 2.7‰ for $\delta^{15}N$. In JBB, the mean isotope values for individual

Table 14.3 Statistical results for analyses assessing isotopic differences among plant species for $\delta^{13}C$ and $\delta^{15}N$ values in JBA and in JBB (sites assessed separately); ANOVA, parametric analysis of variance; KWA, Kruskal–Wallis ANOVA. All comparisons are significant.

Comparison	Site	Parameter	Analysis	*n*	*F*	*H*	*p*
Species	JBA	$\delta^{13}C$	ANOVA	196	18.2		< 0.0001
		$\delta^{15}N$	KWA	196		89.2	< 0.0001
	JBB	$\delta^{13}C$	ANOVA	40	3.6		0.0146
		$\delta^{15}N$	ANOVA	40	12.7		< 0.0001

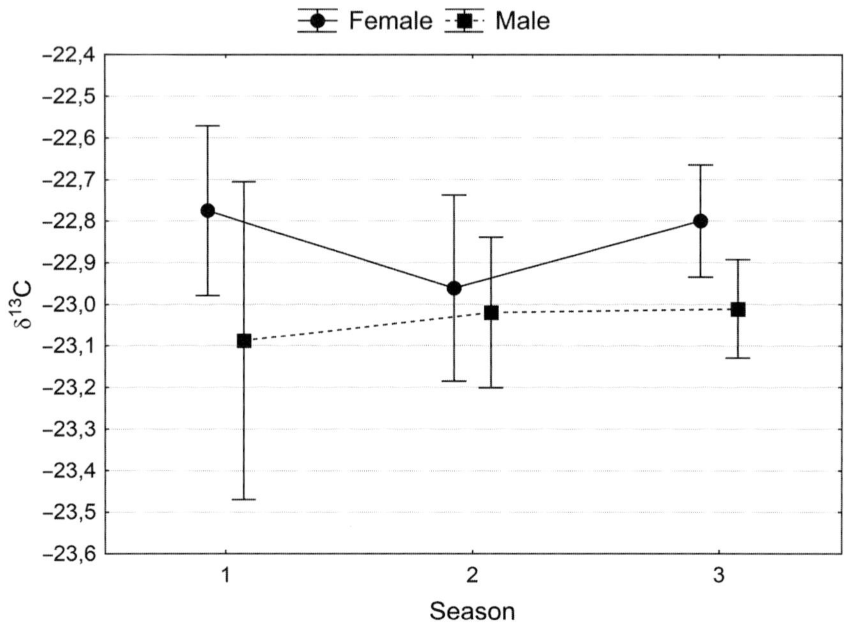

Figure 14.2 Variations in $\delta^{13}C$ values for *M. murinus* in seasons 1, 2, and 3 separated by sex (male/female); means and 0.95 confidence intervals are shown.

species varied from −29.9 ± 1.3‰ to −31.5 ± 1.1‰ for $\delta^{13}C$, and 1.6 ± 0.9‰ and 6.4 ± 0.5‰ for $\delta^{15}N$. Significant differences in both $\delta^{13}C$ and $\delta^{15}N$ values existed between plant species within both sites (Table 14.3).

Isotopic variation in *M. murinus* fur

Carbon isotope values for gray mouse lemurs were influenced by sex and body mass. Carbon isotope values did not differ significantly among seasons. Overall, female gray mouse lemurs have significantly higher $\delta^{13}C$ values than males (Table 14.4). However, these differences were not significant when the three seasons were tested separately with a post-hoc Tukey test (all p-values > 0.05, Figure 14.2). Body mass had a significant influence on the $\delta^{13}C$ values of gray mouse lemur fur (Table 14.4). A subsequent Pearson correlation revealed a significant positive relationship between body mass and $\delta^{13}C$ values ($n = 61$, $r_p = 0.46$, $p < 0.05$). Nitrogen

Table 14.4 Determinants of intraspecific variation in $\delta^{13}C$ and $\delta^{15}N$ values (results of intraspecific GLMs). Significant results are presented in bold.

Species	Comparison	Parameter	*n*	*F*	*p*
M. murinus	Body mass	$\delta^{13}C$	62	17.5	**0.0001**
	Sex			4.5	**0.0375**
	Season			0.5	0.6167
	Sex × Season			0.6	0.5440
	Body mass	$\delta^{15}N$		0.0	0.9724
	Sex			0.1	0.7018
	Season			1.2	0.3197
	Sex × Season			0.6	0.5616
M. ravelobensis	Body mass	$\delta^{13}C$	170	25.1	**<0.0001**
	Site			163.6	**<0.0001**
	Sex			2.6	0.1076
	Season			10.8	**0.0014**
	Site × Sex			1.7	0.1970
	Site × Season			1.1	0.2918
	Sex × Season			0.4	0.5413
	Site × Sex × Season			0.04	0.8360
	Body mass	$\delta^{15}N$	170	0.04	0.8447
	Site			414.4	**<0.0001**
	Sex			0.4	0.5321
	Season			0.8	0.3787
	Site × Sex			1.3	0.2575
	Site × Season			10.6	**0.0014**
	Sex × Season			0.3	0.5696
	Site × Sex × Season			0.07	0.7982

isotope values for fur did not vary significantly between sexes, seasons or with body mass variations (Table 14.4). None of the interaction parameters were significant for either GLM.

Isotopic variation in *M. ravelobensis* fur

Carbon isotope values for golden-brown mouse lemurs were influenced by site, season, and body mass (Table 14.4, Figure 14.3). There were no differences in $\delta^{13}C$ values between sexes. Individuals from JBA had significantly higher $\delta^{13}C$ values than JBB in all seasons (Tukey test: all p values < 0.0001, Figure 14.3), and $\delta^{13}C$ values increased significantly from season 2 to season 3 in the site-independent post-hoc comparison (Tukey test: $p < 0.0001$). As was the case for *M. murinus*, body mass had a significant influence on fur $\delta^{13}C$ values (Table 14.4). A Pearson correlation revealed a significant, positive relationship between body mass and $\delta^{13}C$ values (Table 14.4; Pearson correlation: $n = 170$, $r_p = 0.25$, $p < 0.05$). None of the interaction parameters for either GLM were significant.

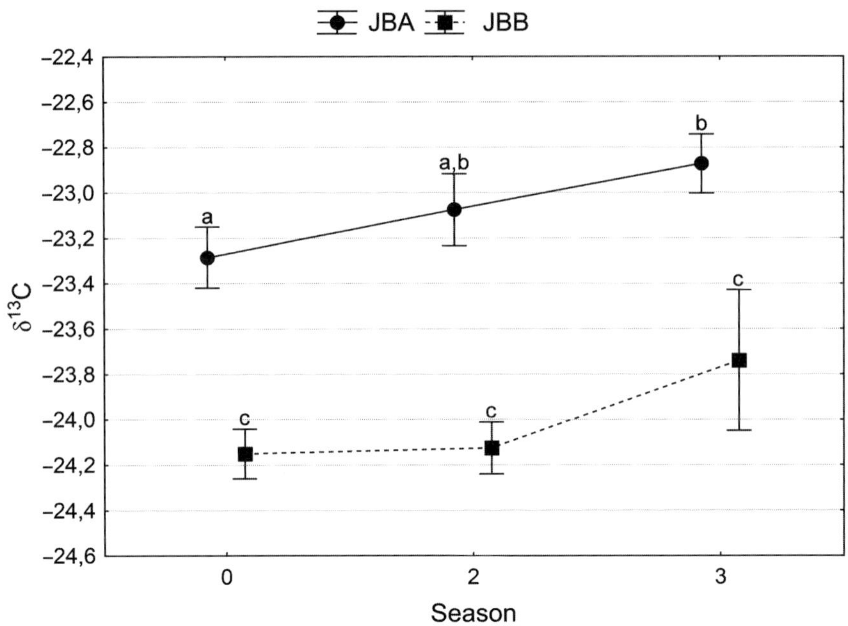

Figure 14.3 Variations in $\delta^{13}C$ values for *M. ravelobensis* in seasons 0, 2, and 3 separated by site (JBA/JBB); means and 0.95 confidence intervals are shown. Small letters indicate statistical distinction with different letters indicating a statistical difference of at least $p < 0.05$ (details in text).

Site and the interaction between site and season had a significant influence on $\delta^{15}N$ values. Golden-brown mouse lemurs from JBA had significantly higher $\delta^{15}N$ values than those from JBB in seasons 2 and 3 (Tukey test: $p < 0.00001$). However, whereas the $\delta^{15}N$ values from JBA declined significantly towards season 3 (Tukey test: $p < 0.05$, Table 14.4), those from JBB remained more or less stable (Figure 14.4). No other variable or interactions had a significant effect on $\delta^{15}N$ values for golden-brown mouse lemurs.

Interspecific comparison of isotopic values

Neither season nor species had a significant effect on $\delta^{13}C$ values, although there was a statistical trend for season ($p < 0.1$), with $\delta^{13}C$ values being smallest in season 2 (Table 14.5, Figure 14.5). The lack of interspecific differences was probably due to high intraspecific isotopic variability (Figure 14.5).

Similar to the intraspecific analyses (see above), $\delta^{13}C$ values for fur were significantly influenced by body mass in the overall GLM (Table 14.5). A separate ANOVA revealed that season, species and their interaction had a significant influence on body mass (Figure 14.6, Table 14.5). *Microcebus ravelobensis* was significantly heavier than *M. murinus* only during season 1 (Tukey test: $p < 0.001$). Body mass for *M. ravelobensis* decreased significantly from season 1 to seasons 2 and 3 (Tukey test: $p < 0.01$), whereas the body mass of *M. murinus* remained more or less stable across seasons (Tukey $p > 0.05$).

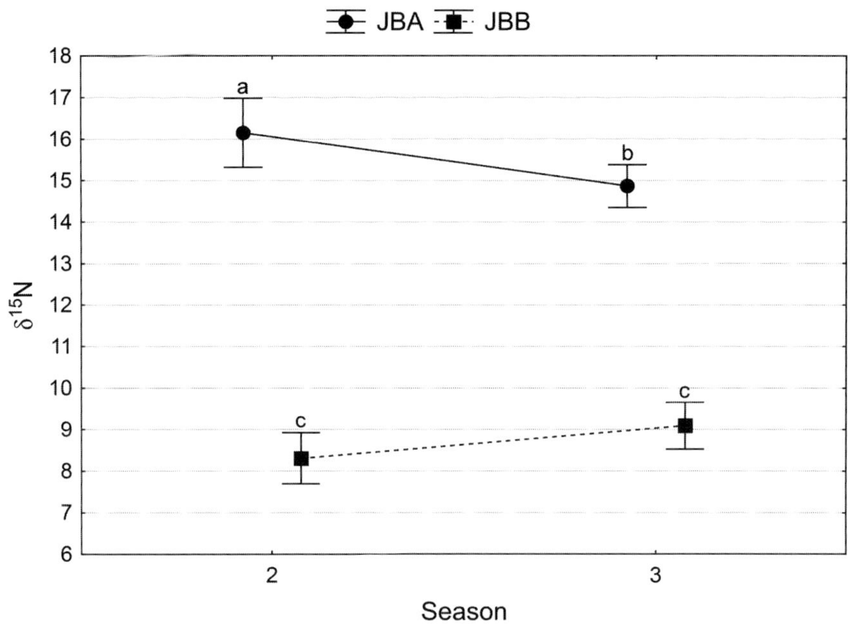

Figure 14.4 Variations in $\delta^{15}N$ values for *M. ravelobensis* in seasons 2 and 3 separated by site (JBA/JBB); means and 0.95 confidence intervals are shown. Small letters indicate statistical distinction with different letters indicating a statistical difference of at least $p < 0.05$ (details in text).

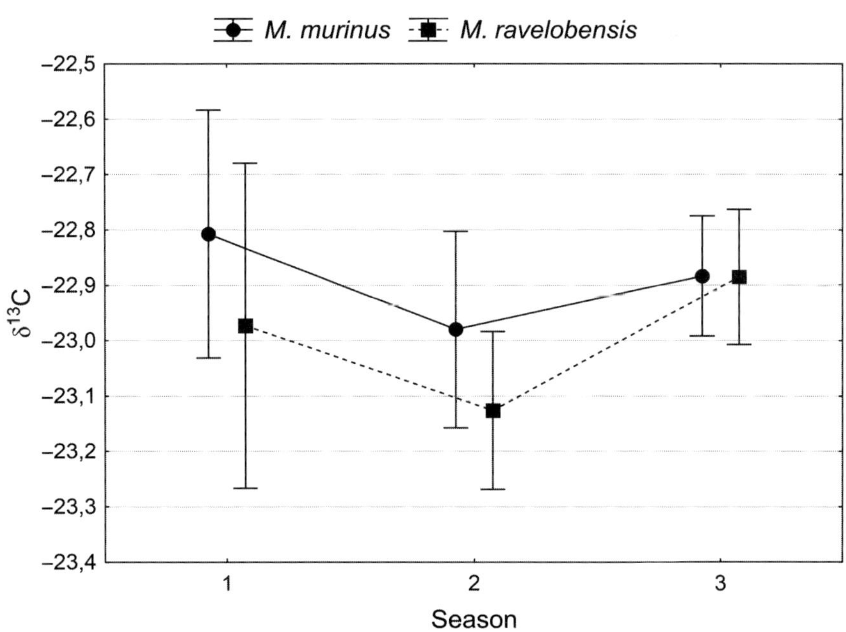

Figure 14.5 Seasonal variations in $\delta^{13}C$ values for *M. ravelobensis* and *M. murinus* in JBA; means and 0.95 confidence intervals are shown.

Table 14.5 Determinants of interspecific variation in $\delta^{13}C$ and $\delta^{15}N$ values (results of interspecific GLMs) and results of two-way ANOVA (body mass). Significant results are presented in bold.

Type of analysis	Comparison	Parameter	*n*	*F*	*p*
Two-way ANOVA	Season	Body mass	120	5.6	**0.0047**
	Species			32.7	**<0.0001**
	Season × Species			6.2	**0.0027**
GLM	Body mass	$\delta^{13}C$	120	12.9	**0.0005**
	Season			3.0	0.0538
	Species			1.6	0.2093
	Season × Species			0.7	0.5122
	Body mass	$\delta^{15}N$		0.03	0.8650
	Season			4.7	**0.0111**
	Species			2.3	0.1319
	Season × Species			0.8	0.4378

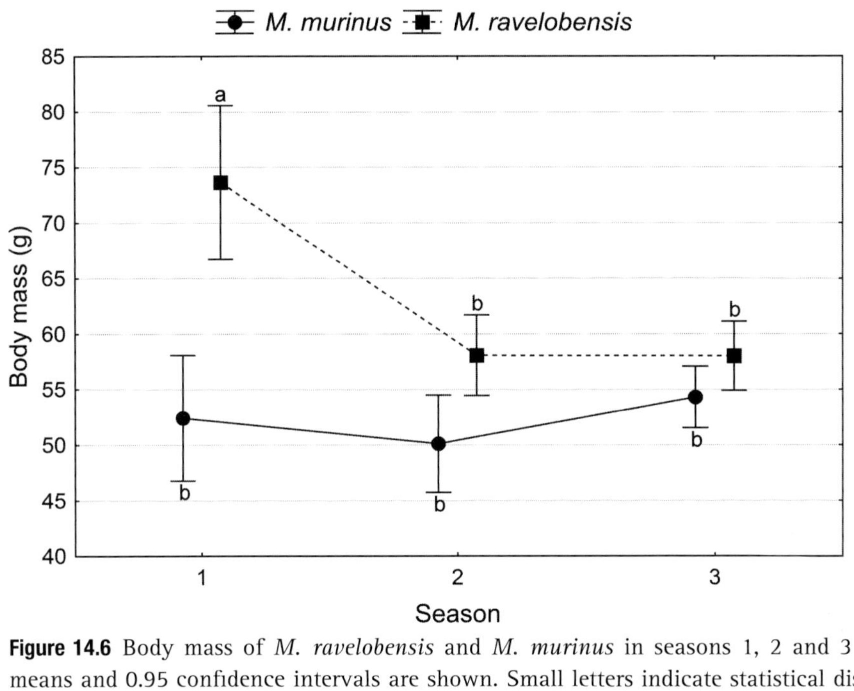

Figure 14.6 Body mass of *M. ravelobensis* and *M. murinus* in seasons 1, 2 and 3 in JBA; means and 0.95 confidence intervals are shown. Small letters indicate statistical distinction with different letters indicating a statistical difference of at least $p < 0.05$ (details in text).

GLM results revealed significant differences in $\delta^{15}N$ values among seasons. None of the other variables or interactions (species, body mass, species × season) had a significant influence on $\delta^{15}N$ values. Average nitrogen isotope values decreased from seasons 1 and 2 to season 3 for both species, but the decrease was larger for *M. ravelobensis* (Figure 14.7, Table 14.5). As a consequence, $\delta^{15}N$ values for both species differed significantly in season 3.

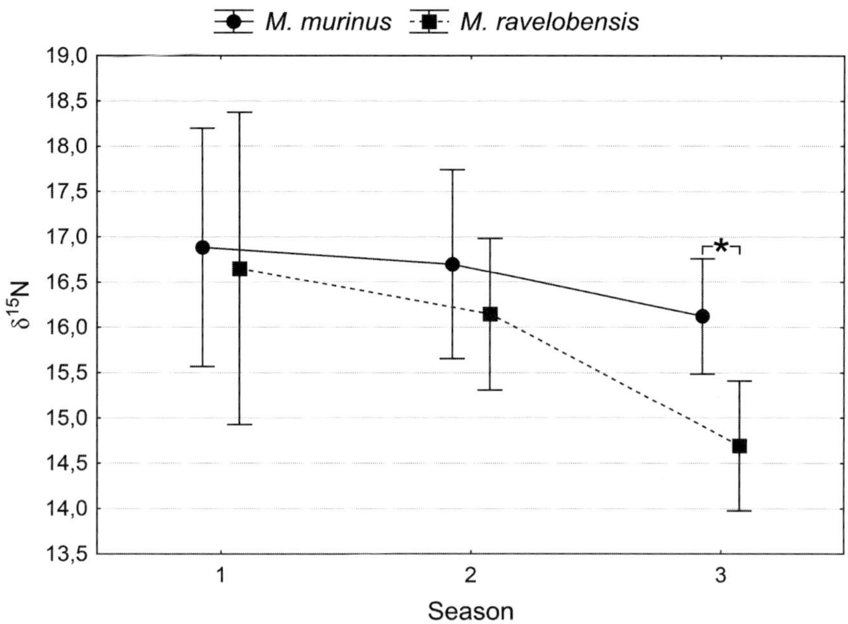

Figure 14.7 Seasonal variations in $\delta^{15}N$ values for *M. ravelobensis* and *M. murinus* in JBA; means and 0.95 confidence intervals are shown. $^*p < 0.05$.

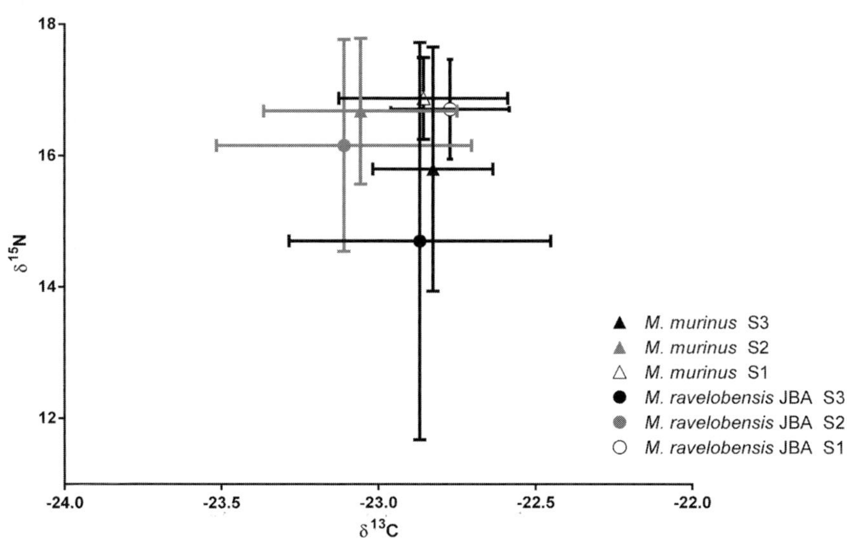

Figure 14.8 Mean (± 1 SD) $\delta^{13}C$ and $\delta^{15}N$ values for *M. ravelobensis* and *M. murinus* in JBA for each season. S1–S3: season 1–season 3.

To integrate the results from both isotopes at JBA, mean isotope values (± 1 SD) for each mouse lemur species were plotted for each season separately (Figure 14.8). This allowed us to visualize qualitative seasonal and species-specific isotope shifts. During season 1 (= rainy season), $\delta^{13}C$ values differed between species. Carbon isotope values decreased for both species in season 2 (= early dry season), but increased and converged in season 3 (= late dry season), with *M. ravelobensis* showing a larger increase in carbon isotope values than *M. murinus*.

During season 1, $\delta^{15}N$ values were rather similar for the two species. However, nitrogen isotope values for the two species diverged in season 2 and even more in season 3. This increasing divergence reflects a larger seasonal shift in isotope values for *M. ravelobensis* than *M. murinus*. Nitrogen isotope values for both species decreased during these seasons, but those for *M. ravelobensis* showed a more pronounced decrease than those for *M. murinus*. Taken together, isotopic differences between the two species were most pronounced during season 1 for carbon and during season 3 for nitrogen (Figures 14.5, 14.7). Season 2 was mainly characterized by lower $\delta^{13}C$ values compared to seasons 1 or 3 (Figure 14.5). Nitrogen isotope values were highest for both species during season 1 and lowest in season 3 (Figure 14.7).

Discussion

Sources of variation in the isotopic content of plant tissues

Isotope values for plants were expected to vary with canopy height, plant part, plant species, and study site due to variations in sunlight exposure, assimilation of CO_2, soil characteristics and species-specific physiological and biochemical characteristics (van der Merwe and Medina, 1989; Crowley *et al.*, 2012). In accordance with expectations, canopy samples (both leaves and fruits) showed higher $\delta^{13}C$ values than the samples collected close to the ground. Although this effect is well known for leaves, it is not well documented for fruits. As expected, the results did not indicate any "canopy effect" for $\delta^{15}N$ values (Cerling *et al.*, 2004; Crowley *et al.*, 2012).

Isotope values for fruits and leaves did not differ, at least not in the four plant species for which both sample types were available. Although isotopic differences among plant parts have been documented previously, these differences are typically small and inconsistent (e.g. Codron *et al.*, 2005; Crowley *et al.*, 2011). Small differences in isotope values for fruits and leaves may be swamped by other sources of isotopic heterogeneity such as plant species or soil characteristics. Alternatively, leaves and fruits at Ankarafantsika may in general have a more similar chemical composition than those plant parts that were compared in other studies (e.g., seeds and fruits: Badeck *et al.*, 2005; fruits and leaves: Crowley *et al.*, 2011; leaves, bark, stems, flowers and fruits: Codron *et al.*, 2005).

In agreement with expectations, plant samples from more mesic JBB had lower $\delta^{13}C$ and $\delta^{15}N$ values than those from drier JBA. These patterns likely reflect differences in soil composition and water-holding capacity between the two sites (Muzuka, 1999; Sehen *et al.*, 2010; Crowley *et al.*, 2012).

As expected, we also found significant isotopic differences between plant species both in JBA and JBB despite some degree of intraspecific variation (Table 14.2). Average $\delta^{13}C$ values for plant species differed by up to 3‰ in JBA and 1.6‰ in JBB. These differences among species are similar to those that have been previously reported (e.g., Ehleringer *et al.*, 1987; Zhang and Cregg, 1996), although Murray *et al.*

(1998) provided evidence that different plant species of the genus *Shorea* can even vary by up to 6.1‰ in their $\delta^{13}C$ values. Average $\delta^{15}N$ values for plant species differed by 5.5‰ in JBA and 4.8‰ in JBB. Interestingly, isotope values of the two species of the Fabaceae (legumes: *Baudouinia fluggeiformis* and Rohimena), which could be expected to have the lowest $\delta^{15}N$ values due to their presumed ability to fix nitrogen directly from the atmosphere by means of symbiotic bacteria (Ambrose, 1991; Schmidt and Stewart, 2003), fell well within the range of the other species (Table 14.2). Because $\delta^{15}N$ values in plants can be influenced by the source of nitrogen (e.g., NH_3, NO, N_2O, NO_3^-, NH_4^+) as well as species-specific symbiotic mycorrhiza (Muzuka, 1999; Adams and Grierson, 2001; Evans, 2007), explanations for differences in $\delta^{15}N$ values among species are generally complex, and beyond the scope of this chapter.

Sources of variation in the isotopic content of mouse lemur fur

The isotope values of an animal's fur can be influenced by many factors, including body mass, sex, species, habitat type and diet (e.g., Rau *et al.*, 1981; Crowley *et al.*, 2011; Rakotondranary *et al.*, 2011). Before discussing in more detail the signals of seasonal dietary changes in *M. murinus* and *M. ravelobensis* at Ankarafantsika, the influences of body mass, sex, and study site shall be evaluated.

First, we expected that mouse lemur body mass would be positively correlated with $\delta^{15}N$ values as has been previously shown in wild dover sole (Rau *et al.*, 1981) and in earthworms (Schmidt *et al.*, 2004). However, we did not find this relationship. Instead, a significant positive correlation between body mass and fur $\delta^{13}C$ values was found for both mouse lemur species. Dammhahn and Kappeler (2010) reported a significant positive correlation between body mass and both $\delta^{13}C$ and $\delta^{15}N$ values in *M. berthae* males, but no further explanation was offered. Slightly higher $\delta^{13}C$ values (about 1‰) in heavier individuals suggest more consumption of foods with higher $\delta^{13}C$ values (e.g., exudates; Dammhahn and Kappeler, 2010; Crowley *et al.*, 2014), although body mass-dependent dietary habits have not yet been reported for mouse lemurs. Another explanation might be that larger individuals could forage more in the canopy than smaller individuals. However, it was previously observed that *M. murinus* and *M. ravelobensis* use the same vertical space at Ankarafantsika (Radespiel *et al.*, 2006). Further studies would be needed to investigate such potential size-dependent feeding strategies in mouse lemurs in more detail.

Second, we expected to find isotopic differences between males and females, particularly for *M. murinus*. Gray mouse lemurs exhibit female dominance (Radespiel and Zimmermann, 2001), which might lead to a monopolization of energy-rich resources by females across the year. Our results confirmed that female *M. murinus* had higher $\delta^{13}C$ values than males, but no significant differences were found between the sexes for $\delta^{15}N$ values. Carbon isotope values for females were especially elevated in the rainy season (season 1). Unfortunately, neither comparative seasonal observational feeding data for both sexes nor seasonal samples from all possible food sources are available to evaluate this hypothesis. However, because females were not significantly heavier than males ($p > 0.05$, results not shown), we can exclude the possibility that differences in body mass alone are driving this pattern. No differences in $\delta^{13}C$ or $\delta^{15}N$ values among sexes were found in *M. ravelobensis*. This might be explained by the lack of female dominance in *M. ravelobensis* (Eichmueller *et al.*, 2013).

Third, we expected that isotope values for *M. ravelobensis* fur would differ between study sites because the significant differences in plant isotope values between JBA and JBB should influence the isotope values of primary and secondary consumers. As expected, individuals from JBA had higher $\delta^{13}C$ and $\delta^{15}N$ values than those from JBB. On average, isotope values for *M. ravelobensis* fur were +6.6‰ higher in carbon and +4.8‰ higher in nitrogen than those of plants from JBA. Similarly, isotope values for *M. ravelobensis* fur were +6.6‰ higher in carbon and +5.2‰ higher in nitrogen than those of plants in JBB. These enrichment values are similar to those previously reported for mouse lemurs (carbon 6.3–6.4‰, nitrogen 4.0–5.8‰: Dammhahn and Kappeler, 2010; carbon 5.0‰, nitrogen 5.9‰: Crowley *et al.*, 2011) and suggest that *M. ravelobensis* consumed a similar amount of animal matter at both study sites. In accordance with this finding, Meusel (2009) found the same relative amount of arthropod remains in a small number of *M. ravelobensis* fecal samples from JBA and JBB.

The influences of species and season on isotope values are evaluated together. No consistent isotopic differences between species were detected across seasons in JBA. However, some seasonal changes were apparent and were not completely identical for both species. Seasonal changes in diet have previously been reported based on observational data for both *M. murinus* and *M. ravelobensis* at Ankarafantsika (Thorén *et al.*, 2011). Therefore, it was predicted that there would be isotopic differences among fur samples collected during different seasons. A statistical trend between season and $\delta^{13}C$ values (i.e., higher values in season 1 and 3 vs. lower values in season 2), and significant differences in $\delta^{15}N$ values among seasons (i.e., low values in season 3), were, indeed, detected in the interspecific comparisons (Figures 14.5, 14.7). Similar results were also visible in the intraspecific comparison for *Microcebus ravelobensis*. However, there were no isotopic differences in $\delta^{13}C$ or $\delta^{15}N$ values among seasons for the intraspecific comparison for *M. murinus*. The lack of significant differences among seasons for *M. murinus* might have been caused by differences in sample composition or high intraspecific variability. The small sample size for *M. murinus* and the unequal sample distribution caused by temporal (season 1: $n = 9$, season 2: $n = 15$, season 3: $n = 38$) and spatial effects might have hidden seasonal influences.

Isotope values in fur should reflect the food that was consumed by the mouse lemurs over the preceding three months (e.g. Rakotondranary *et al.*, 2011; Crowley *et al.*, 2014). Therefore, the seasonal decline in $\delta^{15}N$ values described above suggests a dietary shift from a more omnivorous diet (e.g., arthropods) to a more plant-based diet over the course of the dry season for both species. Overall, the isotopic signatures of both species were rather similar in the rainy season (season 1), but increasingly diverged as the dry season progressed (season 2 and 3; Figure 14.8). This shift is the result of larger changes in the isotopic signature for *M. ravelobensis* than for *M. murinus*. On average, $\delta^{15}N$ values were slightly higher for *M. murinus* than *M. ravelobensis* in seasons 1 and 2 (nitrogen difference < 1.5‰) and significantly higher in season 3 (Figure 14.7). The divergence in nitrogen isotope values between the two species during season 3 most likely indicates the existence of different realized feeding niches in the form of a slightly higher trophic level for *M. murinus* than *M. ravelobensis* (Dammhahn and Kappeler, 2010; Crowley *et al.*, 2011; Rakotondranary *et al.*, 2011). Season 3 coincides with the period of increased food scarcity (late dry season). Given the seasonal environmental challenges that act on mouse lemurs in dry deciduous forests, ecological plasticity may reduce the potential for interspecific competition in times of limited food supply. Such flexibility may be a key facilitating factor in the coexistence of these two mouse lemur species in the forests of northwestern Madagascar (Hutchinson, 1957; Thorén *et al.*, 2011; Chapter 26).

Observational data lend some support to dietary plasticity in these two species. However, in contradiction to the isotope results from this study, previously published observational data suggest that *M. ravelobensis* feeds at a slightly higher trophic level than *M. murinus* (Thorén *et al.*, 2011). These authors reported that the percentage of dietary overlap between the two species was smallest during the late dry season and early rainy season, which correspond partly with season 3 (late dry season) from this study. *Microcebus ravelobensis* consumed a higher amount of fruits (~43%) from February to April (season 1) and switched to gum and insect secretions (~43% each) from May to October (seasons 2 and 3). This species complemented its diet with arthropods and small vertebrates from August to April (10–35% during seasons 1 and 3) but not from May to July (season 2; Thorén *et al.*, 2011). *Microcebus murinus* consumed a high amount of gum (~65–80%) from February to October (seasons 1–3; Thorén *et al.*, 2011). It supplemented this diet with arthropods (~2–8% throughout the year), and insect secretions (~18%) from May to October (seasons 2 and 3), but these amounts were likely too small to cause significant shifts in isotope values among seasons.

Because isotope values in fur predominantly reflect dietary protein, it is likely that the consumption of animal matter is the driving force behind isotopic differences between species and among seasons. Consumption of fruits, exudates, and arthropod secretions may have a negligible effect on fur isotope values (Crowley *et al.*, 2014). The literature states that some exudates from tree species are protein rich (Hladik *et al.*, 1980; Dammhahn and Kappeler, 2010; Nussinovitch, 2010). However, this is not the case for all tree species (Porter *et al.*, 2009). Therefore, it is unknown how gum consumption might have influenced the isotopic composition of mouse lemur fur in this study. Our results suggest that both species were more omnivorous than observational data suggest and that *M. murinus* consumed more arthropods and small vertebrates than has been observed. These opposing results might reflect a small sample size for the observational study combined with difficulties in observing small, nocturnal, arboreal animals.

Because the two mouse lemur species at Ankarafantsika consume foods from multiple plant species (Radespiel *et al.*, 2006; Thorén *et al.*, 2011), and complement these foods with a variable amount of animal matter, one could ask whether the isotopic variability among different plant species, plant parts, and arthropod and small vertebrate prey are able to provide enough isotopic distinctiveness to allow us to reconstruct the omnivorous feeding niches of mouse lemurs in any more detail. Although the mean isotope values for the different plant species in this study varied, they still revealed a substantial amount of interspecific overlap and intraspecific variability (Table 14.2). Fruits and leaves at Ankarafantsika cannot be distinguished isotopically. There are expected differences in isotope values between plants and animal matter (reviewed in Crowley *et al.*, 2014). However, we do not have isotope data for mouse lemur prey items, and therefore cannot evaluate the degree to which the isotopic composition of arthropod and vertebrate prey items may vary. In addition, such a reconstruction would be hampered by current methodological limitations. The number of potential food sources that can be processed by the algorithms of available computer programs such as IsoSource 1.3.1 (Version 1.3, EPA, Washington, DC, USA) is limited. Therefore, it is highly questionable that a fine-grained feeding budget could be reconstructed for either of these species without additional data.

Nevertheless, this study has demonstrated that stable isotopes are useful for investigating ecological differences between sexes and species, across space, and among seasons. Although the results on isotopic variation in sympatric gray and golden-brown mouse lemurs hint towards feeding niche differentiation between two ecologically similar species, the temporal and dietary resolution of isotope values in fur for these omnivores remains limited (Dammhahn and Kappeler, 2008a; Rakotondranary *et al.*, 2011; Crowley *et al.*, 2014). Further research with a broader sampling scheme is needed to determine the precise enrichment and the detailed dietary composition of the mouse lemurs at Ankarafantsika National Park.

Acknowledgments

We would like to thank Lalandy Chanu and Herisoa V. Razakanirina for collecting the plant samples used in this study. We would also like to thank the Département des Eaux et Forêts, CAFF/CORE, and Madagascar National Parks (MNP) for the permission to work in the Ankarafantsika National Park, as well as the staff of the National Park for their support. In addition, we thank Blanchard Randrianambinina, Solofonirina Rasoloharijaona, Romule Rakotondravony, as well as Elke Zimmermann for their continuous support. All field handling and sampling procedures adhered to the legal requirements of Madagascar and were approved by the Ministry of Environment and Forests. We have complied with the ethical standards for the treatment of primates and with the national laws and research rules formulated by the Malagasy authorities. We finally thank Dyke Andreasen, Sean Edgerton, Paul Koch, Hong Li, and Uli Wortmann for logistical and technical assistance during sample preparation and analysis.

References

Adams MA, Grierson PF. 2001. Stable isotopes at natural abundance in terrestrial plant ecology and ecophysiology: an update. *Plant Biology* 3:299–310.

Ambrose SH. 1991. Effects of diet, climate and physiology on nitrogen isotope abundances in terrestrial foodwebs. *Journal of Archaeological Science* 18:293–317.

Amundson R, Austin AT, Schuur EAG, *et al.* 2003. Global patterns of the isotopic composition of soil and plant nitrogen. *Global Biogeochemical Cycles* 17:1031.

Atsalis S. 1999. Seasonal fluctuations in body fat and activity levels in a rain-forest species of mouse lemur, *Microcebus rufus*. *International Journal of Primatology* 20:883–910.

Badeck F-W, Tcherkez G, Nogués S, Piel C, Ghashghaie J. 2005. Post-photosynthetic fractionation of stable carbon isotopes between plant organs – a widespread phenomenon. *Rapid Communications in Mass Spectrometry* 19:1381–1391.

Ben-David M, Flynn RW, Schell DM. 1997. Annual and seasonal changes in diets of martens: evidence from stable isotope analysis. *Oecologia* 111:280–291.

Caut S, Angulo E, Courchamp F. 2008. Discrimination factors (δ15N and δ13C) in an omnivorous consumer: effect of diet isotopic ratio. *Functional Ecology* 22(2):255–263.

Cerling TE, Hart JA, Hart TB. 2004. Stable isotope ecology in the Ituri Forest. *Oecologia* 138:5–12.

Cernusak LA, Tcherkez G, Keitel C, *et al.* 2009. Why are non-photosynthetic tissues generally 13C enriched compared with leaves in C3 plants? Review and synthesis of current hypotheses. *Functional Plant Biology* 36:199–213.

Chanu L, Goetze D, Rajeriarison C, *et al.* 2012. Can differences in floristic composition explain variation in the abundance of two sympatric mouse lemur species (*Microcebus*) in the Ankarafantsika National Park, northwestern Madagascar? *Malagasy Nature* 6:83–102.

Codron J, Codron D, Lee-Thorp JA, *et al.* R. 2005. Taxonomic, anatomical, and spatio-temporal variations in the stable carbon and nitrogen isotopic compositions of plants from an African savanna. *Journal of Archaeological Science* 32:1757–1772.

Crowley BE, Thorén S, Rasoazanabary E, *et al.* 2011. Explaining geographical variation in the isotope composition of mouse lemurs (*Microcebus*). *Journal of Biogeography* 38:2106–2121.

Crowley BE, McGoogan KC, Lehman SM. 2012. Edge effects on foliar stable isotope values in a Madagascan tropical dry forest. *PLoS ONE* 7(9):e44538.

Crowley BE, Blanco MB, Arrigo-Nelson SJ, Irwin MT. 2013. Stable isotopes document resource partitioning and effects of forest disturbance on sympatric cheirogaleid lemurs. *Naturwissenschaften* 100:943–956

Crowley BE, Rasoazanabary E, Godfrey LR. 2014. Stable isotopes complement focal individual observations and confirm dietary variability in reddish-gray mouse lemurs (*Microcebus griseorufus*) from southwestern Madagascar. *American Journal of Physical Anthropology* 155:77–90.

Dammhahn M, Kappeler PM. 2005. Social system of *Microcebus berthae*, the world's smallest primate. *International Journal of Primatology* 26:407–435.

Dammhahn M, Kappeler PM. 2008a. Comparative feeding ecology of sympatric *Microcebus berthae* and *M. murinus*. *International Journal of Primatology* 29:1567–1589.

Dammhahn M, Kappeler PM. 2008b. Small-scale coexistence of two mouse lemur species (*Microcebus berthae* and *M. murinus*) within a homogeneous competitive environment. *Oecologia* 157:473–483.

Dammhahn M, Kappeler PM. 2010. Scramble or contest competition over food in solitarily foraging mouse lemurs (*Microcebus* spp.): New insights from stable isotopes. *American Journal of Physical Anthropology* 141:181–189.

Dammhahn M, Kappeler PM. 2014. Stable isotope analyses reveal dense trophic species packing and clear niche differentiation in a Malagasy primate community. *American Journal of Physical Anthropology* 153:249–259.

Dausmann KH, Glos J, Ganzhorn JU, Heldmaier G. 2004. Physiology: hibernation in a tropical primate. *Nature* 429:825–826.

DeNiro MJ, Epstein S. 1978. Influence of diet on the distribution of carbon isotopes in animals. *Geochimica et Cosmochimica Acta* 42:495–506.

DeNiro MJ, Epstein S. 1981. Influence of diet on the distribution of nitrogen isotopes in animals. *Geochimica et Cosmochimica Acta* 45:341–351.

Donovan LA, Ehleringer JR. 1992. Contrasting water-use patterns among size and life history classes of a semiarid shrub. *Functional Ecology* 6:482–488.

Ehleringer JR, Lin ZF, Field CB, Sun GC, Kuo CY. 1987. Leaf carbon isotope ratios of plants from a subtropical monsoon forest. *Oecologia* 72:109–114

Eichmueller P, Thorén S, Radespiel U. 2013. The lack of female dominance in golden-brown mouse lemurs suggests alternative routes in lemur social evolution. *American Journal of Physical Anthropology* 150:158–164.

Evans RD. 2007. Soil nitrogen isotope composition. In RM Michener, K Lajtha (eds.), *Stable Isotopes in Ecology and Environmental Science* (pp. 83–98). Blackwell Publishing Ltd, Oxford.

Génin F. 2003. Female dominance in competition for gum trees in the grey mouse lemur *Microcebus murinus. Revue d'Écologie* 58:14.

Génin F. 2008. Life in unpredictable environments: first investigation of the natural history of *Microcebus griseorufus. International Journal of Primatology* 29(2):303–321.

Gligor M, Ganzhorn JU, Rakotondravony D, *et al.* 2009. Hybridization between mouse lemurs in an ecological transition zone in southern Madagascar. *Molecular Ecology* 18:520–33.

Handley LL, Austin AT, Stewart GR, *et al.* 1999. The ^{15}N natural abundance (δ^{15}N) of ecosystem samples reflects measures of water availability. *Functional Plant Biology* 26:185–199.

Heaton TH. 1999. Spatial, species, and temporal variations in the ^{13}C/^{12}C ratios of C_3 plants: implications for palaeodiet studies. *Journal of Archaeological Science* 26:637–649.

Hladik CM, Charles-Dominique P, Petter J-J. 1980. Feeding strategies of five nocturnal prosimians in the dry forest of the west coast of Madagascar. In P Charles-Dominique, HM Cooper, A Hladik, *et al.* (eds.), *Nocturnal Malagasy Primates: Ecology, Physiology, and Behavior* (pp. 41–73). Academic Press, New York.

Hobson KA. 1995. Reconstructing avian diets using stable-carbon and nitrogen isotope analysis of egg components: patterns of isotopic fractionation and turnover. *Condor* 97:752–762.

Hutchinson GE. 1957. Concluding remarks. *Cold Spring Harbor Symposia on Quantitative Biology* 22:415–427.

Irwin MT, Raharison JL, Raubenheimer D, Chapman CA, Rothman JM. 2014. Nutritional correlates of the "lean season": effects of seasonality and frugivory on the nutritional ecology of diademed sifakas. *American Journal of Physical Anthropology* 153:78–91.

Jury MR. 2003. The climate of Madagascar. In SM Goodman, JP Benstead (eds.), *The Natural History of Madagascar* (pp. 75–87). University of Chicago Press, Chicago.

Kelly JF. 2000. Stable isotopes of carbon and nitrogen in the study of avian and mammalian trophic ecology. *Canadian Journal of Zoology* 78:1–27.

Lahann P. 2006. Feeding ecology and seed dispersal of sympatric cheirogaleid lemurs (*Microcebus murinus, Cheirogaleus medius, Cheirogaleus major*) in the littoral rainforest of south-east Madagascar. *Journal of Zoology* 271:88–98.

Marshall JD, Brooks JR, Lajtha K. 2007. Sources of variation in the stable isotopic composition of plants. In RM Michener, K Lajtha (eds.), *Stable Isotopes in Ecology and Environmental Science* (pp. 22–60). Wiley-Blackwell, Boston.

Meusel C. 2009. Vergleichende Untersuchung zum saisonalen Nahrungsspektrum von zwei sympatrischen und allopatrischen Mausmakiarten (*Microcebus murinus, M. ravelobensis*) – Ergebnisse von Kotanalysen. BSc dissertation, Gottfried Wilhelm Leibniz Universität, Hannover.

Mittermeier RA, Louis EE, Richardson MS, *et al.* 2010. *Lemurs of Madagascar.* Conservation International, Arlington.

Murray AP, Edwards D, Hope JM, *et al.* 1998. Carbon isotope biogeochemistry of plant resins and derived hydrocarbons. *Organic Geochemistry* 29:1199–1214.

Muzuka ANN. 1999. Isotopic compositions of tropical East African flora and their potential as source indicators of organic matter in coastal marine sediments. *Journal of African Earth Sciences* 28:757–766.

Nussey DH, Wilson AJ, Brommer JE. 2007. The evolutionary ecology of individual phenotypic plasticity in wild populations. *Journal of Evolutionary Biology* 20:831–844.

Nussinovitch A. 2010. *Plant Gum Exudates of the World: Sources, Distribution, Properties, and Applications.* CRC Press, Boca Raton.

O'Leary M. 1981. Carbon isotope fractionation in plants. *Phytochemistry* 20:553–567.

Popa-Lisseanu AG, Delgado-Huertas A, Forero MG, *et al.* 2007. Bats' conquest of a formidable foraging niche: the myriads of nocturnally migrating songbirds. *PLoS ONE* 2:e205.

Porter LM, Garber PA, Nacimento E. 2009. Exudates as a fallback food for *Callimico goeldii. American Journal of Primatology* 71:120–129.

Radespiel U, Zimmermann E. 2001. Female dominance in captive gray mouse lemurs (*Microcebus murinus*). *American Journal of Primatology* 54:181–192.

Radespiel U, Sarikaya Z, Zimmermann E, Bruford MW. 2001. Sociogenetic structure in a free-living nocturnal primate population: sex-specific differences in the grey mouse lemur (*Microcebus murinus*). *Behavioral Ecology and Sociobiology* 50:493–502.

Radespiel U, Lutermann H, Schmelting B, Bruford MW, Zimmermann E. 2003. Patterns and dynamics of sex-biased dispersal in a nocturnal primate, the grey mouse lemur, *Microcebus murinus. Animal Behaviour* 65:709–719.

Radespiel U, Reimann W, Rahelinirina M, Zimmermann E. 2006. Feeding ecology of sympatric mouse lemur species in northwestern Madagascar. *International Journal of Primatology* 27:311–321.

Radespiel U, Jurić M, Zimmerman E. 2009. Sociogenetic structures, dispersal and the risk of inbreeding in a small nocturnal lemur, the golden-brown mouse lemur (*Microcebus ravelobensis*). *Behaviour* 146:607–628.

Radespiel U, Ratsimbazafy J, Rasoloharijaona S, *et al.* 2012. First indications of a highland specialist among mouse lemurs (*Microcebus* spp.) and evidence for a new mouse lemur species from eastern Madagascar. *Primates* 53:157–170.

Rakotondravony R, Radespiel U. 2009. Varying patterns of coexistence of two mouse lemur species (*Microcebus ravelobensis* and *M. murinus*) in a heterogeneous landscape. *American Journal of Primatology* 71:928–938.

Rakotondranary S, Ratovonamana YR, Ganzhorn J. 2010. Distributions et caractéristiques des microhabitats de *Microcebus griseorufus* (Cheirogaleidae) dans le Parc National de Tsimanampetsotsa (Sud-ouest de Madagascar). *Malagasy Nature* 4:55–64.

Rakotondranary SJ, Struck U, Knoblauch C, Ganzhorn JU. 2011. Regional, seasonal and interspecific variation in 15N and 13C in sympatric mouse lemurs. *Naturwissenschaften* 98:909–917.

Ramanankirahina R, Joly M, Zimmermann E. 2012. Seasonal effects on sleeping site ecology in a nocturnal pair-living lemur (*Avahi occidentalis*). *International Journal of Primatology* 33:428–439.

Randrianambinina B, Rakotondravony D, Radespiel U, Zimmermann E. 2003. Seasonal changes in general activity, body mass and reproduction of two small nocturnal primates: a comparison of the golden brown mouse lemur (*Microcebus ravelobensis*) in Northwestern Madagascar and the brown mouse lemur (*Microcebus rufus*) in Eastern Madagascar. *Primates* 44:321–331.

Rasoloarison RM, Weisrock DW, Yoder AD, Rakotondravony D, Kappeler PM. 2013. Two new species of mouse lemurs (Cheirogaleidae: *Microcebus*) from eastern Madagascar. *International Journal of Primatology* 34:455–469.

Rau G, Sweeney R, Kaplan I, Mearns A, Young D. 1981. Differences in animal 13C, 15N and D abundance between a polluted and an unpolluted coastal site: likely indicators of sewage uptake by a marine food web. *Estuarine, Coastal and Shelf Science* 13:701–707.

Rendigs A, Radespiel U, Wrogemann D, Zimmermann E. 2003. Relationship between microhabitat structure and distribution of mouse lemurs (*Microcebus* spp.) in northwestern Madagascar. *International Journal of Primatology* 24:47–64.

Schmelting B. 2000. Saisonale Aktivität und Reproduktionsbiologie von Grauen Mausmaki-Männchen (*Microcebus murinus*, J.F. Miller 1777) in Nordwest-Madagaskar. Dissertation, University of Veterinary Medicine, Hannover.

Schmid J, Kappeler PM. 1998. Fluctuating sexual dimorphism and differential hibernation by sex in a primate, the gray mouse lemur (*Microcebus murinus*). *Behavioral Ecology and Sociobiology* 43:125–132.

Schmidt O, Curry JP, Dyckmans J, Rota E, Scrimgeour CM. 2004. Dual stable isotope analysis (δ13C and δ15N) of soil invertebrates and their food sources. *Pedobiologia* 48:171–180.

Schmidt S, Stewart GR. 2003. d15N values of tropical savanna and monsoon forest species reflect root specialisations and soil nitrogen status. *Oecologia* 134: 569–577.

Sehen L, Goetze D, Rajeriarison C, *et al.* 2010. Structural and floristic traits of habitats with differing relative abundance of the lemurs *Microcebus murinus* and *M. ravelobensis* in northwestern Madagascar. *Ecotropica* 16:15–30.

Sorg J, Rohner U. 1996. Climate and tree phenology of the dry deciduous forest of the Kirindy Forest. *Primate Report* 46:57–80.

Swap RJ, Aranibar JN, Dowty PR, Gilhooly WP, Macko SA. 2004. Natural abundance of 13C and 15N in C3 and C4 vegetation of southern Africa: patterns and implications. *Global Change Biology* 10:350–358.

Szepanski MM, Ben-David M, Van Ballenberghe V. 1999. Assessment of anadromous salmon resources in the diet of the Alexander Archipelago wolf using stable isotope analysis. *Oecologia* 120:327–335.

Thorén S, Quietzsch F, Radespiel U. 2010. Leaf nest use and construction in the golden-brown mouse lemur (*Microcebus ravelobensis*) in the Ankarafantsika National Park. *American Journal of Primatology* 72:48–55.

Thorén S, Quietzsch F, Schwochow D, *et al.* 2011. Seasonal changes in feeding ecology and activity patterns of two sympatric mouse lemur species, the gray mouse lemur (*Microcebus murinus*) and the golden-brown mouse lemur (*M. ravelobensis*), in northwestern Madagascar. *International Journal of Primatology* 32:566–586.

van der Merwe NJ, Medina E. 1989. Photosynthesis and 13C/12C ratios in Amazonian rain forests. *Geochimica et Cosmochimica Acta* 53:1091–1094.

Winter K, Holtum JAM, Edwards GE, O'Leary MH. 1982. Effect of low relative humidity on δ13C value in two C3 grasses and in *Panicum milioides*, a C3–C4 intermediate species. *Journal of Experimental Botany* 33:88–91.

Wright PC. 1999. Lemur traits and Madagascar ecology: coping with an island environment. *American Journal of Physical Anthropology* 110(S29):31–72.

Zhang JW, Cregg BM. 1996. Variation in stable carbon isotope discrimination among and within exotic conifer species grown in eastern Nebraska, USA. *Forest Ecology and Management* 83:181–187.

Zimmermann E, Cepok S, Rakotoarison N, Radespiel U. 1998. Sympatric mouse lemurs in North-West Madagascar: a new rufous mouse lemur species (*Microcebus ravelobensis*). *Folia Primatologica* 69:106–114.

15 Your food, my food: patterns of resource use in two sympatric mouse lemur species

Sandra Thorén, Kate F. Carstens, Doreen Schwochow, and Ute Radespiel

Introduction

Competition over resources, both between individuals of the same species and between different species, plays a crucial role in natural selection (Danchin *et al.*, 2008). Competition is believed to play a major role in regulating the local coexistence of individuals and species (Gause, 1934); therefore, adaptations should have evolved to reduce intra- as well as interspecific resource competition (review in Gause 1934; Brown and Wilson 1956; Hardin 1960; Amarasekare, 2003; Chase and Leibold, 2003). In principle, competition can be avoided through resource partitioning (arthropods: Behmer and Joern, 2008; invertebrates: Pianka, 1973; birds: Garcia and Arroyo, 2005; mammals: Azevedo *et al.*, 2006; Sushma and Singh, 2006). Partitioning of food resources (of focus in this chapter) can be expressed in different ways. First, a food resource might be used by several individuals and/or species in different proportions. For instance, even though three sympatric primate species in Bolivia (*Callimico goeldii*, *Saguinus labiatus* and *S. fuscicollis*) show dietary overlaps, the relative proportion of shared food resources in their total diet differ between species (Porter, 2001). Second, food resources might be temporally partitioned: shared resources may be used at different times of the day and/or during different seasons. This is illustrated by the diurnal versus the nocturnal feeding patterns of different lemur species in Madagascar (Petter, 1962), as well as the varying levels of nocturnal activity between two sympatric species of foxes in Brazil (Vieira and Port, 2007). Third, food resources might be spatially partitioned. The same resources can then be used by competitors who occupy different areas or microhabitats. Spatial separation of resources can occur on a local scale as well as on a larger geographic scale. One example is given by two mouse lemur species, *Microcebus murinus* and *M. berthae*, in western Madagascar. These solitary foragers show a high degree of dietary overlap, but their coexistence appears to be facilitated by spatial separation on a local scale (Dammhahn and Kappeler, 2008a, 2008b).

The Dwarf and Mouse Lemurs of Madagascar: Biology, Behavior and Conservation Biogeography of the Cheirogaleidae, eds. S.M. Lehman, U. Radespiel, E. Zimmermann. Published by Cambridge University Press. © Cambridge University Press 2016.

On an intraspecific level, spatial separation is most prominent in territorial species where individuals exclude each other from access to the resources within their territories, and reduce direct competitive interactions by marking them with olfactory cues (Paquet, 1991; Fawcett *et al.*, 2012), acoustic signals (Chivers and MacKinnon, 1977; Robinson, 1981) or visual displays (Peek, 1972).

In this study, we investigate the nature of competition over shared food resources between two ecologically similar mouse lemur species, the gray mouse lemur (*M. murinus*) and the golden-brown mouse lemur (*M. ravelobensis*). The former has the widest distribution of all mouse lemurs, and occurs from the south to the northwest of the island. In contrast, *M. ravelobensis* has a restricted distribution and occurs only in the area between the Betsiboka and the Mahajamba rivers in the northwest of the country (Olivieri *et al.*, 2007). Their ranges overlap in northwestern Madagascar where they exist in zones of both sympatry and allopatry (Rendigs *et al.*, 2003; Olivieri *et al.*, 2007; Rakotondravony and Radespiel, 2009). The two species do not differ significantly in body length (*M. murinus*: 83.3 ± 0.9 mm, *M. ravelobensis*: 81.3 ± 1.9 g), body mass (*M. murinus*: 53.9 ± 0.9 g, *M. ravelobensis*: 56.2 ± 1.8 g; Zimmermann *et al.*, 1998) or seasonal reproduction (Schmelting *et al.*, 2000). They do, however, differ in their choice of sleeping sites and the composition of sleeping groups. *M. murinus* uses mostly tree holes for sleeping, whereas *M. ravelobensis* uses branches, lianas, leaves, and self-constructed spherical leaf nests (Weidt, 2001; Radespiel *et al.*, 2003; Thorén *et al.*, 2009). *M. murinus* and *M. ravelobensis* are both nocturnal solitary foragers (Bearder, 1987), but often sleep together in female groups (*M. murinus*) or mixed-sex groups (*M. ravelobensis*) during the day (Radespiel, 2000; Weidt *et al.*, 2004). Despite their solitary nocturnal habits, the home ranges of sleeping group members do overlap (Radespiel, 2000; Weidt *et al.*, 2004; Lutermann *et al.*, 2006). Overlap also occurs between the diet of the two mouse lemur species (Thorén *et al.*, 2011; Chapter 14), and consists partly of predictable stationary food resources (e.g. gum, insect secretions, nectar, and fruits), and partly of unpredictable food resources such as arthropods (Reimann, 2002; Radespiel *et al.*, 2006; Joly and Zimmermann, 2007; Thorén *et al.*, 2011).

In this study we specifically want to investigate if individuals of the two mouse lemur species, *M. murinus* and *M. ravelobensis*, use temporal or spatial partitioning of food resources to avoid competition. Nightly observations were conducted at known stationary food resources (SFRs) and the following questions will be addressed. (1) Are SFRs utilized by more than one individual mouse lemur? (2) Are they used by individuals of one or both mouse lemur species? (3) Do social encounters occur at SFRs or are the resources only used in succession? (4) In case of social encounters, what qualities do they have (affiliative, neutral or agonistic)?

Methods

Study site and trapping

We conducted this study from May to November 2007 in Ankarafantsika National Park, northwestern Madagascar (135,000 ha). The climate in this region is characterized by a cooler dry season from May to October and a warmer wet season from November to April, with peak rainfall in January and February. Data were collected from Jardin Botanique A (JBA) study site; a dry deciduous forest located near the Ampijoroa Forestry Station (16°19′S, 46°48′E). The site contains a 30.6-ha grid system where *M. ravelobensis* and *M. murinus* live in sympatry (Rendigs *et al.*, 2003).

We trapped mouse lemurs of both species during 1–4 capture nights per month using 100 Sherman live traps baited with banana at the intersections of the trails (about every 50 m). Individuals were sexed and measured according to Hafen *et al.* (1998). New individuals were ear-marked (1–2 small triangular cuts of approximately 2 mm^2 in the outer part of the pinna) and injected with a unique transponder chip (Trovan Small Animal Marking System, Telinject®) under the dorsal skin. We distinctively trimmed tail fur of all captured lemurs to facilitate determination of sex and species during the nocturnal observations. Females were given a circular 1.5–2.0 cm wide cut around the middle of the tail, whereas males toward the tip of the tail. To distinguish the two mouse lemur species, individuals of *M. ravelobensis* individuals were all trimmed at the base of the tail.

At the same time as our study, a feeding study was conducted on 14 female mouse lemurs (*M. murinus*: 7, *M. ravelobensis*: 7; Thorén *et al.*, 2011). For identification, these individuals were equipped with radiocollars (TW4-button cell tags, Biotrack) that weighed 2.5–3.0 g each (max. 5% of body mass). Collars were removed at the end of the study from those individuals that could be recaptured.

Observations at stationary food resources

A total of 108 SFRs were identified during focal observations of the radiocollared female mouse lemurs (n = 14). Fourteen of these resources, scattered across several sleeping group ranges, were included in this study (insect secretion: 2, gum: 9, fruit: 2, water: 1). Of these, 11 SFRs were consumed exclusively by one species and not the other (*M. murinus*: 7 SFRs, *M. ravelobensis*: 4 SFRs) and three were consumed by both lemur species. We conducted continuous observations at SFRs from 17:30 to 23:00 h during 16 evenings, providing a total of 4632 minutes of observation (range: 215–321 min/night). Start times were adjusted seasonally to changing light conditions. All observations were conducted using a flash light (Maglite, USA), a head lamp (Petzl MYO 5 or Petzl tikka plus, France), and a telemetry receiver (Telonics TR4, USA).

Data collection

We recorded all mouse lemur activities at each SFR (n = 14) as well as within its proximity (Prox ≤ 10 m away). We recorded species, sex and identity of the visiting mouse lemurs where possible. To assess the number of visitors, we defined visiting bouts for each individual that visited the SFR or entered the proximity. If the animal left the proximity for more than 3 min, the visiting bout was defined as finished. Bouts were separated into "stationary food resource (SFR) bouts" and "proximity

(Prox) bouts". The former required an individual entering the SFR, whereas the latter was only noted for individuals that entered the proximity of the SFR. We estimated the minimum and maximum number of different visitors at each SFR based on individual identification and interindividual overlap. For instance, if we observed two unknown visitors at two different occasions during one night, the number of visiting individuals could either be one (minimum) or two (maximum). However, if we simultaneously observed two unknown individuals, we would be able to conclude that it must be two different individuals.

The behaviors of visiting animals were categorized as feeding-related behaviors (feeding, drinking or foraging), moving or resting (Thorén *et al.*, 2011). In cases of uncertainty in distinguishing between foraging and moving, moving was recorded. Mouse lemurs were assumed to be drinking when sticking their heads inside a hole in the tree trunk that was filled with water.

When a minimum of two mouse lemurs were at SFR and/or Prox simultaneously, we considered this to be a social encounter. All social encounters at the food resources and within their proximity were registered and categorized as according to Thorén *et al.* (2011): approach (decrease of distance to another individual), avoid (leave the SFR or Prox quickly within 10 s without being chased), chase (rapidly follow and displace the other individual), flee (rapidly increase distance to the partner as a response to being chased), allogrooming (two individuals grooming each other). Any social interaction that contained agonistic behaviors (aggressive or submissive) was termed a conflict. A conflict was defined as decided if one individual showed a submissive response to an approach or chase, and undecided if none of the participants showed submissive behaviors. The probability of encountering another individual at the feeding site (SFR and Prox) was calculated as the number of events with two or more visitors divided by the total number of events with visitors. We used the program STATISTICA 6 (StatSoft, Inc., 2004) to provide descriptive data for our data set. Due to insufficient data, we were not able to make statistical comparisons of the data between different types of food items.

Results

Visitors at the food resources

During 4632 min of observations, we recorded 99 mouse lemur visiting bouts (median: 1.2 visiting bouts/h, range: 0.2–4.7 visiting bouts/h, $N_{\text{resources}} = 14$, $N_{\text{evenings}} = 16$; Figure 15.1). Of these, 77% were visits in the actual SFR (median: 0.8 visiting bouts/h, range: 0–4.7 visiting bouts/h), whereas 23% were visits in Prox (median: 0.0 visiting bouts/hour, range: 0–1.2 visiting bouts/h). We observed the highest frequency of visiting bouts (SFR and Prox) in one of the two insect secretion colonies (median: 2.8 visiting bouts/h, range: 1–4.7 visiting bouts/h, $N_{\text{evenings}} = 2$, $N_{\text{bouts}} = 27$, Figure 15.1) and at the drinking tree (3.1 visiting bouts/h, $N_{\text{evenings}} = 1$, $N_{\text{bouts}} = 15$, Figure 15.1). Fewer visiting bouts were observed at the gum

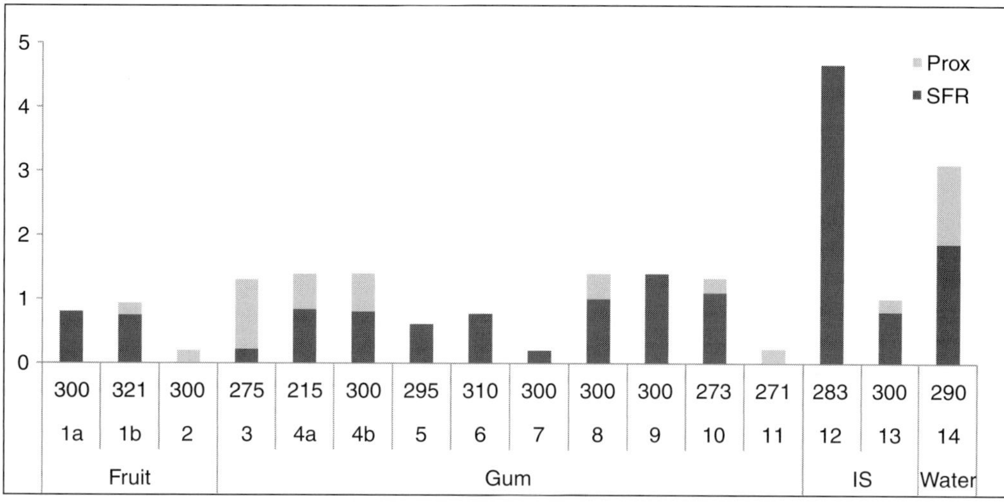

Figure 15.1 The number of visiting bouts/h registered at the food resources (1a–14), separated into type of food item; SFR, in the stationary food resource; Prox, ≤ 10 m away of the resource; IS, insect secretion; a/b, first/second observation occasion. The total minutes of observations (215–321) are provided below each column for each observation occasion.

trees (median: 1.3 visiting bouts/h, range: 0.2–2.6 visiting bouts/h, $N_{evenings}$ = 10, N_{bouts} = 47), and fewest at the fruit trees (median: 0.8 visiting bouts/h, range: 0.2–0.9 visiting bouts/h, $N_{evenings}$ = 3, N_{bouts} = 10).

For the majority of the visiting mouse lemurs (n = 57, 57.6%), we could not identify the species. Of those visiting bouts where species could be determined, about half were individuals of *M. murinus* (n = 22, 22.2%) and half were *M. ravelobensis* (n = 20, 20.2%). The majority of the stationary food resources (87.5%) had at least one unidentified visitor. Five of the 14 food resources (35.7%) had confirmed visitors from both *M. murinus* and *M. ravelobensis*. Three of these were already known to be food resources of both mouse lemur species by the start of this study (gum: n = 2, insect colony: n = 1), whereas the two others were previously only known as resources of *M. ravelobensis* (insect colony: 1, water: 1).

The minimum estimate of individuals visiting food resources (SFR and Prox) ranged from 1 to 5 different mouse lemurs (median: 2.0), whereas the maximum estimate of individuals visiting the food resources ranged from 1 to 22 (median: 5.0; Figure 15.2). We observed the highest median of individuals at the drinking tree (min: 5 individuals, max: 14 individuals), followed by median of individuals at the insect colonies ($median_{min}$: 3.5 individuals, range: 2–5; $median_{max}$: 13.5 individuals, range: 5–22), whereas the lowest median of individuals were observed in the gum trees ($median_{min}$: 2.0 individuals, range: 1–3; $median_{max}$: 4.0 individuals, range: 1–10) and fruit trees ($median_{min}$: 1.5 individuals, range: 1–2; $median_{max}$: 5.0 individuals, range: 1–9).

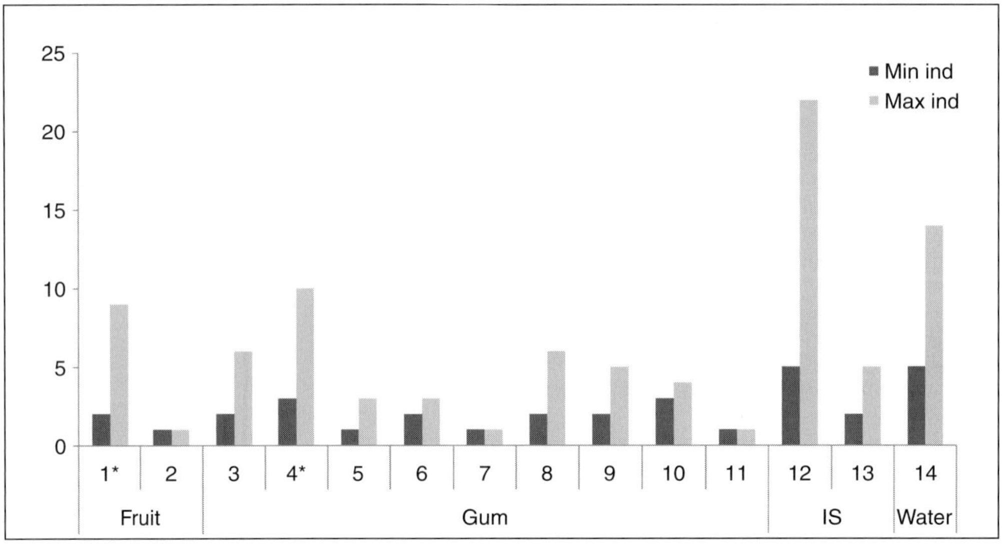

Figure 15.2 The minimum (min) and maximum (max) number of mouse lemur individuals that visited the food resources (1a–14; SFR and Prox), separated into type of food item; IS, insect secretion; *data pooled from two observation evenings.

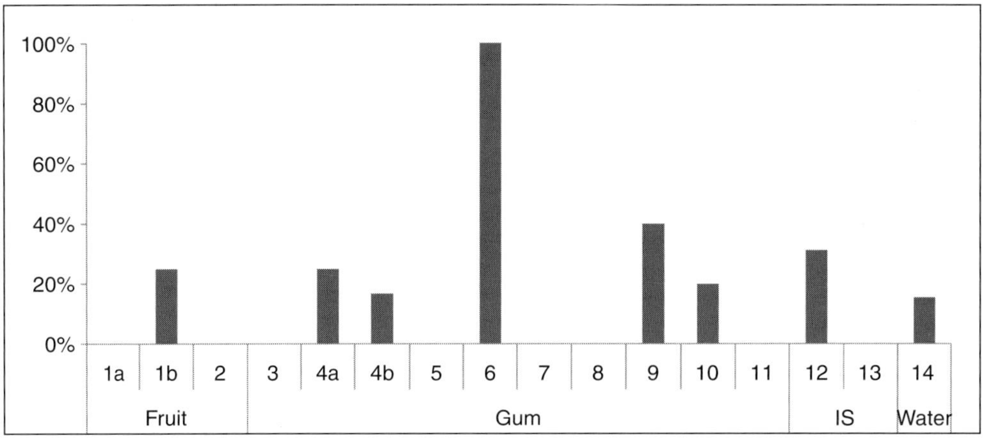

Figure 15.3 The probability of encountering another mouse lemur individual at the food resources (1a–14; SFR and Prox), separated into type of food item; IS, insect secretion; a/b, first/second observation occasion.

Social encounters at the food resources

We observed a total of 16 social encounters, recorded at all 4 different food categories and at 7 of the 14 food resources (median: 0.1/h, range: 0–1.1/h). The median probability for a visitor to encounter another mouse lemur in SFR or in Prox was 7.7% (median range: 0–100%; Figure 15.3). The majority of the encounters (n = 13, 81.3%) involved at least one individual of unknown species. In the remaining three

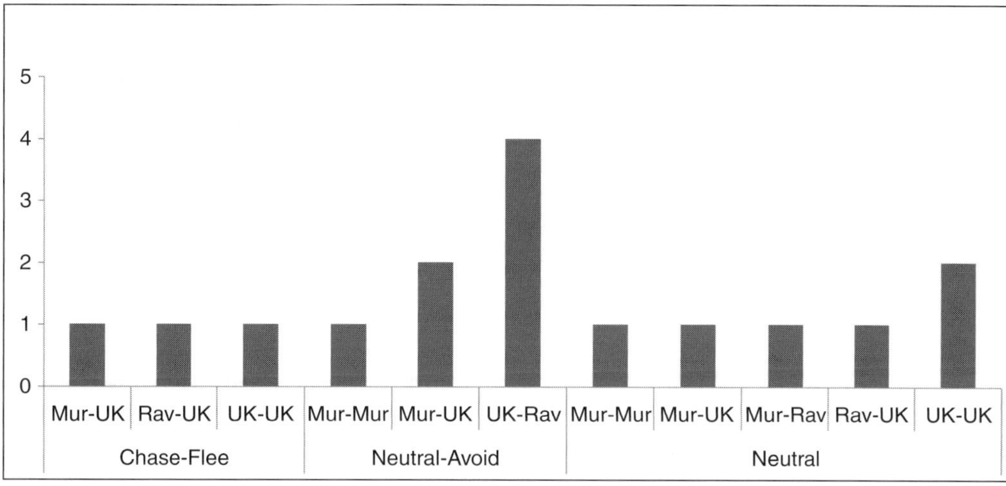

Figure 15.4 Outcomes of the social encounters between individuals. Mur, *M. murinus*; Rav, *M. ravelobensis*; UK, unknown species.

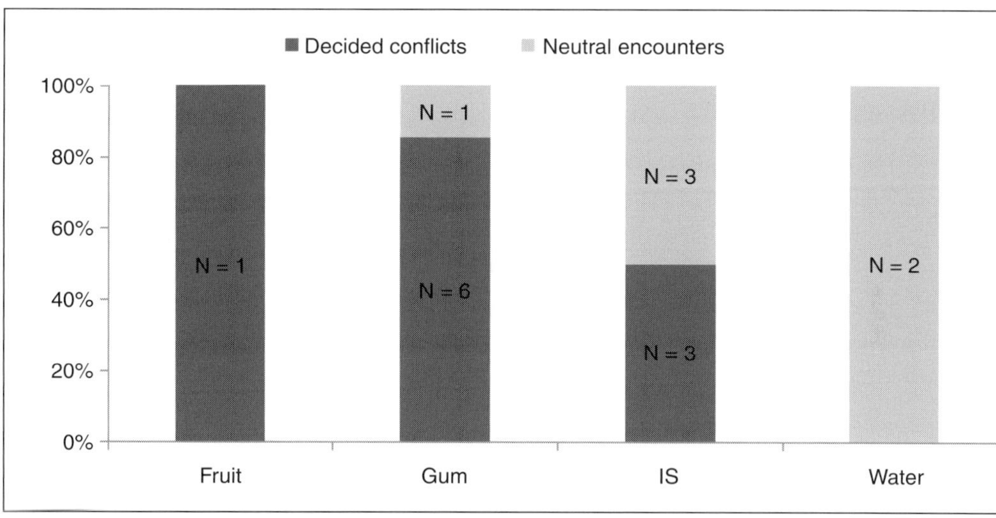

Figure 15.5 The proportion of the social interactions that resulted in decided conflicts, separated into type of food item. IS, insect secretion.

encounters, we were able to confirm the interspecific (N = 1) and intraspecific (Mur–Mur, N = 2) nature of the interactions (Figure 15.4).

Six encounters (37.5%) were non-agonistic (Figure 15.4). Of these, five were mutually neutral (Mur–Rav: 1, Rav–UK: 1, Mur–UK: 1, UK–UK: 2) and one was an allogrooming encounter between two *M. murinus* individuals. The proportion of the social interactions resulting in conflicts ($N_{encounters}$ = 10) was highest in fruit trees and close to gum secretions, whereas conflicts were never observed at the drinking tree (Figure 15.5). All conflicts were of decided outcome. Typically, one individual

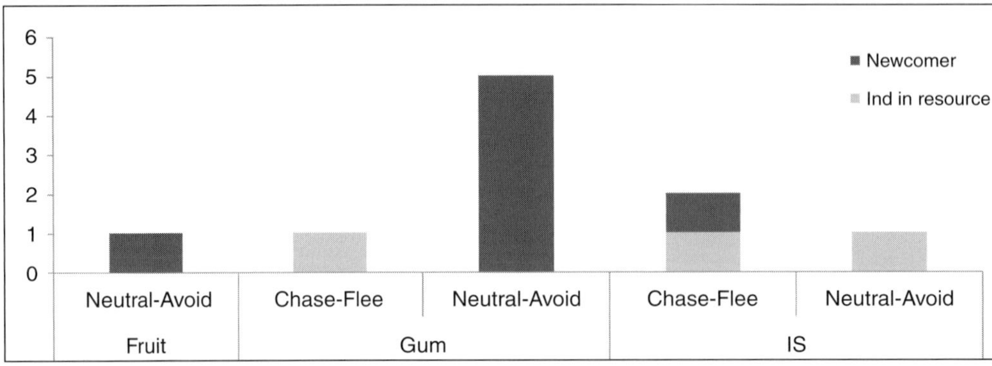

Figure 15.6 The total number of conflicts won by the individual already in the resource (Ind in resource: light gray) versus the newcoming individual (Newcomer: dark gray), separated into type of interaction and type of food item. IS, insect secretion.

avoided the neutral approach of another individual (70.0%). Usually, the individual already in the food resource when the conflict was initiated was displaced by the newcomer (70.0%; Figure 15.6). Aggressive behaviour was observed infrequently (30.0%; Figure 15.4).

Discussion

In this study, we investigated whether intra- as well as interspecific competition over food resources between *M. murinus* and *M. ravelobensis* exists, and if so, how it is avoided. It has been shown that (1) the diets of *M. murinus* and *M. ravelobensis* partially overlap (Thorén *et al.*, 2011), and (2) individuals of *M. murinus* and *M. ravelobensis* have overlapping home ranges intra- as well as interspecifically (Ehresmann, 2000; Radespiel, 2000; Rendigs *et al.*, 2003). However, it was not clear whether the same food resources are typically used by several different mouse lemurs.

Based on minimum visitor estimation in our study, most SFRs (71.4%) received visits from at least two mouse lemur individuals, and about one fourth (28.5%) from more than three visitors. Based on the maximum visitor calculation, the majority of the resources (81.3%) had visits from more than three visitors (median ≤ 4.5 individuals). Although we were unable to provide an exact number of how many different individuals used a certain resource, this study revealed that SFRs are typically shared by multiple individuals of mouse lemurs.

Despite trimming the tail fur distinctively of all captured lemurs to assist in the identification of the sex and species during focal observations, poor visibility and visits from many unmarked individuals prevented species and sex determination of many individuals. Consequently, we were only able to confirm interspecific visits from both *M. murinus* and *M. ravelobensis* in 5 of the 14 food resources (35.7%). Because most SFRs (87.5%) had at least one unidentified visitor, it is likely that the number of resources used by both species is much higher in reality than we were able to confirm in this study. Therefore, it appears that spatial resource partitioning among these two species is rather limited, in contrast to findings in studies of other sympatric *Microcebus* species pairs in western and southern Madagascar (Dammhahn and Kappeler, 2008a, 2008b; Rakotondranary *et al.*, 2011). However, resource partitioning can also be achieved on a temporal scale (Petter, 1962; Vieira and Port, 2007). In this study, only every thirteenth visit in or within the proximity of a food resource resulted in an encounter with another mouse lemur (median: 7.7%). Thus, food resources were most commonly used in succession rather than simultaneously by two or more individuals, and this seems to hold for inter- as well as intraspecific encounters. To conclude, mouse lemurs in northwestern Madagascar most likely avoid competing over shared food resources by partitioning resources on a temporal scale. This could explain (1) how individuals of the same species are able to maintain partially overlapping home ranges, and (2) how the two sympatric mouse lemur species, *M. murinus* and *M. ravelobensis*, are able to coexist in this region.

We observed 16 social encounters during our study. During some encounters (37.5%), the participants were either mutually neutral (n = 5) or mutually affiliative (n = 1). The remaining encounters (62.5%, n = 10) resulted in conflicts with decided outcomes. Aggressive behaviours in our study were only observed in the minority of the all social encounters (18.8%). Instead, conflicts were mostly resolved by one individual avoiding the other. It should be noted that our data set included encounters of individuals of both *M. murinus* and *M. ravelobensis*, but the proportion of aggressive behaviours was similar to that observed in *M. ravelobensis* in the wild (16.8%; Weidt *et al.*, 2004), despite it occurring in a competitive context at food resources. In contrast, focal observations on free-living *M. murinus* revealed a larger proportion of aggressive behaviours during intraspecific encounters (29.4%; Radespiel, 2000). There should be trade-offs between relative costs and benefits of fighting. Unless an animal has a sure indication that it will win without injury, or the resources are valuable enough for the risk of death, animals usually avoid fighting (Barrows, 2011). In our study, individuals already in the resources typically avoided the approaching newcomers without any fight, which might be explained by the fact that a hungry newcomer has more to win by gaining access to the resource than the individuals who have already spent some time feeding. Mouse lemurs are cooperative breeders (Eberle and Kappeler, 2003, 2006), and alternatively, the newcomer could have been a member of the same sleeping group as the individual in the resource. If so, sharing the food resources could also be explained by kin selection (Eberle and Kappeler, 2006; Lutermann *et al.*, 2006).

When looking at visitor rates, we found that the drinking hole had the most visitors and fruit trees had the fewest. We therefore expected the highest competition and likelihood of conflicts to be at the drinking hole. Interestingly, no conflicts were observed at this resource. We found that social encounters resulting in conflicts were proportionally highest in the fruit trees, followed by gum trees then insect secretion colonies. It appears that conflict rates may be better explained by differential resource quality or restricted food availability. For instance, compared to fruits, both insect secretions and gum are renewable and available for longer periods of time. The amount of water in the drinking hole, however, might have been sufficient for many individuals. In this study, we could not show convincingly whether conflicts are indeed more likely in certain food resources than in others, and if so, what factors can explain these differences. Therefore, future studies would certainly assist in investigating this interesting matter.

Acknowledgments

We thank the Malagasy National Parks (MNP), the Dèpartment des Eaux et Forêts (DEF) and the members of *CAFF/CORE* for approving our research proposal and issuing the research authorization. We furthermore thank the University of Antananarivo (D. Rakotondravony, and the late O. Ramilijaona) for continuous support and we are particularly indebted to Solofo Rasoloharijaona and Blanchard Randrianambinina for practical help during the beginning of the study. We also highly appreciate the continuous support from the local staff of the National Park. The DFG (Ra 502/9-1) supported the study financially.

Ethical standards

The study was approved by *CAFF/CORE*, the Dèpartment des Eaux et Forêts (DEF) and MNP. All procedures adhered to the legal requirements of Madagascar and were approved by the Malagasy Ministry of Environment, Water and Forests. We have complied to the international ethical standards concerning the treatment of primates and with the national laws and research rules formulated by the Malagasy authorities.

References

Amarasekare P. 2003. Competitive coexistence in spatially structured environments: a synthesis. *Ecology Letters* 6:1109–1122.

Azevedo FCC, Lester V, Gorsuch W, *et al.* 2006. Dietary breadth and overlap among five sympatric prairie carnivores. *Journal of Zoology* 269:127–135.

Barrows, EM. 2011. *Animal Behavior Desk Reference: A Dictionary of Animal Behavior, Ecology, and Evolution.* CRC Press, Boca Raton.

Bearder SK. 1987. Lorises, bushbabies and tarsiers: diverse societies in solitary foragers. In D Cheney, R Seyfarth, B Smuts, T Struhsaker, R Wrangham (eds.), *Primate Societies* (pp. 11–24). University of Chicago Press, Chicago.

Behmer ST, Joern A. 2008. Coexisting generalist herbivores occupy unique nutritional feeding niches. *Proceedings of the National Academy of Sciences* 105:1977–1982.

Brown Jr WL, Wilson EO. 1956. Character displacement. *Systematic Zoology* 49–64.

Chase JM, Leibold MA. 2003. *Ecological Niches: Linking Classical and Contemporary Approaches.* University of Chicago Press, Chicago.

Chivers DJ, MacKinnon J. 1977. On the behaviour of siamang after playback of their calls. *Primates* 18:943–948.

Dammhahn M, Kappeler PM. 2008a. Small-scale coexistence of two mouse lemur species (*Microcebus berthae* and *M. murinus*) within a homogeneous competitive environment. *Oecologia* 157:473–483.

Dammhahn M, Kappeler PM. 2008b. Comparative feeding ecology of sympatric *Microcebus berthae* and *M. murinus*. *International Journal of Primatology* 29:1567–1589.

Danchin E, Giraldeau L, Cezilly F. 2008. *Behavioural Ecology*. Oxford University Press, New York.

Eberle M, Kappeler PM. 2003. Cooperative breeding in grey mouse lemurs (*Microcebus murinus*). *Folia Primatologica* 74:367.

Eberle M, Kappeler PM. 2006. Family insurance: kin selection and cooperative breeding in a solitary primate (*Microcebus murinus*). *Behavioral Ecology and Sociobiology* 60:582–588.

Ehresmann P. 2000. Ökologische Differenzierung von zwei sympatrischen Mausmaki-Arten (*Microcebus murinus* und *M. ravelobensis*) im Trockenwald Nordwest-Madagaskars. Unpublished PhD thesis, University of Hannover.

Fawcett JK, Fawcett JM, Soulsbury CD. 2012. Seasonal and sex differences in urine marking rates of wild red foxes *Vulpes vulpes*. *Journal of Ethology* 31:41–47.

Garcia JT, Arroyo BE. 2005. Food-niche differentiation in sympatric Hen *Circus cyaneus* and Montagu's Harriers *Circus pygargus*. *Ibis* 147:144–154.

Gause GF. 1934. *The Struggle for Existence*. Williams & Wilkins, Baltimore, MD.

Hafen T, Neveu H, Rumpler Y, Wilden I, Zimmermann E. 1998. Acoustically dimorphic advertisement calls separate morphologically and genetically homogenous populations of the grey mouse lemur (*Microcebus murinus*). *Folia Primatologica* 69:342–356.

Hardin G. 1960. The competitive exclusion principle. *Science* 131:1292–1297.

Joly M, Zimmermann E. 2007. First evidence for relocation of stationary food resources during foraging in a strepsirrhine primate (*Microcebus murinus*). *American Journal of Primatology* 69:1045–1052.

Lutermann H, Schmelting B, Radespiel U, Ehresmann P, Zimmermann E. 2006. The role of survival for the evolution of female philopatry in a solitary forager, the grey mouse lemur (*Microcebus murinus*). *Proceedings of the Royal Society B: Biological Sciences* 273:2527–2533.

Olivieri G, Zimmermann E, Randrianambinina B, *et al.* 2007. The ever-increasing diversity in mouse lemurs: three new species in north and northwestern Madagascar. *Molecular Phylogenetics and Evolution* 43:309–327.

Paquet PC. 1991. Scent-marking behavior of sympatric wolves (*Canis lupus*) and coyotes (*C. latrans*) in Riding Mountain National Park. *Canadian Journal of Zoology* 69:1721–1727.

Peek FW. 1972. An experimental study of the territorial function of vocal and visual display in the male red-winged blackbird (*Agelaius phoeniceus*). *Animal Behaviour* 20:112–118.

Petter JJ. 1962. Ecological and behavioural studies of Madagascar lemurs in the field. *Annals of the New York Academy of Sciences* 102:267–281.

Pianka ER. 1973. The structure of lizard communities. *Annual Review of Ecology, Evolution and Systematics* 4:53–74.

Porter LM. 2001. Dietary differences among sympatric Callitrichinae in northern Bolivia: *Callimico goeldii*, *Saguinus fuscicollis* and *S. labiatus*. *International Journal of Primatology* 22:961–992.

Radespiel U. 2000. Sociality in the gray mouse lemur (*Microcebus murinus*) in northwestern Madagascar. *American Journal of Primatology* 51: 21–40.

Radespiel U, Ehresmann P, Zimmermann E. 2003. Species-specific usage of sleeping sites in two sympatric mouse lemur species (*Microcebus murinus* and *M. ravelobensis*) in northwestern Madagascar. *American Journal of Primatology* 59:139–151.

Rakotondranary SJ, Hapke A, Ganzhorm JU. 2011. Distribution and morphological variation of *Microcebus* spp. along an environmental gradient in southeastern Madagascar. *International Journal of Primatology* 32:1037–1057.

Rakotondravony R, Radespiel U. 2009. Varying patterns of coexistence of two mouse lemur species (*Microcebus ravelobensis* and *M. murinus*) in a heterogeneous landscape. *American Journal of Primatology* 71:928–938.

Reimann WE. 2002. Koexistenz und Nahrungsökologie von Weibchen des grauen und goldbraunen Mausmakis (*Microcebus murinus* und *M. ravelobensis*) in Nordwest-Madagaskar. Doctoral dissertation, University of Veterinary Medicine, Hannover.

Rendigs A, Radespiel U, Wrogemann D, Zimmermann E. 2003. Relationship between microhabitat structure and distribution of mouse lemurs (*Microcebus* spp.) in Northwestern Madagascar. *International Journal of Primatology* 24:47–64.

Robinson JG. 1981. Vocal regulation of inter- and intragroup spacing during boundary encounters in the titi monkey, *Callicebus moloch*. *Primates* 22:161–172.

Schmelting B, Ehresmann P, Lutermann H, Randrianambinina B, Zimmermann E. 2000. Reproduction of two sympatric mouse lemur species (*Microcebus murinus* and *M. ravelobensis*) in north-west Madagascar: first results of a long term study. In WR Lourenço, SM Goodman (eds.), *Diversity and Endemism in Madagascar* (pp. 165–175). Société de Biogéographie, Paris.

Sushma HS, Singh M. 2006. Resource partitioning and interspecific interactions among sympatric rain forest arboreal mammals of the Western Ghats, India. *Behavioral Ecology* 17:479.

Thorén S, Quietzsch F, Radespiel U. 2009. Leaf nest use and construction in the golden-brown mouse lemur (*Microcebus ravelobensis*) in the Ankarafantsika National Park. *American Journal of Primatology* 71:1–8.

Thorén S, Quietzsch F, Schwochow D, *et al.* 2011. Seasonal changes in feeding ecology and activity patterns of two sympatric mouse lemur species, the gray mouse lemur (*Microcebus murinus*) and the golden-brown mouse lemur *(M. ravelobensis*), in northwestern Madagascar. *International Journal of Primatology* 32:566–586.

Vieira EM, Port D. 2007. Niche overlap and resource partitioning between two sympatric fox species in southern Brazil. *Journal of Zoology* 272:57–63.

Weidt A. 2001. Ökologie und Sozialbiologie von Weibchen des goldbraunen Mausmakis (*Microcebus ravelobensis*) während der Trockenzeit in Nordwest Madagaskar. Diploma thesis, Göttingen University, Göttingen, Germany.

Weidt A, Hagenah N, Randrianambinina B, Radespiel U, Zimmermann. E. 2004. Social organization of the golden brown mouse lemur (*Microcebus ravelobensis*). *American Journal of Physical Anthropology* 123:40–51.

Zimmermann E, Cepok S, Rakotoarison N, Zietemann V, Radespiel U. 1998. Sympatric mouse lemurs in north-west Madagascar: a new rufous mouse lemur species (*Microcebus ravelobensis*). *Folia Primatologica* 69:106–114.

16 The physiology of phyletic dwarfism in Cheirogaleidae

Fabien Génin and Judith C. Masters

Introduction

Hill's (1953, p. 28) multivolume monograph, *Primates: Comparative Anatomy and Taxonomy*, presented a phylogenetic tree of living primates in which cheirogaleids occupied a position at the base of the Malagasy "lemuroid" radiation, 40–45 Ma if the time scale is read literally. Some years later, Charles-Dominique and Martin (1970) sought to provide evidence to support this view of cheirogaleid archaism. They pooled their expert knowledge of dwarf galagos and mouse lemurs to derive an extensive list of characters shared by the smallest representatives of the Afro-Asian and Malagasy strepsirrhine lineages, which they then interpreted as plesiomorphic retentions from the common strepsirrhine ancestor. The list included small size, nocturnal habit, occupation of a "fine branch and creeper" microhabitat, omnivorous but partly insectivorous diet, use of woven, globular nests and oral transport for the protection of altricial infants, and a "noyau" social system mediated largely by urine washing and other scent marking behaviours. Cartmill (1974) elaborated his model of the adaptive origins of the primate clade and its diagnostic characters (forward-facing and convergent orbits; grasping hands and feet; claws transformed into nails) from a similar scenario. In Cartmill's view, the primate ancestor was primarily a predator of flying insects, anchoring itself to a branch with its grasping feet, and launching its body with grasping hands outstretched to capture passing prey, its manual dexterity guided by its forward-facing eyes and consequent stereoscopic vision. The scenario found ready acceptance among many primate evolutionary biologists, and other evidence in its support was rapidly adduced. Mouse lemurs, with their large numbers of small acrocentric chromosomes (2n = 66), were perfect ancestors for a model of karyotypic evolution that favoured Robertsonian fusions (Dutrillaux, 1979). Small fossil teeth from Eocene and Oligocene sediments were declared to resemble those of cheirogaleids (Marivaux *et al.*, 2001; Godinot, 2006). Models of the evolution of primate sociality were grounded in cheirogaleid (especially *Microcebus*) social systems (Kappeler, 1998).

Cheirogaleids have an additional feature, unique among primates, to recommend them as model ancestors: they use energy-saving strategies similar to those of other

The Dwarf and Mouse Lemurs of Madagascar: Biology, Behavior and Conservation Biogeography of the Cheirogaleidae, eds. S.M. Lehman, U. Radespiel, E. Zimmermann. Published by Cambridge University Press. © Cambridge University Press 2016.

small mammals like rodents and insectivores. Cheirogaleids accumulate fat stores which they draw on in periods of food scarcity. This energetic facility offered an appealing resolution to the conundrum of early primate biogeography. Primates are currently distributed on various landmasses which are separated by substantial tracts of ocean. The most sophisticated models we have of continental break-up and drift imply that these landmasses separated long before the emergence of the primate clade (see Masters *et al.*, 2007, for a review). A small mammal with high metabolic requirements is unlikely to survive a period of transoceanic rafting, say, from Africa to Madagascar, but an ancestor with the capacity to reduce its metabolism makes a far more likely candidate for successful island colonisation (Martin, 1972; Yoder, 1996; Kappeler, 2000).

The development of sophisticated methods for phylogenetic and ancestral state reconstruction has presented a major challenge to this interpretation of evolutionary polarity. No molecular phylogeny has reconstructed the Cheirogaleidae as a particularly ancient family; this distinction goes to the Daubentoniidae, which diverged at least 10 Ma before the other extant families. All recent phylogenies support the inclusion of *Lepilemur* within the cheirogaleid clade (Roos *et al.*, 2004; Horvath *et al.*, 2008; Chatterjee *et al.*, 2009; Fabre *et al.*, 2009; Perelman *et al.*, 2011; Springer *et al.*, 2012; Masters *et al.*, 2013), and ancestral state reconstructions indicate that the small-bodied cheirogaleids shared a common ancestor at least as large as living *Lepilemur* (Génin and Masters, 2011; Masters *et al.*, 2013, 2014; Chapter 4). On the basis of striking similarities with juvenile *Lepilemur*, Masters *et al.* (2014) proposed that the smallest cheirogaleid taxa (*Microcebus*, *Allocebus* and *Cheirogaleus medius s.l.*) are the paedomorphic products of parallel episodes of phyletic dwarfism, probably as a result of truncated ontogenies (progenesis). Their interpretation followed Gould's (1977) view that shifts in body size are not generally driven by direct selection on size *per se*, but are pleiotropic by-products of alterations in life history. In cheirogaleids, body size reduction was driven by environmental hypervariability, defined as a combination of extreme seasonality (intra-annual variability) and unpredictability (high interannual variability) in terms of rainfall (Dewar and Richard, 2007; Génin, 2008; Génin *et al.*, 2010; Génin and Masters, 2011; Masters *et al.*, 2014). Environmental hypervariability is particularly pronounced in regions subject to El Niño-related droughts, and in primates is associated with a number of convergent adaptations including dwarfing (Génin and Masters, 2011; Masters *et al.*, 2014), hypothermy (Génin and Perret, 2003) and gummivory (Génin *et al.*, 2010; Chapter 4).

This alternative framework implies that many cheirogaleid peculiarities, previously viewed as archaic, may in fact be aspects of a suite of life-history adaptations to the hypervariable environmental conditions that became established in the Afro-tropical region in the Pliocene (5–2 Ma) (Bobe, 2006). In addition to the list of characters provided by Charles-Dominique and Martin (1970), we would include: shortened gestation periods, polyestry, larger litters of altricial infants, heterothermy, and seasonal fattening (Génin and Perret, 2003; Génin, 2008; Génin *et al.*, 2010). We are not the first to seek explanations for the distinguishing features of lemuriforms in the hypervariability and/or low productivity of Malagasy habitats, which have been invoked as drivers of lemur energy-saving strategies, of their unusual life histories, and even of the frequent occurrence of monogamy and female dominance (Hildwein and Goffart, 1975; Müller, 1979; McCormick, 1980; Daniels, 1984; Jolly, 1984; Richard and Nicoll, 1987; Young *et al.*, 1990; Tilden and Oftedal, 1995; Wright, 1999; Génin and Perret, 2003; Dewar and Richard, 2007; Génin, 2008; but see: Harcourt, 2008; Simmen *et al.*, 2010). Additionally, aspects of the "island rule" may apply to Madagascar, particularly with respect to body size shifts, relative brain size and heterothermy (Weston and Lister, 2009; Masters *et al.*, 2014).

In this chapter we re-evaluate current knowledge of cheirogaleid ecophysiology from the perspective of recent phylogenetic insights, assessing the relative likelihood that these unique features are (a) primitive retentions, or (b) adaptations to environmental hypervariability. We combine the elements of a critical review with a re-analysis of some published work and the presentation of previously unpublished data on body composition in sympatric cheirogaleids. To accommodate this rather chimeric mix without losing clarity, we divide our report into two (not unrelated) sections. In the first section, we review and re-interpret information regarding the phenomenon of seasonal fattening in cheirogaleids, and its relation to their reproduction; we compare our field data on cheirogaleid body composition with published data taken from other lemurs. In the second section, we interrogate the concept of hypometabolism in strepsirrhines in general by re-examining an earlier study purporting to demonstrate that their low basal metabolism is a phylogenetic constraint, or primitive retention.

Unique features of cheirogaleid ecophysiology

Seasonal rhythms of reproduction and fattening

Physiological rhythms may be endogenous, in which case they are maintained in captivity under free-running conditions (i.e. in the absence of environmental triggers), or they may be entrained by day length (i.e. photoperiod). Endogenous and entrained rhythms can only be distinguished in the laboratory under controlled conditions. Long-term captive studies have shown that seasonal reproduction in grey mouse lemurs (*Microcebus murinus*) is an endogenous circannual rhythm (i.e. approximately one-year periodicity), maintained under constant photoperiod, but synchronised by photoperiodic changes (Perret and Aujard, 2001). By contrast, the changes in body composition in mouse lemurs are seasonal but not endogenous. Captive *M. murinus* fatten after exposure to short photoperiods (10 h daylight); Génin and Perret, 2000; Génin *et al.*, 2003b, 2005), but this rhythm disappears if animals are artificially maintained under long-day exposure (14 h daylight; Perret and Aujard, 2001). Photoperiodic entrainment allows more flexibility than circannual rhythms, and hence facilitates adaptation to unpredictable environmental conditions (Génin, 2007; Canale *et al.*, 2010).

Circannual rhythms are strongly influenced by body size. Large-bodied mammals that combine reproductive seasonality with long gestation periods are usually "short-day breeders": they mate in autumn (photoperiod < 12 h daylight) and give birth in spring (photoperiod > 12 h daylight). Small-bodied, seasonal mammals are often "long-day breeders" because their short gestations allow them to accomplish their reproduction in spring and summer (photoperiod > 12 h daylight). *M. murinus* (and probably other small cheirogaleids) are the only long-day breeders among primates (Masters *et al.*, 2014), but in their case the transition from short- to long-day breeding appears incomplete. Only the females evince an endogenous circannual rhythm synchronised by long photoperiod; males exhibit photo-refractoriness and start their testicular recrudescence well before long-day exposure (Perret and Aujard, 2001; Figure 16.1).

The postulate that dwarf cheirogaleids have evolved small body size by reducing the length of their gestation (i.e. progenesis; Masters *et al.*, 2014) provides a logical explanation for their anomalous breeding schedule. Like the diurnal lemurs, the larger-bodied members of the lepilemurid–cheirogaleid clade, *Lepilemur* and *Phaner*, mate in the dry season (albeit later), and have unusually long gestations for their body sizes (Masters *et al.*, 2014; see Figure 16.1). Further reduction of gestation duration in the very small-bodied species led to a shift from a short- to a

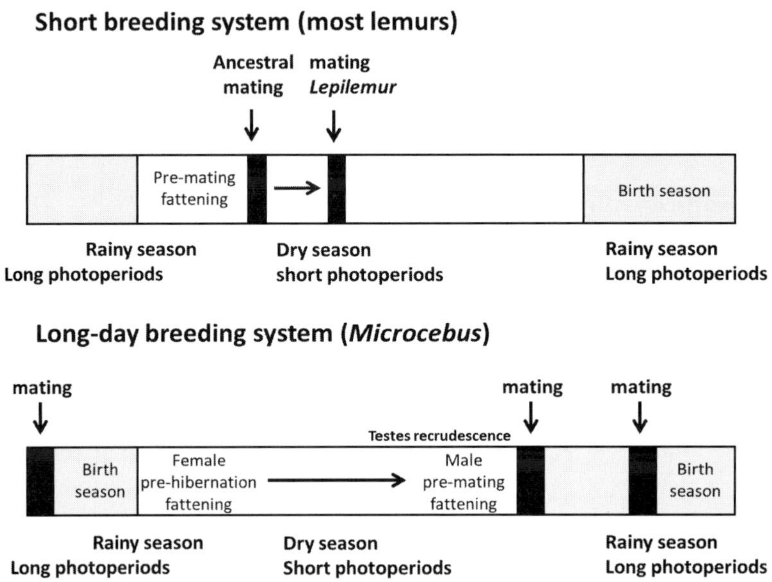

Figure 16.1 The annual rhythm of *Microcebus* (bottom) probably derived from a short-day breeding system similar to that of living Lemuridae (top). In *Lepilemur* and *Phaner*, the mating period is delayed (arrow) but still controlled by short photoperiods, leading to unusually long gestations relative to body size. In *Microcebus*, short photoperiod-induced fattening is delayed in males (arrow) and associated with photo-refractoriness and testes recrudescence, and ancestral pre-mating fattening. In contrast, females evolved a long-day circannual reproduction, allowing three mating periods. Although individual females usually only give birth once or twice per year, they may replace a lost litter with a new litter. The rainy season (November–March) is indicated in grey while the dry season (April–October) is in white.

long-day breeding system, at least in *M. murinus*. The proposal is supported by the non-endogenous rhythms of fattening in cheirogaleids.

In Table 16.1 we present measurements of lean mass taken from 68 sympatric cheirogaleid individuals of four species (see Génin *et al.*, 2005 and Schülke, 2005 for capture methods), in the western deciduous forest of Kirindy in March–April 1999, i.e. the onset of the short photoperiod in advance of the austral winter. The sample included 56 *Microcebus murinus* (28 males and 28 females), 8 *Cheirogaleus medius* (5 males and 3 females), 2 (2 females), and 2 *Phaner pMirza coquerelialles-cens* (1 male and 1 female). Data were collected using a small animal body composition TOBEC analyser (SA-2, EMSCAN, Springfield, IL, USA; van Loan and Mayclin, 1987) on anaesthetised animals (ketamine 500 mg, 2 mg/100 g; Génin *et al.*, 2005). The large mouse lemur sample (N = 56) was analysed further. The regression of body mass (dependent factor) against the E-TOBEC value (independent factor) was significant ($F1/54$ = 20.0, P < 0.001), but lean mass only contributed 27% of the variance of body mass, indicating that 73% of the variation in body mass can be attributed to variation in fat deposits. Because the E-TOBEC value is expressed in arbitrary units, we used the residuals of the body mass/E regression as an estimate of fat mass, assuming that the leanest individual captured in Kirindy (minimum residual = –21 g) had a fat mass < 1 g (accuracy of body mass measurements). This calibration allowed us to calculate a correction factor as follows: if fat mass is equal to zero, then body mass = lean mass = E – 10 g.

Diurnal lemurs, like ring-tailed lemurs (*Lemur catta*) and brown lemurs (*Eulemur fulvus*), fatten in April before the mating season (Simmen *et al.*, 2010; Figure 16.1). Likewise, female mouse lemurs (and some older resident males) fatten in April, which enables them to sustain periods of torpor when food is scarce (Génin, 2007; Canale *et al.*, 2010); i.e. they undergo pre-hibernation fattening. Most males only fatten later, at the end of the dry season, when their testes recrudesce. This is unusual among mammals; for example, in rodents, fattening is inhibited by testosterone. Hence, when Schmid and Kappeler (1998) measured an increase in body mass > 50 g in *M. murinus* males in September, prior to the mating season, they interpreted it as an increase in muscle/lean mass. Génin *et al.* (2005), however, demonstrated that seasonal fattening can occur in mouse lemurs at any time after exposure to short photoperiods, and is not inhibited by testes recrudescence (as is well known in rodent species; e.g. Bartness, 1996). In fact, testes only develop in males that have been exposed at least once to the short photoperiods (Perret and Aujard, 2001), providing strong evidence that mouse lemur ancestors had a short-day breeding rhythm, like other lemurs. Males of *M. murinus* (Génin, 2007) and *M. griseorufus* (Génin, 2008) fatten in September – late in the dry season – and lose a great deal of weight during the mating season, implying that mouse lemurs show pre-mating fattening like other lemurs, but its timing is delayed by a period of five months.

Hence, like reproductive rhythms, fattening rhythms are desynchronised between male and female grey mouse lemurs: males continue to develop a delayed pre-mating fattening, while females exhibit pre-hibernation fattening that has retained the

Table 16.1 Body composition prior to the dry season (March–April: end of fattening period) and environmental description in six lemur species found in the west and south of Madagascar, including four sympatric cheirogaleids measured in Kirindy (west).

	BM (g)	% Fat	*N*	Thermoregulation[1]	Diet[2]
Microcebus murinus	73±2	29±2	56	DT, FH	Fr/In/Gu/Ne
Cheirogaleus medius	196±16	25±5	8	OH	Fr/Ne
Mirza coquereli	225**	6**	2	N	Fr/In/Gu/Ne
Phaner pallescens	313**	5**	2	N	Gu/In/Ne
**Eulemur rufifrons*	1801±77	18±1	14	N	Fr/Le/Ne
**Lemur catta*	2299±80	19±1	11	N	Fr/Le

[1]N, normothermic at all times; DT, daily torpor; FH, facultative hibernation; OH, obligatory hibernation; [2]Le, leaves; Fr, fruits; Ne, nectar; Gu, gum; In, insects and other prey; *Lemuridae, Berenty Reserve (southeast); data from Simmen *et al.* (2010). **Small samples.

timing of ancestral pre-mating fattening (Figure 16.1). In male mouse lemurs, the testes regress during short photoperiod exposure, possibly also as a consequence of their truncated ontogeny. Progenesis implies growth is suspended prior to the ancestral period of sexual maturation; the annual rhythm of testes regression/recrudescence in mouse lemurs may hence be a paedomorphic trait, as males repeat the final phase of their sexual maturation (and natal dispersal) every year.

Body size and hypometabolism

An animal's basal metabolic rate (BMR) is the rate at which it needs to expend energy to maintain a constant body temperature (endothermy), excluding all other energy-consuming activities. Interspecific comparisons of BMR are complicated by its allometric relationship to body mass. Harcourt (2008) compiled average BMRs for 40 primate species (24 haplorhines and 16 strepsirrhines), and analysed them using two different methods, both of which led him to conclude that strepsirrhines have lower average BMRs than haplorhines. First he compared residuals calculated from Kleiber's formula, often employed for comparisons across mammal groups to identify analogous adaptations. Second, he calculated his own regression residuals of primate BMR, a more appropriate method for phylogenetic comparisons (Harcourt, 2008). He attributed his observation of lower BMRs in strepsirrhine species to the retention of ancestral hypometabolism by members of the suborder, rather than as adaptations to the unpredictable ecology of Madagascar, because low BMRs occur in Afro-Asian strepsirrhines as well as in lemurs.

We repeated Harcourt's second analysis, using 95% confidence intervals of body mass-independent residuals to identify cases of hypo- and hypermetabolism. While Harcourt conducted his analysis at the generic level, we examined the data at a specific level, which allowed us both to increase the sample size and to detect species-level deviations. We compared the allometric slopes of the regressions calculated from the haplorhine and strepsirrhine samples. Instead of direct comparisons of the values of BMR, made independent of body mass by the use of residuals, we compared the influence of body mass on BMR in the two suborders, using *t*-tests and the general formula: $t = (s_1 - s_2)$/standard deviation of the compared slopes s_1 and s_2, with $df = n - 2$.

Our analysis (Figure 16.2) agreed with Harcourt's in that strepsirrhines did indeed have lower average BMRs than haplorhines, as is evident from the significantly lower allometric exponent of the regression of BMR against BM in strepsirrhines than in haplorhines ($t_{38} = 2.6$, $P = 0.013$). In strepsirrhines, the allometric exponent of the relationship was 0.665 ± 0.013 ($N = 16$), as against 0.822 ± 0.060 ($N = 24$) in haplorhines. However, somewhat counter-intuitively, the major differences in BMR were found in larger species. The small cheirogaleid species that undergo periods of hypothermy to reduce their energetic requirements (*Microcebus murinus* and *Cheirogaleus medius*) do not have low BMRs relative to their body sizes (Figure 16.2; Table 16.2). The difference in allometric exponents may be a consequence of the frequency distribution of body sizes in our haplorhine sample (right-skewed) and strepsirrhine sample (left-skewed; Figure 16.2). Thus, the low BMRs observed in strepsirrhines might be a consequence of methodological bias, due to the classical U-shape of the relationship of BMR with body size (Brown *et al.*, 1996).

Furthermore, case-by-case analyses reveal a number of exceptions that suggest that hypometabolism is adaptive and convergent, rather than a constraint, and this is supported by its sporadic occurrence in different strepsirrhine families. For example, Harcourt's (2008) data set includes lorisids, in which low BMRs are linked to slow, chameleon- or sloth-like quadrupedal locomotion (Hildwein and Goffart, 1975; Müller, 1979). Such slow climbing is in turn associated with spectacular antipredatory displays (*Arctocebus* spp., *Nycticebus* spp.), or mechanical (*Perodicticus potto*) or even chemical armour (e.g. repellent odours and toxicity). Lorises are specialised in the consumption of slow-moving and/or toxic prey, such as ants and caterpillars, and are among the rare cases of venomous mammals (documented for *Nycticebus* spp.: Grow *et al.*, 2015). The effects of diet (McNab, 1986), locomotion, and armour on BMR have been well studied in other mammals (Lovegrove, 2001, 2004). Economic locomotion can also be interpreted as an energy-saving strategy, suggesting that low BMR may be an effect of slow locomotion, rather than the other way around (Lovegrove, 2001, 2004; McNab, 2002). The reduction of expensive locomotor behaviour is a characteristic that falls under the island rule (McNab, 2002, p. 337). Sportive lemurs (*Lepilemur* spp.) have been shown to exhibit spectacular energy-saving strategies such as reduced locomotor activity and use of caecotrophy (Hladik *et al.*, 1971; Chapter 4). Such slow locomotion may also have characterised large-bodied subfossil lemurs *Megaladapis*, *Babakotia* and *Paleopropithecus* (Godfrey *et al.*, 1997). Terrestrial and semiterrestrial primates such as baboons and patas monkeys also have low BMRs compared with other primates (Table 16.2), although their locomotion can hardly be described as slow. Small size and hypometabolism are also characteristic of island faunas (Lomolino, 1985, 2005) and of those inhabiting

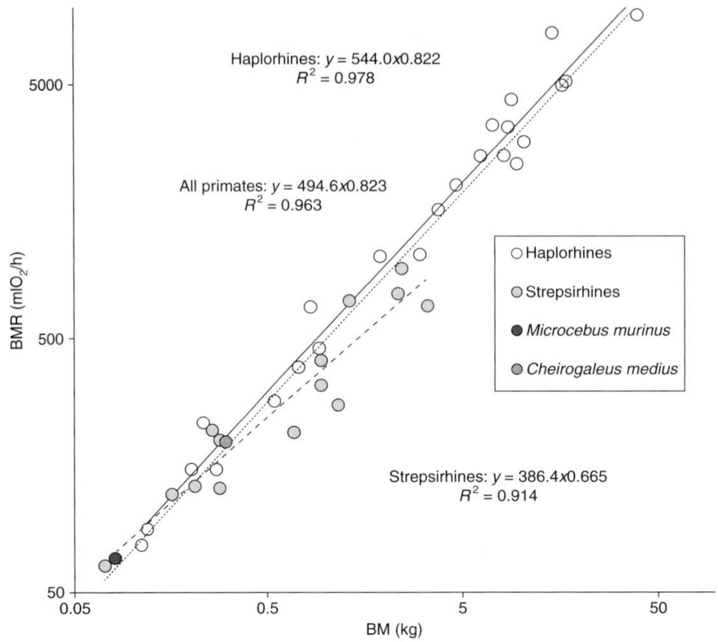

Figure 16.2 Relation between body mass and basal metabolic rate in haplorhine and strepsirhine primates showing different allometric exponents, and the two species of Cheirogaleidae included in the sample (data from Harcourt, 2008).

island-like environments, like fragmented forests for arboreal animals (Whittaker and Fernández-Palacios, 2007).

Hypometabolism is often observed in primates occupying regions that have unpredictable rainfall regimes: Madagascar (Lemuriformes), southeastern Africa (Galagidae), South America (Callitrichinae) and South-East Asia (Lorisidae) are all affected by El Niño-related unpredictable droughts (Génin *et al.*, 2010; Masters *et al.*, 2014). Such environmental unpredictability offers potential explanations for the low BMRs observed in *Otolemur crassicaudatus* (Müller and Jaksche, 1980), Asian lorises (in particular *Nycticebus coucang*: Müller, 1979), and callitrichines, notably *Callithrix jacchus* (Petry *et al.*, 1986; see Table 16.2). Other examples of hypometabolism, including the torpor observed in Malagasy tenrecs (Lovegrove and Génin, 2008) and Australian sugar gliders (Körtner and Geiser, 2000), are also likely to be linked to the fact that they occur in regions subjected to El Niño-related droughts. In contrast, several species of galagids that occur in more climatically predictable regions have high BMRs: i.e. *Galago moholi*, *G. senegalensis*, *Galagoides demidoff* and *Euoticus elegantulus* (Table 16.2).

Hence, we conclude that the low BMRs observed in some strepsirrhines are unlikely to be examples of incompletely evolved endothermy, but adaptations to lifestyles and habitats that place strict limitations on available energy.

Table 16.2 Exceptions to the rule: extreme basal metabolic rates (BMR) in primates, using 95% confidence intervals on body mass-independent residuals (hypo- and hypermetabolism in **bold;** data from Harcourt, 2008).

Species	BM (kg)	BMR (mlO_2/h)	Residuals
Nycticebus coucang	**0.064**	**2.436**	**−0.311**
Propithecus verreauxi	**0.519**	**2.829**	**−0.292**
Lepilemur ruficaudatus	**−0.161**	**2.328**	**−0.233**
Perodicticus potto	**−0.018**	**2.515**	**−0.165**
Loris tardigradus	**−0.553**	**2.107**	**−0.132**
Eulemur fulvus (rufifrons)	**0.367**	**2.873**	**−0.124**
Papio anubis	**0.978**	**3.387**	**−0.113**
Otolemur crassicaudatus	**−0.022**	**2.614**	**−0.062**
Colobus guereza	**1.019**	**3.474**	**−0.060**
Erythrocebus patas	**0.477**	**3.029**	**−0.059**
Callithrix jacchus	**−0.569**	**2.185**	**−0.041**
Daubentonia madagascariensis	**0.391**	**2.976**	**−0.041**
Pan troglodytes	1.594	3.973	−0.035
Callimico goeldii	−0.260	2.453	−0.027
Macaca mulatta	0.908	3.418	−0.024
Arctocebus calabarensis	−0.678	2.117	−0.019
Tarsius syrichta	−0.959	1.886	−0.018
Aotus trivirgatus	−0.032	2.659	−0.009
Leontopithecus rosalia	−0.137	2.582	0.000
Papio ursinus	1.228	3.712	0.006
Pongo pygmaeus	1.210	3.697	0.006
Cebuella pygmaea	−0.921	1.944	0.009
Cheirogaleus medius	−0.523	2.290	0.026
Galago moholi	**−0.796**	**2.079**	**0.041**
Cercocebus torquatus	**0.574**	**3.208**	**0.041**
Microcebus murinus	**−1.097**	**1.833**	**0.042**
Galagoides demidoff	**−1.155**	**1.799**	**0.056**
Alouatta palliata	**0.669**	**3.302**	**0.057**
Otolemur garnettii	**0.117**	**2.848**	**0.057**
Galago senegalensis	**−0.553**	**2.297**	**0.058**
Tarsius spectrum	**−0.699**	**2.182**	**0.063**
Papio papio	**0.794**	**3.417**	**0.068**
Cercopithecus mitis	**0.929**	**3.530**	**0.070**
Hylobates lar	**0.279**	**3.021**	**0.097**
Euoticus elegantulus	**−0.585**	**2.334**	**0.122**
Macaca fascicularis	**0.851**	**3.539**	**0.143**
Macaca fuscata	**0.954**	**3.639**	**0.158**
Saimiri sciureus	**−0.081**	**2.822**	**0.194**
Saguinus geoffroyi	**−0.638**	**2.369**	**0.201**

Discussion: physiological evidence of phyletic dwarfing in cheirogaleids

Mouse and dwarf lemurs bear superficial resemblances to small mammals of other orders, leading them to be viewed as "primitive". The details of their physiologies, however, reveal that these similar features have very different mechanisms of causation, indicating ecological convergence. The progenesis hypothesis provides an alternative interpretive framework for the physiological peculiarities of Cheirogaleidae, and offers a number of new directions for future research to physiologists, biogeographers and socioecologists alike.

Dwarfing and the evolution of adaptive hypothermy

Harcourt's (2008) claim that basal metabolic rate is uniformly lower in strepsirrhines than in haplorhines was not sustained in our re-analysis. Finer comparisons revealed that some strepsirrhines have high metabolism while some haplorhines have low metabolism; and the low BMRs seen in some strepsirrhines are more satisfactorily explained as adaptations to environmental hypervariability than as phylogenetic constraints. Paradoxically, the occurrence of hypothermy may be associated with relatively high BMRs, as observed in *Microcebus* and *Cheirogaleus*, although it too is a form of hypometabolism associated with environmental hypervariability (Lovegrove, 2001, 2011) and insularity (Whittaker and Fernández-Palacios, 2007).

Heterotherms decrease their metabolic requirements by giving up endothermy during torpor bouts, the duration of which depends on body size and fat storage. Small heterotherms usually remain torpid for under 24 h (daily hypothermy), while larger species engage in longer torpor bouts (hibernation). Cheirogaleids follow this rule: the smallest species are daily heterotherms (*Microcebus berthae*, 30 g); larger species are obligate hibernators (*Cheirogaleus* spp., 150–400 g), while *M. murinus* individuals, of intermediate size (60–100 g), may use both daily torpor and short bouts of facultative hibernation (Wright and Martin, 1995; Génin and Perret, 2000, 2003; Schmid, 2000; Schmid *et al.*, 2000; Dausmann *et al.*, 2004; Génin, 2007; Blanco and Rahalinarivo, 2010). In cheirogaleids, as in other groups of mammals, body size is a good predictor of the form of hypothermy used by different species, but does not predict the actual occurrence of hypothermy, which has been observed in the smallest (*Microcebus* spp.) and some larger (*C. major*) species, but probably does not occur in *Mirza* and *Phaner* spp. Therefore, the advantages deriving from hypothermy are unlikely to have driven the evolution of body size in cheirogaleids. Rather, hypothermy evolved as a by-product of dwarfing, in species feeding on seasonal (*Cheirogaleus* spp.) or unpredictable (*Microcebus* spp.) foods.

The historical view of hypothermy as primitive, "imperfect" endothermy (Kayser, 1962) was recently resurrected by Lovegrove (2012), along with a typology of endothermy corresponding to phylogenetically reconstructed "evolutionary grades". Unfortunately, he used an outdated phylogeny for primates that did not reflect the dwarfing episodes revealed by recent studies. Nevertheless, the phylogenetic constraints he identified indicate a high level of homoplasy in the evolution of mammalian thermoregulation, arguing against plesiomorphic retention of heterothermy, and in favour of lineage-specific adaptive scenarios subject to ecological convergence (Geiser and Ruf, 1995; McNab, 2002).

An example of such convergence is found in the mechanism that small hypometabolic mammals employ to warm up actively after torpor bouts: i.e. non-shivering thermogenesis. This process takes place in specialised but diffuse tissues called brown adipose tissue, or brown fat (Génin *et al.*, 2003a). While all heterothermic eutherians have brown fat, the distribution of these tissues in the grey mouse lemur, *M. murinus*, is highly unusual and suggests an independent evolutionary development. Brown adipose tissues are commonly observed in the early developmental stages of mammals, but regress in adults. In large-bodied species such as humans, functional brown fat is only adaptive in newborns, because they have higher heat loss than adults (Nicholls and Locke, 1984). In mouse lemurs, the retention of functional brown fat in the adult stage may be another effect of paedomorphosis (Masters *et al.*, 2014), but is also likely to be adaptive for cold acclimatisation and may occur in all cheirogaleids, including species that do not use hypothermy. Non-shivering thermogenesis is also practiced by small-bodied callitrichines (Rothwell and Stock, 1985, Thompson *et al.*, 1994; Genoud *et al.*, 1997). Rothwell and Stock (1985) interpreted this process in marmosets as a mechanism for dissipating excess energy, but they may have underestimated the challenge of colder temperatures for small animals living in shaded and moist environments with seasonally reduced food availability. Indeed, many characteristics of callitrichines have been interpreted as responses to increased energetic demands (Hrdy, 1977); and cold acclimatisation is probably the most common function of non-shivering thermogenesis. In mouse lemurs, brief and moderate cold exposure rapidly induces the production of the uncoupling protein (UCP) directly involved in heat production in brown adipose tissue (Génin *et al.*, 2003a).

As discussed above, mouse lemurs exhibit atypical seasonal fattening entrained by short photoperiod exposure. Fat deposition results from increased appetite and decreased metabolic rate during the phase of shallow daily hypothermy, which is associated with increased melatonin and decreased thyroxin levels (Génin and Perret, 2003; Génin *et al.*, 2003b). Short photoperiod exposure appears to induce thyroid regression in *Microcebus murinus* (Génin and Perret, 2000; Génin *et al.*, 2003b). Thyroid hormones are involved both in growth and in thermoregulation, including non-shivering thermogenesis, and may have played an important role in the recurrent evolution of endothermy (McNab, 2002).

Biogeography and insular dwarfing

One of the most profound implications of the dwarfing perspective is that it renders less likely the popular biogeographic scenario of the colonisation of Madagascar by transoceanic rafting with the assistance of heterothermy: large ancestors are unlikely to have been hibernators, severely compromising their survival at sea for several weeks (Masters *et al.*, 2007). Other scenarios, particularly with respect to geological evolution of the Indian Ocean, must be examined more seriously.

A second important implication is that if small body size in cheirogaleids is not an ancestral retention, recurrent episodes of dwarfing in the cheirogaleid clade require explanation. Dwarfing is a common characteristic among island faunas and floras, along with a suite of other attributes often grouped under the umbrella of the island rule. Madagascar has generally been viewed as too large to be subject to island effects, but several seem to apply to its biota: specifically, insular dwarfism and gigantism, the use of energy-saving strategies including hypometabolism, increased digestive efficiency, slow locomotion, the presence of near-flightless birds, predominance of woody plants (insular woodiness), and unusual interspecific associations (Wallace, 1892; Carlquist, 1974; Weston and Lister, 2009; Masters *et al.*, 2014; Chapter 4). Many hypotheses explaining these observations are explicitly based on characteristics of young oceanic islands: isolation requiring colonisation, moderate climates and small surface areas (Carlquist, 1974). For example, Lomolino's (1985) rule of convergence – whereby members of island faunas are hypothesised to evolve towards a common "optimal body size" – is interpreted as the result of relaxation of interspecific competition.

Gould (1977) proposed an alternative explanation for the paradoxical phenomena of insular dwarfism and gigantism, which integrates and clarifies Lomolino's (1985, 2005) rule of convergence. Indeed, vertebrate body size tends to converge towards medium sizes on islands; insular animals tend to dwarf when they derive from large ancestors and become giants when they derive from small ancestors (Lomolino, 1985, 2005). Gould (1977) used the example of Malagasy tenrecs to show that developmental constraints may allow species to increase their fecundity by dwarfing (like mouse lemurs) or by becoming giants (like the largest of the Malagasy tenrecs, *Tenrec ecaudatus*). Because ancestral female mouse lemurs gave birth to one young, increased litter size required considerable decrease in body size. In contrast, tenrecs probably evolved from small ancestors with a small litter size, and larger litter sizes required an increase in body size, maximized in *T. ecaudatus*, which has the largest litter size recorded for eutherian mammals of comparable size (>30 offspring for an adult mass of about 1 kg).

Dwarfing and polyandry

Hypervariability has also been invoked to explain the high frequency of monogamy and polyandry in lemurs, as a result of female dominance (Jolly, 1984; Richard and Nicoll, 1987; Young *et al.*, 1990; Tilden and Oftedal, 1995; Wright, 1999; Génin, 2013). A striking convergence between callitrichines and cheirogaleids is the evolution of polyandry and increased litter size in the smallest forms, which coevolved with phyletic dwarfing. In both cases, polyandry evolved from monogamy. *Lepilemur* and the largest cheirogaleids are pair-living (Schülke, 2005; Dröscher and Kappeler, 2013), while the smallest species show a system of dispersed serial polyandry (Huchard *et al.*, 2011). Cheirogaleid polyandry and monogamy have puzzled socioecologists, because they cannot be explained by paternal care which is absent in the lepilemurid–cheirogaleid clade. Mouse lemurs have two or three mating periods per year (Figure 16.1), but females are only receptive for one night per mating period (Eberle *et al.*, 2007). Females live in dispersed matriarchal groups, socialising mainly in their sleeping sites, and associated females are generally receptive on different nights. Further, they do not give birth during the same mating period, allowing parental care to be shared by associated females (Génin, 2008). Moreover, females give birth to 2–4 young per reproductive period, and may raise a second litter or replace a lost litter (Génin, 2008).

As discussed in detail above, male and female reproduction preparedness and seasonal fattening are largely desynchronised in mouse lemurs (Génin, 2007). Females and a few males fatten in April (typical pre-hibernation fattening), while most males only fatten in August (pre-mating fattening; Schmid and Kappeler, 1998; Génin, 2007). In captivity, no sex-specific difference is observed if males and females are housed separately, and fattening is not inhibited by testes recrudescence (Génin *et al.*, 2005). This desynchronisation is explained by social factors: males fatten when most females are still hibernating and their territorial defence is relaxed (Génin *et al.*, 2005; Génin, 2007). The delayed fattening and the nomadic behaviour of most males during the dry season are probably paedomorphic traits corresponding to a prolonged phase of dispersal, resulting in a loss of male territoriality, which is only observed in a few old resident males (Génin, 2007).

Female mouse lemurs typically mate with many males during their night of receptivity (up to 19 in *M. griseorufus*, pers. obs.). Polyandry in mouse lemurs has been interpreted in terms of sperm competition, but sperm competition is logically an effect rather than the cause of polyandry, and tests have failed to confirm several predictions of the theory (Aslam *et al.*, 2002; Eberle *et al.*, 2007). A more likely explanation is a mechanism of induced ovulation, i.e. induced by repeated copulations. Eberle *et al.* (2007) observed that the fathers of most young were those males that had copulated about 40–60 min after the first copulation with another male. Although these authors interpret their results in terms of “optimal timing”, induced ovulation is further suggested by prolonged copulation, the presence of penile spines, the formation of vaginal plugs, and the low motility of sperm (Aslam *et al.*, 2002). Induced ovulation, also observed in shrews, voles, rabbits, cats, and ferrets (Bakker *et al.*, 2001), explains the secondary evolution of polyestry and multiple ovulations in mouse lemurs. In captive grey mouse lemurs, female reproductive success is higher in the presence of several males (Perret, pers. comm.), suggesting a system of induced ovulation.

Conclusion

The superficial similarities of mouse lemurs with other small, nocturnal primates are more likely to result from ecological convergence than from primitive retention. Because such ecological conditions arose recently (13–2 Ma), this suite of adaptations to hypervariability is unlikely to have any equivalent in deep primate ancestry (Génin and Masters, 2011; Masters *et al.*, 2014). Our preferred interpretation follows the reasoning of Gould (1977), who proposed that decreased body size may occur as a by-product of ontogenetic acceleration (progenesis and true acceleration) directly entrained by hypervariable conditions. We suggest that acceleration was a response to increased seasonality in the Miocene (< 13 Ma), and progenesis was driven by the unpredictability that became a permanent fixture of the early Pliocene (5 Ma), resulting in Madagascar's present hypervariable environment (Bobe, 2006; Fedorov *et al.*, 2010).

Acknowledgements

We thank Shawn Lehman, Ute Radespiel and Elke Zimmermann for inviting us to contribute to this book, and for their patience. We also thank Peter Kappeler for welcoming us to Kirindy, and FG thanks Oliver Schülke and Kathrin Dausmann for allowing him to measure individuals of *Phaner pallescens* and *Cheirogaleus medius* they captured in 1999. This is A.P.I.E.S. publication no. 5.

References

Aslam H, Schneiders A, Perret M, Weinbauer GF, Hodges JK. 2002. Quantitative assessment of testicular germ cell production and kinematic and morphological parameters of ejaculated spermatozoa in the grey mouse lemur, *Microcebus murinus*. *Reproduction* 123:323–332.

Bakker J, Kelliher KR, Baum MJ. 2001. Mating induces gonadotropin-releasing hormone neuronal activation in anosmic female ferrets. *Biology of Reproduction* 64:1100–1105.

Bartness TJ. 1996. Photoperiod, sex, gonadal steroids, and housing density affect body fat in hamsters. *Physiology and Behavior* 60:517–529.

Blanco MB, Rahalinarivo V. 2010. First direct evidence of hibernation in an eastern dwarf lemur species (*Cheirogaleus crossleyi*) from the high-altitude forest of Tsinjoarivo, central-eastern Madagascar. *Naturwissenschaften* 97:945–950.

Bobe R. 2006. The evolution of arid ecosystems in eastern Africa. *Journal of Arid Environments* 66:564–584.

Brown JH, Marquet PA, Taper, ML. 1993. Evolution of body size: consequences of an energetic definition of fitness. *The American Naturalist* 142(4):573–584.

Canale CI, Perret M, Théry, HP. 2010. Physiological flexibility and acclimation to food shortage in a heterothermic primate. *Journal of Experimental Biology* 214:551–560.

Carlquist S. 1974. *Island Biology*. Columbia University Press, New York.

Cartmill M. 1974. Rethinking primate origins. *Science* 184: 436–443.

Charles-Dominique P, Martin RD. 1970. Evolution of lemurs and lorises. *Nature* 227:257–260.

Chatterjee HJ, Ho SYW, Barnes I, Groves C. 2009. Estimating the phylogeny and divergence times of primates using a supermatrix approach. *BMC Evolutionary Biology* 9:259 doi: 10.1186/1471-2148-9-259.

Daniels HL. 1984. Oxygen consumption in *Lemur fulvus*: deviation from the ideal model. *Journal of Mammalogy* 65:584–592.

Dausmann KH, Glos J, Ganzhorn JU, Heldmaier G. 2004. Hibernation in a tropical primate. *Nature* 429:825–826.

Dewar RE, Richard AF. 2007. Evolution in the hypervariable environment of Madagascar. *Proceedings of the National Academy of Sciences of the United States of America* 104:13723–13727.

Dröscher I, Kappeler PM. 2013. Defining the low end of primate social complexity: the social organization of the nocturnal white-footed sportive lemur (*Lepilemur leucopus*). *International Journal of Primatology* doi: 10.1007/s10764-013-9735-3.

Dutrillaux B. 1979. Chromosomal evolution in primates: tentative phylogeny from *Microcebus murinus* (prosimian) to man. *Human Genetics* 48:251–314.

Eberle M, Perret M, Kappeler PM. 2007. Sperm competition and optimal timing of matings in *Microcebus murinus*. *International Journal of Primatology* 28:1267–1278.

Fabre P-H, Rodrigues A, Douzery EJP. 2009. Patterns of macroevolution among Primates inferred from a supermatrix of mitochondrial and nuclear DNA. *Molecular Phylogenetics and Evolution* 53:808–825.

Fedorov AV, Brierley CM, Emanuel K. 2010. Tropical cyclones and permanent El Niño in the early Pliocene epoch. *Nature* 463:1066–1070.

Geiser F, Ruf T. 1995. Hibernation versus daily torpor in mammals and birds: physiological variables and classification of torpor patterns. *Physiological Zoology* 68:935–966.

Génin F. 2007. Energy-dependent plasticity of grey mouse lemur social systems: lessons from field and captive studies. *Revue d'Écologie (La Terre et la Vie)* 62:245–256.

Génin F. 2008. Life in unpredictable environments: first investigation of the natural history of *Microcebus griseorufus*. *International Journal of Primatology* 29:303–321.

Génin F. 2013. Venus in fur: female power in mouse lemurs *Microcebus murinus* and *M. griseorufus*. In JC Masters, M Gamba, F Génin (eds.), *Leaping Ahead: Advances in Prosimian Biology* (pp. 121–126). Springer, New York.

Génin F, Masters JC. 2011. Le mythe du microcèbe primitif. *Revue de Primatologie* 3:1–17.

Génin, F, Perret M. 2000. Photoperiod-induced changes in energy balance in gray mouse lemurs. *Physiology and Behavior* 71:315–321.

Génin F, Perret M. 2003. Daily hypothermia in captive grey mouse lemurs (*Microcebus murinus*): effects of photoperiod and food restriction. *Comparative Biochemistry and Physiology B* 136:71–81.

Génin F, Nibbelink M, Galand M, Perret M, Ambid L. 2003a. Brown fat and nonshivering thermogenesis in the gray mouse lemur (*Microcebus murinus*). *American of Journal of Physiology* 284:R811–R818.

Génin F, Schilling A, Claustrat B. 2003b. Melatonin and methimazole mimic short day-induced fattening in gray mouse lemurs. *Physiology and Behavior* 79:553–559.

Génin F, Schilling A, Perret M. 2005. Social inhibition on seasonal fattening in wild and captive grey mouse lemurs. *Physiology and Behavior* 86:185–194.

Génin F, Masters JC, Ganzhorn JU. 2010. Gummivory in cheirogaleids: primitive retention or adaptation to hypervariable environments? In AM Burrows, LT Nash (eds.), *The Evolution of Exudativory in Primates. Developments in Primatology: Progress and Prospects* (pp. 123–140). Springer, New York.

Genoud M, Martin RD, Glaser D. 1997. Rate of metabolism in the smallest simian primate, the pygmy marmoset (*Cebuella pygmaea*). *American Journal of Primatology* 41:229–245.

Godfrey LR, Jungers WL, Reed KE, Simons EL, Chatrath PS. 1997. Subfossil lemurs. In SM Goodman, BD Patterson (eds.), *Natural Change and Human Impact in Madagascar* (pp. 218–256). Smithsonian Institution Press, Washington, DC.

Godinot M. 2006. Lemuriform origins as viewed from the fossil record. *Folia Primatologica* 77:446–464.

Gould SJ. 1977. *Ontogeny and Phylogeny*. Belknap Press, Cambridge, MA.

Grow NB, Wirdateti, Nekaris KAI. 2015. Does toxic defence in *Nycticebus* spp. relate to ectoparasites? The lethal effects of slow loris venom on arthropods. *Toxicon* 95:1–5.

Harcourt AH. 2008. Are lemurs' low basal metabolic rates an adaptation to Madagascar's unpredictable climate? *Primates* 49:292–294.

Hildwein G, Goffart M. 1975. Standard metabolism and thermoregulation in a prosimian *Perodicticus potto*. *Comparative Biochemistry and Physiology A* 50:201–213.

Hill WCO. 1953. *Primates: Comparative Anatomy and Taxonomy. I – Strepsirhini.* Edinburgh University Press, Edinburgh.

Hladik CM, Charles-Dominique P, Valdebouze P, Delort-Laval J, Flanzy J. 1971. La caecotrophie chez un Primate phyllophage du genre *Lepilemur* et les corrélations avec les particularités de son appareil digestif. *Compte Rendus de l'Académie des Sciences de Paris* 272:3191–3194.

Hladik CM, Charles-Dominique P, Petter JJ. 1980. Feeding strategies of five nocturnal prosimians in the dry forest of the west coast of Madagascar. In P Charles-Dominique, HM Cooper, A Hladik, *et al.* (eds.), *Nocturnal Malagasy Primates: Ecology, Physiology, and Behavior* (pp. 41–73). Academic Press, New York.

Horvath JE, Weisrock DW, Embry SL, *et al.* 2008. Development and application of a phylogenomic toolkit: resolving the evolutionary history of Madagascar's lemurs. *Genome Research* 18: 489–499.

Hrdy SB. 1977. *The Langurs of Abu: Female and Male Strategies of Reproduction.* Harvard University Press, Cambridge, MA.

Huchard E, Canale CI, Le Gros C, *et al.* 2011. Convenience polyandry or convenience polygyny? Costly sex under female control in a promiscuous primate. *Proceedings of the Royal Society B* 279:1371–1379.

Jolly A. 1984. The puzzle of female feeding priority. In M Small (ed.), *Female Primates: Studies by Women Primatologists* (pp. 197–216). Alan R Liss, New York.

Kappeler PM. 1998. Nests, tree holes, and the evolution of primate life histories. *American Journal of Primatology* 46:7–33.

Kappeler PM. 2000. Lemur origins: rafting by groups of hibernators? *Folia Primatologica* 71:422–425.

Kayser C. 1962. *The Physiology of Natural Hibernation.* Pergamon Press, New York.

Körtner G, Geiser F. 2000. Torpor and activity patterns in free-ranging sugar gliders *Petaurus breviceps* (Marsupiala). *Oecologia* 123:350–357.

Lomolino MV. 1985. Body size of mammals on islands: the Island Rule reexamined. *The American Naturalist* 125:310–316.

Lomolino MV. 2005. Body size evolution in insular vertebrates: generality of the island rule. *Journal of Biogeography* 32:1683–1699.

Lovegrove BG. 2001. The evolution of body armor in mammals: plantigrade constraints of large body size. *Evolution* 55:1464–1473.

Lovegrove BG 2004. Locomotor mode, maximum running speed and basal metabolic rate in placental mammals. *Physiology and Biochemical Zoology* 77:916–928.

Lovegrove BG 2012. The evolution of endothermy in Cenozoic mammals: a plesiomorphic–apomorphic continuum. *Biological Reviews* 87(1):128–162.

Lovegrove BG, Génin F. 2007. Torpor and hibernation in a basal placental mammal, the lesser hedgehog tenrec *Echinops telfairi. Journal of Comparative Physiology B* 178:691–698.

Marivaux L, Welcomme J-L, Antoine P-O, *et al.* 2001. A fossil lemur from the Oligocene of Pakistan. *Science* 294:587–591.

Martin RD. 1972. Adaptive radiation and behaviour of the Malagasy lemurs. *Philosophical Transactions of the Royal Society of London B* 264:295–352.

Masters JC, Lovegrove BG, de Wit MJ. 2007. Eyes wide shut: can hypometabolism really explain the primate colonization of Madagascar? *Journal of Biogeography* 34:21–37.

Masters JC, Silvestro D, Génin F, DelPero M. 2013. Seeing the wood through the trees: the current state of higher systematics in the Strepsirhini. *Folia Primatologica* 84:201–219.

Masters JC, Génin F, Silvestro D, Lister A, DelPero M. 2014. The red island and the seven dwarfs: body size reduction in Cheirogaleidae. *Journal of Biogeography* 41:1833-1847.

McCormick SA. 1980. Oxygen consumption and torpor in the fat-tailed dwarf lemur (*Cheirogaleus medius*): rethinking prosimian metabolism. *Comparative Biochemistry and Physiology A* 68:605–610.

McNab BK. 1986. The influence of food habits on the energetics of eutherian mammals. *Ecological Monographs* 56:1–19.
McNab BK. 2002. *The Physiological Ecology of Vertebrates. A View from Energetics.* Cornell University Press, Ithaca.
Müller EF. 1979. Energy metabolism, thermoregulation and water budget in the slow loris (*Nycticebus coucang*, Boddaert 1785). *Comparative Biochemistry and Physiology A* 64:109–119.
Müller EF, Jaksche H. 1980. Thermoregulation, oxygen consumption, heart rate and evaporative water loss in the thick-tailed bushbaby (*Galago crassicaudatus*, Geoffroy, 1812). *Zeitschrift für Säugetierkunde* 45:269–278.
Nicholls DG, Locke RM. 1984. Thermogenic mechanisms in brown fat. *Physiological Reviews* 64:1–64.
Perelman P, Johnson WE, Roos C, *et al.* 2011. A molecular phylogeny of living primates. *PLoS Genetics* 7:e11342.
Perret M, Aujard F. 2001. Regulation by photoperiod of seasonal changes in body mass and reproductive function in gray mouse lemurs (*Microcebus murinus*): differential responses by sex. *International Journal of Primatology* 22:5–24.
Petry VH, Riehl I, Zucker H. 1986. Energieumsatzmessungen an weissbüscheläffchen (*Callithrix jacchus*). *Journal of Animal Physiology and Animal Nutrition* 55:214–224.
Richard AF, Nicoll ME. 1987. Female social dominance and basal metabolism in a Malagasy primate, *Propithecus verreauxi. American Journal of Primatology* 12:309–314.
Roos C, Schmitz J, Zischler H. 2004. Primate jumping genes elucidate strepsirrhine phylogeny. *Proceedings of the National Academy of Sciences of the United States of America* 101:10650–10654.
Rothwell NJ, Stock MJ. 1985. Thermogenic capacity and brown adipose tissue activity in the common marmoset. *Comparative Biochemistry and Physiology A* 81:683–686.
Schmid J. 2000. Daily torpor in the gray mouse lemur (*Microcebus murinus*) in Madagascar: energetic consequences and biological significance. *Oecologia* 123:175–183.
Schmid J, Kappeler PM. 1998. Fluctuating sexual dimorphism and differential hibernation by sex in a primate, the grey mouse lemur (*Microcebus murinus*). *Behavioral Ecology and Sociobiology* 43:125–132.
Schmid J, Ruf T, Heldmaier G. 2000. Metabolism and temperature regulation during daily torpor in the smallest primate, the pygmy mouse lemur (*Microcebus myoxinus*) in Madagascar. *Journal of Comparative Physiology B* 170:59–68.
Schülke O. 2005. Evolution of pair-living in *Phaner furcifer. International Journal of Primatology* 26:903–919.
Simmen B, Bayart F, Rasamimanana H, Zahariev A, Blanc S, Pasquet P. 2010. Total energy expenditure and body composition in two free-living sympatric lemurs. *PloS ONE* 5:e9860 doi: 10.1371/journal.pone.0009860.
Springer MS, Meredith RW, Gatesy J, *et al.* 2012. Macroevolutionary dynamics and historical biogeography of primate diversification inferred from a species supermatrix. *PLoS ONE* 7:e49521.
Thompson SD, Power ML, Rutledge, CE, Kleiman DG. 1994. Energy metabolism and thermoregulation in the golden lion tamarin (*Leontopithecus rosalia*). *Folia Primatologica* 63:131–143.
Tilden DC, Oftedal OT. 1995. The bioenergetics of reproduction in prosimian primates: is it related to female dominance? In L Alterman, GA Doyle, MK Izard (eds.), *Creatures of the Dark: The Nocturnal Prosimians* (pp. 119–131). Plenum Publishing Corporation, New York.
Van Loan M, Mayclin P. 1987. A new TOBEC instrument and procedure for the assessment of body composition: use of the Fourier coefficients to predict lean body mass and total body water. *American Journal of Clinical Nutrition* 45:131–137.
Wallace AR. 1892. *Island Life.* Macmillan and Co., London.

Weston EM, Lister AM. 2009. Insular dwarfism in hippos and a model for brain size reduction in *Homo floresiensis*. *Nature* 459:85–88.

Whittaker RJ, Fernández-Palacios JM. 2007. *Island Biogeography: Ecology, Evolution, and Conservation*, 2nd edition. Oxford University Press, Oxford.

Wright PC. 1999. Lemur traits and Madagascar ecology: coping with an island environment. *Yearbook of Physical Anthropology* 42:31–72.

Wright PC, Martin LB. 1995. Predation, pollination and torpor in two nocturnal prosimians: *Cheirogaleus major* and *Microcebus rufus* in the rainforest of Madagascar. In L Alterman, GA Doyle, MK Izard (eds.), *Creatures of the Dark: The Nocturnal Prosimians* (pp. 45–60). Plenum Press, New York.

Yoder AD. 1996. The use of phylogeny for reconstructing lemuriform biogeography. In WR Lourenço (ed.), *Biogéographie de Madagascar* (pp. 245–258). Editions de l'ORSTOM, Paris.

Young AL, Richard AF, Aiello LC. 1990. Female dominance and maternal investment in strepsirrhine primates. *The American Naturalist* 135:473–488.

17 Possible causes and consequences of different hibernation patterns in *Cheirogaleus* species – Mitovy fatsy sahala

Kathrin H. Dausmann and Marina B. Blanco

Heterothermy in lemurs – flexibility is key

All dwarf lemurs (*Cheirogaleus*) of Madagascar are obligate hibernators in their natural environments (Petter, 1962, 1978; Petter *et al.*, 1977; Hladik *et al.*, 1980; Petter-Rousseaux *et al.*, 1980; Wright and Martin, 1995; Dausmann, 2008, 2014). Hibernation is an actively suppressed metabolic state of endothermic species to overcome predictable, usually seasonal, metabolic energy and water crises. It is typified by dramatically reduced bodily functions and expressed by lowered metabolic rates and core body temperature (T_b), prolonged pauses between breaths, and lowered heart rates (Carey *et al.*, 2003; Geiser, 2004; Heldmaier *et al.*, 2004). This strategy is especially advantageous for small species, due to the allometric relationship between metabolism and body size (higher mass-specific metabolic rates and greater heat losses over their relatively larger surface areas; Heldmaier *et al.*, 2004; White and Seymour, 2005). The energy needed to sustain vital functions during hibernation is stored as fat deposits or food caches at the end of the active season. Although the ultimate aim – decreasing energy and water demands drastically by using hibernation to survive the harsh conditions of the Malagasy winter – is the same for all *Cheirogaleus* species, proximate *modi operandi* differ notably between species inhabiting different habitats.

Biomes in Madagascar can roughly be divided between western dry forests and eastern rainforests (Dufils, 2003). Degree of seasonality (i.e., changes in occurrence and intensity of rainfall, and variation in ambient temperature, T_a) and resource predictability differ markedly between these two regions due to a combination of geographic history (mountains as physical barriers running north to south along central Madagascar), wind systems, and oceanic currents (Donque, 1972; Dufils, 2003; Jury, 2003). In addition, altitudinal gradients also influence environmental conditions and are important in explaining geographic distributions of certain

The Dwarf and Mouse Lemurs of Madagascar: Biology, Behavior and Conservation Biogeography of the Cheirogaleidae, eds. S.M. Lehman, U. Radespiel, E. Zimmermann. Published by Cambridge University Press. © Cambridge University Press 2016.

taxa (Raxworthy and Nussbaum, 1994). These differences in environmental conditions not only shape Malagasy taxonomy, but also influence species' physiological processes.

In lemurs, heterothermy has so far only been confirmed in the Cheirogaleidae, a family of rather small, omnivorous, night-active species. Within this family, individuals have proven flexible in regard to their use and patterns of heterothermy, even within populations inhabiting the same forest areas, or within individuals between different years, depending on environmental seasonality (mainly changes in temperature and precipitation), unpredictability of environmental factors, and individual body condition (Dausmann, 2008, 2014). In several *Microcebus* species, individuals can alternate between occasional torpor, daily torpor, prolonged torpor, and months-long hibernation both within and between winter seasons (Schmid, 1999; Kobbe and Dausmann, 2009; Schmid and Ganzhorn, 2009; Kobbe *et al.*, 2011). Yet, there are distinct differences between the general constraints linked with specific habitats for all species occupying them (Dufils, 2003). Because the Cheirogaleidae can adjust their thermoregulatory behavior flexibly to current ambient and bodily conditions, this enables them to inhabit a wide variety of habitats, including some that are not only seasonal, but also very unpredictable.

In contrast to this flexible deployment of heterothermy in *Microcebus* species, all *Cheirogaleus* species enter continuous, months-long hibernation during the winter (Dausmann, 2008, 2014). This strategy requires obligate preparation before the dry season (winter); for example, building up fat stores to sustain hibernation or modification of the digestive and reproductive system. Additionally, it relies on sufficient availability of resources for recovery when hibernation is terminated and reproduction needs to commence promptly. Thus, it can only be expressed in predictable habitats or when body condition is sufficient to sustain hibernation. Nevertheless, hibernation is expressed in very different habitats by the family Cheirogaleidae, from the western dry and spiny forests to the eastern rainforests.

In this context, we want to discuss differences and commonalities, and the underlying constraints, between the hibernation of western and eastern/highland *Cheirogaleus* species. Does hibernaculum choice vary with environmental conditions and how does this affect hibernation patterns? Which could be the ancestral *modus operandi*? Could one physiological adaptation have served as a preadaptation to colonize another habitat?

Study sites

All data discussed in this paper were collected in the field in Madagascar. *C. medius* were studied in the Kirindy forest/CNFEREF (Centre National de Formation, d'Etudes et de Recherche en Environnement et Foresterie Morondava; 20°03′S, 44°39′E, 40 m), a dry deciduous forest near the west coast of Madagascar; *C. crossleyi* and *C. sibreei* in a forest fragment at Tsinjoarivo, a high-altitude rainforest in central-eastern Madagascar (19°41′S, 47°46′E, 1660 m). The Kirindy forest is characterized by a marked dry season during the austral winter with virtually no precipitation from

Table 17.1 Climatic characteristics (mean ± SD) of different *Cheirogaleus* habitats in Madagascar (coldest months (lowest mean $T_{a\ min}$) differ between years: Ranomafana and Tsinjoarivo: June–August, Kirindy: June and July).

Climatic variables	Eastern montane rainforest (Ranomafana) Data from 5 years (2004–2008)	High-altitude rainforest (Tsinjoarivo) Data from 3 years (2011–2013)	Dry forest (Kirindy CNFEREF forest) Data from 3 years (1999–2001)
Mean $T_{a\ min}$ during the coldest months (°C)	11.8 ± 1.6	9.1 ± 0.8	11.1 ± 2.6
$T_{a\ min}$ absolute (°C)	9.5	1.0	5.4
Mean $T_{a\ max}$ during the coldest months (°C)	17.4 ± 0.9	16.8 ± 2.7	30.9 ± 1.5
Approximate rainfall per rainy season (mm)	2000	1500	1000
Months without rain (≤ 3 mm)	0	0	6.7

April through October. The coldest months are June and July, with temperatures dropping below 10°C at night, but always rising to over 28°C, usually over 30°C, during the day, resulting in strong diurnal fluctuations in temperature (Table 17.1; Dausmann *et al.*, 2005). During the dry winter months, there is no standing water available. Tsinjoarivo is one of the coldest environments in Madagascar, with temperatures dropping below 0°C on occasion during the winter, and always remaining well below 30°C during June and July (mean daily maximal temperature during these months is around 17°C (Table 17.1; Blanco and Rahalinarivo, 2010). Rainfall occurs throughout the year, although it is less pronounced during winter.

Trapping

Lemurs were caught in Sherman and Tomahawk live traps, baited with banana pieces. Trapped individuals were sexed, measured, weighed (300 g and 600 g spring balance, Pesola, Switzerland), and individually marked with subcutaneously injected passive identification transponders (Trovan, EURO I.D. Usling GmbH, Weilerswist, Germany; AVID Identification Systems, Inc., CA, USA). Tail circumference was measured with a string at the fattest part of the tail. All animals were released in the late afternoon on the day of capture at the capture site.

Temperature measurements

Skin temperature of the individuals during the hibernation period was recorded hourly using temperature-sensitive collar transmitters that change pulse rate with temperature (*C. medius*; 4 g TW-4 button cell tags, Biotrack Ltd, Wareham, Dorset, UK) or collar temperature loggers (*C. crossleyi* and *C. sibreei*; 10 g ARC 400, Advanced Telemetry Systems, Isanti, MN, USA). These methods are accurate for assessing T_b, especially during daily sleeping phases and hibernation, when the animals curl up and cover the transmitter firmly inside (Dausmann, 2005). Wherever possible, relevant T_a and temperature of the hibernaculum were measured together with skin temperatures, using temperature loggers (TinyTalks, Orion Components, Chichester, UK; thermochron i-buttons, Dallas Semiconductor, USA). For temperature of the hibernaculum, the logger was placed inside the hibernaculum, about 10 cm from the individual; for T_a, the logger was placed at 2 m height on the shady side of the tree.

Measurements of metabolic rate

We measured energy expenditure/metabolic rate (MR) as rate of oxygen consumption (VO_2) of single individuals with open-flow respirometry. The hibernacula served as metabolic chambers, with the entrances serving as air inlets. In the case of underground hibernacula, individuals were placed in plastic containers at their hibernation site for the duration of the measurements. Portable gas analyzers (MagicBox, designed and constructed by G. Heldmaier and T. Ruf, Philipps-University Marburg, and OxBox, designed and constructed by T. Ruf and T. Paumann, FIWI, University of Veterinary Medicine Vienna) were connected to the relevant hibernaculum by an airtight tube (Tygon®, Saint-Gobain Performance Plastics, France). The air was drawn at a flow rate of 50–80 l/h from the hibernacula (depending on size) and filtered from particles and dried with silica gel orange (Carl Roth®, Germany) before entering the gas analyzer. Oxygen content was determined every minute. To control for any drift of the oxygen sensor, we sampled the reference air (baseline) of the ambient environment outside the hibernacula for five minutes every hour and adjusted values accordingly. VO_2 (ml O_2/h) was calculated with the equation given by Heldmaier and Ruf (1992) as: VO_2 = 10 flow (l/h) Δvol% O_2 and corrected for the CO_2 effect. Daily energy expenditure (kJ/day or watt) was calculated assuming an average respiratory quotient (RQ) of 0.85 (combustion of 50% fat and 50% carbohydrates) during the activity period, and 0.7 (mainly combustion of fat) during hibernation. For these specifications, the caloric equivalent, which gives the amount of energy released per liter O_2 consumption, equals 20.37 and 19.7 kJ/l O_2, respectively.

For details regarding study sites, trapping, and measurements of temperature and metabolic rate, see Dausmann *et al.* (2005) and Blanco *et al.* (2013).

Different habitats, different hibernacula

Of the six currently recognized *Cheirogaleus* species (Thiele *et al.*, 2013), only three have been studied in regard to their thermoregulatory physiology so far: *C. medius*, *C. crossleyi*, and *C. sibreei*. *C. medius* live in the dry forests along the west coast of Madagascar; *C. crossleyi* occur in eastern rainforests from lowlands to the highlands running north–south along the center of the island (Thiele *et al.*, 2013). *C. sibreei* are only found in a few of the remaining high-altitude forest patches (above 1400 m) on the highlands, sometimes in sympatry with *C. crossleyi* (Blanco *et al.*, 2009). Although

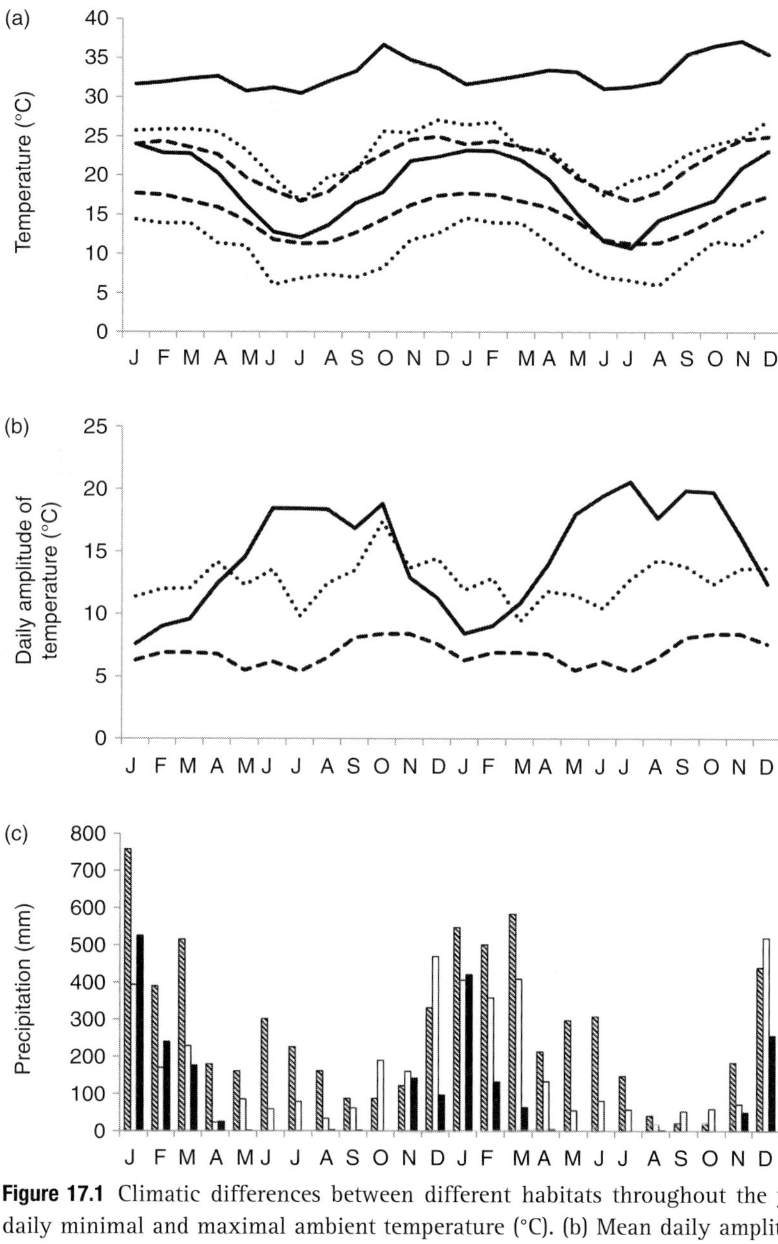

Figure 17.1 Climatic differences between different habitats throughout the years. (a) Mean daily minimal and maximal ambient temperature (°C). (b) Mean daily amplitude of ambient temperature (°C). (c) Precipitation (mm). Dry forest (Kirindy CNFEREF forest): solid lines, black bars; eastern montane rainforest (Ranomafana): dashed lines, shaded bars; high-altitude rainforest (Tsinjoarivo): dotted lines, white bars; letters indicate the months.

all of these habitats obviously share general characteristics of tropical/subtropical areas, there are distinct differences between them (Figure 17.1; Table 17.1). The western dry forests experience a long, pronounced dry season during the austral winter with no precipitation for 5–8 months. In contrast, rainfall occurs throughout the year

Table 17.2 Changes in body parameters (mean ± SD) over the hibernation season and savings of daily energy expenditure (DEE) during hibernation in *Cheirogaleus* species.

	C. medius	*n*	*C. sibreei*	*n*	*C. crossleyi*	*n*
Body mass (g)						
before hibernation	232.1 ± 20.2	25	385.7 ± 70.7	7	429.0 + 51.8	13
after hibernation	130.5 ± 16.2	95	282.8+38.2	12	351.6 + 28.0	8
Loss in body mass	44%		26.7%		28.5%	
Tail circumference (mm)						
before hibernation	65 ± 7	15	52.1 ± 10.4	7	55.4 ± 6.5	13
after hibernation	38 ± 4	59	40.5 ± 7.3	12	52.1 ± 10.4	8
Loss in tail circumference	42%		20.3%		6%	
DEE savings during hibernation (%)	72	11	89		86	

in the eastern rainforests and the highlands, although it is more pronounced during the austral summer (December–March). Temperatures during the coldest months (June–August) are lowest in the highlands (sometimes dropping below 0°C at night), intermediate in the dry forests and highest in the lowland rainforests (although still cool at night during winter: < 15°C). These climatic differences are reflected in food availability. Flowers, fruits, and insects are most abundant during the summer in all habitats, but scarcer during winter, with decreasing availability from the lowland and montane rainforest, to the high-altitude rainforest, to the dry forests. In the last, there is also a crucial lack of standing water during the dry season. All *Cheirogaleus* species store fat in their tails before the winter to fuel hibernation (in *C. medius* almost doubling their body masses; Table 17.2), and do not build up food caches (Fietz and Ganzhorn, 1999; Fietz and Dausmann, 2006).

During the active season all *Cheirogaleus* species spend the regular daily sleeping phases in tree hollows or more or less elaborate nests on tree forks (Fietz, 1999; Lahann, 2007; Dausmann, 2013; Blanco and Godfrey, 2014). When the hibernation season approaches, however, the species occupying highland rainforests retreat about 10–20 cm below ground, whereas in dry forests individuals remain in tree hollows of varying insulation capacities for the entire hibernation phase (Petter *et al.*, 1977; Dausmann *et al.*, 2004; Blanco *et al.*, 2013).

Hibernaculum choice determines hibernation patterns

These strikingly different strategies of hibernaculum choice have extensive consequences on hibernation parameters and patterns. During hibernation bouts, T_b is not actively maintained, but mainly exogenously determined (Dausmann *et al.*, 2004, 2005). Dwarf lemurs hibernating underground, or in very large, thick-walled trees, thus show stable T_b, due to the substantial insulation capacities of their hibernacula (Figure 17.2a, c). In the dry forests (tree hollows), this level is comparatively high, at about 22–25°C (Dausmann *et al.*, 2005), whereas in the highlands (underground), it is much lower, around 12–14°C (Blanco and Rahalinarivo, 2010; Blanco *et al.*,

2013). However, this is still not nearly as low as in temperate underground hibernators, such as marmots, ground squirrels, or dormice that hibernate at T_b of around 0°C (Barnes, 1989; Arnold *et al.*, 1991; Heldmaier and Ruf, 1992). T_b of dwarf lemurs hibernating in less-insulated tree hollows, on the other hand, tracks T_a to a lesser or greater extent, depending on the specific insulation properties, in some cases mirroring daily fluctuations of up to 25°C (Dausmann *et al.*, 2005; Figure 17.2d).

The choice of hibernacula not only influences the pattern of T_b during hibernation bouts, but the patterns of the hibernation bouts themselves are also strongly affected. All temperate and Arctic hibernators have to enter spontaneous arousals and initiate active heat production periodically (every 1–2 weeks; Willis, 1982; Wang, 1989), with the notable exception of black bears, which hibernate at significantly higher T_b (above 30°C; Tøien *et al.*, 2011). Around 70–80% of all the energy needed during the hibernation period is utilized during the periodic arousals (Wang, 1989). The ultimate factors necessitating these energy-demanding phases of euthermy remain debated, but they seem to be indispensable to maintain vital body functions in hibernators maintaining continuous low T_b (Humphries *et al.*, 2003). However, *C. medius* individuals hibernating in poorly insulated tree hollows, where T_b is raised above 30°C passively at least occasionally, dismiss the regular active arousals and conform T_b without displaying active thermogenesis over several months. Indeed, together with the black bear, *C. medius* and some *Microcebus* species (and possibly Malagasy tenrecs) are the only hibernators known to be able to abandon the necessity of arousals during hibernation and stay in a hypometabolic state for many months (even at $T_a > 35$°C). Obviously, the maximum level of T_b reached during hibernation is a crucial factor regulating arousals, regardless of whether it is reached actively or passively. Thus, the choice of hibernaculum determines the physiological realization of hibernation. The choice of tree hollow, however, does not seem to be ultimately decisive in terms of energy expenditure. Higher metabolic rates maintained by individuals being passively heated during the hot midday hours inside poorly insulated tree hollows are counterbalanced by lower metabolic rates during the cool hours of the night (Dausmann *et al.*, 2009). However, T_a in the highlands does not exceed 25°C during the winter days (Table 17.1), preventing passive heating to levels that render arousals dispensable and thus producing a "classical" torpor bout–arousal pattern. Frequency of arousals, on the other hand, seems to be dependent on T_a. Whereas they occur on average weekly in *C. medius* in the dry forests (Dausmann *et al.*, 2005), intervals are distinctly longer in the colder highlands (7–14 days in *C. sibreei*; Blanco *et al.*, 2013; Blanco and Godfrey, 2014). This agrees with temperate and Arctic hibernators that also show increasing hibernation bout lengths with decreasing T_a (Carey *et al.*, 2003). All three species change hibernacula occasionally during the hibernation season; thus, there seems to be an active choice that can be adjusted (but probably more for social than for physiological reasons, or following a predator attack; Dausmann and Glos, 2015).

Observations of *C. crossleyi* undergoing daily torpor bouts inside nest structures at the beginning and at the end (Figure 17.2b) of the winter season suggest the possibility that they may, in fact, hibernate in non-underground settings (e.g., tree

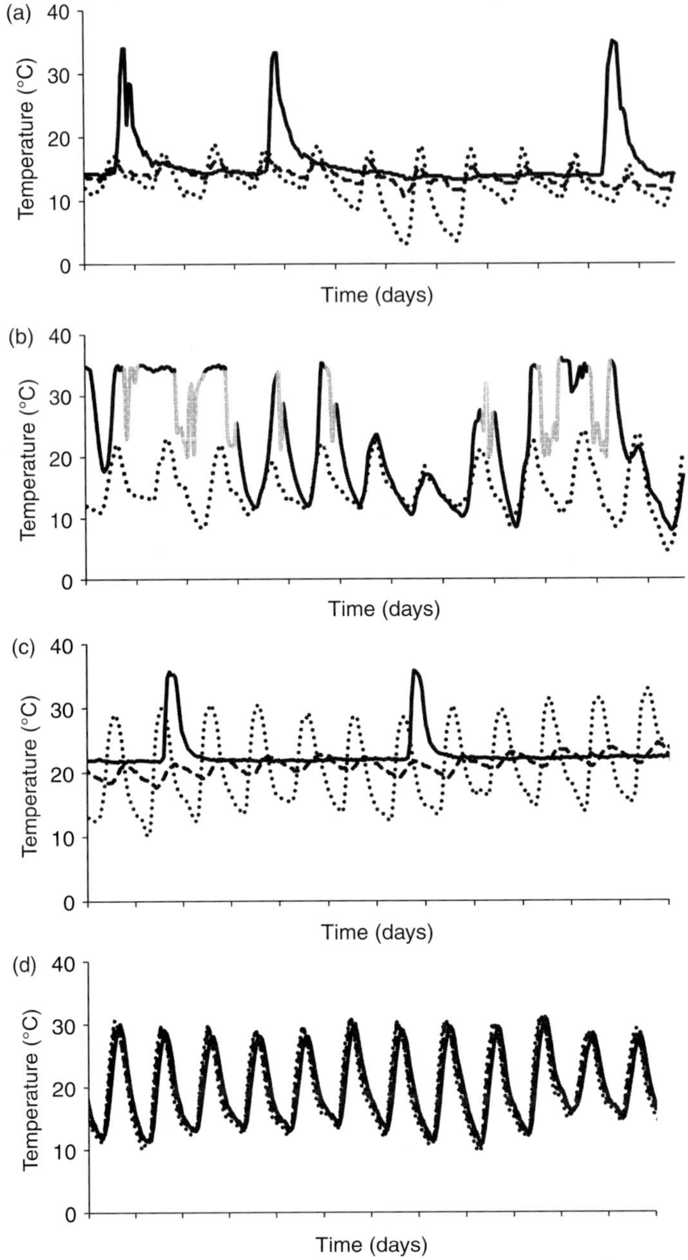

Figure 17.2 Hibernation patterns of *Cheirogaleus crossleyi* and *C. medius*. (a) *C. crossleyi* during the core of the hibernation period (June/July). (b) *C. crossleyi* soon after emergence from underground hibernaculum (end of August). Note important insulation properties of soil vs. nest; (c) *C. medius* hibernating in a well-insulated tree hollow. (d) *C. medius* hibernating in a poorly insulated tree hole. Skin temperature, soil/tree hollow temperature, ambient temperature; gray indicates phases where the animals were actively moving and thus values deviate substantially from true skin temperature.

hollows, nests) when inhabiting lower-altitude, warmer habitats where there is no risk of freezing. The fact that no *C. sibreei* to date has been reported to undergo torpor bouts in tree hollows or nests points to high reliance of underground hibernation in this species, possibly related to their geographic restriction to high, cold environments.

Hibernation duration varies between species and habitats

The utilization of hibernation in lemurs, as in other mammalian hibernators, varies depending on T_a (which determines how much energy is required to maintain euthermy and how much energy can be saved under heterothermy) and/or the degree of resource seasonality and predictability (whether food and water sources are limited to short seasons consistently; Figure 17.3), and is affected by body mass (Geiser, 2004; Heldmaier *et al.*, 2004). According to expectations, dwarf lemurs living in more seasonal habitats with a longer dry season hibernate for longer time periods (6–8 months in *C. medius* compared to 4.5–7 months *C. sibreei* and 3.5–5.5 months in *C. crossleyi*; Blanco and Godfrey, 2014). However, surprisingly, one adult female *C. sibreei* started to hibernate at the end of February and two others at the beginning of March in the highland rainforest site, suggesting that the onset of the dry season *per se* clearly is not the exclusive factor determining the timing or duration of hibernation. It also demonstrates how flexible dwarf lemurs can be in expressing hibernation, depending on ambient and/or bodily conditions and constraints. In the more seasonal dry forest (*C. medius*), on the other hand, the

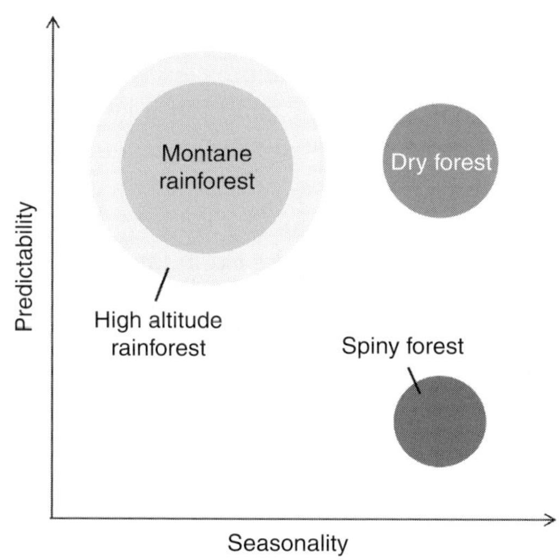

Figure 17.3 Hibernation hyperspace: schematic diagram of the profitability of hibernation in Malagasy Cheirogaleidae in relation to the seasonality and climatic predictability of their habitats. The darker the color of the circles, the warmer the habitat is during winter (mean ambient temperature, see Table 17.1; for spiny forest see Dausmann *et al.*, 2012); the larger the circles, the more profitable hibernation in this habitat is.

Table 17.3 Duration (D) of hibernation phase (mean ± SD) in days of sympatric *Cheirogaleus* species at Tsinjoarivo, a highland rainforest.

Species	*n*	Mean duration (days)	Max duration (days)	Min duration (days)	Wilcoxon *p*
Cheirogaleus crossleyi	10	124.3 ±35.7	182	70	
Cheirogaleus sibreei	8	171 ± 33.9	213	128	0.03

beginning and termination of hibernation is much more synchronized within 2–4 weeks, the exception being young from the previous rainy season, which may continue to be active and feed for another 4 weeks.

C. crossleyi and *C. sibreei* are also distinctly heavier than *C. medius* (Table 17.2). Due to the fact that mass-specific metabolic rate increases with decreasing body mass during euthermy but is comparable across species during hibernation, hibernation is energetically most profitable for small-bodied mammals, as long as they are able to store sufficient amounts of fat to survive several months without food intake, albeit at low metabolic rates. Thus, its significance decreases with increasing body mass (Geiser, 2004; Heldmaier *et al.*, 2004). This is consistent with decreasing hibernation periods from *C. medius* to *C. sibreei* and *C. crossleyi*, but could be superimposed by the climatic effects. However, as sympatric *C. sibreei* and *C. crossleyi* also have significantly different hibernation duration (with the smaller *C. sibreei* generally hibernating longer than *C. crossleyi*; Table 17.3), there clearly also are species-specific differences, independent from climatic variables. It had been suggested that the expression of tropical hibernation may be restricted by body mass to individuals that are around or below 200 g, because the energetic savings are lower than in temperate and Arctic hibernators, and more closely resemble temperate and Arctic daily torpor, which seems to be restricted to species of less than 200 g (Dausmann *et al.*, 2009). Although indeed none of the larger lemurs (> 1 kg) have been shown to employ torpor, it is a favorable strategy displayed by the *Cheirogaleus* species, regardless of weight. Although there are no detailed physiological studies, behavioral observations also suggest hibernation in *C. major* (around 350 g during the activity period; Lahann, 2007; Blanco, pers. obs.) and *C. lavasoensis* (around 300 g during the activity period; Thiele *et al.*, 2013). *C. minusculus* is only known from one museum specimen, and thus conclusions on physiology are not possible (and its status as a species is debated).

Within a species, individuals adjust hibernation characteristics (such as timing and duration of hibernation) in different habitats, depending on the length of the period of scarcity and climatic parameters. For example, in the western dry forests, individuals of *C. medius* hibernate for about 6–8 months, whereas only for 4–5 months in the much less seasonal eastern littoral rainforests (Lahann and Dausmann, 2011). However, the rainforest individuals generally have a shorter life expectancy due to higher mortality rates. They seem to compensate for this by higher reproductive rates (larger litter sizes, greater number of litters). Home ranges and social organization, on the other hand, do not vary between habitats (Lahann and Dausmann, 2011).

Energy saveaholics in the highlands

During hibernation bouts, energy savings increase with decreasing T_a, and thus T_b, because the rate of biochemical reactions decrease with decreasing temperature (Q_{10} temperature coefficient). As T_b approximates temperature in the hibernaculum during hibernation, hibernation in colder habitats, such as the highlands, should be energetically more advantageous, and, indeed, *C. medius* in the dry forest are able to save less energy by hibernation (total daily energy expenditure compared between the active and the hibernation period: 72%) and lose more weight (44.0%) than *C. crossleyi* (86%; 26.7%) and *C. sibreei* (89%; 28.5%) living in colder habitats and experiencing lower average T_b during hibernation (Figure 17.3, Table 17.2). This might be reflected in the pattern of reproduction: whereas *C. crossleyi* and *C. sibreei* produce offspring every year, *C. medius* in western dry deciduous forests only do so every second year, which is surprisingly infrequent for a small mammal. As the habitat of *C. medius* is the most challenging of the *Cheirogaleus* species, maybe the duration of hibernation needs to be extended in comparison to other *Cheirogaleus* species in order to emerge when environmental conditions have sufficiently increased again to secure a body condition sufficient to enable successful reproduction. Even then, regeneration, at least for the females, seems to take two years, unless prolongation of the time span between reproductive efforts is necessary to enable the offspring successful emigration and establishment in their own territories. This might also be a key to explain why hibernation is prevalent in some Cheirogaleidae, but not in others: if the interval between reproductive events cannot be shortened because of some external factor (such as limited availability of territories), it may be advantageous to hibernate (e.g., to escape predation), instead of remaining active.

Did the ancestral dwarf lemur hibernate underground or in tree hollows?

One assumption is that hibernation in lemurs evolved or was sustained (if the ancestral lemurs arriving in Madagascar already possessed this trait; Kappeler, 2000; Masters *et al.*, 2007; Nowack and Dausmann, 2015) in the highland habitats of Madagascar (Blanco and Godfrey, 2014). A retreat into the highlands during the Quaternary, a period of colder, drier climate in Madagascar (watershed hypothesis: Wilmé *et al.*, 2006), could possibly have promoted this process. As this environment becomes significantly cold during winter (below 0°C on occasion), hibernation in well-buffered underground hibernacula would be advantageous to reduce the risk of freezing in poorly insulated tree hollows, especially as thick large trees, such as baobabs (*Adansonia* spp.) and arofys (*Commiphora* spp.) are not available in this habitat. Even if temperatures below 0°C are only reached infrequently, this would be a substantial evolutionary force. However, when hibernating species spread to the dry forests of the west coast of Madagascar, burrowing below ground might not have been a possibility, given the hard and compacted soil, especially during the cool dry season, making the switch to hibernacula in tree holes necessary. Thus, the ability to hibernate in an ancestral *Cheirogaleus* (in the "classical" *modus operandi* of hibernation below ground, with stable, fairly low T_b for most of the time) could have been a preadaptation to cope with the highly seasonal habitat of the dry forests. Interestingly, the tree hollows used during the hibernation season are usually different from those used during the activity season. They are generally located in larger trees, providing more stable hibernation patterns (and maybe reducing predation risk; Dausmann, 2013), suggesting that more stable environments in the hibernacula are preferred, and possibly underground hibernacula, like in their temperate and Arctic counterparts, represent the ancestral condition and were retained in the rainforests (including habitats that are not life-threatingly cold, but have soft soil, such as the eastern rainforests). Moreover, *C. sibreei*, which was never observed showing torpor bouts in tree hollows, branches off first in the *Cheirogaleus* tree and holds a basal position within the clade (Groeneveld *et al.*, 2010). Thus, it is conceivable that this species may still have retained the ancestral condition of underground hibernation.

Alternatively, hibernation in tree holes could have been the original *modus operandi*, but was abandoned in highland habitats because of the risk of freezing. The occurrence of torpor bouts within poorly insulated nests or tree hollows at the beginning and the at end of the hibernation season in *C. crossleyi*, when T_a is more moderate, indicates hibernation can occur despite temperature fluctuations in sleeping sites and might point to a preference of tree shelters, if possible. Predation during hibernation seems to be low in both settings, precluding inferences. However, this is not surprising, given that any strategy with substantial losses due to predation would not have been evolutionarily stable.

Concluding remarks

The ability to enter hibernation can give species a selective physiological advantage and help them master energy crises, whether they are induced by seasonal changes in environmental conditions, unpredictable events, or anthropogenic changes. If we are to fully understand the ecology and physiology behind heterothermic Cheirogaleidae (e.g., hibernating dwarf lemurs), it is important to survey a variety of (Madagascar's) environments under different temperature regimes, thus including the warmer lowlands – dry or wet – and the colder highlands, where temperatures can drop to below 0°C during the winter.

Acknowledgments

Our studies were carried out under the "Accord de Collaboration" between Madagascar National Parks (MNP), the University of Antananarivo, and the University of Hamburg and all procedures comply with the current laws of Madagascar. We thank

MNP, C. Andrianarivo, J. Rakotomala, D. Rakotomalala, the late O. Ramilijaona, and D. Rakotondravony for their collaboration and support and the German Primate Centre (DPZ), P. Kappeler, and M. Irwin for the opportunity to work at the field station in Kirindy and in Tsinjoarivo. Thanks also to N. Razanadrakoto, N. Rakotoniaina, J. Rafalimanatsoa, and N. Domoinaharivelo for their assistance at Tsinjoarivo and R. Rasaolarison, L. Razafimanatsoa, J. Glos, Alpha de Marofandilia, François le Jeunesse, Jean-Claude de Beroboka, and T. Stahl at Kirindy. We acknowledge the authorization and support of this study by the Ministère de l'Environnement et des Forêts, MNP, the University of Antananarivo and MICET/ICTE. Our studies were financed by the "Deutsche Forschungsgemeinschaft" DFG (Ga 342/14; DA 1013/6-1) and the "Deutscher Akademischer Austauschdienst" DAAD.

References

Arnold W, Heldmaier G, Ortmann S, *et al.* 1991. Ambient temperatures in hibernacula and their energetic consequences for alpine marmots (*Marmota marmota*). *Journal of Thermal Biology* 16:223–226.

Barnes BM. 1989. Freeze avoidance in a mammal: body temperatures below 0 degrees C in an Arctic hibernator. *Science* 244:1593–1595.

Blanco MB, Godfrey LR. 2014. Hibernation patterns of dwarf lemurs in the high altitude forest of eastern Madagascar. In NB Grow, S Gursky-Doyen, A Krzton (eds.), *High Altitude Primates. Developments in Primatology: Progress and Prospects* (pp. 23–42). Springer, New York.

Blanco MB, Rahalinarivo V. 2010. First direct evidence of hibernation in an eastern dwarf lemur species (*Cheirogaleus crossleyi*) from the high-altitude forest of Tsinjoarivo, central-eastern Madagascar. *Naturwissenschaften* 97:945–950.

Blanco MB, Godfrey LR, Rakotondratsima M, *et al.* 2009. Discovery of sympatric *Cheirogaleus* species in the high-altitude rainforest of Tsinjoarivo, eastern central Madagascar: implications for biogeography and conservation. *Folia Primatologica* 80:1–17.

Blanco MB, Dausmann KH, Ranaivoarisoa JF, Yoder AD. 2013. Underground hibernation in a primate. *Scientific Reports* 3: doi10.1038/srep01768.

Carey HV, Andrews MT, Martin SL. 2003. Mammalian hibernation: cellular and molecular responses to depressed metabolism and low temperature. *Physiological Reviews* 83:1153–1181.

Dausmann KH. 2005. Measuring body temperature in the field – evaluation of external vs. implanted transmitters in a small mammal. *Journal of Thermal Biology* 30:195–202.

Dausmann KH. 2008. Hypometabolism in primates: torpor and hibernation. In BG Lovegrove, AE McKechnie (eds.), *Hypometabolism in Animals: Torpor, Hibernation and Cryobiology* (pp. 327–336). University of KwaZulu-Natal, Pietermaritzburg.

Dausmann KH. 2013. Spoilt for choice: selection of hibernacula by *Cheirogaleus medius*. In J Masters, M Gamba, F Génin (eds.), *Leaping Ahead: Advances in Prosimian Biology* (pp. 205–214). Springer, New York.

Dausmann KH. 2014. Flexible patterns in energy savings: heterothermy in primates. *Journal of Zoology* 292:101–111.

Dausmann KH, Glos J. 2015. No energetic benefits from sociality in tropical hibernation. *Functional Ecology* 29:498–505.

Dausmann KH, Glos J, Ganzhorn JU, Heldmaier G. 2004. Hibernation in a tropical primate. *Nature* 429:825–826.

Dausmann KH, Glos J, Ganzhorn JU, Heldmaier G. 2005. Hibernation in the tropics: lessons from a primate. *Journal of Comparative Physiology B Biochemical Systemic and Environmental Physiology* 175:147–155.

Dausmann KH, Glos J, Heldmaier G. 2009. Energetics of tropical hibernation. *Journal of Comparative Physiology B Biochemical Systemic and Environmental Physiology* 179:345–357.

Dausmann KH, Nowack J, Kobbe S. 2012. Afrotropical heterothermy: a continuum of possibilities. In T Ruf, C Bieber, W Arnold, E Millesi (eds.), *Living in a Seasonal World: Thermoregulatory and Metabolic Adaptations* (pp. 13–27). Springer, Berlin.

Donque G. 1972. The climatology of Madagascar. In R Battistini, G Richard-Vindard (eds.), *Biogeography and Ecology in Madagascar* (pp. 87–144). Springer, Dordrecht.

Dufils J-M. 2003. Forest ecology. In SM Goodman, JP Benstead (eds.), *The Natural History of Madagascar* (pp. 88–96). The University of Chicago Press, Chicago.

Fietz J. 1999. Monogamy as a rule rather than exception in nocturnal lemurs: the case of the fat-tailed dwarf lemur, *Cheirogaleus medius*. *Ethology* 105:255–272.

Fietz J, Dausmann KH. 2006. Big is beautiful – fat storage and hibernation as a strategy to cope with marked seasonality in the fat-tailed dwarf lemur (*Cheirogaleus medius*). In L Gould, ML Sauther (eds.), *Lemurs: Ecology and Adaptation* (pp. 97–111). Springer, Berlin.

Fietz J, Ganzhorn JU. 1999. Feeding ecology of the hibernating primate *Cheirogaleus medius*: how does it get so fat? *Oecologia* 121:157–164.

Geiser F. 2004. Metabolic rate and body temperature reduction during hibernation and daily torpor. *Annual Review of Physiology* 66:239–274.

Groeneveld LF, Blanco MB, Raharison J-L, *et al.* 2010. MtDNA and nDNA corroborate existence of sympatric dwarf lemur species at Tsinjoarivo, eastern Madagascar. *Molecular Phylogenetics and Evolution* 55(3):833–845.

Heldmaier G, Ruf T. 1992. Body temperature and metabolic rate during natural hypothermia in endotherms. *Journal of Comparative Physiology B Biochemical Systemic and Environmental Physiology* 162:696.

Heldmaier G, Ortmann S, Elvert R. 2004. Natural hypometabolism during hibernation and daily torpor in mammals. *Respiratory Physiology & Neurobiology* 141:317–329.

Hladik CM, Charles-Dominique P, Petter JJ. 1980. Feeding strategies of five nocturnal prosimians in the dry forest of the west coast of Madagascar. In CM Hladik, P Charles-Dominique, JJ Petter, *et al.* (eds.), *Nocturnal Malagasy Primates: Ecology, Physiology, and Behavior* (pp. 41–73). Academic Press, New York.

Humphries MM, Kramer DL, Thomas DW. 2003. The role of energy availability in mammalian hibernation: an experimental test in free-ranging eastern chipmunks. *Physiological and Biochemical Zoology* 76:180–186.

Jury MR. 2003 Climate. In SM Goodman, JP Benstead (eds.), *The Natural History of Madagascar* (pp. 75–87). The University of Chicago Press, Chicago.

Kappeler PM. 2000. Lemur origins: Rafting by groups of hibernators? *Folia Primatologica* 71:422–425.

Kobbe S, Dausmann KH. 2009. Hibernation in Malagasy mouse lemurs as a strategy to counter environmental challenge. *Naturwissenschaften* 96:1221–1227.

Kobbe S, Ganzhorn JU, Dausmann KH. 2011. Extreme individual flexibility of heterothermy in free-ranging Malagasy mouse lemurs (*Microcebus griseorufus*). *Journal of Comparative Physiology B Biochemical Systemic and Environmental Physiology* 181:165–173.

Lahann P. 2007. Biology of *Cheirogaleus major* in a littoral rain forest in southeast Madagascar. *International Journal of Primatology* 28:895–905.

Lahann P, Dausmann KH. 2011. Live fast, die young: flexibility of life history traits in the fat-tailed dwarf lemur (*Cheirogaleus medius*). *Behavioral Ecology and Sociobiology* 65:381–390.

Masters JC, Lovegrove BG, de Wit MJ. 2007. Eyes wide shut: can hypometabolism really explain the primate colonization of Madagascar? *Journal of Biogeography* 34:21–37.

Nowack J, Dausmann KH. 2015. Can heterothermy help the colonization of new habitats? *Mammalian Review* 45:117–127.

Petter JJ. 1962. Recherches sur l'écologie et l'éthologie des Lémuriens malagaches. *Mémoires du Muséum National d'Histoire Naturelle, Sér A, Zoologie, Paris* XXVII:1–146.

Petter JJ. 1978. Ecological and physiological adaptations of five sympatric nocturnal lemurs to seasonal variations in food production. In DJ Chivers, J Herbert (eds.), *Recent Advances in Primatology* (pp. 211–223). Academic Press, New York.

Petter JJ, Albignac R, Rumpler Y. 1977. *Mammifères Lémuriens (Primates Prosimiens). Faune de Madagascar 44*. ORSTOM-CNRS, Paris.

Petter-Rousseaux A, Charles-Dominique P, Cooper HM, *et al.* 1980. Seasonal activity rhythms, reproduction, and body weight variations in five sympatric nocturnal prosimians, in simulated light and climatic conditions. In *Nocturnal Malagasy Primates: Ecology, Physiology, and Behavior*. Academic Press, New York.

Raxworthy CJ, Nussbaum RA. 1994. A rainforest survey of amphibians, reptiles and small mammals at Montagne d'Ambre, Madagascar. *Biological Conservation* 69:65–73.

Schmid J. 1999. Sex-specific differences in activity patterns and fattening in the gray mouse lemur (*Microcebus murinus*) in Madagascar. *Journal of Mammalogy* 80:749–757.

Schmid J, Ganzhorn JU. 2009. Optional strategies for reduced metabolism in gray mouse lemurs. *Naturwissenschaften* 96:737–741.

Thiele D, Razafimahatratra E, Hapke A. 2013. Discrepant partitioning of genetic diversity in mouse lemurs and dwarf lemurs – biological reality or taxonomic bias? *Molecular Phylogenetics and Evolution* 69:593–609.

Tøien O, Blake J, Edgar DM, *et al.* 2011. Hibernation in black bears: Independence of metabolic suppression from body temperature. *Science* 331:906–909.

Wang LCH (ed.) 1989. Ecological, physiological, and biochemical aspects of torpor in mammals and birds. In *Advances in Comparative and Environmental Physiology*. Springer, Berlin.

White CR, Seymour RS. 2005. Allometric scaling of mammalian metabolism. *Journal of Experimental Biology* 208:1611–1619.

Willis JS. 1982. The mystery of the periodic arousal. In CP Lyman, JS Willis, A Malan, L Wang (eds.), *Hibernation and Torpor in Mammals and Birds* (pp. 92–103). Academic Press, New York.

Wilmé L, Goodman SM, Ganzhorn JU. 2006. Biogeographic evolution of Madagascar's microendemic biota. *Science* 312:1063–1065.

Wright PC, Martin LB. 1995. Predation, pollination and torpor in two nocturnal prosimians: *Cheirogaleus major* and *Microcebus rufus* in the rain forest of Madagascar. In L Alterman, GA Doyle, MK Izard (eds.), *Creatures of the Dark: The Nocturnal Prosimians* (pp. 45–60). Plenum Press, New York.

Part IV

Cheirogaleidae: sensory ecology, communication, and cognition

18 Seed dispersal by mouse lemurs: do *Microcebus* represent a unique frugivorous guild?

Kim Valenta and Shawn M. Lehman

Introduction

Seed dispersal by animals

Plants vary in the mechanisms by which they disperse their seeds. Plant dispersal mechanisms are either abiotic (wind, water and ballistic dispersal), or biotic (endozoochory and epizoochory) (Howe and Smallwood, 1982). Epizoochory refers to dispersal of seeds via adhesion to animal fur, while endozoochory refers to the consumption of fleshy fruits and seeds by frugivores, and their subsequent deposition, through either spitting or defecation (Couvreur *et al.*, 2005). While the latter mechanism is relatively rare in temperate forests, up to 90% of tropical tree species depend on endozoochory, relying on animals as dispersal vectors (Howe and Smallwood, 1982). Different animal dispersers affect seed dispersal and seed viability via seed handling (Lambert, 1999), gut passage (Knogge *et al.*, 2003; Valenta and Fedigan, 2009) and the spatial location of seed deposition (Valenta and Fedigan, 2010).

The effect of gut passage on seeds is a potentially crucial determinant of plant reproductive success. In some cases, studies have shown that gut passage neither helps nor hinders a seed's chance of germinating (Smith, 2004). However, the majority of research on the effect of gut passage suggests that it increases the potential for germination for numerous species (Knogge *et al.*, 2003). For some plant species, passage through a specific animal dispersal vector is necessary for successful seed germination (Temple, 1977; Chapman *et al.*, 1992).

Other processes influencing the success of seed dispersal are secondary dispersal and predation (Nathan and Mueller-Landau, 2000; Valenta and Fedigan, 2009). The deposition of a seed is not necessarily the final step in the plant–animal interaction, because the forest floor is home to invertebrates and vertebrates that variously prey on, or safely redeposit, dispersed seeds. One study of primate-dispersed seeds in Costa Rica found that approximately 80% (N = 793) of seeds were removed from animal feces a mere week after initial deposition (Chapman, 1989). A study of

The Dwarf and Mouse Lemurs of Madagascar: Biology, Behavior and Conservation Biogeography of the Cheirogaleidae, eds. S.M. Lehman, U. Radespiel, E. Zimmermann. Published by Cambridge University Press. © Cambridge University Press 2016.

howler monkey (*Alouatta palliata*)-dispersed seeds in Mexico found that a majority of the dung beetle species in the area were overwhelmingly attracted to howler dung, steering clear of the feces of other, more folivorous, herbivores, resulting in the massive redeposition of defecated seeds (Estrada and Coates-Estrada, 1991). In Madagascar, secondary dispersal by ants (*Aphaenogaster swammerdami*) has been found to result in increased rates of seedling establishment in fragments relative to seeds dispersed by Vasa parrots (*Coracopsis* spp.) in continuous forests. This was attributed to both successful secondary dispersal by ants, and the absence of rodent seed predators in fragments (Böhning-Gaese *et al.*, 1999). Secondary dispersal and predation are thus crucial processes in seed dispersal systems.

Seed shadows describe the density and distance of seeds relative to the trees that produced them (Howe and Smallwood, 1982). The spatial parameters of seed shadows generated by dispersal agents are another variable affecting seed and seedling survival and recruitment. Many hypotheses for the evolution of seed dispersal systems rely on the spatial pattern of seed dispersal as the critical variable in seed survival, and in the resultant structure of forests (Janzen, 1970b; Connell, 1971; Chapman and Chapman, 1996). The Janzen–Connell model, or seed escape hypothesis, addresses the negative correlation between seed mortality and distance from parent trees (Janzen, 1970a; Connell, 1971). According to the Janzen–Connell model, seeds that are deposited underneath the canopy of the parent tree will attract and are more likely to fall victim to a host of species-specific seed predators, fungal pathogens and herbivores at a higher rate than those that are deposited further from adult trees. Thus, seed predators may be density-responsive by concentrating their destruction near adult fruiting trees, where initial seed fall will be highest. Also, seedlings that establish closer to parent trees are expected to suffer higher rates of competition with conspecifics (Connell, 1971). Forest diversity is thus maintained, and a mix of different species evenly spaced from one another should prevail, as nonconspecifics successfully establish themselves adjacent to other species that attract seed and seedling predators that are host-specific to their neighbors and competing with each other. Both density- and distance-dependent seed mortality have been well supported empirically (Harms *et al.*, 2000; Russo and Augspurger, 2004; Valenta and Fedigan, 2009).

Variation in seed dispersal strategies (biotic versus abiotic) and the density and diversity of dispersal vectors can result in variation in tree distributions and in variation in species persistence in the absence of key disperser species. The extent of variation in tree distribution and persistence, in turn, depends on seed treatment, the effect of gut passage, secondary dispersal and predation, and distance- and density-dependent effects. In the case of biotic dispersal vectors, fruit and seed morphology, and their effects on fruit choice by animals, such as primates, can also play an important role.

Studies of the loss of frugivores on plant communities in forest fragments and islands have documented significant changes in forest composition and regeneration. Leigh *et al.* (1993) noted a significant loss of plant species diversity on islands lacking native mammalian disperser fauna after 67 years of isolation, which they attribute to the loss of "keystone animals" necessary to the maintenance of plant diversity. In Uganda, Chapman and Onderdonk (1998) documented reduced seedling density and diversity, especially of larger-seeded species, in fragments with reduced primate diversity and abundance. In Tanzania, Cordeiro and Howe (2001) documented a 40-fold reduction in recruitment of endemic, animal-dispersed taxa in small fragments relative to continuous forest, which they attribute to the local absence of primates and birds in these fragments. In the Colombian Amazon, the presence of large-bodied ateline dispersal agents in fragments was identified as the strongest predictor of large-seeded plant species persistence in fragments (Stevenson and Aldana, 2008). In Panama, Wright and Duber (2001) have documented significant spatial shifts in the recruitment of palms along a gradient of anthropogenic disturbance and fragmentation and concluded that the loss of plant diversity in fragments may result from increased spatial aggregation of seedlings around parent trees due to a loss of large-bodied dispersers, followed by increased density-dependent seed predation. In South America, Galetti *et al.* (2006) found that successful seed removal and dispersal of an endemic palm collapsed in small fragments, resulting in low densities of seedlings and juveniles due to the absence of large-bodied dispersers. Also in South America, Rodriguez-Cabal *et al.* (2007) documented the complete local extinction of an endemic mistletoe species in fragments lacking a key dispersal agent. The question arises then as to how plant–animal interactions persist in areas, such as the island of Madagascar, which have experienced massive anthropogenic conversion of forests to grasslands (Harper *et al.*, 2007)? Despite this massive alteration of forested habitats, Madagascar maintains remarkably high levels of plant and animal species richness and endemism (e.g., Mittermeier *et al.*, 2008; Vieites *et al.*, 2009; Schwitzer *et al.*, 2014). This ecological incongruence between high species richness on the one hand and massive forest conversion on the other provides for an ideal but largely unexplored natural laboratory for research on seed dispersal.

In this chapter, we review the patterns and processes associated with seed dispersal by animals in general, and nocturnal mouse lemurs in particular. Our review is divided into two sections. First, we deal with the main hypotheses and models on seed dispersal, with a strong focus on the evolutionary ecology of this complex relationship and its conservation implications. In the second section, we review existing information on seed dispersal by lemurs in the genus *Microcebus*. Within this second section, we also predict conservation implications of vanishing lemur populations on forest growth and regeneration in Madagascar.

Fruit selection by animals – a case for syndromes?

Up to 90% of woody plants in tropical forests rely on birds and mammals as seed dispersers (Howe and Smallwood, 1982). Animal-dispersed plants produce numerous fruit and seed phenotypes, and are often exploited by a wide variety of frugivorous taxa (Fischer and Chapman, 1993). Due to the high variation in the sensory and dispersal capabilities of tropical frugivorous fauna, several researchers have hypothesized the existence of "dispersal syndromes" or "fruit syndromes," i.e., sets of co-occurring fruit and seed morphological traits that are uniquely attractive to different types of frugivores based on the frugivore's behavior, physiology and sensory abilities (Janson, 1983; Lomáscolo *et al.*, 2010).

Although the fruit syndrome hypothesis has been tested many times at many different scales, the evidence for it is mixed. Numerous species of animals with a wide variety of sensory and mechanical capabilities can disperse the seeds of the same fruiting species (Lambert and Garber, 1998). Moreover, given the relatively slow evolutionary rates of plants relative to animals, the extensive gene flow in plant populations and the weak selective pressures on the disperser (reviewed in Lomascolo *et al.*, 2010), it may be unlikely that specific frugivores are driving the evolution of fruit traits. Several empirical studies, especially those using phylogenetic methods, that matched frugivore assemblages to fruit traits failed to detect clear syndromes (Herrera, 1992; Fischer and Chapman, 1993). The lack of a tight match between fruit traits and frugivore assemblages is thought to result primarily from diffuse coevolution driven by high dietary overlap. Because many different species of birds and mammals can disperse the seeds of the same plant species, there should be little selective pressure for a plant to evolve specialized fruit morphology to increase detection by just one disperser species (Fleming *et al.*, 1987). Additionally, when more than one species offers equivalent ecological services to an angiosperm species, then the loss of one of those species should be offset by the presence of the remaining species because of their ecological equivalency to that plant (Loiselle *et al.*, 2007). In these cases of so-called "ecological redundancy" plants can be expected to exhibit adaptations and strategies favoring dispersal by multiple species (Loiselle *et al.*, 2007).

Despite evidence against the existence of fruit syndromes, there is compelling evidence that frugivores and fruiting trees drive morphological adaptations in one another. Plants are sessile organisms that rely on the dispersal of their seeds for successful reproduction, and they have been found to evolve mechanisms to increase the likelihood of attracting effective dispersal vectors, including odorants and color (Lomáscolo *et al.*, 2010). Frugivores affect plant fitness (Howe and Smallwood, 1982), and frugivores show morphological and behavioral adaptations that enable them to successfully exploit fruit, such as color vision and olfaction (Garber and Lambert, 1998; Lomáscolo and Schaefer, 2010). Additionally, frugivores' fruit preferences for certain suites of fruit traits have been documented (Stevenson, 2004), and fruit and seed morphology has been convincingly linked to specific guilds of frugivores (Lomáscolo *et al.*, 2010). Several parameters of fruit and seed morphology have been found to affect fruit choice by frugivorous animals. These include aril:seed ratio (Russo, 2003), exocarp hardness (Janson, 1983), ripe fruit quantity (Russo, 2003), phenological stage (Russo, 2003), chemical fruit traits (Voigt *et al.*, 2004), fruit odor (Hodgkison *et al.*, 2007; Valenta *et al.*, 2013), pulp richness and seed size (McKey, 1975; Snow, 1981), fruit size (Janson, 1983) and, most commonly, fruit color (Smith *et al.*, 2003; Melin *et al.*, 2009; Valenta *et al.*, 2013).

Early studies of fruit syndromes relied on subjective assessments of fruit coloration – fruit color was described according to a series of human categories like red, yellow and blue (e.g., Janson, 1983). Technological advances in the measurement of reflectance spectra mean that fruit color can now be measured quantitatively, and compared to the actual visual capabilities of frugivores (Stevens *et al.*, 2007, 2009). The few studies that have quantified fruit coloration in relation to animal feeding choice have found that fruit color impacts the foraging rates of frugivores (Hiramatsu *et al.*, 2008; Melin *et al.*, 2009). The only study that has applied quantitative measures of fruit coloration to the question of dispersal syndromes within a phylogenetically controlled analysis has provided support for the fruit syndrome hypothesis by demonstrating the existence of bird and bat syndromes in neotropical *Ficus* species (Lomáscolo *et al.*, 2010). Traits significantly correlated with key disperser taxa are fruit size, fruit hardness, visual conspicuousness, odor, and contrast against substrate, all of which are convincingly linked to quantitative measures of the sensory and mechanical capabilities of disperser species in the study ecosystem (Lomáscolo *et al.*, 2010).

The ability to quantitatively measure both disperser sensory abilities (e.g., color vision) and fruit morphological characteristics (e.g. color) enables researchers to measure one axis of the mutualism between seed dispersing animals and fruiting plants. However, the applicability of these methods within habitats experiencing anthropogenic disturbance is a critical, yet largely unexplored, topic in ecology. For example, field studies have revealed that light intensity and wind speed vary due to deforestation and edge effects, which can then alter seed dispersal and biochemical properties, respectively, of plants living in forests undergoing anthropogenic disturbance (Holl and Lulow, 1997; Cadenasso and Pickett, 2001). Thus, regions subject to extreme spatial and temporal patterns of anthropogenic disturbance may be experiencing stochastic changes that can affect the mutualism between seed dispersers and plants, particularly those plants species intolerant of edge habitats in the tropics (Clark *et al.*, 1999).

Since the approximately 2000 years since human settlement, Madagascar has potentially lost numerous dispersers (Dewar and Richard, 2007), with unknown effects on seed dispersal mutualisms. Anthropogenic disturbance is particularly relevant to the forested regions of Madagascar, where it is estimated that the combined effects of logging and slash-and-burn agriculture have resulted in the conversion of 70–90% of the original forest into grasslands and agricultural fields (Harper *et al.*, 2007). Few mammals exploit these non-forest habitats. Moreover, the remaining forest is found in increasingly fragmented landscapes that are also prone to edge effects. Although recent studies indicate that edge effects influence spatial variations in foliar stable isotopes (Crowley *et al.*, 2012), there are few data on how anthropogenic factors influence other ecological processes, such as seed dispersal. In the context of the increasing forest fragmentation and defaunation in Madagascar, understanding the extent to which plant–animal mutualisms affect what remains of Madagascar's forests requires understanding the dynamics of forests in the absence of dispersers.

Plant–animal mutualisms in Madagascar

In most tropical forest systems, disperser guilds include a high number and wide variety of species capable of dispersing the seeds of any given tree species (Howe and Miriti, 2004). For this reason, the loss of any given disperser species should not adversely affect the viability of a tree species, as other birds or mammals capable of dispersing that fruit are able to fill the niche left by the extirpated species (Fischer and Chapman, 1993). However, this general pattern of multiple seed dispersers is quite different in Madagascar. While Malagasy forests are floristically amongst the most speciose on earth (Barthlott *et al.*, 1996), these forests support a surprisingly depauperate frugivore community (Goodman *et al.*, 1997). Preliminary studies of seed dispersal in Madagascar indicate that although birds contribute to seed dispersal of some species (Böhning-Gaese *et al.*, 1995; Bollen *et al.*, 2004), lemurs appear to be primarily responsible, and in some cases essential, for the dispersal of many plant taxa (Ganzhorn *et al.*, 1999).

Tight coevolutionary relationships are considered rare, or limited to cases where seed size constrains a potential disperser assemblage, or to areas with a low diversity of disperser species (Loiselle *et al.*, 2007). The tropical dry forests of Madagascar represent an ideal environment within which to test the effects of dispersers on plant evolution and persistence because of the following: they maintain a depauperate Malagasy frugivore community (Bollen *et al.*, 2004), lack all but a very few large-bodied dispersers (Ganzhorn *et al.*, 1999), lack seedling recruitment under parent trees for at least some endemic species (Deleporte *et al.*, 1996), support a high degree of plant and animal endemism (Goodman and Benstead, 2005), and the urgency of conserving what remains of the forest, which is experiencing severe anthropogenic disturbance.

Fruit selection by *Microcebus* spp.

While little is known about the role of seed handling, gut passage and dispersal distance for *Microcebus*, there are considerable data on fruit choice by *Microcebus* species, which is an important first step in understanding the seed dispersal process. Here, we review what is known about the role of *Microcebus* as frugivores in the forests of Madagascar.

Previous studies of *Microcebus* spp. have found that, despite their small body mass of 30.6–77.8 g which predicts a high degree of insectivory (Atsalis, 1999), many species of this genus throughout Madagascar are highly frugivorous (Atsalis, 1999; Lahann, 2006; Dammhahn and Kappeler, 2008; Thoren *et al.*, 2011). Published estimates of percent fruit in the *Microcebus* diet range up to 83% during periods of overall high fruit availability (Thoren *et al.*, 2011), and all extant studies that have included fecal analysis have found intact seeds in feces (Atsalis, 1999; Lahann, 2006; Dammhahn and Kappeler, 2008; Thoren *et al.*, 2011; Valenta *et al.*, 2013). Thus, despite their small body mass, *Microcebus* spp. can be highly frugivorous and pass seeds of several species intact in their feces.

Because of their small mass, it is intuitive that species of the genus *Microcebus* are limited in terms of the size of seeds they can ingest and disperse. Indeed, several studies have measured the size of seeds dispersed by *Microcebus* and all showed that the seed sizes that *Microcebus* spp. disperse tend to be quite small. One study found that the mean seed width consumed by Cheirogalidae in southeast Madagascar is 6 mm (Lahann, 2006). Maximum seed size varies between species, but is as large as 12.5 mm in maximum diameter for *Cheirogaleus* spp. (Lahann, 2006). Thus, unlike sympatric frugivorous primates that can disperse large-seeded species (e.g., *Eulemur* spp., *Varecia* spp.; Wright *et al.*, 2011), *Microcebus* spp. are limited to small-seeded species that might otherwise be thought of as bird-dispersed species – animals with opposite foraging patterns and sensory phenotypes (Valenta *et al.*, 2013). This size-related pattern of seed ingestion has interesting implications for the evolution of small-seeded Malagasy fruits, because while avian seed dispersers are tetrachromatic and diurnal, and rely heavily on visual cues during foraging, *Microcebus* spp. are nocturnal, and dichromatic, with highly retained olfactory capabilities (Barton *et al.*, 1995; Hohenbrink *et al.*, 2013). The difference in small fruit morphology that might be expected based on this coexistence of two small-seeded disperser guilds with different sensory abilities raises an interesting question: do *Microcebus* disperse the same seeds as sympatric avian frugivores, or do they represent their own guild in the uniquely depauperate frugivore communities of Madagascar? While extensive comparisons of bird- and *Microcebus*-dispersed fruits are not available, measures of *Microcebus* fruit choice as a function of fruit morphology and nutritional rewards can offer clues as to whether *Microcebus* spp. share a disperser niche with avian frugivores, or inhabit their own.

In terms of the effect of morphological variation of fleshy fruits on fruit choice by *Microcebus* species, two studies have documented the important role of olfaction in fruit detection and choice (Siemers *et al.*, 2007; Valenta *et al.*, 2013). One experimental study found that, controlling for other sensory cues, gray mouse lemurs (*Microcebus murinus*) were able to detect fruit based on smell alone, without the aid of any visual stimuli (Siemers *et al.*, 2007). Another recent study documented the role of the overall volatile organic compounds (odors) present in fruits dispersed and not dispersed by two species of *Microcebus* and found that dispersed fruits emitted significantly higher odor signals than non-dispersed fruits (Valenta *et al.*, 2013). Unlike birds, *Microcebus* spp. have retained extensive neuroanatomical structures associated with enhanced olfactory discrimination, including moist rhinaria and large olfactory bulbs (Barton *et al.*, 1995; Siemers *et al.*, 2007; Piep *et al.*, 2008). The combination of highly retained olfactory ability in *Microcebus* spp. and its absence in most avian taxa may indicate functional separation of these frugivorous guilds.

Visual signals – specifically chromaticity, or color, and luminance, or brightness – have been shown to be important in finding fruit (Hiramatsu *et al.*, 2008). Unlike some other nocturnal primates, *Microcebus* spp. have retained dichromatic cone function, allowing them to distinguish between blues and yellows (Tan and Li, 1999). Dichromatic animals have also been found to have an increased ability to notice variation in luminance relative to trichromats (Morgan *et al.*, 1992).

Therefore, it is reasonable to posit that both fruit chromaticity and luminance might act as important fruit signals for *Microcebus* spp. One recent study measured fruit chromaticity and luminance, and found that *Microcebus*-dispersed fruits showed significantly higher chromatic contrasts than non-dispersed fruits, although the relationship did not hold for luminance (Valenta *et al.*, 2013). More data on the role of fruit chromaticity and luminance under different illuminants and for different species would help resolve the role of plant visual signals in attracting *Microcebus* spp.

In addition to fruit signals, studies have documented the nutritional rewards of *Microcebus*-consumed fruits in order to identify nutritional preferences (Atsalis, 1999). In addition to fruit color, odor and seed size, there is evidence that at least one species of *Microcebus* preferentially eats foods based on nutrient values; Atsalis (1999) discovered that *Microcebus rufus* in Ranomafana prefer lipid-rich fruits, while avoiding fruits with a high degree of tannins. Typically, mammals prefer sugar-rich fruits, while birds prefer fruits rich in lipids and protein (Fleming *et al.*, 1987; Debussche and Isenmann, 1989; Galetti, 2000).

Conservation implications

The effect of frugivore loss on forests and forest fragments can fall anywhere along a spectrum from increased seedling survival and growth (recruitment) of certain species (e.g., Böhning-Gaese *et al.*, 1999) to local extinction (e.g., Chapman *et al.*, 1992). Studies in Madagascar and elsewhere have documented seedling recruitment and plant species persistence in the absence of some, or all, of the local frugivore guild. Dausmann *et al.* (2008) found increased seedling establishment of one Malagasy species, *Commiphora guillaumini*, in fragments versus continuous forest. This plant species is primarily dispersed by lesser Vasa parrots (*Coracopsis nigra*) and secondarily dispersed by the ant *Aphaenogaster swammerdami*. Using seed exclosure experiments, the authors conclude that the absence in fragments, but not in continuous forests, both of native secondary seed predators (rodents including *Hypogeomys antimena*, *Eliurus myoxinus* and *Macrotarsomys bastardi*) and of invasive seed predators (e.g. *Rattus rattus*) resulted in increased seedling recruitment of this species in fragments. Ganzhorn *et al.* (1999) identified only one vertebrate species, the red-fronted brown lemur (*Eulemur rufifrons*), that could ingest seeds > 11 mm long, and that in forest fragments without this species, fewer large-seeded species regenerated. This finding was echoed by Dausmann *et al.* (2008), who found that at least one-tenth of tree species in a Malagasy tropical dry forest depend exclusively on the red-fronted brown lemur (*Eulemur rufifrons*) for primary dispersal and successful regeneration.

Throughout tropical forests, the absence of animal dispersal agents in forest fragments results in variation in seedling recruitment of animal-dispersed species, specifically larger-seeded species wherein only one or a few larger-bodied animals are able to successfully disperse seeds. Woody plant population persistence requires successful seed dispersal and seedling establishment, both of which can be disrupted due to anthropogenic disturbance with potentially devastating consequences for endemic species.

Conclusion

Understanding the nature and extent to which primates influence the processes of forest regeneration and maintenance is critical to understanding primates in the context of the ecological communities of which they are a part; it is also critical to understanding the potentially enormous implications that their behavior exerts on the survival of entire ecosystems. Anthropogenic and climatic forest disturbance is projected to continue well into the future in Madagascar (Myers *et al.*, 2000) and understanding the capacity for ecosystem resiliency in the face of disturbance is essential. This is especially true in the case of lemurs, which are important dispersers of seeds in an otherwise uniquely depauperate frugivore community (Ganzhorn *et al.*, 1999).

While numerous studies have documented the variables driving fruit choice by *Microcebus*, few studies have quantified or qualified actual seed dispersal variables, namely seed handling, gut passage effects and the spatial parameters of seed dispersal by *Microcebus*. These basic details of the natural history of *Microcebus* would greatly increase our understanding of their role in forest regeneration. Additionally, to truly measure the extent to which *Microcebus* spp. might represent their own frugivorous guild in Madagascar, studies comparing the diets of frugivorous Malagasy birds and *Microcebus* are needed.

Acknowledgments

For helpful commentary and suggestions, we thank S.K. Monckton, K.T. Firth-Evans, the reviewer, and Elke Zimmermann. For funding we thank Sigma Xi, GM Women in Science (K.V.), the University of Toronto (K.V.) and Natural Sciences and Engineering Research Council of Canada (K.V., S.M.L.).

References

Atsalis S. 1999. Diet of the brown mouse lemur (*Microcebus rufus*) in Ranomafana National Park, Madagascar. *International Journal of Primatology* 20:193–229.

Barthlott W, Lauer W, Placke A. 1996. Global distribution of species diversity in vascular plants: towards a world map of phytodiversity. *Erdkunde* 50:317–327.

Barton RA, Purvis A, Harvey PH. 1995. Evolutionary radiation of visual and olfactory brain systems in primates, bats and insectivores. *Philosophical Transactions of the Royal Society of London B* 348:381–392.

Böhning-Gaese K, Gaese B, Rabemanantsoa S. 1995. Seed dispersal by frugivorous tree visitors in the Malagasy tree species *Commiphora guillaumini*. *Ecotropica* 1:41–50.

Böhning-Gaese K, Gaese BH, Rabemanantsoa SB. 1999. Importance of primary and secondary seed dispersal in the Malagasy tree *Commiphora guillaumini*. *Ecology* 80(3):821–832.

Bollen A, Elsacker LV, Ganzhorn JU. 2004. Tree dispersal strategies in the littoral forest of Sainte Luce (SE-Madagascar). *Oecologia* 139(4):604–616.

Cadenasso ML, Pickett ST. 2001. Effect of edge structure on the flux of species into forest interiors. *Conservation Biology* 15(1):91–97.

Chapman CA. 1989. Primate seed dispersal: the fate of dispersed seeds. *Biotropica* 21(2):148–154.

Chapman CA, Chapman LJ. 1996. Frugivory and the fate of dispersed and non-dispersed seeds of six African tree species. *Journal of Tropical Ecology* 12(04):491–504.

Chapman CA, Onderdonk DA. 1998. Forests without primates: primate/plant codependency. *American Journal of Primatology* 45:127–141.

Chapman CA, Chapman LJ, Wrangham R, *et al.* 1992. Estimators of fruit abundance of tropical trees. *Biotropica* 24(4):527–531.

Clark JS, Silman M, Kern R, Macklin E, HilleRisLambers J. 1999. Seed dispersal near and far: patterns across temperate and tropical forests. *Ecology* 80(5):1475–1494.

Connell JH. 1971. On the role of natural enemies in preventing competitive exclusion in some marine animals and in rain forest trees. In P van Boer, G Gradwell (eds.), *Dynamics of Populations* (pp. 298–312). Centre for Agricultural Publishing and Documentation, Wageningen.

Cordeiro N, Howe H. 2001. Low recruitment of trees dispersed by animals in African forest fragments. *Conservation Biology* 15(6):1733–1741.

Couvreur M, Cosyns E, Hermy M, Hoffmann M. 2005. Complementarity of epi- and endozoochory of plant seeds by free ranging donkeys. *Ecography* 28(1):37–48.

Crowley BE, McGoogan KC, Lehman SM. 2012. Edge effects on foliar stable isotope values in a Madagascan tropical dry forest. *PLoS ONE* 7(9):1–9.

Dammhahn M, Kappeler PM. 2008. Comparative feeding ecology of sympatric *Microcebus berthae* and *M. murinus*. *International Journal of Primatology* 29:1567–1589.

Dausmann K, Glos J, Linsenmair K, Ganzhorn J. 2008. Improved recruitment of a lemur-dispersed tree in Malagasy dry forests after the demise of vertebrates in forest fragments. *Oecologia* 157(2):307–316.

Debussche M, Isenmann P. 1989. Fleshy fruit characters and the choices of bird and mammal seed dispersers in a Mediterranean region. *Oikos* 56:327–338.

Deleporte P, Randrianasolo J, Rakotonirina G. 1996. Sylviculture in the dry dense forest of western Madagascar. In JU Ganzhorn, JP Sorg (eds.), *Ecology and Economy of a Tropical Dry Forest in Madagascar, Primate Report* (pp. 89–116). Goltze, Gottingen.

Dewar RE, Richard AF. 2007. Evolution in the hypervariable environment of Madagascar. *Proceedings of the National Academy of Sciences* 104(34):13723–13727.

Estrada A, Coates-Estrada R. 1991. Howler monkeys (*Alouatta palliata*), dung beetles (*Scarabaeidae*) and seed dispersal: ecological interactions in the tropical rain forest of Los Tuxtlas, Mexico. *Journal of Tropical Ecology* 7(4):459–474.

Fischer K, Chapman C. 1993. Frugivores and fruit syndromes: differences in patterns at the genus and species level. *Oikos* 66(3):472–482.

Fleming T, Breitwisch R, Whitesides G. 1987. Patterns of tropical vertebrate frugivore diversity. *Annual Review of Ecology and Systematics* 18:91–109.

Galetti M. 2000. Frugivory by toucans (Rhampastidae) at two altitudes in the Atlantic Forest of Brazil. *Biotropica* 32:842–850.

Galetti M, Donatti C, Pires A, Guimaraes P, Jordano P. 2006. The palms: seed survival and dispersal of an endemic Atlantic forest palm: the combined effects of defaunation and forest fragmentation. *Botanical Journal of the Linnean Society* 151(1):141–149.

Ganzhorn JU, Fietz J, Rakotovao E, Schwab D, Zinner D. 1999. Lemurs and the regeneration of dry deciduous forest in Madagascar. *Conservation Biology* 13(4):794–804.

Garber PA, Lambert JE. 1998. Primates as seed dispersers: ecological processes and directions for future research. *American Journal of Primatology* 45:3–8.

Goodman SM, Benstead JP. 2005. Updated estimates of biotic diversity and endemism for Madagascar. *Oryx* 39(01):73–77.

Goodman SM, Ganzhorn JU, Wilme L. 1997. Observations at a *Ficus* tree in Malagasy humid forest. *Biotropica* 29(4):480–488.

Harms K, Wright S, Calderón O, Hernández A, Herre E. 2000. Pervasive density-dependent recruitment enhances seedling diversity in a tropical forest. *Nature* 404(6777):493–495.

Harper G, Steininger M, Tucker C, Juhn D, Hawkins F. 2007. Fifty years of deforestation and forest fragmentation in Madagascar. *Environmental Conservation* 34(04):325–333.

Herrera C. 1992. Interspecific variation in fruit shape: allometry, phylogeny, and adaptation to dispersal agents. *Ecology* 73(5):1832–1841.

Hiramatsu C, Melin AD, Aureli F, *et al.* 2008. Importance of achromatic contrast in short-range fruit foraging of primates. *PLoS ONE* 3(10):e3356.

Hodgkison R, Ayasse M, Kalko E, *et al.* 2007. Chemical ecology of fruit bat foraging behavior in relation to the fruit odors of two species of paleotropical bat-dispersed figs (*Ficus hispida* and *Ficus scortechinii*). *Journal of Chemical Ecology* 33(11):2097–2110.

Hohenbrink P, Mundy NI, Zimmerman E, Radespiel U. 2013. First evidence for functional comeronasal 2 receptor genes in primates. *Biology Letters* 9(1):20121006.

Holl KD, Lulow ME. 1997. Effects of species, habitat, and distance from edge on post-dispersal seed predation in a tropical rainforest. *Biotropica* 29(4):459–468.

Howe H, Smallwood J. 1982. Ecology of seed dispersal. *Annual Review of Ecology and Systematics* 13:201–228.

Howe HF, Miriti MN. 2004. When seed dispersal matters. *BioScience* 54(7):651–660.

Janson CH. 1983. Adaptation of fruit morphology to dispersal agents in a neotropical forest. *Science* 219(4581):187–189.

Janzen DH. 1970a. Herbivores and the number of tree species in tropical forests. *The American Naturalist* 104(940):501.

Janzen DH. 1970b. Herbivores and the number of tree species in tropical forests. *American Naturalist* 104:501.

Knogge C, Herrera ERT, Heymann EW. 2003. Effects of passage through tamarin guts on the germination potential of dispersed seeds. *International Journal of Primatology* 24(5):1121–1128.

Lahann P. 2006. Feeding ecology and seed dispersal of sympatric cheirogaleid lemurs (*Microcebus murinus*, *Cheirogaleus medius*, *Cheirogaleus major*) in the littoral rainforest of south-east Madagascar. *Journal of Zoology* 271(1):88–98.

Lambert JE. 1999. Seed handling in chimpanzees (*Pan troglodytes*) and redtail monkeys (*Cercopithecus ascanius*): implications for understanding hominoid and cercopithecine fruit-processing strategies and seed dispersal. *American Journal of Physical Anthropology* 109(3):365–386.

Lambert JE, Garber PA. 1998. Evolutionary and ecological implications of primate seed dispersal. *American Journal of Primatology* 45(1):9–28.

Leigh EG, Wright SJ, Herre EA, Putz FE. 1993. The decline of tree diversity on newly isolated tropical islands: a test of a null hypothesis and some implications. *Evolutionary Ecology* 7(1):76–102.

Loiselle BA, Blendinger PG, Blake JG, Ryder TB. 2007. Ecological redundancy in seed dispersal systems: a comparison between manakins (Aves: Pipridae) in two tropical forests. In AJ Dennis, EW Schupp, R Green, DW Westcott (eds.), *Seed Dispersal: Theory and its Application in a Changing World* (pp. 178–196). CABI Publishing, Wallingford.

Lomáscolo S, Schaefer H. 2010. Signal convergence in fruits: a result of selection by frugivores? *Journal of Evolutionary Biology* 23:614–624.

Lomáscolo S, Levey D, Kimball R, Bolker B, Alborn H. 2010. Dispersers shape fruit diversity in *Ficus* (Moraceae). *Proceedings of the National Academy of Sciences* 107(33):14668–14672.

McKey D. 1975. The ecology of coevolved seed dispersal systems. In LE Gilbert, PH Raven (eds.), *Coevolution of Animals and Plants* (pp. 155–191). University of Texas Press, Austin.

Melin AD, Fedigan LM, Hiramatsu C, *et al.* 2009. Fig foraging by dichromatic and trichromatic *Cebus capucinus* in a tropical dry forest. *International Journal of Primatology* 30(6):753–775.

Mittermeier R, Ganzhorn J, Konstant W, *et al.* 2008. Lemur diversity in Madagascar. *International Journal of Primatology* 29(6):1607–1656.

Morgan MJ, Adam A, Mollon JD. 1992. Dichromats detect colour-camouflaged objects that are not detected by trichromats. *Proceedings: Biological Sciences* 248(1323):291–295.

Myers N, Mittermeier RA, Mittermeier CG, da Fonseca GAB, Kent J. 2000. Biodiversity hotspots for conservation priorities. *Nature* 403(6772):853–858.

Nathan R, Mueller-Landau HC. 2000. Spatial patterns of seed dispersal, their determinants and consequences for recruitment. *Trends in Ecology and Evolution* 15:278–285.

Piep M, Radespiel U, Zimmermann E, Schmidt S, Siemers BM. 2008. The sensory basis of prey detection in captive-born grey mouse lemurs, *Microcebus murinus. Animal Behaviour* 75(3):871–878.

Rodríguez-Cabal M, Aizen M, Novaro A. 2007. Habitat fragmentation disrupts a plant-disperser mutualism in the temperate forest of South America. *Biological Conservation* 139(1–2):195–202.

Russo S. 2003. Responses of dispersal agents to tree and fruit traits in *Virola calophylla* (Myristicaceae): implications for selection. *Oecologia* 136(1):80–87.

Russo SE, Augspurger CK. 2004. Aggregated seed dispersal by spider monkeys limits recruitment to clumped patterns in *Virola calophylla. Ecology Letters* 7:1058–1067.

Schwitzer C, Mittermeier RA, Johnson SE, *et al.* 2014. Averting lemur extinctions amid Madagascar's political crisis. *Science* 343(6173):842–843.

Siemers BM, Goerlitz HR, Robsomanitrandrasana E, *et al.* 2007. Sensory basis of food detection in wild *Microcebus murinus. International Journal of Primatology* 28:291–304.

Smith AC, Buchanan-Smith HM, Surridge AK, Osorio D, Mundy NI. 2003. The effect of colour vision status on the detection and selection of fruits by tamarins (*Saguinus* spp.). *The Journal of Experimental Biology* 206(Pt 18):3159–3165.

Smith S. 2004. A preliminary study of seed dispersal by white-faced capuchins (*Cebus capucinus*) and mantled howlers (*Alouatta palliata*) in Costa Rica. *American Journal of Physical Anthropology* 123(S1):184.

Snow DW. 1981. Tropical frugivorous birds and their food plants: a world survey. *Biotropica* 13:1–14.

Stevens M, Paraga CA, Cuthill IC, Partridge JC, Troscianko TOMS. 2007. Using digital photography to study animal coloration. *Biological Journal of the Linnean Society* 90(2):211–237.

Stevens M, Stoddard MC, Higham JP. 2009. Studying primate color: towards visual system-dependent methods. *International Journal of Primatology* 30(6):893–917.

Stevenson P, Aldana A. 2008. Potential effects of ateline extinction and forest fragmentation on plant diversity and composition in the western Orinoco Basin, Colombia. *International Journal of Primatology* 29(2):365–377.

Stevenson PR. 2004. Fruit choice by woolly monkeys in Tinigua National Park, Colombia. *International Journal of Primatology* 25(2):367–381.

Tan Y, Li WH. 1999. Trichromatic vision in prosimians. *Nature* 402:36.

Temple SA. 1977. Plant-animal mutualism: coevolution with dodo leads to near extinction of plant. *Science* 197(4306):885–886.

Thoren S, Quietzsch F, Schwochow D, *et al.* 2011. Seasonal changes in feeding ecology and activity patterns of two sympatric mouse lemur species, the gray mouse lemur (*Microcebus murinus*) and the golden-brown mouse lemur (*M. ravelobensis*) in northwestern Madagascar. *International Journal of Primatology* 32:566–586.

Valenta K, Fedigan LM. 2009. Effects of gut passage, feces, and seed handling on latency and rate of germination in seeds consumed by capuchins (*Cebus capucinus*). *American Journal of Physical Anthropology* 13:486–492.

Valenta K, Fedigan LM. 2010. Spatial patterns of seed dispersal by white-faced capuchins in Costa Rica: evaluating distant-dependent seed mortality. *Biotropica* 42(2):223–228.

Valenta K, Burke RJ, Styler SA, *et al.* 2013. Colour and odour drive fruit selection and seed dispersal by mouse lemurs. *Scientific Reports* 3(2424):1–5.

Vieites DR, Wollenberg KC, Andreone F, *et al.* 2009. Vast underestimation of Madagascar's biodiversity evidenced by an integrative amphibian inventory. *Proceedings of the National Academy of Sciences, USA* 106(20):8267–8272.

Voigt F, Bleher B, Fietz J, *et al.* 2004. A comparison of morphological and chemical fruit traits between two sites with different frugivore assemblages. *Oecologia* 141(1):94–104.

Wright PC, Tecot SR, Erhart EM, *et al.* 2011. Frugivory in four sympatric lemurs: implications for the future of Madagascar's forests. *American Journal of Primatology* 73:585–602.

Wright S, Duber H. 2001. Poachers and forest fragmentation alter seed dispersal, seed survival, and seedling recruitment in the palm *Attalea butyraceae*, with implications for tropical tree diversity. *Biotropica* 33(4):583–595.

19 Predation in the dark: antipredator strategies of Cheirogaleidae and other nocturnal primates

Claudia Fichtel

Introduction

Predation can incur extreme fitness costs and can therefore drive the evolution and maintenance of elaborate and costly antipredator behavior in prey species (Caro, 2005; Fichtel, 2012). Although about one-third of all primates are nocturnal, small-bodied, and presumably facing a high predation risk (van Noordwijk *et al.*, 1993; Isbell, 1994; Janson, 2003; Hart, 2007; Scheumann *et al.*, 2007), few data exist regarding their alarm-calling behavior and other antipredator strategies (Fichtel, 2012). In general, predation occurs quickly, is rarely observed, and is difficult to study; even more so in nocturnal primates due to poor visibility at night. Small body size and nocturnality have been suggested to represent adaptations to reduce predation risk (Clutton-Brock and Harvey, 1977). Accordingly, nocturnal and diurnal species have been suggested to differ in their antipredator strategies; whereas nocturnal species may depend on reduced detectability and crypsis, diurnal species rely on early detection to ward off predators (Terborgh and Janson, 1986). However, recent studies revealed that not all nocturnal primates necessarily rely on reduced detectability and crypsis; some also mob predators and appear to rely on a combination of alarm calls and escape responses that are adapted to the different hunting tactics of predators (Fichtel, 2012).

Because little is known about antipredator behavior in cheirogaleids, in this chapter I will first summarize the current knowledge on antipredator behavior in nocturnal primates and then portray what is known about antipredator behavior in cheirogaleids. I begin with summarizing the kind of animals that prey upon cheirogaleids and which cues nocturnal primates use to recognize predators. Because predation imposes direct costs of mortality and chronic indirect costs of employing behaviors that reduce mortality risk, antipredator behavior can be broadly classified into tactics used to avoid encountering predators and those employed after predator

The Dwarf and Mouse Lemurs of Madagascar: Biology, Behavior and Conservation Biogeography of the Cheirogaleidae, eds. S.M. Lehman, U. Radespiel, E. Zimmermann. Published by Cambridge University Press. © Cambridge University Press 2016.

detection (Stanford, 2002). Therefore, I will secondly summarize tactics taken before encountering predators such as predator-sensitive foraging, appropriate sleeping site selection, increased vigilance, and the formation of mixed-species associations. Thirdly, I will address tactics that take effect during encounters such as alarm-calling and mobbing as well as various escape and defense tactics. Moreover, I will briefly discuss the evolution of grouping effects. Finally, I conclude this chapter by discussing future aspects of research on antipredator behavior in nocturnal primates.

Cheirogaleid predators

Predation occurs rarely, unpredictably, and is therefore difficult to observe, in particular at night. In addition, predators are typically wary of human observers, who are equipped with headlamps and torches, thereby presumably deterring predatory attacks, resulting in only anecdotal observations of predation on nocturnal primates.

Existing information indicates that all cheirogaleids are preyed upon both by nocturnal and diurnal birds of prey (Table 19.1). During their active period at night, cheirogaleids are preyed upon by barn owls (*Tyto alba*, Scopoli, 1786) that use acoustic cues to locate and capture their prey in the darkness (Payne, 1971). During their inactive period, cheirogaleids also face predation at their resting sites in tree holes: the Madagascar harrier hawk (*Polyboroides radiateus*, Scopoli, 1786) has specialized morphological adaptations that allows it to bend its legs to probe into narrow tree cavities and to extract cheirogaleids (Goodman *et al.*, 1993). Even hook-billed vangas (*Vanga curvirostris*, Linnaeus, 1766) have been observed to prey upon mouse lemurs.

Carnivores also prey upon cheirogaleids. The fosa (*Cryptoprocta ferox*, Bennet, 1833) and mongooses (*Mungotictis decemlineata*, A. Grandidier, 1866; *Galidia elegans*, Saint Hilaire, 1837) prey upon all cheirogaleids by either hunting by stealth, pursuing them, or extracting them out of their tree holes (Table 19.1). In addition, stray dogs (*Canis lupus familiaris*, Linnaeus, 1758) have been observed to prey upon mouse lemurs. Snakes (*Sanzinia madagascariensis*, Gray, 1849; *Ithycyphus miniatus*, Schlegel, 1867) prey upon *Cheirogaleus* and *Microcebus*, whereas no observation of snake predation has been documented so far in members of other genera (Table 19.1). Observations of joint snake mobbing by fork-marked lemurs and Coquerel's dwarf lemurs (*Mirza coquereli*, A. Grandidier, 1866) of a boa (*Boa manditra*, Kluge, 1991), however, suggest that they were also preyed upon by snakes (Schülke, 2001).

Some primates also prey upon each other (reviewed in Fichtel, 2012). Coquerel's dwarf lemurs have been observed preying upon gray mouse lemurs (*Microcebus murinus*, Miller 1777; Schliehe-Diecks *et al.*, 2010). A case of cannibalism has been documented in gray mouse lemurs, but it is not clear whether it was preceded by predation (Hämalainen, 2012). These observations suggest that cheirogaleids face a high predation risk by raptors, carnivores, snakes, and other cheirogaleids during both their active and inactive periods. Predation risk inferred from the number

Table 19.1 Predators of cheirogaleids.

Genus	Raptors	Carnivores	Reptiles	Primates	References
Microcebus	*Accipiter henstii* *Asio madagascariensis* *Polyboroides radiatus* *Tyto alba* *Tyto soumagnei* *Vanga curvirostris*	*Canis lupus* *Cryptoprocta ferox* *Galidia elegans* *Mungotictis decemlineata*	*Sanzinia madagascariensis* *Ithycyphus miniatus*	*Mirza coquereli*	1, 2, 3, 6, 7, 8, 9, 10, 11, 12, 14, 15, 16, 17, 19
Cheirogaleus	*Accipiter henstii* *Asio madagascariensis* *Buteo brachypterus* *Polyboroides radiatus* *Tyto alba*	*Cryptoprocta ferox* *Galidea elegans* *Mungotictis decemlineata*	*Sanzinia madagascariensis*		1, 2, 3, 4, 5, 12, 13, 14
Mirza	*Asio madagascariensis* *Buteo brachypterus*	*Cryptoprocta ferox* *Mungotictis decemlineata*			3, 4, 18
Phaner	*Asio madagascariensis* *Buteo brachypterus*	*Cryptoprocta ferox*			3

[1]Rand, 1935; [2]Rasoloarison *et al.*, 1995; [3]Goodman *et al.*, 1993; [4]Goodman and Langrand, 1996; [5]Gilbert and Tingay, 2001; [6]Richard, 1978; [7]Sterling and McFadden, 2000; [8]Karpanty and Goodman, 1999; [9]Thorstrom and La Marca, 2000; [10]Langrand, 1990; [11]Goodman *et al.*, 1998; [12]Wright and Martin, 1995; [13]Goodman *et al.*, 1997; [14]Goodman, 2003; [15]Albignac, 1976; [16]Petter *et al.*, 1977; [17]Deppe *et al.*, 2008; [18]Kappeler, 1997; [19]Schliehe-Diecks *et al.*, 2010.

of predators that prey upon a genus appears to be highest for *Microcebus* and *Cheirogaleus*, which are preyed upon by 9–13 predator species, and intermediate (4–8 predator species) for *Mirza* and *Phaner* (Scheumann *et al.*, 2007).

Predator recognition

In order to understand the evolution of antipredator behavior, it is important to understand which cues primates use to recognize predators. In this context, it is important to consider perceptual modalities separately. Primates can use pelage and skin patterns to recognize carnivores and snakes, but also the shape of silhouettes to recognize raptors (Fichtel, 2012). In nocturnal primates, little is known about the use of visual cues to recognize predators. In cheirogaleids, a study on captive gray and Goodman's mouse lemurs (previously *Microcebus rufus*, now *Microcebus lehilahytsara*, Kappeler *et al.*, 2005) showed that mouse lemurs produced alarm calls (whistles) in response to unknown or potentially dangerous objects (empty bottle, living python or python dummy, and barn owl dummy or barn owl silhouette; Zimmermann *et al.*, 2000), suggesting that experience may shape visual predator recognition. Similarly, wild brown mouse lemurs (*Microcebus rufus*, É. Geoffroy, 1834) responded more strongly to the skin of a local than a non-local snake, suggesting that visual snake recognition might not be innate (Deppe, 2006). Finally, in caged gray mouse lemurs in the wild, presentation of carnivore and snake models elicited monitoring behavior, probably to assess the potential risk that emanates from the predator. In response to raptor models, mouse lemurs behaved cryptically and exhibited freezing behavior, suggesting that they rely on visual cues to recognize predators (Rahlfs and Fichtel, 2010).

Primates may identify predators by acoustic cues alone, and in areas where predators regularly occur, calls of raptors or carnivores elicit antipredator responses (Hauser and Wrangham, 1990; Zuberbühler *et al.*, 1999; Karpanty and Grella, 2001; Fichtel and Kappeler, 2002; Arnold and Zuberbühler, 2006). Among nocturnal lemurs, recognition of predator calls – i.e., the association of predator vocalizations with a specific escape response – has so far only been demonstrated in sportive lemurs (*Lepilemur ruficaudatus*, Grandidier, 1867; *L. sahamalazensis*, Andriaholinirina *et al.*, 2006). Both species showed predator-specific escape responses to calls of aerial and terrestrial predators (Fichtel, 2007; Seiler *et al.*, 2013a). After playback experiments with vocalizations of aerial predator calls, red-tailed sportive lemurs behaved cryptically and showed freezing behavior, but after playback experiments with terrestrial predator calls they climbed up the tree they were currently in and scanned the ground (Fichtel, 2007).

The importance of paying attention to acoustic cues in predator avoidance is also reflected by the large number of animals that respond in adaptive ways to alarm calls emitted by other sympatrically living species (Caro, 2005; Fichtel, 2012). In nocturnal primates, recognition of heterospecific alarm calls has been documented in Sahamalaza sportive lemurs: during active periods at diurnal sleeping sites, lemurs increased their vigilance after playback experiments with alarm calls of sympatric birds or aerial predator alarm calls of blue-eyed lemurs (*Eulemur flavifrons*, Gray, 1867), but not after blue-eyed lemurs' terrestrial predator alarm calls (Seiler *et al.*, 2013b). In cheirogaleids, one study on acoustic-predator recognition in captive predator-naïve gray mouse lemurs suggested that acoustic predator recognition is influenced by experience (Bunkus *et al.*, 2005). In this study, mouse lemurs getting a reward at a fixed location were exposed to one of four acoustic stimuli: historical predator, introduced predator, historical non-predator, and a control. Mouse lemurs showed avoidance behaviors after presentation of unknown stimuli, suggesting that experience may shape acoustic recognition of their environment. However, it is currently unknown whether cheirogaleids in the wild use acoustic cues to recognize predators or pay attention to alarm calls of sympatric species. Nevertheless, these studies tentatively suggest that, as in diurnal primates, visual and acoustic recognition of predators seem to be based on experience and learning (Fichtel, 2012).

Primates also use their sense of smell to recognize predators (Fichtel, 2012). A study in captive gray mouse lemurs indicated that they possess an innate ability to recognize the scents of predators from their scat (Sündermann *et al.*, 2008). Similarly, wild gray and golden-brown mouse lemurs (*Microcebus ravelobenis*) in a two-choice experimental design avoided olfactory stimuli (fecal samples) of carnivores (fosa) but interestingly not raptors (barn owl) (Kappel *et al.*, 2011). The lack of avoidance of raptor stimuli might be explained by fewer or no aversive metabolites in owl feces, or by a lower biological significance for olfactory recognition of raptors (Kappel *et al.*, 2011). Moreover, wild brown mouse lemurs also responded more strongly to olfactory stimuli of local and novel predators compared to olfactory stimuli of non-feared non-predator or new odors, suggesting that olfactory recognition of predators has an innate component (Deppe *et al.*, 2007). As suggested for mammals in general (reviewed in Apfelbach *et al.*, 2005), potential prey may recognize predators through their urine and feces via metabolites produced by the digestion of meat. The production of a consistent olfactory cue may lead to selection favoring the ability of prey species to recognize predators through such cues. Because seeing or hearing a predator also provides additional cues that allow primates to classify an animal as a predator, smell alone rarely does so. Hence, recognition of predators by smell should be independent of any learning.

Behavioral tactics to avoid predation

The antipredator behavior of individuals can be divided into tactics employed before and after encountering predators. Tactics employed before encountering predators include predator-sensitive foraging, choosing safe sleeping sites, vigilance, and forming mixed-species associations. Tactics employed during or after encountering predators include emitting alarm calls, fleeing, and confronting and/or mobbing predators. Because the shift from an originally solitary to a gregarious lifestyle has been considered to be one of the major evolutionary transitions (Maynard Smith and Szathmáry, 1995), some of the tactics such as selfish herding, confusion, and dilution effects might not be available for nocturnal primates, because they roam mostly solitarily during their active period (Kappeler, 2012).

Tactics employed before encountering predators

Predator-sensitive foraging

Because predators move unpredictably in space and time (Sih, 1997), animals face a trade-off between finding food and avoiding becoming food for others (Lima and Dill, 1990). Predator- or threat-sensitive foraging has been documented in many diurnal primates (reviewed in Miller, 2002; Fichtel, 2012). Patterns of habitat use appear to reflect the trade-off between foraging demands and the constant risk of predation. Primates spend less time feeding in habitats with higher predation risk, or adjust their time spent feeding to the party size of conspecifics (van Schaik *et al.*, 1983; Cowlishaw, 1997a, 1997b; Miller, 2002).

Risk-sensitive foraging has so far only been documented in one nocturnal primate, the gray mouse lemur (Dammhahn and Almeling, 2012). Mouse lemurs' readiness to accept risks during foraging has been studied experimentally by presenting feeding platforms at a height of 1.5 m and on the ground, assuming that platforms on the ground pose a higher risk. Mouse lemurs entered low-risk feeding platforms faster than high-risk platforms and commenced feeding faster on low-risk platforms, suggesting that they adapt their foraging tactics to perceived predation risk.

In addition, lunar phase influences activity patterns of nocturnal primates, perhaps because it affects perceived predation risk. Nocturnal white-footed sportive lemurs (*Lepilemur leucopus*, Major, 1894), slender and pygmy lorises (*Loris lydekkerianus*, Cabrera, 1908; *Nycticebus pygmaeus*, Bonhote, 1907), spectral tarsiers (*Tarsius spectrum*, Erxleben, 1777), and Colombian night monkeys (*Aotus lemurinus*, Geoffroy, 1846) are relatively active during moonlit nights (Fernandez-Duque and Erkert, 2006; Nash, 2007; Starr *et al.*, 2012). Moreover, southern lesser galagos (*Galago moholi*, A. Smith, 1836) avoid traveling on the ground during nights without moonlight, presumably because small carnivores represent a significant risk to them (Bearder *et al.*, 2002). Thus, bright nights may enhance foraging efficiency to capture insects or ripe fruits but also detectability by predators. However, whether cheirogaleids also adjust their ranging behavior strategically as a means to avoid predation or adjust their activity to the lunar phases has not been studied so far.

Selecting safe sleeping sites

Individual primates spend over half of their lives sleeping, and deciding where to sleep therefore represents a significant problem. Sleeping sites must be carefully chosen to avoid predators and to deal with thermoregulation, parasites, and territory defense (Anderson, 1998; Kappeler, 1998). Physical characteristics of sleeping sites, their patterns of use, the timing of retirement, and pre-retirement behavior could all influence predation risk, and diurnal primates have evolved corresponding behavioral adaptations (Fichtel, 2012). Rapid movements toward sleeping sites, cryptic behavior near them, increased vigilance when departing the shelter, and avoidance of fixed sleeping sites may reduce the vulnerability of individuals to predation.

Nocturnal primates, and in particular cheirogaleids, usually sleep and rest in tree holes or nests in dense vegetation (reviewed in Kappeler, 1998). These may serve to protect them from diurnal predators (Goodman *et al.*, 1993; Schülke and Ostner, 2001), but they may also perform other functions, including thermoregulation, especially for young infants (Martin, 1972; Kappeler, 1998; Schmid and Kappeler, 1998; Dausmann *et al.*, 2004). In general, sleeping holes located higher up in a tree, with smaller entrances and thicker walls, may provide better protection against predation (Fietz, 1998; Radespiel *et al.*, 1998; Lahann, 2008). Nocturnal primates also switch between preferred areas frequently (Wright, 1994; Radespiel *et al.*, 1998, 2003; Schmid and Kappeler, 1998; Schülke and Kappeler, 2003; Rasoloharijaona *et al.*, 2003, 2008; Zinner *et al.*, 2003; Dammhahn and Kappeler, 2005; Eberle and Kappeler, 2006; Lahann, 2008; Seiler *et al.*, 2013c).

Some nocturnal primates sleep in dense vegetation, an observation consistent with the hypothesis that the use of tree holes and nests has functions other than to mitigate predation (Kappeler, 1998; Bearder *et al.*, 2003; Dammhahn and Kappeler, 2005). In cheirogaleids, *Mirza* rest in nests (Kappeler, 1997), *Phaner* in both nests and tree holes (Schülke and Kappeler, 2003), and *Cheirogaleus* usually rest in tree holes (Dausmann *et al.*, 2004; Lahann, 2008). In gray mouse lemurs, females usually rest in tree holes with other related females, whereas males most often rest alone in open vegetation (Radespiel *et al.*, 1998, 2003; Schmid and Kappeler, 1998; Lahann, 2008). This sex difference in sleeping site choice might be due to the fact that females monopolize high-quality sleeping holes (Radespiel *et al.*, 1998, 2001; Lahann, 2008). Males may thus prefer to rest in open vegetation rather than in low-quality tree holes to mitigate predation (Lahann, 2008).

Reduced activity has also been suggested as a behavioral strategy to avoid predators (Caro, 2005). Gray mouse lemurs in Kirindy forest showed sex differences in activity patterns: females fall into torpor during the dry season, whereas males remain active. These sex differences in activity have been suggested to reflect a female strategy to mitigate predation during periods of low food availability and high predation risk (Rasoloarison *et al.*, 1995; Schmid and Kappeler, 1998; Schmid, 1999). A 10-year capture–mark–recapture study revealed, however, that females and males survive equally well during the dry season, suggesting that reduced activity does not provide direct benefits against predation (Kraus *et al.*, 2008; see also Lutermann *et al.*, 2006; Schmelting *et al.*, 2007 for similar results at the study site Ampijoroa in northwestern Madagascar). In summary, tree holes and nests likely serve multiple functions, one of them being protection against predators. However, behavioral adaptations such as cryptic behavior, vigilance, and movements around sleeping sites prior to retiring to mitigate predation have not been studied systematically in cheirogaleids and other nocturnal primates.

Vigilance

Early detection of approaching predators substantially enhances the survival chances of prey because it gives them time to observe the predator, to hide, to alert other conspecifics, or to deter attacks. Antipredator vigilance has been referred to as scanning for predators, sometimes precluding other activities such as foraging. Many studies clearly show that diurnal primates increase their levels of vigilance after detecting predators or predator models, or after hearing predator calls or alarm calls (van Schaik *et al.*, 1989; Baldellou and Henzi, 1992; König, 1998; Hirsch, 2002; Fichtel, 2004, 2007). Several ecological and social factors, such as characteristics of the habitat and the degree of canopy cover, as well as an individual's activity, sex, age, dominance rank, spatial position, or the presence of infants can influence individual levels of vigilance (reviewed in Fichtel, 2012).

It has been suggested that vigilance is not a viable antipredator behavior in nocturnal primates due to poor visibility at night (Terborgh and Janson, 1986). However, playback experiments with pair-living red-tailed and Sahamalazenis sportive lemurs in the wild, as well as solitarily foraging gray mouse lemurs in the wild, revealed that they increase their scanning rates after playbacks of either species-typical alarm calls or predator calls/models compared to control calls/models (Fichtel, 2007; Rahlfs and Fichtel, 2010; Seiler *et al.*, 2013a). Red-tailed sportive lemurs also increased their scanning rates after hearing calls of their pair-mate (Fichtel and Hilgartner, 2012), indicating that social vigilance also affects scanning behavior. Thus, there is evidence that nocturnal primates, including some cheirogaleids, also scan for predators. However, auditory vigilance, which is focused on acoustic cues that may indicate the presence of predators, such as wing flaps or footfalls, has not yet been investigated in any primate species.

Mixed-species associations

Some diurnal primates associate with other species, resulting in increased effective group size (reviewed in Fichtel, 2012). Such mixed-species groups usually consist of members of related taxa – i.e., the same genus or family (but see Newton, 1989). The formation of mixed-species associations, however, has not been documented in nocturnal primates, presumably because the benefits of early predator detection do not accrue.

Tactics employed after predator detection

Once a predator has been detected, prey can respond in several ways: they can give acoustic warning signals that might be directed to both predators and conspecifics, or they might flee or even confront predators (Scheumann *et al.*, 2007; Fichtel, 2012). Diurnal primates adapt their escape strategies and associated behaviors to the different hunting strategies employed by predators and adjust their antipredator behavior flexibly to habitat types, varying sets of predators, and perceived predation risk (Fichtel, 2012).

Alarm calls, mobbing, and escape responses in nocturnal primates

It has been suggested that diurnal and nocturnal species may differ in their antipredator strategies; while the former rely on early detection to ward off predators, nocturnal species may depend on reduced detectability and crypsis (Clutton-Brock and Harvey, 1977; van Schaik, 1983; Terborgh and Janson, 1986). Recent studies, however, revealed that nocturnal primates, including some cheirogaleids, also produce alarm calls, show predator-specific escape strategies, and mob predators (Schülke, 2001; Bearder *et al.*, 2002; Gursky, 2005; Fichtel, 2007; Scheumann *et al.*, 2007; Eberle and Kappeler, 2008), thus challenging earlier notions about a premium on crypsis.

Among nocturnal primates, spectral tarsiers (*Tarsius tarsier*), which are secondarily nocturnal (Martin and Ross, 2005), have been suggested to produce two different predator-specific alarm calls in response to models of aerial and terrestrial predators to warn conspecifics (Gursky, 2006, 2007). In other nocturnal primates, alarm calls appear to be directed at predators but also conspecifics. For example, alarm calls of large-eared greater galagos (*Otolemur crassicaudatus*, É. Geoffroy, 1812) and southern lesser galagos have been suggested to reflect the perceived risk of the caller and may serve to deter predators (Bearder, 2007). Red-tailed and Milne

Edwards' sportive lemurs do not alarm-call to alert conspecifics to the presence of predators, but instead direct these calls to the predators themselves (Fichtel, 2007; Scheumann *et al.*, 2007). These so-called general alarm calls are also given during aggressive interactions with conspecifics (Fichtel and Kappeler, 2002; Fichtel, 2007), a phenomenon that has also been found in other nocturnal primates such as Senegal bushbabies, and mouse lemurs (Zimmermann, 1985a; Scheumann *et al.*, 2007; Rahlfs and Fichtel, 2010). In cheirogaleids, gray mouse lemurs have been observed to mob snakes (Eberle and Kappeler, 2008). Alarm calls of an individual caught by a boa recruited conspecifics that mobbed and attacked the snake until the caught individual could escape (Eberle and Kappeler, 2008). Alarm-calling here might have served to recruit conspecifics to join the mobbing (Rahlfs and Fichtel, 2010). Interestingly other cheirogaleids – i.e., fork-marked and Coquerel's dwarf lemurs – mobbed a boa jointly, although they also compete over food (Schülke, 2001).

Little is known about escape responses in nocturnal primates. However, the few studies that examined antipredator behavior in nocturnal primates in more detail suggest that their escape responses also correspond to the hunting strategies of the corresponding predators. Spectral tarsiers, for example, sometimes froze in response to raptor models and sometimes mobbed them, whereas carnivore and snake models elicited upwards climbing and mobbing (Gursky, 2006, 2007). Playback experiments with calls of predators showed that red-tailed sportive lemurs climb up after presentation of terrestrial predator calls and show freezing behavior in response to aerial predator calls (Fichtel, 2007). Similarly, Sahamalazensis sportive lemurs at diurnal resting sites showed adaptive escape responses corresponding to different types of predators (Seiler *et al.*, 2013a). In cheirogaleids, escape responses have been studied in more detail only in gray mouse lemurs (Rahlfs and Fichtel, 2010). In response to visual models of raptors, which rely on acoustic cues to locate prey, they showed freezing behavior, but they remained vigilant in confrontations with ambush-hunting terrestrial predator and snake models (Rahlfs and Fichtel, 2010). Although studies in captivity showed that *Mirza* also produce alarm calls (Macedonia and Stanger, 1994; Stanger, 1995), little is known about the functions of these calls and associated escape behavior. In particular, studies in *Mirza* would be of interest because they also prey upon smaller primates (Schliehe-Diecks *et al.*, 2010).

Slow lorises lack alarm calls (Zimmermann, 1985b), but have brachial glands that produce oily exudates with a repelling odor, which they spread over their body while autogrooming. Although these exudates do not contain toxic compounds, they have been suggested to play a role in deterring predators (Hagey *et al.*, 2007). Thus, solitary foraging and crypsis remain effective predator avoidance strategies for many nocturnal primates. However, they have evolved specialized antipredator behaviors that are well adapted to the different hunting styles of their predators. Alarm calls, however, may primarily serve to communicate with predators.

Most diurnal primates live in groups (van Schaik, 1983), reducing thereby the likelihood of being preyed upon after detecting predators (Hamilton, 1971). Individuals benefit thereby from the dilution, confusion, and the selfish herd effect (Miller, 1922; Hamilton, 1971). Because many nocturnal primates are solitary or organized into dispersed pairs in which pair partners roam solitarily during their active period (Kappeler, 2012), these grouping effects may not be beneficial for nocturnal primates. Instead, it has been suggested that small body size and nocturnality represent adaptations to reduce predation risk (Clutton-Brock and Harvey, 1977). Predation risk estimated by the number of predatory species per genus across lemurs, however, did not differ between nocturnal and diurnal/cathemeral species, and also did not correlate with group size but with body mass, suggesting that lemurs may not benefit from grouping effects and that particularly small-bodied lemurs face a high predation risk (Scheumann *et al.*, 2007). In contrast, Hill and Lee (1998) suggested in a comparative study across cercopithecoid primates that predation risk, estimated by predator presence, density, and attack rates, is positively correlated with group size. Because reliable data on predator presence, density, and predator attack rates are quite often not available, estimating predation risk across species is challenging. However, the transition from a solitary to a pair-living lifestyle provides theoretically the most benefits of grouping, because for each individual the probability that it will be preyed upon is reduced by 50%. Grouping effects should be most beneficial in nocturnal pair-living primates which coordinate their activities, as for example woolly lemurs (*Avahi* ssp.). Accordingly, studies on vigilance behavior or alarm calls and escape responses would provide important information on benefits of group-living.

In conclusion, cheirogaleids are preyed upon by both nocturnal and diurnal raptors, carnivores, and snakes, suggesting that they face a high predation risk (Goodman *et al.*, 1993; Hart, 2007; Scheumann *et al.*, 2007). In general, nocturnal primates and in particular cheirogaleids employ tactics before encountering predators including predator-sensitive foraging, choosing safe sleeping sites, and vigilance to reduce predation risk. In addition, nocturnal primates also eavesdrop on heterospecific alarm calls of sympatric species. Tactics employed during or after encountering predators including the emission of alarm calls, fleeing, and confronting or mobbing predators have also been described in nocturnal primates and in some cheirogaleids. These studies indicate that solitary foraging and crypsis remain effective predator avoidance strategies for many nocturnal primates. In addition, they have, similarly to diurnal primates, evolved specialized antipredator behaviors that are well adapted to the different hunting styles of their predators. Alarm calls, however, may primarily serve to communicate with predators. Because nocturnal strepsirrhines are the most evolutionarily basal primates living today, their alarm-calling behavior might represent the ancestral form for the entire order (Fichtel, 2007).

In cheirogaleids only a few aspects of antipredator behavior have been studied in detail. For example, little is known about which cues they use to recognize predators (Bunkus *et al.*, 2005; Scheumann *et al.*, 2007; Sündermann *et al.*, 2008; Rahlfs and Fichtel, 2010; Kappel *et al.*, 2011). Moreover, studies on tactics to avoid predation before predator encounters, such as vigilance and coordination at sleeping sites or predator-sensitive foraging, might provide insights into the strategic habitat use in cheirogaleids. Finally, research on functions of alarm calls in solitary (*Microcebus*, *Mirza*) and pair-living species (*Cheirogaleus*, *Phaner*) may provide important insights into the evolution of alarm calls – i.e., whether alarm calls are directed at predators or whether alarm-calling individuals might receive indirect fitness by doing so (Eberle and Kappeler, 2008; Fichtel, 2012).

Acknowledgments

I am very grateful to Shawn Lehman, Elke Zimmermann, and Ute Radespiel for inviting me to contribute to this book on cheirogaleids. I also thank Peter Kappeler for discussions of predation in nocturnal primates. I thank Anja Deppe, Peter Kappeler, and Elke Zimmermann for constructive comments on this chapter.

References

Albignac R. 1976. L'écologie du *Mungotictis decemlineata* dans les forêts decidues de l'ouest de Madagascar. *Revue d'Écologie (La Terre et la Vie)* 30:347–376.

Anderson J. 1998. Sleep, sleeping sites, and sleep-related activities: awakening to their significance. *American Journal of Primatology* 46:63–75.

Andriaholinirina N, Fausser J-L, Roos C, *et al.* 2006. Molecular phylogeny and taxonomic revision of the sportive lemurs (*Lepilemur*, Primates). *BMC Evolutionary Biology* 6:17.

Apfelbach R, Blanchard C, Blanchard R, Hayes R, McGregor I. 2005. The effects of predator odors in mammalian prey species: a review of field and laboratory studies. *Neuroscience and Biobehavioral Reviews* 29:1123–1144.

Arnold K, Zuberbühler K. 2006. The alarm-calling system of adult male putty-nosed monkeys, *Cercopithecus nictitans martini. Animal Behavior* 72:643–653.

Baldellou M, Henzi SP. 1992. Vigilance, predator detection and the presence of supernumerary males in vervet monkey troops. *Animal Behavior* 43:451–462.

Bearder S. 2007. A comparison of calling patterns in two nocturnal primates, *Otolemur crassicaudatus* and *Galago moholi* as a guide to predation risk. In S Gursky, KAI Nekaris (eds.), *Primate Anti-Predator Strategies* (pp. 206–221). Springer, New York.

Bearder SK, Nekaris KAI, Buzzell CA. 2002. Dangers in the night: are some nocturnal primates afraid of the dark? In L Miller (ed.), *Eat or Be Eaten: Predator Sensitive Foraging Among Primates* (pp. 11–24). Cambridge University Press, Cambridge.

Bearder SK, Ambrose L, Harcourt C, *et al.* 2003. Species-typical patterns of infant contact, sleeping site use and social cohesion among nocturnal primates in Africa. *Folia Primatologica* 74:337–354.

Bunkus E, Scheumann M, Zimmermann E. 2005. Do captive-born grey mouse lemurs (*Microcebus murinus*) recognize their natural predators by acoustic cues alone? *Primate Report* 72:22.

Caro T. 2005. *Antipredator Defenses in Birds and Mammals.* The University of Chicago Press, Chicago.

Clutton-Brock TH, Harvey PH. 1977. Primate ecology and social organization. *Journal of Zoology, London* 183:1–39.

Cowlishaw G. 1997a. Refuge use and predation risk in a desert baboon population. *Animal Behavior* 54:241–253.

Cowlishaw G. 1997b. Trade-offs between foraging and predation risk determine habitat use in a desert baboon population. *Animal Behavior* 53:667–686.

Dammhahn M, Almeling L. 2012. Is risk taking during foraging a personality trait? A field test for cross-context consistency in boldness. *Animal Behavior* 84:1131–1139.

Dammhahn M, Kappeler PM. 2005. Social system of *Microcebus berthae*, the world's smallest primate. *International Journal of Primatology* 26:407–435.

Dausmann K, Glos J, Ganzhorn JU, Heldmaier G. 2004. Hibernation in a tropical primate. *Nature* 429:825–826.

Deppe AM. 2006. Visual snake recognition in wild brown mouse lemurs (*Microcebus rufus*). *American Journal of Primatology* 68(Suppl 1):34.

Deppe AM, Randriamiarisoa M, Wright PC. 2007. Mammalian predator recognition via olfactory cues in wild brown mouse lemurs (*Microcebus rufus*). *American Journal of Physical Anthropology* 44:99.

Deppe AM, Randriamiarisoa M, Kasprak AH, Wright PC. 2008. Predation on the brown mouse lemur (*Microcebus rufus*) by a diurnal carnivore, the ring-tailed mongoose (*Galidia elegans*). *Lemur News* 13:17–18.

Eberle M, Kappeler PM. 2006. Family insurance: kin selection and cooperative breeding in a solitary primate (*Microcebus murinus*). *Behavioral Ecology and Sociobiology* 60:582–588.

Eberle M, Kappeler P. 2008. Mutualism, reciprocity, or kin selection? Cooperative rescue of a conspecific from a boa in a nocturnal solitary forager the gray mouse lemur. *American Journal of Primatology* 70:410–414.

Fernandez-Duque E, Erkert H. 2006. Cathemerality and lunar periodicity of activity rhythms in owl monkeys of the Argentinian Chaco. *Folia Primatologica* 77:123–138.

Fichtel C. 2004. Reciprocal recognition in sifaka (*Propithecus verreauxi verreauxi*) and redfronted lemur (*Eulemur fulvus rufus*) alarm calls. *Animal Cognition* 7:45–52.

Fichtel C. 2007. Avoiding predators at night: antipredator strategies in red-tailed sportive lemurs (*Lepilemur ruficaudatus*). *American Journal of Primatology* 69:611–624.

Fichtel C. 2012. Predation. In JC Mitani, J Call, P Kappeler, R Palombit, J Silk (eds.), *The Evolution of Primate Societies* (pp. 169–194). Chicago University Press, Chicago.

Fichtel C, Hilgartner R. 2012. Noises in the dark: vocal communication in nocturnal pair-living primates. In J Master, M Gamba, F Génin (eds.), *Leaping Ahead: Advances in Prosimian Biology* (pp. 297–304). Springer, New York.

Fichtel C, Kappeler PM. 2002. Anti-predator behavior of group-living Malagasy primates: mixed evidence for a referential alarm call system. *Behavioral Ecology and Sociobiology* 51:262–275.

Fietz J. 1998. Body mass in wild *Microcebus murinus* over the dry season. *Folia Primatologica* 69:183–190.

Gilbert M, Tingay RE. 2001. Predation of a fat-tailed dwarf lemur *Cheirogaleus medius* by a Madagascar harrier-hawk *Polyboroides radiatus*: an incidental observation. *Lemur News* 6:6.

Goodman SM. 2003. Predation on lemurs. In SM Goodman, JP Benstead (eds.), *The Natural History of Madagascar* (pp. 1221–1228). University of Chicago Press, Chicago.

Goodman SM, Langrand O. 1996. Food remains found in a nest of the Madagascar buzzard (*Buteo brachypterus*) in the Vohibasia Forest. *Working Group on Birds in the Madagascar Region Newsletter* 6:13–14.

Goodman SM, O'Connor S, Langrand O. 1993. A review of predation on lemurs: implications for the evolution of social behavior in small, nocturnal primates. In PM Kappeler, JU Ganzhorn (eds.), *Lemur Social Systems and Their Ecological Basis* (pp. 51–66). Plenum Press, New York.

Goodman SM, Langrand O, Rasolonandrasana BPN. 1997. The food habits of *Cryptoprocta ferox* in the high mountain zone of the Andringitra Massif, Madagascar (Carnivora, Viverridae). *Mammalia* 61:185–192.

Goodman SM, de Roland LAR, Thorstrom R. 1998. Predation on the eastern wooly lemur (*Avahi laniger*) and other vertebrates by Henst's goshawk (*Accipiter henstii*). *Lemur News* 3:14–15.

Gursky S. 2003. Predation experiments on infant spectral tariser (*Tarsius spectrum*). *Folia Primatologica* 74:272–284.

Gursky S. 2005. Predator mobbing in *Tarsius spectrum*. *International Journal of Primatology* 26:207–221.

Gursky S. 2006. Function of snake mobbing in spectral tarsiers. *American Journal of Physical Anthropology* 129:601–608.

Gursky S. 2007. The response of spectral tarsiers toward avian and terrestrial predators. In S Gursky, KAI Nekaris (eds.), *Primate Anti-Predator Strategies* (pp. 241–252). Springer, New York.

Hagey L, Fry B, Fitch-Snyder H. 2007. Talking defensively, a dual use for the brachial gland exudate of slow and pygmy lorises. In S Gursky, KAI Nekaris (eds.), *Primate Anti-Predator Strategies* (pp. 253–272). Springer, New York.

Hämäläinen A. 2012. A case of adult cannibalism in the gray mouse lemur, *Microcebus murinus. American Journal of Primatology* 74:783–787.

Hamilton WD. 1971. Geometry of the selfish herd. *Journal of Theoretical Biology* 31:295–311.

Hart D. 2007. Predation on primates: a biogeographical analysis. In S Gursky, KAI Nekaris (eds.), *Primate Anti-Predator Strategies* (pp. 27–59). Springer, New York.

Hauser MD, Wrangham RW. 1990. Recognition of predator and competitor calls in nonhuman primates and birds: a preliminary report. *Ethology* 86:116–130.

Hill R, Lee PC. 1998. Predation risk as an influence on group size in cercopithecoid primates: implications for social structure. *Journal of Zoology, London* 245:447–456.

Hirsch B. 2002. Social monitoring and vigilance behavior in brown capuchin monkeys (*Cebus apella*). *Behavioral Ecology and Sociobiology* 52:458–464.

Isbell L. 1994. Predation on primates: ecological patterns and evolutionary consequences. *Evolutionary Anthropology* 3:61–71.

Janson C. 2003. Puzzles, predation, and primates: using life history to understand selection pressures. In PM Kappeler, ME Pereira (eds.), *Primate Life Histories and Socioecology* (pp. 103–131). University of Chicago Press, Chicago.

Kappel P, Hohenbrink S, Radespiel U. 2011. Experimental evidence for olfactory predator recognition in wild mouse lemurs. *American Journal of Primatology* 73:928–938.

Kappeler P. 1998. Nests, tree holes, and the evolution of primate life histories. *American Journal of Primatology* 46:7–33.

Kappeler PM. 1997. Intrasexual selection in *Mirza coquereli*: evidence for scramble competition polygyny in a solitary primate. *Behavioral Ecology and Sociobiology* 45:115–127.

Kappeler PM. 2012. The behavioral ecology of strepsirrhines and tarsiers. In JC Mitani, J Call, PM Kappeler, R Palombit, J Silk (eds.), *The Evolution of Primate Societies* (pp. 17–42). Chicago University Press, Chicago.

Kappeler PM, Rasoloarison RM, Razafimanantsoa L, Walter L, Roos C. 2005. Morphology, behaviour and molecular evolution of giant mouse lemurs (*Mirza* spp.) Gray, 1970, with description of a new species. *Primate Report* 71:3–26.

Karpanty SM, Goodman SM. 1999. Diet of the Madagascar harrier-hawk, *Polyboroides radiatus*, in southeastern Madagascar. *Journal of Raptor Research* 33(4):313–316.

Karpanty SM, Grella R. 2001. Lemur responses to diurnal raptor calls in Ranomafana National Park, Madagascar. *Folia Primatologica* 172:100–103.

König A. 1998. Visual scanning by common marmosets (*Callithrix jacchus*): functional aspects and the special role of adult males. *Primates* 39:85–90.

Kraus C, Eberle M, Kappeler P. 2008. The costs of risky male behaviour: sex differences in seasonal survival in a small sexually monomorphic primate. *Proceedings of the Royal Society B* 275:1635–44.

Lahann P. 2008. Habitat utilization of three sympatric cheirogaleid lemur species in a littoral rain forest of Southeastern Madagascar. *International Journal of Primatology* 29:117–134.

Langrand O. 1990. *Guide to the Birds of Madagascar.* Yale University Press, New Haven.

Lima SL, Dill LM. 1990. Behavioral decisions made under the risk of predation: a review and prospectus. *Canadian Journal of Zoology* 68:616–640.

Lutermann H, Schmelting B, Radespiel U, Ehresmann P, Zimmermann E. 2006. The role of survival for the evolution of female philopatry in a solitary forager, the grey mouse lemur (*Microcebus murinus*). *Proceedings of the Royal Society B* 273:2527–2533.

Macedonia J, Stanger K. 1994. Phylogeny of the Lemuridae revisited: evidence from communication signals. *Folia Primatologica* 63:1–43.

Martin RD. 1972. Adaptive radiation and behaviour of the Malagasy lemurs. *Philosophical Transactions of the Royal Society, London B* 264:295–352.

Martin RD, Ross CF. 2005. The evolutionary and ecological context of primate vision. In *The Primate Visual System: A Comparative Approach* (pp. 1–36). John Wiley, Chichester.

Maynard Smith J, Szathmáry E. 1995. *The Major Transitions in Evolution.* Oxford University Press, Oxford.

Miller L. 2002. *Eat or Be Eaten: Predator Sensitive Foraging Among Primates.* Cambridge University Press, New York.

Miller RC. 1922. The significance of the gregarious habit. *Ecology* 3:122–126.

Nash L. 2007. Moonlight and behavior in nocturnal and cathemeral primates, especially *Lepilemur leucopus*: illuminating possible anti-predator efforts. In S Gursky, KAI Nekaris (eds.), *Primate Anti-Predator Strategies* (pp. 173–205). Springer, New York.

Newton PN. 1989. Associations between Langur Monkeys (*Presbytis entellus*) and Chital Deer (*Axis axis*): chance encounters or a mutualism? *Ethology* 83:89–120.

Payne R. 1971. Acoustic location of prey by barn owls (*Tyto alba*). *Journal of Experimental Biology* 54:535–573.

Petter J-J, Albignac R, Rumpler Y. 1977. *Mammifères Lémuriens (Primates Prosimiens).* Vol. 44 of Faune de Madagascar. ORSTOM/CNRS, Paris.

Radespiel U, Cepok S, Zietemann V, Zimmermann E. 1998. Sex-specific usage patterns of sleeping sites in grey mouse lemurs (*Microcebus murinus*) in northwestern Madagascar. *American Journal of Primatology* 46:77–84.

Radespiel U, Sarikaya Z, Zimmermann E, Bruford MW. 2001. Sociogenetic structure in a free-living nocturnal primate population: sex-specific differences in the grey mouse lemur (*Microcebus murinus*). *Behavioral Ecology and Sociobiology* 50:493–502.

Radespiel U, Ehresmann P, Zimmermann E. 2003. Species-specific usage of sleeping sites in two sympatric mouse lemur species (*Microcebus murinus* and *M. ravelobensis*) in northwestern Madagascar. *American Journal of Primatology* 59:139–151.

Rahlfs M, Fichtel C. 2010. Anti-predator behaviour in a nocturnal primate, the grey mouse lemur (*Microcebus murinus*). *Ethology* 116:429–439.

Rand AL. 1935. On the habits of some Madagascar mammals. *Journal of Mammalogy* 16:89–104.

Rasoloarison RM, Rasolonandrasana BPN, Ganzhorn JU, Goodman SM. 1995. Predation on vertebrates in the Kirindy forest, Western Madagascar. *Ecotropica* 1:59–65.

Rasoloharijaona S, Rakotosamimanana B, Randrianambinina B, Zimmermann E. 2003. Pair-specific usage of sleeping sites and their implications for social organization in a nocturnal Malagasy primate, the Milne Edwards' sportive lemur (*Lepilemur edwardsi*). *American Journal of Physical Anthropology* 122:251–258.

Rasoloharijaona S, Randrianambinina B, Zimmermann E. 2008. Sleeping site ecology in a rain-forest dwelling nocturnal lemur (*Lepilemur mustelinus*): implications for sociality and conservation. *American Journal of Primatology* 70:247–253.

Richard A. 1978. *Behavioral Variation: Case Study of a Malagasy Lemur.* Bucknell University Press, Lewisburg.

Scheumann M, Rabesandratana, Zimmermann E. 2007. Predation, communication, and cognition in lemurs. In S Gursky, KAI Nekaris (eds.), *Primate Anti-Predator Strategies* (pp. 100–126). Springer, New York.

Schliehe-Diecks S, Markolf M, Huchard E. 2010. When big lemurs swallow up small ones: the Coquerel's dwarf lemur as a predator of grey mouse lemurs and endemic rodents. *Lemur News* 15:13–14.

Schmelting B, Zimmermann E, Berke O, Bruford MW, Radespiel U. 2007. Experience-dependent recapture rates and reproductive success in male grey mouse lemurs (*Microcebus murinus*). *American Journal of Physical Anthropology* 133:743–752.

Schmid J. 1999. Sex-specific differences in activity patterns and fattening in the gray mouse lemur (*Microcebus murinus*) in Madagascar. *Journal of Mammalogy* 80:749–757.

Schmid J, Kappeler PM. 1998. Fluctuating sexual dimorphism and differential hibernation by sex in a primate, the gray mouse lemur (*Microcebus murinus*). *Behavioral Ecology and Sociobiology* 43:125–132.

Schülke O. 2001. Social anti-predator behaviour in a nocturnal lemur. *Folia Primatologica* 172:332–334.

Schülke O, Kappeler PM. 2003. So near and yet so far: territorial pairs but low cohesion between pair-partners in a nocturnal lemur, *Phaner furcifer*. *Animal Behavior* 65:331–343.

Schülke O, Ostner J. 2001. Predation on a lepilemur by a harrier hawk and implications of sleeping site quality. *Lemur News* 6:5.

Seiler M, Schwitzer C, Holderied M. 2013a. Anti-predator behaviour of Sahamalaza sportive lemurs, *Lepilemur sahamalazensis*, at diurnal sleeping sites. *Contributions to Zoology* 82:131–143.

Seiler M, Schwitzer C, Gamba M, Holderied M. 2013b. Interspecific semantic alarm call recognition in the solitary Sahamalaza sportive lemur, *Lepilemur sahamalazensis*. *PLoS ONE* 8:e67397.

Seiler M, Holderied M, Schwitzer C. 2013c. Effects of habitat degradation on sleeping site choice and use in Sahamalaza sportive lemurs (*Lepilemur sahamalazensis*). *International Journal of Primatology* 34:260–280.

Sih A. 1997. To hide or not to hide? Refuge use in a fluctuating environment. *TREE* 12:375–376.

Stanford C. 2002. Avoiding predators: expectations and evidence in primate antipredator behavior. *International Journal of Primatology* 23:741–757.

Stanger K. 1995. Vocalizations of some cheirogaleid prosimians evaluated in a phylogenetic context. In L Alterman, G Doyle, M Izard (eds.), *Creatures of the Dark* (pp. 353–376). Plenum Press, New York.

Starr C, Nekaris KAI, Leung L. 2012. Hiding from the moonlight: luminosity and temperature affect activity of Asian nocturnal primates in a highly seasonal forest. *PLoS ONE* 7:e36396.

Sterling EJ, McFadden K. 2000. Rapid census of lemur populations in the Parc National de Marojejy, Madagascar. *Fieldiana Zoology* 97:265–274.

Sündermann D, Scheumann M, Zimmermann E. 2008. Olfactory predator recognition in predator-naïve gray mouse lemurs (*Microcebus murinus*). *Journal of Comparative Psychology* 122:146–155.

Terborgh J, Janson CH. 1986. The socioecology of primate groups. *Annual Review of Ecological Systems* 17:111–135.

Thorstrom R, La Marca G. 2000. Nesting biology and behavior of the Madagascar harrier-hawk (*Polyboroides radiatus*) in northeastern Madagascar. *Journal of Raptor Research* 34:120–125.

van Noordwijk M, Hemelrijk C, Herremans L, Sterck E. 1993. Spatial position and behavioral sex differences in juvenile long-tailed macaques. In M Pereira, L Fairbanks (eds.), *Juvenile Primates* (pp. 77–85). Oxford University Press, New York.

van Schaik C. 1983. Why are diurnal primates living in groups? *Behaviour* 87:120–144.

van Schaik CP, van Noordwijk MA, Warsono B, Sutriono E. 1983. Party size and early detection of predators in Sumatran forest primates. *Primates* 24:211–221.

Wright P. 1994. The nocturnal niche in the New World. *Journal of Human Evolution* 18:635–658.

Wright P, Martin L. 1995. Predation, pollination and torpor in two nocturnal prosimians: *Cheirogaleus major* and *Microcebus rufus* in the rainforest of Madagascar. In L Alterman, G Doyle, M Izard (eds.), *Creatures of the Dark* (pp. 45–60). Plenum Press, New York.

Zimmermann E. 1985a. The vocal repertoire of the adult Senegal bushbaby (*Galago senegalensis senegalensis*). *Behaviour* 94:211–233.

Zimmermann E. 1985b. Vocalizations and associated behaviours in adult slow loris (*Nycticebus coucang*). *Folia Primatologica* 44:52–64.

Zimmermann E, Vorobieva E, Wrogemann D, Hafen T. 2000. Use of vocal fingerprinting for specific discrimination of gray (*Microcebus murinus*) and rufous mouse lemurs (*Microcebus rufus*). *International Journal of Primatology* 21:837–852.

Zinner D, Hilgartner RD, Kappler PM, Pietsch T, Ganzhorn JU. 2003. Social organization of *Lepilemur ruficaudatus*. *International Journal of Primatology* 24:869–888.

Zuberbühler K, Cheney DL, Seyfarth RM. 1999. Conceptual semantics in a nonhuman primate. *Journal of Comparative Psychology* 113:33–42.

20 The gray mouse lemur (*Microcebus murinus*): a novel cognitive primate brain aging model

Jean-Luc Picq

Introduction

The use of the small cheirogaleid, the gray mouse lemur (*Microcebus murinus*, which will be referred to as "mouse lemur" throughout this chapter), for laboratory studies of the brain and cognitive aging began in the early 1990s (Bons *et al.*, 1991, 1994; Picq, 1992, 1993; Mestre and Bons, 1993). In the decades that followed, several research groups have become interested in *M. murinus* and carried out routine assessment of cognitive capacities on various behavioral tasks (see present review). Therefore, the mouse lemur represents an exception among the prosimians, which were largely ignored for a long time by researchers studying cognition. Indeed, when consulting a textbook, such as *Primate Cognition* (Tomasello and Call, 1997), the following type of remark is commonly encountered: "as usual, we know little about cognitive skills of prosimians ...". In fact, there was a longstanding lack of interest regarding the investigation of cognitive functions in prosimians, which resulted from two major ideas. First, prosimians were generally thought to be untrainable, as they were considered to be nervous, easily distracted, and non-cooperative (Wilkerson and Rumbaugh, 1979). Early attempts to work with these animals were very disappointing. For example, it was reported that "when removed from the home cage, lemurs grew hysterical, lorisoids sulked, and both refused to work" (Jolly, 1964b). Second, prosimians were believed to be less intelligent when compared to other primates. It was claimed that "Old and New World monkeys' intelligence outdistanced that of other mammals, including the prosimian primates" (Jolly, 1966). Two major explanations were proposed to explain the modest cognitive capacity of prosimians. First of all, it was suggested that they have found an ecological niche in which intelligence is not required (i.e., a niche with reduced opportunities for competition between species) (Andrew, 1962; Jolly, 1966). Second, they have been described as primitive, representing "surviving approximations of those ancestral primate forms from whose core components the higher primates evolved" (Wilkerson

The Dwarf and Mouse Lemurs of Madagascar: Biology, Behavior and Conservation Biogeography of the Cheirogaleidae, eds. S.M. Lehman, U. Radespiel, E. Zimmermann. Published by Cambridge University Press. © Cambridge University Press 2016.

and Rumbaugh, 1979). Indeed, according to the view of Le Gros Clark (1959), the primate order is composed of a series of forms that are successive approximations of the evolutionary antecedents of man (characterized by high cognitive abilities), with prosimians at the bottom of the scale. In the case of cheirogaleids, and especially mouse lemurs, this argument might suggest very low intelligence because their major traits (e.g., small body size, heterothermy, and nocturnality) are often viewed as retention of the primitive features of ancestral primates (Kappeler, 2000). Thus, according to Martin (1972), "... the behavior of the ancestral lemurs was probably quite similar to that now exhibited by some Cheirogaleinae (particularly *Microcebus*) ...". Likewise, according to MacLean *et al.* (2008), "Prosimian primates ... are thus our best living model of the ancestral primate mind." Unfortunately, the perception that cheirogaleids represent primitive primates inadvertently conveys notions that they are "archaic, outdated, unelaborated, simple, invariant, inflexible, and hardwired," as noted by Schülke and Ostner (2005). Nevertheless, recent studies have demonstrated great behavioral flexibility in cheirogaleids (Schülke and Ostner, 2005) as well as generally high cognitive abilities in prosimians. In fact, these abilities are comparable sometimes to those demonstrated in monkeys and apes, including the capacity for inferential reasoning (Maille and Roeder, 2012), orienting toward and following the attention of other individuals (Shepherd and Platt, 2008), representing number (Lewis *et al.*, 2005) and hidden objects (Deppe *et al.*, 2009), and reasoning with tools (Santos *et al.*, 2005). Strikingly, these findings (see Fichtel and Kappeler, 2010 for review) are highly discordant with earlier data, such as a report demonstrating that prosimians were incapable of mastering single visual discrimination tasks (Jolly, 1964a). Thus, if prosimians are more intelligent than previously alleged, it might suggest that they are not so primitive (i.e., they are not lower primates when compared to higher primates, such as apes). In this regard, it must be remembered that there are no hierarchical relationships between living species, but only different degrees of relatedness. Notably, Butler and Hodos (1996) emphasized that "... terms like 'higher', 'lower' ... are inappropriate ways of comparing animals and have no place in the vocabulary of evolutionary biology," and "we should only speak of 'primitive' characteristics, not 'primitive' species." Thus, some have questioned whether the characteristics of cheirogaleids, and particularly of mouse lemurs, are primitive as commonly claimed. According to some authors, this is clearly not the case. For example, the small size of cheirogaleids has been suggested to result from an evolutionary episode of dwarfism rather than retention of a primitive feature (Masters *et al.*, 2007; Montgomery and Mundy, 2013). Likewise, heterothermy in mouse lemurs has been viewed as a sophisticated thermoregulatory process (Aujard *et al.*, 1998). Also, there is no evidence to support that mouse lemurs present poor or crude cognitive functions.

Despite potential misconceptions about mouse lemurs, this species became an attractive model for cognitive studies following the preliminary work of Bons *et al.* (1991), which was devoted to brain changes in an aging laboratory population of mouse lemurs. Their investigation revealed that some aged animals presented histological lesions, notably amyloid plaques in the cerebral cortex, which resembled those associated with Alzheimer's disease in man. Such Alzheimer-like amyloid deposits were previously identified in a few other species, including aged monkeys (Struble *et al.*, 1985), bears (Cork *et al.*, 1988), and dogs (Wisniewski *et al.*, 1970). Therefore, interest in the use of mouse lemurs as a laboratory model was aroused following this publication due to the fact that they are primates (i.e., exhibit a close phylogenetic relationship with humans), and that they do not display the practical limitations associated with monkeys. Indeed, the laboratory use of monkeys (usually the rhesus macaque) for aging research is hampered by numerous drawbacks associated with their biological traits (Austad and Fischer, 2011): they exhibit a long lifespan (more than three decades in captivity), low reproductive rate, and large body size. Consequently, the cost of obtaining and housing a sufficient number of subjects for studies is prohibitive, there is low availability of aged animals, and longitudinal studies require very long-term research projects. Moreover, monkeys (e.g., macaques) are group-living species that are sensitive to the psychopathological consequences associated with individual housing (Bayne and Novak, 1998), which is required for some experimental designs and can potentially interfere with behavioral assessments. Finally, both macaques and humans can transmit serious zoonotic diseases (Huemer *et al.*, 2002; Lefaux *et al.*, 2004).

In contrast, the mouse lemur is a small (body length 12 cm, 60–120 g), rapidly maturing (puberty occurs at about 6–8 months), relatively short-lived (8–12 years), and rather solitary primate with a reasonably high fecundity (in captivity, typically one litter a year, 1–3 offspring per litter; Perret, 2005). Furthermore, these animals have been successfully raised in captivity since 1953 in France (in the breeding colony of Brunoy which houses today about 500 mouse lemurs of various ages), and several reproducing colonies have been developed (Austad, 1997). Thus, the mouse lemur appeared to represent a promising and suitable species for the study of aging. This experimental animal model offered a useful compromise between the practicalities and affordability of rodents and the evolutionary proximity of monkeys or apes. Collectively, these theoretical arguments prompted researchers to expand the study of brain aging in *M. murinus*. These investigations revealed similarities in the age-related changes occurring in brains of humans and mouse lemurs, including amyloid plaque formation and neurofibrillary changes (Bons *et al.*, 1991; Mestre-Francés *et al.*, 1996); pathological tau metabolism (Delacourte *et al.*, 1995); neurochemical alterations (Jallageas *et al.*, 1998); neuronal loss in specific cerebral structures (e.g., the nucleus basalis of Meynert; Mestre and Bons, 1993); iron accumulation (Dhenain *et al.*, 1998a); and varying patterns of cerebral atrophy (Dhenain *et al.*, 2000). This last point is of particular interest. Indeed, in humans, age-related cerebral atrophy is closely associated with cognitive impairments observed during neurodegenerative diseases (Deweer *et al.*, 1995; Stoub *et al.*, 2010). However, brain atrophy appears to be very rare or very modest in most animal species, including transgenic mouse models of Alzheimer's disease (Delatour *et al.*, 2006). Sherwood *et al.* (2011), using *in vivo* structural magnetic resonance imaging (MRI), compared age-related brain atrophy in humans with that of their closest relatives, the chimpanzees. Their findings indicated that brain volumetric loss during aging was much less pronounced in chimpanzees when compared to humans. Thus, the authors concluded that humans are uniquely vulnerable to age-related neurodegeneration, which might result from an extended lifespan and enlarged brain. Nevertheless, marked brain shrinkage has also been observed in aged mouse lemurs (Kraska *et al.*, 2011), potentially making *M. murinus* a precious model for the study of mechanisms underlying human neurodegeneration, which involves cognitive impairment. Therefore, it is critical to assess cognitive function in mouse lemurs in order to determine their relevance as a model for normal or pathological cerebral aging. In particular, it is necessary to investigate whether there is an age-related cognitive decline in these animals, and if so, how closely it resembles events occurring in humans. Indeed, neuropsychological studies in humans have indicated that the brain aging process, whether normal or pathological, differentially affects neural networks that subserve specific cognitive processes. In particular, aging has a continuous life-long effect upon the prefrontal lobe systems that mediate critical aspects of executive functions (Dempster, 1992). This age-related impairment in executive function is dramatically amplified in neurodegenerative diseases, such as Parkinson's and Alzheimer's disease (Gabrieli, 1996). Also, it is known that declarative memory experiences a mild decline with normal aging; however, it is profoundly impaired in Alzheimer's disease due to medial temporal neurodegeneration (Deweer *et al.*, 1995). In contrast, procedural memory appears to be well preserved in both normal aging and Alzheimer's disease (Cohen and Squire, 1980; Squire, 1992). If the mouse lemur is to be used as a model of normal or pathological aging, it is important to characterize the cognitive processes that are differentially susceptible to aging in this species. It is also essential to target cognitive components that have shown central diagnostic value in human clinical studies, such as executive functions, declarative memory, and procedural memory. These topics have been addressed by several teams, which have designed appropriate experiments using mouse lemurs to target relevant cognitive functions that possess psychological validity. Here, we review the main findings of the research carried out on this issue. Our goal is to explain the theoretical framework and to provide precise and up-to-date empirical data concerning cognitive aspects of the mouse lemur aging previously summarized by Languille *et al.* (2012).

Experimental evaluation of cognitive aging in the mouse lemur

General considerations

When attempting to assess cognitive functions in a species, every effort has to be made to test them in ways that are appropriate to the behavioral and biological traits of this species, thus enabling the full expression of its cognitive potential. Accordingly, standardized equipment and tasks that are used for research on other animals (e.g., macaques, rats, or mice) cannot be used directly in mouse lemurs. Compared to monkeys, mouse lemurs have short arms and cannot move their fingers independently to achieve a precision grip, resulting in reduced capacity to reach and manipulate objects. Even if they can grasp a food item through a grid with their hand when forced, they prefer to use their mouth (Scheumann *et al.*, 2011) and show less tendency to spontaneously touch and explore objects with a hand than Old World monkeys and as required for the Wisconsin General Test Apparatus (WGTA), a piece of laboratory equipment essential for investigating primate learning. Moreover, mouse lemurs display a more developed olfactory sense than monkeys (Schilling, 1979, 2000; Piep *et al.*, 2008; Hohenbrink *et al.*, 2014). Thus, although covering a cup containing food is sufficient to hide a reward when conducting classical tasks with monkeys (e.g., delayed response task), this does not preclude the perception of a reward in mouse lemurs. In contrast, due to their small size, it is far easier to conduct spatial exploration tasks in mouse lemurs than in monkeys or apes. These types of experiments, which are commonly performed on rodents, are crucial for taxing important cognitive functions. Nevertheless, devices and procedures used in rats or mice require modifications because mouse lemurs behave differently than rodents. For example, in spite of being nocturnal, mouse lemurs rely more heavily on visual perception when compared to rodents. They have very large forward-facing eyes that allow stereoscopic vision, and their retina possesses a tapetum lucidum, which increases sensitivity under dim light conditions (Pariente, 1979). Indeed, vision is critical for a species that occupies a complex three-dimensional ecological fine branch niche in the forest and feeds partly on flying insects. Although in a limited way relative to anthropoid primates, vision is also used for social communication via a very restricted repertoire of facial mimics and body postures (Schilling, 2000). Unlike rodents, mouse lemurs are very powerful jumpers and are capable of vertical leaps, which means that experimental devices need to be completely enclosed to prevent escape. Nevertheless, a major limitation for testing mouse lemurs involves their relative unwillingness to move and explore in an open space or unusual environment. Two reasons may account for this behavior. First, the mouse lemur is naturally subject to high predation pressure from a wide range of carnivores (e.g., birds and snakes), which has resulted in antipredator strategies, including staying still and motionless within dense vegetation (Scheumann *et al.*, 2007). Second, mouse lemurs have spent a relatively short time in captivity, not having been genetically domesticated like laboratory rodents, and are therefore subject to stress-related responses related to husbandry techniques (Perret, 1982). Identifying the best incentive reward to motivate active responses of mouse lemurs is critical, and to this end, access to nest boxes appears to be at least as efficient as food (even following food deprivation). The urge to find their nest box is a very powerful stimulus because the species naturally makes use of tree holes as vital means to evade predation, ensure thermoregulation, and park infants during food foraging (Kappeler, 1998). Considering their behavioral and biological traits, researchers have used devices and tasks specifically designed for mouse lemurs in order to assess cognitive abilities. New devices are still under evaluation, especially devices allowing movements in the 3D spaces that are characteristic of a species with an arboreal lifestyle.

Most studies were cross-sectional and compared aged to young adults. The age categories were classically based on the median survival time (time at which half of the population has died), which is 4.9 years for females and 5.7 years for males at the world's largest breeding colony of Brunoy (Languille *et al.*, 2012), where most of the tests reported in this review were performed. Accordingly, mouse lemurs older than 6 years were generally considered as aged.

Assessment of procedural memory

Theoretical framework

Procedural memory is defined as the implicit acquisition of sensory–motor or cognitive biases, skills, or habits that can be revealed by measures of performance (Cohen and Squire, 1980; Squire, 2004). This kind of learning consists of rigidly associating stimuli with responses during repetition of a series of trials. Acquisition is gradual and relies on positive or negative reinforcement, which either validates or invalidates correct or incorrect responses. This kind of memory is specialized for extracting and retaining invariant features from a series of training episodes (Sherry and Schacter, 1987). Consequently, the tasks used to tax procedural memory are trial-independent with the relationship between stimulus, response, and reward constant across trials. These tasks must also be externally guided in order to allow association between a specific stimulus and a response. Simple discrimination tasks fit these requirements, and many versions have been used in studies involving mouse lemurs (Picq, 1995, 2007; Joly *et al.*, 2014).

Simultaneous spatial or visual simple discrimination tasks

These discrimination tasks were performed in an apparatus (Figure 20.1) specifically designed for mouse lemurs (Picq, 1995, 2007). The apparatus was composed of a starting box that gave access to a work chamber, from which there were six corridors with a circular opening. These corridors led to a reinforcement chamber, which housed the mouse lemur's usual nest box, and in some cases, a food reward. The mouse lemur had to select, by paying attention to visual or spatial characteristics, which corridor enabled free entrance to the reinforcement chamber (i.e., the positive corridor). If the mouse lemur entered a negative corridor, it was confined via a sliding door and then returned to the starting box for another trial. In simple discrimination tasks, only two corridors were used, the opening of the four other corridors being hidden by an opaque door. The spatial location of the corridor (i.e., left or right) was relevant for choosing the positive corridor for spatial discrimination tasks, whereas for visual discrimination tasks, relevance was based on whether the corridor was lit or unlit. Mouse lemurs were given daily sessions of up to 30 trials, and testing continued until they reached a criterion (usually 9 correct choices for 10 consecutive trials). Mouse lemurs quickly mastered such tasks and no differences were noted between young and aged animals (see Table 20.1).

Go/no-go successive discrimination tasks

Go/no-go tasks required an animal to produce a simple motor response to one stimulus while inhibiting reaction to the presence of another stimulus. These tasks

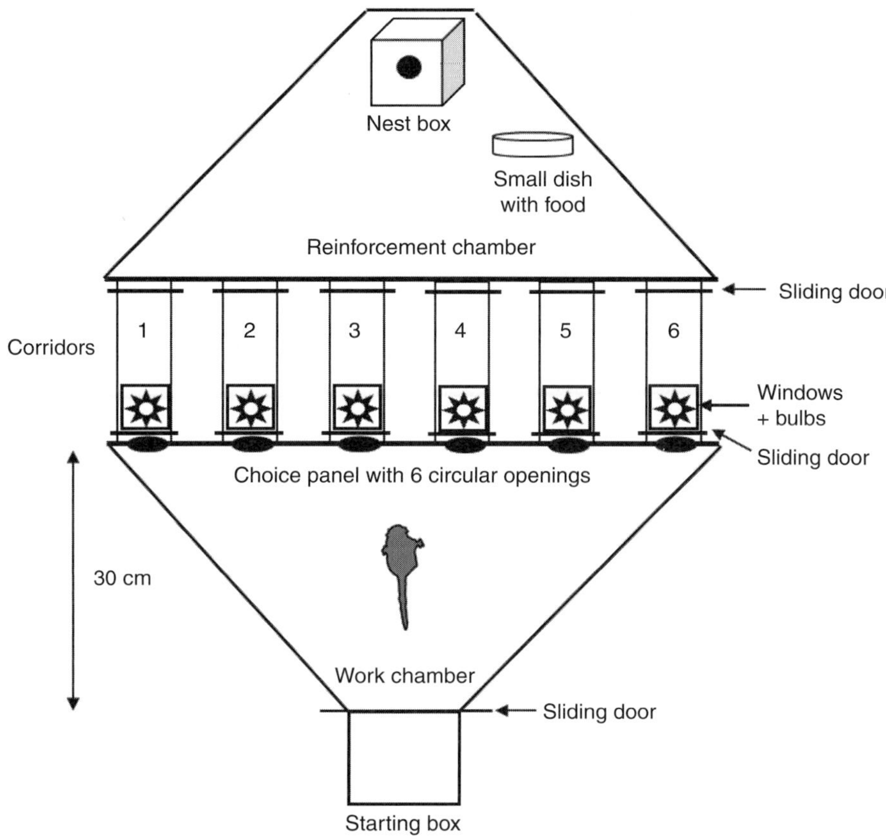

Figure 20.1 Apparatus with corridors used in mouse lemurs for discrimination tasks.

are better than simultaneous discrimination tasks for taxing procedural memory because they facilitate the expression of biases toward single items (i.e., simple stimulus–response associations). A spatial version of this task was performed using the apparatus (Figure 20.1) described above (Picq, 2007). A single corridor, either the positive or the negative one (depending on its spatial position), was presented at each trial. The mouse lemur had 70 seconds to make a choice (i.e., go or no-go) before an opaque screen was lowered, which prevented access to the corridor opening. The mouse lemurs were tested until the ratio between the median latency to enter the negative and the positive corridor was ≥ 2 over the 20 trials of a session. An olfactory version was performed by adapting an automated olfactometer (Craik, 1977), which had been designed for use with rodents (Joly *et al.*, 2004). In this task, the mouse lemur had to insert its snout into an odor delivery port and respond by licking a reinforcement tube upon detection of the positive odor stimulus. Also, no response to the negative odor was scored as a correct reaction. However, responses to the negative odors or no reaction to positive odors were scored as errors. The mouse lemurs had to make at least 33 correct responses during 2 consecutive 40-trial sessions (Joly *et al.*, 2006). These two go/no-go procedures

Table 20.1 Summary of tasks assessing cognitive aging in mouse lemurs.

Tasks	Groups	*N*	Age (years)	Impairment in aged animals	Stat. signif.	Reference
Simultaneous simple discrimination	Young–aged	4* 8**	2–4 for * and **	No for * and ** –	–	*Picq, 1995
		15***	1.1–4.1***	yes for***	–	**Picq, 2007
		4* 9**	9–10* 7–11**		$p < 0.01$	***Joly *et al.*, 2014
		9***	6.9–9.5***			
Go/no-go successive discrimination	Young	8*10**	2–4* 3–4**	No		*Picq, 2007
	Aged	9* 8**	7–11* 6–14**			**Joly *et al.*, 2006
Circular platform	Young	5	2–3	Yes, but only in a subset	$p < 0.05$	Picq *et al.*, 2012
	Aged	11	6–8			
Generalization	Young	8	2–4	Yes	$p < 0.05$	Picq, 2007
	Aged	9	7–11			
Spatial rule-guided discrimination	Young	8	2–4	Yes	$p < 0.05$	Picq, 2007
	Aged	9	7–11			
Delayed non-matching to sample	Young	6	3–4	No for d = 5 s	$p < 0.01$	Dhenain *et al.*, 1998b
	Aged	5	6–10	Yes for d = 40 s		
Free object exploration	Young	6	1–4	No for object exploration,	$p < 0.01$	Picq and Dhenain, 1998
	Aged	6	8–9	yes for response to novelty		
Novel object recognition	Young	13	2–4	No (only a limited subset		Languille *et al.*, 2015
	Middle-aged	16	5–6	impaired)		
	Aged	12	7–11			
Delayed response	Young	4	2–4	Yes	$p < 0.01$	Picq, 1995
	Aged	4	9–10			
Continuous spontaneous alternation	Young	13	2–4	Yes	$p < 0.05$	Languille *et al.*, 2015
	Middle-aged	16	5–6			
	Aged	12	7–11			
Three-panel runway	Young	14	2–3	Yes	Tendency	Trouche *et al.*, 2010
	Aged	15	6–12			
Set-shifting	Young	5	2–3	Yes	$p < 0.01$	Picq *et al.*, 2012
	Aged	11	6–8			

* Indicates to which study data refer.

result in fast learning and no influence of age on performance was detected (see Table 20.1).

The data showed that mouse lemurs are fast learners on discrimination problems and to the extent that such discrimination tasks primarily involve procedural memory, they suggest that this form of memory is preserved during aging in this species.

Assessment of declarative-like memory

Theoretical framework

Declarative memory in humans is defined as the ability to remember facts and events in a manner that can be subjected to verbal report (Cohen and Squire, 1980; Squire and Zola-Morgan, 1991). This definition is not relevant for animals, but, according to Sherry and Schacter (1987), it is possible to define declarative memory in a more fundamental manner. They suggest that it can be considered as a form of memory specializing in the preservation of contextual details, which uniquely mark individual experiences. Thus, declarative memory is capable of preserving variance across episodes, whereas procedural memory preserves invariance. This requires the declarative memory system to have at least three main properties: the capacity to store a conjunction of different items and events that constitute a specific episode and to associate them into a relational network (Eichenbaum *et al.*, 1992; Squire, 1992); flexibility, which is capacity to access and use this stored relational network under conditions that significantly differ from the circumstances of original learning, thus allowing novel responses in new situations (Eichenbaum, 1999); and the ability to support rapid one-trial learning. Thus, tasks requiring flexible use of acquired information or encouraging comparisons between stimuli rather than separate representation of the isolated items are centrally driven by the declarative memory system. Accordingly, variants of discrimination tasks can be used to assess declarative memory when they require flexibility. Likewise, because allocentric spatial learning places strong emphasis on relational representation, spatial navigation tasks can be used to assess hippocampal-dependent declarative memory, as previously described (Eichenbaum *et al.*, 1994). In addition, trial-unique tasks represent another way of assessing declarative memory. Indeed, memory of unique events is dependent on the hippocampal system, and trial-unique tests based on delayed matching to sample (DMS) or delayed non-matching to sample (DNMS) procedures have been developed in monkeys (Mishkin and Delacour, 1975) to assess hippocampal integrity (Malkova *et al.*, 1995). However, as noted by Ennaceur and Delacour (1988), DMS or DNMS tests do not purely represent one-trial tasks, as they are based on learning a rule (matching or non-matching to sample) during a long pre-training stage using repetition of stimulus–response–reward associations. Thus, these tasks require two different memory processes: memory for the rule of the task and memory of the unique stimulus presented before the delay. Simple recognition tasks based on the differentiation of familiar and new objects are considered to be true one-trial tasks, as they do not involve the learning of a rule (i.e., they are based on spontaneous exploration of objects).

Circular platform test (CPT)

The CPT was developed by Barnes (1979) as a spatial learning memory task for rodents and has now been applied to mouse lemurs. This test displays several advantages with respect to the main task, the Morris water maze, which is used to evaluate spatial memory in rats. Indeed, the CPT does not involve submersion into water, which might induce a strong stress response, and does not require high motor ability. As a spatial reference memory task, it requires animals to remember a specific location in relation to extra-maze cues. Therefore, the CPT is conceptually similar to a water maze and is equally sensitive to hippocampal integrity (McLay *et al.*, 1999). Indeed, the animal must learn to walk toward a reinforced hole guided by various distal cues surrounding the maze. This task encourages the comparison of these various cues as well as representation of their spatial relationships (i.e., with each other, the different start positions, and the goal). The apparatus used for mouse lemurs (Figure 20.2) is an adaptation of the device described by Barnes (1979). It consists of a white, circular platform with 12 equally spaced circular holes located near the perimeter. A nest box, which serves as a refuge (goal box), can be inserted or removed beneath each hole. In addition, a Plexiglas ceiling allows the mouse lemurs to see extra-maze visual cues. Also, a small black box made of plywood can be slid beneath the non-goal holes in order to block the passage, allowing only entry of the lemur's head. The starting box was an open-ended dark cylinder positioned in the center of the platform, and each trial was begun by placing the animal within it. After 30 s, the box was lifted to release the animal. When the lemur entered the goal box, the trial was stopped, with the test subject allowed to remain in the goal box for 5 min. For all trials, the position of the goal box was kept constant in the room (i.e., under one specific hole). However, the platform was randomly rotated on its central axis after each trial to avoid the use of intra-maze cues. Performances were assessed by the number of errors prior to reaching the goal box. An error was recorded when the lemur inspected an incorrect hole by inserting its nose. Although the circular platform task was easily mastered by mouse lemurs, it revealed performance differences between young and aged animals (see Table 20.1). In the study by Picq *et al.* (2012), the mean percentage of errors was 1.8 for young adults and 6.3 for an older group. However, this significant difference concealed considerable heterogeneity within the aged group. While some of the older animals performed as well as younger subjects, a subset of aged mouse lemurs (approximately 60%) was severely impaired, which was suggestive of spatial memory deficits. Interestingly, deficiency in this task was correlated with hippocampal atrophy (Picq *et al.*, 2012), strengthening the notion that it crucially relies on integrity of the hippocampus.

Generalization task

The generalization task was performed after a simple simultaneous visual discrimination task, such as the one described above (see "Simultaneous spatial or visual simple discrimination tasks"). The mouse lemurs had to apply a previously acquired rule (e.g., the lit corridor was the positive one and the unlit corridor was

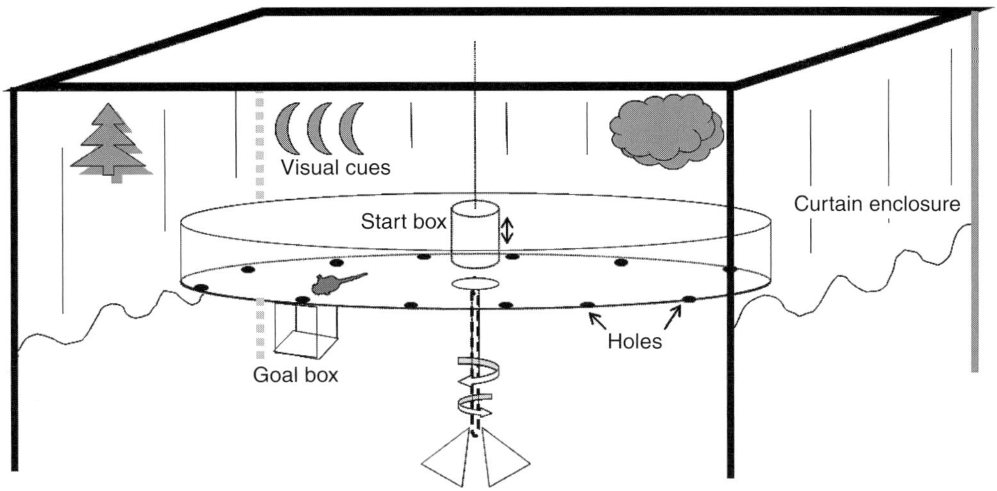

Figure 20.2 Apparatus used for the circular platform test.

the negative one) in a new context, which required new motor responses towards corridors whose location had changed (Picq, 2007). The mouse lemurs had to ignore the change in the irrelevant information (i.e., spatial location of the corridors). Most mouse lemurs did not display difficulties in directly applying a learned rule to a changed context that required new motor responses. However, a subset of aged animals (approximately 45%) was severely impaired in this task, suggesting that they were deficient in the flexible use of acquired memories (see Table 20.1).

Spatial rule-guided discrimination task

This task was also performed in the same corridor apparatus (Figure 20.1; Picq, 2007). For this discrimination task, each corridor was placed in one of six possible locations. In each trial, the mouse lemur was presented with a different pair of corridors, with each individual corridor belonging to several pairs (having an equal chance of positive and negative valence over a session). Therefore, the valence of a given corridor was determined by its spatial position with respect to the other. The task required comparison between items (i.e., the two corridors of a pair), as one single item gave no indication for guiding the choice. Thus, flexibility was required since the response to a given item could be inverted between trials. Each daily session consisted of 18 trials, which included two presentations of each corridor pair according to a pseudo-randomized sequence. The test comprised three sessions, and performance was measured by the total percentage of correct choices.

In general, mouse lemurs performed very well in this task, which required comparison between items and flexibility instead of simply the expression of biases toward single items (e.g., discrimination tasks). On average the percentage of correct responses over the three sessions was 80% for young adults and 60% for an aged group (Picq, 2007). Only a subgroup of aged mouse lemurs (approximately 30%) exhibited severe deficits. Moreover, a strong correlation was observed between low performance in the generalization task and the spatial rule-guided discrimination task (i.e., the same-aged mouse lemurs exhibited deficits in both tests). This may suggest that these two tasks employ the same cognitive abilities.

Delayed non-matching to sample task

The same six-corridor apparatus (Figure 20.1) was used for a spatial version of the DNMS task (Dhenain *et al.*, 1998b). Prior to this task, the mouse lemurs were trained in visual discrimination using lit corridors that had positive valence and unlit corridors with negative valence. The DNMS task was divided into two subtests. Initially, the mouse lemurs had to acquire the rule of "non-matching to sample". For this, one of the six corridors was lit during a presentation phase, and the mouse lemurs had to enter this corridor for access to the reinforcement chamber. Five seconds later, the mouse lemurs were placed back into the starting box for a choice phase, in which two corridors were now lit (the previous corridor and a new one). The mouse lemurs had to enter the new corridor to receive their rewards. For the second subtest, which was performed following learning of the "non-matching to sample" rule, the delay between the presentation and choice phases was increased to 40 s.

Acquisition of the general rule of non-matching was achieved similarly in young and old mouse lemurs, with a very short retention delay of 5 s. In the memory subtest, when the delay between sample and choice was increased to 40 s, the percentage of correct responses remained steady (> 80%) in young adults, but decreased to approximately 65% in the aged group (Dhenain *et al.*, 1998b).

Free object exploration task

The free object exploration test was performed within a five-wall enclosure containing five objects placed in permanent positions (Figure 20.3). During a preceding test 3 weeks earlier, each mouse lemur was allowed to explore the enclosure for 1 h per day (3 days a week) in order to become familiar with the environment and objects. Following this habituation period, the task consisted of 6 tests of 8 min each for 6 consecutive days. On days 1–3, a new object was added cumulatively and placed in a fixed location, with the old objects from previous days remaining in their familiar positions. On days 4–6, one of the old objects was displaced to a new location and replaced by a new item. The various exploratory interactions with objects (touching with hands, touching with mouth, sniffing, looking at) were scored (Picq and Dhenain, 1998).

The results showed that young adult mouse lemurs explored new objects longer than familiar objects and ran directly towards them after having observed the enclosure from the starting platform. However, although their interest in these objects markedly decreased upon second contact (after one day), any change in object location sparked renewed exploratory interest. Thus, mouse lemurs can mentally "map out" a complex and changing environment, memorize it for at least 24 h, and update it with a few minutes of exploration after alteration of the surroundings. Aged mouse lemurs also exhibited preferences for new objects, which was expressed by prolonged periods of exploration. Nevertheless, they did not make an

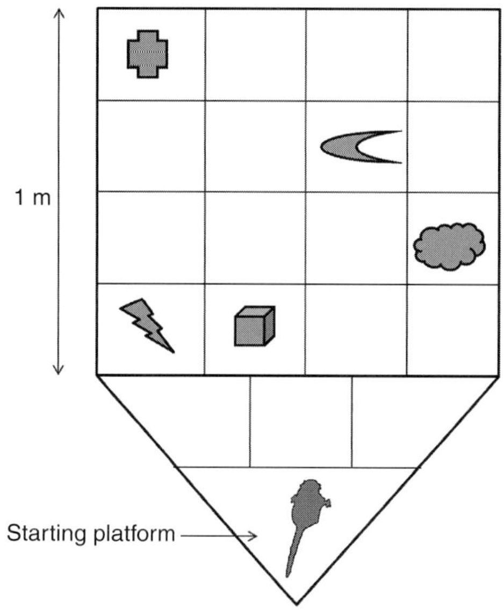

Figure 20.3 Apparatus for the free object exploration task. The location of the five objects present during habituation sessions is represented.

immediate priority visit to novel items, as though the objects were not immediately noticed within the complex environment. Moreover, interest in objects declined gradually over repeated sessions rather than suddenly upon the second presentation. Thus, it appeared that the novel objects were not perfectly memorized in the internal map during the first contact. Finally, older mouse lemurs displayed no reaction to changes in the location of familiar objects, which suggests a deficiency in memorizing spatial positioning (Picq and Dhenain, 1998).

Novel object recognition (NOR) task

The NOR task was performed in a T-maze (Languille *et al.*, 2015). Each session comprised two trials. During the sample trial (T1), two identical objects were placed in symmetrical positions, with each object fixed to a goal arm of the T-maze. After a delay of 5 min, 1 h, or 24 h the choice trial (T2) took place. In T2, a duplicate of the sample object was replaced by a new object. Notably, the position of the sample and the new object was randomly permuted with each session. The mouse lemurs were allowed 10 min of exploration for each trial, and the time spent in investigating each object was recorded. Mouse lemurs of all ages were found to spend more time exploring novel objects rather than familiar ones, irrespective of the retention delay. Interestingly, using a 24-h retention delay, only 1 of 27 young/middle-aged animals displayed difficulty in discriminating between familiar and novel objects, whereas 4 of 11 older animals showed bad discrimination. Therefore, aging does not alter short-term recognition, but instead seems to impair intermediate-term memory (at least in a subset of mouse lemurs). It is worth noting that, according to Eichenbaum *et al.* (1992, 1994), this intermediate-term memory may bridge the gap between short-term and hippocampal-dependent permanent memory.

Assessment of executive function

Theoretical framework

Executive function represents a set of cognitive processes that allow guidance and control of behavior in a flexible manner in relation to internal goals (Logan, 1985). It enables a subject to adapt to changing environmental circumstances and to succeed at non-routine problem solving that requires the generation of novel solution strategies (Logan, 1985; Norman and Shallice, 1986). This function includes various cognitive abilities, including working memory, planning, generating/evaluating strategies, inhibition of prepotent responses, as well as maintenance and shifting of the cognitive set. The most common methods of assessing executive function in animals are set-shifting tasks, which require cognitive flexibility, and working memory tasks. Thus, these tests have been performed with mouse lemurs.

Delayed response (DR) task

In this trial-dependent task, the relationship between stimulus, response, and reward is varied across trials. Information regarding the right choice is provided on the pre-delay phase of each trial, but is no longer available at the moment of response. Consequently, this information must be kept in memory during the delay. Moreover, as relevant information is changed across trials, this short-term memory has to be continuously updated against the interfering background of earlier trials. Thus, the DR task assesses the central feature of working memory, which is the capacity to remember relevant information over short periods of time for guiding ongoing behavior. In mouse lemurs, the DR task was performed in the corridor apparatus described above (Figure 20.1; Picq, 1995). They were first trained on a simple visual discrimination task, where the choice of a lit corridor led to the reward. Then the test was performed via a series of trials in which one of the four corridors was lit, and a transparent screen was used to block access to the corridors. After the mouse lemurs viewed the lit corridor for 4 s, a second opaque screen was lowered, masking all corridors, and the light stimulus was switched off. The two screens were immediately lifted, and the test subjects had to enter the previously lit corridor to receive a reward. In this 0-s delay test, mouse lemurs were trained until they achieved 16 of 20 correct responses for 2 consecutive daily sessions. During testing, the delays (0, 5, 15, 30, or 60 s) between observation and choice phases were given in a randomized order. Also, half of the trials involved a 180° rotation of the work chamber during the delay so that correct choices were not simply made by maintaining position in front of a positive corridor.

The delayed response performance of young mouse lemurs decreased slightly from the 0- to 15-s delay period, remaining stable with an average of 65% correct responses. In contrast, older mouse lemurs displayed severe time-related memory impairment as they performed at chance level (25%) for delays ≥ 30 s (Picq, 1995). Notably, a similar age-related alteration of working memory has been described in both rhesus monkeys (Bartus *et al.*, 1979; Rapp and Amaral, 1992) and humans (Craik, 1977).

Continuous spontaneous alternation

Spontaneous alternation is a measure of exploratory behavior. It requires motivation for exploring and is heavily dependent on both working memory and strategy efficiency. Indeed, optimal exploration requires that recently visited areas not be revisited. Therefore, precise memory of temporal order is required. The test was performed in a plus-maze that harbored four arms ending in 90° left turns. These parts of the arms were not visible from the center of the maze, creating an incentive effect on mouse lemur exploratory behavior. In order to avoid jumps over the maze walls, a one-way mirror was used as a cover. This permitted observation by researchers, while preventing mouse lemurs from seeing extra-maze cues. Intra-maze cues (e.g., pieces of plastic, foam rubber, or cardboard) were placed on the walls of each arm for differentiation. At the beginning of the trial, an animal was placed in the center of the maze with all the four alleys obstructed by opaque doors. After 10 s, the doors were slowly raised, and the mouse lemur was allowed to explore the four arms freely for 20 min. The number and sequence of entries (i.e., all four paws into a given arm) were recorded. Latency before the first entry was also noted. A total alternation was defined as entry into four different arms on overlapping sets of four consecutive choices. A partial alternation occurred when lemurs entered three different arms on overlapping sets of four consecutive choices. According to the recent study by Languille *et al.* (2015), the number of visits to alleys in a cross-maze did not differ significantly with age. However, a negative relationship between spontaneous alternation and age was demonstrated, revealing a significant decrease of spontaneous alternation with age. These findings suggest the occurrence of age-related impairment in working memory and/or exploratory strategy.

Three-panel runway task

The three-panel runway task involved modification of a classical spatial working memory procedure used in rodents (Furuya *et al.*, 1988) so that it could be applied to mouse lemurs. The apparatus was composed of a start box and a goal box, and between these boxes existed an alley with four consecutive choice panels, consisting of three gates. Two of the three gates were locked, and the mouse lemurs had to memorize the sequence of the open gates. These gates remained constant over a session, but changed daily. The animals were required to perform six consecutive trials per daily session (Trouche *et al.*, 2010). Young lemurs showed significantly more perseverative errors in the three-panel runway task, which is likely due to a higher level of anxiety and distress within novel environments. Although aged mouse lemurs displayed more rapid habituation to this task, they showed a tendency to commit more reference and working memory errors.

Set-shifting tasks

Set-shifting tasks were used to assess cognitive flexibility. They require shifts in assignment of affective association to specific stimuli ("affective shifting") or in selective attention from one perceptual dimension of a complex stimulus to another ("attentional set-shifting"). Picq *et al.* (2012) performed set-shifting tasks using the corridor apparatus described above (Figure 20.1) with only two corridors available. After visual discrimination, in which mouse lemurs were trained to prefer the lit corridor to the dark corridor, the reward contingencies were reversed (i.e., they had to choose the unlit corridor to receive a reward). The mouse lemurs were subsequently given an extra-dimensional shift task, where the previously irrelevant dimension of the stimulus became correlated with the reward. The positive stimulus became the right corridor, regardless of luminosity, requiring shift of attentional set from visual characteristics to spatial location. Finally, a spatial reversal discrimination task was conducted so that the left corridor became the correct choice. For each of these tests, the score was represented by the number of errors made before reaching 9 correct choices for 10 consecutive trials. In addition, Joly *et al.* (2006) assessed cognitive flexibility using the olfactory discrimination task, which was described earlier. Once mouse lemurs had achieved the discrimination performance criterion, they were required to perform a reversal task in which the positive odor became the negative one (and the opposite).

As reported in the above paragraphs, performances on simple visual or spatial discrimination tasks appeared to be independent of age. However, older mouse lemurs displayed difficulties in reversing visual or spatial discrimination and in shifting their attentional set from visual to spatial characteristics. Notably, all of the aged lemurs were impaired in set-shifting tasks, with almost no overlap between young and older group scores, suggesting widespread age-related cognitive rigidity in mouse lemurs (Picq *et al.*, 2012). It is worth noting that aged mouse lemurs differ in their pattern errors. Some old mouse lemurs displayed difficulties in disengaging from the previous rule, resulting in a large number of perseverative errors. In contrast, others displayed difficulties in maintaining the new rule, causing a large number of regressive errors (i.e., reverting back to the previously correct rule after having applied a new rule; Picq, 2007). Thus, different forms of executive dysfunction could be distinguished among these aged animals.

A similar conclusion was reported by Joly *et al.* (2006) using the olfactory test. Indeed, it was demonstrated that some aged mouse lemurs (but not the majority of them as in the above test) displayed perseverative behavior, continuing to respond to the previously rewarded stimulus. Recently, the same age-associated cognitive decline in reversal learning was also described in a visual discrimination task using a touch-screen standardized automated system (Joly *et al.*, 2014).

Discussion

A major lesson of this review is that the mouse lemur does not represent an untrainable animal. In fact, by tailoring experimental devices and procedures to the biology of this species, it is possible to assess these animals in a variety of cognitive tasks. However, it is much more challenging to test mouse lemurs than mice or rats, especially because they are not as prone to spontaneously explore novel environments and can remain motionless for a long period before responding to stimuli (for example, the latency of the first movement in an open-field was about 1800 s for the young adults in the study by Languille *et al.*, 2015). Moreover, tasks requiring precise measure of time spent exploring an object, such as the novel object recognition task, are difficult to perform. In contrast to rodents, mouse lemurs can use distance vision from a stationary position as well as direct contact (e.g., sniffing or touching) to inspect objects. In addition, they have a very wide visual field (230°) and can quickly move their eyes without moving their head, making it difficult to determine the target of their visual attention. Moreover, they display a trend to consecutively repeat erroneous choices, which must be taken into account by researchers. For example, in the three-panel runway maze, Trouche *et al.* (2010) noted that "contrarily to rodents, lemurs showed a strong tendency to go back towards closed gates even if they had already tested them." In many cases, especially when failure is due to a physical obstacle, their actions may reflect willfulness or obstinacy rather than an issue with working memory, which represents a possible limitation when interpreting results.

Another important message of this review is that mouse lemurs perform very well in a broad range of cognitive tasks. Although prosimians were originally considered incapable of passing such tests (Jolly, 1964a), they are actually quite capable of mastering spatial or visual discrimination tasks in a very small number of trials (as few as 20 in some experiments). Moreover, they display an excellent long-term memory. Thus, mouse lemurs tested several months after training for a given visual discrimination task can perform perfectly (see, for example Picq, 1995). They also exhibit good working memory, which enables them to keep relevant information in mind during short delays, using it to guide behavior in the absence of external informative stimuli. In addition, they can use this memory to correctly complete tasks in the face of interfering information provided by earlier trials. Moreover, they display cognitive flexibility. For example, mouse lemurs can change strategy when rules are altered or apply learned rules in new contexts. Furthermore, they can memorize the nature and location of many objects in an environment and use allocentric cues to memorize a goal location. Thus, they are capable of using spatial skills for cognitive mapping. Notably, in their natural environment, mouse lemurs have been shown to remember and revisit sites with stationary food resources (e.g., gum or homopteran larvae secretions), showing preferred orientation to these foraging sites when leaving their sleeping site at dusk (Joly and Zimmermann, 2007). These findings strongly suggest that mouse lemurs have remarkable spatial memory ability, which has been confirmed by studies using artificial feeding platforms (Joly *et al.*, 2008; Luhrs *et al.*, 2009), and show evidence for travel route planning (Joly and Zimmermann, 2011). Taken together, these data support the notion that mouse lemurs have higher cognitive competencies than previously claimed. In this regard, mouse lemurs and other prosimians have been described as being incapable of understanding and manipulating objects (Jolly, 1964b). Nevertheless, recent experiments (Scheumann *et al.*, 2011; Schilling, 2013) have shown that mouse lemurs can solve complex manipulation problems in a flexible manner. In fact, these animals are able to open several kinds of baited boxes via distinct methods (e.g., raising a lid, pulling/pushing a drawer, or performing successive and simultaneous handling of different parts of a complex object).

Here, we have also highlighted the core goal of behavioral studies on mouse lemurs, which is to establish the pattern of cognitive aging in this species. In regard to this topic, different conclusions can be drawn. (1) Mouse lemurs display cognitive alterations during aging. As the same-aged animals perfectly succeed in some tasks and fail in others that differ only by a key parameter (for example, delay in DNMS or familiar or non-familiar position of the same corridors in visual discrimination task vs. generalization task), this decline in cognitive task performance is dependent on the cognitive demand of each task and cannot be attributed to general factors (e.g., attention deterioration, low motivation, or sensory–motor alteration) or global cognitive dysfunction. The loss of specific cognitive abilities, along with the preservation of others, suggests selective vulnerability during aging. (2) Not all of the aged subjects are equally affected by cognitive impairment. In a given task, some aged mouse lemurs are profoundly deficient, whereas others perform similarly to younger animals. This age-related interindividual variability has often been reported in other species, including humans (Valdois *et al.*, 1990; Bachevalier *et al.*, 1991; Rapp and Amaral, 1992; Ylikoski *et al.*, 1999). Different subgroups of aged mouse lemurs displaying distinct patterns of cognitive alterations can be identified. Indeed, this provides a unique opportunity to determine the neurobiological correlates of cognitive decline and the possibility to distinguish normal processes from pathological aging. (3) Not all cognitive tasks are equally sensitive to the effects of aging in this species. For example, the ability to learn spatial or visual discrimination is preserved (but see Joly *et al.*, 2014 for diverging results that could be explained, according to the authors, by differences in the difficulty of the discrimination). These tasks encourage simple stimulus–response associations acquired through the repetition of similar trials and chiefly involve procedural memory. Thus, procedural memory appears to be spared in aging mouse lemurs. On the other hand, performance in executive function tasks is severely impaired in most aged mouse lemurs. Indeed, aged animals display greater difficulty than younger animals in performing reversal discrimination tasks, disengaging from a previous rule, or maintaining a new rule on set-shifting tasks. Moreover, aged mouse lemurs are more sensitive to delay lengths during delayed response tasks and make more working memory errors in the three-panel runway and spontaneous alternation tasks. Thus cognitive flexibility and working memory, which are essential parts of executive function, are especially vulnerable to aging in mouse lemurs. This decline in executive function begins as early as middle age, worsening with time as it does in humans (Davis *et al.*, 1990; Gazzaley *et al.*, 2005). Like executive functions, aging can also affect the ability of lemurs to master allocentric spatial memory tasks (the CPT), location/object recognition tasks, or use previously acquired flexible information (i.e., generalization and spatial rule-guided discrimination tasks). However, in the case of these tasks, only subsets of aged mouse lemurs show impairments. Indeed, some aged mouse lemurs perform similarly to younger animals in these tasks. Because rapid learning, flexible use of memories, and ability to compare/contrast items in memory can be viewed as the hallmarks of declarative memory, these findings suggest its impairment in some aged mouse lemurs.

Overall, the behavioral studies reviewed in this chapter have allowed us to establish a pattern of cognitive aging in mouse lemurs, namely preserved procedural memory, progressive/widespread decline in executive function, and impairment in declarative-like memory, which is limited to a subpopulation of aged individuals. Strikingly, this pattern of age-related alterations mimics that described in humans (Gabrieli, 1996). In particular, executive dysfunction (especially cognitive rigidity) has been considered as a fundamental sign of normal cognitive aging (Dempster, 1992), whereas declarative memory undergoes only mild decline during normal aging, although it is profoundly impaired in Alzheimer's disease due to medial temporal degeneration (Stoub *et al.*, 2010). Interestingly, poor performance in the CPT by old mouse lemurs was found to correlate with hippocampal or entorhinal cortex atrophy (Picq *et al.*, 2012). This may suggest a similar neuroanatomical basis for declarative memory dysfunction in humans and mouse lemurs.

Although these findings are very promising, progress must be made to standardize behavioral tests for mouse lemurs. Indeed, comparison of laboratory findings remains difficult because different teams have developed distinct tests to examine cognitive function in mouse lemurs (for example, the Brunoy team used the continuous spontaneous alternation task to evaluate working memory, whereas the Montpellier team used the three-panel runway task). Furthermore, several researchers have reported low adaptation of the mouse lemurs to experimental devices, leading to excessive duration of experimental trials and/or disproportionate elimination of individuals as non-responders (see, for example, Trouche *et al.*, 2010). Although it is true that much work remains in order to streamline the cognitive assessment of mouse lemurs, the evidence presented in this chapter reveals the remarkable cognitive abilities of these animals. Furthermore, recent data have highlighted the proximity of their age-related cognitive profile to that of humans. Thus, the study of mouse lemurs might provide a unique opportunity to understand the mechanisms that drive cognitive impairment in humans and to potentially modulate the aging process. Such scientific challenges require longitudinal follow-up of large cohorts of mouse lemurs and, accordingly, rapid and neuropsychologically relevant cognitive tests that result in minimal learning transfer when repeatedly given. Such studies, especially those evaluating the impact of diet on cognitive aging (Dal-Pan *et al.*, 2011; Vinot *et al.*, 2011), are underway and have already yielded novel and promising results, thus showing that the gray mouse lemur is an animal model of paramount interest still at the dawn of its development.

References

Andrew RJ. 1962. Evolution of intelligence and vocal mimicking: studies of large-brained mammals promise to elucidate some problems of human evolution. *Science* 137(3530):585–589.

Aujard F, Perret M, Vannier G. 1998. Thermoregulatory responses to variations of photoperiod and ambient temperature in the male lesser mouse lemur: a primitive or an advanced adaptive character? *Journal of Comparative Physiology, B* 168(7):540–548.

Austad SN. 1997. Small nonhuman primates as potential models of human aging. *ILAR Journal* 38(3):142–147.

Austad SN, Fischer KE. 2011. The development of small primate models for aging research. *ILAR Journal* 52(1):78–88.

Bachevalier J, Landis LS, Walker LC, *et al.* 1991. Aged monkeys exhibit behavioral deficits indicative of widespread cerebral dysfunction. *Neurobiology of Aging* 12(2):99–111.

Barnes CA. 1979. Memory deficits associated with senescence: a neurophysiological and behavioral study in the rat. *Journal of Comparative and Physiological Psychology* 93(1):74–104.

Bartus RT, Dean RL 3rd, Fleming DL. 1979. Aging in the rhesus monkey: effects on visual discrimination learning and reversal learning. *Journal of Gerontology* 34(2):209–219.

Bayne K, Novak M. 1998. Behavioral disorders. In TB Bennett, CR Abee, R Henrickson (eds.), *Nonhuman Primates in Biomedical Research* (pp. 485–500). Academic Press, San Diego.

Bons N, Mestre N, Petter A. 1991. Senile plaques and neurofibrillary changes in the brain of an aged lemurian primate, *Microcebus murinus. Neurobiology of Aging* 13(1):99–105.

Bons N, Mestre N, Ritchie K, *et al.* 1994. Identification of amyloid beta protein in the brain of the small, short-lived lemurian primate *Microcebus murinus. Neurobiology of Aging* 15(2):215–220.

Butler AB, Hodos W. 1996. *Comparative Vertebrate Neuroanatomy: Evolution and Adaptation.* Wiley-Liss, New York.

Cohen NJ, Squire LR. 1980. Preserved learning and retention of pattern-analyzing skill in amnesia: dissociation of knowing how and knowing that. *Science* 210(4466):207–210.

Cork LC, Powers RE, Selkoe DJ, *et al.* 1988. Neurofibrillary tangles and senile plaques in aged bears. *Journal of Neuropathology and Experimental Neurology* 47(6):629–641.

Craik FIM. 1977. Age differences in human memory. In JE Birren, KW Schaie (eds.), *Handbook of the Psychology of Aging.* Van Nostrand-Reinhold, New York.

Dal-Pan A, Pifferi F, Marchal J, Picq JL, Aujard F. 2011. Cognitive performances are selectively enhanced during chronic caloric restriction or resveratrol supplementation in a primate. *PLoS ONE* 6(1).

Davis HP, Cohen A, Gandy M, *et al.* 1990. Lexical priming deficits as a function of age. *Behavioral Neuroscience* 104(2):288–297.

Delacourte A, Sautiere PE, Wattez A, *et al.* 1995. Biochemical characterization of Tau proteins during cerebral aging of the lemurian primate *Microcebus murinus. Comptes Rendus de l'Académie des Sciences Série III, Sciences de la Vie* 318(1):85–89.

Delatour B, Guegan M, Volk A, Dhenain M. 2006. *In vivo* MRI and histological evaluation of brain atrophy in APP/PS1 transgenic mice. *Neurobiology of Aging* 27(6):835–847.

Dempster FN. 1992. The rise and fall of the inhibitory mechanism: toward a unified theory of cognitive development and aging. *Developmental Review* 12(1):45–75.

Deppe AM, Wright PC, Szelistowski WA. 2009. Object permanence in lemurs. *Animal Cognition* 12(2):381–388.

Deweer B, Lehericy S, Pillon B, *et al.* 1995. Memory disorders in probable Alzheimer's disease: the role of hippocampal atrophy as shown with MRI. *Journal of Neurology, Neurosurgery and Psychiatry* 58(5):590–597.

Dhenain M, Duyckaerts C, Michot JL, *et al.* 1998a. Cerebral T2-weighted signal decrease during aging in the mouse lemur primate reflects iron accumulation. *Neurobiology of Aging* 19(1):65–69.

Dhenain M, Michot J-L, Volk A, *et al.* 1998b. Etude préliminaire du vieillissement cognitif des microcèbes murins dans une version spatiale du test de non appariement retardé (Preliminary study of cerebral aging in mouse lemurs by a spatial version of the delayed non-matching to sample task). *Primatologie* 1:309–331.

Dhenain M, Michot JL, Privat N, *et al.* 2000. MRI description of cerebral atrophy in mouse lemur primates. *Neurobiology of Aging* 21(1):81–88.

Eichenbaum H. 1999. The hippocampus and mechanisms of declarative memory. *Behavioural Brain Research* 103(2):123–133.

Eichenbaum H, Otto T, Cohen NJ. 1992. The hippocampus – what does it do? *Behavioral and Neural Biology* 57(1):2–36.

Eichenbaum H, Otto T, Cohen NJ. 1994. Two functional components of the hippocampal memory system. *Behavioral and Brain Sciences* 17(3):449–472.

Ennaceur A, Delacour J. 1988. A new one-trial test for neurobiological studies of memory in rats. 1: Behavioral data. *Behavioural Brain Research* 31(1):47–59.

Fichtel C, Kappeler PM. 2010. Human universals and primate symplesiomorphies: etablishing the lemur baseline. In PM Kappeler, JB Silk (eds.), *Mind the Gap: Tracing the Origins of Human Universals* (pp. 395–426). Springer, Heidelberg.

Furuya Y, Yamamoto T, Yatsugi S, Ueki S. 1988. A new method for studying working memory by using the three-panel runway apparatus in rats. *Japanese Journal of Pharmacology* 46(2):183–188.

Gabrieli JD. 1996. Memory systems analyses of mnemonic disorders in aging and age-related diseases. *Proceedings of the National Academy of Sciences of the United States of America* 93(24):13534–13540.

Gazzaley A, Cooney JW, Rissman J, D'Esposito M. 2005. Top-down suppression deficit underlies working memory impairment in normal aging. *Nature Neuroscience* 8(10):1298–1300.

Hohenbrink P, Dempewolf S, Zimmermann E, Mundy NI, Radespiel U. 2014. Functional promiscuity in a mammalian chemosensory system: extensive expression of vomeronasal receptors in the main olfactory epithelium of mouse lemurs. *Frontiers in Neuroanatomy* 8:102.

Huemer HP, Larcher C, Czedik-Eysenberg T, Nowotny N, Reifinger M. 2002. Fatal infection of a pet monkey with *Human herpesvirus 1*. *Emerging Infectious Diseases* 8(6):639–642.

Jallageas V, Privat N, Mestre-Francés N, Bons N. 1998. Age-related changes in serotonergic and catecholaminergic brain systems in the lemurian primate *Microcebus murinus*. *Proceedings of the Zoological Society of London* 839:628–630.

Jolly A. 1964a. Choice of cue in prosimian learning. *Animal Behaviour* 12:571–577.

Jolly A. 1964b. Prosimians' manipulation of simple object problems. *Animal Behaviour* 12:560–570.

Jolly A. 1966. Lemur social behavior and primate intelligence. *Science* 153(735):501–506.

Joly M, Zimmermann E. 2007. First evidence for relocation of stationary food resources during foraging in a strepsirrhine primate (*Microcebus murinus*). *American Journal of Primatology* 69(9):1045–1052.

Joly M, Zimmermann E. 2011. Do solitary foraging nocturnal mammals plan their routes? *Biology Letters* 7(4):638–640.

Joly M, Michel B, Deputte B, Verdier JM. 2004. Odor discrimination assessment with an automated olfactometric method in a prosimian primate, *Microcebus murinus*. *Physiology & Behavior* 82(2–3):325–329.

Joly M, Deputte B, Verdier JM. 2006. Age effect on olfactory discrimination in a non-human primate, *Microcebus murinus*. *Neurobiology of Aging* 27(7):1045–1049.

Joly M, Scheumann M, Zimmermann E. 2008. Wild mouse lemurs revisit artificial feeding platforms: implications for field experiments on sensory and cognitive abilities in small primates. *American Journal of Primatology* 70(9):892–896.

Joly M, Ammersdorfer S, Schmidtke D, Zimmermann E. 2014. Touchscreen-based cognitive tasks reveal age-related impairment in a primate aging model, the grey mouse lemur (*Microcebus murinus*). *PLoS ONE* 9(10):e109393.

Kappeler PM. 1998. Nests, tree holes, and the evolution of primate life histories. *American Journal of Primatology* 46(1):7–33.

Kappeler PM. 2000. Lemur origins: rafting by groups of hibernators? *Folia Primatologica* 71(6):422–425.

Kraska A, Dorieux O, Picq JL, *et al.* 2011. Age-associated cerebral atrophy in mouse lemur primates. *Neurobiology of Aging* 32(5):894–906.

Languille S, Blanc S, Blin O, *et al.* 2012. The grey mouse lemur: a non-human primate model for ageing studies. *Ageing Research Reviews* 11(1):150–162.

Languille S, Liévin-Bazin A, Picq JL, *et al.* 2015. Deficits of psychomotor and mnesic functions across aging in mouse lemur primates. *Frontiers in Behavioral Neuroscience* 8:446.

Le Gros Clark WE. 1959. *The Antecedent of Man.* Edinburgh University Press. London.

Lefaux B, Duprez R, Tanguy M, *et al.* 2004. Nonhuman primates might be highly susceptible to cross-species infectivity by human alpha-herpesviruses. *Veterinary Pathology* 41(3):302–304.

Lewis KP, Jaffe S, Brannon EM. 2005. Analog number representations in mongoose lemurs (*Eulemur mongoz*): evidence from a search task. *Animal Cognition* 8(4):247–252.

Logan GD. 1985. Executive control of thought and action. *Acta Psychologica* 60:193–210.

Luhrs ML, Dammhahn M, Kappeler PM, Fichtel C. 2009. Spatial memory in the grey mouse lemur (*Microcebus murinus*). *Animal Cognition* 12(4):599–609.

MacLean EL, Merritt DJ, Brannon EM. 2008. Social complexity predicts transitive reasoning in prosimian primates. *Animal Behaviour* 76(2):479–486.

Maille A, Roeder JJ. 2012. Inferences about the location of food in lemurs (*Eulemur macaco* and *Eulemur fulvus*): a comparison with apes and monkeys. *Animal Cognition* 15(6):1075–1083.

Malkova L, Mishkin M, Bachevalier J. 1995. Long-term effects of selective neonatal temporal lobe lesions on learning and memory in monkeys. *Behavioral Neuroscience* 109(2):212–226.

Martin RD. 1972. Adaptive radiation and behaviour of the Malagasy lemurs. *Philosophical Transactions of the Royal Society of London Series B: Biological Sciences* 264(862):295–352.

Masters JC, de Wit MJ, Lovegrove BG. 2007. Eyes wide shut: can hypometabolism really explain the primate colonization of Madagascar? *Journal of Biogeography* 34(1):21–37.

McLay RN, Freeman SM, Harlan RE, Kastin AJ, Zadina JE. 1999. Tests used to assess the cognitive abilities of aged rats: their relation to each other and to hippocampal morphology and neurotrophin expression. *Gerontology* 45(3):143–155.

Mestre N, Bons N. 1993. Age-related cytological changes and neuronal loss in basal forebrain cholinergic neurons in *Microcebus murinus* (lemurian, primate). *NeuroDegeneration* 2:25–32.

Mestre-Francés N, Silhol S, Bons N. 1996. Evolution of β-amyloid deposits in the cerebral cortex of *Microcebus murinus* lemurian primate. *Alzheimer Research* 2:19–28.

Mishkin M, Delacour J. 1975. An analysis of short-term visual memory in the monkey. *Journal of Experimental Psychology* 1(4):326–334.

Montgomery SH, Mundy NI. 2013. Parallel episodes of phyletic dwarfism in callitrichid and cheirogaleid primates. *Journal of Evolutionary Biology* 26(4):810–819.

Norman DA, Shallice T. 1986. Attention in action: willed and automatic control of behavior. In RJ Davidson, GE Schwartz, D Shapiro (eds.), *Consciouness and Self-Regulation: Advance in Research and Theory* (pp. 1–18). Plenum, New York.

Pariente GF. 1979. The role of vision in prosimian behavior. In GA Doyle, RD Martin (eds.), *The Study of Prosimian Behavior* (pp. 411–459). Academic Press, New York.

Perret M. 1982. Stress-effects in *Microcebus murinus. Folia Primatologica* 39(1–2):63–114.

Perret M. 2005. Relationship between urinary estrogen levels before conception and sex ratio at birth in a primate, the gray mouse lemur. *Human Reproduction* 20(6):1504–1510.

Picq JL. 1992. Aging and social behaviour in captivity in *Microcebus murinus. Folia Primatologica* 59(4):217–220.

Picq JL. 1993. Radial maze performance in young and aged grey mouse lemurs (*Microcebus murinus*). *Primates* 24:223–226.

Picq JL. 1995. Effects of aging upon recent memory in *Microcebus murinus. Aging* 7(1):17–22.

Picq JL. 2007. Aging affects executive functions and memory in mouse lemur primates. *Experimental Gerontology* 42(3):223–232.

Picq JL, Dhenain M. 1998. Reaction to new objects in young and aged grey mouse lemurs (*Microcebus murinus*). *Journal of Experimental Psychology* 51:337–348.

Picq JL, Aujard F, Volk A, Dhenain M. 2012. Age-related cerebral atrophy in nonhuman primates predicts cognitive impairments. *Neurobiology of Aging* 33(6):1096–1109.

Piep M, Radespiel U, Zimmerman E, Schmidt S, Siemers BM. 2008. The sensory basis of prey detection in captive-born grey mouse lemurs. *Animal Behaviour* 75(3):871–878.

Rapp PR, Amaral DG. 1992. Individual differences in the cognitive and neurobiological consequences of normal aging. *Trends in Neurosciences* 15(9):340–345.

Santos LR, Mahajan N, Barnes JL. 2005. How prosimian primates represent tools: experiments with two lemur species (*Eulemur fulvus* and *Lemur catta*). *Journal of Comparative Psychology* 119(4):394–403.

Scheumann M, Rabesandratana A, Zimmerman E. 2007. Predation, communication, and cognition in lemurs. In SL Gursky-Doyen, KAI Nekaris (eds.), *Primate Anti-Predator Strategies* (pp. 100–126). Springer, New York.

Scheumann M, Joly-Radko M, Leliveld L, Zimmermann E. 2011. Does body posture influence hand preference in an ancestral primate model? *BMC Evolutionary Biology* 11:52.

Schilling A. 1979. Olfactory communication in prosimians. In GA Doyle, RD Martin (eds.), *The Study of Prosimian Behavior* (pp. 461–538). Academic Press, New-York.

Schilling A. 2000. Sensory organs and communication in *Microcebus murinus*. *Primatologie* 3:85–143.

Schilling A. 2013. Cognitive capacities of captive gray mouse lemurs as evidenced by object manipulation. In J Masters, M Gamba, F Génin (eds.), *Leaping Ahead: Advances in Prosimian Biology* (pp. 331–340). Springer, New York.

Schülke O, Ostner J. 2005. Big times for dwarfs: social organisation, sexual selection, and cooperation in the Cheirogaleidae. *Evolutionary Anthropology* 14:170–185.

Shepherd SV, Platt ML. 2008. Spontaneous social orienting and gaze following in ringtailed lemurs (*Lemur catta*). *Animal Cognition* 11(1):13–20.

Sherry DF, Schacter DL. 1987. The evolution of multiple memory systems. *Psychological Review* 94(4):439–454.

Sherwood CC, Gordon AD, Allen JS, *et al.* 2011. Aging of the cerebral cortex differs between humans and chimpanzees. *Proceedings of the National Academy of Sciences of the United States of America* 108(32):13029–13034.

Squire LR. 1992. Memory and the hippocampus: a synthesis from findings with rats, monkeys, and humans. *Psychological Review* 99(2):195–231.

Squire LR. 2004. Memory systems of the brain: a brief history and current perspective. *Neurobiology of Learning and Memory* 82(3):171–177.

Squire LR, Zola-Morgan S. 1991. The medial temporal lobe memory system. *Science* 253(5026):1380–1386.

Stoub TR, Rogalski EJ, Leurgans S, Bennett DA, deToledo-Morrell L. 2010. Rate of entorhinal and hippocampal atrophy in incipient and mild AD: relation to memory function. *Neurobiology of Aging* 31(7):1089–1098.

Struble RG, Price DL Jr, Cork LC, Price DL. 1985. Senile plaques in cortex of aged normal monkeys. *Brain Research* 361(1–2):267–275.

Tomasello M, Call J. 1997. *Primate Cognition*. Oxford University Press, New York.

Trouche SG, Maurice T, Rouland S, Verdier JM, Mestre-Francés N. 2010. The three-panel runway maze adapted to *Microcebus murinus* reveals age-related differences in memory and perseverance performances. *Neurobiology of Learning and Memory* 94(1):100–106.

Valdois S, Joanette Y, Poissant A, Ska B, Dehaut F. 1990. Heterogeneity in the cognitive profile of normal elderly. *Journal of Clinical and Experimental Neuropsychology* 12(4):587–596.

Vinot N, Jouin M, Lhomme-Duchadeuil A, *et al.* 2011. Omega-3 fatty acids from fish oil lower anxiety, improve cognitive functions and reduce spontaneous locomotor activity in a non-human primate. *PLoS ONE* 6(6):e20491.

Wilkerson J, Rumbaugh DM. 1979. Learning and intelligence in prosimians. In GA Doyle, RD Martin (eds.), *The Study of Prosimian Behavior* (pp. 207–246). Academic Press, New York.

Wisniewski H, Johnson AB, Raine CS, Kay WJ, Terry RD. 1970. Senile plaques and cerebral amyloidosis in aged dogs. A histochemical and ultrastructural study. *Laboratory Investigation* 23(3):287–296.

Ylikoski R, Ylikoski A, Keskivaara P, *et al.* 1999. Heterogeneity of cognitive profiles in aging: successful aging, normal aging, and individuals at risk for cognitive decline. *European Journal of Neurology* 6(6):645–652.

21 Acoustic divergence in communication of cheirogaleids with special emphasis to mouse lemurs

Elke Zimmermann

Introduction

How and why species emerge, persist, and become extinct is a central and controversially debated topic in evolutionary biology (Martin, 1990; Meyer, 1993; Masters, 1999; Mayr, 1999; see also Chapter 1). Acoustic signaling systems are known to mediate discrimination within and between species and play a major role in speciation and evolution (Wilkins *et al.*, 2013). However, empirical research on this topic for primates is currently in its infancy.

Most research related to this topic comes from various non-mammalian animal groups, such as birds (e.g. Irwin *et al.*, 2008; Podos, 2010), frogs (e.g. Amézquita *et al.*, 2011; Willink *et al.*, 2014), and insects (e.g. McNett and Cocroft, 2008), although singing mice present a rare exception for mammals (Campbell *et al.*, 2010). Current comparative bioacoustic research suggests that three major selective forces drive acoustic divergence within and among species, favoring speciation and evolution (see Figure 21.1): adaptive mechanisms such as ecological and sexual selection as well as neutral mechanisms such as drift. For instance, in Darwin finches (*Geospiza* ssp.), climate constraints shape feeding ecology and thereby bill shape and indirectly the structure of trill calls in their song (Podos, 2010). Ecological selection for beak size results in acoustic and population divergence through assortative mating. Sexual selection following host shifts is strongly supported for treehoppers (McNett and Cocroft, 2008), in which the frequency of a pure tone mating signal favors species divergence. Sexual selection for song complexity along a gradient and drift are proposed to explain song divergence in greenish warblers (Irwin *et al.*, 2008), whereas only drift is suggested to be the driving force for divergence in singing mice (Campbell *et al.*, 2010).

Disentangling the relative contributions of these selective forces for the evolution of acoustic divergence and species diversity is urgently needed (Wilkins *et al.*, 2013). For primates, a first and important step in this direction is to study the extent that

The Dwarf and Mouse Lemurs of Madagascar: Biology, Behavior and Conservation Biogeography of the Cheirogaleidae, eds. S.M. Lehman, U. Radespiel, E. Zimmermann. Published by Cambridge University Press. © Cambridge University Press 2016.

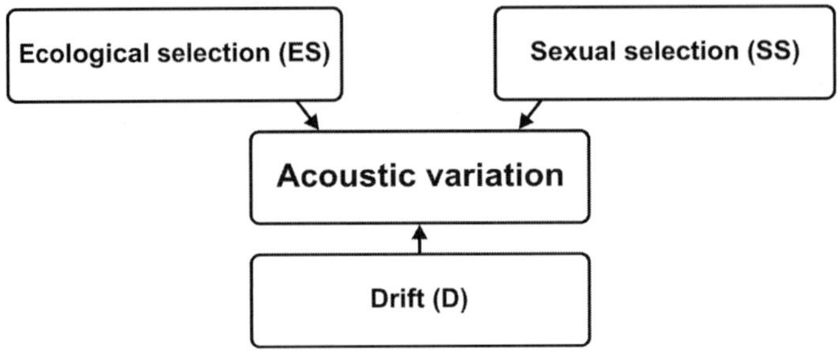

Figure 21.1 Acoustic divergence within and among sound-producing animal taxa is driven by ecological selection, sexual selection, and drift, and favors speciation and evolution.

acoustic divergence in communication is present and biologically relevant. It is also important to examine these aspects in genetically characterized species with known sympatric and allopatric distribution and varying in ranges, habitat, and body mass.

The small and nocturnal mouse lemurs of Madagascar provide a unique primate radiation for exploring the significance of vocal communication for species recognition and discrimination. Mouse lemurs are known to display a high cryptic species diversity of currently 21 genetically characterized species which vary about twofold in body mass (Radespiel *et al.*, 2012; Rasoloarison *et al.*, 2013). The rufous-colored species show local endemism with local to regional distribution in either dry deciduous or rainforest habitats across Madagascar (see Figure 21.2 and Chapter 26), where they are most often limited to one single so-called "inter-river system" (IRS; see also Chapter 26). In contrast, the gray mouse lemur (*Microcebus murinus*) shows a broad range across all dry deciduous forests from the northwest to the southeast encompassing several IRS (see Figure 21.2), often providing the basis for sympatry with one of the rufous-colored species. Thus, current field research shows that genetically defined species vary in their degree of sympatry and allopatry, in body mass as well as in habitat.

Survival of mouse lemurs in the dark of their dense forest environment is strongly dependent on olfaction and audition; vision is environmentally and physiologically more constrained (Charles-Dominique *et al.*, 1980; Pariente, 1980; Siemers *et al.*, 2007; Piep *et al.*, 2008; Valenta *et al.*, 2013; Veilleux *et al.*, 2013; Chapter 18). Mouse lemurs represent the smallest-bodied cheirogaleids and thus face the highest predation risk in primates (Scheumann *et al.*, 2007a). For predator detection and avoidance, they use primarily olfactory and acoustic cues (Bunkus *et al.*, 2005; Sündermann *et al.*, 2008; Rahlfs and Fichtel, 2010; Kappel *et al.*, 2011; Hohenbrink *et al.*, 2012, 2013; see also Chapter 19). These cues are equally important for prey detection, localization, and recognition (Goerlitz and Siemers, 2007; Siemers *et al.*, 2007; Piep *et al.*, 2008; Siemers, 2013) as well as for social communication (Perret, 1995; Buesching *et al.*, 1998; Zimmermann, 2010). In the best-studied mouse lemur species, males and females forage solitarily, whereas they may also form bonded uni-sex or mixed-sex sleeping groups (Radespiel, 2006). Under such a nocturnal, solitary foraging regime, mouse lemurs face major social challenges, such as how to

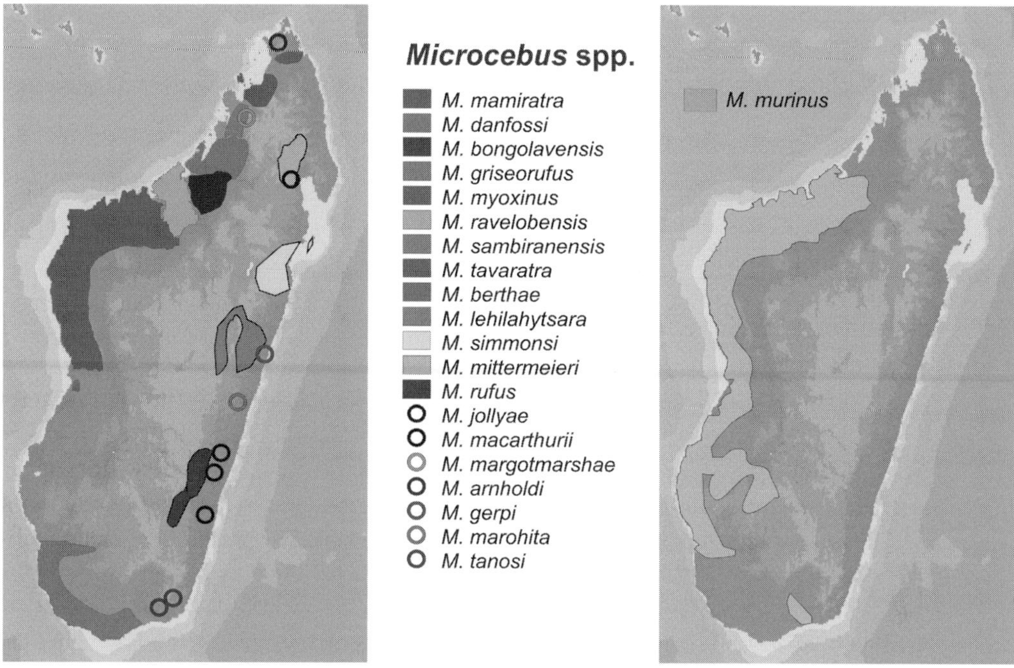

Figure 21.2 Geographical distributions of mouse lemurs in Madagascar (according to Radespiel *et al.*, 2012 and Rasoloarison *et al.*, 2013). (A black and white version of this figure will appear in some formats. For the color version, please refer to the plate section.)

find a suitable mating partner, how to reunite mother and offspring when the latter are parked temporarily in dense vegetation, how to coordinate group movements to and from sleeping sites, and how to avoid rivals or predators. Olfactory cues are known to regulate spacing between groups, and mediate group ownership (Braune *et al.*, 2005), social dominance (Schilling and Perret, 1987; Zimmermann and Lerch, 1993), and reproductive status (Buesching *et al.*, 1998). However, olfactory signals do not allow rapid location-specific communication; for example, to avoid predators or to coordinate pair, mother–offspring or group movement to prepare meeting at a given place at a specific time or reuniting after some time of social separation. For these purposes, mouse lemurs evolved a set of acoustically complex, tonal vocalizations in the audible and/or ultrasonic range conveying indexical and emotional information and governing mating, mother–infant or group reunion, agonistic conflicts, or antipredator strategies (Zimmermann, 2010, 2013; see also Chapter 19). Auditory-evoked brainstem responses of the gray mouse lemur showed that mouse lemurs exhibit a broad hearing range, extending from the low frequency to the near ultrasonic range (Schopf *et al.*, 2014), and coinciding with the range for acoustic communication (Zimmermann, 2010; Leliveld *et al.*, 2011). Mouse lemurs display a best frequency around 8 kHz, matching the frequency of territorial calls of raptors (Schopf *et al.*, 2014) and suggesting that predation was a driving force for the evolution of hearing in small-bodied primates.

Thus the high and genetically defined species diversity, the disparate geographical ranges providing the basis for allopatry and sympatry, the variation in body mass as well as the relevance of acoustic signaling for behavioral decision-making makes mouse lemurs an excellent primate model group in which to explore current hypotheses for acoustic divergence driving speciation and evolution in sound-producing animals. In this chapter I will take the four best-studied species of the mouse lemur radiation in order to examine to what extent the acoustic space for communication is flexible within and between species. I will further explore major hypotheses discussed to drive speciation and evolution via acoustic divergence by the model of mouse lemurs' tonal calls (communication calls with a harmonic structure, meaning that harmonics are spread in equal distances to the fundamental frequency across the frequency range). Finally, I will give perspectives for further research.

The acoustic window as the basis for acoustic divergence

Terrestrial mammals communicating by sound are exposed to major challenges to get their message through in a given environment. The message has to be loud and clear enough to be transmitted over some distance to conspecific recipients without too much distortion. Ambient noise by abiotic (e.g., wind, rain, thunderstorms) or biotic (the sound-producing community; e.g., cicadas, crickets, frogs, birds, mammals) sources and habitat structure (e.g., density of vegetation and resulting reflections and reverberations) may mask or modify the transmitted sound. Likewise, it can also be eavesdropped by prey or predators.

Considering these facts, it is generally postulated that selection drives acoustic signal evolution for a multidimensional effective "acoustic window" to optimize transmission, a notion put forth by the acoustic adaptation hypothesis (Morton, 1975) or the broader Sensory Drive framework (Endler, 1992), describing coevolution of signals, sensory systems, and microhabitat choice (Wilkins *et al.*, 2013). Such an approach outlines the available acoustic variation, open for ecological and sexual selection, and drift in a given habitat. A recent meta-analysis of song structure in birds revealed that habitat structure (open vs. closed) affected peak frequency significantly via sensory drive, while other spectral traits or interval duration remained almost unchanged (Boncoraglio and Saino, 2007). The relatively low effect of sensory drive shown in birds suggests that signal divergence within an acoustic window might be a common phenomenon (Wilkins *et al.*, 2013).

Habitat, predation, body mass, and the acoustic window are known (Figure 21.3) from four mouse lemur species (*M. murinus, M. ravelobensis, M. berthae, and M. lehilahytsara*). Two allopatric rufous-colored species (*M. ravelobensis, M. berthae*) are locally endemic and distributed in the dry deciduous forests of northwestern and western Madagascar, the third locally endemic and allopatric species in the evergreen rainforest of eastern Madagascar (*M. lehilahytsara*). Two studied populations of the gray mouse lemur (*M. murinus*) live in the dry deciduous forests of northwestern and western Madagascar, separated by about 500 km and several IRS (Schmid and Kappeler, 1994; Hafen, 1998; Zimmermann *et al.*, 1998;

Zietemann, 2000; Braune *et al.*, 2005, 2008; Olivieri *et al.*, 2007; see also Chapters 1 and 26). One of the gray mouse lemur populations in the dry deciduous forest lives sympatrically to the locally distributed same-bodied rufous-colored species in northwestern Madagscar (Zimmermann *et al.*, 1998), the other one to the significantly smaller-bodied rufous-colored species in western Madagascar (Schmid and Kappeler, 1994). Ambient noise is likely to differ in habitats of all of the allopatric species and populations due to divergent sound-producing animal communities (which is true even for the lemur community; see Mittermeier *et al.*, 2010) and divergent climatic conditions (such as rainfall and wind in dry vs. rainforest; see also Zietemann, 2000). Predation pressure is high and most likely comparable between rain and dry forest habitats (Scheumann *et al.*, 2007a; see also Chapter 19). The acoustic window of the mouse lemur's communication can be assessed by the lowest and highest peak frequency of the first harmonic of their most commonly used tonal call types (trill call, tsak call, whistle call; see Zimmermann, 2010, 2013). For all tested species and populations the used frequency window for communication coincides in ranging between about 8 and 40 kHz, despite differences in habitat and ambient noise conditions and divergences in body mass and distribution pattern (Figure 21.3). A potential explanation is that mouse lemurs bear a high predation risk because of their small body size (Goodman *et al.*, 1993; Scheumann *et al.*, 2007a). Two major predator categories, raptors and viverrids, either cannot hear high frequencies in this range (raptors), or they may have difficulties in locating their prey by the barely perceived high frequencies (viverrids; Fay, 1988). These findings suggest that predation may act as a major selective force constraining acoustic window variation for communication in small-bodied lemurs, such as mouse lemurs, and support findings in birds that signal divergence within an acoustic window might be a common phenomenon (Wilkins *et al.*, 2013).

	Predation	**Habitat**	**Body mass (g)**	**Acoustic window (kHz)**
M. murinus Ampijoroa Kirindy	raptors, viverrids, snakes	dry deciduous forest	58 58	8 - 40
M. ravelobensis	raptors, viverrids, snakes	dry deciduous forest	58	8 - 40
M. berthae	raptors, viverrids, snakes	dry deciduous forest	34	8 - 40
M. lehilahytsara	raptors, viverrids, snakes	rainforest	42	8 - 40

Figure 21.3 The "acoustic window" in mouse lemurs is uniform and extends from the high frequency to the ultrasonic range across populations and species, irrespective of habitat, distribution, and body mass. This suggests that predation is a major selective force constraining the available acoustic space for tonal communication in this small-bodied primate radiation.

Adaptive and neutral mechanisms of acoustic divergence

Ecological selection hypothesis for the evolution of acoustic divergence

If ecological selection is the primary driver of signal divergence, ecological trait divergence should covary with the degree of acoustic differentiation between sister taxa, while strength for local variants or strength of sexual selection does not (Wilkins *et al.*, 2013). It is known from various sound-producing animal groups that ecological selection for body mass or body size induces divergence in acoustic signals, in particular in signal pitch and duration (Gillooly and Ophir, 2010), and favors speciation. Mass and tension of vocal folds and lung capacity (reflected in air pressure) determine fundamental frequency and duration of a vocal output in mammals (Fitch, 2010). Body mass and bioacoustic information on three tonal call types used in the mating/group reunion (trill and chirp call), alarm/disturbance (whistle call), and agonistic context (tsak call) are available from allopatric rufous-colored mouse lemur species (*M. ravelobensis*, *M. berthae*, and *M. lehilahytsara*) and allopatric populations of the gray mouse lemur (*M. murinus*), separated by about 500 km and several IRS (Schmid and Kappeler, 1994; Hafen, 1998; Zimmermann *et al.*, 1998; Zietemann, 2000; Braune *et al.*, 2005, 2008; see Figures 21.3 and 21.4). One of the gray mouse lemur populations in the dry deciduous forest lives sympatrically to the locally distributed same-bodied rufous-colored species in northwestern Madagascar (Zimmermann *et al.*, 1998), the other to the significantly smaller-bodied rufous-colored species in western Madagascar (Schmid and Kappeler, 1994). Current research indicated that in the mating/group reunion context, there is no significant effect of species on fundamental frequency or bandwidth (Braune *et al.*, 2008), despite variation in body mass. There was no significant effect of population, but of species, on duration found (Zietemann, 2000; Braune *et al.*, 2008). Same-bodied (*M. murinus*, *M. ravelobensis*) and significantly different-bodied species (*M. murinus*, *M. lehilahytsara*) differed in this acoustic trait. In the agonistic context, there is no significant effect of population on the frequency of the first harmonics or bandwidth (Zietemann, 2000), but a significant effect of species on bandwidth in the sympatric species pair (*M. murinus*, *M. ravelobensis*), which do not differ in body mass. In another sympatric species pair (*M. murinus* and *M. berthae*) with a twofold and significant difference in body mass, no significant effect of species was revealed on minimum fundamental frequency, but a significant effect on maximum fundamental frequency and bandwidth. In the alarm/disturbance context, neither fundamental frequency nor duration showed prominent population or species effects (Zimmermann *et al.*, 1998; Zietemann, 2000). Thus, so far, findings make it unlikely that ecological selection for body mass drives acoustic divergence in mouse lemurs. However, ecological selection for morphological traits involved in the differentiation of feeding niches is known to affect vocal production in other sound-producing vertebrate groups. Thus, in Galapagos finches, divergence of beak size for efficient seed extraction is known to have a pleiotropic effect on song trill rate, so that birds with larger bills produce slower-paced trills (Podos, 2001). Divergently beaked morphs respond also

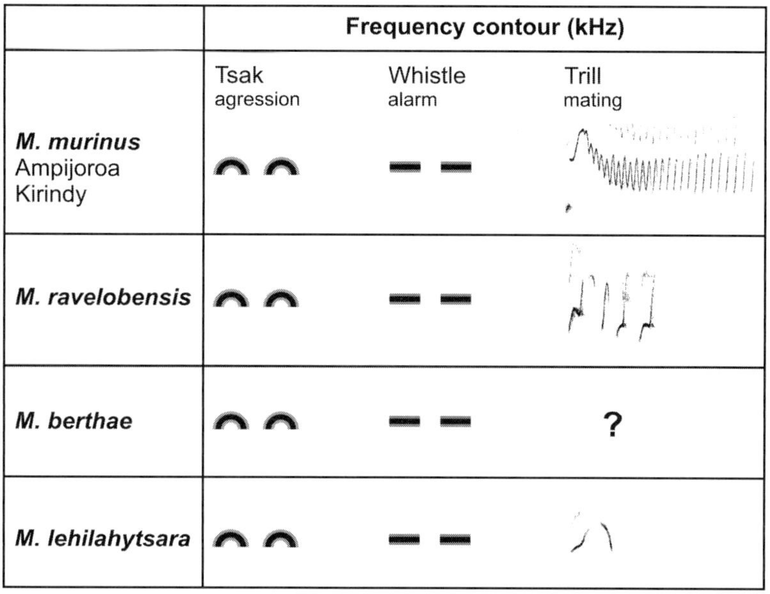

Figure 21.4 Tonal communication calls, given in the agonistic and alarm/disturbance context, show a uniform inverted U-shaped (tsak call) or slightly frequency-modulated frequency contour (short whistle call) across populations and species, irrespective of habitat, distribution, and body mass. Whereas they exhibit subtle statistically significant differences in specific acoustic dimensions between species in the agonistic context, they do not differ species-specifically in the alarm/disturbance context. Tonal communication calls, produced in the mating/group reunion context (trill call), show highest variation between individuals, groups, populations, and species. Species-specificity is displayed in frequency contour and duration.

more strongly to same-type morphs and are consequently leading to assortative mating (Podos, 2010).

More research is necessary on the relation between morphological traits (e.g., snout length and width and related vocal tract length) and vocal emission, production and perception to better understand the role of ecology for acoustic divergence and speciation in mouse lemurs.

Sexual selection hypothesis for the evolution of acoustic divergence

If sexual selection has been the major driver of acoustic divergence, strength of preference for, or intrasexual aggression elicited by, local acoustic signals should be a stronger predictor of acoustic divergence than differentiation in ecological traits (Wilkins *et al.*, 2013). One of the best examples supporting this hypothesis comes from Hawaiian swordtail crickets (Mendelson and Shaw, 2005; Grace and Shaw, 2012). Here, phylogenetic methods revealed that genetic differences between species are associated with differences in pulse rates and playback experiments showed that females preferred mean values of local pulse rates and therefore local mates.

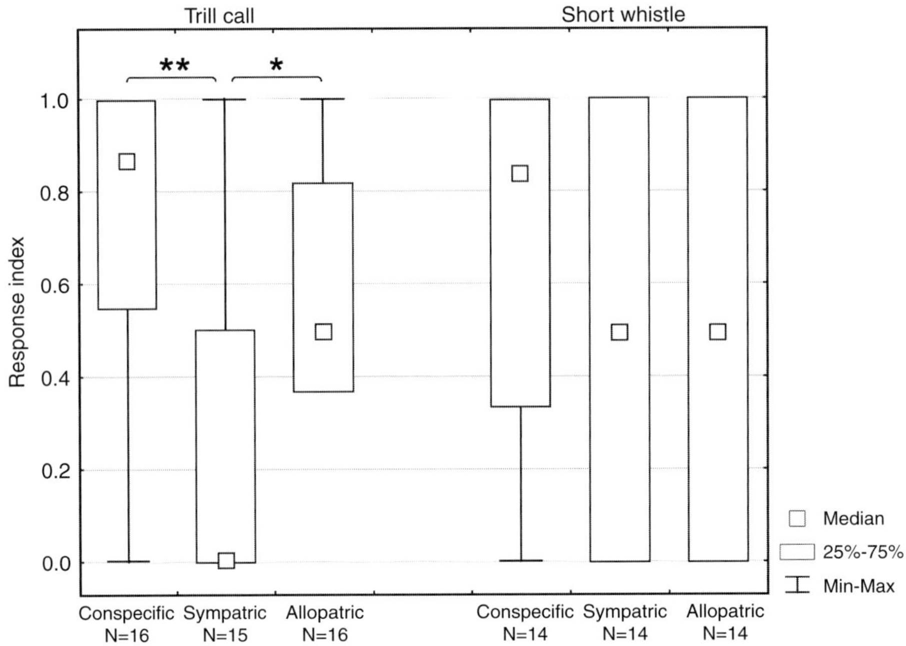

Figure 21.5 Playback experiments were performed in the gray mouse lemur from Ampijoroa, northwestern Madagascar. Individual response indices are shown (N = number of subjects). Subjects displayed stronger attention to the sympatric conspecific mating/group reunion call (trill call) than to the sympatric heterospecifics. The high attention to the allopatric heterospecific call may be due to a novelty effect. Subjects did not differ in their attention toward sympatric conspecific and heterospecific or allopatric heterospecific alarm/disturbance calls (whistle call). (Wilcoxon $^{*}p < 0.05$; $^{**}p < 0.01$; N = number of lemurs.)

Research on genetically characterized species and populations of mouse lemurs revealed quite interesting results. Clear-cut species-specific differences appeared in the frequency contour of the first harmonics of calls used in the context of mating/group reunion (trill call, advertisement call; see Figure 21.4), but not in those used in the context of agonism (tsak call) and alarm/disturbance (short whistle call; Zimmermann *et al.*, 1998; Zietemann, 2000; Braune *et al.*, 2008; Zimmermann, 2013). Playback experiments performed so far with the gray mouse lemur showed that members of this lemur species showed significantly more attention to the conspecific than to the sympatric heterospecific mating/group reunion/advertisement call (trill or chirp call). No such divergence in attention was revealed between sympatric conspecific and heterospecific or allopatric alarm/disturbance calls (short whistle call), which also did not differ species-specifically in call structure (Figure 21.5; Zimmermann *et al.*, 1998; Braune *et al.*, 2008). These findings suggest that sexual selection drives the evolution of acoustic divergence in communication calls of mouse lemurs specifically and thus may have a strong effect on speciation in mouse lemurs. Further studies are needed to examine the predictions of this hypothesis for calls used in agonistic conflicts. Likewise, further research both on the proximate mechanisms of acoustic divergence (divergences in mechanisms of sound production and perception) and its functional consequences (significance of divergences for mate choice, mother–infant and group coordination) is urgently needed to link acoustic divergence to functional morphology, neural control and genome-level data and to explain its relevance for evolution.

Neutral nechanisms

Drift hypothesis – a stochastic basis for acoustic divergence

Theoretically, it is possible that neutral mechanisms rather than ecological or sexual selection drives acoustic divergence between populations and favors speciation. Acoustic trait divergence should then increase linearly with genetic or geographic distance, as shown e.g. for greenish warblers (Irwin *et al.*, 2008), frogs (Amézquita *et al.*, 2011), or singing mice (Campbell *et al.*, 2010), and population-specific assortative mating should occur (Wilkins *et al.*, 2013).

Previous comparative and integrative studies on the gray mouse lemur (*M. murinus*), linking acoustic divergence to genetic distance between populations and taking into account morphological differentiations, yielded fascinating insights (Hafen *et al.*, 1998; Zietemann, 2000; Zimmermann and Hafen, 2001). First of all, two studied populations of the gray mouse lemur of northwestern and western Madagascar, separated macrogeographically by about 500 km and several IRS, are genetically distinct (Yoder *et al.*, 2000; Pastorini *et al.*, 2001; Olivieri *et al.*, 2007). Populations are also acoustically divergent in tonal communication calls, but on different levels for the respective call types. Thus, in call series of the short whistle call, given in alarm/disturbance contexts, and the tsak call, produced during agonistic conflicts, neither measured parameters of the frequency or temporal space are significantly different between populations (Zietemann, 2000). However, the trill call, used by males in the mating context, differed significantly and most prominently in the spectral and temporal domain (Figure 21.6A). Longer whistle duration at the onset of the trill call and a broader bandwidth of modulations separate the population of Kirindy from Ampijoroa (see Figure 21.6B and C; Zietemann, 2000). A further study focusing on the male's trill call in Kirindy (Hafen *et al.*, 1998) revealed that two demes of mouse lemurs, separated on a microgeographic scale by a distance of 1.5 km, belong to a morphologically and genetically homogenous population. Nevertheless, dialects in their calls were revealed in so far as acoustically divergent trill calls were found separating the two demes. Another study focusing on the male's trill call compared three colonies of this species in three European facilities (Brunoy, F; Hannover, D; Kassel, D). It was shown that all three colonies are genetically and acoustically clearly distinct (Zimmermann & Hafen, 2001). However, acoustic divergence was only weakly related to genetic and not to morphological divergence, suggesting that acoustic traits change at a more obvious and rapid pace than morphology in isolated populations. Interestingly, the acoustic dimensions of trill calls in mouse lemur infants, given in the reunion context, are

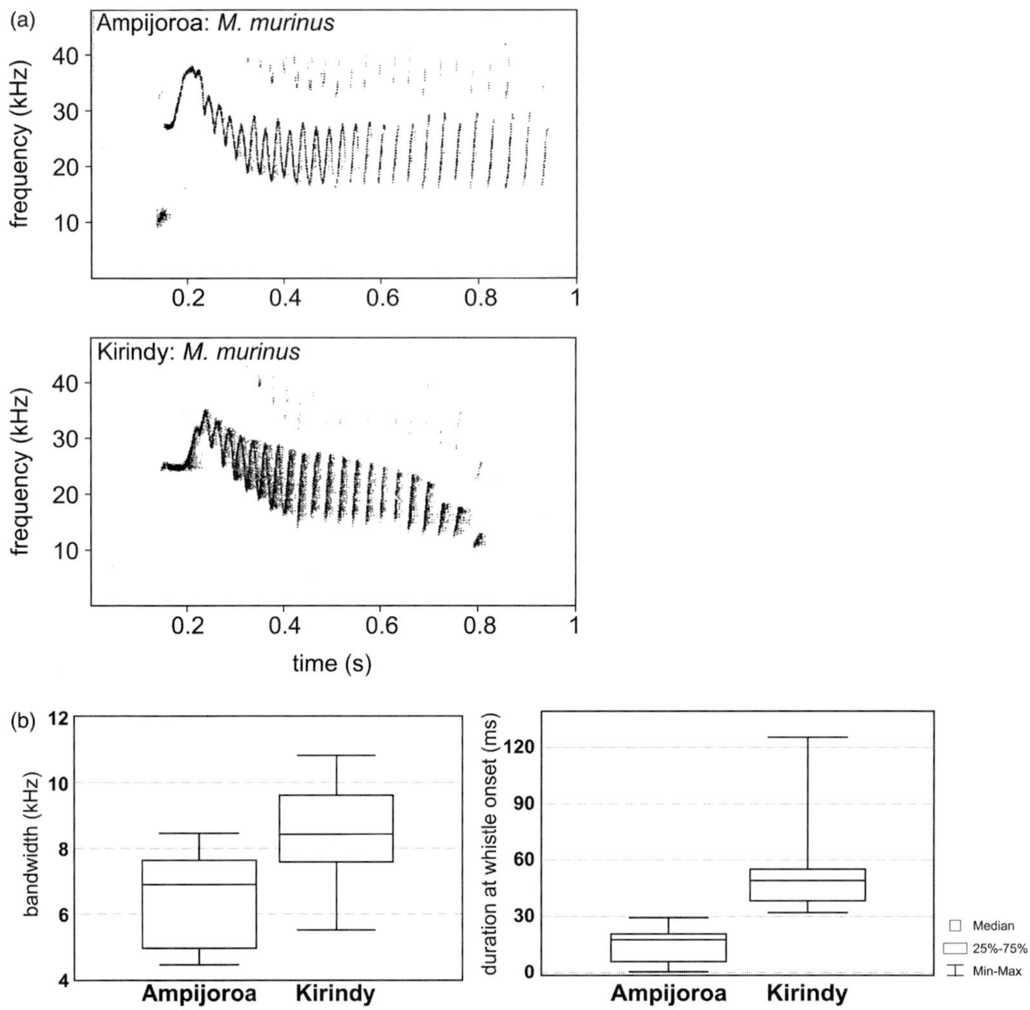

(c)

	% correct	Ampijoroa	Kirindy
Ampijoroa	100	10	0
Kirindy	100	0	18
Total	100	10	18

Figure 21.6 (a) The male's mating call (trill call) diverged between two allopatric populations of the gray mouse lemur. (b) Bandwidth of the seven first modulations (left) and duration at whistle onset (right) separates the population of Ampijoroa from Kirindy. (c) The classification table based on a stepwise discriminant analysis (SDA) with the two most discriminant parameters. The SDA assigned calls (after cross-validation) reliably to the respective population (Zietemann, 2000).

highly plastic (Zimmermann, 1991). Future experimental research on acoustically guided social interactions of genetically known subjects and populations is necessary to illuminate the role of maturation and learning on acoustic divergence. It is further important to assess how neutral and adaptive processes shape acoustic signals over time and contribute to the divergence of populations.

Statistical differences separating populations and species by their calls are not always perceptually salient and thus have to be examined for their biological relevance. From many studies of vertebrates communicating by sound, it is known that not all of the acoustic variation found on the sender's side is relevant for listeners (e.g., Zimmermann *et al.*, 2013). Standardized playback experiments (e.g., Braune *et al.* 2008; Kessler *et al.*, 2012) would allow examination of the extent to which subjects and sexes differ in their attention/reaction towards context-homologous calls varying on the microgeographic or macrogeographic scale by controlling for motivation, familiarity, individuality, kin, and sex. We have already performed pilot playback experiments with subjects from the Ampijoroa population. Findings suggested that subjects give the same amount of attention to short whistle and trill calls of the conspecific sympatric and allopatric population (Braune, Schmidt, Zimmermann, pers. observation). If these findings are supported by a larger sample size, this would mean that it is likely that the above-shown statistical differences in call structure, probably caused by drift, are not perceived and thus are actually of minor relevance for speciation.

Conclusions and perspectives for further research

All in all, the comparative and integrative approach on acoustic divergence in mouse lemurs suggests that multiple factors such as neutral and adaptive processes interact in shaping acoustic divergence and speciation in mouse lemurs.

Selection via predation (and probably ambient noise) and small body size seem to have constrained the acoustic window phylogenetically and thereby shaped the available acoustic space for communication in mouse lemurs. These constraints may have favored the evolution of a uniform high to ultrasonic frequency niche for tonal signal transmission, resulting in a lack of variation in this acoustic trait among the studied populations and species, despite varying habitat and ambient noise conditions. Sexual selection is most likely responsible for driving specific acoustic divergence in mating/group reunion calls (trill call). The lack of sexual selection may be the reason for a missing acoustic divergence in alarm/disturbance calls (short whistle call) between species. The effect of drift is less clear, but may be involved in shaping specific acoustic dimensions across tonal calls separating demes, populations, and species specifically. The relevance of these specific acoustic variations for speciation and evolution is so far largely unstudied and merits further playback experiments, both in the field and under controlled experimental conditions. Disentangling the relative contributions of the different forces and thereby illuminating causes and consequences of acoustic divergence and its significance for the distribution and evolution of mouse lemurs needs further comparative and

Table 21.1 Synopsis of current bioacoustic data on cheirogaleids (data supplemented by sonograms are summarized), related to taxon, habitat, distribution, locomotion, body mass, predation, food, and social organization.

Taxon	Habitat	Distrib.	Loco.	Body mass (g)	Pred.	Food	Social organ.	Vocal repert.	Level of acoustic divergence	References
Allocebus trichotis	ERF	Patchy	AQ	65–90	Potentially high	G, also O	P	Field, partly k	Unknown	8, 10
Cheirogaleus major	ERF	Broad	AQ	250–500	High	F, N, also O	?	Partly k	Unknown	6, 10
Cheirogaleus medius	DDF	Broad	AQ	120–270	High	F, also O	P	Lab	Unknown	2, 6, 10, 12, 13
Microcebus berthae (previously *myoxinus*)	DDF	Local	AQ, VCL	34	High	O, partly G	SR, SS?	Field	Species	13, 14
Microcebus lehilahytsara (previously *rufus*)	ERF	Local	AQ, VCL	42	High	O	SR, SFG, SMA	Field and lab	Species, individual	9, 10, 11, 13
Microcebus murinus	DDF	Broad	AQ, VCL	58	High	O, partly G, IS	SR, SFG, SMA	Field and lab	Species, population, deme, kin, group, individual	2, 3, 4, 5, 6, 7 10, 12, 13, 14
Microcebus ravelobensis	DDF	Local	AQ, VCL	58	High	O, partly G, IS,	SR, SMSG	Field	Species	1, 13, 14
Mirza coquereli	DDF		AQ, VCL	300–320	Medium	O	SR, SS	Field and lab, partly k	Unknown	5, 6, 10, 12
Phaner pallescens (previously *furcifer*)	DDF	Patchy	VCL	327–340	Medium	Mostly tree exudates	DP	Field, partly k	Unknown	6, 10

Notes: Habitat: ERF, evergreen rainforest; DDF, dry deciduous forest; Distrib., distribution; Loc., locomotion: Q, quadrupedal walking and climbing; VCL, vertical clinging and leaping; Pred., predation risk: low, 0–3 predator species; medium, 4–8 predator species; high, more than 9 predator species; Social organ., social organization: P, pair; DP, dispersed pair; SR, solitary ranging; SFG, sleeping in female groups; SS, sleeping solitary; SMSG, sleeping in mixed-sex groups; SMS, sleeping male solitary; Food: F, fruit; G, gum; IS, insect secretions; N, nectar; O, omnivorous; Vocal repert., vocal repertoire: noted if calls were displayed together with sonogram and context information; k, known; Level of acoustic divergence: known statistically affirmed divergences on the level of individual, group, deme, population, or species.

References: 1, Braune *et al.*, 2005, 2008; 2, Cherry *et al.*, 1987; 3, Hafen *et al.*, 1998, Hafen, 1998; 4, Leliveld *et al.*, 2010; 5, Pagès, 1980; 6, Petter and Charles-Dominique, 1979; 7, Rahlfs and Fichtel, 2010; 8, Rakotoarison *et al.*, 1996; 9, Randrianambinina, 2001; 10, Scheumann *et al.*, 2007a, 2007b; 11, Schopf, 2008; 12, Stanger, 1989, 1992; 13, Zimmermann and Lerch, 1993, Zimmermann *et al.*, 2000, Zimmermann, 1995a, 1995b, 2010, 2012; 14, Zietemann, 2000. Information on the biology of all taxa was taken from Mittermeier *et al.*, 2010, Radespiel, 2006 and Schülke and Ostner, 2005; ?, unknown.

integrative approaches. The latter approaches should cover a micro- and macrogeographic scale and more than four species and should combine bioacoustic, behavioral, ecological, functionally morphologic, and biogeographic and genome-level data (see also Zimmermann and Radespiel, 2014). Comparative and integrative approaches are urgently needed to quantify the role of acoustic divergence and adaptation in shaping the evolution and current distribution of species. They are also of the utmost importance for conservation because they may help to develop non-invasive bioacoustic tools for assessing species diversity and distribution. Similar approaches should be applied to other sound-producing nocturnal cheirogaleids which to date have been fairly neglected bioacoustically (Table 21.1). There is almost no information on the small-bodied, rainforest-dwelling and pair-living hairy-eared dwarf lemur (*Allocebus* ssp.), besides a sonogram showing a whistle-like disturbance call (Rakotoarison *et al.*, 1996). The giant mouse lemurs (*Mirza* ssp.), squirrel-sized lemurs living in the dry forest habitats of western Madagascar in comparable social settings as mouse lemurs (Schülke and Ostner, 2005), are largely unstudied with regard to the relevance of acoustic divergence for speciation and evolution, and so are the highly vocal, weasel-sized, pair-living forked-marked lemurs (*Phaner* ssp.) with their patchy distribution in western, northwestern and eastern Madagascar. The same is true for the squirrel-sized, pair-living dwarf lemurs (*Cheirogaleus* ssp.), inhabiting a diversity of habitats across Madagascar (Schülke and Ostner, 2005). Because these groups vary largely in body size, locomotion, feeding habits, sociality, and distribution patterns as well as in predation pressure, comparative and integrative approaches covering these groups will provide unique insight into the extent to which these factors drive acoustic divergence via neutral and adaptive processes in primates and contribute to the current distribution and evolution of cheirogaleids in Madagascar.

Acknowledgments

I thank my colleague Sabine Schmidt for comments on a previous version of this manuscript. I also thank my colleagues Ute Radespiel, Marina Scheumann, Romule Rakotondravony, Blanchard Randrianambinina, and Solofonirina Rasoloharijaona for long-term support and Elisabeth Engelke and Sönke von den Berg for technical support.

References

Amézquita A, Flechas SV, Lima AP, Gasser H, Hödl W. 2011. Acoustic interference and recognition space within a complex assemblage of dendrobatid frogs. *Proceedings of the National Academy of Sciences of the United States of America* 108:17058–17063.

Boncoraglio G, Saino N. 2007. Habitat structure and the evolution of bird song: a meta-analysis of the evidence for the acoustic adaptation hypothesis. *Functional Ecology* 21:134–142.

Braune P, Schmidt S, Zimmermann E. 2005. Spacing and group coordination in a nocturnal primate, the golden brown mouse lemur (*Microcebus ravelobensis*): the role of olfactory and acoustic signals. *Behavioral Ecology and Sociobiology* 58:587–596.

Braune P, Schmidt S, Zimmermann E. 2008. Acoustic divergence in the communication of cryptic species of nocturnal primates (*Microcebus* ssp.). *BMC Biology* 6:19.

Buesching CD, Heistermann M, Hodges JK, Zimmermann E. 1998. Multimodal oestrus advertisement in a small nocturnal prosimian, *Microcebus murinus*. *Folia Primatologica* 69:295–308.

Bunkus E, Scheumann M, Zimmermann E. 2005. Does acoustic predator recognition require experience in ancient primates? 29th International Ethological Conference, Budapest, 20–27 August 2005, p. 39.

Campbell P, Pasch B, Pino JL, *et al.* 2010. Geographic variation in the songs of Neotropical singing mice: testing the relative importance of drift and local adaptation. *Evolution* 64:1955–1972.

Charles-Dominique P, Cooper HM, Hladik A, *et al.* 1980. *Nocturnal Malagasy Primates: Ecology, Physiology, and Behavior.* Academic Press, New York.

Cherry JA, Izard MK, Simons EL. 1987. Description of ultrasonic vocalizations of the mouse lemur (*Microcebus murinus*) and the fat-tailed dwarf lemur (*Cheirogaleus medius*). *American Journal of Primatology* 13:181–185.

Endler JA. 1992. Signals, signal conditions, and the direction of evolution. *American Naturalist* 139:S125–S153.

Fay RR. 1988. Comparative psychoacoustics. *Hearing Research* 34:295–306.

Fitch WT. 2010. *The Evolution of Language.* Cambridge University Press, New York.

Gillooly JF, Ophir AG. 2010. The energetic basis of acoustic communication. *Proceedings Biological Sciences* 277:1325–1331.

Goerlitz HR, Siemers BM. 2007. Sensory ecology of prey rustling sounds: acoustical features and their classification by wild grey mouse lemurs. *Functional Ecology* 21:143–153.

Goodman SM, O'Connor S, Langrand O. 1993. A review of predation on lemurs: implications for the evolution of social behavior in small, norturnal primates. In PM Kappeler, JU Ganzhorn (eds.), *Lemur Social Systems and their Ecological Basis* (pp. 51–66). Plenum Press, New York.

Grace JL, Shaw KL. 2012. Incipient sexual isolation in *Laupala cerasina*: females discriminate population-level divergence in acoustic characters. *Current Zoology* 58:416–425.

Hafen TG. 1998. Dialekte bei Lemuren: bioakustische, morphometrische und molekulargenetische Untersuchungen zur intraspezifischen Variabilität beim grauen Mausmaki (*Microcebus murinus*). Dissertation, Universität Konstanz, Konstanz.

Hafen T, Neveu H, Rumpler Y, Wilden I, Zimmermann E. 1998. Acoustically dimorphic advertisement calls separate morphologically and genetically homogenous populations of the grey mouse lemur (*Microcebus murinus*). *Folia Primatologica* 69:342–356.

Hohenbrink P, Radespiel U, Mundy NI. 2012. Pervasive and ongoing positive selection in the vomeronasal-1 receptor (V1R) repertoire of mouse lemurs. *Molecular Biology and Evolution* 29:3807–3816.

Hohenbrink P, Mundy NI, Zimmermann E, Radespiel U. 2013. First evidence for functional vomeronasal 2 receptor genes in primates. *Biology Letters* 9:20121006.

Irwin DE, Thimgan MP, Irwin JH. 2008. Call divergence is correlated with geographic and genetic distance in greenish warblers (*Phylloscopus trochiloides*): a strong role for stochasticity in signal evolution? *Journal of Evolutionary Biology* 21:435–448.

Kappel P, Hohenbrink S, Radespiel U. 2011. Experimental evidence for olfactory predator recognition in wild mouse lemurs. *American Journal of Primatology* 73:928–938.

Kessler SE, Scheumann M, Nash LT, Zimmermann E. 2012. Paternal kin recognition in the high frequency/ultrasonic range in a solitary foraging mammal. *BMC Ecology* 12:26.

Leliveld LMC, Scheumann M, Zimmermann E. 2010. Effects of caller characteristics on auditory laterality in an early primate (*Microcebus murinus*). *PLoS ONE* 5:e9031.

Leliveld LMC, Scheumann M, Zimmermann E. 2011. Acoustic correlates of individuality in the vocal repertoire of a nocturnal primate (*Microcebus murinus*). *Journal of the Acoustical Society of America* 129:2278–2288.

Martin RD. 1990. *Primate Origins and Evolution: A Phylogenetic Reconstruction*. Chapman and Hall, London.

Masters J. 1999. The role of selection in speciation. In RS Singh, CB Krimbas (eds.), *Evolutionary Genetics: From Molecules to Morphology*, Vol. 1 (pp. 513–531). Cambridge University Press, Cambridge.

Mayr E. 1999. Understanding evolution. *Trends in Ecology and Evolution* 14:372–373.

McNett GD, Cocroft RB. 2008. Host shifts favor vibrational signal divergence in *Enchenopa binotata* treehoppers. *Behavioral Ecology* 19:650–656.

Mendelson TC, Shaw KL. 2005. Rapid speciation in an arthropod. *Nature* 433:375–376.

Meyer A. 1993. Phylogenetic relationships and evolutionary processes in East African cichlid fishes. *Trends in Ecology and Evolution.* 8:279–284.

Mittermeier RA, Louis EE Jr, Richardson M, *et al.* 2010. *Lemurs of Madagascar.* Conservation International, Washington, DC.

Morton ES. 1975. Ecological sources of selection on avian sounds. *American Naturalist* 109:17–34.

Olivieri G, Zimmermann E, Randrianambinina B, *et al.* 2007. The ever-increasing diversity in mouse lemurs: three new species in north and northwestern Madagascar. *Molecular Phylogenetics and Evolution* 43:309–327.

Pagès E. 1980. Ethoecology of *Microcebus coquereli* during the dry season. In P Charles-Dominique, HM Cooper, A Hladik, *et al.* (eds.), *Nocturnal Malagasy Primates: Ecology, Physiology, and Behavior* (pp. 97–116). Academic Press, New York.

Pariente GF. 1980. Quantitative and qualitative study of the light available in the natural biotope of Malagasy prosimians. In P Charles-Dominique, HM Cooper, A Hladik, *et al.* (eds.), *Nocturnal Malagasy Primates: Ecology, Physiology, and Behavior* (pp. 117–134). Academic Press, New York.

Pastorini J, Martin RD, Ehresmann P, Zimmermann E, Forstner MRJ. 2001. Molecular phylogeny of the lemur family Cheirogaleidae (Primates) based on mitochondrial DNA sequences. *Molecular Phylogenetics and Evolution* 19:45–56.

Perret M. 1995. Chemocommunication in the reproductive function of mouse lemurs. In L Alterman, GA Doyle, MK Izard (eds.), *Creatures of the Dark: The Nocturnal Prosimians* (pp. 377–392). Plenum Press, New York.

Petter JJ, Charles-Dominique P. 1979. Vocal communication in prosimians. In GA Doyle, RD Martin (eds.), *The Study of Prosimian Behavior* (pp. 247–305). Academic Press, New York.

Piep M, Radespiel U, Zimmermann E, Schmidt S, Siemers BM. 2008. The sensory basis of prey detection in captive-born grey mouse lemurs, *Microcebus murinus. Animal Behaviour* 75:871–878.

Podos J. 2001. Correlated evolution of morphology and vocal signal structure in Darwin's finches. *Nature* 409:185–188.

Podos J. 2010. Acoustic discrimination of sympatric morphs in Darwin's finches: a behavioural mechanism for assortative mating? *Philosophical Transactions of the Royal Society of London Series B, Biological Sciences* 365:1031–1039.

Radespiel U. 2006. Ecological diversity and seasonal adaptations of mouse lemurs (*Microcebus* spp.). In L Gould, ML Sauther (eds.), *Lemurs: Ecology and Adaptation* (pp. 211–233). Springer, New York.

Radespiel U, Ratsimbazafy JH, Rasoloharijaona S, *et al.* 2012. First indications of a highland specialist among mouse lemurs (*Microcebus* spp.) and evidence for a new mouse lemur species from eastern Madagascar. *Primates* 53:157–170.

Rahlfs M, Fichtel C. 2010. Anti-predator behaviour in a nocturnal primate, the grey mouse lemur (*Microcebus murinus*). *Ethology* 116(5):429–439.

Rakotoarison N, Zimmermann H, Zimmermann E. 1996. Hairy-eared dwarf lemur (*Allocebus trichotis*) discovered in a highland rainforest of eastern Madagascar. In WR Lourenço (ed.), *Biogéographie de Madagascar* (pp. 275–282). OSTROM, Paris.

Randrianambinina B. 2001. Contribution à l'étude comparative de l'écoéthologie de deux microcèbes rouges de Madagascar *Microcebus ravelobensis* (Zimmermann *et al.*, 1998), *Microcebus rufus* (Lesson, 1840). Dissertation, Université d'Antananarivo, Antananarivo, Madagascar.

Rasoloarison RM, Weisrock DW, Yoder AD, Rakotondravony D, Kappeler PM. 2013. Two new species of mouse lemurs (Cheirogaleidae: *Microcebus*) from eastern Madagascar. *International Journal of Primatology* 34:455–469.

Scheumann M, Rabesandratana A, Zimmermann E. 2007a. Predation, communication, and cognition in lemurs. In SL Gursky, KAI Nekaris (eds.), *Primate Anti-Predator Strategies* (pp. 100–126). Springer, New York.

Scheumann M, Zimmermann E, Deichsel G. 2007b. Context-specific calls signal infants' needs in a strepsirrhine primate, the gray mouse lemur (*Microcebus murinus*). *Developmental Psychobiology* 49:708–718.

Schilling A, Perret M. 1987. Chemical signals and reproductive capacity in a male prosimian primate (*Microcebus murinus*). *Chemical Senses* 12:143–158.

Schmid J, Kappeler PM. 1994. Sympatric mouse lemurs (*Microcebus* spp.) in western Madagascar. *Folia Primatologica* 63:162–170.

Schopf C. 2008. Verhaltensexperimentelle und bioakustische Untersuchungen zum akustischen Verhalten und zur Bedeutung von Soziallauten beim Goodman's Mausmaki (*Microcebus lehilahytsara*). Diploma thesis, Leibniz Universität Hannover, Hannover.

Schopf C, Zimmermanna E, Tünsmeyer J, *et al.* 2014. Hearing and age-related changes in the gray mouse lemur. *Journal of the Association for Research in Otolaryngology*. doi: 10.1007/s10162-014-0478-4.

Schülke O, Ostner J. 2005. Big times for dwarfs: social organization, sexual selection, and cooperation in the Cheirogaleidae. *Evolutionary Anthropology* 14:170–185.

Siemers BM. 2013. The sensory ecology of foraging for animal prey. In J Masters, M Gamba, F Génin (eds.), *Leaping Ahead: Advances in Prosimian Biology* (pp. 257–263). Springer, New York.

Siemers BM, Goerlitz HR, Robsomanitrandrasana E, *et al.* 2007. Sensory basis of food detection in wild *Microcebus murinus*. *International Journal of Primatology* 28:291–304.

Stanger K. 1989. A comparative approach to the organization of vocal signals in noctural prosimians. *Human Evolution* 4:181–185.

Stanger KF. 1992. Structure and function of the vocalizations of nocturnal prosimians (Cheirogaleidae). Dissertation, Eberhard-Karls-Universität, Tübingen.

Sündermann D, Scheumann M, Zimmermann E. 2008. Olfactory predator recognition in predator-naïve gray mouse lemurs (*Microcebus murinus*). *Journal of Comparative Psychology* 122:146–155.

Valenta K, Burke RJ, Styler SA, *et al.* 2013. Colour and odour drive fruit selection and seed dispersal by mouse lemurs. *Scientific Reports* 3:2424.

Veilleux CC, Louis EE, Bolnick DA. 2013. Nocturnal light environments influence color vision and signatures of selection on the *OPN1SW* opsin gene in nocturnal lemurs. *Molecular Biology and Evolution* 30(6):1420–1437. doi:10.1093/molbev/mst058.

Wilkins MR, Seddon N, Safran RJ. 2013. Evolutionary divergence in acoustic signals: causes and consequences. *Trends in Ecology and Evolution* 28:156–166.

Willink B, Bolaños F, Pröhl H. 2014. Conspicuous displays in cryptic males of a polytypic poison-dart frog. *Behavioral Ecology and Sociobiology* 68:249–261.

Yoder AD, Rasoloarison RM, Goodman SM, *et al.* 2000. Remarkable species diversity in Malagasy mouse lemurs (primates, *Microcebus*). *Proceedings of the National Academy of Sciences of the United States of America* 97:11325–11330.

Zietemann V. 2000. Artendiversität bei Mausmakis: die Bedeutung der akustischen Kommunikation. Dissertation, Universität Hannover, Hannover.

Zimmermann E. 1991. Ontogeny of acoustic communication in prosimian primates. In A Ehara, O Takenaka, M Iwamoto (eds.), *Primatology Today* (pp. 337–340). Elsevicr, Amsterdam.

Zimmermann E. 1995a. Acoustic communication in nocturnal prosimians. In L Alterman, GA Doyle, MK Izard (eds.), *Creatures of the Dark: The Nocturnal Prosimians* (pp. 311–330). Plenum Press, New York.

Zimmermann E. 1995b. Loud calls in nocturnal prosimians: structure, evolution and ontogeny. In E Zimmermann, JD Newman, U Jürgens (eds.), *Current Topics in Primate Vocal Communication* (pp. 47–72). Plenum Press, New York.

Zimmermann E. 2010. Vocal expression of emotion in a nocturnal prosimian primate group, mouse lemurs. In SM Brudzynski (ed.), *Handbook of Mammalian Vocalization: An Integrative Neuroscience Approach* (pp. 215–226). Academic Press, Oxford.

Zimmermann E. 2013. Primate serenades: call variation, species diversity, and adaptation in nocturnal strepsirrhines. In J Masters, M Gamba, F Génin (eds.), *Leaping Ahead: Advances in Prosimian Biology* (pp. 287–295). Springer, New York.

Zimmermann E, Hafen TG. 2001. Colony specificity in a social call of mouse lemurs (*Microcebus* ssp.). *American Journal of Primatology* 54:129–141.

Zimmermann E, Lerch C. 1993. The complex acoustic design of an advertisement call in male mouse lemurs (*Microcebus murinus*, Prosimii, Primates) and sources of its variation. *Ethology* 93:211–224.

Zimmermann E, Radespiel, U. 2014. Species concepts, diversity, and evolution in primates: lessons to be learned from mouse lemurs. *Evolutionary Anthropology* 23:11–14.

Zimmermann E, Cepok S, Rakotoarison N, Zietemann V, Radespiel U. 1998. Sympatric mouse lemurs in north-west Madagascar: a new rufous mouse lemur species (*Microcebus ravelobensis*). *Folia Primatologica* 69:106–114.

Zimmermann E, Vorobieva E, Wrogemann D, Hafen T. 2000. Use of vocal fingerprinting for specific discrimination of gray (*Microcebus murinus*) and rufous mouse lemurs (*Microcebus rufus*). *International Journal of Primatology* 21(5):837–852.

Zimmermann E, Leliveld LMC, Schehka S. 2013. Toward the evolutionary roots of affective prosody in human acoustic communication: a comparative approach to mammalian voices. In E Altenmüller, S Schmidt, E Zimmermann (eds.), *Evolution of Emotional Communication: From Sounds in Nonhuman Mammals to Speech and Music in Man* (pp. 116–132). Oxford University Press, Oxford.

22 Modeling the origins of primate sociality: social flexibility and kinship in mouse lemurs (*Microcebus* spp.)

Sharon E. Kessler, Ute Radespiel, Leanne T. Nash, and Elke Zimmermann

The dawn of primate social complexity: kin selection in asocial mammals

Since Hamilton's ground-breaking theory of inclusive fitness in 1964, kin-biased behavior has been theorized to have played a crucial role in the evolution of mammalian sociality (Hamilton, 1964; de Waal and Tyack, 2003; Chapais and Berman, 2004). Given the amount of attention given to the topic over the subsequent decades, it is surprising that while group-living and social complexity has evolved multiple times in mammals, we still know very little about how this process occurs (Waser and Jones, 1983; Müller and Thalmann, 2000; de Waal and Tyack, 2003). In this section we review what is known about ancestral mammals and how they were the foundation for the evolution of ancestral primates.

Ancestral mammals are believed to have been asocial, as are many extant mammal species (Waser and Jones, 1983; Müller and Thalmann, 2000). Asocial species forage alone and maintain no relationships outside of the mating and infant-rearing seasons (Charles-Dominique, 1974, 1978; Waser and Jones, 1983; Müller and Thalmann, 2000). Interactions between adults, including adult kin, are marked by avoidance and aggression (Charles-Dominique, 1974; Waser and Jones, 1983; Müller and Thalmann, 2000). This is note-worthy because in many species, females typically disperse shorter distances than males, leading to a spatial clustering of female kin (Waser and Jones, 1983; Stoen *et al.*, 2005; Maher, 2009). For many scientists, it is this spatial clustering of kin which is the first step towards increasing sociality (Waser and Jones, 1983; Perrin and Lehmann, 2001; Kappeler *et al.*, 2002; Lutermann *et al.*, 2006; Meshriy *et al.*, 2011; Messier *et al.*, 2012). The transition to group-living is believed to have occurred through solitary foraging (Müller and Thalmann, 2000). Extant solitary foragers forage alone, but, in contrast to the

The Dwarf and Mouse Lemurs of Madagascar: Biology, Behavior and Conservation Biogeography of the Cheirogaleidae, eds. S.M. Lehman, U. Radespiel, E. Zimmermann. Published by Cambridge University Press. © Cambridge University Press 2016.

asocial mammalian ancestors, maintain year-round social networks, communicating with conspecifics via scent-marks and vocalizations (Charles-Dominique, 1974, 1978; Zimmermann, 1990, 1995a, 1995b, 2010; Müller and Thalmann, 2000; Nash, 2004; see also Chapter 21). Individuals may interact affiliatively during their active periods and sometimes sleep in social groups, often consisting of matrilineal kin, during the inactive periods (e.g., Radespiel *et al.*, 2001b; Eberle and Kappeler 2006; for review, see Müller and Thalmann, 2000). These social networks are believed to have been the foundations from which group-living evolved (Müller and Thalmann, 2000); if these social networks enabled solitary foragers to recognize their kin, they could have facilitated kin selection (the preferential treatment of genetic relatives; Hamilton, 1964; Müller and Thalmann, 2000; Nash, 2004; Kessler *et al.*, 2012). This, in turn, may have led to the formation of kin-based foraging groups and the diversity of complex social systems seen today in mammals (Waser and Jones, 1983; Müller and Thalmann, 2000; Perrin and Lehmann, 2001).

This transition from solitary foraging with kin-based networks to kin-based foraging groups hinges upon the ability of individuals to use these kin networks to benefit themselves and their kin (i.e., kin selection; Hamilton, 1964; Kessler et al., 2012, 2014). In order to better understand how this transition occurred, we focus on Primates, an order where kin-based group-living has evolved multiple times (Müller and Thalmann, 2000). Although kin-biased sociality is not unique to primates (e.g., hyenas: Holekamp *et al.*, 2012; killer whales: Yurk *et al.*, 2002; elephants: McComb *et al.*, 2000; for review, see de Waal and Tyack, 2003), it is argued to have been a particularly vital adaptation in the primate order (Dunbar, 1998; Silk, 2002, 2007, 2009; Chapais and Berman, 2004). Today, the primate order contains diverse social systems including both group-living and dispersed systems (Kappeler, 1997; Müller and, Thalmann 2000; Kappeler and van Schaik, 2002; Strier, 2007). Group-living (gregarious) primates form their groups so that the benefits of group-living (e.g., increased protection from predators, foraging benefits, affiliative social interactions, social learning) are given to themselves and their kin (kin selection; Hamilton, 1964; Whiten *et al.*, 1999; Silk, 2002, 2007, 2009; Chapais and Berman, 2004; Strier, 2007; Perry, 2011; see also Chapter 19). Group-living primates forage in many different types of groups including: (1) multiple males and multiple females, (2) one male and multiple females, (3) one female and multiple males, (4) monogamous pairs, (5) fission–fusion communities that break apart and join together into different subgroups, and (6) multilevel societies with smaller permanent units embedded within them (Kappeler and van Schaik, 2002; Strier, 2007).

Dispersed social systems are also complex (Charles-Dominique, 1974; Bearder, 1987; Müller and Thalmann, 2000; Nash, 2004). Although primates with dispersed social systems usually forage solitarily, they have overlapping ranges and maintain social networks (Charles-Dominique, 1974; Bearder, 1987; Müller and Thalmann, 2000; Nash, 2004). Although less is known about solitary foragers, they may gain a similar advantage by directing the benefits of their dispersed social networks to their kin (Charles-Dominique, 1974; Nash, 2004). The structure of these networks varies across species to include, at a minimum, dispersed multimale/multifemale, dispersed monogamy, and dispersed harem systems (Charles-Dominique, 1974; Bearder, 1987; Müller and Thalmann, 2000; Nash, 2004). They use vocalizations to interact with individuals who are distant in space and leave scent-marks to communicate with individuals who are distant in time (Charles-Dominique, 1974, 1978; Bearder, 1987; Zimmermann, 1995a, 1995b; Müller and Thalmann, 2000; Nash, 2004; see also Chapter 21).

The mouse lemur model of ancestral primates: support and criticism

The last common ancestor of primates is believed to have been a nocturnal solitary forager which diverged from other primitive mammals between 90 and 55 million years ago (Martin, 1993; Gingerich and Uhen, 1994; Müller and Thalmann, 2000; Tavaré *et al.*, 2002; Springer *et al.*, 2003; Bloch *et al.*, 2007; Martin *et al.*, 2007; Strier, 2007; Steiper and Seiffert, 2012). Therefore, studies of extant nocturnal, solitary foragers are of great theoretical value because these primates have social systems and ecology (small brain relative to body size, foraging alone at night) that are thought to be similar to those of the hypothesized last common ancestor of the primate order (Cartmill, 1972, 1974, 1992; Charles-Dominique, 1974, 1978; Rasmussen, 1990, 2002; Sussman, 1991; Kappeler, 1997, 1998; Müller and Thalmann, 2000; Kappeler and van Schaik, 2002; Kappeler *et al.*, 2002; Gebo, 2004; Barton, 2006; Radespiel, 2006; Martin *et al.*, 2007; Ravosa and Dagosto, 2007; Dammhahn and Kappeler, 2008a; Scheumann and Zimmermann, 2008; Zimmermann, 2010; Thorén *et al.*, 2011). Mouse lemurs, in particular, have long been regarded as excellent ancestral primate models because they fit two major theories of primate origins. The first is the visually directed predation theory (Cartmill, 1972, 1974, 1992, 2012), which hypothesizes that the last common ancestor of primates was small-bodied with convergent eyes and grasping hands with which they preyed upon insects in the forest canopy and undergrowth. This is supported by fossils indicating that some of the earliest primates were small-bodied insectivores (i.e., *Teilhardina asiatica* dated to 55 mya; Ni *et al.* 2004, 2005). The second major theory of primate origins is the angiosperm coevolution theory (Sussman and Raven, 1978; Sussman, 1991; Sussman *et al.*, 2013), which predicts that the last common ancestor of primates exploited the fruits, nectar, and flowers (and insects attracted to them) in the thin terminal ends of angiosperm tree branches. This theory is supported by the relatively complete skeleton of the plesiadapiform *Carpolestes simpsoni* dated between 55 and 56 mya (Bloch and Boyer, 2002). This fossil has an opposable hallux (big toe) adapted for grasping, with a nail rather than a claw, and low-crowned molars indicating frugivory, thus supporting the angiosperm coevolution theory (Bloch and Boyer, 2002; Sussman *et al.*, 2013). In addition, *C. simpsoni* has divergent orbits, which are inconsistent with the visual predation hypothesis (Bloch and Boyer, 2002; Sussman *et al.*, 2013). Rasmussen (1990) suggested that these two theories are not necessarily mutually exclusive, meaning that an ancestral omnivore could have fed upon both fruits and insects in a fine branch niche. This view has since been supported by comparative work on marsupials (Lemelin, 1999; Schmitt and Lemelin, 2002) and lorises (Nekaris, 2005). Given that there is a mismatch between the divergence times estimated from the fossil record (first primates ~55 mya; Ni *et al.*, 2004, 2005) and divergence times estimated from molecular clock studies (~82 mya: Tavaré *et al.*, 2002; 76–63 mya: Steiper and Seiffert, 2012), many open questions remain.

Thus, living models for ancestral primates remain an important contribution to this debate, both because they provide comparative data from species in similar ecological niches and because they serve as models for those aspects of primate evolution that do not fossilize or get recorded in the genome, e.g., social behaviour (e.g., Müller and Thalmann, 2000; Gebo, 2004). Among the extant taxa used as ancestral primate models (i.e., shrews, tree-shrews, mouse lemurs, galagos, marsupials, colugos; Rasmussen, 1990; Müller and Thalmann, 2000; Gebo, 2004), mouse lemurs are among the most frequently used (e.g., Charles-Dominique and Martin, 1970; Sussman and Raven, 1978; Müller and Thalmann, 2000; Gebo, 2004), and are compatible with both the visual predation theory and the angiosperm coevolution theory. As the visually directed predation hypothesis (Cartmill, 1972, 1974, 1992, 2012) predicts for ancestral primates, mouse lemurs have convergent eyes and grasping hands with which they prey upon insects in the forest canopy and undergrowth (Charles-Dominique and Martin, 1972). As predicted by the angiosperm coevolution theory (Sussman and Raven, 1978; Sussman, 1991; Sussman *et al.*, 2013), mouse lemurs also forage for fruits and nectar in the thin terminal ends of branches (Charles-Dominique and Martin, 1972; Radespiel, 2006; Siemers *et al.*, 2007; Dammhahn and Kappeler, 2008a, 2009, 2010; Thorén *et al.*, 2011; Toussaint *et al.*, 2013). Furthermore, as *Microcebus* are the smallest primates, this makes them particularly useful models for estimates of ancestral primate body size that put ancestral primates at or smaller than the size of all living primates (e.g., Gebo, 2004; Silcox *et al.*, 2007). For example, based on fossil evidence suggesting that the last common ancestor of primates was likely smaller than extant primates, Gebo (2004) estimated that ancestral primates would have been approximately shrew-sized (10–15 g) and exhibited a mixture of behavioral traits similar to shrews and mouse lemurs.

Within the last 15 years, the idea that the first primates were mouse lemur sized or smaller has come under repeated criticism (e.g., Chapter 4; Soligo and Müller, 1999; Müller and Soligo, 2005; Soligo and Martin, 2006; Montgomery and Mundy, 2013). A series of studies proposed a revised synthesis of the adaptive nature of the origin of primates (Soligo and Müller, 1999; Müller and Soligo, 2005; Soligo and Martin, 2006). These authors argue that the stem lineage of primates increased in body mass leading to an ancestral primate that weighed approximately 1 kg and was largely frugivorous (Soligo and Müller, 1999; Müller and Soligo, 2005; Soligo and Martin, 2006). In a comparative study of rodents, they argue that frugivory is associated with the presence of social networks due to the foraging benefits of information sharing (Müller and Soligo, 2005). More recent work has focused on the dwarfing process that would have then ensued in the cheirogaleids (Montgomery and Mundy, 2013). A comparison of callitrichid primates and cheirogaleids showed that both lineages exhibit reductions in body size, although the dwarfing appears to have occurred through different mechanisms in each lineage (Montgomery and Mundy, 2013). The authors suggest that similarities between cheirogaleids and small-bodied ancestral primates should then be considered to be convergent evolution rather than the retention of primitive traits (Montgomery and Mundy, 2013). Similarly, Andrews *et al.* (Chapter 4) used phylogenetic ancestral state reconstructions of body mass and diet to conclude that the ancestral primate may have weighed as much as 2 kg, and was a slow-moving faunivore, similar to extant lorises. They reconstruct the ancestral lemur as having been approximately 1 kg and folivorous/frugivorous (Chapter 4). They argue that mouse lemurs underwent dwarfing relatively recently (< 8 mya) in association with forest fragmentation and drought, during which the gut adaptations for folivory aided the transition to gummivory/exudativory (Chapter 4).

What is immediately obvious in these scenarios is that, while multiple studies have argued that cheirogaleids have undergone dwarfing (Chapter 4; Soligo and Müller, 1999; Müller and Soligo, 2005; Soligo and Martin, 2006; Montgomery and Mundy, 2013), there is no clear consensus about what the ancestral primates would have been like. The authors differ on exactly how large the ancestral primate would have been, how it would have moved, and what it would have fed on (Chapter 4; Soligo and Müller, 1999; Müller and Soligo, 2005; Soligo and Martin, 2006; Montgomery and Mundy, 2013). Given that the social system is often inferred from this information, based on comparisons with living models (see discussion above), it leaves a significant amount of uncertainty as to what the ancestral primate social system might have been.

In the second half of this review, we propose using a different, yet complementary, strategy. Rather than inferring the social system from body size, diet, etc., we propose using the living models to examine the *diversity of social systems* that might have been possible with a certain suite of characteristics (i.e., body size, diet, etc.). Instead of reconstructing *the* ancestral primate social system, this approach estimates flexibility and the range of possibilities that an ancestral primate social system might have had. It also simultaneously allows identification of those characteristics which are consistent across species and therefore likely to be ancestral.

The mouse lemur radiation is an ideal model for this approach. Over the past several decades, mouse lemur diversity has come to be better understood, such that there are now 21 recognized species (Zimmermann and Radespiel, 2014). This presents an excellent opportunity to examine the variation in kin networks and how they influence social grouping patterns. While the social systems of many of these newly described species are still largely unknown, we are now developing a body of literature that allows us to begin to describe the variation that exists between the species, and occasionally even between populations (e.g., Schülke and Ostner, 2005; Lahann *et al.*, 2006; Radespiel, 2006). In the next section, we review what is known about the variation in kin networks in five of the best-studied mouse lemur species with a focus on how this is reflected in the formation of sleeping group associations (Figure 22.1 and Table 22.1). We focus on sleeping group associations for both theoretical and practical reasons. Theoretical reasons include that if there are benefits associated with sociality (i.e., thermoregulation, information sharing regarding foraging opportunities, cooperative breeding, etc.) then these sleeping

Figure 22.1 A sleeping group of three matrilineally related *M. murinus* in Ankarafantsika National Park. Genetic data presented in full in Kessler *et al.* (2014). Photo credit: Sharon E. Kessler. (A black and white version of this figure will appear in some formats. For the colour version, please refer to the plate section.)

group associations are when they are very likely to occur (e.g., Radespiel *et al.*, 2003; Müller and Soligo, 2005; Eberle and Kappeler, 2006; Génin, 2010). In practice, these are the most visible social contacts (Figure 22.1), thus they are easiest to study and use for assessing sociality patterns (Sterling *et al.*, 2000). Therefore, we focus on studies which radio-collared and tracked wild or free-ranging individuals of the species for which the most information is available: Madam Berthe's mouse lemur (*M. berthae*), the reddish-gray mouse lemur (*M. griseorufus*), Goodman's mouse lemur (*M. lehilahytsara*), the gray mouse lemur (*M. murinus*), and the golden-brown mouse lemur (*M. ravelobensis*) (Table 22.1).

Social flexibility as a crucial characteristic of ancestral primate sociality

In this section, we review the variation that occurs within and across mouse lemur species with the goal of examining the range of flexibility that exists. We examine the interactions of resources, energetic benefits, and reproductive season with sociality.

Resources: Included in Table 22.1 are populations that exist in very different habitats, ranging from rainforests to forests with harsh seasonality in temperature and rainfall including dry deciduous forests and spiny forest. Habitat, including

Table 22.1 Some of the variables thought to influence the co-sleeping behaviors of the best-studied mouse lemurs. Shaded rows show the variables influencing co-sleeping behaviors and non-shaded rows summarize the data available. When possible, data have been pooled from multiple studies and the source of the data is depicted with a superscript. Note that the season in which the data were collected is indicated by the superscripts in row "Season." Thereafter, when that superscript is listed next to corresponding data, it refers to *both* the study listed below and the season in which those data were collected. Numbers shown in the row "Radio-collared" depict the numbers of males and females radio-collared in the superscripted studies. "Solitary" shows the numbers of males and females who always slept alone. "M-M dyads," "F-F dyads," and "M-F dyads" show the number of male–male, female–female, and male–female dyads, respectively. Y or N indicates presence or absence when numbers are not given in the studies. "% Male" and "% Female" indicate what percentage of observations the males and females, respectively, slept in with at least one other lemur (grp), solitarily (sol) or, when given, sleeping group undetermined (unk). For photos of these species, see Zimmermann and Radespiel (2014) and Chapter 13.

Species	*M. berthae*	*M. griseorufus*	*M. lehilahytsara*		*M. murinus*			*M. ravelobensis*
Location	Kirindy	Berenty	Mantadia	Zürich Zoo	Kirindy	Mandena	Ankarafantsika	Ankarafantsika
Forest	Dry deciduous	Spiny forest	Evergreen rainforest	Masoala Hall	Dry deciduous	Littoral rainforest	Dry deciduous	Dry deciduous
Density (ind./ha)	[u]0.36 (or 0.70)	[c]2.2–4.3	[d]3–4	[f]46.4	[u]7.1		[v]0.8–1.3	[m]7.3–10.9
Temp. (°C)	[s]10-37		[s]10–26	[f]18–34	[s]10–37		[s]15–35	[s]15–35
Diet	[o]Om: Is, An	[t]Om: G, Fr		[f]Prov.	[o]Om: Fr, G	[p]Om: Fr	[r]Om: G, Fr	[r]Om: G, Is
Season	[a,b,q]Non	[c]Lact & Non	[d]Lact,[e]Non	[f]Non	[g]Lact, [h,q]Non	[i]Lact	[j,k]Non	[l,m,n]Non
Radio-collared*	[a]M:2, F:2 [b]M:19, F:9 [q]F:18	[c]M:12, F:14	[d]M:3, F:3 [e]M:5, F:2	[f]M:5, F:5	[g]F:33 [h]M:6, F:6 [q]F:17	[i]M:5, F:4	[j]M:15, F:15 [k]M:4, F:5	[l,m]M:10, F:9 [l,n]M:1, F:10, ?:5
Solitary	[a]M:2, F:2	[c]M:6, F:3	[d]F:1 [e]M:4	[f]rare	[g]F:2 [h]M:Y		[j]M:12, F:2	[m]M:2, F:1 [n]?:5
M-M dyads**	[a]0 [b]Y	[c]≤6	[d]3 [e]1 M-?	[f]common	[h]Y	[i]Y	[j]1	[l]5
F-F dyads**	[a]0 [b]Y	[c]≤10	[d]1 [e]1	[f]common	[g]51 [h]Y	[i]Y	[j]13	[l]10

M-F dyads**	[a]0 [b]Y	[c]Y	[d]0 [e]1 M-?	[f]rare	[h]Y	[i]<1% days	[j]2	[l]14
Male (%)	[a]100% sol [b]46% grp	[c]25% grp	[d]55% sol [d]45% grp [e]80% sol [e]20% grp			[i]71% sol [i]9% grp [i]20% unk	[k]100% sol	
Female (%)	[a]100% sol [b]48% grp [q]34% grp	[c]56% grp	[d]33% sol [d]66% grp [e]63% sol [e]37% grp		[q]73% grp	[i]3% sol [i]74% grp [i]23% unk	[k]7% sol [k]84% grp [k]9% unk	
Genetics	[b]3/15 dyads rel.	–	–	[f]Mat.	[g]Mat.		[j]Mat	[l]Mat.

Diet descriptions provide a general category and important food types as determined by the cited references. Om, omnivorous, includes some or all of the following: nectar (N), flowers (fl), gum (G), insects (In), insect secretions (Is), leaves (L), fruits (Fr), animal matter (An). Prov, provisioned. Season: Lact, lactation, Non, non-lactation.

Temp., temperature. [x]Temperatures are the ranges of mean monthly minimums. [f]Typical temperature ranges experienced by animals in the climate controlled Masoala Hall of the Zürich Zoo. Rel., related, Mat., matrilineally related.

[*] Animals collared in multiple years in the same study are counted only once. Animals collared in both lactation and non-lactation seasons in the same study are counted separately for each season. For *M. griseorufus*, *M. murinus* (Ankarafantsika), and *M. ravelobensis* some of the same animals may be included in multiple studies. When this occurs, the radio-collared animals are listed separately in the row "Radio-collared," and the numbers of dyads and percentage of time that the animals spent together is taken, as much as possible, from the most comprehensive data set.

[**] The definition of "dyad" varies across studies as follows: [c]counts include dyads listed as "pairs" not "associations," because "pairs" appear to be more stable, [g]≥50% of days, [j]≥2 days for males, [l]≥3 days of co-sleeping, [n]≥2 days, [o]≥2 days. For all other studies dyad is assumed to mean ≥1 day of co-sleeping.

[a]Schwab, 2000; [b]Dammhahn and Kappeler, 2005; [c]Génin, 2008; [d]table XXXI of Randrianambinina, 2001; [e]table XXX of Randrianambinina, 2001; [f]Jürges *et al.*, 2013; Kitzler, 2011; [g]Eberle and Kappeler, 2006; [h]Rasoazanabary, 2006; [i]Lahann, 2008; [j]Radespiel *et al.*, 2001b; [k]Radespiel *et al.*, 1998; [l]Radespiel *et al.*, 2009; [m]Weidt *et al.*, 2004; [n]Braune, 2007; Braune *et al.*, 2005; [o]Dammhahn and Kappeler, 2008b, 2009; [p]Lahann, 2007; [q]Dammhahn and Kappeler, 2009; [r]Thorén *et al.*, 2011; [s]Schülke and Ostner, 2007; [t]Génin, 2010; [u]Schülke and Ostner, 2005; [v]Radespiel *et al.*, 2001a.

the availability of different types of sleeping sites (e.g., tree holes vs. vegetation tangles) and the type and distribution of food resources in time and space, appears to impact the degree of sociality (Table 22.1). For example, *M. berthae* subsists upon smaller, more dispersed food sources (insect secretions and animal matter), which matched predictions for its lower degree of sociality, relative to the more social gray mouse lemur which has a broader feeding niche and subsists upon larger patches of food, including more fruit (Dammhahn and Kappeler, 2005, 2008a, 2009). Similarly, *M. griseorufus*, which also consumes substantial amounts of fruit, sometimes forages more cohesively and may exclude non-sleeping group members from fruit trees (Génin, 2008, 2010; Chapter 13). Individuals which formed regular sleeping group associations had larger ranges, had more fruit in their diets, and were heavier than more solitary conspecifics (Génin, 2008, 2010; Chapter 13).

Energetic benefits: Energetic benefits of co-sleeping may also have a large influence on sociality (Kappeler, 1998; Perret, 1998; Schmid, 1998; Dammhahn and Kappeler, 2005; Schülke and Ostner, 2005, 2007; Lutermann *et al.*, 2010; Dausmann 2014). Sleeping in groups has been shown to provide energetic benefits in captive *M. murinus* (Perret, 1998) and has been suggested to help aid thermoregulation in infant *M. murinus* (Radespiel *et al.*, 1998; Lutermann *et al.*, 2010) and subadult male *M. ravelobensis* who may delay dispersal until after their first breeding season (Radespiel *et al.*, 2009). Although co-sleeping did not fluctuate seasonally in a study of *M. berthae*, subjects were only radio-collared for a short time (7–50 days per subject outside the infant rearing season) and year-round follow-up studies would be highly interesting (Dammhahn and Kappeler, 2005). Sleeping groups of *M. griseorufus* were better explained by social territoriality than by thermoregulation or predation pressure (Génin, 2010). However, the energetic factors that drive co-sleeping may be different across populations and species and involve interactions of seasonality, food shortages, ambient temperature fluctuations, and the availability of high-quality sleeping sites (Table 22.1; Radespiel *et al.*, 2003; Schülke and Ostner, 2007; Dausmann, 2014; Chapter 13).

Reproductive season: There is also some evidence that some sleeping group associations may result from male mating effort (Table 22.1). At Kirindy, females hibernate for longer periods of time and change sleeping sites less frequently during this period than do males (Schmid and Kappeler, 1998; Rasoazanabary, 2006). The males remain far more active, at times joining the females in their sleeping sites, joining other males, or sleeping alone (Schülke and Ostner, 2005; Rasoazanabary, 2006). However, they were frequently found sleeping near the sites of the females (Rasoazanabary, 2006). This presents two non-mutually exclusive possibilities. The first is that adult males may be monitoring the females' locations in preparation for mating when the females come out of hibernation (Eberle and Kappeler, 2004b; Schülke and Ostner, 2005; Rasoazanabary, 2006). The second possibility is that while subadult male mouse lemurs have not been documented to disperse in groups (Schliehe-Dieck*s et al.*, 2012), those who have not yet dispersed might still obtain benefits from sharing sleeping sites with their maternal kin. This may include co-sleeping with female maternal kin from their natal group or, in the case of all-male groups, subadult male littermates. Additional genetic data are necessary to test these possibilities.

More data exist on the non-lactation season than the lactation season (Table 22.1). This makes it difficult to judge how widespread cooperative breeding and allonursing of foreign young may be. *M. murinus* were filmed allonursing infants inside communal nests at Kirindy (Eberle and Kappeler, 2006) and indirect evidence exists for *M. griseorufus* at Berenty (Génin, 2008). Allonursing among co-sleeping *M. griseorufus* was inferred by the observation of a female who continued lactating longer than necessary for raising only her own offspring, but shorter than would be necessary for her to have raised a second litter of her own (Génin, 2008). While these observations suggest that, at the very least, these two species under the conditions found at Kirindy and Berenty breed cooperatively, it is still unclear whether all mouse lemur populations breed cooperatively (Table 22.1). While all *Microcebus* species have been proposed to have the potential to breed cooperatively (Schülke and Ostner, 2005), more studies are needed to determine under what conditions they may actually do so. In particular, *M. berthae* has been suggested to be unlikely to breed cooperatively (Dammhahn and Kappeler, 2009). As the least social of the mouse lemurs studied to date (Dammhahn and Kappeler, 2005), it would be highly interesting to see whether this species is more social during the lactation season. If so, this would be similar to *M. lehilahytsara*, which also showed comparatively high rates of solitary sleeping, but females slept in groups more frequently during the lactation season than the non-lactation season (Randrianambinina, 2001).

In this section, we have demonstrated that the degree of sociality appears to vary in association with resources, energetic needs, and reproductive season (Table 22.1). This suggests that for an ancestral primate, social flexibility itself may have been a crucial adaptive trait allowing it to colonize new habitats, exist in different climates, and cope with seasonality (see also Chapter 13). The range of social systems possible for ancestral primates may have varied from relatively asocial as seen by *M. berthae* to cooperatively breeding matrilines where individuals may even sometimes forage cohesively and exclude non-sleeping group members from resources as seen in *M. murinus* and *M. griseorufus* (e.g., Dammhahn and Kappeler, 2005; Eberle and Kappeler, 2006; Génin, 2008, 2010). In the next section, we will discuss what aspects of mouse lemur sociality remain stable within and across species.

Stability of matrilineal kin networks

Overall, though the definition of a co-sleeping dyad differs across studies (possibly reflecting differing frequencies of co-sleeping), female mouse lemurs seem to co-sleep more consistently than males do (Table 22.1). Females formed sleeping groups more often than males: *M. murinus* at Ankarafantsika (Radespiel *et al.*, 1998, 2001b) and Mandena (Lahann, 2008), *M. griseorufus* at Berenty

(Génin, 2008), and *M. lehilahytsara* at Mantadia (Randrianambinina, 2001). More males than females were observed to always sleep solitarily in *M. ravelobensis*, *M. murinus* (Ankarafantsika), and in *M. lehilahytsara* (Mantadia; Table 22.1). Additionally, female–female co-sleeping dyads occurred in greater numbers than male–male or male–female co-sleepers in *M. murinus* (Ankarafantsika) and *M. griseorufus* (Table 22.1). In a recent study on *M. lehilahytsara* in the semi-free colony of the Zürich Zoo, females formed stable sleeping groups with the same group members more often than the males, who changed partners much more frequently (Jürges *et al.*, 2013). Given that in the wild, male *M. lehilahytsara* have been reported to mostly sleep solitarily (Randrianambinina 2001), these unstable sleeping groups may have been, in part, a result of the high population density and/or a scarcity of suitable sleeping sites in the zoo enclosure. (It does not exclude the possibility that males may also form sleeping groups when the benefits are greater than the costs.) The only species in which females do not appear to have a greater likelihood of forming sleeping groups is *M. berthae* and this species is, for both sexes, among those with lower rates of co-sleeping (Dammhahn and Kappeler, 2005).

In three of the five populations/species where genetic data were available, stable groups consisted primarily of matrilineal kin (*M. murinus* in Kirindy, *M. murinus* and *M. ravelobensis* in Ankarafantsika; Table 22.1). Although both sexes of *M. ravelobensis* often disperse, female offspring are more likely to remain philopatric and in their natal sleeping group over multiple years (Radespiel *et al.*, 2009). For *M. lehilahytsara* in the Zürich Zoo, females formed stable groups with female matrilineal kin, while males formed unstable groups, but co-slept with related males more often than unrelated males (Kitzler, 2011; Jürges *et al.*, 2013).

Sleeping groups consisting of unrelated dyads were commonly observed in *M. berthae* at Kirindy and male *M. lehilahytsara* in the Zürich Zoo (Dammhahn and Kappeler, 2005; Jürges *et al.*, 2013). Interestingly, these two situations are very different. *M. berthae* in Kirindy have been argued to exist at very low densities due to their high subsistence on insect secretions and animal matter (Dammhahn and Kappeler, 2005, 2008a, 2008b, 2009). They also have a high predation rate which may limit the number of kin available to individuals (Dammhahn and Kappeler, 2005). In *M. berthae* only 3 of 15 co-sleeping dyads were close kin and these were two male–male dyads and one male–female dyad (Dammhahn and Kappeler, 2005). While mtDNA was not investigated and more distant matrilineal relatives may have gone undetected, this is clearly different from the species in which females form matrilineal groups (Dammhahn and Kappeler, 2005; Eberle and Kappeler, 2006; Radespiel *et al.*, 2001b, 2009). Furthermore, the proportion of kin dyads within co-sleeping dyads was not different from the proportion of kin in dyads that could have co-slept but did not (Dammhahn and Kappeler, 2005). Co-sleeping dyads co-slept with varying degrees of stability for an average of sleeping alone every other day (Dammhahn and Kappeler, 2005). Additionally, a follow-up study found that the females co-slept with other individuals only 34% of the time (Dammhahn and Kappeler, 2009). These factors led the authors to conclude that co-sleeping in *M. berthae* is largely opportunistic (Dammhahn and Kappeler, 2005).

In contrast to *M. berthae* (Dammhahn and Kappeler, 2005, 2009), male *M. lehilahytsara* in the Masoala Hall of Zürich Zoo are provisioned, semi-free mouse lemurs living at an extremely high density (Jürges *et al.*, 2013). Because the hall is an enclosed area with relatively young plants, suitable sleeping sites may have been scarce relative to the wild (Jürges *et al.*, 2013). While relatedness did correlate positively with the frequency of co-sleeping, unrelated male dyads still co-slept frequently (up to 30% of observation days; Jürges *et al.*, 2013). Overall, male–male groups were far less stable than the kin groups of female *M. lehilahytsara* (Jürges *et al.*, 2013). What is evident in both *M. berthae* and *M. lehilahytsara* is that unrelated co-sleeping dyads tend to be less stable, suggesting that the identities of the co-sleepers are not as important (Dammhahn and Kappeler, 2005; Jürges *et al.*, 2013). This is a stark contrast to the more stable infant rearing groups consisting of matrilineal kin observed in *M. murinus* (Kirindy: Eberle and Kappeler, 2006; Ankarafantsika: Lutermann *et al.*, 2006; Radespiel *et al.*, 2001b) and *M. ravelobensis* (Ankarafantsika: Radespiel *et al.*, 2009). It suggests that for unrelated dyads sociality may result from mutualism rather than kin selection, especially if, as may be the case for *M. berthae*, the availability of close kin may be limited by a high mortality rate (Dammhahn and Kappeler, 2005).

Overall, the data presented in these studies (Table 22.1) show that across the genus *Microcebus*, matrilineal kin networks are the most pervasive, strongly suggesting that matrilineal networks are likely to have been the foundation for sociality in ancestral primates as well. It also suggests that when reconstructing an ancestral primate social system, this flexibility must be incorporated into the model. This means that, based on interactions with environmental factors, the ancestral primate social system may have been highly flexible in the degree of sociality exhibited within and between populations. We would also argue that this is highly likely to have been the case, regardless of the body size of the ancestral primate. This prediction is supported by a growing body of evidence suggesting substantial flexibility within and between species in the lorises and galagos (for review, see Nekaris and Bearder, 2011). As social flexibility in these species becomes better understood (Nekaris and Bearder, 2011), comparisons with the mouse lemur radiation will be an important test of this model. We expect that they will also show levels of social flexibility similar to the mouse lemurs and a high prevalence of matrilineal kin networks.

Mechanisms for kin recognition in mouse lemurs and ancestral primates

Given the high likelihood that ancestral primates had kin networks, we can then ask what mechanisms these ancestral primates might have used to recognize their kin. Several different mechanisms have been proposed in animals and are relevant for mouse lemurs living in the diverse social systems described above. We will use the following three definitions based on Komdeur and Hatchwell (1999, p. 238, box 2).

1. **Familiarity**: individuals learn who their kin are during a period of familiarization (i.e., maturation, nursing, etc.).
2. **Phenotype matching**: the individual's own phenotype or that of closely related conspecifics is learned and used as a template against which unknown individuals are compared. Kin are recognized based on how well they match the template.
3. **Spatially based recognition**: individuals located in a specific place (i.e., the nest) are treated as kin.

Although there is a commonly discussed fourth mechanism in the literature, recognition alleles (a specific gene produces a recognizable phenotype and the ability to recognize that gene in others), this mechanism is believed to be largely theoretical with very little empirical support, particularly among complex organisms (Komdeur and Hatchwell, 1999). Therefore, we have excluded it here. This leaves three possible recognition mechanisms: familiarity, phenotype matching, and spatially based recognition. We will discuss what is known about kin recognition in mouse lemurs and the evidence that exists for each.

Familiarity is widely accepted to be the mechanism through which mammals recognize maternal kin (Rendall, 2004). Because immature mammals associate closely with their mothers during nursing, they are most familiar with their maternal kin and progressively less familiar with more distant relatives, producing a reliable correlation between maternal kinship and familiarity (Rendall, 2004). This is likely to be the main process by which mouse lemur species form their sleeping group associations with matrilineal kin (Radespiel *et al.*, 2009; Figure 22.1). This is supported by evidence showing that even when a genetic connection between two individuals in a sleeping group could not be found, it could often be found by going back a generation, suggesting that the missing link may have been a matrilineally related lemur with whom they had both previously shared a nest and which had since died (e.g., Radespiel *et al.*, 2001b, 2009). Furthermore, Kirindy *M. murinus* mothers carried only their own infants and not foreign infants when changing sleeping sites (Eberle and Kappeler, 2006). This suggests that the mothers recognized their own young in the nest (Eberle and Kappeler, 2006) and can be explained by the familiarity of the mother with her own offspring acquired during and after birth. That the mother carried only her own young refutes spatially based recognition, because the mother is clearly not using the simple rule of "recognize all young in nest as own young," as has been observed in birds which will feed even chicks of foreign species which are placed in the nest (Komdeur and Hatchwell, 1999; Hatchwell, 2010).

The existence of non-kin-based groups, such as in *M. berthae* (Dammhahn and Kappeler, 2005), should not be taken as evidence that *M. berthae* lacks the capacity for kin recognition. Instead, it is likely that this species has the cognitive ability to form kin-based groups via familiarity, but that given the high mortality rate, larger home ranges, and low densities at which this species has been observed, kin may not always be available to each individual (Dammhahn and Kappeler, 2005, 2008a, 2009). This is supported by the finding that the frequency of kin-based groups approximately matches the frequency of home-range overlaps with kin and thus the availability of kin (Dammhahn and Kappeler, 2005).

In contrast to the recognition of maternal kin, where the existence of clear matrilineal kin networks offer the strongest evidence of matrilineal kin recognition, the effects of patrilineal kin recognition are more difficult to observe. *M. murinus* male mating calls, but not alarm calls, were found to contain patrilineal signatures (Kessler *et al.*, 2012). Furthermore, female mouse lemurs paid more attention to mating calls from non-kin (control males) than to those of their fathers (Kessler *et al.*, 2012). No difference was found in the females' responses to alarm calls from their fathers and non-kin control males (Kessler *et al.*, 2012). Because the females were housed so as to be equally familiar with their fathers and their control males, this suggests that the recognition may have occurred via phenotype matching (Kessler *et al.*, 2012). Pfefferle and colleagues (2014) argue against phenotype matching as the underlying mechanism in this study, saying that because the females were familiar with the males, they may have linked caller identity to relatedness information driven by other cues (i.e., scent, visual). However, we think that this does not exclude phenotype matching as a possible mechanism for kin recognition. Because the females were equally familiar with their fathers and their control males, familiarity alone could not explain the findings (Kessler *et al.*, 2012). If the listeners linked the call characteristics with other individual-based cues (like a memory of the caller's scent), which then correlated with relatedness based upon comparisons to the qualities (i.e., scent) of the listener herself (or known relatives), this would still be phenotype matching.

Mouse lemurs are a particularly interesting genus for investigating phenotype matching, because, in contrast to more gregarious species, mouse lemurs cannot rely on familiarity to recognize paternal kin (Kessler *et al.*, 2012). While the prevalence of paternal half-siblings as familiar playmates in age-based cohorts has been argued to facilitate patrilineal kin recognition in group-living species with a high male reproductive skew and short male breeding tenure (Widdig, 2013), it cannot explain patrilineal kin recognition in mouse lemurs (Kessler *et al.*, 2012). As species where males obtain matings primarily through scramble competition and paternal half-siblings are not likely to grow up together (e.g., Radespiel *et al.*, 2001a; Eberle and Kappeler, 2004a, 2004b), phenotype matching based on inherited traits would be necessary (Kessler *et al.*, 2012).

Phenotype matching is predicted in species which do not provide paternal care, have multiple paternity litters, and/or nest communally (Mateo, 2004). Among mouse lemurs, at least the gray mouse lemur (and possibly other mouse lemur species) fulfils these criteria (Kessler *et al.*, 2012). First, *M. murinus* males do not provide paternal care, co-nest, or co-forage with their mates or with their young (Radespiel *et al.*, 2001b; Eberle and Kappeler, 2006). This strongly limits the effectiveness of the familiarity-based mechanisms often seen in more gregarious species with cohesive foraging groups (e.g., Rendall, 2004; Widdig, 2007, 2013). Second, gray mouse lemur litters can have multiple paternities within the same litter (Radespiel *et al.*, 2002; Eberle and Kappeler, 2004a), thus infant mouse lemurs could be predicted to evolve self-referential phenotype matching to distinguish between full-siblings and maternal half-siblings (Kessler *et al.*, 2012). Finally, given that multiple females may breed in the same nest (Eberle and Kappeler, 2006), infant mouse lemurs could encounter paternal half-siblings within the other mother's litter and use self-referential phenotype matching to recognize them (Kessler *et al.*, 2012). Self-referential phenotype matching via olfactory cues has been observed in ground squirrels which, after emerging from hibernation, recognized their siblings but not foster mates (Mateo, 2010). Given that some individuals in *M. murinus* populations (as well as other cheirogaleids) have been shown to hibernate (e.g., Rasoazanabary, 2006; Lahann and Dausmann, 2011; Dausmann, 2014), further studies on mouse lemur kin recognition mechanisms and the maintenance of kin networks across hibernation seasons would be a highly interesting comparison to the squirrel studies.

The extent to which other mouse lemur species fulfill these criteria which predict the evolution of phenotype matching remains to be seen, but it seems likely that many of them will. Several species require additional data to be collected during the lactation season to verify a lack of paternal care and communal nesting during the infant rearing period (Table 22.1). In addition, genetic data are needed to verify the presence of multiple paternities within litters (Table 22.1). Intriguingly, some of the genetic data suggest that these kin recognition mechanisms are not perfect (Radespiel *et al.*, 2009). Two cases of first-degree inbreeding have been documented in *M. ravelobensis* (Radespiel *et al.*, 2009); however, this is not likely to indicate that kin recognition does not exist in this species. It is more likely to simply indicate that the existing mechanisms are not perfect.

Implications for the evolution of sociality

In the above sections, we used the mouse lemur to (1) demonstrate the importance of using a comparative perspective to model social flexibility into our understanding of ancestral primate sociality, and (2) show that matrilineal kin networks were likely to be the foundation from which more complex forms of kin-based sociality evolved in primates.

By looking across species, we start to see the range of possibilities that may have been key to the success of ancestral primates across different habitats (Table 22.1). For example, in habitats where ancestral primates existed at low densities, had large ranges in which they foraged widely for insects, and/or where the need and ability (availability) to share sleeping sites was lower, their social systems were probably similar to the less social *M. berthae* today (Schwab, 2000; Dammhahn and Kappeler, 2005, 2008a, 2009). In contrast, in habitats where individuals could form larger sleeping groups to maintain larger ranges and access to fruit trees, individuals may have been increasingly social, sometimes even foraging together as seen in *M. griseorufus* (Génin, 2008, 2010) and/or forming large matrilineal family groups and communally rearing infants as documented extensively in *M. murinus*

(Eberle and Kappeler, 2006). Similar levels of social flexibility appear to exist in the closely related giant mouse lemurs (*Mirza* spp.: Kappeler *et al.*, 2005; Rode *et al.*, 2013), suggesting that if ancestral primates were larger than extant *Microcebus* (e.g., Chapter 4; Soligo and Müller, 1999; Müller and Soligo, 2005; Soligo and Martin, 2006; Montgomery and Mundy, 2013), the observed flexibility in kin-based sociality may still hold true.

In addition, we showed that matrilineal kin networks are the most stable and most pervasive (Table 22.1), suggesting that ancestral primates probably formed matrilineal kin networks through familiarity learned during infancy. These matrilineal networks were probably the most prominent and stable kin networks shaping sleeping group associations and dispersal patterns, though phenotype matching may have also played a role in the recognition of patrilineal kin and inbreeding avoidance (Kessler *et al.*, 2012; Kessler, 2014). Although kin recognition has not been determined in many mouse lemur species (e.g., Eberle and Kappeler, 2006; Kessler *et al.*, 2012, 2014), it is highly likely that other closely related mouse lemur species also possess similar kin recognition mechanisms. That kin recognition was found based on the mating call, but not the alarm call, does not suggest that mouse lemurs do not have the cognitive ability to distinguish kin via alarm calls (Kessler *et al.*, 2012). The more parsimonious explanation is that the cognitive structures are similar regardless of call type, and that not all call types are necessarily involved in kin recognition in all contexts (Kessler, 2014). In the case of alarm calls, it may even be advantageous not to discriminate, but instead to react to all alarm calls. Alternatively, alarm calls may not contain the acoustic variation necessary to make distinguishing possible (Kessler *et al.*, 2012). Comparative tests of kin recognition across mouse lemur species are needed to verify this.

An additional avenue for future work would be to study the influence of additional cue types (e.g., Ghazanfar, 2013; Kulahci *et al.*, 2014). Given that the perception of pheromones is expected to be highly important for mouse lemurs (Hohenbrink *et al.*, 2012, 2013; Yoder *et al.*, 2014), olfactory signals are expected to be rich in social information which could complement acoustic cues. A combination of imperfect information from multiple sensory channels may be sufficient to produce sufficiently reliable information for kin recognition and selection to occur (Kessler, 2014). We encourage future work to incorporate such a multimodal approach.

Surprisingly, there is still some debate as to the importance of kin recognition in the evolution of mammalian social complexity and we will consider to what extent the data presented in this chapter support the two possibilities: (1) kin recognition aside from spatially based recognition is not required (cryptic kin selection: Hatchwell, 2010), or (2) kin recognition was crucial to the evolution of mammalian social complexity (social network model: Müller and Thalmann, 2000; Kessler *et al.*, 2012, 2014).

The cryptic kin selection model (Hatchwell, 2010) predicts that kin-based sociality could evolve in a population where kin are spatially clustered when animals behave altruistically towards their neighbors (who will then, on average, be kin). This model does not require any kin recognition more precise than recognition based on spatial cues (Hatchwell, 2010). Thus, recognition based upon familiarity and phenotype matching is not required. This explanation for the evolution of kin-based sociality is not supported by the data on mouse lemur social organization. The strongest evidence against this model is the comparative data suggesting that across mouse lemur species social groups based on kinship appear to be far more stable than those which are not based upon kinship (Table 22.1). For example, the least social species, *M. berthae*, forms groups which are the most unstable and has a frequency of forming sleeping groups with kin that reflects the availability of kin (number of kin with home-range overlaps), suggesting that they are neither actively selecting kin nor actively selecting particular non-kin individuals (i.e., "friends"; Dammhahn and Kappeler, 2005). This implies that ancestral primates are not likely to have evolved kin-based sociality simply by chance due to having kin for neighbors. It suggests that when individuals do not actively select group members, the result is unstable groups with unrelated members. In contrast, other species such as *M. murinus* and *M. ravelobensis* form groups with matrilineal kin that are stable (Radespiel *et al.*, 2001b, 2009; Eberle and Kappeler, 2006). When the group changes the sleeping site, all members typically reconvene at the new site using gathering calls, which have not yet been tested for kin recognition (Radespiel *et al.*, 2001b, 2009; Braune *et al.*, 2005; Eberle and Kappeler, 2006; Kessler, 2014), thus they are repeatedly choosing to associate with kin (likely via familiarity). These more stable, kin-based social groups are consistent with the alternative model: the social network model (Müller and Thalmann, 2000; Kessler *et al.*, 2012, 2014).

The social network model predicts that the social networks of solitary foragers provided the foundation for the evolution of more complex forms of sociality (Müller and Thalmann, 2000). It follows from this model that if solitary foragers were able to use these social networks to recognize kin, that these social networks may have then facilitated the kin selection that drove the evolution of kin-based social groups (Kessler *et al.*, 2012, 2014). The social network model is compatible with evidence suggesting that the neurophysiological structures involved in mammalian mother–infant recognition and bonding were the evolutionary basis from which other forms of mammalian social bonding evolved, including the various forms of primate social complexity (Müller and Thalmann, 2000; Broad *et al.*, 2006), and with models arguing that gregarious sociality evolved via kin recognition and nepotism in the philopatric sex (e.g., Waser and Jones, 1983; Perrin and Lehmann, 2001).

An important limitation of our review is that it is not, at this time, possible to differentiate between different models predicting which forms of gregarious sociality evolved first in primates. Various models include (1) the socio-ecological model, which has been predicted to produce an "unstructured" model in which all transitions are equally likely (i.e., pairs to multimale, to harems), and are produced by individual reactions to resource availability (Sterck *et al.*, 1997; Eberle and Kappeler, 2002; Kappeler and van Schaik, 2002; Schülke and Ostner, 2005; Shultz *et al.*, 2011), or (2) various step-wise models which argue that certain transitions are more likely than others (e.g., Müller and Thalmann, 2000; Shultz *et al.*, 2011). Prior work (Schülke and Ostner, 2005; Dammhahn and Kappeler, 2009) suggesting that mouse lemurs, like more gregarious primates, do conform to the predictions of the socio-ecological model suggests that, at the very least, this is a good starting point. The flexibility in the kin-based sociality within and across mouse lemur species is broadly consistent with the socio-ecological model (Table 22.1). However, we do not yet have the data necessary to definitively reject the step-wise models by showing that some transitions are more likely than others. The existing data do suggest that whatever the pattern of transitions was, it was structured, at least in part, by kin relations. We say "at least in part" because transitions from nocturnality to diurnality and the corresponding changes in predator avoidance are also believed to have been significant (Müller and Thalmann, 2000; Shultz *et al.*, 2011; see also Chapter 19). Activity period models are not necessarily mutually exclusive with models arguing for the importance of kin relations and dispersal patterns (e.g., Waser and Jones, 1983; Perrin and Lehmann, 2001).

Synthesis and perspectives

In this review, we have used the mouse lemur model of primate origins to illustrate the importance of incorporating social flexibility into reconstructions of ancestral primate sociality. We have also showed the pervasiveness of matrilineal kin networks across species and we argue that this suggests that these networks were the backbone of ancestral primate sociality and the foundation from which more complex forms of sociality evolved in primates. This is supported by a growing body of research suggesting that other strepsirrhine species are also likely to be highly socially flexible and have strong matrilineal kin networks (e.g., Nekaris and Bearder, 2011). Future comparisons between mouse lemurs, galagos, and lorises will be powerful tests of this model.

Acknowledgments

We would like to thank Shawn Lehman for the invitation to participate in this volume and Marina Scheumann for helpful discussions on this manuscript. Parts of this chapter are from SEK's dissertation which was funded by an NSF Dissertation Improvement Grant (#0961779), PEO Scholar Award, the Animal Behavior Society, Lewis and Clark Fund of the American Philosophical Society, German Academic Exchange Service (#A/09/81743), American Society of Primatologists, Sigma Xi (National Chapter, grants #G200810150592, #G2009101504), Arizona State University (ASU) chapter of Sigma Xi, ASU Graduate and Professional Student Association, ASU School of Human Evolution and Social Change, an ASU Graduate College Dissertation Writing Fellowship, a postdoctoral fellowship from the Fonds de Recherche du Québec – Nature et Technologies, and by the University of Veterinary Medicine Hanover and the German Research Foundation (DFG FOR 499 to EZ).

References

Barton RA. 2006. Primate brain evolution: integrating comparative, neurophysiological, and ethological data. *Evolutionary Anthropology* 15(6):224–236.

Bearder SK. 1987. Lorises, bushbabies, and tarsiers: diverse societies in solitary foragers. In BB Smuts, DL Cheney, RM Seyfarth, R Wrangham, TT Struhsaker (eds.), *Primate Societies* (pp. 11–24). University of Chicago Press, Chicago.

Bloch JI, Boyer DM. 2002. Grasping primate origins. *Science* 298(5598):1606–1610.

Bloch JI, Silcox MT, Boyer DM, Sargis EJ. 2007. New Paleocene skeletons and the relationship of plesiadapiforms to crown-clade primates. *Proceedings of the National Academy of Sciences of the United States of America* 104(4):1159–1164.

Braune P. 2007. *Acoustic Variability and its Biological Significance in Nocturnal Lemurs.* Gottfried Wilhelm Leibniz Universität Hannover, Hannover.

Braune P, Schmidt S, Zimmermann E. 2005. Spacing and group coordination in a nocturnal primate, the golden brown mouse lemur (*Microcebus ravelobensis*): the role of olfactory and acoustic signals. *Behavioral Ecology and Sociobiology* 58(6):587–596.

Broad KD, Curley JP, Keverne EB. 2006. Mother–infant bonding and the evolution of mammalian social relationships. *Philosophical Transactions of the Royal Society B Biological Sciences* 361(1476):2199–2214.

Cartmill M. 1972. Arboreal adaptations and the origin of the order Primates. In RH Tuttle (ed.), *The Functional and Evolutionary Biology of Primates* (pp. 97–122). Aldine-Atherton, Chicago.

Cartmill M. 1974. Rethinking primate origins. *Science* 184(4135):436–443.

Cartmill M. 1992. New views on primate origins. *Evolutionary Anthropology* 1:105–111.

Cartmill M. 2012. Primate origins, human origins, and the end of higher taxa. *Evolutionary Anthropology* 21(6):208–220.

Chapais B, Berman CM (eds.) 2004. *Kinship and Behavior in Primates.* Oxford University Press, New York.

Charles-Dominique P. 1974. Aggression and territoriality in nocturnal prosimians. In RL Holloway (ed.), *Primate Aggression, Territoriality, and Xenophobia* (pp. 31–48). Academic Press, New York.

Charles-Dominique P. 1978. Nocturnality and diurnality: an ecological interpretation of these two modes of life by an analysis of the higher vertebrate fauna in tropical forest ecosystems. In WP Luckett, FS Szalay (eds.), *Phylogeny of the Primates* (pp. 69–88). Plenum Press, New York.

Charles-Dominique P, Martin RD. 1970. Evolution of lorises and lemurs. *Nature* 227(5255):257–260.

Charles-Dominique P, Martin RD. 1972. *Behaviour and Ecology of Nocturnal Prosimians: Field Studies in Gabon and Madagascar.* P. Parey, Berlin.

Dammhahn M, Kappeler PM. 2005. Social system of *Microcebus berthae*, the world's smallest primate. *International Journal of Primatology* 26(2):407–435.

Dammhahn M, Kappeler P. 2008a. Comparative feeding ecology of sympatric *Microcebus berthae* and *M. murinus*. *International Journal of Primatology* 29(6):1567–1589.

Dammhahn M, Kappeler P. 2008b. Small-scale coexistence of two mouse lemur species (*Microcebus berthae* and *M. murinus*) within a homogeneous competitive environment. *Oecologia* 157(3):473–483.

Dammhahn M, Kappeler PM. 2009. Females go where the food is: does the socio-ecological model explain variation in social organisation of solitary foragers? *Behavioral Ecology and Sociobiology* 63(6):939–952.

Dammhahn M, Kappeler PM. 2010. Scramble or contest competition over food in solitarily foraging mouse lemurs (*Microcebus* spp.): new insights from stable isotopes. *American Journal of Physical Anthropology* 141(2):181–189.

Dausmann KH. 2014. Flexible patterns in energy savings: heterothermy in primates. *Journal of Zoology* 292(2):101–111.

de Waal F, Tyack PL (eds.) 2003. *Animal Social Complexity: Intelligence, Culture, and Individualized Societies*. Harvard University Press, Cambridge.

Dunbar R. 1998. The social brain hypothesis. *Evolutionary Anthropology* 6(5):178–190.

Eberle M, Kappeler PM. 2002. Mouse lemurs in space and time: a test of the socioecological model. *Behavioral Ecology and Sociobiology* 51(2):131–139.

Eberle M, Kappeler PM. 2004a. Selected polyandry: female choice and inter-sexual conflict in a small nocturnal solitary primate (*Microcebus murinus*). *Behavioral Ecology and Sociobiology* 57(1):91–100.

Eberle M, Kappeler PM. 2004b. Sex in the dark: determinants and consequences of mixed male mating tactics in *Microcebus murinus*, a small solitary nocturnal primate. *Behavioral Ecology and Sociobiology* 57(1):77–90.

Eberle M, Kappeler PM. 2006. Family insurance: kin selection and cooperative breeding in a solitary primate (*Microcebus murinus*). *Behavioral Ecology and Sociobiology* 60(4):582–588.

Gebo DL. 2004. A shrew-sized origin for primates. *Yearbook of Physical Anthropology* 47:40–62.

Génin F. 2008. Life in unpredictable environments: first investigation of the natural history of *Microcebus griseorufus*. *International Journal of Primatology* 29:303–321.

Génin F. 2010. Who sleeps with whom? Sleeping association and socio-territoriality in *Microcebus griseorufus*. *Journal of Mammalogy* 91(4):942–951.

Ghazanfar AA. 2013. Multisensory vocal communication in primates and the evolution of rhythmic speech. *Behavioral Ecology and Sociobiology* 67(9):1441–1448.

Gingerich PD, Uhen MD. 1994. Time of origin of primates. *Journal of Human Evolution* 27(5):443–445.

Hamilton WD. 1964. The genetical evolution of social behavior. I and II. *Journal of Theoretical Biology* 7:1–52.

Hatchwell BJ. 2010. Cryptic kin selection: kin structure in vertebrate populations and opportunities for kin-directed cooperation. *Ethology* 116(3):203–216.

Hohenbrink P, Radespiel U, Mundy NI. 2012. Pervasive and ongoing positive selection in the vomeronasal-1 receptor (V1R) repertoire of mouse lemurs. *Molecular Biology and Evolution* 29(12):3807–3816.

Hohenbrink P, Mundy NI, Zimmermann E, Radespiel U. 2013. First evidence for functional vomeronasal 2 receptor genes in primates. *Biology Letters* 9(1):20121006.

Holekamp KE, Smith JE, Strelioff CC, Van Horn RC, Watts HE. 2012. Society, demography and genetic structure in the spotted hyena. *Molecular Ecology* 21(3):613–632.

Jürges V, Kitzler J, Zingg R, Radespiel U. 2013. First insights into the social organization of Goodman's mouse lemur (*Microcebus lehilahytsara*) – testing predictions from socio-ecological hypotheses in the Masoala Hall of Zurich Zoo. *Folia Primatologica* 84:32–48.

Kappeler P, Rasoloarison R, Razafimanantsoa L, Walter L, Roos C. 2005. Morphology, behavior and molecular evolution of giant mouse lemurs (*Mirza* spp.), Gray 1970, with a description of a new species. *Primate Report* 71:3–26.

Kappeler PM. 1997. Determinants of primate social organization: comparative evidence and new insights from Malagasy lemurs. *Biological Reviews of the Cambridge Philosophical Society* 72(1):111–151.

Kappeler PM. 1998. Nests, tree holes, and the evolution of primate life histories. *American Journal of Primatology* 46(1):7–33.

Kappeler PM, van Schaik CP. 2002. Evolution of primate social systems. *International Journal of Primatology* 23(4):707–740.

Kappeler PM, Wimmer B, Zinner D, Tautz D. 2002. The hidden matrilineal structure of a solitary lemur: implications for primate social evolution. *Proceedings of the Royal Society of London Series B Biological Sciences* 269(1502):1755–1763.

Kessler SE. 2014. *Modeling the Origins of Primate Sociality: Kin Recognition in Mouse Lemurs.* Arizona State University, Tempe.

Kessler SE, Scheumann M, Nash LT, Zimmermann E. 2012. Paternal kin recognition in the high frequency/ultrasonic range in a solitary foraging mammal. *BMC Ecology* 12:26.

Kessler SE, Radespiel U, Hasiniaina A, *et al.* 2014. Modeling the origins of mammalian sociality: moderate evidence for matrilineal signatures in mouse lemur vocalizations. *Frontiers in Zoology* 11:14.

Kitzler J. 2011. Genetische struckturen in der population von Goodman mausmakis in der Masoala Halle, Zoo Zürich. Bachelorarbeit, Gottfried Wilhelm Leibniz Universität Hannover, Hannover.

Komdeur J, Hatchwell BJ. 1999. Kin recognition: function and mechanism in avian societies. *Trends in Ecology and Evolution* 14:237–241.

Kulahci IG, Drea CM, Rubenstein DI, Ghazanfar AA. 2014. Individual recognition through olfactory–auditory matching in lemurs. *Proceedings of the Royal Society B Biological Sciences* 281(1784):6.

Lahann P. 2007. Feeding ecology and seed dispersal of sympatric cheirogaleid lemurs (*Microcebus murinus, Cheirogaleus medius, Cheirogaleus major*) in the littoral rainforest of south-east Madagascar. *Journal of Zoology* 271(1):88–98.

Lahann P. 2008. Habitat utilization of three sympatric cheirogaleid lemur species in a littoral rain forest of southeastern Madagascar. *International Journal of Primatology* 29(1):117–134.

Lahann P, Dausmann KH. 2011. Live fast, die young: flexibility of life history traits in the fat-tailed dwarf lemur (*Cheirogaleus medius*). *Behavioral Ecology and Sociobiology* 65(2):381–390.

Lahann P, Schmid J, Ganzhorn JU. 2006. Geographic variation in populations of *Microcebus murinus* in Madagascar: resource seasonality or Bergmann's rule? *International Journal of Primatology* 27(4):983–999.

Lemelin P. 1999. Morphological correlates of substrate use in didelphid marsupials: implications for primate origins. *Journal of Zoology* 247:165–175.

Lutermann H, Schmelting B, Radespiel U, Ehresmann P, Zimmermann E. 2006. The role of survival for the evolution of female philopatry in a solitary forager, the grey mouse lemur (*Microcebus murinus*). *Proceedings of the Royal Society B Biological Sciences* 273(1600):2527–2533.

Lutermann H, Verburgt L, Rendigs A. 2010. Resting and nesting in a small mammal: sleeping sites as a limiting resource for female grey mouse lemurs. *Animal Behaviour* 79(6):1211–1219.

Maher CR. 2009. Genetic relatedness and space use in a behaviorally flexible species of marmot, the woodchuck (*Marmota monax*). *Behavioral Ecology and Sociobiology* 63(6):857–868.

Martin RD. 1993. Primate origins – plugging the gaps. *Nature* 363(6426):223–234.

Martin RD, Soligo C, Tavare S. 2007. Primate origins: implications of a cretaceous ancestry. *Folia Primatologica* 78(5–6):277–296.

Mateo JM. 2004. Recognition systems and biological organization: the perception component of social recognition. *Annales Zoologici Fennici* 41(6):729–745.

Mateo JM. 2010. Self-referent phenotype matching and long-term maintenance of kin recognition. *Animal Behaviour* 80(5):929–935.

McComb K, Moss C, Sayialel S, Baker L. 2000. Unusually extensive networks of vocal recognition in African elephants. *Animal Behaviour* 59(6):1103–1109.

Meshriy MG, Randall JA, Parra L. 2011. Kinship associations of a solitary rodent, *Dipodomys ingens*, at fluctuating population densities. *Animal Behaviour* 82(4):643–650.

Messier GD, Garant D, Bergeron P, Reale D. 2012. Environmental conditions affect spatial genetic structures and dispersal patterns in a solitary rodent. *Molecular Ecology* 21(21):5363–5373.

Montgomery SH, Mundy N. 2013. Parallel espidoes of phyletic dwarfism in callitrichid and cheirogaleid primates. *Journal of Evolutionary Biology* 26:810–819.

Müller A, Soligo C. 2005. Primate sociality in evolutionary context. *American Journal of Physical Anthropology* 128(2):399–414.

Müller AE, Thalmann U. 2000. Origin and evolution of primate social organisation: a reconstruction. *Biological Review* 75:405–435.

Nash LT. 2004. Kinship and behavior among nongregarious nocturnal prosimians: what do we really know? In B Chapais, CM Berman (eds.), *Kinship and Behavior in Primates* (pp. 200–222). Oxford University Press, New York.

Nekaris KAI. 2005. Foraging behaviour of the slender loris (*Loris lydekkerianus lydekkerianus*): implications for theories of primate origins. *Journal of Human Evolution* 49(3):289–300.

Nekaris KAI, Bearder S. 2011. The strepsirrhine primates of Asia and Mainland Africa: diversity shrouded in darkness. In C Campbell, A Fuentes, K MacKinnon, S Bearder, R Stumpf (eds.), *Primates in Perspective*, 2nd edition. Oxford University Press, Oxford.

Ni XJ, Wang YQ, Hu YM, Li CK. 2004. A euprimate skull from the early Eocene of China. *Nature* 427(6969):65–68.

Ni XJ, Hu YM, Wang YQ, Li CK. 2005. A clue to the Asian origin of euprimates. *Anthropological Science* 113(1):3–9.

Perret M. 1998. Energetic advantage of nest-sharing in a solitary primate, the lesser mouse lemur (*Microcebus murinus*). *Journal of Mammalogy* 79(4):1093–1102.

Perrin N, Lehmann L. 2001. Is sociality driven by the costs of dispersal or the benefits of philopatry? A role for kin-discrimination mechanisms. *American Naturalist* 158(5):471–483.

Perry S. 2011. Social traditions and social learning in capuchin monkeys (*Cebus*). *Philosophical Transactions of the Royal Society B Biological Sciences* 366(1567):988–996.

Pfefferle D, Ruiz Lambides AV, Widdig A. 2014. Female rhesus macaques discriminate unfamiliar paternal sisters in playback experiments: support for acoustic phenotype matching. *Proceedings of the Royal Society B* 281(1774):20131628.

Radespiel U. 2006. Ecological diversity and seasonal adaptations of mouse lemurs (*Microcebus* spp.). In L Gould, ML Sauther (eds.), *Lemurs: Ecology and Adaptation* (pp. 211–234). Springer, New York.

Radespiel U, Cepok S, Zietemann V, Zimmermann E. 1998. Sex-specific usage patterns of sleeping sites in grey mouse lemurs (*Microcebus murinus*) in northwestern Madagascar. *American Journal of Primatology* 46(1):77–84.

Radespiel U, Ehresmann P, Zimmermann E. 2001a. Contest versus scramble competition for mates: the composition and spatial structure of a population of gray mouse lemurs (*Microcebus murinus*) in northwest Madagascar. *Primates* 42(3):207–220.

Radespiel U, Sarikaya Z, Zimmermann E, Bruford MW. 2001b. Sociogenetic structure in a free-living nocturnal primate population: sex-specific differences in the grey mouse lemur (*Microcebus murinus*). *Behavioral Ecology and Sociobiology* 50(6):493–502.

Radespiel U, Dal Secco V, Drogemuller C, *et al.* 2002. Sexual selection, multiple mating and paternity in grey mouse lemurs, *Microcebus murinus*. *Animal Behaviour* 63:259–268.

Radespiel U, Ehresmann P, Zimmermann E. 2003. Species-specific usage of sleeping sites in two sympatric mouse lemur species (*Microcebus murinus* and *M. ravelobensis*) in northwestern Madagascar. *American Journal of Primatology* 59(4):139–151.

Radespiel U, Juric M, Zimmermann E. 2009. Sociogenetic structures, dispersal and the risk of inbreeding in a small nocturnal lemur, the golden-brown mouse lemur (*Microcebus ravelobensis*). *Behaviour* 146:607–628.

Randrianambinina B. 2001. Contribution à l'étude comparative de l'écoéthologie de deux microcèbes rouges de Madagascar: *Microcebus ravelobensis* (Zimmermann et al., 1998), *Microcebus rufus* (Lesson, 1840). These de Doctorat 3ième cycle, University of Antananarivo, Antananarivo.

Rasmussen DT. 1990. Primate origins – lessons from a neotropical marsupial. *American Journal of Primatology* 22(4):263–277.

Rasmussen DT. 2002. The origin of primates. In W Hartwig (ed.), *The Primate Fossil Record* (pp. 5–9). Cambridge University Press, Cambridge.

Rasoazanabary E. 2006. Male and female activity patterns in *Microcebus murinus* during the dry season at Kirindy Forest, western Madagascar. *International Journal of Primatology* 27(2):437–464.

Ravosa MJ, Dagosto M (eds). 2007. *Primate Origins: Adaptations and Evolution*. Springer, Chicago.

Rendall D. 2004. Recognizing kin: mechanisms, media, minds, modules, and muddles. In B Chapais, CM Berman (eds.), *Kinship and Behavior in Primates* (pp. 295–316). Oxford University Press, New York.

Rode JE, Nekaris A, Markolf M, *et al.* 2013. Social organisation of the northern giant mouse lemur *Mirza zaza* in Sahamalaza, north western Madagascar, inferred from nest group composition and genetic relatedness. *Contributions to Zoology* 82(2):71–83.

Scheumann M, Zimmermann E. 2008. Sex-specific asymmetries in communication sound perception are not related to hand preference in an early primate. *BMC Biology* 6(3).

Schliehe-Diecks S, Eberle M, Kappeler PM. 2012. Walk the line – dispersal movements of gray mouse lemurs (*Microcebus murinus*). *Behavioral Ecology and Sociobiology* 66(8):1175–1185.

Schmid J. 1998. Tree holes used for resting by gray mouse lemurs (*Microcebus murinus*) in Madagascar: insulation capacities and energetic consequences. *International Journal of Primatology* 19(5):797–809.

Schmid J, Kappeler PM. 1998. Fluctuating sexual dimorphism and differential hibernation by sex in a primate, the gray mouse lemur (*Microcebus murinus*). *Behavioral Ecology and Sociobiology* 43(2):125–132.

Schmitt D, Lemelin P. 2002. Origins of primate locomotion: gait mechanics of the woolly opossum. *American Journal of Physical Anthropology* 118(3):231–238.

Schülke O, Ostner J. 2005. Big times for dwarfs: social organization, sexual selection, and cooperation in the Cheirogaleidae. *Evolutionary Anthropology* 14(5):170–185.

Schülke O, Ostner J. 2007. Physiological ecology of cheirogaleid primates: variation in hibernation and torpor. *Acta Ethologica* 10(1):13–21.

Schwab D. 2000. A preliminary study of spatial distribution and mating system of pygmy mouse lemurs (*Microcebus cf myoxinus*). *American Journal of Primatology* 51(1):41–60.

Shultz S, Opie C, Atkinson QD. 2011. Stepwise evolution of stable sociality in primates. *Nature* 479(7372):219–224.

Siemers BM, Goerlitz HR, Robsomanitrandrasana E, *et al.* 2007. Sensory basis of food detection in wild *Microcebus murinus*. *International Journal of Primatology* 28(2):291–304.

Silcox MT, Boyer DM, Bloch JI, Sargis EJ. 2007. Revisiting the adaptive origins of primates (again). *Journal of Human Evolution* 53(3):321–324.

Silk JB. 2002. Kin selection in primate groups. *International Journal of Primatology* 23(4):849–875.

Silk JB. 2007. Social components of fitness in primate groups. *Science* 317(5843):1347–1351.

Silk JB. 2009. Nepotistic cooperation in non-human primate groups. *Philosophical Transactions of the Royal Society B Biological Sciences* 364(1533):3243–3254.

Soligo C, Martin RD. 2006. Adaptive origins of primates revisited. *Journal of Human Evolution* 50(4):414–430.

Soligo C, Müller AE. 1999. Nails and claws in primate evolution. *Journal of Human Evolution* 36(1):97–114.

Springer MS, Murphy WJ, Eizirik E, O'Brien SJ. 2003. Placental mammal diversification and the Cretaceous–Tertiary boundary. *Proceedings of the National Academy of Sciences of the United States of America* 100(3):1056–1061.

Steiper ME, Seiffert ER. 2012. Evidence for a convergent slowdown in primate molecular rates and its implications for the timing of early primate evolution. *Proceedings of the National Academy of Sciences of the United States of America* 109(16):6006–6011.

Sterck EHM, Watts DP, van Schaik CP. 1997. The evolution of female social relationships in nonhuman primates. *Behavioral Ecology and Sociobiology* 41:291–309.

Sterling EJ, Nguyen N, Fashing PJ. 2000. Spatial patterning in nocturnal prosimians: a review of methods and relevance to studies of sociality. *American Journal of Primatology* 51(1):3–19.

Stoen OG, Bellemain E, Saebo S, Swenson JE. 2005. Kin-related spatial structure in brown bears *Ursus arctos*. *Behavioral Ecology and Sociobiology* 59(2):191–197.

Strier KB. 2007. *Primate Behavioral Ecology*. Allyn and Bacon, San Francisco.

Sussman RW. 1991. Primate origins and the evolution of angiosperms. *American Journal of Primatology* 23(4):209–223.

Sussman RW, Raven PH. 1978. Pollination by lemurs and marsupials – archaic coevolutionary system. *Science* 200(4343):731–736.

Sussman RW, Rasmussen DT, Raven PH. 2013. Rethinking primate origins again. *American Journal of Primatology* 75(2):95–106.

Tavaré S, Marshall CR, Will O, Soligo C, Martin RD. 2002. Using the fossil record to estimate the age of the last common ancestor of extant primates. *Nature* 416:726–729.

Thorén S, Quietzsch F, Schwochow D, *et al.* 2011. Seasonal changes in feeding ecology and activity patterns of two sympatric mouse lemur species, the gray mouse lemur (*Microcebus murinus*) and the golden-brown mouse lemur (*M. ravelobensis*), in northwestern Madagascar. *International Journal of Primatology* 32(3):566–586.

Toussaint S, Reghem E, Chotard H, *et al.* 2013. Food acquisition on arboreal substrates by the grey mouse lemur: implication for primate grasping evolution. *Journal of Zoology* 291(4):235–242.

Waser PM, Jones WT. 1983. Natal philopatry among solitary mammals. *Quarterly Review of Biology* 58(3):355–390.

Weidt A, Hagenah N, Randrianambinina B, Radespiel U, Zimmermann E. 2004. Social organization of the golden brown mouse lemur (*Microcebus ravelobensis*). *American Journal of Physical Anthropology* 123(1):40–51.

Whiten A, Goodall J, McGrew WC, *et al.* 1999. Cultures in chimpanzees. *Nature* 399(6737):682–685.

Widdig A. 2007. Paternal kin discrimination: the evidence and likely mechanisms. *Biological Review* 82(2):319–334.

Widdig A. 2013. The impact of male reproductive skew on kin structure and sociality in multi-male groups. *Evolutionary Anthropology* 22(5):239–250.

Yoder AD, Chan LM, dos Reis M, *et al.* 2014. Molecular evolutionary characterization of a V1R subfamily unique to strepsirrhine primates. *Genome Biology and Evolution* 6(1):213–227.

Yurk H, Barrett-Lennard L, Ford JKB, Matkin CO. 2002. Cultural transmission within maternal lineages: vocal clans in resident killer whales in southern Alaska. *Animal Behaviour* 63:1103–1119.

Zimmermann E. 1990. Differentiation of vocalizations in bushbabies (Galaginae, Prosimiae, Primates) and the significance for assessing phylogenetic relationships. *Zeitschrift Fur Zoologische Systematik Und Evolutionsforschung* 28(3):217–239.

Zimmermann E. 1995a. Acoustic communication in nocturnal prosimians. In L Alterman, GA Doyle, MK Izard (eds.), *Creatures of the Dark* (pp. 311–330). Plenum Press, New York.

Zimmermann E. 1995b. Loud calls in nocturnal prosimians: structure, evolution and ontogeny. In E Zimmermann, JD Newman, U Jurgens (eds.), *Current Topics in Primate Vocal Communication* (pp. 47–72). Plenum Press, New York.

Zimmermann E. 2010. Vocal expression of emotion in a nocturnal prosimian primate group, mouse lemurs. In SM Brudzynski (ed.), *Handbook of Mammalian Vocalization: An Integrative Neuroscience Approach* (pp. 215–225). Academic Press, Oxford.

Zimmermann E, Radespiel U. 2014. Species concepts, diversity, and evolution in primates: lessons to be learned from mouse lemurs. *Evolutionary Anthropology* 23:11–14.

Part V

Cheirogaleidae: conservation biogeography

23 Ecological niche modeling of mouse lemurs (*Microcebus* spp.) and its implications for their species diversity and biogeography

Jason M. Kamilar, Marina B. Blanco, and Kathleen M. Muldoon

Introduction

Recent studies of lemur species diversity have particularly focused on mouse lemurs (Cheirogaleidae: *Microcebus*) – small, nocturnal, and superficially monotypic creatures (Radespiel *et al.*, 2012; Zimmermann and Radespiel, 2014). *Microcebus* are widespread across the diverse habitats of Madagascar, and are present in primary, secondary, and disturbed forest types where suitable area remains (Mittermeier *et al.*, 2010). Until the end of the last century, it was assumed that diversity in mouse lemurs consisted only of two morphologically and geographically distinct species – a larger, grayish western morph, *Microcebus murinus*, and a smaller, reddish eastern species, *M. rufus* (Mittermeier *et al.*, 1994). Each of these species was thought to have a broad distribution, encompassing a relatively diverse array of climates and habitats. In addition, mouse lemurs were believed to have largely allopatric distributions, with very little, if any, sympatry observed. However, the number of recognized mouse lemur species on Madagascar has vastly increased in recent years with an emphasis on methods to delineate cryptic species in the field (Mittermeier *et al.*, 2010). There are currently 21 formally described species of mouse lemur, including some taxa that have sympatric distributions (Zimmermann and Radespiel, 2014).

Some controversy exists over the taxonomic validity over the newly named species, associated in large part with problems of sample size and geographic coverage, and difficulty in distinguishing clinal variation from distinct phylogenetic species (Tattersall, 2007; Markolf *et al.*, 2011). Although cryptic species are often differentiated solely on the basis of genetics (Pastorini *et al.*, 2001; Olivieri *et al.*, 2007; Horvath *et al.*, 2008; Groeneveld *et al.*, 2009; Weisrock *et al.*, 2012), integrative studies of primate taxonomy have recently used diet, social system,

The Dwarf and Mouse Lemurs of Madagascar: Biology, Behavior and Conservation Biogeography of the Cheirogaleidae, eds. S.M. Lehman, U. Radespiel, E. Zimmermann. Published by Cambridge University Press. © Cambridge University Press 2016.

communication signals, sleeping site ecology, and reproductive behavior to delineate species boundaries (Kamilar, 2006; Radespiel *et al.*, 2006; Zimmermann, 2013; Zimmermann and Radespiel, 2014). Such studies have demonstrated that sympatric species of *Microcebus* coexist in several regions of Madagascar (Yoder *et al.*, 2005; Weisrock *et al.*, 2010; Rasoloarison *et al.*, 2013). Thus, mouse lemurs provide an excellent opportunity to investigate the evolutionary and ecological mechanisms that allow species to coexist.

One relatively new tool to investigate species diversity and species delineation is ecological niche modeling (i.e., species distribution modeling; Elith *et al.*, 2006; Phillips *et al.*, 2006). Ecological niche models can be applied to questions of species delineation because this method can provide new information regarding the distribution and ecological niche separation across a set of taxa. Raxworthy *et al.* (2007) used this approach to investigate the biogeography and taxonomy of Malagasy day geckos (genus *Phelsuma*). Although the taxa exhibited some degree of genetic and morphological disparity, ecological niche models performed significantly better when taxa were divided into separate species, rather than subspecies. These findings supported the elevation of three species to full species status, as well as a new species being named. A recent paper by Blair *et al.* (2013) used ecological niche models to explicitly test whether closely related *Eulemur* taxa exhibited significantly divergent niche spaces. They found that some sister species displayed niches no different than expected by chance, implying an allopatric mode of speciation because a failure to adapt to new environments might be hypothesized when populations are spatially isolated from each other. In other words, spatial isolation may be facilitated by ecological niche conservatism (Wiens, 2004). In contrast, one pair of species displayed significantly different ecological niches, suggesting a parapatric model of speciation.

While it is not our intent to evaluate the validity of newly named species of *Microcebus*, an analysis of ecological niche space can be useful for testing hypotheses about species diversity in mouse lemurs, and evaluating species divergence between sympatric sister species. A wide range of approaches to construct quantitative models of species' ecological niches have been developed, commonly termed ecological niche models (ENMs; Franklin, 2009; Peterson *et al.*, 2011). Studies of the degree to which ecological niches vary across a phylogeny are crucial for understanding how and when environmental factors can influence natural selection and generate biological diversity (Wiens, 2004; Kamilar and Muldoon, 2010; Wiens *et al.*, 2010; MacColl, 2011; Kamilar and Cooper, 2013). The goal of our chapter is thus to evaluate the species diversity of mouse lemurs by using an ecological niche modeling approach. Specifically, we quantified mouse lemur climatic niches and tested the degree of niche overlap among species. Generating niche models can inform us about the potential distribution of taxa, and consequently the possibility of species co-occurrence. In addition, niche models with good predictive power indicate that a species distribution is strongly related to climatic factors. In contrast, species with poorly performing models may indicate that they are ecological generalists and not well connected to climatic factors and/or that multiple species may actually exist. Finally, examining the degree of niche overlap among taxa, especially sister species, can provide insights into speciation mechanisms.

In this chapter, we have the following goals: (1) generate ENMs for six mouse lemur species to determine how well climatic variables predict their distributions, (2) examine the degree of niche overlap among these species and discuss these results in the context of their geographic distribution and taxonomy. Ideally, we would have examined the ecological niche space of all mouse lemur species, though the available data limited us to examining six species, two of which are sister species (*M. griseorufus* and *M. murinus*) according to the phylogeny presented in several recent sources (Springer *et al.*, 2012; Weisrock *et al.*, 2012; Thiele *et al.*, 2013).

Methods

Data collection

We used known locality data from unpublished as well as published sources (Kamilar and Muldoon, 2010). The data from Kamilar and Muldoon (2010) were originally sourced from Wilmé *et al.* (2006) and supplemented with published and unpublished data provided by KMM, Shawn Lehman, and other field researchers (e.g., Muldoon and Goodman, 2010). Our analyses required a minimum of 10 known localities per species. Based on this criterion and the available data, we included six mouse lemur species in our study: *Microcebus griseorufus*, *M. murinus*, *M. myoxinus*, *M. ravelobensis*, *M. rufus*, and *M. tavaratra*. We used a total of 125 localities, with the number of localities per species presented in Table 23.1 and visually displayed in Figure 23.1.

Our ecological niche models were defined by six climate variables that characterized various aspects of rainfall and temperature: (1) isothermality, (2) temperature seasonality, (3) minimum temperature of coldest month, (4) temperature annual range, (5) precipitation of driest quarter, and (6) precipitation of coldest quarter. These and similar climatic variables are good proxies for the various habitat types that exist in Madagascar (Kamilar and Muldoon, 2010) and are likely related to lemur physiology and life history (Wright, 1999; Dewar and Richard, 2007). In addition, these measures of climatic variation were recently used in a niche modeling study of *Eulemur* species (Blair *et al.*, 2013). All climate data were obtained from the WorldClim database (Hijmans *et al.*, 2005), with an approximate 1 km resolution.

Data analysis

We used Maxent (version 3.3.3k; Phillips *et al.*, 2006; Phillips and Dudık, 2008) to construct climate-based distribution models for each species. Maxent employs a machine learning algorithm to predict the distribution of species using known occurrences. In the past five years, Maxent has become the most commonly used method to construct species distribution models. Previous research evaluating a variety of niche modeling methods found that Maxent performs as well as or better

Table 23.1 Maxent results using climatic variables to model the distribution of mouse lemur species.

Species	*N**	Mean test AUC	Test AUC SD	Omission error –fold 1**	Omission error –fold 2**	Omission error –fold 3**	Omission error –fold 4**
Microcebus griseorufus	32	0.926	0.018	0, $P < 0.001$	0, $P < 0.001$	0.125, $P < 0.001$	0, $P < 0.001$
Microcebus murinus	34	0.680	0.123	0.111, $P = 0.521$	0, $P = 0.056$	0, $P = 0.103$	0.125, $P = 0.568$
Microcebus myoxinus	12	0.890	0.021	0, $P = 0.008$	0, $P = 0.008$	0, $P = 0.008$	0.333, $P = 0.101$
Microcebus ravelobensis	10	0.987	0.007	0, $P < 0.001$	0, $P < 0.001$	0, $P = 0.002$	0.5, $P = 0.032$
Microcebus rufus	23	0.919	0.027	0.167, $P = 0.001$	0.167, $P = 0.004$	0, $P = 0.004$	0, $P = 0.001$
Microcebus tavaratra	14	0.980	0.003	0, $P < 0.001$	0.500, $P = 0.003$	0, $P < 0.001$	0, $P < 0.001$

*Number of localities used to build model. **Under a minimum training presence threshold.

than other software packages (Elith *et al.*, 2006). In addition, one advantage of Maxent is that it only requires known presence data (as opposed to known absences as well). This is particularly advantageous for our study because mouse lemurs are cryptic animals. Their nocturnal lifestyle, small body size, and relatively dispersed social structure (Zohdy *et al.*, 2012) make it difficult to be certain that species are absent from a particular location.

For each species, we used 75% of their known localities as training data and set the remaining localities as test data. We used a regularization multiplier equal to one for all species. As lower values may yield overfitted models (Radosavljevic and Anderson, 2013), we also examined models using regularization multipliers equal to two and three. Yet, using these latter regularization multipliers produced models that were similar or performed less well. One way we judged model performance is based on area under the curve (AUC) values of the receiver operating curve plots. An AUC value of 0.5 indicates that the model is no better than random at predicting the presence of a species at a locale. An AUC value greater than 0.5 suggests improved model performance, with values of 1.0 indicating a model with perfect predictive ability. However, AUC values are known to be influenced by the total niche space occupied by a species. Taxa occupying narrow environmental niches (relative to the total available niche space) tend to have higher AUC values compared to other species. In addition, we followed the default recommendations from the Maxent developers for the other model options (Phillips *et al.*, 2006; Phillips and Dudık, 2008).

As recommended and performed by recent studies (Peterson *et al.*, 2011; Blair *et al.*, 2013), we implemented a fourfold cross-validation procedure that randomly splits the occurrence data into equal sized groups. This approach is advantageous because it uses all data for validation, which is especially useful for small data sets.

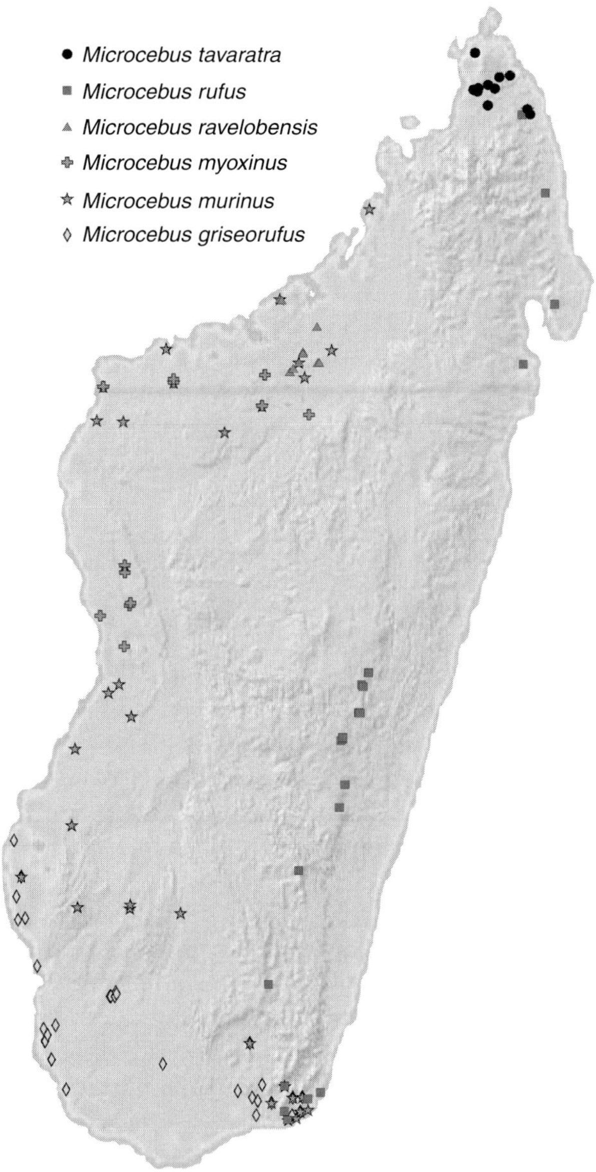

Figure 23.1 Species localities included in our analyses. (A black and white version of this figure will appear in some formats. For the color version, please refer to the plate section.)

This is important considering the relatively small sample size for most of our species. Using fourfold partitioning yields four models per species, with AUC values produced for each model. We present the mean test AUC for each species, as well as the standard deviation for the models. In addition, we used a binomial test of omission to calculate the statistical significance of each model's prediction (Anderson *et al.*, 2002). We present the detailed results produced by each model for each species.

We calculated the niche overlap based on the distribution models for all pairwise-species using Schoener's D (Schoener, 1968) and Hellinger's I (Warren *et al.*, 2008). Both metrics vary from zero to one, indicating no niche overlap to complete niche overlap, respectively. These values were quantified using the ENMTools software (Warren *et al.*, 2010). In addition, we tested whether species exhibited significantly different niches using the identity test feature in ENMTools. This procedure pools all the locality data for each species-pair and then randomly assigns localities to two new "pseudo" species. We repeated this procedure 100 times to create a distribution of random species-pairs with associated niche overlap values, and then compared the niche overlap of the real species-pair to this hypothetical distribution. We considered the ecological niches of species-pairs to be significantly different if their overlap value was lower than the five lowest values from the randomized distribution. This is equivalent to a one-tailed test with an accepted alpha level of 0.05.

Results

Maxent predictive models performed well for five of the six mouse lemur species examined, with the exception being *Microcebus murinus* (Table 23.1 and Figure 23.2A–F). For the five species that exhibited niche models that were well predicted, their mean test AUC values ranged from 0.890 to 0.987. Four of these five species exhibited statistically significant binomial tests for all of the four folds (p values ranged from < 0.001 to 0.032 and most omission errors were 0, and ranged from 0 to 0.5). The remaining species, *M. myoxinus*, exhibited a significant binomial test for three of four folds ($p = 0.008–0.101$ and omission errors from 0 to 0.333). For *M. murinus*, omission errors ranged from 0 to 0.125 and p values ranged from 0.056 to 0.568.

The niche overlap values for species-pairs ranged from 0.022 (*M. rufus–M. ravelobensis*) to 0.779 (*M. murinus–M. griseorufus*) for Hellinger's I and 0.002 (*M. rufus–M. ravelobensis*) to 0.554 (*M. myoxinus–M. murinus*) for Schoener's D (Tables 23.2 and 23.3). We found significantly low overlap values for all pairwise comparisons and for both overlap indices. In fact, the lowest possible p value was found in all cases ($p < 0.01$). *M. murinus* exhibited the highest niche overlap values on average, which is in accordance with their broadly predicted niche distribution (Figure 23.2B).

Discussion

Our study supports the hypothesis that climate plays a significant role in determining mouse lemur distribution in Madagascar (Pearson and Raxworthy, 2009; Kamilar and Muldoon, 2010; Blair *et al.*, 2013). Our climate-based ENMs demonstrate significant predictive ability of occurrence records for all species except for *Microcebus murinus*. This is not surprising, given that genetic sampling has demonstrated fine-scale patterns of lineage divergence within regional populations of this geographically widespread species. The taxonomic grouping of *M. murinus* (*sensu lato*) likely encompasses multiple cryptic novel species (Weisrock *et al.*, 2010).

Table 23.2 Niche overlap results using Hellinger's I. Higher I values indicate greater niche overlap between species.

Species	*M. griseorufus*	*M. murinus*	*M. myoxinus*	*M. ravelobensis*	*M. rufus*	*M. tavaratra*
Microcebus griseorufus	×	0.779	0.471	0.171	0.499	0.174
Microcebus murinus	< 0.01	×	0.778	0.502	0.593	0.454
Microcebus myoxinus	< 0.01	< 0.01	×	0.773	0.068	0.447
Microcebus ravelobensis	< 0.01	< 0.01	< 0.01	×	0.022	0.417
Microcebus rufus	< 0.01	< 0.01	< 0.01	< 0.01	×	0.147
Microcebus tavaratra	< 0.01	< 0.01	< 0.01	< 0.01	< 0.01	×

I value above diagonal; *p* value below diagonal.
Significant *p* values indicate species occupy different niche spaces.
Hellinger's I (Warren *et al.*, 2008)

Table 23.3 Niche overlap results using Schoener's D. Higher D values indicate greater niche overlap between species.

Species	*M. griseorufus*	*M. murinus*	*M. myoxinus*	*M. ravelobensis*	*M. rufus*	*M. tavaratra*
Microcebus griseorufus	×	0.476	0.221	0.056	0.260	0.043
Microcebus murinus	< 0.01	×	0.554	0.236	0.303	0.210
Microcebus myoxinus	< 0.01	< 0.01	×	0.489	0.011	0.250
Microcebus ravelobensis	< 0.01	< 0.01	< 0.01	×	0.002	0.235
Microcebus rufus	< 0.01	< 0.01	< 0.01	< 0.01	×	0.038
Microcebus tavaratra	< 0.01	< 0.01	< 0.01	< 0.01	< 0.01	×

D value above diagonal; *p* value below diagonal.
Significant *p* values indicate species occupy different niche spaces.
Schoener's D (Schoener, 1968).

Our climate-based ENMs show possible overprediction in some areas in comparison with known ranges. Species may be absent in areas predicted by the model because there are geographic barriers that limit their dispersal into these areas. All *Microcebus* species are small-bodied primates (Kamilar *et al.*, 2012), and small taxa often have limited geographic ranges due to reduced dispersal ability across geographic barriers (Ayres and Clutton-Brock, 1992; Harcourt and Wood, 2012). Mouse

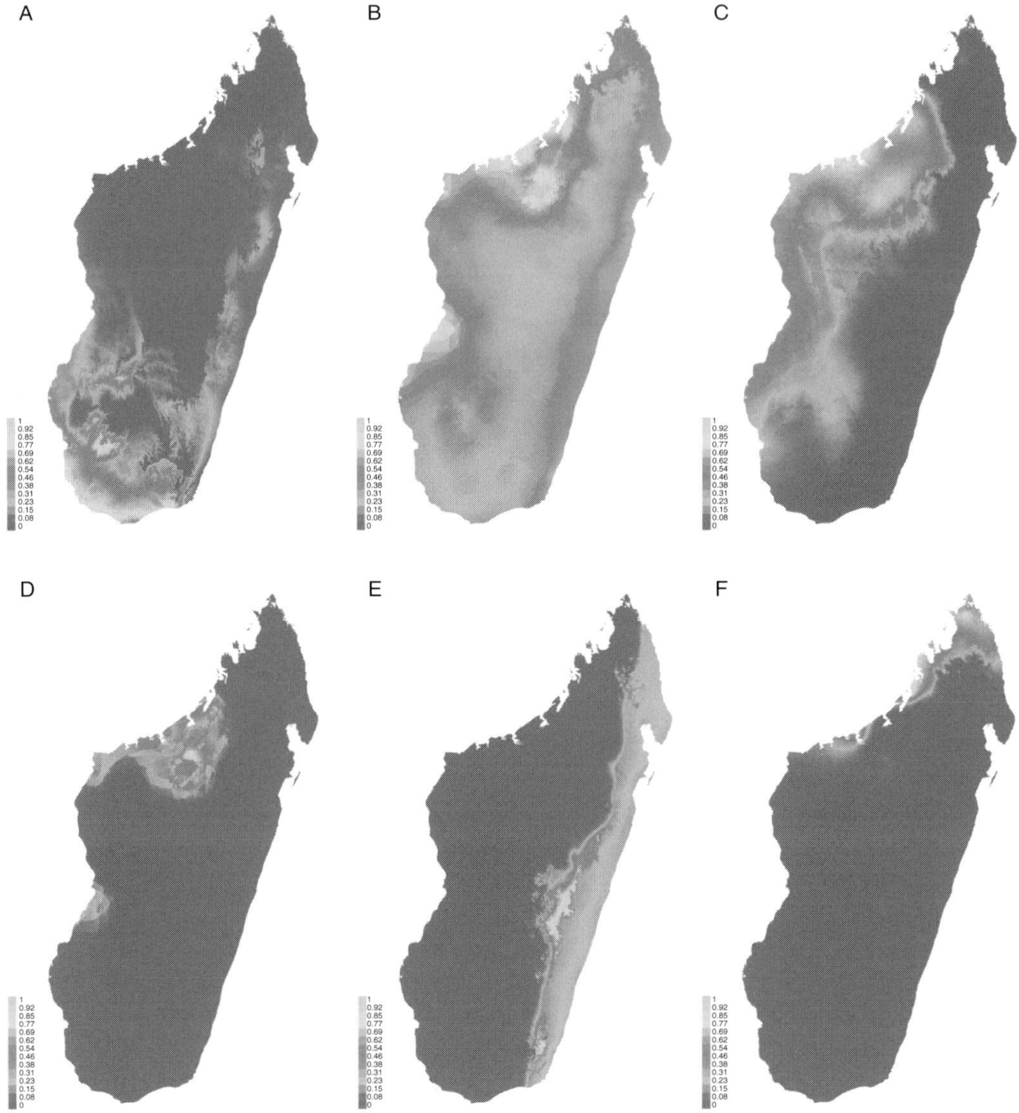

Figure 23.2 Graphical representation of the Maxent models for mouse lemur species. Each map represents the point-wise mean of the four replicate runs for each species. Warmer colors (e.g., orange and red) indicate areas that are more suitable for species to occur, based on climate variables. Cooler colors (e.g., dark and light blue) indicate regions with poorly predicted conditions. Species include: (A) *Microcebus griseorufus*, (B) *M. murinus*, (C) *M. myoxinus*, (D) *M. ravelobensis*, (E) *M. rufus*, and (F) *M. tavaratra*. (A black and white version of this figure will appear in some formats. For the color version, please refer to the plate section.)

lemurs in particular have been found to exhibit significant genetic differentiation at riverine boundaries (Olivieri *et al.*, 2007), supporting the idea that individuals dispersing across this type of barrier is limited. However, it should be noted that our data support the presence of *M. rufus* in the northeast (Figure 23.2E), where it is sympatric with other newly described, distinct species that were excluded from our analysis based on minimum locality requirements.

Non-mutually exclusive phenomena may be at work. For instance, some mouse lemur species may actually have a broader distribution than currently known, as their small, nocturnal, and cryptic nature does not lend itself to easy identification. Our predictive models may provide direction for future surveys to focus on extending the geographic areas likely to be occupied by mouse lemur species.

In addition, species may be prevented from living in suitable habitat due to competitive exclusion from ecologically similar species. Competitive exclusion has been argued as an important influence of primate distributions at broad spatial scales (Ganzhorn, 1997; Lehman, 2006; Kamilar and Ledogar, 2011; Beaudrot *et al.*, 2013). Yet, this may also be an important factor at finer spatial scales within Madagascar. For example, *Microcebus murinus* and *M. griseorufus* occur sympatrically at sites within Andohahela Parcel 2 (xeric forest). *Microcebus murinus* also occurs in the ecotone that separates Parcel 2 and Parcel 1, but field researchers have not captured *M. murinus* within the boundaries of Parcel 1 (mesic forest) at Andohahela. The pattern of *Microcebus* spp. turnover along the ecotonal gradient could be due to a series of competitive exclusions: *M. griseorufus* may be best adapted to xeric vegetation formations; as conditions become more mesic, it is replaced by *M. murinus*, which in turn is replaced by *M. rufus* at the most humid end of the gradient (Rakotondranary *et al.*, 2011). Each species seems to have an evolutionary advantage within their respective vegetation type. However, recent studies have demonstrated that there is high intraspecific variation in morphology and physiology within mouse lemurs (Lahann *et al.*, 2006; Atsalis, 2007; Génin, 2008; Kobbe et al., 2011). Interestingly, there is evidence of hybridization between *M. murinus* and *M griseorufus* at Andohahela despite clear ecological preferences (xeric vs. gallery forest; Hapke *et al.*, 2011). Hence, the ecological data currently available for these species do not provide any explanation for a selective advantage of one species over the other in different vegetation formations.

Two possible explanations exist for why *Microcebus murinus* was associated with a relatively poor ENM. This species may be an ecological generalist, being able to persist in a relatively broad range of climates and habitat types (Kamilar and Muldoon, 2010). If *M. murinus* is a generalist, then ENMs will not be able to predict their distribution with a high degree of confidence because their known occurrences are found in a wide range of environments. Assuming the broad niche of *M. murinus* is reflective of their biology, then this may have important implications for mouse lemur conservation. The relatively high level of discordance between their occurrence and the local climate and habitat may make this species better able to adapt to future shifts in climate (Dufresne *et al.*, 2013). Conversely, the relatively strong connection between climate and the distribution of other mouse lemur species suggest that these taxa will be susceptible to extinction or geographic range shifts under future models of climate change. Alternatively, the poor performing ENM for *M. murinus* may indicate that populations from this species should actually be elevated to full species. Molecular analyses using mtDNA and nDNA found clear genetic differentiation with *M. murinus*. In particular, *M. murinus* populations in southeastern Madagascar are completely distinct from western populations, with no evidence of gene flow (Weisrock *et al.*, 2010).

Finally, our study clearly demonstrates a significant amount of ecological (i.e., climatic) niche separation among all mouse lemurs examined. If the niche space occupied is indicative of species distinctness, then our results support the idea that each of these six species are in fact truly separate species. Our findings also support previous work showing divergent climatic niches among mouse lemurs (as well as among many other lemur species) in a phylogenetic context (Kamilar and Muldoon, 2010). In addition, our results are concordant with a recent study demonstrating that rainfall and temperature were poor predictors of body mass variation in lemurs, including eight mouse lemur species (Kamilar *et al.*, 2012). The lemurs in our study are similar in body size, but the lack of niche overlap suggests that the particular environment in which they live does not significantly impact their body size. The lack of connection between climatic factors and body mass may at least be partly explained by use of daily torpor or hibernation that is known to occur in some cheirogaleid species (Blanco *et al.*, 2013; e.g., Blanco and Rahalinarivo, 2010). The use of these metabolic strategies may act as a buffer to the energetic constraints that mouse lemurs may have to face in different habitats.

In conclusion, we found that ecological niche models performed well for predicting the occurrence of mouse lemur species, with the exception of *M. murinus*. These results suggest that *M. murinus* is an ecological generalist or actually multiple species. In addition, we found that all of the mouse lemur species we examined exhibit little niche overlap, adding further support to recent genetic studies that have named new species.

Acknowledgments

We thank the editors for inviting us to contribute to this volume. We appreciate the helpful comments from two anonymous reviewers that improved an earlier version of this chapter. Finally, we thank Steve Goodman for many discussions about lemur biogeography.

References

Anderson RP, Gomez-Laverde M, Peterson AT. 2002. Geographical distributions of spiny pocket mice in South America: insights from predictive models. *Global Ecology and Biogeography* 11:131–141.

Atsalis S. 2007. *A Natural History Of The Brown Mouse Lemur.* Routledge, Abingdon.

Ayres JM, Clutton-Brock TH. 1992. River boundaries and species range size in Amazonian primates. *American Naturalist* 140:531–537.

Beaudrot L, Struebig MJ, Meijaard E, *et al.* 2013. Interspecific interactions between primates, birds, bats and squirrels may affect community composition on Borneo. *American Journal of Primatology* 75:170–185.

Blair ME, Sterling EJ, Dusch M, Raxworthy CJ, Pearson RG. 2013. Ecological divergence and speciation between lemur (*Eulemur*) sister species in Madagascar. *Journal of Evolutionary Biology* 26:1790–1801.

Blanco MB, Rahalinarivo V. 2010. First direct evidence of hibernation in an eastern dwarf lemur species (*Cheirogaleus crossleyi*) from the high-altitude forest of Tsinjoarivo, central-eastern Madagascar. *Naturwissenschaften* 97:945–950.

Blanco MB, Dausmann KH, Ranaivoarisoa JF, Yoder AD. 2013. Underground hibernation in a primate. *Scientific Reports* 3:1768. doi:10.1038/srep01768.

Dewar RE, Richard AF. 2007. Evolution in the hypervariable environment of Madagascar. *Proceedings of the National Academy of Sciences, USA* 104:13723–13727.

Dufresne JL, Foujols MA, Denvil S, *et al.* 2013. Climate change projections using the IPSL-CM5 Earth System Model: from CMIP3 to CMIP5. *Climate Dynamics* 40:2123–2165.

Elith J, Graham CH, Anderson RP, *et al.* 2006. Novel methods improve prediction of species' distributions from occurrence data. *Ecography* 29:129–151.

Franklin J. 2009. *Mapping Species Distributions.* Cambridge University Press, Cambridge.

Ganzhorn JU. 1997. Test of Fox's assembly rule for functional groups in lemur communities in Madagascar. *Journal of Zoology (London)* 241:533–542.

Génin F. 2008. Life in unpredictable environments: first investigation of the natural history of *Microcebus griseorufus. International Journal of Primatology* 29:303–321.

Groeneveld LF, Weisrock DW, Rasoloarison RM, Yoder AD, Kappeler PM. 2009. Species delimitation in lemurs: multiple genetic loci reveal low levels of species diversity in the genus *Cheirogaleus. BMC Evolutionary Biology* 9:30.

Hapke A, Gligor M, Rakotondranary SJ, Rosenkranz D, Zupke O. 2011. Hybridization of mouse lemurs: different patterns under different ecological conditions. *BMC Evolutionary Biology* 11:297.

Harcourt AH, Wood MA. 2012. Rivers as barriers to primate distributions in Africa. *International Journal of Primatology* 33:168–183.

Hijmans RJ, Cameron SE, Parra JL, Jones PG, Jarvis A. 2005. Very high resolution interpolated climate surfaces for global land areas. *International Journal of Climatology* 25:1965–1978.

Horvath JE, Weisrock DW, Embry SL, *et al.* 2008. Development and application of a phylogenomic toolkit: resolving the evolutionary history of Madagascar's lemurs. *Genome Research* 18:489–499.

Kamilar JM. 2006. Geographic variation in savanna baboon (*Papio*) ecology and its taxonomic and evolutionary implications. In SM Lehman, JG Fleagle (eds.), *Primate Biogeography* (pp. 169–200). Springer, New York.

Kamilar JM, Cooper N. 2013. Phylogenetic signal in primate behaviour, ecology, and life history. *Philosophical Transactions of the Royal Society London B Biological Sciences* 368:20120341.

Kamilar JM, Ledogar JA. 2011. Species co-occurrence patterns and dietary resource competition in primates. *American Journal of Physical Anthropology* 144:131–139.

Kamilar JM, Muldoon KM. 2010. The climatic niche diversity of Malagasy primates: a phylogenetic approach. *PLoS ONE* 5:e11073. doi:10.1371/journal.pone.0011073.

Kamilar JM, Muldoon KM, Lehman SM, Herrera JP. 2012. Testing Bergmann's rule and the resource seasonality hypothesis in Malagasy primates using GIS-based climate data. *American Journal of Physical Anthropology* 147:401–408.

Kobbe S, Ganzhorn JU, Dausmann KH. 2011. Extreme individual flexibility of heterothermy in free-ranging Malagasy mouse lemurs (*Microcebus griseorufus*). *Journal of Comparative Physiology B* 181:165–173.

Lahann P, Schmid J, Ganzhorn JU. 2006. Geographic variation in populations of *Microcebus murinus* in Madagascar: resource seasonality or Bergmann's rule? *International Journal of Primatology* 27:983–999.

Lehman SM. 2006. Nested distribution patterns and the historical biogeography of the primates of Guyana. In SM Lehman, JG Fleagle (eds.), *Primate Biogeography* (pp. 63–80). Springer, New York.

MacColl A. 2011. The ecological causes of evolution. *Trends in Ecology & Evolution* 26:514–522.

Markolf M, Brameier M, Kappeler PM. 2011. On species delimitation: yet another lemur species or just genetic variation? *BMC Evolutionary Biology* 11:216.

Mittermeier RA, Tattersall I, Konstant WR, Meyers DM, Mast RB. 1994. *Lemurs of Madagascar.* Conservation International, Washington, DC.

Mittermeier RA, Louis EE, Richardson M, *et al.* 2010. *Lemurs of Madagascar*, 3rd edition. Conservation International, Washington, DC.

Muldoon K, Goodman S. 2010. Ecological biogeography of Malagasy non-volant mammals: community structure is correlated with habitat. *Journal of Biogeography* 37:1144–1159.

Olivieri G, Zimmermann E, Randrianambinina B, *et al.* 2007. The ever-increasing diversity in mouse lemurs: three new species in north and northwestern Madagascar. *Molecular Phylogenetics and Evolution* 43:309–327.

Pastorini J, Forstner MRJ, Martin RD. 2001. Phylogenetic history of sifakas (*Propithecus*: Lemuriformes) derived from mtDNA sequences. *American Journal of Primatology* 53:1–17.

Pearson RG, Raxworthy CJ. 2009. The evolution of local endemism in Madagascar: watershed versus climatic gradient hypothesis evalauted by null biogeographic models. *Evolution* 63:959–967.

Peterson AT, Soberon J, Pearson RG, *et al.* 2011. *Ecological Niches and Geographic Distributions.* Princeton University Press, Princeton, NJ.

Phillips SJ, Dudık M. 2008. Modeling of species distributions with Maxent: new extensions and a comprehensive evaluation. *Ecography* 31:161–175.

Phillips SJ, Anderson RP, Schapire RE. 2006. Maximum entropy modeling of species geographic distributions. *Ecological Modelling* 190:231–259.

Radespiel U, Reimann W, Rahelinirina M, Zimmermann E. 2006. Feeding ecology of sympatric mouse lemur species in northwestern Madagascar. *International Journal of Primatology* 27:311–321.

Radespiel U, Ratsimbazafy JH, Rasoloharijaona S, *et al.* 2012. First indications of a highland specialist among mouse lemurs (*Microcebus* spp.) and evidence for a new mouse lemur species from eastern Madagascar. *Primates* 53:157–170.

Radosavljevic A, Anderson RP. 2013. Making better Maxent models of species distributions: complexity, overfitting and evaluation. *Journal of Biogeography* 41:629–643.

Rakotondranary SJ, Hapke A, Ganzhorn JU. 2011. Distribution and morphological variation of *Microcebus* spp. along an environmental gradient in southeastern Madagascar. *International Journal Of Primatology* 32:1037–1057.

Rasoloarison RM, Weisrock DW, Yoder AD, Rakotondravony D, Kappeler PM. 2013. Two new species of mouse lemurs (Cheirogaleidae: *Microcebus*) from eastern Madagascar. *International Journal of Primatology* 34:455–469.

Raxworthy CJ, Ingram CM, Rabibisoa N, Pearson RG. 2007. Applications of ecological niche modeling for species delimitation: a review and empirical evaluation using day geckos (*Phelsuma*) from Madagascar. *Systematic Biology* 56:907–923.

Schoener TW. 1968. Size of feeding territories among birds. *Ecology* 49:123–141.

Springer MS, Meredith RW, Gatesy J, *et al.* 2012. Macroevolutionary dynamics and historical biogeography of primate diversification inferred from a species supermatrix. *PLoS ONE* 7:e49521.

Tattersall I. 2007. Madagascar's lemurs: cryptic diversity or taxonomic inflation? *Evolutionary Anthropology* 16:12–23.

Thiele D, Razafimahatratra E, Hapke A. 2013. Discrepant partitioning of genetic diversity in mouse lemurs and dwarf lemurs – biological reality or taxonomic bias? *Molecular Phylogenetics and Evolution* 69:593–609.

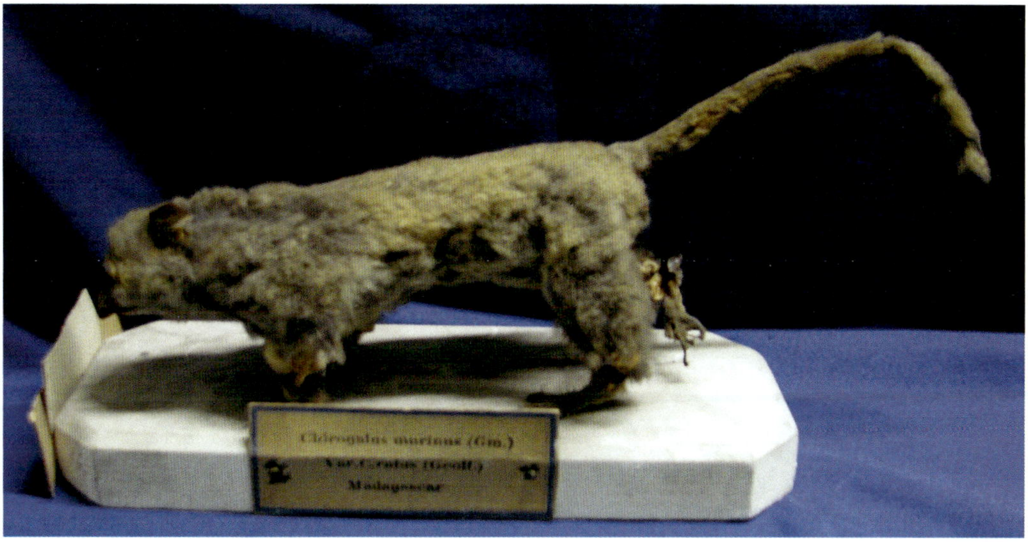

Figure 2.1 "Le Microcèbe roux".

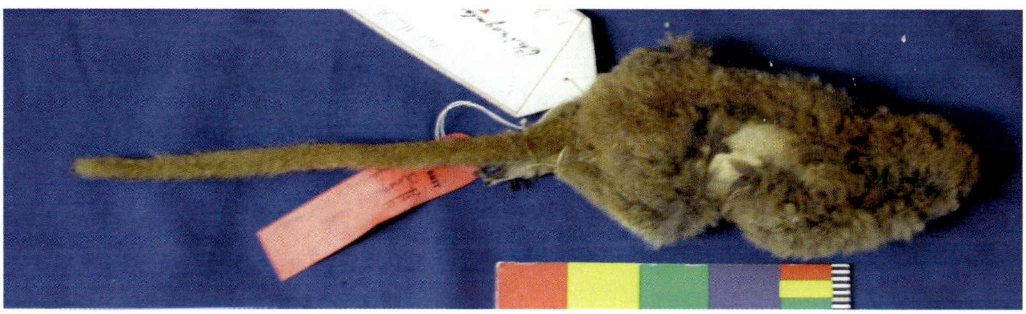

Figure 2.2a British Museum (Natural History), type of *Microcebus smithii* (Gray, 1842); (a) skin.

Figure 2.3a Holotype of *Cheirogaleus milii*, neotype of *Cheirogaleus major* MNHNP no.148: (a) skin.

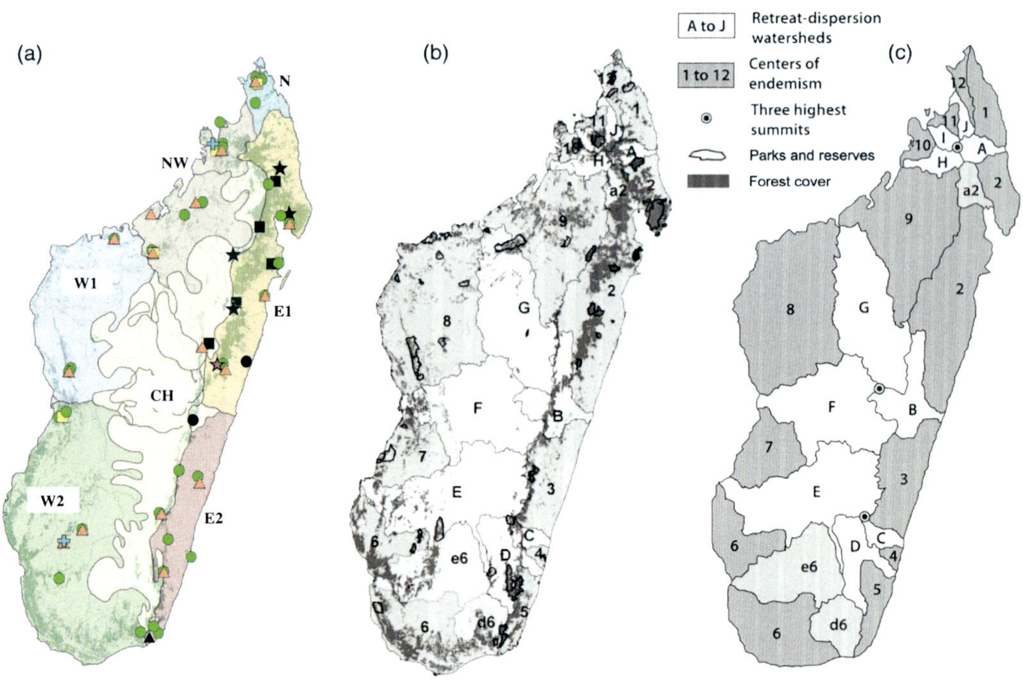

Figure 3.4 Biogeographic regions based on hypotheses proposed to explain biodiversity patterns in Madagascar. (a) Modified map presenting the seven biogeographic regions of lemur distributions based on phytogeographic criteria and similarities of lemur communities as proposed by Martin (1972, 1995); Thalmann (2000) and Ganzhorn *et al.* (2006). (b) Biogeographic map presenting "retreat-dispersion watersheds" (RWD) and Centers of Endemism (COE) modified from Wilmé *et al.* (2006). (c) Modified map presented in Townsend *et al.* (2009) showing RWD and COE for leaf chameleons (*Brookesia*).

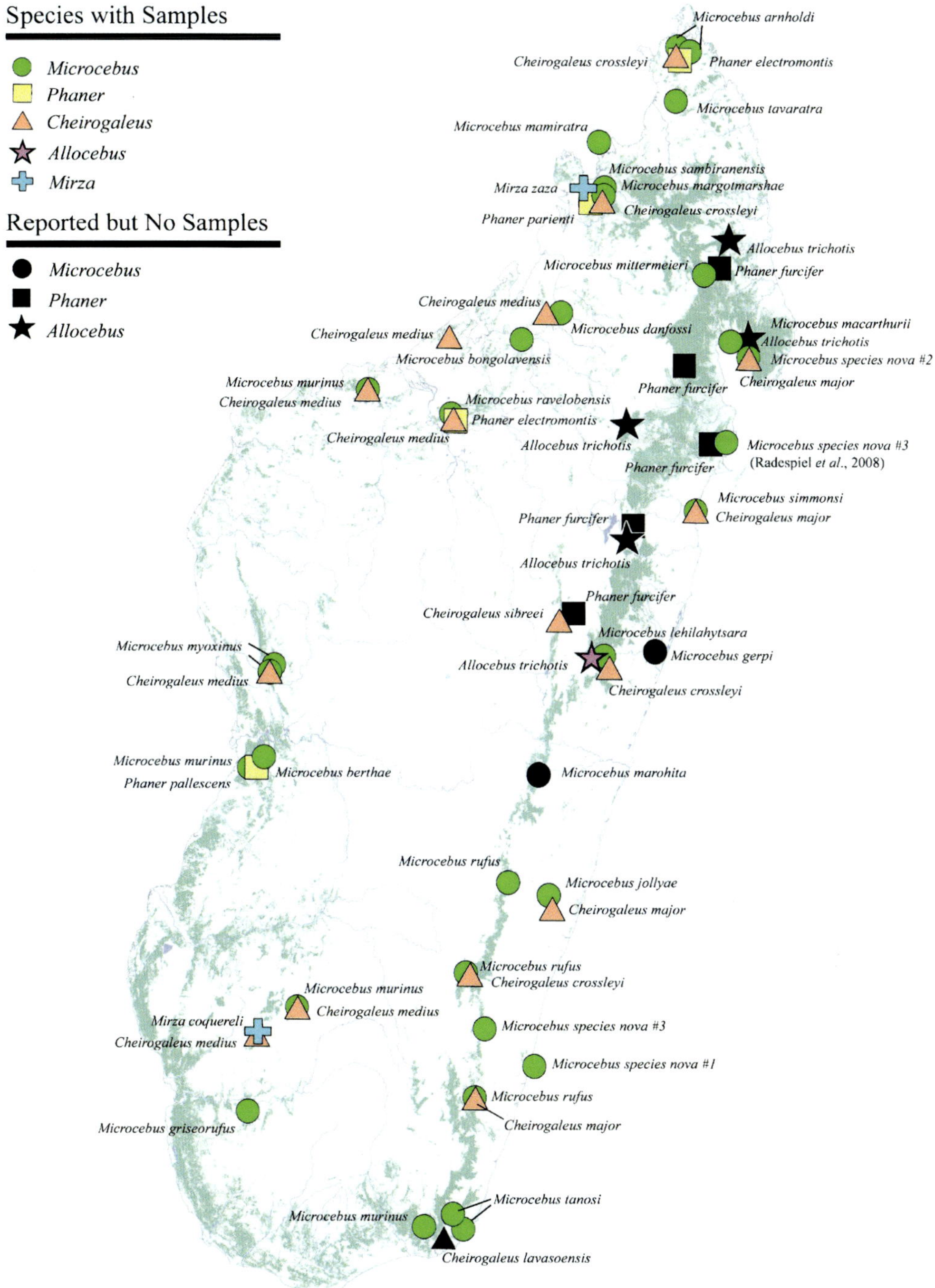

Figure 3.1 Map of sampling localities of the cheirogaleid lemurs of Madagascar. Circles, square, triangle, star, and cross represent the genera *Microcebus*, *Phaner*, *Cheirogaleus*, *Allocebus*, and *Mirza*, respectively. Solid black symbols denote published localities for different species with no molecular data available for this study. Detailed information for locality sites is presented in Table 3.1 or from Meier and Albignac (1989, 1991), Rakotoarison *et al.* (1996), Rakotoarison (1998), Schütz and Goodman (1998), Garbutt (2000), Goodman and Raselimanana (2002), Mittermeier *et al.* (2010), Radespiel *et al.* (2012), and Rasoloarison *et al.* (2013).

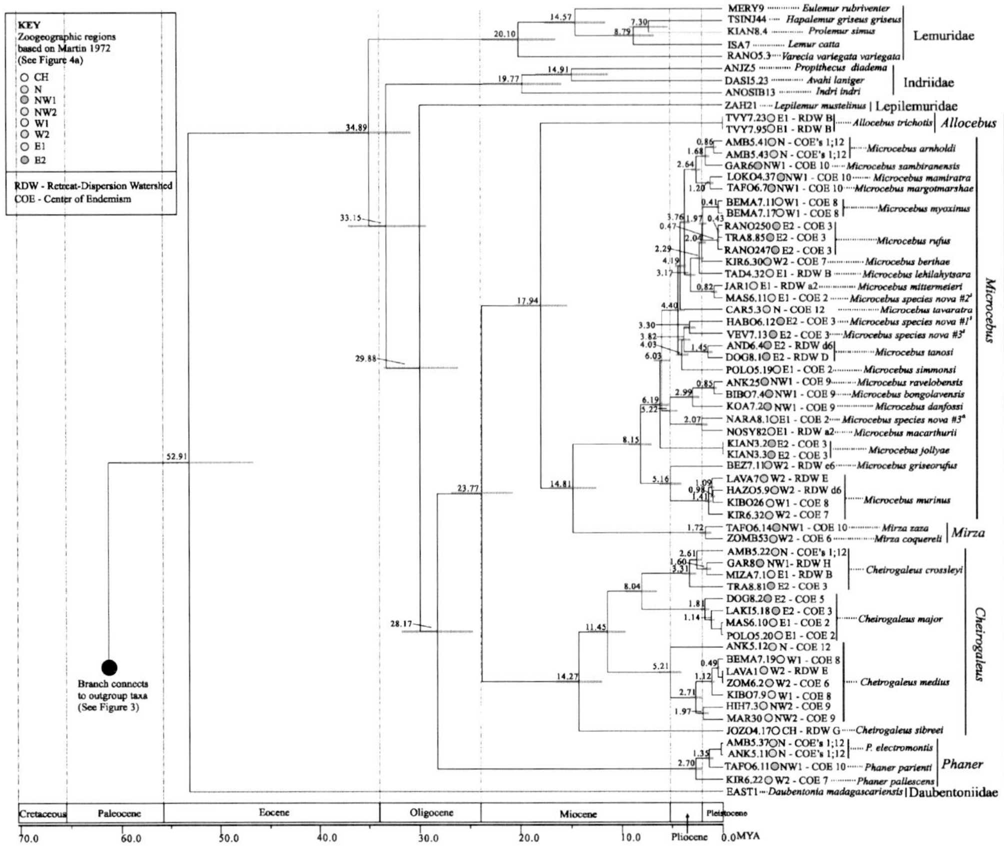

Figure 3.5 Truncated view of Figure 3.3 of estimated divergence dates for the Family Cheirogaleidae in Lemuriformes based on the mtDNA genome heavy-stranded coding region sequences of the 56 Cheirogaleidae individuals and the 9 outgroup lemur species utilizing a Bayesian relaxed molecular clock analysis with BEAST (Drummond and Rambaut, 2007). The gray bars display the 95% HPD (highest probability density) interval of node ages. Colored circles following taxa designation correspond to biogeographic regions in Figure 3.4a. Abbreviations designate specific regions in Figure 3.4a and COE and RDW for each lemur taxon located in Figures 3.4b and 3.4c.

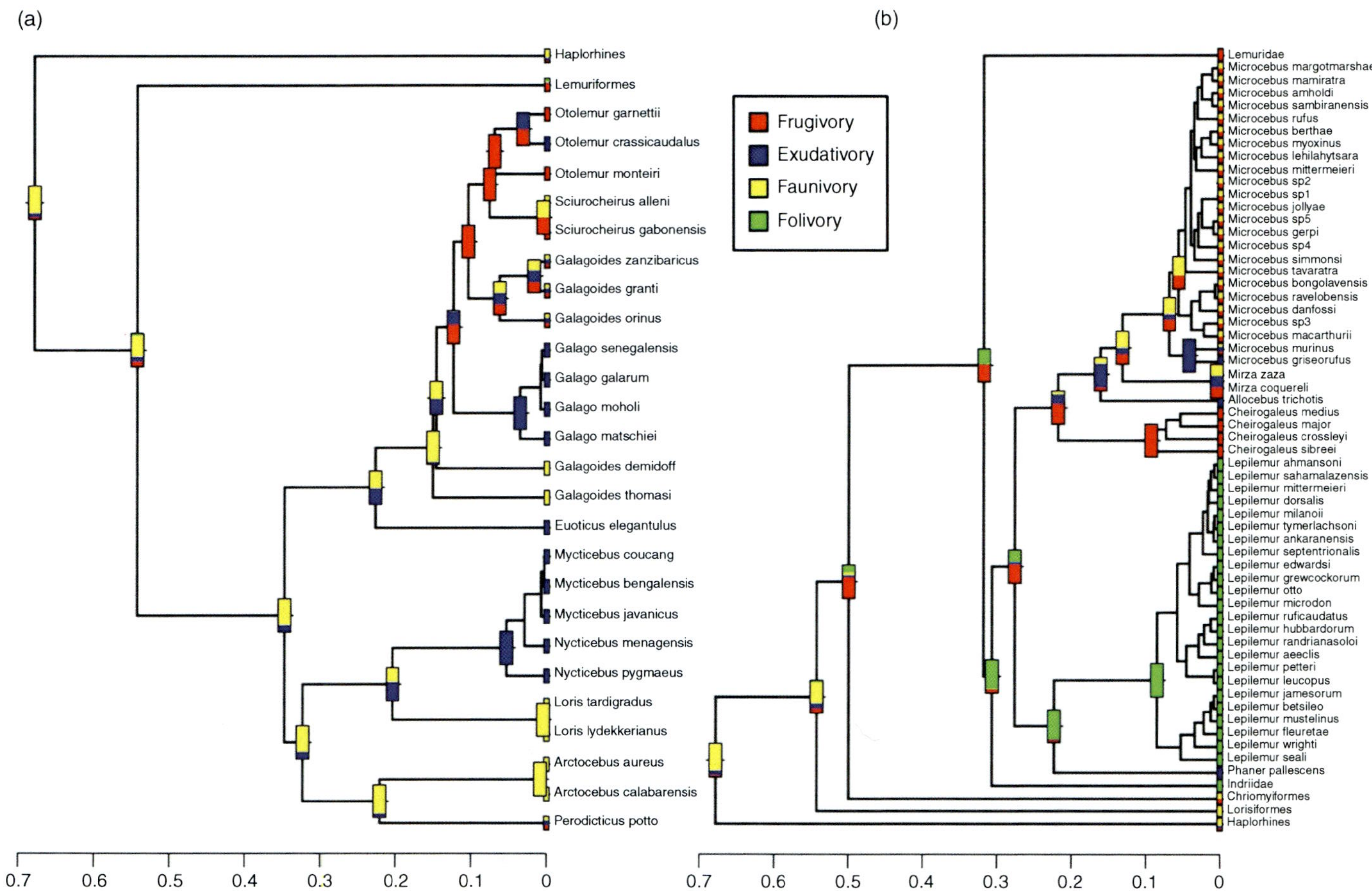

Figure 4.2 Phylogenetic reconstruction of the diets of (a) Lorisoidea, and (b) Cheirogaleidae. The time scales are expressed in 100 Ma. The color codes indicate the probabilities of the reconstructed diets.

Figure 13.1 The two forest habitats at Beza Mahafaly. Top row: mesic (gallery) forest, from left to right, wet season and dry season (Emilienne Rasoazanabary placing trap). Bottom: xerophytic (spiny) forest, from left to right, wet season and dry season. (Photo credits: top left, Emilienne Rasoazanabary; top right, Brooke Crowley; bottom left, Brooke Crowley; bottom right, Laurie Godfrey.)

Figure 5.2 Photographs displaying the morphological variation in *Cheirogaleus crossleyi* (A) and *C. sibreei* (B) faces (a), dorsum and ventrum (b), and female genital morphology (c).

Figure 6.2 Custom-made induction chamber with a small port for fresh gas inlet and a large port for connecting to an active, negative pressure scavenging system. The translucent material (polypropylene, SAMLA box, IKEA) allows observation of the animal during induction.

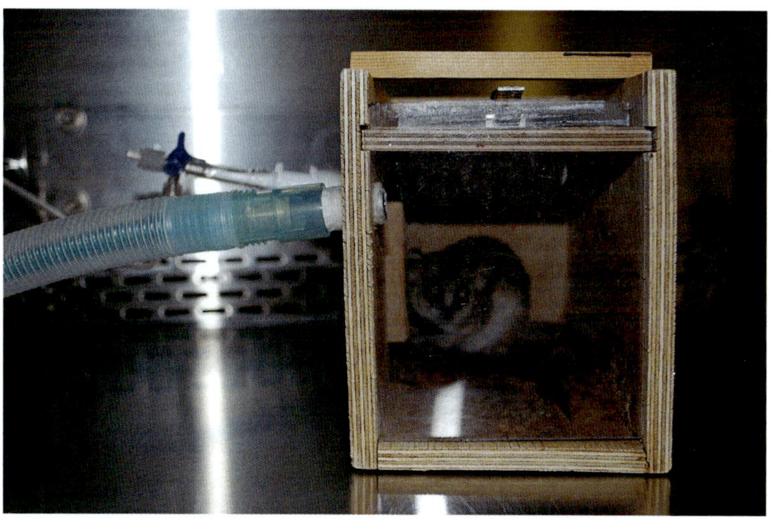

Figure 6.1b Set-up for anesthesia induction using the sleeping boxes of mouse lemurs. Side openings of the box are connected to the anesthetic system supplying the anesthetic gas mixture. A second opening in the box is connected to an active charcoal canister to scavenge the waste gas. Detail of the connection of a Bain breathing system to the sleeping box. The animal can be observed during induction.

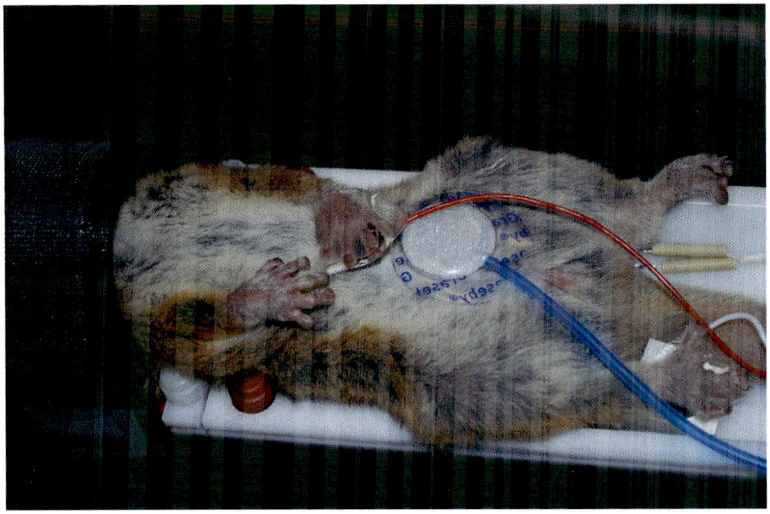

Figure 6.7a Pediatric ECG surface electrodes on one hand and one foot of a gray mouse lemur. An air cushion transmits respiratory excursions to pressure changes. The lemur is equipped for MRI scans.

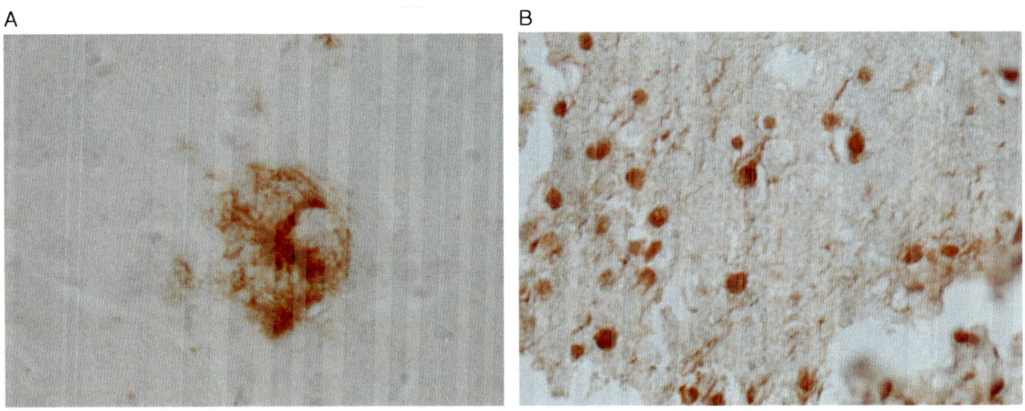

Figure 7.1 Aβ plaques in the occipital cortex (A) and tau-PHF neurons in the piriform cortex (B) of a mouse lemur brain.

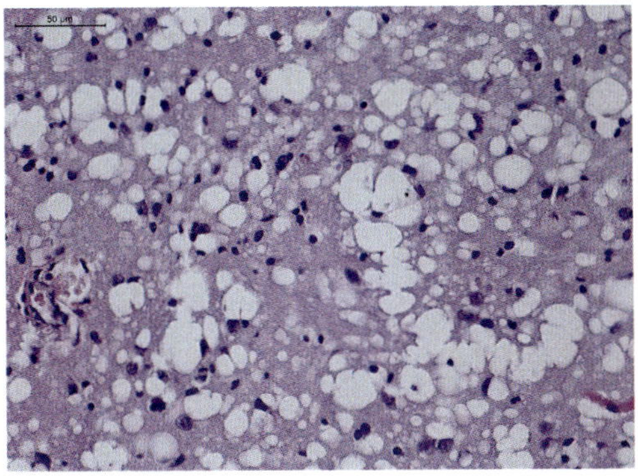

Figure 7.2 Spongiosis in intracerebrally inoculated animals showing large confluent vacuoles.

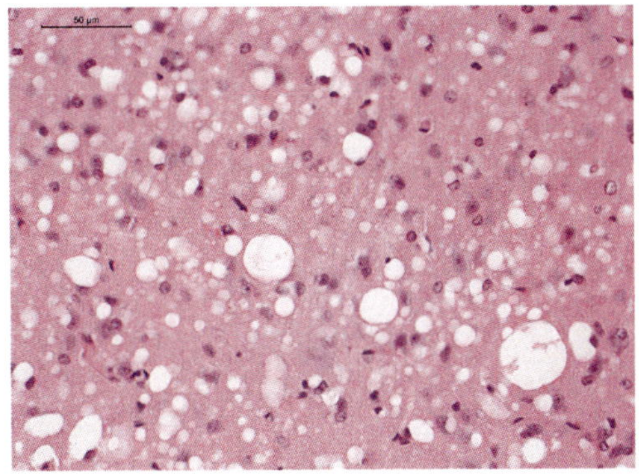

Figure 7.3 Spongiosis in orally infected animals.

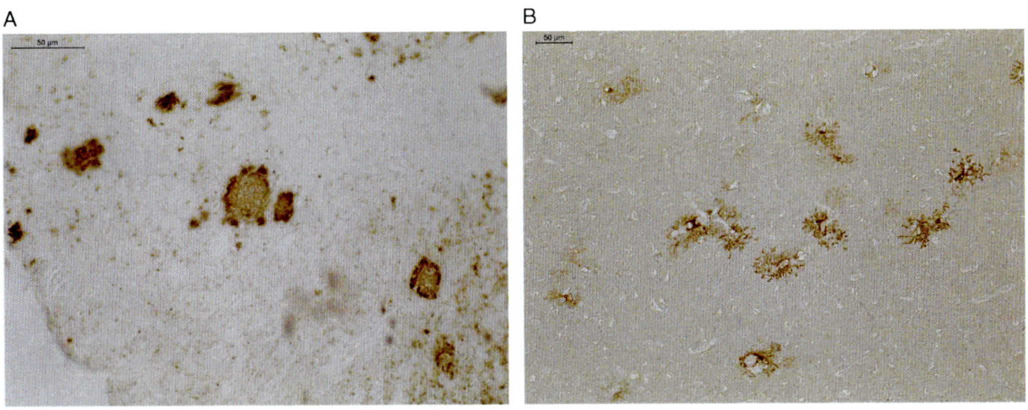

Figure 7.6 Immunodetection of PrP^{res} in plaques with 3F4 antibody (A). GFAP antibody labeling shows reactive astrocytic gliosis (B).

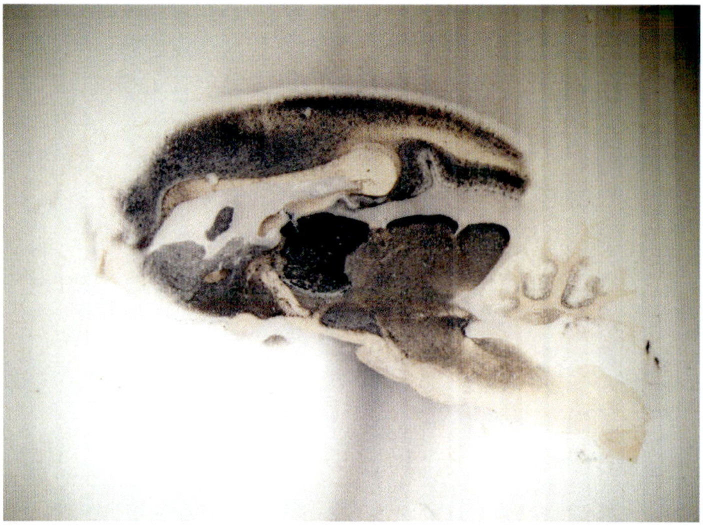

Figure 7.4 PET-blot showing PrP^{res} distribution in the whole brain of IC-inoculated mouse lemurs.

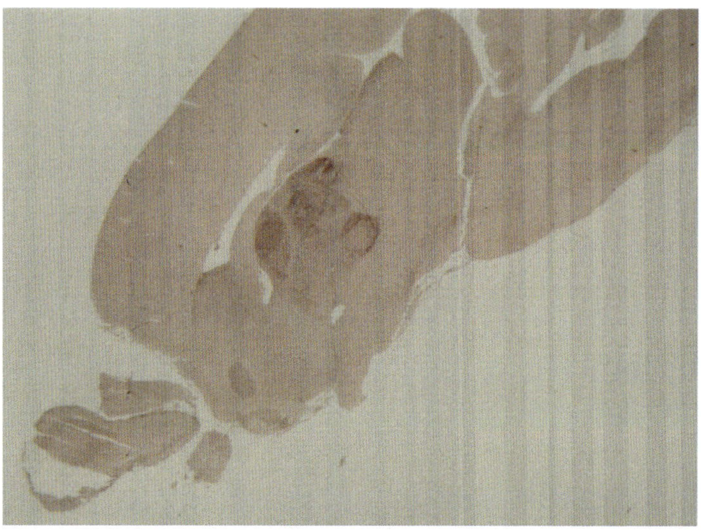

Figure 7.5 PET-blot showing PrP^{res} distribution in the whole brain of orally inoculated mouse lemurs.

Figure 8.8 Phenotypic changes during aging in the gray mouse lemur. Displayed is a gray mouse lemur infant of a few days (photo: M. Scheumann), a young adult (2 years) (photo: E. Zimmermann), and an old (10 years) gray mouse lemur (photo: E. Zimmermann).

Figure 12.3 Animals of *Mirza zaza* were observed licking homopteran Flatidae larvae (left) and their sugary excretions at Ankarafa, Sahamalaza.

Figure 22.1 A sleeping group of three matrilineally related *M. murinus* in Ankarafantsika National Park. Genetic data presented in full in Kessler *et al.* (2014). Photo credit: Sharon E. Kessler.

Figure 13.3 Comparison of nest types. Left: tree hole in a dead tree (*Alantsilodendron humbertii*). Right: open nest of tangled vegetation in a famata tree (*Euphorbia tirucalli*). Photo credits: Emilienne Rasoazanabary.

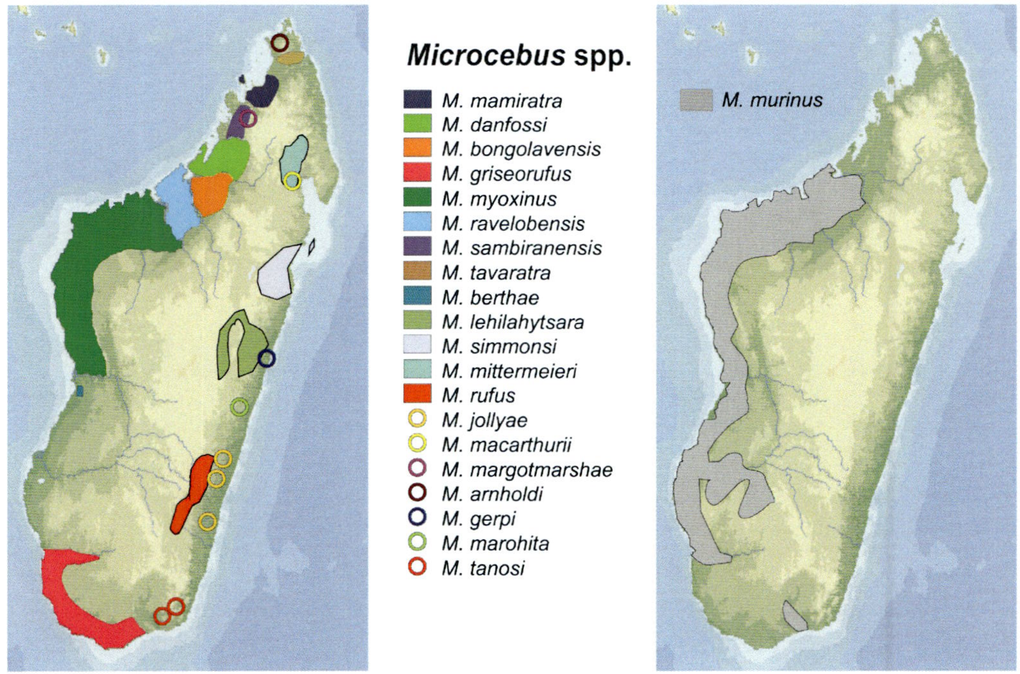

Figure 21.2 Geographical distributions of mouse lemurs in Madagascar (according to Radespiel *et al.*, 2012 and Rasoloarison *et al.*, 2013).

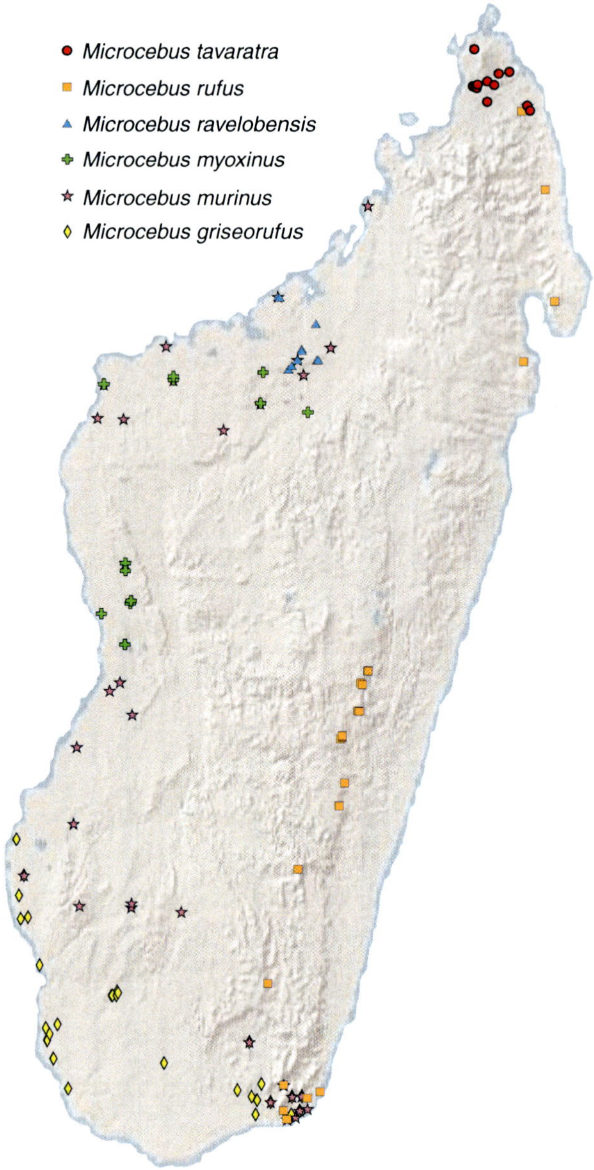

Figure 23.1 Species localities included in our analyses.

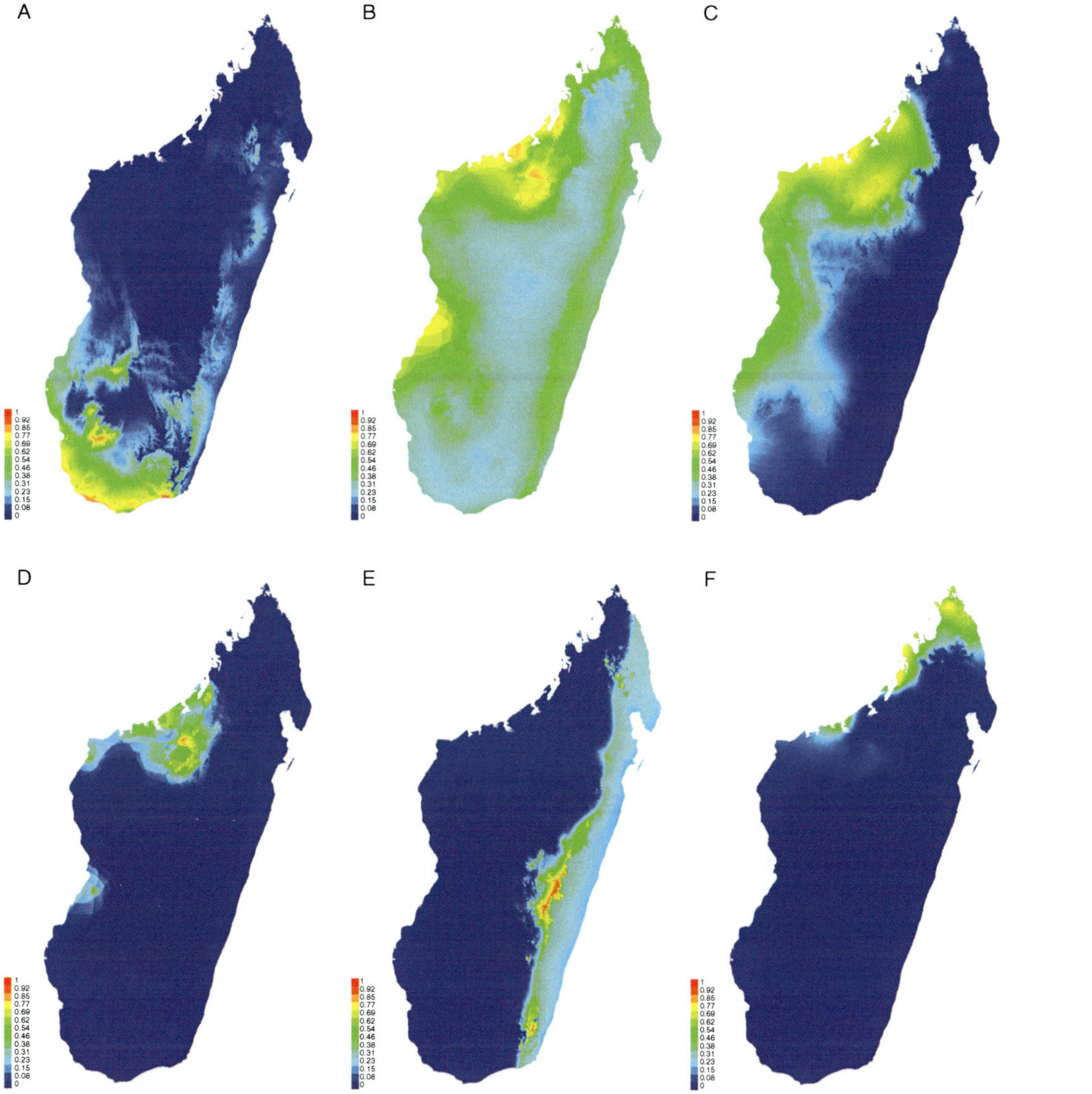

Figure 23.2 Graphical representation of the Maxent models for mouse lemur species. Each map represents the point-wise mean of the four replicate runs for each species. Warmer colors (e.g., orange and red) indicate areas that are more suitable for species to occur, based on climate variables. Cooler colors (e.g., dark and light blue) indicate regions with poorly predicted conditions. Species include: (A) *Microcebus griseorufus*, (B) *M. murinus*, (C) *M. myoxinus*, (D) *M. ravelobensis*, (E) *M. rufus*, and (F) *M. tavaratra*.

Figure 24.1 Overview of interior forest in Ampijoroa.

Figure 24.2 Forest edge and matrix at edge effects research site in Ampijoroa.

Warren D, Glor R, Turelli M. 2010. ENMTools: a toolbox for comparative studies of environmental niche models. *Ecography* 33:607–611.

Warren DL, Glor RE, Turelli M. 2008. Environmental niche equivalency versus conservatism: quantitative approaches to niche evolution. *Evolution* 62:2868–2883.

Weisrock DW, Rasoloarison RM, Fiorentino I, *et al.* 2010. Delimiting species without nuclear monophyly in Madagascar's mouse lemurs. *PLoS ONE* 5:e9883. doi:10.1371/journal.pone.0009883.

Weisrock DW, Smith SD, Chan LM, *et al.* 2012. Concatenation and concordance in the reconstruction of mouse lemur phylogeny: an empirical demonstration of the effect of allele sampling in phylogenetics. *Molecular Biology and Evolution* 29:1615–1630.

Wiens JJ. 2004. Speciation and ecology revisited: phylogenetic niche conservatism and the origin of species. *Evolution* 58:193–197.

Wiens JJ, Ackerly DD, Allen AP, *et al.* 2010. Niche conservatism as an emerging principle in ecology and conservation biology. *Ecology Letters* 13:1310–1324.

Wilmé L, Goodman SM, Ganzhorn JU. 2006. Biogeographic evolution of Madagascar's microendemic biota. *Science* 312:1063–1065.

Wright PC. 1999. Lemur traits and Madagascar ecology: coping with an island environment. *Yearbook of Physical Anthropology* 42:31–72.

Yoder AD, Olson LE, Hanley C, *et al.* 2005. A multidimensional approach for detecting species patterns in Malagasy vertebrates. *Proceedings of the National Academy of Sciences of the United States of America* 102:6587–6594.

Zimmermann E. 2013. Primate serenades: call variation, species diversity, and adaptation in nocturnal strepsirrhines. In J Masters, M Gamba, F Génin (eds.), *Leaping Ahead* (pp. 287–295). Springer, New York.

Zimmermann E, Radespiel U. 2014. Species concepts, diversity, and evolution in primates: lessons to be learned from mouse lemurs. *Evolutionary Anthropology* 23:11–14.

Zohdy S, Kemp AD, Durden LA, Wright PC, Jernvall J. 2012. Mapping the social network: tracking lice in a wild primate (*Microcebus rufus*) population to infer social contacts and vector potential. *BMC Ecology* 12:4.

24 Edge effects on tree dendrometrics, abiotics, and mouse lemur densities in western dry forests in Madagascar

Shawn M. Lehman

Introduction

Anthropogenic disturbance of forests leads to fragmentation and habitat loss, which strongly influence genetic diversity, extirpations, and extinction of many species (e.g., Fahrig and Merriam, 1994; Gibson *et al.*, 2013; Radespiel and Bruford, 2014). One of the most significant consequences of fragmentation and habitat loss is an increase in the amount of edge effects and edge habitat (Chen *et al.*, 1992). Edge effects represent the penetration, to varying depths and intensities, of biotic and abiotic conditions from the surrounding environment (matrix) into the forest interior (Malcolm, 1994). Murcia (1995) described three ecological consequences of edge effects: (1) abiotic effects, (2) direct biological effects, and (3) indirect biological effects. Abiotic effects occur as the result of the penetration of environmental factors, such as temperature and light levels, from the matrix into the forest interior. Direct biological effects are changes in the abundance and distribution of organisms due to physical conditions near the forest edge. For example, Laurance *et al.* (1997) documented that wind damage led to a drastic drop in tree biomass within 100 m of the forest edge in South America. Indirect biological effects involve changes in species interactions (e.g., herbivory, frugivory), such that species that avoid the edge may be doing so due to the scarcity of preferred food items in these stochastic habitats (Mills, 1995). The combination of these three ecological processes results in dynamic, multidimensional edge habitats that differ from the matrix and forest interior. Although edge effects have been shown to alter forest structure and animal communities (e.g., Lovejoy *et al.*, 1986; Laurance and Yensen, 1991; Laurance *et al.*, 1997), they have rarely been studied directly in regards to primate ecology and biogeography.

Edge effects are particularly relevant to ecological and biogeographic studies in Madagascar. Madagascar has an unrivaled level of plant endemicity and diversity: of the estimated 12,000 species of plants in Madagascar, 81% are endemic to

The Dwarf and Mouse Lemurs of Madagascar: Biology, Behavior and Conservation Biogeography of the Cheirogaleidae, eds. S.M. Lehman, U. Radespiel, E. Zimmermann. Published by Cambridge University Press. © Cambridge University Press 2016.

this country (e.g., Phillipson, 1994, 1996; Du Puy *et al.*, 1999). This plant diversity is remarkable given that much of the original forest cover has been lost to human perturbations, such as slash-and-burn agriculture, logging, mining, and associated erosion (Green and Sussman, 1990; Du Puy and Moat, 1998). Deforestation continues at as estimated rate of approximately 0.9% (105,543 ha) of total forest cover (11,727,000 ha) per year (FAO, 2003), resulting in highly fragmented forest landscapes with poorly studied abiotic and biotic edge effects on local plants and animals (Ganzhorn *et al.*, 1997; Lowry *et al.*, 1997; Ramanamanjato, 2001; Ramanamanjato and Ganzhorn, 2001; Schatz, 2001; Lehtinen *et al.*, 2003).

Although anthropogenic disturbance has likely produced systemic edge effects in Madagascar, most data on these ecological processes come from a few sites in eastern humid forests and SE littoral forests (Jenkins *et al.*, 2003; Lehtinen *et al.*, 2003; Watson *et al.*, 2004; Lehman *et al.*, 2006b; Irwin, 2008). In eastern Madagascar, the structure and composition of forest habitats vary largely as a function of altitude, from sea level up to 1300 m above sea level, along a steep escarpment that maintains most of the remaining humid forest (Du Puy and Moat, 1996). These altitudinally complex forests are characterized by high species richness of plants and animals, high humidity, temperature seasonality, high levels of rainfall (ca. 2000–4500 mm/year), predominantly evergreen trees, and a well-developed, continuous canopy (Lehman, 2013). However, anthropogenic deforestation over the last few hundreds years in combination with recent intensified logging of hardwood trees has produced consistent patterns of reduced species richness and tree dendrometrics in forest fragments (Brown *et al.*, 2013; Schwitzer et al., 2014). For example, Lehman (2007) documented significantly lower mean tree height and DBH (diameter at breast height) in edge habitats compared to interior habitats in a large forest fragment in eastern Madagascar. Similar tree abundance and dendrometrics patterns have been reported for littoral forests in SE Madagascar (Lehtinen *et al.*, 2003). Abiotic edge effects on ambient temperature, humidity, light levels, and wind speed were consistently higher in edge than interior habitats (Lehman, unpublished data).

It should not be assumed that edge effects are monotonic throughout Madagascar. Regional variations in climate, forest cover, and plant biology likely result in concomitant but largely unexplored differences in edge effects between eastern humid and western dry forests (Randrianambinina *et al.*, 2003). Western Madagascar is characterized by hot, arid climatic conditions, with limited rainfall of 750–1500 mm/year during the typhoon season (ca. January–March) due to the rain shadow caused by the high, central plateau (Elmqvist *et al.*, 2007). The remaining forest is composed primarily of deciduous, tropical dry forest characterized by a low, broken canopy that exists predominantly in isolated fragments of varying sizes and shapes surrounded by large tracts of grasslands (Ganzhorn *et al.*, 1999). Although there are few data on abiotic and biotic edge effects in western dry forests, the local climate (hot, dry), deciduous plant biology (i.e., seasonal loss of leaves), and reduced size of trees (compared to tropical rain forest) likely results in strong depth and intensity of edge effects in the remaining forest fragments. In one of the few studies of its kind in western dry forests, Crowley *et al.* (2012) found that foliar $\delta^{15}N$ values (the product of partitioning of isotopes of the same element with differing masses in leaves) were significantly lower near edge habitats, which likely reflected spatial variations in nutrients, N-sources, or water availability. Moreover, some biotic edge effects, such as the distribution of food trees preferred by *Propithecus coquereli*, can penetrate up to 500 m into the interior of dry forest (McGoogan, 2011), which exceeds the 100–125 m distances measured at other tropical forest sites (Harper *et al.*, 2005). Thus, the combination of hot, dry climate with seasonal opening of the canopy should result in strong abiotic effects in western dry forests. Specifically, conditions in the open, arid matrix should result in higher wind speed, temperature, and light levels as well as reduced humidity in edge habitats (Gehlhausen *et al.*, 2000).

There are few data on animal responses to edge effects in Madagascar, and most of them are for species in humid or littoral forests. The sensitivity of some reptiles and frugivorous birds to edge habitats indicates increased extinction probabilities in littoral forests (Lehtinen *et al.*, 2003; Watson *et al.*, 2004). Data on lemur responses to edge effects comes predominantly from studies of large-bodied, diurnal species (Lehman *et al.*, 2006b; Lehman, 2007; Schwitzer *et al.*, 2007; Irwin, 2008; Quéméré *et al.*, 2010; McGoogan, 2011; Brenneman *et al.*, 2012). For example, Quéméré *et al.* (2010) documented cryptic genetic clusters in golden-crowned sifakas (*Propithecus tattersalli*) that were associated with individuals exploiting forest edges, which may indicate that edge tolerance is a key feature for successful dispersal between fragments. Of the two cheirogaleids (*Cheirogaleus major* and *Microcebus rufus*) studied to date for their density responses to forest edges in eastern humid forests, *C. major* exhibited reduced densities in edge habitats, whereas *M. rufus* had significantly higher densities in edge habitats (Lehman *et al.*, 2006b, 2006c). Consequently, *C. major* was classified as having a negative edge response and *M. rufus* having a positive edge response. These differential density responses likely reflected spatial variations in the abundance of preferred food items for frugivores (*C. major*) and omnivores that supplement their diet with insects (*M. rufus*) between edge and interior habitats (Ries and Sisk, 2004). The question arises then as to responses of nocturnal lemurs to edge effects in tropical dry forests in Madagascar.

There are some data on the effects of habitat variation on cheirogaleid biogeography in western dry forests (Radespiel and Raveloson, 2001; Rakotondravony and Radespiel, 2009; Burke and Lehman, 2014; Radespiel and Bruford, 2014). For example, Burke and Lehman (2014) conducted mark–recapture experiments on sympatric mouse lemurs (*Microcebus murinus* and *M. ravelobensis*) trapped in edge and interior forest habitats in NW Madagascar. In this study, each transect ran parallel to the forest edge, which represents a novel approach to edge detection methods in primate research because most studies have used transects perpendicular to the forest edge. Of the 34 *M. murinus* captured during their study, 82% ($N = 28$) were trapped in interior habitats whereas 72% ($N = 47$) of *M. ravelobensis* were captured in edge habitats. These preliminary data indicate that *M. murinus* is edge-intolerant, despite this species having a large geographic range across a highly fragmented forest landscape subject to presumably extensive edge effects (Rakotondravony and Radespiel, 2009). Conversely, *M. ravelobensis* has a considerably smaller geographic range size, prefers sites near standing bodies of water, and is listed by the IUCN as endangered (Rakotondravony and Radespiel, 2009), yet this mouse lemur may be an edge-tolerant species. Put differently, these are seemingly incongruous results of an edge-intolerant but wide-ranging *M. murinus* and edge-tolerant but rarer *M. ravelobensis* in the fragmented dry forests in NW Madagascar. If edge tolerance provided some benefit in terms of dispersing between fragments whose perimeter provided the preferred habitat for *M. ravelobensis*, then we would expect this species to have a broader distribution across the fragmented landscapes within the park. The question arises then as to differences in estimated edge responses by lemurs depending on whether researchers use transects perpendicular or parallel to the forest edge.

In this chapter, I present data on edge effects on tree dendrometrics, abiotic factors, and mouse lemur densities along edge and interior transects in western dry forests in Madagascar. Based on previous studies in other fragmented landscapes and in Madagascar, I predicted that:

(1) all dendrometrics variables will be higher in interior habitats than in edge habitat;
(2) wind speed, temperature, and light levels will be higher and humidity lower in edge habitats than in the forest interior; and
(3) *M. murinus* will exhibit a negative edge response and *M. ravelobensis* will exhibit a positive edge response.

Methods

Data were collected June–October 2007, where the continuous forest bordered a savannah at the Ampijoroa field station in Ankarafantsika National Park in NW Madagascar (46°48′E, 16°19′S). Ankarafantsika has a highly seasonal climate, with most of the average annual rainfall of 1100–1600 mm per year concentrated during the monsoon season in January and February (Rendigs *et al.*, 2003). The climate is mostly dry with mean yearly rainfall of 1000–1500 mm occurring mostly in the rainy season between November and April (McGoogan, 2011). Soils underlying forests near the station are sandy, resulting in a short canopy of 10–12 m. The plant families represented in the canopy include Bignoniaceae, Euphorbiaceae, Sapindaceae, and Anacardiaceae (Figure 24.1). In the scrub layer, there is a diversity of liana species (Asclepiadaceae) and shrubs (Leguminosae and Rubiaceae). The savannah is dominated by various grass species (Figure 24.2), particularly *Aristida barbicollis* (Bloesch, 1999), with only a few isolated trees, such as *Acridocarpus excelsus*, *Hyphaene shatan*, *Ziziphus mauritiana* and *Strychnos spinosa* (Bloesch, 1999). Although the history of the savannah grassland is poorly understood, annual fires to promote "greencut" for livestock grazing occurred annually until 1997, when the area became designated a National Park and grazing was prohibited (Bloesch, 1999). Because dry forest can degrade to savannah grasslands with a burning period of 3–4 sessions over 15 years, it is possible that the savannah in Ankarafantsika was previously contiguous dry forest habitat.

Figure 24.1 Overview of interior forest in Ampijoroa. (A black and white version of this figure will appear in some formats. For the color version, please refer to the plate section.)

Tree, abiotic, and lemur surveys followed methods used to study edge effects in SE Madagascar (Lehman *et al.*, 2006a). Five 400-m edge transects (0–400 m from edge) and five 400-m interior transects (850–1250 m from edge) were set up for lemur and botanical surveys. Interior transects were set up at the same orientation and at least 450 m from the terminus of edge transects. Thus, interior transects started at least 850 m from the forest edge, which exceeds penetration distances for edge effects on *Propithecus coquereli* food tree dendrometrics and densities measured to date at Ampijoroa (McGoogan, 2011). Each transect was marked in 10-m increments with numbered flagging tape and each ran perpendicular from the forest edge into the forest interior; three of these transects run east–west and the remaining two transects run north–south.

Abiotic data on ambient temperature (°C), percent humidity, wind speed (m/s), and light intensity (LUX) were collected between 0930 and 1430 hours at 26 locations along each transect no more than once per day except for days with rain. Following Chen *et al.* (1995), abiotic and altitude data were collected at set distances from the edge into the forest (0, 5, 10, 15, 30, 45, 60, 90, 120, 150, 180, 210, 240, 270, 300, 360, 420, 480, 540, 600, 720, 840, 960, 1080, 1200, and 1250 m). A Kestrel 4000 Weather Tracker (Nielsen-Kellerman, Nelson, PA 19013) was used to collect data on temperature, percent humidity, wind speed, and altitude. Light

Figure 24.2 Forest edge and matrix at edge effects research site in Ampijoroa. (A black and white version of this figure will appear in some formats. For the color version, please refer to the plate section.)

levels (LUX) were collected using a Cole Parmer DLM2 digital light meter (Cole Parmer, Anjou, Quebec H1J 2H9). All climate data were collected by one team member holding each instrument at breast height (ca. 140 cm), facing in the cardinal direction of the transect, and waiting until readings had stopped fluctuating. Team members always stayed at least 3 m behind the instruments during readings.

Nocturnal surveys were conducted slowly (0.5–1.0 km/h) during 1800–2100 hours, with starting points for all surveys being rotated between each end of a transect to ensure that the data were not biased. Identification of *M. murinus* required clear observations of gray/brown fur and a short tail that was thick at the base. *M. ravelobensis* identification involved animals having rufus fur and a long tail that was thin at the base. Any ambiguity in the identification of *Microcebus* species resulted in identification only to the genus level. The following data were collected whenever lemurs were seen: date, time, transect number, researcher name, distance along trail from the animal, species, sighting distance from trail at 90°, height (m) of animal, and method of detection (eye-shine, branch movement, body sighting). Botanical surveys were conducted along both sides of each transect to a depth of 1 m, for a total area sampled of 1.00 ha. For all trees over 10 cm DBH ($N = 805$), data were collected on height (m), DBH (cm), canopy height (m), canopy width (m), local name, and distance to forest edge (m).

Due to small sample sizes (Buckland *et al.*, 2010), lemur densities (# individuals/ha) were obtained by dividing the number of individuals surveyed by the total survey area in each habitat (Whitesides *et al.*, 1988). Specific-specific sighting widths for each transect were estimated using the perpendicular distance (m) from the individual or center of the group to the transect and the histogram inspection technique, with a 50% criterion for fall-off distance (Whitesides *et al.*, 1988).

General linear models (GLMs) were used to explore biotic data variations within edge and interior habitats and repeated measures GLMs were used to explore abiotic data variations within edge and interior habitats. Thus, GLMs investigated spatial variations from edge into the interior, with expectations of a linear combination of a set of observed values, with the predictors being distance from the edge within each habitat type (edge and interior). Generalized linear models (GeLMs) were used to determine the effects of habitat variation (i.e., edge vs. interior transects) on spatial variations in tree dendrometrics (DBH, tree height, canopy height, canopy width, canopy volume) and abiotic data (wind speed, temperature, humidity, and light). GeLMs were chosen because of the common pattern of differences in the error distribution (i.e., not a normal distribution) between the stochastic parameters in edge habitats compared to the more monotonic patterns found in interior habitats (Fortin and Dale, 2005). For example, many abiotic conditions, such as wind speed, vary considerably near the edge, but then become increasingly monotonic moving into the interior, resulting in a stronger correlation in edge habitats and no correlation in interior habitats. Consequently, comparing standard linear models between edge and interior habitats can increase chances of type I errors (rejection of a true null hypothesis) when careful considerations are not of the non-normal data distributions within and between each population (edge or interior). One fixed factor (edge or interior habitat) representing the dependent variable and one random covariate (transect ID), as a diagnostic test for spatial patterns other than those associated with edge effects, were used in full models of all abiotic and biotic parameters. I assumed that edge effects penetrated 500 m into the forest interior, following analyses of plot data on plant diversity and dendrometrics undertaken by McGoogan (2011) in the same area. Mann–Whitney *U* tests were used to determine if there were any differences in detection distances between habitats (i.e., edge vs. interior) for each species as well as overall perpendicular detection distances between the three mouse lemur categories (*M. murinus* and *Microcebus* sp. as well as *M. ravelobensis* and *Microcebus* sp.). The rationale for this last test was to determine whether failure to determine species determinations (or lack thereof) was a result of increasing perpendicular distances (i.e., animals further from transect were less likely to be identified to the species level). All analyses were conducted using SPSS ver. 22 (IBM) and the alpha level was set at $P < 0.05$.

Results

A total of 51 surveys were conducted at Ampijoroa, resulting in higher stem numbers in interior habitats ($N = 523$) than in edge habitats ($N = 282$). Diagnostic GLMs

Table 24.1 General linear models comparing tree dendrometrics and abiotic data within edge and interior transects at Ampijoroa. For each variable, *F* refers to the ANOVA value and *P* for the associated significance value in each GLM model.

		Edge transects (N = 5)		Interior transects (N = 5)	
Category	Variable	F	P	F	P
Dendrometric	DBH (cm)	6.14	0.014	0.37	0.546
	Height (m)	21.76	0.000	0.01	0.912
	Canopy height (m)	14.88	0.000	1.91	0.168
	Canopy width (m)	39.93	0.000	9.77	0.002
	Canopy volume (m^3)	12.56	0.000	6.98	0.009
Abiotic	Wind speed (m/s)	6.64	0.010	2.54	0.111
	Temperature (°C)	24.14	0.000	1.52	0.218
	Humidity (%)	9.02	0.003	3.68	0.055
	Light (LUX)	100.58	0.000	0.11	0.744

indicated marked intrahabitat variations in tree dendrometrics and abiotic data for edge transects at Ampijoroa (Table 24.1). Specifically, all dendrometric and abiotic data varied significantly among the five edge transects. Conversely, all but two dendrometrics variables (canopy width and volume in interior habitats) did not display intertransect variations in interior habitats.

GeLMs indicated that six of the nine dendrometric and abiotic variables differed between edge and interior habitats (Table 24.2). For dendrometrics, DBH did not differ between edge and interior habitats, although tree height was significantly greater in interior habitats. Canopy width and canopy volume were greater in edge habitats than in interior habitats. Thus, although there were more, taller trees along interior transects, trees had generally larger canopies along edge transects. For abiotic data, wind speed and light levels were significantly higher in interior habitats than in edge habitats. Conversely, humidity was significantly higher in edge habitats than interior habitats. There were no temperature differences between edge and interior habitats.

Of the 116 sightings made of *Microcebus* along the transects, 41.3% (N = 48) were of *M. murinus*, 12.0% (N = 14) were of *M. ravelobensis*, and 46.5% (N = 54) were made on individuals that could not be described to the species level (Table 24.3). There were no significant differences in perpendicular detection distances between edge and interior transects for *M. murinus* (U = 267.0, Z = –0.275, P = 0.78) or *M. ravelobensis* (U = 27.0, Z = 1.05, P = 0.35). Moreover, there were no significant differences in detection distance between *M. murinus* and *Microcebus* sp. (U = 3260.0, Z = 1.73, P = 0.08) or between *M. ravelobensis* and *Microcebus* sp. (U = 879.0, Z = 0.984, P = 0.32). Thus, the survey data were not artifacts of differential detection probabilities between edge and interior habitats or due to perpendicular distance effects on species identification. Overall density for all *Microcebus* sighted during surveys was 17.2 ind./km^2 in the interior and 17.6 ind./km^2 in the edge. The density of *M. murinus* was lower for interior transects (7.8 ind./km^2) than

Table 24.2 Descriptive statistics and generalized linear models of differences in dendrometrics and abiotic data between edge and interior habitats.

		Edge		Interior		Habitat effect	
Category	Variable	Mean	SD	Mean	SD	χ^2	*P*
Dendrometric	DBH (cm)	15.81	6.97	15.23	5.49	0.05	0.818
	Height (m)	7.86	2.64	8.96	2.64	15.88	0.015
	Canopy height (m)	2.72	0.96	2.65	0.93	0.48	0.487
	Canopy width (m)	3.60	1.33	3.38	1.33	9.17	0.002
	Canopy volume (m^3)	49.27	56.28	41.99	46.17	5.29	0.021
Abiotic	Wind speed (m/s)	0.23	0.26	0.26	0.24	8.83	0.003
	Temperature (°C)	26.22	2.12	26.79	1.30	2.72	0.099
	Humidity (%)	51.42	7.03	48.45	5.35	4.92	0.026
	Light (LUX)	3.53	5.70	7.17	14.48	24.44	0.000

Table 24.3 Median detection distance and density estimates for two species of mouse lemurs between edge and interior habitats.

	N		Median detection distance ± 1 SD (m)		Density estimate (ind./km^2)	
Species	Interior	Edge	Interior	Edge	Interior	Edge
M. murinus	20	28	3.0 ± 3.5	3.0 ± 2.3	7.8	9.8
M. ravelobensis	5	9	3.0 ± 1.8	3.0 ± 1.8	3.2	2.6
All *Microcebus*[a]	61	55	5.0 ± 3.4	4.0 ± 2.9	17.2	17.6

[a] Includes all individuals seen even in the case where species designations could not be determined in the field.

edge transects (9.8 ind./km^2). For *M. ravelobensis*, density estimates were slightly higher on interior transects (3.2 ind./km^2) than on edge transects (2.6 ind./km^2).

Discussion

My first objective was to test the prediction that tree dendrometrics would be higher in interior habitats at Ampijoroa. Although trees were more numerous and significantly taller in interior than in edge habitats, canopy width and volume were higher in edge habitats. Moreover, DBH measures did not differ between edge and interior habitats. Thus, only two of the five tree variables conformed to predicted edge effects. These results for tree height generally confirm prevailing models on edge effects, in which matrix conditions are hypothesized to have a negative influence on plant biology and tree growth near the edge (Esseen and Renhorn, 1998; Didham and Lawton, 1999; Mesquita *et al.*, 1999). For example, cyclonic winds can and do destroy large trees in Madagascar (Ratsimbazafy, 2002). Lehman *et al.* (2006c) reported similar results in that trees were significantly shorter near forest edges in eastern humid forests. However, the lack of an edge response for tree DBH is difficult to explain, as this result differs from comparable studies in tropical and temperate forests (Donovan *et al.*, 1997; Lehman *et al.*, 2006c; Irwin, 2008). One possible explanation is that the broad definition of edge habitat extending from 0 to 400 m into the forest may obscure subtle spatial variations in tree dendrometrics. Causal factors relating to edge effects on the canopy differences between edge and interior habitats are difficult to rationalize, as there are few data on these variables for any edge effects studies, let alone those in western dry forests. Thus, causal factors influencing edge effects on tree dendrometrics requires further research in western dry forests.

My second objective was to test the prediction of higher wind speed, temperature, and light levels as well as lower humidity in edge habitats than in the forest interior at Ampijoroa. I did not confirm this prediction because abiotic patterns showed more spatial variability than tree dendrometrics, with humidity being higher in the edge and two other variables (wind speed and light levels) being relatively lower in the edge than in the forest interior. Temperature showed no variation between edge and interior transects. Although it would be ideal to seek correlates of tree dendrometrics to the abiotic data, the collection methods are not comparable, as the abiotic data were not collected continuously along each transect. In general, the abiotic patterns I documented in edge habitats are similar to those collected in edge habitats that bordered on open woodlands, where the proximity of trees mitigated edge effects in nearby continuous forests (Gascon *et al.*, 1999; Debinski, 2006; Nascimento *et al.*, 2006). Matrix composition can greatly influence edge effects in adjacent edge habitats by altering wind shear and abiotic conditions, providing rapidly changing microhabitats due to plant colonization, supplementary food resources for animals (particularly for agricultural plots abutting forest edges), and novel predation pressures and seed dispersal patterns (Debinski, 2006). Matrix conditions are particularly relevant to changing abiotic and biotic edge conditions when the matrix is composed of rotating patterns of agriculture and then fallow fields dominated by rapidly growing colonizing plant species, such as those seen and studied in eastern humid forests in Madagascar (Lehman *et al.*, 2006b). However, the matrix is composed of open grassland near forest edges at Ampijoroa (Figure 24.2), which should have resulted in a monotonic interaction with edge habitats. These results are all the more enigmatic given that studies in eastern humid forests and littoral forests in SE Madagascar produced abiotic data matching predictions from prevailing edge models (Lehtinen *et al.*, 2003; Lehman *et al.*, 2006b). For example, Lehtinen *et al.* (2003) documented significantly higher temperature and wind speed as well as lower relative humidity in the edge compared to 50 m into the forest in littoral forest fragments at Mandena. Thus, dendrometric and abiotic edge effects are clearly stochastic ecological phenomena during the time period of my study.

My final objective was to determine if *M murinus* would exhibit a negative edge response and *M. ravelobensis* would exhibit a positive edge response. Due to the small sample sizes for both species, particularly for *M. ravelobensis*, any conclusions on their edge responses should be viewed with caution. In general, however, density estimates for both species did not support predictions based on previous studies of conspecifics by Burke and Lehman (2014) using transects parallel to the forest edge. Moreover, Burke and Lehman (2014) were able to amass a larger data set for both species with fewer transects in a shorter time period, leading to a preliminary conclusion of greater research efficiency when using transects running parallel to the forest edge. Another possible explanation for the differences in edge responses between the two studies is that temporal differences in local ecology influenced the results. The data reported here were collected in 2007 whereas those for Burke and Lehman (2014) were collected in 2012, resulting in data sets representing ecologies before and after the 2009 political crisis in Madagascar. This political crisis resulted in the suspension of numerous conservation projects, which then led to widespread exploitation of endangered lemurs and hardwoods (Schwitzer *et al.*, 2014). For example, presumed increases in lemur hunting pressures have led to the disappearance of many large-bodied lemurs, such as *P. coquereli* and *Eulemur mongoz*, in edge habitats at Ampijoroa (Valenta, Lehman, Ross, personal observations). The resulting disruption in resource competition for food resources, such as fruit, shared among the mouse lemurs and larger lemurs may have created a dynamic ecological system, similar to that reported as density compensation by Peres and Dolman (2000) in Amazonia and the Guianas. In this hypothesis, the loss of large-bodied competitors results in a biomass release for smaller-bodied taxa as they gained greater, non-competitive access to fruits.

The consistent, pervasive variations in biotic and abiotic data within edge transects created an unexpected analytical issue complicating my study of edge effects at Ampijoroa. These edge habitats are highly stochastic and thus provide some compelling, preliminary support for investigation of additive edge effects. Put differently, edge habitats are likely experiencing what amounts to additivity from nearby edges (Malcolm, 1994; Fletcher, 2005; Ewers *et al.*, 2007), which implies that edge effects are exacerbated through the proximity of multiple nearby edges (Malcolm, 2001). Under this model of edge additivity, undulating edge borders and changes in fragment shape can result in complicated edge effects. For example, Fernandez *et al.* (2002) found it difficult to find identical gradients of nearest-edge relationships in irregularly shaped fragments. Testing for edge additivity could be done in a more fragmented landscape, such as those found in the border regions of Ankarafantsika National Park. If these results are consistent across the seasons and repeatable in future studies, then researchers should recognize the possibility that edge effects and the landscape patterns that produce these ecological phenomena may be site- or region-specific and thus highly variable between eastern humid and western dry forests in Madagascar.

Acknowledgments

I thank the Madagascar National Parks, Ministère de L'Eau et de Forêt, L'Office National pour l'Environnement à Madagascar, and University of Antananarivo for permission to conduct my research in Madagascar. I am extremely grateful to my former Master's student, Courtney Sendall, for kindly allowing me to use data she collected in part to complete the field component of her degree. I thank Dr. Patricia Wright, Dr. Benjamin Andriamahaja as well as the staff at ICTE and MICET for their support, advice, and hospitality. I also thank the reviewers and editor for their insightful comments on a draft manuscript. I gratefully acknowledge Prof. Drs. Elke Zimmermann, Ute Radespiel, and their students for collegial support of research conducted by my students and me at Ampijoroa. This research was supported in part by a NSERC Discovery Grant, NSERC CGS-M, Connaught Foundation, and a Conservation Grant from the American Society of Primatology.

References

Bloesch U. 1999. Fire as a tool in the management of a savanna/dry forest reserve in Madagascar. *Applied Vegetation Science* 2(1):117–124.

Brenneman RA, Johnson SE, Bailey CA, *et al.* 2012. Population genetics and abundance of the Endangered grey-headed lemur *Eulemur cinereiceps* in south-east Madagascar: assessing risks for fragmented and continuous populations. *Oryx* 46(02):298–307.

Brown KA, Johnson SE, Parks KE, *et al.* 2013. Use of provisioning ecosystem services drives loss of functional traits across land use intensification gradients in tropical forests in Madagascar. *Biological Conservation* 161:118–127.

Buckland S, Plumptre A, Thomas L, Rexstad E. 2010. Design and analysis of line transect surveys for primates. *International Journal of Primatology* 31(5):833–847.

Burke RJ, Lehman S. 2014. Edge effects on morphometrics and body mass in two sympatric species of mouse lemurs in Madagascar. *Folia Primatologica* 85:277–291.

Chen J, Franklin JF, Spies TA. 1992. Vegetation responses to edge environments in old-growth Douglas-fir forests. *Ecological Applications* 2(4):387–396.

Chen J, Franklin JF, Spies TA. 1995. Growing-season microclimatic gradients from clearcut edges into old-growth Douglas-fir forests. *Ecological Applications* 5:74–86.

Crowley BE, McGoogan KC, Lehman SM. 2012. Edge effects on foliar stable isotope values in a Madagascan tropical dry forest. *PLoS ONE* 7(9):e44538.

Debinski DM. 2006. Forest fragmentation and matrix effects: the matrix does matter. *Journal of Biogeography* 33(10):1791–1792.

Didham RK, Lawton JH. 1999. Edge structure determines the magnitude of changes in microclimate and vegetation structure in tropical forest fragments. *Biotropica* 31(1):17–30.

Donovan TM, Jones PW, Annand EM, Thompson III FR. 1997. Variation in local-scale edge effects: mechanisms and landscape context. *Ecology* 78(7):2064–2075.

Du Puy D, Moat J. 1998. Vegetation mapping and classification in Madagascar (using GIS): implications and recommendations for the conservation of biodiversity. In CR Huxley, JM Lock, DF Cutler (eds.), *Chorology, Taxonomy & Ecology of the Floras of Africa and Madagascar* (pp. 97–117). Royal Botanic Gardens, Kew.

Du Puy D, Cribb P, Bosser J, Hermans J, Hermans C. 1999. *The Orchids of Madagascar.* Royal Botanic Gardens, Kew.

Du Puy DJ, Moat J. 1996. A refined classification of the primary vegetation of Madagascar based on the underlying geology: using GIS to map its distribution and to assess its conservation status. In WR Lourenço (ed.), *Proceedings of the International Symposium on the Biogeography of Madagascar* (pp. 205–218). Editions de l'ORSTOM, Paris.

Elmqvist T, Pyykönen M, Tengö M, *et al.* 2007. Patterns of loss and regeneration of tropical dry forest in Madagascar: the social institutional context. *PLoS ONE* 2(5):e402.

Esseen P-A, Renhorn K-E. 1998. Edge effects on an epiphytic lichen in fragmented forests. *Conservation Biology* 12(6):1307–1317.

Ewers RM, Thorpe S, Didham RK. 2007. Synergistic interactions between edge and area effects in a heavily fragmented landscape. *Ecology* 88(1):96–106.

Fahrig L, Merriam G. 1994. Conservation of fragmented populations. *Conservation Biology* 8(1):50–59.

FAO. 2003. *The State of the World's Forests*. Food and Agricultural Organization, United Nations, Rome.

Fernandez C, Acosta FJ, Abella G, Lopez F, Diaz M. 2002. Complex edge effect fields as additive processes in patches of ecological systems. *Ecological Modeling* 149:273–283.

Fletcher RJ. 2005. Multiple edge effects and their implications in fragmented landscapes. *Journal of Animal Ecology* 74(2):342–352.

Fortin M-J, Dale, MRT. 2005. *Spatial Analysis: A Guide for Ecologists*. Cambridge University Press, Cambridge.

Ganzhorn JU, Malcolmber A, Adrianantoanina O, Goodman SM. 1997. Habitat characteristics and lemur species richness in Madagascar. *Biotropica* 29(3):331–343.

Ganzhorn JU, Wright PC, Ratsimbazafy HJ. 1999. Primate communities: Madagascar. In JG Fleagle, CH Janson, K Reed (eds.), *Primate Communities* (pp. 75–89). Cambridge University Press, Cambridge.

Gascon C, Lovejoy TE, Bierregaard RO, *et al.* 1999. Matrix habitat and species richness in tropical forest remnants. *Biological Conservation* 91:223–229.

Gehlhausen SM, Schwartz MW, Augspurger CK. 2000. Vegetation and microclimatic edge effects in two mixed-mesophytic forest fragments. *Plant Ecology* 147:21–35.

Gibson L, Lynam AJ, Bradshaw CJA, *et al.* 2013. Near-complete extinction of native small mammal fauna 25 years after forest fragmentation. *Science* 341(6153):1508–1510.

Green GM, Sussman RW. 1990. Deforestation history of the eastern rain forests of Madagascar from satellite images. *Science* 248(4952):212–215.

Harper KA, Macdonald SE, Burton P, *et al.* 2005. Edge influence on forest structure and composition in fragmented landscapes. *Conservation Biology* 19(3):1–15.

Irwin MT. 2008. Diademed sifaka (*Propithecus diadema*) ranging and habitat use in continuous and fragmented forest: higher density but lower viability in fragments? *Biotropica* 40(2):231–240.

Jenkins RKB, Brady LD, Bisoac M, Rabearivony J, Griffiths RA. 2003. Forest disturbance and river proximity influence chameleon abundance in Madagascar. *Biological Conservation* 109:407–415.

Laurance WF, Yensen E. 1991. Predicting the impacts of edge effects in fragmented habitats. *Biological Conservation* 57(2):205–219.

Laurance WF, Laurance SG, Ferreira LV, *et al.* 1997. Biomass collapse in Amazonian forest fragments. *Science* 278:1117–1118.

Lehman SM. 2007. Spatial variations in *Eulemur fulvus rufus* and *Lepilemur mustelinus* densities in Madagascar. *Folia Primatologica* 78(1):46–55.

Lehman SM. 2013. Effects of altitude on the conservation biogeography of lemurs in south east Madagascar. In S Gursky, A Krzton, N Grows (eds.), *High Altitude Primates* (pp. 3–22). Springer Press, New York.

Lehman SM, Rajaonson A, Day S. 2006a. Lemur responses to edge effects in the Vohibola III Classified Forest, Madagascar. *American Journal of Primatology* 68(3):293–299.

Lehman SM, Rajoanson A, Day S. 2006b. Edge effects and their influence on lemur distribution and density in southeast Madagascar. *American Journal of Physical Anthropology* 129(2):232–241.

Lehman SM, Rajoanson A, Day S. 2006c. Edge effects on the density of *Cheirogaleus major*. *International Journal of Primatology* 27(6):1569–1588.

Lehtinen RM, Ramanamanjato J-B, Raveloarison JG. 2003. Edge effects and extinction proneness in a herpetofauna from Madagascar. *Biodiversity & Conservation* 12(7):1357–1370.

Lovejoy TE, Bierregaard RO, Rylands AB, *et al.* 1986. Edge and other effects of isolation on Amazon forest fragments. In ME Soulé (ed.), *Conservation Biology: the Science of Scarcity and Diversity* (pp. 257–285). Sinauer Associates, Sunderland, MA.

Lowry PP, Schatz GE, Phillipson PB. 1997. The classification of natural and anthropogenic vegetation in Madagascar. In SM Goodman, BD Patterson (eds.), *Natural Change and Human Impact in Madagascar* (pp. 93–123). Smithsonian Institution Press, Washington, DC.

Malcolm JR. 1994. Edge effects in central Amazonian forest fragments. *Ecology* 75(8):2438–2445.

Malcolm JR. 2001. Extending models of edge effects to diverse landscape configurations, with a test case from the Neotropics. In RO Bierregaard, C Gascon, TE Lovejoy, RM Mesquita (eds.), *Lessons from Amazonia: The Ecology and Conservation of a Fragmented Forest* (pp. 346–357). Yale University Press, New Haven.

McGoogan K. 2011. Edge effects on Coquerel's sifaka (*Propithecus coquereli*) in northwest Madagascar. Unpublished PhD, University of Toronto.

Mesquita RM, Delamônica P, Laurance WF. 1999. Effect of surrounding vegetation on edge-related tree mortality in Amazonian forest fragments. *Biological Conservation* 91(2–3):129–134.

Mills LS. 1995. Edge effects and isolation: red-backed voles on forest remnants. *Conservation Biology* 9(2):395–403.

Murcia C. 1995. Edge effects in fragmented forests: implications for conservation. *Trends in Ecology and Evolution* 10(2):58–62.

Nascimento HE, Andrade AC, Camargo JL, *et al.* 2006. Effects of the surrounding matrix on tree recruitment in Amazonian forest fragments. *Conservation Biology* 20(3):853–860.

Peres CA, Dolman PM. 2000. Density compensation in neotropical primate communities: evidence from 56 hunted and nonhunted Amazonian forests of varying productivity. *Oecologia* 122(2):175–189.

Phillipson PB. 1994. Madagascar. In D Davis, VH Heywood, AC Hamilton (eds.), *Centres of Plant Diversity* (pp. 271–281). WWF and IUCN, Gland.

Phillipson PB. 1996. Endemism and non-endemism in the flora of south-west Madagascar. In WR Lourenço (ed.), *Biogeography of Madagascar* (pp. 125–136). ORSTROM, Paris.

Quéméré E, Louis E, Ribéron A, Chikhi L, Crouau-Roy B. 2010. Non-invasive conservation genetics of the critically endangered golden-crowned sifaka (*Propithecus tattersalli*): high diversity and significant genetic differentiation over a small range. *Conservation Genetics* 11(3):675–687.

Radespiel U, Bruford MW. 2014. Fragmentation genetics of rainforest animals: insights from recent studies. *Conservation Genetics* 15(2):245–260.

Radespiel U, Raveloson H. 2001. Preliminary study on the lemur communities at three sites of dry deciduous forest in the Réserve Naturelle d'Ankarafantsika. *Lemur News* 6:22.

Rakotondravony R, Radespiel U. 2009. Varying patterns of coexistence of two mouse lemur species (*Microcebus ravelobensis* and *M. murinus*) in a heterogeneous landscape. *American Journal of Primatology* 71(11):928–938.

Ramanamanjato JB. 2001. Fragmentation effects on reptile and amphibian diversity in the littoral forest of southeastern Madagascar. In G Rheinwald (ed.), *Isolated Vertebrate Communities in the Tropics* (pp. 299–310). Museum Alexander Koenig, Bonn.

Ramanamanjato JB, Ganzhorn JU. 2001. Effects of forest fragmentation, introduced *Rattus rattus* and the role of exotic tree plantations and secondary vegetation for the conservation of an endemic rodent and a small lemur in littoral forests of southeastern Madagascar. *Animal Conservation* 4(2):175–183.

Randrianambinina B, Rakotondravony D, Radespiel U, Zimmermann E. 2003. Seasonal changes in general activity, body mass and reproduction of two small nocturnal primates: a comparison of the golden brown mouse lemur (*Microcebus ravelobensis*) in Northwestern Madagascar and the brown mouse lemur (*Microcebus rufus*) in Eastern Madagascar. *Primates* 44(4):321–331.

Ratsimbazafy J. 2002. On the brink of extinction and the process of recovery: responses of black-and-white ruffed lemurs (*Varecia variegata variegata*) to disturbance in Manombo Forest, Madagascar. Unpublished PhD dissertation, Stony Brook University, Stony Brook, NY.

Rendigs A, Radespiel U, Wrogemann D, Zimmermann E. 2003. Relationship between microhabitat structure and distribution of mouse lemurs (*Microcebus* spp.) in Northwestern Madagascar. *International Journal of Primatology* 24(1):47–64.

Ries L, Sisk TD. 2004. A predictive model of edge effects. *Ecology* 85(11):2917–2926.

Schatz GE. 2001. *Generic Tree Flora of Madagascar.* Royal Botanic Gardens and Missouri Botanical Garden Press, Kew and St. Louis.

Schwitzer C, Mittermeier RA, Johnson SE, *et al.* 2014. Averting lemur extinctions amid Madagascar's political crisis. *Science* 343(6173):842–843.

Schwitzer N, Randriatahina GH, Kaumanns W, Hoffmeister D, Schwitzer C. 2007. Habitat utilization of blue-eyed black lemurs, *Eulemur macaco flavifrons* (Gray, 1867), in primary and altered forest fragments. *Primate Conservation* 22:79–87.

Watson JEM, Whittaker RJ, Dawson TP. 2004. Habitat structure and proximity to forest edge affect the abundance and distribution of forest-dependent birds in tropical coastal forests of southeastern Madagascar. *Biological Conservation* 120(3):311–327.

Whitesides GH, Oates JF, Green SM, Kluberdanz RP. 1988. Estimating primate densities from transects in a West African rain forest: a comparison of techniques. *Journal of Animal Ecology* 57:345–367.

25 Factors determining *Microcebus* abundance in a fragmented landscape in Ankarafantsika National Park, Madagascar

Travis S. Steffens and Shawn M. Lehman

Introduction

A fundamental issue in conservation biogeography is determining how and why population density (the number of individuals per unit area) and abundance (number of individual animals) vary across the geographic range of plant and animal species (Whittaker *et al.*, 2005). Understanding these relationships is critical because they provide information on population dynamics and extinction probabilities (Wilcox and Murphy, 1985; Lima and Zollner, 1996). For example, Davidson *et al.* (2009) conducted a meta-analysis of 4500 mammal species and found that population density was one of the main predictors of extinction risk. Consequently, researchers have explored numerous covariates to primate density and abundance, including food quality and amount (Stevenson, 2001; Chapman *et al.*, 2004), temporal distribution of key food resources (Terborgh, 1986), habitat structure such as basal area, stem density, tree height and size (Rovero and Struhsaker, 2007; Pozo-Montuy *et al.*, 2011; Grow *et al.*, 2013), habitat quality such as plant productivity (Janson and Chapman, 1999), rainfall and temperature (Pinto *et al.*, 2009), distribution (Harcourt and Doherty, 2005), fragment area and isolation (Anzures-Dadda and Manson, 2007), and anthropogenic disturbance (Peres, 1990; Ganzhorn and Schmid, 1998; Anzures-Dadda and Manson, 2007; Peres and Palacios, 2007). Despite the wide range of species and habitats studied, few consistent patterns have emerged that explain spatial variations in primate density and abundance. Moreover, those patterns that have found statistical support tend to have been focused on large-bodied, diurnal species, leaving many unanswered questions on the conservation biogeography of small-bodied, nocturnal taxa (McGoogan *et al.*, 2007).

The Dwarf and Mouse Lemurs of Madagascar: Biology, Behavior and Conservation Biogeography of the Cheirogaleidae, eds. S.M. Lehman, U. Radespiel, E. Zimmermann. Published by Cambridge University Press. © Cambridge University Press 2016.

Food availability

Food availability typically refers to temporal and spatial variations in food quantity and quality, and has long been considered one of the main drivers of primate abundance and density. Despite extensive investigations into the relationship between food quality and amount and primate density and abundance (Hanya and Chapman, 2012), it is often difficult to determine quantitatively rigorous measures of food availability. This issue arises due to the high plant diversity at many primate research sites in the tropics, as well as stochastic variations in the temporal availability of food resources. Many researchers have employed plant dendrometrics as proxies for food availability, such as diameter at breast height (DBH), basal area, and stem density (e.g., Wieczkowski, 2004; Anzures-Dadda and Manson, 2007; Grow *et al.*, 2013). However, comparative ecological studies have found contrasting results in terms of how these proxies relate to primate biogeography. For example, basal area has been both negatively and positively correlated with colobine and mangabey abundance (Wieczkowski, 2004; Rovero and Struhsaker, 2007), while Rovero and Struhsaker (2007) found no relationship in Syke's monkeys (*Cercopithecus albogularis*) in Africa. In the Neotropics, the density of understory trees was negatively associated with the abundance (mean detections per day) of *Capucinus melamura*, while no relationship was observed with *C. donacophilus*. Tree density was associated with increased abundance in mangabeys (Wieczkowski, 2004) and pygmy tarsier density (Grow *et al.*, 2013). Canopy height has been correlated with howler monkey abundance in fragments (Anzures-Dadda and Manson, 2007) and within matrix elements (Pozo-Montuy *et al.*, 2011). Grow *et al.* (2013) found that pygmy tarsier (*Tarsius pumilus*) density was lower in high-altitude versus low-altitude forest and that this was associated with reduced tree size. Other researchers have found no relationship between primate density and abundance and forest structure in Africa or SE Asia (Teelen, 2007; Paciulli, 2010).

The most consistent pattern that has arisen in studies of the effects of food availability on primate density and distribution has been in folivores and, more recently, insectivorous primates (Hanya and Chapman, 2012). Specifically, folivore density seems to be strongly affected by protein–fiber ratios (Chapman *et al.*, 2004), with greater density levels in areas that provide animals with leaves containing more protein and less fiber. Few studies have looked at how food quality and abundance affect non-folivores and frugivores, although the density of some frugivores is strongly affected by increased fruit production and abundance of preferred plant foods (Stevenson, 2001). The abundance and density of insectivorous primates may be related to food abundance and quality. For example, Grow *et al.* (2013) found that the distribution of *T. pumilus* was associated with higher insect abundance along forest edges than in forest interior.

Habitat loss and fragmentation

Habitat loss and fragmentation represent two of the greatest threats to biodiversity maintenance in the tropics. Although there is a strong link between habitat loss and habitat fragmentation (Fahrig, 2003), the impact of habitat loss and fragmentation on primate species is poorly understood (Arroyo-Rodriguez and Dias, 2010). Biogeographers often invoke the abundance–area relationship to describe how reduced habitat area results in a decrease in population in many different species including birds, mammals, and insects (Connor *et al.*, 2000). Primate density has been shown to be influenced by habitat area. For example, Chiarello and de Melo (2001) documented that primate density was higher in larger than in smaller fragments. Habitat fragmentation can also increase abiotic and biotic edge effects that can secondarily alter the structure and composition of forest edges along habitat patches (Murcia, 1995; Harper *et al.*, 2005). The resulting edge effects change the structural composition of forest patches by impacting the size and height of trees, stem density, and basal area (Murcia, 1995; Lehman, 2006). These changes in forest composition influence food availability, which can alter species diversity and abundance (Stevenson, 2001; Wieczkowski, 2004; Hanya and Chapman, 2012).

Many researchers invoke Island Biogeography Theory when formulating hypotheses on animal responses to changes in fragment size and isolation. Island biogeography theory states that as islands (fragments) of habitat become more isolated the potential for a species to colonize decreases (MacArthur, 1967). Therefore, increasing isolation should reduce the number of individuals that are capable of immigrating into a habitat patch affecting both primate density and abundance. Conversely, small but less-isolated habitat patches may suffer from overcrowding. For example, *Alouatta* spp. have been found at high population densities in small fragments and in some cases within individual trees (Anzures-Dadda and Manson, 2007; Pozo-Montuy *et al.*, 2011). In a fragmented landscape in Mexico, Pozo-Montuy *et al.* (2011) found that isolated matrix elements closer to fragments of forest (less isolated) had higher *Alouatta pigra* density than matrix elements further from fragments of forest (more isolated). How long a patch has been isolated can impact primate density and abundance. For example, Michalski and Peres (2007) found that patch isolation age had a negative effect on the abundance of *Callicebus moloch* (relative density).

Anthropogenic disturbance

Anthropogenic disturbance can impact primate abundance and density. Deforestation and agricultural activities are widely cited as drivers of habitat loss and fragmentation (Skole and Tucker, 1993; Gade, 1996; Laurance *et al.*, 2011). Further issues arise in areas where habitats are altered by logging or fire. Primate responses to these anthropogenic disturbances are complex and often population-specific, although there are some general patterns that emerge as a function of primate body size. In the Neotropics, the density and distribution of large-bodied primates, such as *Ateles* spp., are negatively impacted by hunting (Peres, 1990; Peres and Palacios, 2007). Conversely, smaller primate species, such as *Saguinus* spp., may show positive associations with hunting (Peres and Palacios, 2007), which may be the result of density compensation by small to medium primate species when larger species are preferentially removed from an ecosystem (Peres and Dolman, 2000). The impact of logging has been associated with both positive and negative affects on primate density and abundance. In the Budongo Forest Reserve, Uganda, Plumptre and Reynolds (1994) found higher densities of *Cercopithecus mitis*, *C. ascanius* and *Colobus guereza* in logged rather than unlogged areas. In another example, Remis and Robsinson (2012) found that the abundance of monkeys was significantly related to logging and not hunting in 2002 and then the opposite relationship in 2009 in the Congo Basin forest.

Density and abundance in mouse lemurs

Microcebus spp. are the smallest primates in the world, with adult body mass ranging from 30.6 to 77.8 g, and tend to occur at high population abundance and densities relative to sympatric, diurnal lemurs (Robinson and Redford, 1986; Fa and Purvis, 1997; Smith and Jungers, 1997). *Microcebus* spp. consume fruits, flowers, arthropods, insect secretions, gum, small vertebrates, and leaves (Corbin and Schmid, 1995; Atsalis, 1999; Lahann, 2007; Dammhahn and Kappeler, 2008). Although they are highly frugivorous, it is most appropriate to classify *Microcebus* as fauni-frugivorous because of the degree to which they consume insects and small vertebrates (Lahann, 2007). For example, the availability of insect secretions influences the ranging behavior of *M. murinus* (Corbin and Schmid, 1995). Although we would expect *Microcebus* density and abundance to be related to fruit production in similar ways as in other frugivores/insectivores, their reliance on insects, insect secretions, and gum may buffer them when fruit production is low (Dammhan and Kappeler, 2008).

Spatial and temporal variations in *Microcebus* biogeography have been linked to habitat characteristics such as forest type (Ganzhorn and Schmid, 1998) and microhabitat structure (Rendigs *et al.*, 2003). Population density of *M. murinus* is often lower in secondary forest than primary forest; moreover, *Microcebus* spp. tend to have reduced body mass in secondary forests (Ganzhorn and Schmid, 1998). Ganzhorn and Schmid (1998) also found that fewer *M. murinus* females go into energy-saving torpor in secondary forests than primary forests and that *M. murinus* occurred at lower densities in secondary versus primary forest. Randrianambinina *et al.* (2010) determined that there was a higher density of *Microcebus danfossi* in fragments with greater disturbance. They argued that the higher density of *M. danfossi* was related to crowding or edge effects within the smaller fragments.

Microcebus murinus and *Microcebus ravelobensis*

M. murinus ranges through much of the remaining forest in western Madagascar (Lahann *et al.*, 2006). Although these lemurs prefer deciduous dry forest, they have been recorded in littoral rainforest in the southern portion of the island (Lahann *et al.*, 2006). *M. ravelobensis* is a recently described species found only in the discontinuous forest between the Betsiboka and Maevarano rivers of northwest

Madagascar (Zimmermann *et al.*, 1998; Guschanski *et al.*, 2007). Both species are broadly sympatric in the forested areas in and around Ankarafantsika National Park. According to Zimmerman *et al.* (1998), *M. murinus* and *M. ravelobensis* are similar in body mass (39–69 g), but each species can be distinguished based on pelage color (grayish brown in *M. murinus*, golden brown in *M. ravelobensis*), and tail morphology (longer, thinner tail in *M. ravelobensis*). There are also physiological differences between the two species: *M. murinus* stores fat in its tail and exhibits behavioral torpor during the dry season while *M. ravelobensis* does not store fat in its tail and is more active during the dry season (Zimmermann *et al.*, 1998; Thorén *et al.*, 2011).

However, much of the data regarding the conservation biogeography of *M. murinus* and *M. ravelobensis* comes from studies in continuous forest sites. Thus, a fundamental but largely unexplored question is to determine whether or not the ecological patterns observed for *M. murinus* and *M. ravelobensis* in continuous forests will hold for conspecifics in fragmented forests. Our specific aim was to investigate how the abundance and density of *M. murinus* and *M. ravelobensis* varied across a fragmented landscape in Ankarafantsika National Park in NW Madagascar. The objective of our research was to determine the causal factors that influenced *Microcebus* abundance and density.

Methods

We collected our biogeographic data between June and November of 2011 in Ankarafantsika National Park, Madagascar. Ankarafantsika National Park is approximately 135,800 ha, and consists of a mosaic of approximately 72,670 ha of dry deciduous forest (Alonso *et al.*, 2002; Razafy Fara, 2003). The surrounding matrix is a relatively homogeneous matrix of grassland savannah mostly of the species *Aristida barbicollis* (Bloesch, 1999) containing some structural elements such as trees, such as *Strychnos spinosa* (Loganiaceae). The climate is mostly dry with mean yearly rainfall of 1000–1500 mm occurring mostly in the rainy season between November and April (Alonso *et al.*, 2002). Data were collected during the dry season and early part of the wet season (June–November) to facilitate access to the survey landscape, and because there was increased visibility due to reduced foliage.

We defined abundance as the number of total individuals within a sample and density as the number of individuals per unit area. The difference between these two measures may not be immediately obvious, as they appear similar. Measures of abundance include the number of individuals, total biomass, or population size whereas measures of density are always made with reference to some unit of area; for example, individuals per square kilometer. In some studies, relative density, such as observations (individuals or groups) per kilometer of survey or time, is used as a proxy of abundance (Michalski and Peres, 2007; Rovero and Struhsaker, 2007; Teelen, 2007; Pyritz *et al.*, 2010; Randrianambinina *et al.*, 2010; Remis and Robinson, 2012). Using density as a proxy for abundance in fragmented landscapes is problematic because it is entirely possible that a small fragment may have a high density but low abundance and the inverse may be true for large fragments. We acknowledge that density and abundance are area related and that mechanisms that predict each may be similar, but we feel it is important to separate density and abundance in order to fully understand what determines each. For example, if a species occurs at low abundance but high density within a habitat fragment, the low abundance may be explained by fragment area while the high density may be explained by other factors, such as high food availability. In our brief review of the potential impacts on primate density and abundance we use the terms abundance and density as defined above. All mention of relative density will be associated with abundance and not density.

We surveyed *Microcebus* in 42 forest fragments in the landscape; however, we report results only from the 38 fragments that we were able to collect habitat characteristics for (Figure 25.1). We defined habitat fragments as patches of forest with a connected canopy, including primary and secondary vegetation, that were separated from other fragments by a clear gap in the canopy. The history of fragmentation within the landscape was not known, but investigation of topographic maps from the 1950s showed a similar spatial arrangement of fragments. Current land-use management in the landscape allows for seasonal burning and subsequent grazing by cattle of the savannah. Although there are no permanent residents within the landscape, local people create temporary settlements to graze their cattle during the wet season. Despite prohibitions against hunting lemurs, there is evidence in other areas of the park that lemurs are hunted (García & Goodman, 2003; Rabesandratana *et al.*, 2012), and hunting attempts by local people on *Eulemur* sp. and *Lepilemur* have been observed by other researchers in the vicinity of the main Ampijoroa research station (McGoogan, Ross, and Valenta, pers. comms.) and within the habitat fragments (Steffens, pers. obs.). However, there is no evidence of hunting of *Microcebus* either at the main station or in the fragmented landscape.

To determine the population density of *Microcebus* spp., a total team of 10 persons conducted early and late nocturnal surveys in each fragment. All team members received the same training on survey methods and identifying species. Early nocturnal surveys took place between 18:00 and 21:27 hours (N = 299) and late nocturnal surveys took place between 02:17 and 5:55 hours (N = 265). To ensure temporal independence for each survey, the research team only conducted one survey along each transect per 24-h period. In total we conducted 11–21 surveys in each fragment (a total of 564 surveys). To ensure there was no spatial bias we alternated which direction we surveyed each transect. Following Rakotondravony and Radespiel (2009), two researchers walked slowly together during all nocturnal surveys and each researcher focused on one side of the transect for the entire duration of the survey. Each team member used high-powered flashlights and headlamps during nocturnal surveys to observe eye shine of each species. During survey walks, each researcher walked slowly (approximately 1 km/h), visually scanning and listening for all *Microcebus* spp. When members of the survey team observed a group or individual *Microcebus* spp. they spent up to 15 min determining species

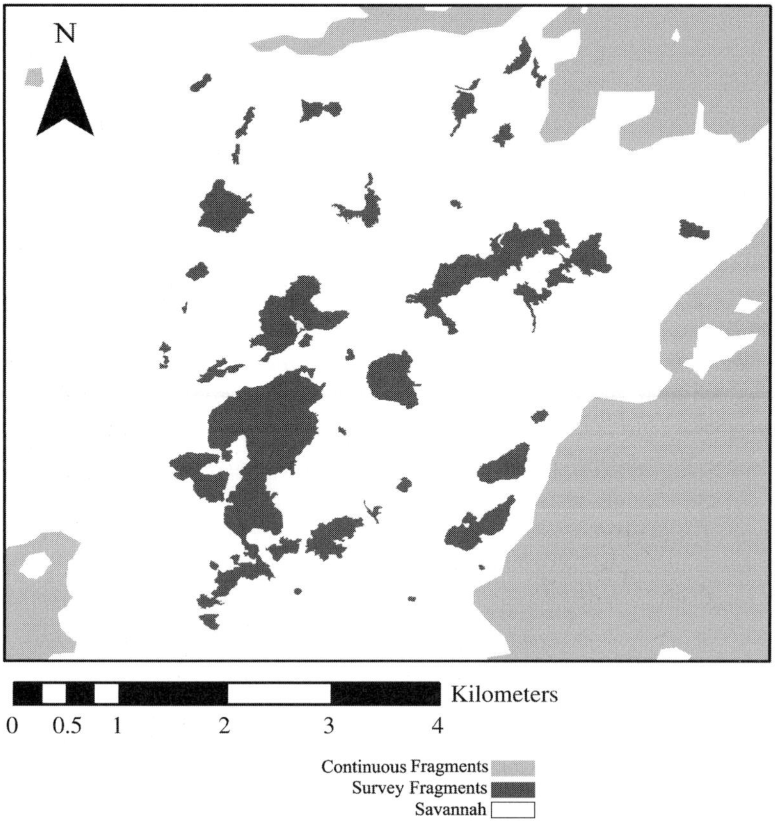

Figure 25.1 Study site landscape containing habitat fragments.

identity. Similar to Rakatonadrony and Radespiel (2009), we found that it was easy to spot both species of *Microcebus*, but difficult to systematically identify all individuals of each of the two *Microcebus* species. We used a suite of characteristics to determine the species identity of the two *Microcebus* species. A positive identification of a *M. murinus* individual required the following characteristics: gray/brown fur and a short tail that was thick at the base. A positive identification of *M. ravelobensis* required confirmation of the following characteristics: rufus fur and long tail that was thin at the base. If there was any ambiguity in the identification of *Microcebus* species to the species level, then we assigned identification to the genus level. During each *Microcebus* sighting we also recorded the following information: distance of animals from observer, angle of animal from transect, perpendicular distance of animal from the transect, height of animal, tree height where the animal was spotted, start time of sighting, end time of sighting, which researcher first observed the animal, which side of the transect the animal was observed, and activity of the animal. For each survey, we recorded the start and end time, the type of survey (diurnal or nocturnal), which side each researcher was surveying, direction of survey, and general weather.

To facilitate comparisons with existing studies, we calculated *Microcebus* population density according to Müller *et al.* (2000). Müller *et al.* (2000) calculates density as the mean number of individuals observed within the survey area ($\bar{n}$). The survey area is calculated as two times the "effective width" ($\hat{w}$) multiplied by the length of each transect. The "effective width" is the distance below which individuals can be reliably observed during surveys and this metric is determined by creating a histogram of the observer to animal distances recorded during surveys and determining at which distance there is a substantial fall-off. Based on Müller *et al.* (2000) and Rakotondravony and Radespiel (2009), we used 2-m distance classes and determined the "effective width" as the upper end of the distance class where the number of sightings fell by 2/3 of the previous distance class. We found no statistical difference between observer to animal distances among transects ($N = 38$, $F = 3.66$, $p = 0.06$). Thus, we calculated the "effective width" using pooled sightings among all transects. Population density (D) for each fragment was determined as the mean number of individuals observed along each transect for all survey walks ($\bar{n}$) divided by survey area ($2l\hat{w}$):

$$D = \frac{\bar{n}}{2l\hat{w}}.$$

To determine *Microcebus* abundance we first determined *Microcebus* density for each fragment and then multiplied *Microcebus* density by the area (A) of each fragment.

$$Ab = D * A$$

Biogeographic variables affecting *Microcebus* abundance

We measured the area of each fragment by recording a track along the perimeter of a fragment using a handheld global positioning device (Garmin GPS map 60csx) and inputting the results into QGIS (2012; $N = 37$). If obstructions prevented a complete perimeter walk ($N = 1$), we traced the perimeter on a high-resolution DigitalGlobe™ satellite image taken during the study (10/8/2011) from Google Earth™ and input the polygon in QGIS to determine its area. We placed primate survey transects along the longest axis of the fragment while going through the center of the fragment except in one fragment (12A) where the line transect was placed along the longest axis of the largest portion of the fragment.

Within ArcMap we determined isolation metrics by measuring the nearest distance from the edge of each fragment to the nearest edge of continuous forest (DCF) and by measuring the distance from the center of each fragment to the center of the nearest neighboring fragment (DNN) using the ArcMap Spatial Join tool.

Habitat variables affecting *Microcebus* abundance

We used vegetation surveys to measure habitat characteristics that may impact *Microcebus* spp. abundance within a fragment. Surveys were conducted along each primate survey transect, recording the diameter at breast height (DBH) of all trees > 5 cm DBH within 1 m of either side of the transect. We recorded the location of each tree using a handheld GPS device. We used these data to calculate four habitat variables: stem density, basal area, mean DBH, and mean tree height.

Human disturbance variables affecting *Microcebus* abundance

We used two measures of human disturbance: total human disturbance within a fragment, and the distance from each fragment to the nearest temporary settlement. We measured total human disturbance within a fragment by recording with a GPS unit the locations of all human activities within 1 m of either side of the primate survey transects: the number of cut trees, holes dug to collect *maciba* (*Dioscoreaceae maciba*) tubers for human consumption, zebu (*Bos primigenius*) dung, hunting traps, or other human disturbance variables (trails, artifacts). We did not include fire damage as a variable because of the stochastic pattern of this natural and anthropogenic parameter across the landscape. For analytical purposes, we treated all observed incidences of human disturbance as equivalent and combined them into one single human disturbance score (totaldist). We measured the distance between each fragment and the nearest temporary settlement (DNS) using ArcMap Spatial Join tool. We used the perimeter:area ratio as a proxy for the amount of edge within a fragment. Perimeter:area ratio was measured using the ArcMap Calculate Geometry tool.

Analysis

We tested for normality in our data using a Shapiro–Wilk test for normality (Shapiro and Wilk, 1965) from the "stats" package in R statistical software (R Development Core Team, 2014). We found that many variables were not normally distributed. We normalized the independent variables using natural log (ln) and square root transformations (sqrt) re-testing with a Shapiro–Wilk test to confirm normality. Our dependent variables, density and abundance, were not normally distributed, and were best described by a Poisson distribution; therefore, for regression analysis we used a least squares generalized linear model (GLM) with a Poisson (log) link function. To test for correlations between density and abundance we used a Spearman's Rho test. To compare density and abundance values between *M. murinus* and *M. ravelobensis* among fragments we used a Student's *t*-test from the "stats" package in R statistical software (R Development Core Team, 2014).

We conducted a principal component analysis (PCA) on all the independent variables as a data reduction technique (Quinn and Keough, 2002). Prior to conducting a PCA, we tested for bivariate correlations among the normally distributed independent variables using a Spearman's Rho test. We found that there was statistically significant collinearity among many of the independent variables. We then used all the components that had eigenvalues greater than one as inputs into our GLM. We then ran principal component regression analysis (PCR) with a GLM using a Poisson link function of both *Microcebus* species density and abundance against the four dominant principal components.

For any dependent variables (species-specific abundance and density) that did not perform well in the PCR analysis, we performed a hierarchical partitioning procedure using the "hier.part" package (version 1.0–4; Mac Nally, 2000) in R statistical software (R Development Core Team, 2014). This procedure does not create a predictive model; rather, it determines causal relationships between our independent and dependent variables. Hierarchical partitioning allows for the analysis of causal factors of a dependent variable in a multiple regression while accounting for issues of multicollinearity among independent variables (Olea *et al.*, 2010). Hierarchical partitioning uses least squares estimate for model fitting (Chevan and Sutherland, 1991; Mac Nally, 1996, 2000, 2002) and we chose to use the R^2 measure of goodness of fit within the hier.part package. All mean values are listed with ± 1 standard deviation.

Results

Within the survey landscape, fragments ranged in size from 0.23 to 117.7 ha with a mean size of 8.82 ± 19.6 ha. The distance from each fragment to continuous forest ranged from 17.06 to 2481.39 m with a mean distance of 994.86 ± 759.03 m. The distance between each fragment and the nearest neighbor fragment ranged from 151.54 to 1089.28 m with a mean distance of 477.52 ± 195.60 m. The distance from each fragment to nearest temporary settlement ranged from 116.84 to 2525.35 m with an overall mean distance of 1295.56 ± 694.06 m.

We measured a total of 4042 trees within the fragments. Tree abundance ranged from a low of 9 trees in fragment 29 (area = 0.71 ha) to a high of 583 trees in fragment 3 (area = 117.70 ha). There was a significant positive correlation between the number of trees measured and fragment area ($r = 0.90$, $p < 0.01$). Basal area within each fragment ranged from 1.67 to 109.53 m^2/ha with a mean of 20.33 ± 23.03 m^2/ha. Mean tree height within each fragment ranged from 4.04 to 7.81 m. Stem density within each fragment ranged from 238.1 to 3196.35/ha with a mean of 1321.65 ± 599.06/ha. Mean DBH within each fragment ranged from 5.53 to 17.89 cm (mean of 10.49 ± 2.67 cm). We documented a total of 1626 indicators of human disturbance within the fragments. The number of indicators ranged from 1 to 172 with an overall mean of 42.79 ± 41.02 indicators across the fragmented landscape.

Of the 933 *Microcebus* spp. sightings, 29% ($N = 269$) were classified as *M. murinus*, 30% ($N = 283$) as *M. ravelobensis*, and 41% ($N = 381$) were unidentified individuals. To create species-specific density and abundance estimates we calculated the proportion of the unidentified individuals that were either *M. murinus* or *M. ravelobensis* based on the proportion of known sightings by transect. We therefore assumed that the proportion of individuals we did not identify followed the same proportion of individuals we did identify within each fragment.

The abundance and density of each species varied across the landscape. The abundance of *M. murinus* ranged from 0 to 91 individuals and of *M. ravelobensis* ranged

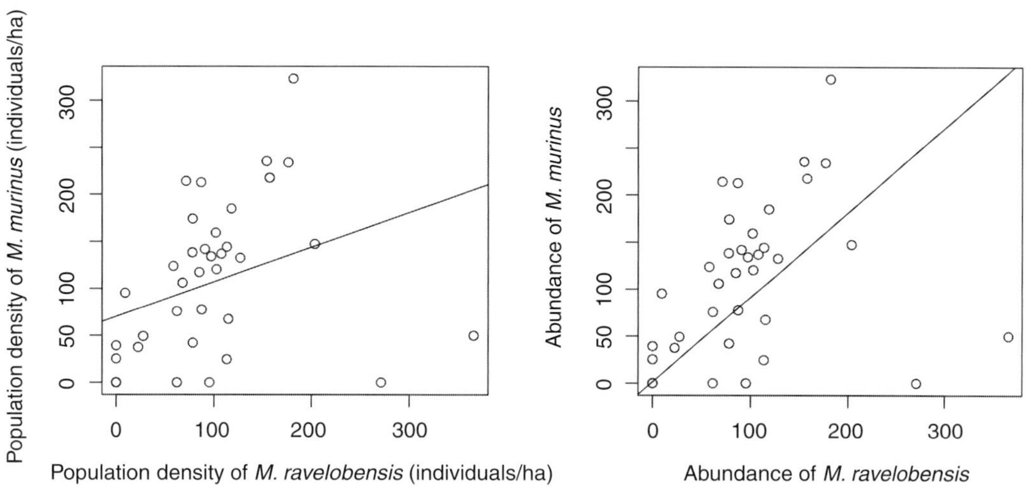

Figure 25.2 Correlation of density and abundance of conspecific mouse lemurs in a fragmented landscape.

from 0 to 103 within the fragments. There was no significant difference in mean abundance of *M. murinus* (9.41 ± 17.94) and *M. ravelobensis* (8.93 ± 19.36) in the fragments (t = 0.11, p = 0.91). The density of *Microcebus* spp. in the fragments ranged from 0–3.23 ind./ha for *M. murinus* to 0–3.66 ind./ha for *M. ravelobensis*. There was no significant difference in mean density of *M. murinus* (1.05 ± 0.83 ind./ha) and *M. ravelobensis* (0.94 ± 0.77 ind./ha) in the fragmented landscape (t = 0.60, p = 0.55). The proportion of occupied patches was greater for *M. ravelobensis* (0.92) than *M. murinus* (0.82). There were significant, positive correlations between *M. murinus* and *M. ravelobensis* for both their density (rs = 0.53, $p < 0.01$) and abundance estimates (rs = 0.87, $p < 0.01$; Figure 25.2).

The first four axes of the landscape PCAs explained 83.6% of the variation in the original variables (Table 25.1). The following plant dendrometric variables displayed negative loading on PCA axis one: basal area, mean tree height, stand density, and mean DBH. Three spatial aspects of the fragments loaded on the second PCA axis: area (+), perimeter:area ratio (–), and total disturbance (+). Distance to nearest fragment loaded heavily on the third PCA axis. Distance to continuous forest (–) and distance to nearest temporary settlement (+) were the variables most strongly correlated with PCA axis 4.

Our PCR models on the biogeography of each species performed well for abundance but not density (Table 25.2). For abundance, there was a consistent effect of PCAs 1 (plant dendrometrics), 2 (fragment area and size), and 4 (distance) on abundance in *M. murinus* (pseudo r^2 = 0.85) and *M. ravelobensis* (pseudo r^2 = 0.82). Conversely, none of the PCAs explained spatial variations in density estimates for *M. murinus* (pseudo r^2 = 0.09) or *M. ravelobensis* (pseudo r^2 = 0.09). Consequently, we undertook hierarchical partitioning analysis of the landscape and *Microcebus* spp. density data.

Table 25.1 Results of a principal component analysis of 10 independent variables and the cumulative variance of each component.

Variables	C1	C2	C3	C4	C5	C6	C7	C8	C9	C10
lnArea	−0.19	0.52	−0.17	0.19		0.29	−0.24	−0.18	0.57	0.35
lnBasal area	−0.44	−0.24		−0.27	0.11	0.30	−0.32		0.30	−0.62
lnMean DBH	−0.34	−0.31	0.25	−0.27	−0.52		−0.32		−0.22	0.47
Stem density	−0.38	−0.13	−0.34	−0.21	0.67		0.10		−0.21	0.41
Mean tree height	−0.40		−0.33	−0.17	−0.40	−0.17	0.68	−0.16	0.13	−0.11
lnTotal disturbance	−0.27	0.41	0.20	−0.16		−0.52	−0.12	0.63		
sqrtContinuous forest distance	0.29	0.18	−0.18	−0.66		−0.36	−0.26	−0.45		
Distance nearest neighbor	−0.23	0.12	0.76		0.31		0.28	−0.41	0.10	
Nearest settlement distance	−0.29	−0.26	−0.15	0.52		−0.59	−0.30	−0.32		
lnPerimeter:area	0.24	−0.53		−0.11		−0.17	0.15	0.22	0.68	0.28
Cumulative variance	0.36	0.62	0.73	0.84	0.89	0.93	0.97	0.99	1.00	1.00

C, component.

Table 25.2 GLM models of *M. murinus* and *M. ravelobensis* density and abundance as function of four principal components.

GLM models	Estimate	Standard error	*z*-value	*p*-value	AIC	Pseudo R^2
M. murinus density = C1+C2+C3+C4					2847.8	0.09
Intercept	4.62	0.02	282.94	< 0.01		
Component 1	−0.10	0.01	−12.38	< 0.01		
Component 2	−0.03	0.01	−3.01	< 0.01		
Component 3	−0.09	0.01	−6.00	< 0.01		
Component 4	0.10	0.02	6.55	< 0.01		
M. ravelobensis density = C1+C2+C3+C4					2432.8	0.09
Intercept	4.51	0.02	261.90	< 0.01		
Component 1	−0.10	0.01	−11.58	< 0.01		
Component 2	−0.08	0.01	−8.24	< 0.01		
Component 3	−0.05	0.02	−3.22	< 0.01		
Component 4	0.02	0.02	1.37	0.17		
M. murinus abundance = C1+C2+C3+C4					220.78	0.85

Table 25.2 *(cont.)*

GLM models	Estimate	Standard error	*z*-value	*p*-value	AIC	Pseudo R^2
Intercept	0.90	0.14	6.65	< 0.01		
Component 1	−0.37	0.05	−7.27	< 0.01		
Component 2	0.99	0.07	14.00	< 0.01		
Component 3	−0.04	0.07	−0.68	0.50		
Component 4	0.35	0.10	3.44	< 0.01		
M. ravelobensis abundance = C1+C2+C3+C4					250.33	0.82
Intercept	0.73	0.15	5.00	< 0.01		
Component 1	−0.33	0.05	−6.21	< 0.01		
Component 2	1.08	0.08	14.08	< 0.01		
Component 3	−0.08	0.07	−1.09	0.28		
Component 4	0.26	0.11	2.38	0.02		

C, component; AIC, Akaike's Information Criterion.

Table 25.3 Hierarchical partitioning of independent and joint effects of 10 variables on *M. murinus* and *M. ravelobensis* density.

	Microcebus murinus			*Microcebus ravelobensis*		
Variables	Independent effects	Joint effects	%	Independent effects	Joint effects	%
lnArea	0.01	0.01	5.06	0.01	−0.01	3.26
lnBasal area	0.02	0.04	16.59	0.04	0.07	19.31
lnMean DBH	0.01	0.00	5.17	0.02	0.04	9.19
Stem density	0.03	0.07	27.23	0.01	0.02	6.55
Mean tree height	0.01	0.02	5.39	0.04	0.05	18.65
lnTotal Disturbance	0.01	0.00	4.96	0.04	0.00	20.66
sqrtContinuous forest distance	0.02	0.05	18.88	0.03	0.05	12.91
Distance nearest neighbor	0.00	0.00	0.64	0.00	0.00	1.64
Nearest settlement distance	0.01	0.03	7.79	0.01	0.00	2.70
lnPerimeter:area	0.01	0.00	8.28	0.01	−0.01	5.13

The hierarchical partitioning models indicated that spatial variations in density for *M. murinus* were due to basal area (16.59%), stem density (27.23%), and distance from continuous forest (18.88%). No other variables contributed to greater than 10% of the variation in *M. murinus* density (Table 25.3). The density of *M. ravelobensis* showed a different relationship with the independent variables. For the spatial variation in density for *M. ravelobensis* we found the following four variables contributed to greater than 10% of the variation: basal area (19.31%), mean tree height (18.65%), distance from the continuous forest (12.91%), and total disturbance (20.66%; Table 25.3).

Discussion

Our research objective was to determine the causal factors that influenced spatial variations in abundance and density of *M. murinus* and *M. ravelobensis* in the fragmented landscape. One of the most striking results of our research was that we found a positive relationship between *M. murinus* and *M. ravelobensis* density and abundance in the landscape. These results contrast with research by Rakotondravony and Radespiel (2009) of conspecifics in continuous forest sites within Ankarafantsika, where there was a negative relationship between *M. murinus* and *M. ravelobensis* densities. Although the underlying causalities of these spatial differences in interspecific associations are difficult to determine due to differences in methodologies between the two respective studies, our results provide further support for the hypothesis that ecological patterns of both nocturnal and diurnal lemurs differ between fragmented and continuous forests. Ganzhorn and Schmid (1998) found that *M. murinus* existed at lower densities in secondary versus primary forest in Kirindy. Randrianambinina *et al.* (2010) noted that density estimates of *M. danfossi* were higher in fragments with greater disturbance, a pattern these researchers argued resulted from crowding or edge effects within the smaller fragments. Diurnal lemurs display similar patterns of ecological differences between disturbed and undisturbed forests (e.g., Irwin, 2008a, 2008b; Gerber *et al.*, 2012). For example, Irwin (2008a, 2008b) studied the feeding ecology of *Propithecus diadema* in fragmented and continuous forests in eastern Madagascar, and noted striking differences in ranging patterns, dietary diversity, and habitat use between the forest types. Therefore, the landscape ecology of *M. murinus* and *M. ravelobensis* in fragmented habitats is just as varied as that seen in large-bodied, diurnal lemurs.

We found that abundance and, to a lesser extent, density estimates of *M. murinus* and *M. ravelobensis* reflected spatial variations in plant dendrometrics, fragment area and size, and proximity to continuous forests and human settlements. Rakotondravony and Radespiel (2009) found that neither species of *Microcebus* was influenced by spatial variations in plant dendrometrics or plant species diversity across the continuous forest sites. However, Burke and Lehman (2014) found that *M. murinus* preferred interior habitats and *M. ravelobensis* were more abundant in edge habitats. Plant dendrometrics as a determinant of *M. murinus* abundance and density in our study likely reflected sleeping site requirements and dietary preferences. Previous research indicates that individual *M. murinus* sleep mainly in holes of living or dead trees and that females use tree holes of better quality with respect to thermoregulation and predator protection than males in continuous forest (Radespiel *et al.*, 1998, 2003; Rakotondravony and Radespiel, 2009), whereas *M. ravelobensis* are less dependent on tree holes, because they have been shown to use self-constructed nests made out of small branches and leaves as sleeping sites (Thorén *et al.*, 2010). In fact, high-quality tree holes are such a critical, restricted resource for *M. murinus* females that Eichmueller *et al.* (2013) suggested that intersexual competition for access to these sites is a driving force of female dominance in *Microcebus*. Plant dendrometrics may have also impacted the feeding ecology of

M. murinus. This species of *Microcebus* preferentially exploits tree exudates and fruit as food resources (Thorén *et al.*, 2011), and increased fruit production and availability is often linked to habitats containing trees with larger diameter and overall size (e.g., Ganzhorn *et al.*, 1999; Balko and Underwood, 2005; Wright *et al.*, 2005). In our landscape, fragments that contained trees with larger plant dendrometric measurements may have provided *M. murinus* with more fruit feeding opportunities. It is important to note, however, that the specific patterns of fruiting abundance and exploitation by lemurs are not well studied in fragmented landscapes in Madagascar, as most relevant research has been conducted in continuous forests. For example, Rakotondravony and Radespiel (2009) found that none of their 19 vegetation parameters, 8 of which involved dendrometrics measures, explained spatial variation in the density of *M. murinus* across 10 research sites in the park, a result that likely reflected spatial monotony of the parameters. Instead, *M. murinus* densities peaked at higher elevation sites that were further away from water (Rakotondravony and Radespiel, 2009). Although the fragments had elevations ranging from 81 to 215 m, our site is embedded in a very dry landscape with no standing bodies of water outside the short rainy season. Therefore, we suggest that spatial variations in the abundance and density of *M. murinus* may be affected by sleeping site and dietary preferences in a highly fragmented forest landscape.

Although the PCR and partitioning results are statistically relevant for understanding spatial patterns of density and abundance in *M. ravelobensis*, they are less clear in terms of underlying ecological processes. This *Microcebus* species prefers to eat insect secretions (i.e., sugary secretion produced by flatid leaf bug nymphs) and arthropods, the distribution and abundance of which can not be easily related to the data we collected. Moreover, *M. ravelobensis* sleep in groups with variable sex composition in relatively open sleeping sites, like lianas or leaf nests (Rendigs *et al.*, 2003). This life-history trait differs from *M. murinus*, where solitary individuals sleep mainly in holes of living or dead trees (Radespiel *et al.*, 2003). Furthermore, these results are difficult to place within the existing literature on the biogeography of *M. ravelobensis*, in which this species exists at higher densities in humid forests and in areas in which *M. murinus* was absent (Rakotondravony and Radespiel, 2009). Neither of these parameters match our study site or community structure.

We suggest that edge effects may explain the spatial patterns in abundance and density for *M. ravelobensis*. In a recent study of edge effects on *Microcebus* in nearby forests in Ankarafantsika, Burke and Lehman (2014) found that *M. ravelobensis* were captured more often in edge habitats, and that these habitats also contained a greater abundance of preferred insect prey and insect secretions. The question arises as to why our perimeter:area ratio, which can be seen as a crude metric on relative edge habitat, did not factor more importantly in either the PCR or partitioning models. Although this perimeter:area ratio has been widely used in landscape ecology, due largely to its ease of implementation over remote imaging, the very nature of its uniform edge ratio rarely works well for the varying depths and intensities of edge effects. Pervasive edge effects mean that small fragments have no forest interior habitat, which has important implications for the conservation biogeography of animal populations (Ries and Sisk, 2004). The orientation of the fragments relative to topography and climate can alter complex patterns of seed dispersal, frugivory, and species diversity. These ecological patterns likely mean that *M. ravelobensis* are responding to complex processes of multiple edge effects in each forest patch within a fragmented landscape (Fernández *et al.*, 2002). Thus, determining if and how edge effects influence the density of *M. ravelobensis* will require fragment-level metrics on edge direction, intensity, and depth.

Understanding the distribution and abundance of animals in a heterogeneous landscape is critical to a better understanding of their population dynamics and biogeography. Madagascar is often cited as representing one of the most fragmented forest landscapes in the tropics, with concomitantly high extinction risks for many of the endemic species (Schwitzer *et al.*, 2014). Despite the need for data on lemur responses to anthropogenic disturbance, few studies have conducted longitudinal research in fragmented landscapes. Those studies that have been conducted have typically focused on large-bodied, diurnal lemurs whose relatively larger ranging patterns preclude large sample sizes. Put differently, researchers can typically only study a few groups of diurnal species that range across large areas. However, most forest-dwelling mammals are closer in body size to *Microcebus* than to the relatively large-bodied diurnal lemurs (Isaac *et al.*, 2005). Moreover, *Microcebus*, such as the two species we studied, are ideal study animals for research on primate responses to deforestation processes because these lemurs have small home ranges, can exist at high densities, display seasonal and habitat-related variations in body mass, and are amenable to capture–mark–recapture methods (Radespiel and Bruford, 2014). For example, applied conservation efforts would benefit from a greater understanding of how protection of larger fragments would extend population persistence of species in anthropogencially disturbed fragments, such as those we studied in Ankarafantsika. This research could be done utilizing metapopulation modeling of varying conservation strategies for mouse lemurs in the fragments. Further studies should seek to experimentally determine if the addition of wooden nest boxes would increase the abundance of *M. murinus* in small, isolated fragments that lack high-quality tree holes (i.e., those with optimal thermoregulation and predator protection). Thus, studies of how these small-bodied, nocturnal lemurs respond to natural and anthropogenic disturbances are critical to a better understanding of the conservation biogeography of mammals.

Acknowledgments

We would like to thank Madagascar National Parks and the University of Antananarivo for permission to conduct our research in Madagascar. For their support, advice, and hospitality, we thank Benjamin Andriamahaja and the staff at ICTE/MICET, Fanja Raoelinirina, and the staff at La Maison du Pyla. We would like to thank the many people who have assisted us with our research in Madagascar, particularly Mamy Razafitsalama, Rindra Rakotoarivony, Jean Paul, Velontsara,

Alpha, Yuri Fraser, Keriann McGoogan, and Kim Valenta. We especially would like to thank the residents of the villages of Andranohobaka and Maevatanbari for their hospitality and essential support for our project. We would like to thank the reviewers for their excellent comments and feedback which greatly improved our manuscript. Our research was supported in part by the Explorers Club Exploration Fund (TS), University of Toronto School of Graduate Studies Travel Grant (TS), the American Society of Primatologists Conservation Committee Small Grant (TS), and a Discovery grant to SML from the Natural Sciences and Engineering Research Council of Canada.

References

Alonso LE, Schulenberg TS, Radilofe S, Missa O. 2002. *A Biological Assessment of the Réserve Naturelle Intégrale d'Ankarafantsika, Madagascar. RAP Bulletin of Biological Assessment*, Vol. 23. Conservation International, Washington, DC.

Anzures-Dadda A, Manson RH. 2007. Patch- and landscape-scale effects on howler monkey distribution and abundance in rainforest fragments. *Animal Conservation* 10(1):69–76.

Arroyo-Rodríguez V, Dias PAD. 2010. Effects of habitat fragmentation and disturbance on howler monkeys: a review. *American Journal of Primatology* 72(1):1–16.

Atsalis S. 1999. Diet of the brown mouse lemur (*Microcebus rufus*) in Ranomafana National Park, Madagascar. *International Journal of Primatology* 20(2):193–229.

Balko EA, Underwood HB. 2005. Effects of forest structure and composition on food availability for *Varecia variegata* at Ranomafana National Park, Madagascar. *Journal of Primatology* 66(1):45–70.

Bloesch, U. 1999. Fire as a tool in the management of a savanna/dry forest reserve in Madagascar. *Applied Vegetation Science* 2(1):117–124.

Burke RJ, Lehman SM. 2014. Edge effects on morphometrics and body mass in two sympatric species of mouse lemurs in Madagascar. *Folia Primatologica* 85(5):277–291.

Chiarello A, de Melo FR. 2001. Primate population densities and sizes in Atlantic forest remnants of northern Espirito Santo, Brazil. *International Journal of Primatology* 22(3):379–396.

Chapman C, Chapman L, Naughton-Treves L, Lawes MJ, McDowell LR. 2004. Predicting folivorous primate abundance: validation of a nutritional model. *American Journal of Primatology* 62:55.

Chevan A, Sutherland M. 1991. Hierarchical partitioning. *The American Statistician* 45(2):90–96.

Connor E, Courtney A, Yoder J. 2000. Individuals–area relationships: the relationship between animal population density and area. *Ecology* 81(3):734–748.

Corbin GD, Schmid J. 1995. Insect secretions determine habitat use patterns by a female lesser mouse lemur (*Microcebus murinus*). *American Journal of Primatology* 37(4):317–324.

Dammhahn M, Kappeler PM. 2008. Small-scale coexistence of two mouse lemur species (*Microcebus berthae* and *M. murinus*) within a homogeneous competitive environment. *Oecologia* 157(3):473–483.

Davidson AD, Hamilton MJ, Boyer AG, Brown JH, Ceballos G. 2009. Multiple ecological pathways to extinction in mammals. *Proceedings of the National Academy of Sciences* 106(26):10702–10705.

Eichmueller P, Thorén S, Radespiel U. 2013. The lack of female dominance in golden-brown mouse lemurs suggests alternative routes in lemur social evolution. *American Journal of Physical Anthropology* 150(1):158–164.

Fa JE, Purvis A. 1997. Body size, diet and population density in afrotropical forest mammals: a comparison with neotropical species. *Journal of Animal Ecology* 66(1):98–112.

Fahrig L. 2003. Effects of habitat fragmentation on biodiversity. *Annual Review of Ecology, Evolution, and Systematics* 34:487–515.

Fernández C, Acosta FJ, Abellá G, López F, Díaz M. 2002. Complex edge effect fields as additive processes in patches of ecological systems. *Ecological Modelling* 149:273–283.

Gade DW. 1996. Deforestation and its effects in highland Madagascar. *Mountain Research and Development* 16(2):101–116.

Ganzhorn JU, Schmid J. 1998. Different population dynamics of *Microcebus murinus* in primary and secondary deciduous dry forests of Madagascar. *International Journal of Primatology* 19(5):785–796.

Ganzhorn JU, Fietz J, Rakotovao E, Schwab D, Zinner D. 1999. Lemurs and the regeneration of dry deciduous forest in Madagascar. *Conservation Biology* 13(4):794–804.

García G, Goodman SM. 2003. Hunting of protected animals in the Parc National d'Ankarafantsika, north-western Madagascar. *Oryx* 37(1):115–118.

Gerber BD, Arrigo-Nelson S, Karpanty SM, Kotschwar M, Wright PC. 2012. Spatial ecology of the endangered Milne-Edwards' sifaka (*Propithecus edwardsi*): do logging and season affect home range and daily ranging patterns? *International Journal of Primatology* 33(2):305–321.

Grow N, Gursky S, Duma Y. 2013. Altitude and forest edges influence the density and distribution of pygmy tarsiers (*Tarsius pumilus*). *American Journal of Primatology* 75(5):464–477.

Guschanski K, Olivieri GL, Funk SM, Radespiel U. 2007. MtDNA reveals strong genetic differentiation among geographically isolated populations of the golden brown mouse lemur, *Microcebus ravelobensis*. *Conservation Genetics* 8(4):809–821.

Hanya G, Chapman CA. 2012. Linking feeding ecology and population abundance: a review of food resource limitation on primates. *Ecological Research* 28(2):183–190.

Harcourt AH, Doherty DA. 2005. Species–area relationships of primates in tropical forest fragments: a global analysis. *Journal of Applied Ecology* 42(4):630–637.

Harper KA, Macdonald SE, Burton P, *et al.* 2005. Edge influence on forest structure and composition in fragmented landscapes. *Conservation Biology* 19(3):768–782.

Irwin MT. 2008a. Diademed sifaka (*Propithecus diadema*) ranging and habitat use in continuous and fragmented forest: higher density but lower viability in fragments? *Biotropica* 40(2):231–240.

Irwin MT. 2008b. Feeding ecology of *Propithecus diadema* in forest fragments and continuous forest. *International Journal of Primatology* 29(1):95–115.

Isaac NJB, Jones KE, Gittleman JL, Purvis A. 2005. Correlates of species richness in mammals: body size, life history, and ecology. *The American Naturalist* 165(5):600–607.

Janson CH, Chapman CA. 1999. Resources and primate community structure. In JG Fleagle, CH Janson, KE Reed (eds.), *Primate Communities* (pp. 237–267). Cambridge University Press, Cambridge.

Lahann P. 2007. Feeding ecology and seed dispersal of sympatric cheirogaleid lemurs (*Microcebus murinus, Cheirogaleus medius, Cheirogaleus major*) in the littoral rainforest of south-east Madagascar. *Journal of Zoology* 271(1):88–98.

Lahann P, Schmid J, Ganzhorn JU. 2006. Geographic variation in populations of *Microcebus murinus* in Madagascar: resource seasonality or Bergmann's rule? *International Journal of Primatology* 27(4):983–999.

Laurance WF, Camargo JLC, Luizão RCC, *et al.* 2011. The fate of Amazonian forest fragments: a 32-year investigation. *Biological Conservation* 144(1):56–67.

Lehman SM. 2006. Conservation biology of Malagasy strepsirrhines: a phylogenetic approach. *American Journal of Physical Anthropology* 130(2):238–253.

Lima SL, Zollner PA. 1996. Towards a behavioral ecology of ecological landscapes. *Trends in Ecology & Evolution* 11(3):131–135.

MacArthur RH. 1967. *The Theory of Island Biogeography*, Vol. 1. Princeton University Press, Chicago.

Mac Nally R. 1996. Hierarchical partitioning as an interpretative tool in multivariate inference. *Australian Journal of Ecology* 21(2):224–228.

Mac Nally R. 2000. Regression and model-building in conservation biology, biogeography and ecology: the distinction between – and reconciliation of – 'predictive' and 'explanatory' models. *Biodiversity and Conservation* 9(5):655–671.

Mac Nally R. 2002. Multiple regression and inference in ecology and conservation biology: further comments on identifying important predictor variables. *Biodiversity and Conservation* 11(8):1397–1401.

McGoogan K, Kivell T, Hutchison M. 2007. Phylogenetic diversity and the conservation biogeography of African primates. *Journal of Biogeography* 34(11):1962–1974.

Michalski F, Peres CA. 2007. Disturbance-mediated mammal persistence and abundance–area relationships in Amazonian forest fragments. *Conservation Biology* 21(6):1–15.

Müller P, Velo A, Raheliarisoa EO. 2000. Surveys of sympatric lemurs at Anjamena, north-west Madagascar. *African Journal of Ecology* 38:248–257.

Murcia C. 1995. Edge effects in fragmented forests: implications for conservation. *Trends in Ecology & Evolution* 10(2):58–62.

Olea PP, Mateo-Tomás P, de Frutos A. 2010. Estimating and modelling bias of the hierarchical partitioning public-domain software: implications in environmental management and conservation. *PLoS ONE* 5(7):e11698.

Paciulli L. 2010. The relationship between nonhuman primate densities and vegetation on the Pagai, Mentawai Islands, Indonesia. In S Gursky-Doyen, J Supriatna (eds.), *Indonesian Primates*. Springer, New York.

Peres CA. 1990. Effects of hunting on western Amazonian primate communities. *Biological Conservation* 54:47–59.

Peres CA, Dolman PM. 2000. Density compensation in neotropical primate communities: evidence from 56 hunted and nonhunted Amazonian forests of varying productivity. *Oecologia* 122(2):175–189.

Peres CA, Palacios E. 2007. Basin-wide effects of game harvest on vertebrate population densities in amazonian forest: implications for animal-mediated seed dispersal. *Biotropica* 39(3):304–315.

Pinto N, Lasky J, Bueno R, Keitt TH, Galetti M. 2009. Primate densities in the Atlantic forest of southeast Brazil: the role of habitat quality and anthropogenic disturbance. In *South American Primates* (pp. 413–431). Springer, New York.

Plumptre AJ, Reynolds V. 1994. The effect of selective logging on the primate populations in the Budongo Forest Reserve, Uganda. *Journal of Applied Ecology* 31(4):631–641.

Pozo-Montuy G, Serio-Silva JC, Bonilla-Sánchez YM. 2011. Influence of the landscape matrix on the abundance of arboreal primates in fragmented landscapes. *Primates* 52(2):139–147.

Pyritz L, Büntge A, Herzog S. 2010. Effects of habitat structure and fragmentation on diversity and abundance of primates in tropical deciduous forests in Bolivia. *International Journal of Primatology* 31:796–812.

Quinn GP, Keough MJ. 2002 *Experimental Design and Data Analysis for Biologists*. Cambridge University Press, Cambridge.

R Core Team. 2014. *R: A Language and Environment for Statistical Computing*. R Foundation for Statistical Computing, Vienna, Austria.

Rabesandratana AZ, Rakotondravony R, Zimmermann E. 2012. Distribution and abundance of Milne Edwards' sportive lemur *Lepilemur edwardsi* in the Ankarafantsika National Park, northwestern Madagascar. *Lemur News* 16:57–59.

Radespiel U, Bruford MW. 2014. Fragmentation genetics of rainforest animals: insights from recent studies. *Conservation Genetics* 15(2):245–260.

Radespiel U, Cepok S, Zietemann V, Zimmermann E. 1998. Sex-specific usage patterns of sleeping sites in grey mouse lemurs (*Microcebus murinus*) in northwestern Madagascar. *American Journal of Primatology* 46:77–84.

Radespiel U, Ehresmann P, Zimmermann E. 2003. Species-specific usage of sleeping sites in two sympatric mouse lemur species (*Microcebus murinus* and *M. ravelobensis*) in northwestern Madagascar. *American Journal of Primatology* 59(4):139–151.

Rakotondravony R, Radespiel U. 2009. Varying patterns of coexistence of two mouse lemur species (*Microcebus ravelobensis* and *M. murinus*) in a heterogeneous landscape. *American Journal of Primatology* 71(11):928–938.

Randrianambinina B, Rasoloharijaona S, Rakotondravony R, Zimmermann E, Radespiel U. 2010. Abundance and conservation status of two newly described lemur species in northwestern Madagascar (*Microcebus danfossi, Lepilemur grewcockorum*). *Madagascar Conservation & Development* 5(2):95–102.

Razafy Fara L. 2003. Rapport sur l'actualisation de la carte de végétation du parc national à Ankarafantsika. ANGAP, GFA, Antananarivo, Madagascar. p. 26.

Remis MJ, Robinson CAJ. 2012. Reductions in primate abundance and diversity in a multiuse protected area: synergistic impacts of hunting and logging in a Congo Basin forest. *American Journal of Primatology* 74(7):602–612.

Rendigs A, Radespiel U, Wrogemann D, Zimmermann E. 2003. Relationship between microhabitat structure and distribution of mouse lemurs (*Microcebus* spp.) in northwestern Madagascar. *International Journal of Primatology* 24(1):47–64.

Ries L, Sisk T. 2004. A predictive model of edge effects. *Ecology* 85(11):2917–2926.

Robinson JG, Redford KH. 1986. Body size, diet, and population density of neotropical forest mammals. *The American Naturalist* 128(5):665–680.

Rovero F, Struhsaker TT. 2007. Vegetative predictors of primate abundance: utility and limitations of a fine-scale analysis. *American Journal of Primatology* 69(11):1242–1256.

Schwitzer C, Mittermeier RA, Johnson SE, *et al.* 2014. Conservation. Averting lemur extinctions amid Madagascar's political crisis. *Science* 343(6173):842–843.

Shapiro S, Wilk M. 1965. An analysis of variance test for normality (complete samples). *Biometrika* 52(3–4):591.

Skole D, Tucker C. 1993. Tropical deforestation and habitat fragmentation in the Amazon: satellite data from 1978 to 1988. *Science* 260(5116):1905.

Smith RJ, Jungers WL. 1997. Body mass in comparative primatology. *Journal of Human Evolution* 32(6):523–559.

Stevenson PR. 2001. The relationship between fruit production and primate abundance in Neotropical communities. *Biological Journal of the Linnean Society* 72:161–178.

Teelen S. 2007. Primate abundance along five transect lines at Ngogo, Kibale National Park, Uganda. *American Journal of Primatology* 69(9):1030–1044.

Terborgh, J. 1986. Keystone plant resources in the tropical forest. In ME Soule, BA Wilcox (eds.), *Conservation Biology: an evolutionary-ecological perspective* (pp. 330–344). Sinauer Associates, Sunderland, MA.

Thorén S, Quietzsch F, Radespiel U. 2010 Leaf nest use and construction in the golden-brown mouse lemur (*Microcebus ravelobensis*) in the Ankarafantsika National Park. *American Journal of Primatology* 72(1):48–55.

Thorén S, Quietzsch F, Schwochow D, *et al.* 2011. Seasonal changes in feeding ecology and activity patterns of two sympatric mouse lemur species, the gray mouse lemur (*Microcebus murinus*) and the golden-brown mouse lemur (*M. ravelobensis*), in northwestern Madagascar. *International Journal of Primatology* 32: 566–586.

Whittaker RJ, Araújo MB, Paul J, *et al.* 2005. Conservation biogeography: assessment and prospect. *Diversity and Distributions* 11(1):3–23.

Wieczkowski J. 2004. Ecological correlates of abundance in the Tana mangabey (*Cercocebus galeritus*). *American Journal of Primatology* 63(3):125–138.

Wilcox BA, Murphy DD. 1985. Conservation strategy: the effects of fragmentation on extinction. *The American Naturalist* 125(6):879–887.

Wright PC, Razafindratsita VR, Pochron ST, Jernvall J. 2005. The key to Madagascar frugivores. In J Dew, J Boubli (eds.), *Tropical Fruits and Frugivores: The Search for Strong Interactors* (pp. 121–138). Springer, Dordrecht.

Zimmermann E, Cepok S, Rakotoarison N, Zietemann V, Radespiel U. 1998. Sympatric mouse lemurs in north-west Madagascar: a new rufous mouse lemur species (*Microcebus ravelobensis*). *Folia Primatologica* 69:106–114.

26 Can behavioral ecology help to understand the divergent geographic range sizes of mouse lemurs?

Ute Radespiel

Introduction

The understanding of the size, shape, overlap, and constraints of geographic ranges of organisms forms one important foundation of the very active field of biogeography, and is tightly connected to the understanding of evolutionary processes that give rise to the diverse distribution patterns that can be found today throughout all terrestrial and aquatic biomes (Cox and Moore, 2010; Lomolino *et al.*, 2010). Madagascar is well known for its hypervariable environments (Dewar and Richard, 2007) and its extraordinary diversity in fauna and flora, most of which evolved endemically during the long period of isolation of Madagascar from other landmasses (Yoder *et al.*, 1996; Poux *et al.*, 2005; Crottini *et al.*, 2012; Samonds *et al.*, 2012). Some meta-analyses and reviews have recently aimed to investigate the evolutionary processes that led to the endemic radiations in Malagasy vertebrates and emphasized the role of climatic factors and physical barriers as drivers of allopatric speciation (e.g., Goodman and Ganzhorn, 2004; Yoder and Nowak, 2006; Wilmé *et al.*, 2006; Pearson and Raxworthy, 2009; Vences *et al.*, 2009). In particular, the role of large rivers and mountain ranges in disrupting gene flow and thereby leading to the evolution of local and regional endemics has been emphasized for various vertebrate clades (e.g., Wilmé & Goodman, 2003; Olson *et al.*, 2004; Wilmé *et al.*, 2006; Olivieri *et al.*, 2007; Townsend *et al.*, 2009). It should be mentioned that many geographic range sizes of Malagasy lemurs are only inferred from a rather small amount of point location data rather than known from extensive survey work which is known to limit the power of testing biogeographic hypotheses (Gaston, 1994). Nevertheless, none of the available biogeographic models for Madagascar can explain why some phylogenetic lineages contain closely related species with quite variable geographic range sizes. Mouse lemurs (*Microcebus* spp.) form one such lineage among lemurs and it shall be explored in this chapter whether ecological and

The Dwarf and Mouse Lemurs of Madagascar: Biology, Behavior and Conservation Biogeography of the Cheirogaleidae, eds. S.M. Lehman, U. Radespiel, E. Zimmermann. Published by Cambridge University Press. © Cambridge University Press 2016.

behavioral traits may explain the existence of such divergent distribution patterns within a single genus.

Mouse lemurs comprise a total of 21 described species (Mittermeier *et al.*, 2010; Radespiel *et al.*, 2012; Rasoloarison *et al.*, 2013) and can be found in all forest habitats of the island. All mouse lemurs are small, nocturnal, arboreal, solitary foragers and have probably evolved from a common ancestor within the last 8–11 Ma (Yoder and Yang, 2004; Thiele *et al.*, 2013), although the exact sequence of speciation events is still not known. Whereas most species seem to have very limited geographic ranges (n = 13, Figure 26.1) or are known only from single locations so far (n = 5), two species show a regional distribution (*M. griseorufus*, *M. myoxinus*), encompassing several so-called "Inter-River Systems," and one species, the gray mouse lemur (*M. murinus*), occurs throughout large parts of western Madagascar. Due to these disparate geographic range sizes, up to two mouse lemur species may coexist in certain areas of the island. In western Madagascar, this is the widely distributed *M. murinus* co-occurring with *M. griseorufus*, *M. berthae*, *M. myoxinus*, *M. ravelobensis*, and *M. bongolavensis*, respectively. From eastern Madagascar, only one such case is known, and this is the coexistence of *M. mittermeieri* and *M. macarthurii* in the Makira region (Radespiel *et al.*, 2008a). In order to explore possible explanations for these divergent distribution patterns and to simplify the line of arguments in the subsequent parts of this chapter, *M. murinus* will be compared and opposed to all other mouse lemur species that have much smaller geographic ranges.

Biogeographic hypotheses (incl. hypotheses 1–5)

At least nine different hypotheses have been suggested to explain divergent geographic range sizes in the biogeographic literature (Table 26.1). However, five of these nine biogeographic hypotheses are not suitable to explain variation in *Microcebus* range sizes: Rapoport's rule, i.e., the tendency for range sizes to be smaller with decreasing latitude (hypothesis 1: Rapoport, 1982; Stevens, 1989), and a varying seasonality in rainfall (hypothesis 2: Cowlishaw and Hacker, 1997) cannot explain divergent geographic range sizes in mouse lemurs, as the coexisting species pairs inhabit the same geographic regions and are subject to the same seasonal climatic variations. Although mouse lemurs can generally be grouped into larger-bodied species (> 50 g body mass, comprising *M. murinus*, *M. tavaratra*, *M. mamiratra*, *M. danfossi*, *M. bongolavensis*, *M. ravelobensis*, *M. griseorufus*, *M. macarthurii*, *M. simmonsi*, *M. gerpi*, *M. jollyae*, *M. marohita*, *M. tanosi*) and smaller-bodied species (< 50 g body mass, comprising *M. mittermeieri*, *M. lehilahytsara*, *M. rufus*, *M. arnholdi*, *M. margotmarshae*, *M. sambiranensis*, *M. myoxinus*, *M. berthae*) (summarized in Radespiel *et al.*, 2012; Rasoloarison *et al.*, 2013), there is no divergence in geographic range size between these two groups (hypothesis 3: compare Figure 26.1). Likewise, the reproductive potential of *M. murinus* (Schmelting *et al.*, 2000; Eberle and Kappeler, 2004, 2006; Lahann *et al.*, 2006) does not seem to be systematically higher than that of its congeners (hypothesis 4), although reproductive

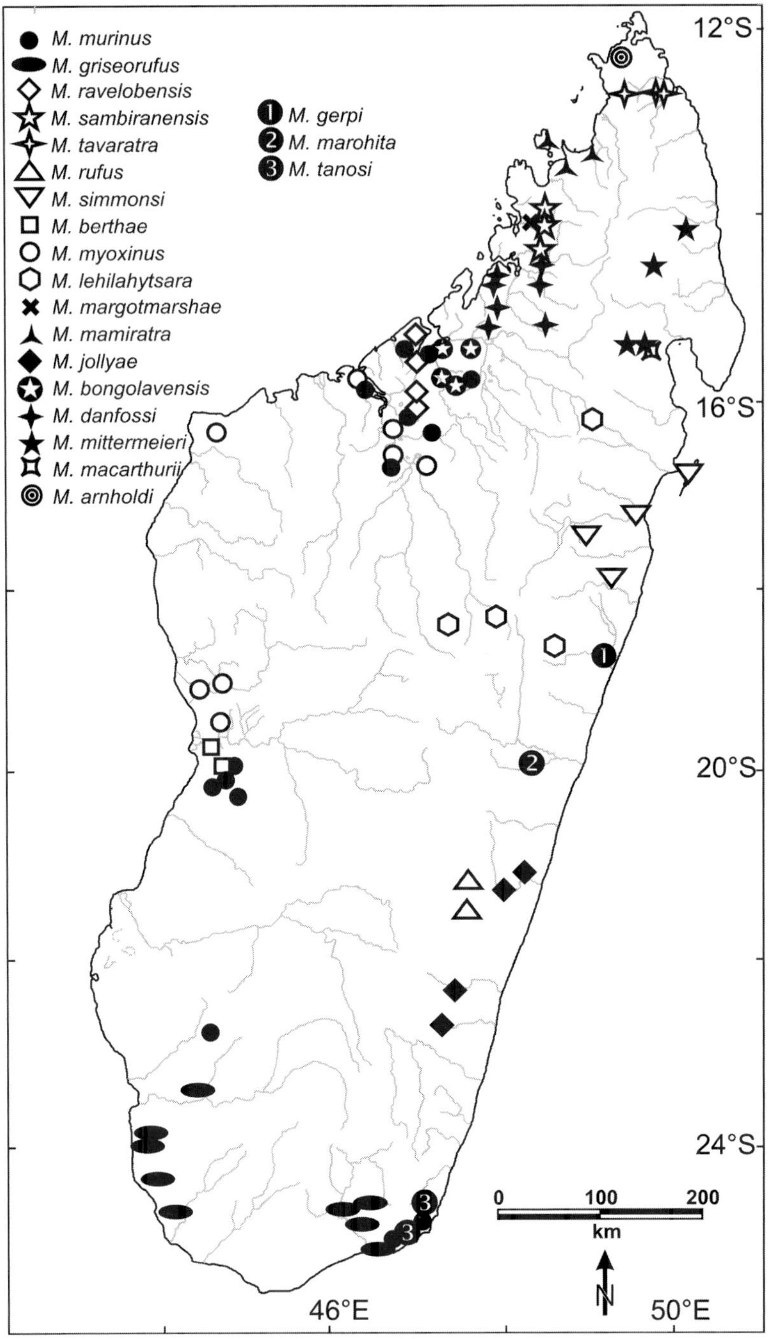

Figure 26.1 Map of Madagascar with confirmed occurrences of the 21 described mouse lemur species. Each species is indicated by a different symbol (modified and extended from Radespiel *et al.*, 2012).

Table 26.1 Predictions derived from nine hypotheses explaining divergent geographic range sizes.

No.	Hypotheses	Predictions	
		Large geographic range size (= *M. murinus*)	Limited geographic range size (all congeners)
1	Rapoport's rule	Further away from equator	Closer to equator
2	Seasonality in rainfall	High seasonality	Low seasonality
3	Body mass	Higher	Lower
4	Reproduction	Faster	Slower
5	Phylogeography	Longer history of expansion (= "older species")	Shorter history of expansion (= "younger species")
6	Habitat use	Generalist	Specialist
7	Resource use	Generalist	Specialist
8	Interspecific competition	Higher competitive potential	Lower competitive potential
9	Vagility	High	Low

records are only available from five other species, *M. ravelobensis* (Schmelting *et al.*, 2000; Randrianambinina *et al.*, 2003), *M. berthae* (Dammhahn and Kappeler, 2005), *M. griseorufus* (Génin, 2008), *M. lehilahytsara* (Randrianambinina *et al.*, 2003), and *M. rufus* (Blanco, 2008, 2011; Blanco and Meyer, 2009). The available literature suggests that all studied mouse lemur species are seasonal breeders, and females can have one or two litters (with an average of two offspring per litter, reviewed in Zimmermann and Radespiel, 2013) per year under natural conditions, depending on geographic location and probably the length and productivity of the rainy season during which infants are reared (Lahann *et al.*, 2006). A differential length of their phylogenetic history is also not likely to explain the divergent geographic range sizes (hypothesis 5), because (1) previous first attempts to date the tip nodes in the mouse lemur phylogeny do not suggest an older age for the most recent common ancestor of *M. murinus* than for species with smaller geographic range sizes (e.g., Kappeler *et al.*, 2005; Thiele *et al.*, 2013); (2) intraspecific sequence divergence is not necessarily larger in *M. murinus* than in other species, but rather seems to depend on sample size (Radespiel *et al.*, 2012; Weisrock *et al.*, 2012; Thiele *et al.*, 2013); and (3) one study suggests a recent expansion of *M. murinus* into northwestern Madagascar during the late Pleistocene (Schneider *et al.*, 2010), indicating that the large range size of this species may be a secondary and more recent phenomenon, possibly not directly coupled to its speciation.

The remaining four hypotheses 6–9 imply certain behavioral and ecological differences between *M. murinus* and its congeners. In the subsequent parts of this chapter, each of them shall be reviewed and evaluated in more detail for those species pairs for which the respective information is available.

Hypothesis 6: habitat use

It is well established in the macroecological literature that species that are ecologically plastic – i.e., possess a broad fundamental niche and can therefore be termed ecological generalists – should be able to colonize a wide spectrum of habitat types resulting in larger geographic range sizes than species that are less plastic and more specialized (e.g., Brown, 1984; Hanski and Gyllenberg, 1997). This relationship has also gained some support for primates (Eeley and Foley, 1999; Harcourt *et al.*, 2002), but has so far not been systematically investigated in Malagasy lemurs. This hypothesis would predict that the widely distributed *M. murinus* should be a habitat generalist, whereas its congeners with smaller geographic ranges should be habitat specialists. This hypothesis is evidently supported if a complete list of habitat types ever reported to be used by *M. murinus* across the island was compared to a list for each of the other species with limited geographic range size (reviewed in Radespiel, 2006). Beyond this obvious plasticity on the species level, however, it can be asked whether this habitat plasticity is also visible on the local level reflecting the ecologic plasticity that may govern the daily patterns of habitat use by individuals. In other words, it should be explored whether *M. murinus* is also a local habitat generalist compared to other mouse lemur species. Such a comparison is so far feasible only for the species pairs *M. murinus*/*M. griseorufus* in the Mandena/Andohahela region (southern Madagascar), *M. murinus*/*M. berthae* in the Kirindy region (western Madagascar), and *M. murinus*/*M. ravelobensis* in the Ankarafantsika National Park (northwestern Madagascar), and the respective findings are summarized in Table 26.2.

The comparison of published evidence suggests that *M. murinus* is a local habitat generalist only in southern Madagascar where it has been found in various different habitat types ranging from rainforests to dry spiny forests in contrast to its congener *M. griseorufus* which seems to be rather specialized in this region and confined to the dry spiny forests (Table 26.2). However, *M. griseorufus* has previously been found in other habitat types as well (e.g., gallery forest, xerophytic shrubs, dry forest, degraded forest; Rakotondranary *et al.*, 2010), whenever *M. murinus* was absent (Bohr *et al.*, 2011). This finding indicates that its absence, for example from the gallery forests in southern Madagascar, may be due rather to competitive exclusion by *M. murinus* than to a high degree of habitat specialization (Rakotondranary *et al.*, 2011a).

In contrast to the expectation, *M. murinus* and *M. berthae* in western Madagascar occupy quite similar habitat types and have not been reported to show a different degree of habitat specialization (Table 26.2). The main difference seems to be that *M. murinus* has been documented to be quite resilient to anthropogenic habitat disturbances (but see Ganzhorn and Schmid, 1998) and was found in degraded and secondary forests and even plantations, whereas this was not the case for *M. berthae* (Ganzhorn, 1994, 1995; Schwab and Ganzhorn, 2004).

Finally, *M. murinus* and *M. ravelobensis* in northwestern Madagascar do contradict the expected pattern of habitat use. A regional study on the relative abundance

Table 26.2 Local habitat plasticity of *M. murinus* compared to its co-occurring congeners.

Species pair	Species	Region	Habitat use	Habitat preference	Reference
M. murinus /*M. ravelobensis*	*M. murinus*	NW	Dry deciduous forest[1,2,4,5] Rare or absent in mesic forest types[3–5] Disturbed forests: yes[2]	Abundance increases with increasing aridity[4]	[1]Rendigs *et al.*, 2003 [2]Olivieri *et al.*, 2005 [3]Rakotondravony and Radespiel, 2007
	M. ravelobensis		Gallery forest[2–4] Alluvial forest[1,3,4,5] Dry deciduous forest[1–5] Disturbed forests: yes[2]	Highest abundance in mesic habitats[4] Present in all habitat types[1–4]	[4]Rakotondravony and Radespiel, 2009 [5]Chanu *et al.*, 2012
M. murinus /*M. berthae*	*M. murinus*	W	Dense dry deciduous forest[6–9] Disturbed forests: yes[7,8]	Not detected so far[6]	[6]Dammhahn and Kappeler, 2008a
	M. berthae		Dense dry deciduous forest[6,9] Disturbed forests: no[9]	Not detected so far[6]	[7]Ganzhorn, 1995 [8]Ganzhorn and Schmid, 1998 [9]Schwab and Ganzhorn, 2004
M. murinus /*M. griseorufus*	*M. murinus*	S	Rainforest[12] Gallery forest[10–12] Dry spiny forest[11,12] Littoral forests[10]	Gallery forests likely to be preferred[11,12]	[10]Lahann *et al.*, 2006 [11]Gligor *et al.*, 2009 [12]Rakotondranary *et al.*, 2011a [13]Rakotondranary and
	M. griseorufus		Dry spiny forest[11–13]	Not detected so far[13]	Ganzhorn, 2011

and distribution of both species in the Ankarafantsika National Park showed that *M. murinus* was more restricted in its regional distribution and habitat use (occurred only in 64% of the study sites) than its congener, and that this restriction was probably based on a preference for more xeric habitat types (Rakotondravony and Radespiel, 2009). *M. ravelobensis*, on the other hand, was found in 86% of the study sites and in all habitat types that were present in the study area. Moreover, both species are known to occur in areas of human disturbances (Olivieri *et al.*, 2005). Taken together, there is only limited evidence for a higher degree of local habitat specialization in mouse lemur species with limited geographic range sizes compared to *M. murinus*.

Hypothesis 7: resource use

The resource use hypothesis predicts that species that have broad ecological niches and are therefore generalists in terms of their resource requirements should have a larger geographic range size than species that are highly specialized in their resource requirements (Brown, 1984, 1995). It should be noted that there are most likely some direct links to the previous hypothesis on habitat plasticity, in that, for example, a dietary generalist is probably able to use more habitat types than a dietary specialist. However, two clear predictions can be derived from this hypothesis for mouse lemurs that are omnivorous and regularly use some sort of daytime shelter for sleep, torpor, or rearing infants (reviewed in Radespiel, 2006). It can be predicted that *M. murinus* should have a broader feeding niche and/or use a wider variety of sleeping sites than its congeners with a more limited geographic distribution. Due to a lack of published data on the comparative feeding ecology and sleeping site ecology on the other sympatric species pairs, comparisons are possible for the same three species pairs as before: *M. murinus*/*M. griseorufus* in the Mandena/Andohahela region (southern Madagascar), *M. murinus*/*M. berthae* from the Kirindy region (western Madagascar), and *M. murinus*/*M. ravelobensis* from the Ankarafantsika National Park (northwestern Madagascar, Table 26.3).

Feeding niche breadth

In all three species pairs there is evidence that the seasonal changes in dietary composition may be larger in *M. murinus* than in its local congener (Table 26.3). These findings indicate that the plasticity in feeding habits may indeed be larger in the widely distributed *M. murinus* than in the species with restricted geographic range size. This statement is further supported by the available data on dietary diversity (*M. murinus*/*M. ravelobensis*) and feeding niche breadth (*M. murinus*/*M. berthae*), which seem to be higher or broader, respectively, in the species with larger geographic ranges (Table 26.3). Comparative data are not yet available for *M. murinus*/*M. griseorufus* on these aspects. It should be mentioned, however, that all relevant studies also suggested that species-specific metabolic changes and different energy-saving strategies (e.g., frequency, type and length of torpor, activity)

Table 26.3 Local plasticity in resource use of *M. murinus* compared to its sympatric congeners.

Species pair	Species	Region	Feeding niche breadth	Variability in daytime shelter	Reference
M. murinus /*M. ravelobensis*	*M. murinus*	NW	More diverse diet[1] Larger seasonal changes in diet[1]	86% tree holes[4] 2% branches[4] 12% other sites[4] Tree holes used for infant rearing[6]	[1]Thorén *et al.*, 2011a [4]Radespiel *et al.*, 2003a [5]Thorén *et al.*, 2010 [6]Lutermann *et al.*, 2010
	M. ravelobensis		Less diverse diet[1] Smaller seasonal changes in diet[1]	46% tree holes[4] 28% branches[4] 26% other sites[4] Leaf nest use high during period of infant rearing[5]	
M. murinus /*M. berthae*	*M. murinus*	W	Season 1: 0.56* (wet to dry)[2] Season 2: 0.41* (dry)[2] Season 3: 0.31* (dry to wet)[2] Larger seasonal changes in diet[2,3]	98% tree holes[8] 2% foliage[8]	[2]Dammhahn and Kappeler, 2008b [3]Dammhahn and Kappeler, 2010 [7]Schwab, 2000 [8]Schmid, 1998
	M. berthae		Season 1: 0.32* (wet to dry)[2] Season 2: 0.07* (dry)[2] Season 3: 0.19* (dry to wet)[2] Smaller seasonal changes in diet[2,3]	11% tree holes[7] 79% branches[7] 5% nest[7] 5% rolled bark[7]	
M. murinus /*M. griseorufus*	*M. murinus*	S	Larger seasonal changes in δ15N No seasonality in δ13C	n.a.	Rakotondranary *et al.*, 2011b
	M. griseorufus		Smaller seasonal changes in δ15N No seasonality in δ13C	n.a.	

* Levin's standardized index for niche breadth (Krebs, 1998).

in particular during the lean dry season may partly explain the differences between the species (Dammhahn and Kappeler, 2010; Rakotondranary *et al.*, 2011b; Thorén *et al.*, 2011a).

Plasticity in sleeping habits

The comparison of the variability in the types of shelters used for daytime resting does not support the prediction of a higher plasticity in *M. murinus* than in its sympatric local congeners (Table 26.3). In contrast, *M. murinus* seems to be a tree hole specialist which has previously been pointed out as a possible limiting factor for this species (e.g., Ganzhorn and Schmid, 1998; Radespiel *et al.*, 2003a; Rendigs *et al.*, 2003; Lutermann *et al.*, 2010). *M. murinus* uses predominantly tree holes for daily resting as well as for rearing infants. On the other hand, *M. ravelobensis* and *M. berthae*, the locally restricted congeners, seem to be much more opportunistic with regard to their choice of sleeping sites, and not only when *M. murinus* shares the same habitat and could potentially monopolize high-quality tree holes (Radespiel *et al.*, 2003a). In conclusion, it is highly unlikely that a differential use of shelters can explain the divergent geographic range sizes in mouse lemurs. However, it has been stated that predation rates on *M. berthae* are larger than for *M. murinus* (Dammhahn and Kappeler, 2008a), which may be partially explained by a lower quality of their sleeping sites. If so, such higher species-specific mortality rates may in the long term limit population growth and the potential for range expansion.

Hypothesis 8: interspecific competition

It can be predicted that a species with relatively high competitive potential may be able to expand its geographic range size despite the possible presence of other ecologically similar congeners (Case *et al.*, 2005; Cox and Moore, 2010). In contrast, a species with a relatively low competitive potential may not be able to expand into areas where its ecological niche may already be filled by another species (Darwin, 1859; Hanski, 1982; Wilson and Keddy, 1986; Walck *et al.*, 1999). How can relative competitive potential be established in the case of mouse lemurs? Direct accounts of interspecific encounters and interactions are rare in mouse lemurs under natural conditions. This can be explained partly by their solitary mode of foraging and partly by limited contact times during behavioral observations. Contact times are usually constrained by (1) the small body size of mouse lemurs, (2) their general phenotypic similarity during nighttime observations, and (3) limited visibility in the dense vegetation at night. Therefore, the relative competitive potential of sympatric mouse lemurs is typically either inferred from indirect evidence (e.g., patterns of habitat or resource use) or from controlled encounter experiments that can be conducted in a cage setting. The former approach has been taken in the species pairs *M. murinus*/*M. griseorufus* in the Mandena/Andohahela region (southern Madagascar) and *M. murinus*/*M. berthae* from the Kirindy region (western Madagascar), whereas the latter approach has been chosen in the case of the species pair *M. murinus*/*M. ravelobensis* in the Ankarafantsika National Park (northwestern Madagascar). In all three cases, evidence suggests that *M. murinus* indeed has a higher competitive potential than its sympatric congener with smaller geographic range size. In the case of *M. murinus*/*M. griseorufus* this conclusion was based on the observation that *M. griseorufus* is absent from more mesic sites in southeastern Madagascar whenever *M. murinus* is present (Rakotondranary *et al.*, 2011a), but inhabits such habitat types at even higher population densities than those established in drier vegetation formations in southwestern Madagascar where *M. murinus* is absent (Rakotondranary *et al.*, 2010; Bohr *et al.*, 2011). In the case of *M. murinus*/*M. berthae* the conclusion is based on (1) the large difference in body mass between the small *M. berthae* (30.6 g ± 0.57 g SE) and *M. murinus* with almost double the size (57.9 ± 1.24 g SE; Schmid and Kappeler, 1994); (2) occasional behavioral observations of interspecific interactions at feeding sites (Dammhahn and Kappeler, 2008a); and (3) higher population densities in *M. murinus* than in *M. berthae* (Dammhahn and Kappeler, 2008a). Body mass, however, is not systematically different between *M. murinus* and *M. ravelobensis* in northwestern Madagascar (Zimmermann *et al.*, 1998), suggesting that physical strength should in this case not be automatically higher in one species than in the other. Systematic encounter experiments were therefore conducted between 14 female pairs of one *M. murinus* and one *M. ravelobensis* that were kept in cages for about one week (Thorén *et al.*, 2011b). In this experimental setting *M. murinus* won significantly more conflicts than *M. ravelobensis*. Moreover, *M. murinus* could be called dominant over their *M. ravelobensis* cage-mates in 87.5% of the decided dyads when applying the same methodological tools that are used to establish female dominance in lemur species (Thorén *et al.*, 2011b). These findings provide strong support for an asymmetry in agonistic power in both species with a higher competitive potential in *M. murinus* compared to the regional endemic *M. ravelobensis*. Taken together, these studies support the prediction derived from hypothesis 8, i.e., that the species with the larger geographic range size (*M. murinus*) should have a higher competitive potential than its congeners with smaller geographic range size.

Hypothesis 9: vagility

This hypothesis predicts that the species with a larger geographic range size, i.e., *M. murinus*, should have a higher vagility (translating into a larger colonization ability) than its congeners with limited geographical range sizes (Lomolino *et al.*, 2010). Such a higher vagility could be expressed in (1) higher migration rates (higher gene flow between populations), (2) higher dispersal rates or larger dispersal distances, and/or (3) a better ability to cross physical barriers. The first and last prediction can only be tested with the help of molecular data. However, such data sets are not yet available for all sympatric mouse lemur species pairs.

Previous studies on the species pair *M. murinus*/*M. ravelobensis* indeed indicate that populations of *M. murinus* in northwestern Madagascar seem to be more connected and less differentiated than those of *M. ravelobensis* (Table 26.4). This higher connectivity is expressed in lower F_{ST} values between populations, lower rates of correct assignment of individuals to their source population, and a lower disruptive effect of roads and savannahs on gene flow than in their local congener (details in Table 26.4). These results suggest that *M. murinus* may be able to maintain gene flow at higher rates and/or longer distances than its local congener, even in view of some degree of habitat disturbance or discontinuity. Unfortunately, comparable data on absolute dispersal distances and on the rate with which dispersal leads to successful settlement and breeding in the new environment are not available so far. Published studies suggest, however, that young male *M. murinus* disperse in larger numbers earlier in the season (before their first mating season; Radespiel *et al.*, 2003b) than young *M. ravelobensis* males who seem to delay their dispersal for at least several months (Radespiel *et al.*, 2009). Because it is known that predation rates on mouse lemurs are generally high (Goodman *et al.*, 1993; Goodman, 2003; Scheumann *et al.*, 2007) and yearly survival rates are relatively low (Schmelting *et al.*, 2007; Kraus *et al.*, 2008), it is possible that a 6-month difference in the timing of dispersal may already translate into species differences in settlement rates and long-term breeding success of dispersers which are prerequisites for gene flow between populations. However, future studies are needed to evaluate this hypothesis in more detail. Currently available data suggest that neither species is maintaining gene flow across the large rivers (Betsiboka, Mahajamba) that limit this inter-river system to the south and to the north, respectively (Guschanski *et al.*, 2007; Schneider *et al.*, 2010). Nevertheless, the populations of *M. murinus* from adjacent inter-river systems have not yet undergone speciation (Olivieri *et al.*, 2007) which has been suggested to be the result of a rather recent colonization (late Pleistocene) of northwestern Madagascar by *M. murinus* (Olivieri *et al.*, 2007; Schneider *et al.*, 2010) via the highland habitats of central Madagascar and possibly the riverine vegetation along the high elevation headwaters of large rivers (Wilmé *et al.*, 2006).

Although a very high proportion of young males disperse in western *M. murinus* (Kirindy) over distances that are comparable to those known from northwestern Madagascar (Schliehe-Diecks *et al.*, 2012), genetic differentiation between adjacent social neighbourhoods was nevertheless significant and not in agreement with a very high degree of connectivity and gene flow (Fredsted *et al.*, 2005; Table 26.4). This relatively strong local genetic structure has been attributed to the heterogeneous distribution of matrilinear female aggregations in the forest that are separated from each other by stretches (several 100 m) of "empty" forest (Wimmer *et al.*, 2002; Fredsted *et al.*, 2005; Dammhahn and Kappeler, 2008a). Although the reasons for this uneven distribution remain unclear, it is plausible that such a gap in distribution may indicate some unfavorable ecological conditions which may have a filtering effect, thereby reducing gene flow and migratory activities across this gap beyond the effect of distance alone. Such a discontinuity in distribution has so far not been reported from other sites. Unfortunately, comparable data from the sympatric *M. berthae* are not yet available beyond preliminary indications of male-biased dispersal and a lack of matrilinear female aggregations in this species

Table 26.4 Vagility of *M. murinus* compared to its sympatric congeners.

Species pair	Species	Region	Migration rates and gene flow	Dispersal pattern and dispersal distances	Ability to cross physical barriers (road, savannah, river, mountain)	References
M. murinus /*M. ravelobensis*	*M. murinus*	NW	F_{ST}: 67.9% sign. values[1] 45.7% of indiv. correctly assigned to source population[1]	Male-biased dispersal: 88.5% of young males disappear before 1st mating season[3] Dispersal distances: up to 1000 m[3]	No disruption of gene flow by road[1] Moderate disruption of gene flow by savannah[1,5] No gene flow across large rivers[5]	[1]von der Wall, 2008 [2]Radespiel *et al.*, 2008b [3]Radespiel *et al.*, 2003b [4]Radespiel *et al.*, 2009 [5]Schneider *et al.*, 2010 [6]Guschanski *et al.*, 2007 [7]Olivieri *et al.*, 2007
	M. ravelobensis		F_{ST}: 88.5% sign. values[2] 92.5% of indiv. correctly assigned to source population[2]	Male-biased dispersal but 52.6% of young males still philopatric in 1st mating season[4] Dispersal of young males delayed[4] Dispersal distances: n.a.	Moderate disruption of gene flow by road[2] Strong disruption of gene flow by savannah[2,6] No gene flow across large rivers[6,7]	

(continued)

Table 26.4 *(cont.)*

Species pair	Species	Region	Migration rates and gene flow	Dispersal pattern and dispersal distances	Ability to cross physical barriers (road, savannah, river, mountain)	References
M. murinus /*M. berthae*	*M. murinus*	W	F_{ST}: values significant even at local scale (area spanning 12.3 km^2)[13]	Male-biased dispersal: - 83.3% of young males disperse before/within 1st mating season[10] - Males further away from relatives than females[12] - Large no. of unique haplotypes among adult males[14] Dispersal distances: 180–960 m[10]	Relatively small neighborhoods consisting of matrilineal female aggregations separated by long stretches of "empty" forest → uneven distribution promotes genetic drift/inbreeding and limits connectivity between subpopulations[13,14]	[10]Schliehe-Diecks *et al.*, 2012 [11]Dammhahn and Kappeler, 2005 [12]Fredsted *et al.*, 2004 [13]Fredsted *et al.*, 2005 [14]Wimmer *et al.*, 2002
	M. berthae		n.a.	Male-biased dispersal: - females more closely related than males[11] - females have more relatives than males[11]	n.a.	
M. murinus /*M. griseorufus*	*M. murinus* *M. griseorufus*	S	F_{ST}: similar values in both species on small spatial scale[8] F_{ST}: values higher in *M. murinus* on larger spatial scale[9]	n.a. n.a.	n.a. n.a.	[8]Hapke *et al.*, 2011 [9]Gligor *et al.*, 2009

(Dammhahn and Kappeler, 2005). Taken together, a direct comparison of the vagility of both sympatric species is presently not possible.

In the case of *M. murinus*/*M. griseorufus* only some first-population genetics data but no dispersal data are yet available (Gligor *et al.*, 2009; Hapke *et al.*, 2011). These preliminary data (F_{ST} values) do not support a higher connectivity of subpopulations of *M. murinus* compared to *M. griseorufus* (Table 26.4) but a rather similar degree of genetic differentiation among subpopulations of comparable spatial settings. Both species do hybridize along their contact zones (Gligor *et al.*, 2009; Hapke *et al.*, 2011), but the patterns of introgression are rather symmetric and do not allow us to conclude that one species has a higher vagility than the other.

Taken together, the available evidence provides partial support (mostly from northwestern Madagascar) for the hypothesis of a higher vagility in *M. murinus* compared to its sympatric congeners with limited geographic range size.

Synthesis

The comparative evaluation of the explanatory power of available biogeographic hypotheses suggests that a combination of several behavioral and ecological factors may have facilitated the spatial expansion of *M. murinus* into most of western Madagascar. More specifically, it appears that a high seasonal plasticity in feeding habits that may be coupled with highly efficient metabolic energy-saving strategies (e.g., the ability for prolonged (seasonal) torpor or lower locomotor activity) during the lean season, a relatively high vagility, and a relatively high competitive potential may best explain the high colonization ability of *M. murinus*. Based on the reviewed evidence and some biogeographic background information from Madagascar, the following evolutionary scenario is proposed.

It is generally believed that periods of glaciation in the temperate zones were paralleled by relatively cold and arid conditions in the tropics that caused tropical forests to contract into Pleistocene refugia (e.g., Anthony *et al.*, 2007; Tosi, 2008). Subsequently, these forests could spatially expand when climatic conditions became favorable again. Wilmé *et al.* (2006) explicitly incorporated Pleistocene vegetation changes into a "watershed model" and hypothesized that 13 rivers with high-elevation headwaters across Madagascar have served as palaeo-refugia, maintaining riparian forests, to where forest-dwelling species may have retreated during the driest periods. As conditions became wetter, riparian habitats expanded into lower altitudes again, permitting range expansions and dispersal processes to happen. In between watersheds, the model proposes the existence of isolated centers of endemism, in which vicariant events led to extensive allopatric speciation.

Within such a general biogeographic scenario for Madagascar, divergent range sizes of different mouse lemur species are most likely the result of differences in the altitudinal plasticity of species that may or may not have enabled them to use high altitude habitats as migratory pathways into other regions of the island ("restricted gene flow hypothesis"; Craul *et al.*, 2008). Schneider *et al.* (2010) already provided some evidence for a large-scale spatial range expansion of *M. murinus* in the course of the late Pleistocene into northwestern Madagascar. Such a recent colonization history was probably facilitated by traits that enable this species to survive in unpredictable, dry and harsh, seasonal environmental conditions, i.e., those that it may have encountered in the highland area of Madagascar during these times of climatic changes. The ability to modify and adapt dietary habits to seasonally and spatially changing food availability (see Hypothesis 7) was probably highly beneficial for *M. murinus* under such circumstances. It is even possible that *M. murinus* evolved in isolation somewhere in such highly seasonal, high-altitude habitats in the course of this period of major climate dynamics. The last common ancestor of western and southern *M. murinus* was recently dated to a median age of 670,000 years (343,000–1,066,000 years, 95% highest posterior density; Thiele *et al.*, 2013) which would lend support to such a succession of events. Largely in congruence, another recent study detected signals of demographic expansion for western and southern *M. murinus* in the Quaternary around 1 Ma (Blair *et al.*, 2014). The ability to adaptively reduce metabolic rates and to save energy in times of seasonal food scarcity may have been another important trait, although it seems that the ability to undergo prolonged periods of torpor depends mainly on air temperature, body size, and individual condition, and is not an autapomorphy of *M. murinus*, but exists in several larger-bodied mouse lemur species and may therefore rather be a symplesiomorphic trait of mouse lemurs (*M. murinus*: Schmid and Kappeler, 1998; Vuarin *et al.*, 2013; *M. rufus*: Atsalis, 1999; *M. lehilahytsara*: Randrianambinina *et al.*, 2003; *M. griseorufus*: Kobbe *et al.*, 2011).

A relatively high vagility would have further facilitated expansion and colonization activities of *M. murinus*. Indicators of a high vagility stem so far mainly from northwestern Madagascar (see Hypothesis 9, Table 26.4), the area where this species possibly arrived last (Schneider *et al.*, 2010). Whether the signals of high gene flow and connectivity between the subpopulations of *M. murinus* in this area are indeed the result of higher migratory potential, or rather the result of a shorter period of time over which genetic drift may have affected the populations, cannot yet be completely answered (Schneider *et al.*, 2010). This second explanation is plausible, because the molecular data sets from the species pairs in southern and western Madagascar do not presently support a strong difference regarding the vagility of *M. murinus* and *M. berthae* or *M. griseorufus*, respectively. However, much more data will be needed to decide the relative importance of these two hypotheses.

What seems clear, however, is that once *M. murinus* colonized a certain area, this species was probably not negatively affected by the presence of another mouse lemur species in local forests: the competitive potential of *M. murinus* seems to be superior to any of the other well-known sympatric congeners (see Hypothesis 8) and its distribution and expansion was therefore probably constrained rather by its own ecological requirements (e.g., of high-quality sleeping sites) than by interspecific competition with local sympatric congeners (Radespiel *et al.*, 2003a; Rakotondravony and Radespiel, 2009; Rakotondranary *et al.*, 2011a). This higher competitive potential may have facilitated expansion into areas where parts of its fundamental niche were already occupied by ecologically similar congeners (Thorén *et al.*, 2011b). It cannot be excluded, however, that the expanding *M. murinus* experienced some degree of interspecific competition by other species, such as the larger dwarf lemurs (*Cheirogaleus* spp.; Schwab and Ganzhorn, 2004) or some of the larger and partly arboreal endemic rodents (Garbutt, 2007; Muldoon and Goodman, 2010).

This review provides evidence for some behavioral and ecological traits that may have enabled ancestral *M. murinus* populations to expand (feeding plasticity, vagility) and to colonize (high competitive potential) new areas beyond the immediate region in which they evolved. Knowledge of the plasticity in the species-specific behavioral ecology of these small and cryptic lemur species therefore enables us to better understand their biogeography and may also enable us in the long-term to better predict and understand their potential to cope with environmental challenges such as future climate change (Hannah *et al.*, 2008; Busch *et al.*, 2012) and increasing anthropogenic disturbances and habitat fragmentation (Green and Sussman, 1990; Allnutt *et al.*, 2008; Irwin *et al.*, 2010). This review, however, also shows that we are still far from fully understanding these complex questions. Our present view of mouse lemurs is mostly based on a preliminary data set collected on a handful of species. Many mouse lemur species are still known only from single or very few locations (Chapter 27). Consequently, a fundamental parameter, their geographic range size, is unknown (Gaston, 1994), and data on their ecological plasticity and behavioral adaptations are not yet available. In conclusion, our knowledge is still largely incomplete and urgently deserves further attention for many more years to come. Inter- and intraspecific comparisons from many areas of the island would be useful and even essential to fully understand the evolution of these small, intriguing "creatures of the dark" (Doyle and Izard, 1995).

Acknowledgments

I thank Shawn Lehman for proposing the idea of this book to me and for advancing this project with much energy and dedication. I furthermore thank Elke Zimmermann and the Institute of Zoology for continuous support and all my present and former students for contributing to the growing knowledge on mouse lemur behavioral ecology. Special thanks go to our Malagasy collaborators, and in particular to Blanchard Randrianambinina, Solofonirina Rasoloharijaona, and Romule Rakotondravony, without whom our long-term work on mouse lemurs would have never been possible. Finally, thanks go to two anonymous reviewers whose comments improved the manuscript.

References

Allnutt TF, Ferrier S, Manion G, *et al.* 2008. A method for quantifying biodiversity loss and its application to a 50-year record of deforestation across Madagascar. *Conservation Letters* 1:173–181.

Anthony NM, Johnson-Bawe M, Jeffery K, *et al.* 2007. The role of Pleistocene refugia and rivers in shaping gorilla genetic diversity in central Africa. *Proceedings of the National Academy of Sciences of the United States of America* 104:20432–20436.

Atsalis S. 1999. Seasonal fluctuations in body fat and activity levels in a rain-forest species of mouse lemur, *Microcebus rufus*. *International Journal of Primatology* 20:883–910.

Blair C, Heckman KL, Russell AL, Yoder AD. 2014. Multilocus coalescent analyses reveal the demographic history and speciation patterns of mouse lemur sister species. *BMC Evolutionary Biology* 14:57.

Blanco MB. 2008. Reproductive schedules of female *Microcebus rufus* at Ranomafana National Park, Madagascar. *International Journal of Primatology* 29:323–338.

Blanco MB. 2011. Timely estrus in wild brown mouse lemur females at Ranomafana National Park, southeastern Madagascar. *American Journal of Physical Anthropology* 145:311–317.

Blanco MB, Meyer JS. 2009. Assessing reproductive profiles in female brown mouse lemurs (*Microcebus rufus*) from Ranomafana National Park, southeast Madagascar, using fecal hormone analysis. *American Journal of Primatology* 71:439–446.

Bohr YE-MB, Giertz P, Ratovonamana YR, Ganzhorn JU. 2011. Gray-brown mouse lemurs (*Microcebus griseorufus*) as an example of distributional constraints through increasing desertification. *International Journal of Primatology* 32:901–913.

Brown JH. 1984. On the relationship between abundance and distribution of species. *American Naturalist* 124:255–279.

Brown JH. 1995. *Macroecology*. University of Chicago Press, Chicago.

Busch J, Dave R, Hannah L, *et al.* 2012. Climate change and the cost of conserving species in Madagascar. *Conservation Biology* 26:408–419.

Case TJ, Holt RD, McPeek MA, Keitt TH. 2005. The community context of species' borders: ecological and evolutionary perspectives. *Oikos* 108:28–46.

Chanu L, Goetze D, Rajeriarison C, *et al.* 2012. Can differences in floristic composition explain variation in the abundance of two sympatric mouse lemur species (*Microcebus*) in the Ankarafantsika National Park, northwestern Madagascar? *Malagasy Nature* 6:83–102.

Cowlishaw G, Hacker JE. 1997. Distribution, diversity, and latitude in African primates. *American Naturalist* 150:505–512.

Cox CB, Moore PD. 2010. *Biogeography: An Ecological and Evolutionary Approach*, 8th edition. John Wiley & Sons, Chichester.

Craul M, Radespiel U, Rasolofoson DW, *et al.* 2008. Large rivers do not always act as species barriers for *Lepilemur* sp. *Primates* 49:211–218.

Crottini A, Madsen O, Poux C, *et al.* 2012. Vertebrate time-tree elucidates the biogeographic pattern of a major biotic change around the K-T boundary in Madagascar. *Proceedings of the National Academy of Sciences of the United States of America* 109:5358–5363.

Dammhahn M, Kappeler PM. 2005. Social system of *Microcebus berthae*, the world's smallest primate. *International Journal of Primatology* 26:407–435.

Dammhahn M, Kappeler PM. 2008a. Small-scale coexistence of two mouse lemur species (*Microcebus berthae* and *M. murinus*) within a homogeneous competitive environment. *Oecologia* 157:473–483.

Dammhahn M, Kappeler PM. 2008b. Comparative feeding ecology of sympatric *Microcebus berthae* and *M. murinus*. *International Journal of Primatology* 29:1567–1589.

Dammhahn M, Kappeler PM. 2010. Scramble or contest competition over food in solitarily foraging mouse lemurs (*Microcebus* spp.): new insights from stable isotopes. *American Journal of Physical Anthropology* 141:181–189.

Darwin C. 1859. *On the Origin of Species by means of Natural Selection, or the Preservation of Favoured Races in the Struggle for Life*, 1st edition. John Murray, London.

Dewar RE, Richard AF. 2007. Evolution in the hypervariable environment of Madagascar. *Proceedings of the National Academy of Sciences of the United States of America* 104:13723–12727.

Doyle GA, Izard MK (eds.) 1995. *Creatures of the Dark: the Nocturnal Prosimians*. Plenum Press, New York.

Eberle M, Kappeler PM. 2004. Selected polyandry: female choice and inter-sexual conflict in a small nocturnal solitary primate (*Microcebus murinus*). *Behavioral Ecology and Sociobiology* 57:91–100.

Eberle M, Kappeler PM. 2006. Family insurance: kin selection and cooperative breeding in a solitary primate (*Microcebus murinus*). *Behavioral Ecology and Sociobiology* 60:582–588.

Eeley HAC, Foley RA. 1999. Species richness, species range size and ecological specialisation among African primates: geographical patterns and conservation implications. *Biodiversity and Conservation* 8:1033–1056.

Fredsted T, Pertoldi C, Olesen JM, Eberle M, Kappeler PM. 2004. Microgeographic heterogeneity in spatial distribution and mtDNA variability of gray mouse lemurs (*Microcebus murinus*, Primates: Cheirogaleidae). *Behavioral Ecology and Sociobiology* 56:393–403.

Fredsted T, Pertoldi C, Schierup MH, Kappeler PM. 2005. Microsatellite analyses reveal fine-scale genetic structure in grey mouse lemurs (*Microcebus murinus*). *Molecular Ecology* 14:2363–2372.

Ganzhorn JU. 1994. Lemurs as indicators for habitat change. In B Thierry, JR Anderson, JJ Roeder, N Herrenschmidt (eds.), *Current Primatology. Vol.1. Ecology and Evolution* (pp. 51–56). Universite Louis Pasteur, Strasbourg.

Ganzhorn JU. 1995. Low-level forest disturbance effects on primary production, leaf chemistry, and lemur populations. *Ecology* 76:2084–2096.

Ganzhorn JU, Schmid J. 1998. Different population dynamics of *Microcebus murinus* in primary and secondary deciduous dry forest of Madagascar. *International Journal of Primatology* 19:785–796.

Garbutt N. 2007. *Mammals of Madagascar: A Complete Guide*. Yale University Press, New Haven.

Gaston KJ. 1994. Measuring geographic range sizes. *Ecography* 17:198–205.

Génin F. 2008. Life in unpredictable environments: first investigation of the natural history of *Microcebus griseorufus*. *International Journal of Primatology* 29:303–321.

Gligor M, Ganzhorn JU, Rakotondravony D, *et al.* 2009. Hybridization between mouse lemurs in an ecological transition zone in southern Madagascar. *Molecular Ecology* 18:520–533.

Goodman SM. 2003. Predation on lemurs. In SM Goodman, JP Benstead (eds.), *The Natural History of Madagascar* (pp. 1221–1228). University of Chicago Press, Chicago.

Goodman SM, Ganzhorn JU. 2004. Biogeography of lemurs in the humid forests of Madagascar: the role of elevational distribution and rivers. *Journal of Biogeography* 31:47–55.

Goodman SM, O'Connor S, Langrand O. 1993. A review of predation on lemurs: Implications for the evolution of social behavior in small, nocturnal primates. In PM Kappeler, JU Ganzhorn (eds.), *Lemur Social Systems and Their Ecological Basis* (pp. 51–66). Plenum Press, New York.

Green GM, Sussman RW. 1990. Deforestation history of the eastern rain forests of Madagascar from satellite images. *Science* 248:212–215.

Guschanski K, Olivieri G, Funk SM, Radespiel U. 2007. MtDNA reveals strong genetic differentiation among geographically isolated populations of the golden brown mouse lemur, *Microcebus ravelobensis*. *Conservation Genetics* 8:809–821.

Hannah L, Dave R, Lowry PP II, *et al.* 2008. Climate change adaptation for conservation in Madagascar. *Biology Letters* 4:590–594.

Hanski I. 1982. Dynamics of regional distribution: the core and satellite species hypothesis. *Oikos* 38:210–221.

Hanski I, Gyllenberg M. 1997. Uniting two general patterns in the distribution of species. *Science* 275:397–400.

Hapke A, Gligor M, Rakotondranary SJ. 2011. Hybridization of mouse lemurs: different patterns under different ecological conditions. *BMC Evolutionary Biology* 11:297.

Harcourt AH, Coppeto SA, Parks SA. 2002. Rarity, specialization and extinction in primates. *Journal of Biogeography* 29:445–456.

Irwin MT, Wright PC, Birkinshaw C, *et al.* 2010. Patterns of species change in anthropogenically disturbed forests of Madagascar. *Biological Conservation* 143:2351–2362.

Kappeler PM, Rasoloarison RM, Razafimanantsoa L, Walter L, Roos C. 2005. Morphology, behaviour and molecular evolution of giant mouse lemurs (*Mirza* spp.) Gray, 1870, with description of a new species. *Primate Report* 71:3–26.

Kobbe S, Ganzhorn JU, Dausmann KH. 2011. Extreme individual flexibility of heterothermy in free-ranging Malagasy mouse lemurs (*Microcebus griseorufus*). *Journal of Comparative Physiology B – Biochemical Systemic and Environmental Physiology* 181:165–173.

Kraus C, Eberle M, Kappeler PM. 2008. The costs of risky male behaviour: sex differences in seasonal survival in a small sexually monomorphic primate. *Proceedings of the Royal Society B, Biological Sciences* 275:1635–1644.

Krebs CJ. 1998. *Ecological Methodology*. Addison-Wesley Educational Publishers, Menlo Park, CA.

Lahann P, Schmid J, Ganzhorn JU. 2006. Geographic variation in populations of *Microcebus murinus* in Madagascar: resource seasonality or Bergmann's rule? *International Journal of Primatology* 27:983–999.

Lomolino MV, Riddle BR, Whittaker RJ, Brown JH. 2010. *Biogeography*, 4th edition. Sinauer Associates, Inc., Sunderland, MA.

Lutermann H, Verburgt L, Rendigs A. 2010. Resting and nesting in a small mammal: sleeping sites as a limiting resource for female grey mouse lemurs. *Animal Behaviour* 79:1211–1219.

Mittermeier RA, Louis EE Jr, Richardson M, *et al.* 2010. *Lemurs of Madagascar*, 3rd edition. Conservation International, Arlington.

Muldoon KM, Goodman SM. 2010. Ecological biogeography of Malagasy non-volant mammals: community structure is correlated with habitat. *Journal of Biogeography* 37:1144–1159.

Olivieri G, Craul M, Radespiel U. 2005. Inventaire des lémuriens dans 15 fragments de forêt de la province de Mahajanga. *Lemur News* 10:11–16.

Olivieri G, Zimmermann E, Randrianambinina B, *et al.* 2007. The ever-increasing diversity in mouse lemurs: three new species in north and northwestern Madagascar. *Molecular Phylogenetics and Evolution* 43:309–327.

Olson LE, Goodman SM, Yoder AD. 2004. Illumination of cryptic species boundaries in long-tailed shrew tenrecs (Mammalia: Tenrecidae; *Microgale*), with new insights into geographic variation and distributional constraints. *Biological Journal of the Linnean Society* 83:1–22.

Pearson RG, Raxworthy CJ. 2009. The evolution of local endemism in Madagascar: watershed versus climatic gradient hypotheses evaluated by null biogeographic models. *Evolution* 63:959–967.

Poux C, Madsen O, Marquard E, *et al.* 2005. Asynchronous colonization of Madagascar by the four endemic clades of primates, tenrecs, carnivores, and rodents as inferred from nuclear genes. *Systematic Biology* 54:719–730.

Radespiel U. 2006. Ecological diversity and seasonal adaptations of mouse lemurs (*Microcebus* spp.). In L Gould, ML Sauther (eds.), *Lemurs: Ecology and Adaptation* (pp. 211–233). Springer, New York.

Radespiel U, Ehresmann P, Zimmermann E. 2003a. Species-specific usage of sleeping sites in two sympatric mouse lemur species (*Microcebus murinus* and *M. ravelobensis*) in northwestern Madagascar. *American Journal of Primatology* 59:139–151.

Radespiel U, Lutermann H, Schmelting B, Bruford MW, Zimmermann E. 2003b. Patterns and dynamics of sex-biased dispersal in a nocturnal primate, the grey mouse lemur, *Microcebus murinus*. *Animal Behaviour* 65:709–719.

Radespiel U, Olivieri G, Rasolofoson DW, *et al.* 2008a. Exceptional diversity of mouse lemurs (*Microcebus* spp.) in the Makira region with the description of one new species. *American Journal of Primatology* 70:1033–1046.

Radespiel U, Rakotondravony R, Chikhi L. 2008b. Natural and anthropogenic determinants of genetic structure in the largest remaining population of the endangered golden-brown mouse lemur, *Microcebus ravelobensis*. *American Journal of Primatology* 70:860–870.

Radespiel U, Jurić M, Zimmermann E. 2009. Sociogenetic structures, dispersal and the risk of inbreeding in a small nocturnal lemur, the golden-brown mouse lemur (*Microcebus ravelobensis*). *Behaviour* 146:607–628.

Radespiel U, Ratsimbazafy JH, Rasoloharijaona S, *et al.* 2012. First indications of a highland specialist among mouse lemurs (*Microcebus* spp.) and evidence for a new mouse lemur species from eastern Madagascar. *Primates* 53:157–170.

Rakotondranary SJ, Ganzhorn JU. 2011. Habitat separation of sympatric *Microcebus* spp. in the dry spiny forest of south-eastern Madagascar. *Folia Primatologica* 82:212–223.

Rakotondravony R, Radespiel U. 2007. Variation de la distribution de deux espèces de microcèbes dans le Parc National Ankarafantsika. *Lemur News* 12:31–34.

Rakotondravony R, Radespiel U. 2009. Varying patterns of coexistence of two mouse lemur species (*Microcebus ravelobensis* and *M. murinus*) in a heterogeneous landscape. *American Journal of Primatology* 71:928–938.

Rakotondranary J, Ratovonamana RY, Ganzhorn JU. 2010. Distributions et caractéristiques des microhabitats de *Microcebus griseorufus* (Cheirogaleidae) dans le Parc National de Tsimanampetsotsa (Sud-ouest de Madagascar). *Malagasy Nature* 4:55–64.

Rakotondranary SJ, Hapke A, Ganzhorn JU. 2011a. Distribution and morphological variation of *Microcebus* spp. along an environmental gradient in southeastern Madagascar. *International Journal of Primatology* 32:1037–1057.

Rakotondranary SJ, Struck U, Knoblauch C, Ganzhorn JU. 2011b. Regional, seasonal and interspecific variation in ^{15}N and ^{13}C in sympatric mouse lemurs. *Naturwissenschaften* 98:909–917.

Randrianambinina B, Rakotondravony D, Radespiel U, Zimmermann E. 2003. Seasonal changes in general activity, body mass and reproduction of two small nocturnal primates: a comparison of the golden brown mouse lemur (*Microcebus ravelobensis*) in northwestern Madagascar and the brown mouse lemur (*Microcebus rufus*) in eastern Madagascar. *Primates* 44:321–331.

Rapoport EH. 1982. *Areography: Geographical Strategies of Species*. Pergamon Press, New York.

Rasoloarison RM, Weisrock DW, Yoder AD, *et al.* 2013. Two new species of mouse lemurs (Cheirogaleidae: *Microcebus*) from eastern Madagascar. *International Journal of Primatology* 34:455–469.

Rendigs A, Radespiel U, Wrogemann D, Zimmermann E. 2003. Relationship between microhabitat structure and distribution of mouse lemurs (*Microcebus* spp.) in northwestern Madagascar. *International Journal of Primatology* 24:47–64.

Samonds KE, Godfrey LR, Ali JR, *et al.* 2012. Spatial and temporal arrival patterns of Madagascar's vertebrate fauna explained by distance, ocean currents, and ancestor type. *Proceedings of the National Academy of Sciences, USA* 109:5352–5357.

Scheumann M, Rabesandratana A, Zimmermann E. 2007. Predation, communication, and cognition in lemurs. In SL Gursky, KAI Nekaris (eds.), *Primate Anti-Predator Strategies* (pp. 100–126). Springer, Oxford.

Schliehe-Diecks S, Eberle M, Kappeler PM. 2012. Walk the line – dispersal movements of gray mouse lemurs (*Microcebus murinus*). *Behavioral Ecology and Sociobiology* 66:1175–1185.

Schmelting B, Ehresmann P, Lutermann H, Randrianambinina B, Zimmermann E. 2000. Reproduction of two sympatric mouse lemur species (*Microcebus murinus* and *M. ravelobensis*) in north-west Madagascar: first results of a long term study. In WR Lourenço, SM Goodman (eds.), *Diversité et Endémisme à Madagascar* (pp. 165–175). Société de Biogéographie, Paris.

Schmelting B, Zimmermann E, Berke O, Bruford MW, Radespiel U. 2007. Experience-dependent recapture rates and reproductive success in male grey mouse lemurs (*Microcebus murinus*). *American Journal of Physical Anthropology* 133:743–752.

Schmid J. 1998. Tree holes used for resting by gray mouse lemurs (*Microcebus murinus*) in Madagascar: insulation capacities and energetic consequences. *International Journal of Primatology* 19:797–809.

Schmid J, Kappeler PM. 1994. Sympatric mouse lemurs (*Microcebus* spp.) in western Madagascar. *Folia Primatologica* 63:162–170.

Schmid J, Kappeler PM. 1998. Fluctuating sexual dimorphism and differential hibernation by sex in a primate, the gray mouse lemur (*Microcebus murinus*). *Behavioral Ecology and Sociobiology* 43:125–132.

Schneider N, Chikhi L, Currat M, Radespiel U. 2010. Signals of recent spatial expansions in the grey mouse lemur (*Microcebus murinus*). *BMC Evolutionary Biology* 10:105.

Schwab D. 2000. A preliminary study of spatial distribution and mating system of pygmy mouse lemurs (*Microcebus cf myoxinus*). *American Journal of Primatology* 51:41–60.

Schwab D, Ganzhorn JU. 2004. Distribution, population structure and habitat use of *Microcebus berthae* compared to those of other sympatric cheirogalids. *International Journal of Primatology* 25:307–330.

Stevens GC. 1989. The latitudinal gradient in geographical range: how so many species coexist in the tropics. *American Naturalist* 133:240–256.

Thiele D, Razafimahatratra E, Hapke A. 2013. Discrepant partitioning of genetic diversity in mouse lemurs and dwarf lemurs – biological reality or taxonomic bias? *Molecular Phylogenetics & Evolution* 69:593–609.

Thorén S, Quietzsch F, Radespiel U. 2010. Leaf nest use and construction in the golden-brown mouse lemur (*Microcebus ravelobensis*) in the Ankarafantsika National Park. *American Journal of Primatology* 72:48–55.

Thorén S, Quietzsch F, Schwochow D, *et al.* 2011a. Seasonal changes in feeding ecology and activity patterns of two sympatric mouse lemur species, the gray mouse lemur (*Microcebus murinus*) and the golden-brown mouse lemur (*M. ravelobensis*), in northwestern Madagascar. *International Journal of Primatology* 32:566–586.

Thorén S, Linnenbrink M, Radespiel U. 2011b. Different competitive potential in two coexisting mouse lemur species in northwestern Madagascar. *American Journal of Physical Anthropology* 145(1):156–162.

Tosi AJ. 2008. Forest monkeys and Pleistocene refugia: a phylogeographic window onto the disjunct distribution of the *Chlorocebus lhoesti* species group. *Zoological Journal of the Linnean Society* 154:408–418.

Townsend TM, Vieites DR, Glaw F, Vences M. 2009. Testing species-level diversification hypotheses in Madagascar: the case of microendemic *Brookesia* leaf chameleons. *Systematic Biology* 58:641–656.

Vences M, Wollenberg KC, Vieites DR, Lees DC. 2009. Madagascar as a model region of species diversification. *Trends in Ecology and Evolution* 24:456–465.

von der Wall D. 2008. Der Einfluss von Habitatfragmentierung auf die genetischen Strukturen von *Microcebus murinus* (Grauer Mausmaki). Diploma thesis, Leibniz University Hanover. 74 p.

Vuarin P, Dammhahn M, Henry PY. 2013. Individual flexibility in energy saving: body size and condition constrain torpor use. *Functional Ecology* 27:793–799.

Walck JL, Baskin JM, Baskin CC. 1999. Relative competitive abilities and growth characteristics of a narrowly endemic and a geographically widespread *Solidago* species (Asteraceae). *American Journal of Botany* 86:820–828.

Weisrock DW, Smith SD, Chan LM, *et al.* 2012. Concatenation and concordance in the reconstruction of mouse lemur phylogeny: an empirical demonstration of the effect of allele sampling in phylogenetics. *Molecular Biology and Evolution* 29:1615–1630.

Wilmé L, Goodman SM. 2003. Biogeography, guild structure, and elevational variation of Madagascar forest birds. In SM Goodman, JP Benstead (eds.), *The Natural History of Madagascar* (pp. 1045–1058). University of Chicago Press, Chicago.

Wilmé L, Goodman SM, Ganzhorn JU. 2006. Biogeographic evolution of Madagascar's microendemic biota. *Science* 312:1063–1065.

Wilson SD, Keddy PA. 1986. Species competitive ability and position along a natural stress/disturbance gradient. *Ecology* 67:1236–1242.

Wimmer B, Tautz D, Kappeler PM. 2002. The genetic population structure of the gray mouse lemur (*Microcebus murinus*), a basal primate from Madagascar. *Behavioral Ecology and Sociobiology* 52:166–175.

Yoder AD, Nowak MD. 2006. Has vicariance or dispersal been the predominant biogeographic force in Madagascar? Only time will tell. *Annual Review of Ecology, Evolution, and Systematics* 37:405–431.

Yoder AD, Yang Z. 2004. Divergence dates for Malagasy lemurs estimated from multiple gene loci: geological and evolutionary context. *Molecular Ecology* 13:757–773.

Yoder AD, Cartmill M, Ruvolo M, Smith K, Vilgalys R. 1996. Ancient single origin for Malagasy primates. *Proceedings of the National Academy of Sciences*, USA 93:5122–5126.

Zimmermann E, Radespiel U. 2013. Life histories in Primates. In W Henke, I Tattersall (eds.), *Handbook of Paleoanthropology*. Springer-Verlag, Berlin.

Zimmermann E, Cepok S, Rakotoarison N, Zietemann V, Radespiel U. 1998. Sympatric mouse lemurs in North-West Madagascar: a new rufous mouse lemur species (*Microcebus ravelobensis*). *Folia Primatologica* 69:106–114.

27 Conservation biology of the Cheirogaleidae: future research directions

Shawn M. Lehman, Ute Radespiel, and Elke Zimmermann

Introduction

Many conservation biologists consider the lemurs of Madagascar to be one of the world's highest conservation priorities (Rylands and Konstant, 2000; Allnutt *et al.*, 2008; Kremen *et al.*, 2008). Of the 103 extant lemurs species monitored under the International Union for Conservation of Nature (IUCN) Red List, 70.8% (N = 73 species) are endangered or critically endangered. Moreover, 6 lemurs are on the most recent list of the 25 most endangered primates: *Eulemur flavifrons*, *Lepilemur septentrionalis*, *Propithecus candidus*, *Microcebus berthae*, *Varecia rubra*, and *Indri indri* (Mittermeier *et al.*, 2012). Considerable research and conservation efforts have been applied to studies of some threatened lemur species and their forest habitats (e.g., Kremen *et al.*, 1999; Seddon et al., 2000; Andrianandrasana *et al.*, 2005). As a result of these efforts, there is a rich literature on the status of some rare lemur species and the need for concerted efforts to integrate local communities with conservation biology programs in the establishment and maintenance of protected areas (Schwitzer *et al.*, 2014). However, most of these efforts have been applied to diurnal, large-bodied taxa, such as species in the genus *Propithecus* (Banks *et al.*, 2007; Patel and Andrianandrasana, 2008; Irwin *et al.*, 2009; Quéméré *et al.*, 2010).

Although the Cheirogaleidae represents the most speciose lemur family, with 35 species described to date (Chapter 2; IUCN), there are relatively few data on the conservation assessments and threats experienced by taxa in this family compared to large-bodied, diurnal taxa in Madagascar. This lack of conservation attention arises, in part, from suggestions that small-bodied, nocturnal taxa have relatively low extinction risks, are ubiquitous in their geographic distribution, suffer least from anthropogenic disturbance (e.g., hunting and deforestation), and have physiological adaptations such as torpor and hibernation in many species that mitigate extinction threats (Cardillo, 2003; Liow *et al.*, 2009; Hanna and Cardillo, 2014). The most endangered cheirogaleids have not benefitted from conservation campaigns

The Dwarf and Mouse Lemurs of Madagascar: Biology, Behavior and Conservation Biogeography of the Cheirogaleidae, eds. S.M. Lehman, U. Radespiel, E. Zimmermann. Published by Cambridge University Press. © Cambridge University Press 2016.

focusing on charismatic species, due largely to their nocturnality and concomitant difficulties in visual access for tourists. Recent assessments of extinction risks in lemurs indicate increased conservation concerns for many cheirogaleids, particularly in light of studies indicating the critical role of some species as seed dispersers for endemic fruiting plants (Valenta *et al.*, 2013), and their high levels of phylogenetic diversity (Lehman, 2006). Thus, a holistic approach to the conservation biology of cheirogaleids needs to consider their current conservation status in existing protected areas and where associated data are deficient for mitigating extinction threats.

Conservation threats for primates result primarily from varying kinds and intensities of habitat modification (deforestation, fire, logging), live capture, and hunting (Chapman and Peres, 2001). For cheirogaleids, the most pressing conservation threat relates directly to the massive anthropogenic conversion that has affected forest habitats on Madagascar. Although there is no consensus on the actual amount of anthropogenic deforestation that has occurred since humans arrived on the island (McConnell and Kull, 2014), landscape analyses of remote imaging indicate that since the 1950s approximately half of the forest cover has been converted to either agricultural fields, particularly in lowland areas ideally suited for rice culture, or deforested grasslands (Green and Sussman, 1990; Harper et al., 2007; Allnutt *et al.*, 2008). Afforestation projects have added to forest cover on the island, but most of these projects involve monocrops of pine and eucalyptus plantations to produce firewood (McConnell, 2002), representing exotic habitats with unknown value to endemic lemurs (Ganzhorn, 1987).

There is marked regional variation in deforestation patterns between western and eastern Madagascar. In western Madagascar, the landscape is composed primarily of vast tracts of grasslands surrounding fragments of tropical dry forest, both of which slope down from the central plateau to the western shore (Ganzhorn *et al.*, 1999). As of the year 2000, anthropogenic deforestation had resulted in the loss of approximately 97% of the original western dry forest cover (Whitmore and Sayer, 1992; Ganzhorn *et al.*, 2001; Allnutt *et al.*, 2008; but see Bond *et al.*, 2008; Willis *et al.*, 2008). In eastern Madagascar, the structure and composition of forest habitats vary largely as a function of altitude along a steep escarpment that maintains most of the remaining humid forest (Du Puy and Moat, 1996). Deforestation patterns are not as extensive in eastern Madagascar as those in the west, although recent political issues have accelerated the illegal exploitation of hardwood trees in eastern humid forests (Schwitzer *et al.*, 2014). Whether in the western dry or eastern humid forests, these deforestation patterns have resulted in the creation of smaller, more isolated fragments, which also suffer from varying intensities of edge effects. Despite these well-documented patterns of habitat modification on the island, there are few data on the effects of forest loss, fragmentation, and edges on Cheirogaleidae species, let alone populations. Most research has taken place in continuous forests inside protected areas (Mittermeier *et al.*, 2008). Moreover, many cheirogaleids exhibit disjunct distribution patterns across their geographic ranges, which complicate conservation assessments and management systems in existing and future protected areas strategies (Ganzhorn *et al.*, 2006).

In addition to anthropogenic disturbance of their forest habitats, cheirogaleids also experience some hunting pressures (Jenkins *et al.*, 2011; Ravaloharimanitra *et al.*, 2011; Razafimanahaka *et al.*, 2012; Golden *et al.*, 2013, 2014). Unlike hunters in some other primate-source regions (Hill *et al.*, 2003), Malagasy hunters have until recently been reticent to discuss hunting practices, let alone enable participant data collection by researchers on their hunting patterns. New data collected from interviews with local people on prey preferences indicate that cheirogaleids are hunted and eaten, but due to their small body size, nocturnality, and varying patterns of hibernation and torpor, are less preferred than some sympatric larger-bodied, diurnal species (Golden, 2009; Jenkins *et al.*, 2011; Razafimanahaka *et al.*, 2012). Live capture of cheirogaleids for the pet trade is also a seemingly rare occurrence in Madagascar (Mutschler and Feistner, 1995), although there is a growing concern that recent movie and celebrity fixation on lemurs as pets may increase the illegal trade in charismatic species. For example, a simple internet search for "lemur pets for sale" produces numerous websites and associated businesses dealing in the sale and purchase of lemurs as pets. At any rate, models of harvesting rates indicate that none of the diurnal lemurs hunted by local people are done so in a sustainable manner (Razafimanahaka *et al.*, 2012).

In this chapter, we describe the conservation biology of cheirogaleids. Our specific goals are to review cheirogaleid conservation status and biogeography in protected areas, assess the literature for relevant extinction threats, and provide future research directions to fill gaps in data on the conservation biology of species in this lemur family.

Conservation status

The conservation classification of cheirogaleids is based primarily on status, criteria, and threat assessments made by "volunteer experts" associated with the Species Survival Commission of the IUCN. Each species is classified into one of nine IUCN Red List categories (IUCN, 2014). The alphabetized criteria used to determine these categories focus on population dynamics (A, C, D: size, decline, level of restriction), geographic range (B: occurrence or occupancy), and quantitative analysis of extinction probability (E). For example, a species listed as Endangered A1 must experience a 70%–90% population reduction where the causes of the reduction are clearly reversible, understood, and have ceased (IUCN, 2014). Threats are based on a complex assessment system that considers everything from residential and commercial development through natural system modifications to climate change and severe weather. Greater emphasis is placed on what are termed major threats, which refer to high-impact, continuing threats, such as large-scale, intentional logging. It is important to note that each species' status reflects a dynamic process of criteria and threats, which is updated by assessors every 5–10 years.

A simple comparison of Red List classification between 2008 and 2012 is illustrative of changes in conservation status for Cheirogaleidae. In 2008, the IUCN Red List classifications for lemurs provided assessments for 29 species of Cheirogaleidae

(Table 27.1). Remarkably, 51.7% (N = 14) of the cheirogaleid species were listed as data deficient, which reflected the lack of data about many newly described species, such as *Microcebus tanosi* and *M. margotmarshae* (Rasoloarison *et al.*, 2013). The remaining 14 species were listed as least concern (N = 7), vulnerable (N = 3), and endangered (N = 4). None were assessed as critically endangered. The lack of information on extinction risks was further highlighted by the fact that only nine species had associated threat classifications, with the main threats coming from agriculture (N = 7 species), logging (N = 6 species), quality loss (N = 4 species), and exploitation (N = 1 species). Recognizing the need for updated assessments, in 2012 the IUCN Primate Specialist Group organized a red-listing and conservation planning workshop for lemurs (Schwitzer *et al.*, 2013). The goal of this workshop was to have participants and stakeholders create a common strategy for the conservation of lemurs. Of the 35 cheirogaleids that were assessed in the workshop, 3 were not evaluated, 6 were listed as data deficient or least concern, 6 as vulnerable, and 16 as endangered. Four species (*Cheirogaleus sibreei*, *M. gerpi*, *M. mamiratra*, and *M. marohita*) were categorized as critically endangered. Although recent research efforts have improved our understanding of the conservation biogeography of most of the currently recognized cheirogaleids, comparing the 2008 IUCN Red List results to the 2012 conference results indicate that cheirogaleids have over the time period experienced deleterious changes in their conservation status. What is not clear, however, is whether these changes represent increased information or actual deleterious changes in their conservation status.

The most recent IUCN Red List assessments list threats for all but two species of cheirogaleids (Table 27.1). Threat assessments have been made for the four critically endangered taxa: *C. sibreei* (agro-farming and shifting agriculture/small holder farming), *M. gerpi* (hunting, logging and wood harvesting, and shifting agriculture/small holder farming), *M. mamiratra* (shifting agriculture/small holder farming), and *M. marohita* (logging and shifting agriculture/small holder farming). Threat assessments are available for 16 endangered cheirogaleids, all of which, except for one species (*P. electromontis*), are experiencing deleterious changes in the conservation status due to agro-farming or shifting agriculture/small holder farming. Eight of these endangered taxa are also experiencing hunting pressures. Of the six vulnerable taxa assessed for threats, five are threatened by agro-farming or shifting agriculture/small holder farming. Threats profiles for *M. tavaratra* (vulnerable) are unique, resulting from a combination of fire and fire suppression, logging and wood harvesting, and mining and quarrying. Criteria for Red List categorization were assigned to 27 species, with 20 species suffering from reductions in geographic range size and 7 from population size reduction. Based on these assessments, extinction threats relate predominantly to anthropogenic farming and are most prominent for species with a small geographic range size.

The recent descriptions of many new cheirogaleid species in this already speciose lemur family highlight perhaps the most fundamental question for their conservation biology: how many species of cheirogaleids are there in Madagascar? The application of molecular techniques to species designations has revolutionized and

Table 27.1 Forest habitat use, geographic distribution, and IUCN Red List categories and threats for 35 species of Cheirogaleidae.

Genus	Species	Forest type(s)[a]	Distribution[b]	2008 Red List[c]	2012 Red List[c]	Threats[d]	Category criteria[e]	No. in protected areas
Allocebus	*trichotis*	Humid	Continuous	DD	VU	A+E	PSR	9
Cheirogaleus	*adipicaudatus*	Dry	Disjunct	DD	NE			4
	crossleyi	Humid	Continuous	DD	DD	S		3
	major	Variety	Disjunct	LC	DD	S		17
	medius	Variety	Disjunct	LC	LC	A+H		12
	minusculus	Humid	Continuous	DD	DD	D+S		0
	ravus	Humid	Continuous	DD	NE			2
	sibreei	Humid	Disjunct	DD	CR	A+S	GR	2
Microcebus	*arnholdi*	Humid	Continuous	NE	EN	H+L+S		1
	berthae	Dry	Continuous	EN	EN	A+H+L	GR	2
	bongolavensis	Dry	Continuous	DD	EN	A+H+L+S	GR	0
	danfossi	Dry	Continuous	DD	EN	H+L+S	GR	1
	gerpi	Humid	Unknown	NE	CR	H+L+S	GR	0
	griseorufus	Spiny/gallery	Continuous	LC	LC	A+M		4
	jollyae	Humid	Disjunct	DD	EN	H+S	GR	1
	lehilahytsara	Humid	Continuous	DD	VU	S	GR	2
	macarthurii	Humid	Unknown	NE	EN	S	GR	1
	mamiratra	Humid	Continuous	DD	CR	S	GR	1
	margotmarshae	Dry	Continuous	NE	EN	S	GR	1
	marohita	Humid	Unknown	NE	CR	L+S	GR	0
	mittermeieri	Humid	Continuous	DD	EN	S	GR	2

	murinus	Dry	Disjunct	LC	LC	H+M+S		14
	myoxinus	Dry	Disjunct	DD	VU	S	PSR	2
	ravelobensis	Dry	Disjunct	EN	EN	F+S	GR	1
	rufus	Humid	Disjunct	LC	VU	S	GR	5
	sambiranensis	Humid/dry	Continuous	EN	EN	A+L	GR	1
	simmonsi	Humid	Continuous	DD	EN	S	GR	1
	tanosi	Dry	Unknown	NE	NE			0
	tavaratra	Dry	Disjunct	EN	VU	F+L+M	GR	3
Mirza	*coquereli*	Dry	Disjunct	DD	EN	F+H+L+S	PSR	5
	zaza	Humid/dry	Disjunct	VU	EN	S	PSR	1
Phaner	*electromontis*	Humid/dry	Disjunct	VU	EN	F+M	GR	1
	furcifer	Humid	Disjunct	LC	VU	L+S	PSR	8
	pallescens	Dry	Disjunct	LC	EN	A+S	PSR	6
	parienti	Humid	Continuous	VU	EN	L+S	GR	2

[a] Determined from comparisons of IUCN (2014) distribution maps with data from the CEPF Madagascar Vegetation Mapping Project (http://www.vegmad.org/).

[b] Data from IUCN (2014).

[c] NE, Not Evaluated; DD, Data Deficient; LC, Least Concern; VU, Vulnerable; EN, Endangered; CR, Critically Endangered. Based on IUCN (2014).

[d] A, Agro-industry Farming; D, Housing and Urban Areas; F, Fire and Fire Suppression; H, Hunting; L, Logging and Wood Harvesting; M, Mining and Quarrying; S, Shifting Agriculture/Small Holder Farming. Data from IUCN (2014).

[e] PSR, Population Size Reduction; GR, Geographic Range. Data from IUCN (2014).

accelerated our understanding of the evolution and phylogenetic systematics of mammals, but this work has also complicated field research on the conservation biology and biogeography of these new species (Ceballos and Ehrlich, 2009). For example, data once applied to a wide-ranging, single species, such as *M. rufus* throughout eastern Madagascar, now reflect interspecific variations across widely separated, small geographic areas (Louis Jr *et al.*, 2008; Radespiel *et al.*, 2008, 2012; Rasoloarison *et al.*, 2013). These reassessments of species richness complicate comparative studies for two reasons: the new species are often described based on very limited collecting localities, and it is difficult to determine which species are represented in data sets spanning longitudinal research. There is also a growing concern among some researchers that genetic studies supporting new species designations lack equally intensive behavioral and ecological data (Heckman *et al.*, 2006; Tattersall, 2007), which can complicate our understanding on the conservation biology of a species (Heckman *et al.*, 2006). This issue is not inconsequential, as most conservation recommendations are made at the species level (Isaac *et al.*, 2004). Therefore, the first, most critical issue for the conservation biology of cheirogaleids is to undertake holistic studies of species richness, with special attention being made to collect complementary data on the behavior and ecology of local animals and for sampling more widely and intensively across the range of purported new species (Zimmermann and Radespiel, 2014).

Presence in protected areas

We reviewed presence/absence data on 35 cheirogaleid species in 47 protected areas in Madagascar (Table 27.2). In terms of maximizing cheirogaleid species richness, one protected area contains the most species (Tsingy de Namoroka National Park, $N = 5$ species). Eight other protected areas contain at least four cheirogaleid species: Andohahela National Park, Andranomena Special Reserve, Betampona Integrated Nature Reserve, Beza-Mahafaly Special Reserve, Kirindy Mitea National Park, Makira National Park, Montagne d'Ambre, and Zombitse Vohibasia National Park. Although only three cheirogaleids are listed as least concern (*C. medius*, *M. griseorufus*, and *M. murinus*), taxa in this category account for the majority of species in the existing protected area system (Figure 27.1). Of the four critically endangered cheirogaleids, two species are not found in the existing protected area network (*M. gerpi* and *M. marohita*). For the 16 endangered taxa, all but one species (*M. bongolavensis*) range into at least one protected area. The apparent failure of the existing protected area system to encompass some of the range of these three critically endangered or endangered cheirogaleids is, again, likely because these taxa have been described only within the last 15 years. Thus, it can be stated that existing protected areas are not maximizing protection of cheirogaleids most at risk of extinction, although the complicated political and financial processes of creating a new protected area can take far longer than the time it takes to locate, collect and analyze data on, and describe new biological species (Joppa *et al.*, 2008). These analytics and the results from recent surveys indicate that presence of a species

Table 27.2 Abiotic and biotic characteristics, including cheirogaleid species richness, for 47 protected areas in Madagascar (data from Hijmans *et al.*, 2005; Stevens and O'Connor, 2006; Moat and Smith, 2007; Kamilar, 2009; Steffens and Lehman, accepted; IUCN, 2014).

Name	Type[a]	Area (ha)	Latitude	Min alt (m)	Max alt (m)	Alt range (m)	Annual rainfall (mm)	Hi temp (°C)	Lo temp (°C)	Temp range (°C)	Species richness
Ambatovaky	RS	60050	15.3	300	500	200	3300	30	18	12	1
Ambohitantely	RS	5600	18.1	1448	1662	214	1678	30	9	21	0
Ambohijanahary	RS	24750	23.4	800	1600	800	1618	32	11	21	0
Analabe	PR	2000		1	50	50	900	34	13	19	2
Analamerana	RS	34700	12.4	10	608	598	1250	30	19	11	1
Andasibe Analamazaotra	RS	15480	18.2	930	1040	110	1700	24	14	10	3
Andohahela	PN	76020	24.5	90	1972	1882	1700	36	9	26	4
Andranomena	RS	6420	20.1	1	100	100	1200	35	14	21	4
Andringitra	PN	31160	22	650	2008	1958	2000	30	5	25	2
Anjanaharibe-Sud	RS	32090	14.4	500	2064	1564	2000	30	14	16	3
Ankarafantiska	PN	135000	16.1	80	333	253	1250	35	19	16	3
Ankarana	PN	18225	12.5	50	409	359	1890	30	19	11	2
Antafondro	RS		14.2								1
Baie de Baly	PN	57142	16	1	400	400	1000	33	19	14	2
Bemaraha	PN	157710	18.5	150	750	600	980	35	19	16	3
Bemarivo	RS	11570	16.9	50	120	70	1248	34	17	17	2
Berenty	PR	200	25	4	10	6	500	40	10	30	2
Betampona	RNI	2228	17	300	400	100	3000	30	18	12	4
Beza-Mahafaly	RS	600	23.4	100	200	100	550	35	14	21	4
Bora	RS	4841	14.75	115	411	296	1800	35	19	16	1
Cap Sainte Marie	RS	1750	25.3	110	199	89	350	35	14	21	2
Isalo	PN	81540	22	514	1268	754	850	30	14	16	2
Ivohibe	RS	3453	22.3	750	2060	1310	2000				2

(*continued*)

Table 27.2 *(cont.)*

Name	Type[a]	Area (ha)	Latitude	Min alt (m)	Max alt (m)	Alt range (m)	Annual rainfall (mm)	Hi temp (°C)	Lo temp (°C)	Temp range (°C)	Species richness
Kalambatritra	RS	28250	23.2	1300	1500	200	2000	30	9	16	1
Kasijy	RS	19800	17	124	454	330	1679	34	15.8	18	3
Kirindy Mitea	PN	72200	20.8	1	300	300	700	33	14	19	4
Lokobe	RNI	740	13.2	1	430	430	2250	35	19	20	1
Makira	PN	372470	15.6				4800	32	20	8	4
Mananara Nord	PN	144000	16.2	1	400	400	3000	24	18	6	3
Maningoza	RS	7900	17	1	800	800	1300	32	18	14	2
Manombo	RS	5320	23	1	137	137	2500	30	14	16	1
Manongarivo	RS	32735	13.75	155	1876	1721	1750	35	19	16	3
Marojejy	PN	60050	14	75	2133	2058	3000	30	19	11	3
Mantadia Andasibe	PN	15480	18.8	900	1250	350	1890	27	11	16	3
Marotandrano	RS	42200	16.1				1314	31	16	15	2
Masoala	PN	240520	15.3	1	1300	1300	4800	32	20	12	3
Midongy Befotaka	PN	192198	23	689	1679	990					2
Montagne d'Ambre	PN	23010	12.3	850	1475	625	2378	25	19	6	4
Tsingy de Namoroka	PN	21742	16.2	71	227	156	1160	35	19	16	5
Pic d'Ivohibe	RS	3453	22.3	775	2060	1285	3000	28	9	19	2
Ranomafana	PN	41601	21.1	800	1200	400	2600	30	14	16	2
Sahamalaza	PN	7483	14								3
Tamboketsa	RS	17150	17	100	1181	1081	1750				1
Tsaratanana	RNI	48622	13.75	227	2876	2649	1704	25	9	16	2
Tsimanampetsotsa	PN	43200	24.1	38	114	76	350	35	14	21	2
Zahamena	PN	64370	17	750	1512	762	1750	30	14	16	3
Zombitse Vohibasia	PN	36308	22.8	300	825	525	750	30	14	16	4

[a] RS, Réserve Spéciale; PN, Parc National; RNI, Réserve Naturelle Intégrale.

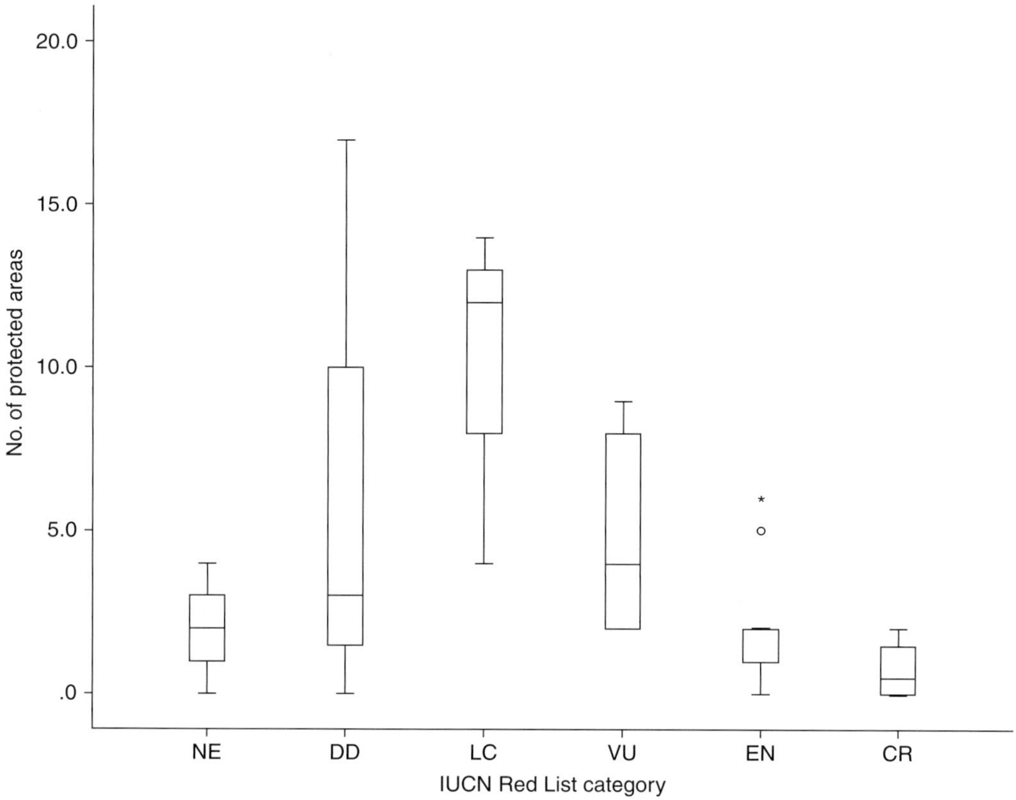

Figure 27.1 Box-and-whisker plots of the number of protected areas containing cheirogaleids in six IUCN Red List categories. See Table 27.1 for Red List category descriptions. Dark horizontal line is the median value, bottom and top of the closed box refer to the 25th percentile and 75th percentile, respectively, t-bars refer to the 95% range of data, and points are the outliers.

in protected areas is no guarantee of population persistence (Meyler *et al.*, 2012). For example, Meyler *et al.* (2012) noted that endangered *M. tavaratra* have higher density estimates outside than inside of a protected area in northern Madagascar.

Conservation biogeography

We assessed abiotic and biotic determinants of cheirogaleid species richness in protected areas. In particular, we were interested in determining how the area of these protected areas related to species richness because numerous biogeographic studies indicate a strong, positive relationship between area and species richness for many mammalian taxonomic groups, including diurnal lemurs (Fleagle and Reed, 1996; Harcourt, 1999; Stevens and O'Connor, 2006). We collected data on biotic factors (cheirogaleid species richness, IUCN criteria, and protected area type/size) and abiotic factors (latitude, altitude, rainfall, and temperature) from published sources

Table 27.3 Results of general linear models of effects of abiotic and biotic factors on species richness of cheirogaleids in 47 protected areas in Madagascar.

	Species richness			EN and CR richness[b]			Least Threatened richness[c]		
Variables[a]	Mean square	F	P	Mean square	F	P	Mean square	F	P
Protected area type	0.09	2.13	0.13	0.77	1.17	0.34	1.45	1.25	0.32
Region	0.08	2.07	0.13	0.85	1.30	0.30	1.95	1.67	0.19
$\log_{10}$ area (km^2)	0.05	1.31	0.27	0.02	0.02	0.88	1.38	1.18	0.29
$\log_{10}$ latitude (°)	0.10	2.40	0.14	2.33	3.56	0.07	2.51	2.15	0.16
$\log_{10}$ minimum altitude (m)	0.08	1.85	0.19	1.65	2.52	0.13	0.69	0.59	0.45
$\log_{10}$ maximum altitude (m)	0.00	0.01	0.94	0.63	0.96	0.34	0.84	0.72	0.41
$\log_{10}$ altitude range (m)	0.02	0.44	0.51	0.66	1.00	0.33	0.61	0.52	0.48
$\log_{10}$ annual rainfall (mm)	0.00	0.08	0.78	0.03	0.04	0.84	0.00	0.00	0.97
$\log_{10}$ maximum temperature (°C)	0.00	0.00	0.96	0.03	0.04	0.84	0.18	0.16	0.70
$\log_{10}$ maximum temperature (°C)	0.02	0.42	0.53	0.72	1.10	0.31	0.47	0.40	0.53
$\log_{10}$ temperature range (°C)	0.02	0.41	0.53	0.05	0.07	0.79	0.05	0.04	0.84
Adjusted R^2		0.03			−0.07			−0.05	

[a] Protected area type: Réserve Spéciale, Parc National, Réserve Naturelle Intégrale; Region: North, South, East, West, and Central.

[b] EN, endangered; CR, critically endangered, based on IUCN Red List criteria in Schwitzer *et al.* (2013).

[c] All taxa not listed as either EN or CR in Schwitzer *et al.* (2013).

and from the WorldClim database (Hijmans *et al.*, 2005; Mittermeier *et al.*, 2006, 2008; Stevens and O'Connor, 2006; Schwitzer *et al.*, 2013; IUCN, 2014). Following Kamilar *et al.* (2012), we extracted the climate data for each protected area using the "Extract-by-point" operation in ArcGIS 9 (ESRI). We then used general linear models to explore the effects of biotic and abiotic variables on overall species richness, high-risk taxa (endangered and critically endangered), and low-risk species (those taxa assessed below endangered) of cheirogaleids in existing protected areas. These analyses were conducted using SPSS ver. 22 (IBM) and the alpha level was set at < 0.05. Contrary to earlier studies on biotic and abiotic correlates to the diversity of diurnal lemurs in protected areas (Stevens and O'Connor, 2006), we found no significant relationship between $\log_{10}$ species richness and $\log_{10}$ protected area for cheirogaleids (Table 27.3). Moreover, the size of protected areas did not influence the richness of either the most-threatened taxa (Critically Endangered and Endangered) or least-threatened taxa (Data Deficient, Least Concern, and Vulnerable). No abiotic or biotic factors explained the species richness of the most threatened or least threatened cheirogaleids across the protected areas in Madagascar (Table 27.3). Thus, the size and abiotic characteristics of the current protected areas system does not support protection of overall cheirogaleid species richness.

Predicted extinction threats

Given poor model performance for determining abiotic and biotic correlates to species richness in protect areas, we reviewed the literature for possible predictors of extinction risk for taxa in this family. Studies of mammalian extinction threats consistently find that small-bodied species (< 3–5 kg) that undergo hibernation or torpor, such as most cheirogaleids, have different functional extinction rates and correlates compared to large-bodied species (> 3–5 kg) in an ecological system (Cardillo, 2003; Liow et al., 2009; Schnell *et al.*, 2013). This difference is consistently related back to the complex interactions of anthropogenic effects with animal life-history characteristics (e.g., litter size) and biogeography (Gaston and Blackburn, 1996). There is the added threat associated with the deleterious effects of dietary specialization on extinction probability (Jernvall and Wright, 1998), although few cheirogaleids would qualify as dietary specialists (exception: *Phaner* spp.).

Two meta-analysis studies of extinction risks in mammals provide insights into threats that may be of value for conservation authorities focusing on cheirogaleids (Cardillo, 2003; Davidson *et al.*, 2009). Cardillo *et al.* (2005) conducted one of the most comprehensive studies on environmental, ecological, and life-history predictors of extinction risk in mammals. Within this work, they refined their analysis to include 1207 species of small-bodied mammals (< 3 kg body mass), finding that extinction risk in these species was significantly influenced by geographic range size, latitude, and human population density. After we reviewed the statistical results within the body mass range of cheirogaleids (ca. 33–500 g), only small geographic range size was a significant correlate of extinction risk. Put differently, in mammals < 3 kg in body mass, extinction risk was correlated with multiple factors; however, small geographic range size was the only significant predictor of extinction risk in mammals with the same body mass range of cheirogaleids (ca. 33–500 g). What, then, is the critical range size below which a cheirogaleid species' extinction risk increases? A study by Davidson *et al.* (2009) on ecological correlates to extinction in mammals provides a threshold that might be tested for applicability in cheirogaleids. Their analysis of data on 4079 mammals indicated increased extinction risks for those taxa with a body mass of < 304 g and a geographic range size of < 28,054 km^2. Therefore, cheirogaleid taxa with geographic range sizes below this threshold should be candidates for prioritized conservation research.

It is important to note that the process of determining geographic range size is methodologically and theoretically complicated (Chapter 26), particularly for taxa with disjunct distribution patterns and those with wide variations in the scale of their ranging patterns (Gaston and Fuller, 2009; Jiménez-Alfaro *et al.*, 2012). There are, in fact, two kinds of geographic range size: extent of occurrence (EOO) and area of occupancy (AOO). Estimations of EOO use data on known, inferred, and projected localities for determining the area contained within the geographic limits of the occurrence of a species (Gaston, 1994a). The IUCN Red List uses a minimum convex polygon method to report EOO for most mammals, including cheirogaleids. Conversely, the AOO is estimated based on known localities of species distributions and also provides estimates of data quality and spatial resolution (Gaston, 1994a). Maps based on the EOO method tend to be of lower predictive strength when the distribution and scale of a species and its habitats are poorly known and quantified. For example, *M. murinus* is perhaps the best-studied cheirogaleid and has one of the largest geographic range sizes, which we estimated from the IUCN EOO map to be 91,160 km^2. However, this EOO map contains extensive areas of non-forested habitats, such as grasslands and human settlements, resulting in over-inflated range estimates that will reduce the power of statistical tests. Many two-dimensional EOO distribution maps also fail to incorporate distribution demarcations due to elevational range limits, which are an issue of particular importance for cheirogaleids in the elevationally complex forests in east Madagascar (Lehman, 2013). For example, the recently described *C. sibreei* likely has complex elevational range limits in eastern humid forests (Blanco *et al.*, 2009; Groeneveld *et al.*, 2010; Chapter 5), which unless quantified can result in erroneous estimates of both AOO and EOO. Thus, we require accurate data on the actual geographic range size of the most threatened cheirogaleids (Chapter 26), particularly the four species that have been listed as critically endangered (*C. sibreei*, *M. gerpi*, *M. mamiratra*, and *M. marohita*).

Improved data on each species' geographic range is also critical to gaining a better understanding of their rarity, which is often described as a precursor to extinction (Gaston, 1994b). Rarity is implicit in most political and scientific discussions of conservation priorities, yet the term is difficult to quantify because it encompasses varying aspects of extinction threats, ecology, and biogeography (Harcourt, 2004). An important initial process for applying rarity to cheirogaleid conservation biology is to assume that it is multidimensional rather than one-dimensional, and therefore we need to consider more than just geographic range size in deciding on extinction threats and conservation priorities. Rabinowitz's (1981) widely used forms of rarity look at species in terms of their multidimensional patterns of geographic range (small or large), habitat breadth (narrow or wide), and abundance (rare or common). The rarest species will have small geographic ranges, narrow habitat breadth, and rare abundance. A key issue that needs consideration involves conflation of density estimates with abundance in primatology; they are not synonymous and can provide differing results in terms of the conservation biogeography of species (Chapter 25). For example, even some of the most intensely surveyed protected areas, such as Ranomafana National Park, have incomplete assessments of cheirogaleid diversity, range, habitat use, and abundance, which further complicates estimates of rarity and thus extinction threats. Therefore, the conservation biology of cheirogaleids requires biogeographic data on the habitat use and abundance of each species.

Deforestation effects on Cheirogaleidae

Despite the potential value of meta-analysis studies for identifying some predictors of rarity and extinction risk in cheirogaleids, there are few biogeographic data on how populations of many species respond to varying kinds and intensities of forest loss and forest fragmentation. A fundamental issue is that in theory, forest loss and fragmentation represent independent effects (Trzcinski *et al.*, 1999), where forest loss refers to the removal of habitat in a forest patch and fragmentation involves the break-up of a patch into isolated fragments. In practice, however, forest fragmentation typically results in concomitant loss of available habitat, making it difficult to determine which process is affecting animals in forests (Tscharntke *et al.*, 2012). These issues can be resolved by having either longitudinal data on a forest patch undergoing anthropogenic modifications or cross-sectional data from multiple patches of varying size and shapes. Irrespective of these theoretical and methodological issues, fragmentation results in an increase in the number of fragments, a decrease in fragment size, or an increase in fragment isolation (Fahrig, 2003). For example, a recent review by Radespiel and Bruford (2014) of fragmentation genetics in animals noted that of the four cheirogaleid species studied to date, each showed varying fragmentation effects manifested in genetic diversity, genetic population divergence, gene flow, and signals of genetic bottleneck. These complex genetic responses to fragmentation are best studied at the landscape level, which encompasses spatially heterogeneous geographic areas (Jackson and Fahrig, 2012). Thus, a comprehensive analysis of cheirogaleid responses to fragmentation requires landscape-level studies of how populations respond to changes in the number, size, and isolation of fragments.

Forest fragmentation also results in the creation of forest edges and subsequent edge effects (Murcia, 1995). Edge effects represent the penetration, to varying depths and intensities, of abiotic and biotic conditions from the surrounding matrix into the forest interior (Murcia, 1995; Ries and Sisk, 2004). These edge effects can have strong primary influences on the structure and composition of plant communities, which in turn cause secondary effects on animal distribution and abundance (Harper *et al.*, 2005). Research on how and why cheirogaleids respond to edge effects needs to take into consideration the fact that these effects are confounded with area effects (Malcolm, 1994; Laurance, 2000). Small fragments will have large perimeter:area ratios, which means that edges will comprise an increasingly greater proportion of habitat as fragments become smaller. Moreover, long, narrow fragments may already be composed entirely of edge habitats, which in turn increases local extinction probabilities for plants and animals that prefer interior forest habitats (MacArthur and Wilson, 1967). These pervasive, dynamic edge:area effects may dominate ecological and biogeographic processes in fragmented landscapes (Woodroffe and Ginsberg, 1998; Harrison & Bruna, 1999; Ewers *et al.*, 2010), which is a potentially major issue in Madagascar (Harper *et al.*, 2007). For example, biogeographic studies of lemur densities revealed that *M. rufus* had a strong positive response and *C. major* had a strong negative response to edges in eastern humid forests (Lehman *et al.*, 2006a, 2006b). In western dry forests, preliminary data indicate that the abundance of sympatric *M. ravelobensis* and *M. murinus* differ between edge and interior forest habitats, with *M. ravelobensis* being more abundant in edge habitats and *M. murinus* preferring interior forest habitats (Burke and Lehman, 2014). These varying edge responses have been related to interspecific differences in feeding ecology in both eastern humid and western dry forests (Corbin and Schmid, 1995; Lehman *et al.*, 2006b; Burke and Lehman, 2014). Thus, we recommend further research on how edge effects influence the biogeography and feeding ecology of cheirogaleids.

Spatially and temporally varying patterns of deforestation can result in dynamic fragmentation and edge effects unique to each landscape, making it difficult to extrapolate responses of conspecific populations at one site or landscape in different parts of their geographic range. For example, *M. ravelobensis* and *M. murinus* are sympatric through much of the range of *M. ravelobensis* in the dry forests of NW Madagascar. Rakotondravony and Radespiel (2009) found a negative relationship between *M. murinus* and *M. ravelobensis* densities in continuous forest sites within Ankarafantsika, whereas Steffens and Lehman (Chapter 25) found a positive relationship between *M. murinus* and *M. ravelobensis* density and abundance in a fragmented forest landscape in Ankarafantsika. However, differing data collection methods between these two studies complicated more refined analyses of how and why spatial variations in abiotic and biotic parameters resulted in the differing patterns of abundance in *M. murinus* and *M. ravelobensis*. Researchers can deal with some of these spatial issues by providing complete, detailed descriptions of the fragments within their study site (e.g., size, shape, orientation, elevation ranges, slopes, and isolation), recording and analyzing any historical information on deforestation processes, and collecting comparative data in both nearby continuous forests and fragments. Thus, researchers seeking to incorporate these spatial data into databases for meta-analysis research on multiple species should first run diagnostic tests on intraspecific variation before proceeding to interspecific analyses (Gauch, 1982). A further issue here is that researchers should exercise caution when combining data on a species from multiple, widely separated sites to create species-specific or even community-level patterns of biogeography (Gauch, 1982).

Summary of recommendations

We make the following five recommendations in regard to the conservation biology of cheirogaleids.

(1) Researchers should undertake holistic studies of species richness, with special attention being made to supplement genetic data with information on the behavior and ecology of local animals and to sample all data sets more widely and intensively across the range of each new or existing species. Because various kinds of agriculture are key drivers of extinction risk in cheirogaleids, we suggest that researchers make greater efforts to determine how local human populations influence forest degradation and lemur hunting in Madagascar. Moreover, researchers

should strive to understand the problems and needs of local people, which can then be used to make more informed recommendations on how to alleviate anthropogenic pressures on local cheirogaleid populations (Dolins *et al.*, 2010).

(2) Researchers should seek to standardize data collection methods, or at least incorporate data using methods already published on the research question(s) and species (Murcia, 1995). Doing so will facilitate meta-analysis studies testing hypotheses on underlying factors influencing species richness. This comparative research will be particularly important for tracking temporal and spatial changes in cheirogaleid conservation biology.

(3) We require data on the geographic range size, habitat breadth, and abundance of all cheirogaleids, with priority for the rarest species with a projected geographic range $< 28{,}054$ km^2. A previous study estimated that this may apply to at least 13 mouse lemur species (Radespiel *et al.*, 2012) that are often confined to a single inter-river system (area between two adjacent large rivers: Olivieri *et al.*, 2007) which may not exceed 9500 km^2. Among the other cheirogaleids, this criterion may apply to *C. sibreei* and two or three *Phaner* species. Moreover, efforts must be made to collect survey data across multiple sites, resulting in distribution data amenable to the production of maps on area of occupancy rather than extent of occurrence.

(4) Biogeographic research on cheirogaleids should focus on landscape studies of how multiple populations of a species respond to changes in the number, size, and isolation of fragments. Additionally, we recommend further research on how edge effects influence the biogeography and feeding ecology of cheirogaleids.

(5) Researchers creating species-specific and community-level data sets for meta-analysis research should be cautious of the potential for considerable intraspecific variation in the conservation biogeography of cheirogaleids. Lack of consideration for this potential variation can result in reduced power of statistical models.

Acknowledgments

We thank Martin Griffith and the editorial staff at Cambridge University Press for their enthusiastic support of this edited volume. We also thank the two reviewers for their very helpful comments and suggestions on our manuscript. We gratefully acknowledge the many people who have assisted us with our research in Madagascar. We thank the following students from the University of Toronto who greatly assisted with editorial work on the book manuscript: Feben Alemu, Rebecca Antonacci, Chey Cameron, Megan Joyce, Aniqa Khan, Samantha Lucchetta, and Veeda Tariq.

References

Allnutt T, Ferrier S, Manion G, *et al.* 2008. A method for quantifying biodiversity loss and its application to a 50-year record of deforestation across Madagascar. *Conservation Letters* 1(4):173–181.

Andrianandrasana HT, Randriamahefasoa J, Durbin J, Lewis RE, Ratsimbazafy JH. 2005. Participatory ecological monitoring of the Alaotra wetlands in Madagascar. *Biodiversity and Conservation* 14(11):2757–2774.

Banks MA, Ellis ER, Antonio, Wright PC. 2007. Global population size of a critically endangered lemur, Perrier's sifaka. *Animal Conservation* 10(2):254–262.

Blanco MB, Godfrey LR, Rakotondratsima M, *et al.* 2009. Discovery of sympatric dwarf lemur species in the high-altitude rain forest of Tsinjoarivo, eastern Madagascar: implications for biogeography and conservation. *Folia Primatologica* 80(1):1–17.

Bond W, Silander Jr J, Ranaivonasy J, Ratsirarson J. 2008. The antiquity of Madagascar's grasslands and the rise of C4 grassy biomes. *Journal of Biogeography* 35(10):1743–1758.

Burke RJ, Lehman SM. 2014. Edge effects on morphometrics and body mass in two sympatric species of mouse lemurs in Madagascar. *Folia Primatologica* 85(5):277–291

Cardillo M. 2003. Biological determinants of extinction risk: why are smaller species less vulnerable? *Animal Conservation* 6(1):63–69.

Cardillo M, Mace GM, Jones KE, *et al.* 2005. Multiple causes of high extinction risk in large mammal species. *Science* 309(5738):1239–1241.

Ceballos G, Ehrlich PR. 2009. Discoveries of new mammal species and their implications for conservation and ecosystem services. *Proceedings of the National Academy of Sciences* 106(10):3841–3846.

Chapman CA, Peres CA. 2001. Primate conservation in the new millennium: the role of scientists. *Evolutionary Anthropology* 10(1):16–33.

Corbin GD, Schmid J. 1995. Insect secretions determine habitat use patterns by a female lesser mouse lemur (*Microcebus murinus*). *American Journal of Primatology* 37(4):317–324.

Davidson AD, Hamilton MJ, Boyer AG, Brown JH, Ceballos G. 2009. Multiple ecological pathways to extinction in mammals. *Proceedings of the National Academy of Sciences*, USA 106(26):10702–10705.

Dolins FL, Jolly A, Rasamimanana H, *et al.* 2010. Conservation education in Madagascar: three case studies in the biologically diverse island-continent. *American Journal of Primatology* 72(5):391–406.

Du Puy DJ, Moat J. 1996. A refined classification of the primary vegetation of Madagascar based on the underlying geology: using GIS to map its distribution and to assess its conservation status. In WR Lourenço (ed.), *Proceedings of the International Symposium on the Biogeography of Madagascar* (pp. 205–218). Editions de l'ORSTOM, Paris.

Ewers RM, Marsh CJ, Wearn OR. 2010. Making statistics biologically relevant in fragmented landscapes. *Trends in Ecology & Evolution* 25(12):699–704.

Fahrig L. 2003. Effects of habitat fragmentation on biodiversity. *Annual Review of Ecology, Evolution, and Systematics* 34(1):487–515.

Fleagle JG, Reed KE. 1996. Comparing primate communities: a multivariate approach. *Journal of Human Evolution* 30:489–510.

Ganzhorn JU. 1987. A possible role of plantations for primate conservation in Madagascar. *American Journal of Primatology* 12(2):205–215.

Ganzhorn JU, Wright PC, Ratsimbazafy HJ. 1999. Primate communities: Madagascar. In JG Fleagle, CH Janson, K Reed (eds.), *Primate Communities* (pp. 75–89). Cambridge University Press, Cambridge.

Ganzhorn JU, Lowry PP, Schatz GE, Sommer S. 2001. The biodiversity of Madagascar: one of the world's hottest hotspots on its way out. *Oryx* 35(4):346–348.

Ganzhorn JU, Goodman SM, Nash S, Thalmann U. 2006. Lemur biogeography. In SM Lehman, JG Fleagle (eds.), *Primate Biogeography* (pp. 229–254). Plenum/Kluwer Press, New York.

Gaston KJ. 1994a. Measuring geographic range sizes. *Ecography* 17(2):198–205.

Gaston KJ. 1994b. What is rarity? In WE Kunin, K Gaston (eds.), *The Biology of Rarity: Causes and Consequences of Rare – Common Differences* (pp. 1–21). Springer, Netherlands.

Gaston KJ, Blackburn TM. 1996. Rarity and body size: importance of generality. *Conservation Biology* 10:1295–1298.

Gaston KJ, Fuller RA. 2009. The sizes of species' geographic ranges. *Journal of Applied Ecology* 46(1):1–9.

Gauch HG. 1982. *Multivariate Analysis in Community Ecology*. Cambridge University Press, Cambridge.

Golden CD. 2009. Bushmeat hunting and use in the Makira Forest, north-eastern Madagascar: a conservation and livelihoods issue. *Oryx* 43(3):386–392.

Golden CD, Wrangham RW, Brashares JS. 2013. Assessing the accuracy of interviewed recall for rare, highly seasonal events: the case of wildlife consumption in Madagascar. *Animal Conservation* 16(6):597–603.

Golden CD, Bonds MH, Brashares JS, Rodolph Rasolofoniaina BJ, Kremen C. 2014. Economic valuation of subsistence harvest of wildlife in Madagascar. *Conservation Biology* 28(1):234–243.

Green GM, Sussman RW. 1990. Deforestation history of the eastern rain forests of Madagascar from satellite images. *Science* 248(4952):212–215.

Groeneveld LF, Blanco MB, Raharison J-L, *et al.* 2010. MtDNA and nDNA corroborate existence of sympatric dwarf lemur species at Tsinjoarivo, eastern Madagascar. *Molecular Phylogenetics and Evolution* 55(3):833–845.

Hanna E, Cardillo M. 2014. Clarifying the relationship between torpor and anthropogenic extinction risk in mammals. *Journal of Zoology* 293(3):211–217.

Harcourt AH. 1999. Biogeographic relationships of primates on South-East Asian islands. *Global Ecology and Biogeography* 8:55–61.

Harcourt AH. 2004. Are rare primate taxa specialists or simply less studied? *Journal of Biogeography* 31:57–61.

Harper GJ, Steininger MK, Tucker CJ, Juhn D, Hawkins F. 2007. Fifty years of deforestation and forest fragmentation in Madagascar. *Environmental Conservation* 34(4):325–333.

Harper KA, Macdonald SE, Burton P, *et al.* 2005. Edge influence on forest structure and composition in fragmented landscapes. *Conservation Biology* 19(3):1–15.

Harrison S, Bruna E. 1999. Habitat fragmentation and large-scale conservation: what do we know for sure? *Ecography* 22(3):225–232.

Heckman KL, Rasoazanabary E, Machlin E, Godfrey LR, Yoder AD. 2006. Incongruence between genetic and morphological diversity in *Microcebus griseorufus* of Beza Mahafaly. *BMC Evolutionary Biology* 6:98 (online 1–11).

Hijmans RJ, Cameron SE, Parra JL, Jones PG, Jarvis A. 2005. Very high resolution interpolated climate surfaces for global land areas. *International Journal of Climatology* 25(15):1965–1978.

Hill K, McMillan G, Farina R. 2003. Hunting-related changes in game encounter rates from 1994 to 2001 in the Mbaracayu Reserve, Paraguay. *Conservation Biology* 17(5):1312–1323.

Irwin MT, Raharison J-L, Wright PC. 2009. Spatial and temporal variability in predation on rainforest primates: do forest fragmentation and predation act synergistically? *Animal Conservation* 12(3):220–230.

Isaac NJB, Mallet J, Mace GM. 2004. Taxonomic inflation: its influence on macroecology and conservation. *Trends in Ecology & Evolution* 19:464–469.

IUCN. 2014. *IUCN Red List of Threatened Species*. IUCN Species Survival Commission, Gland.

Jackson HB, Fahrig L. 2012. What size is a biologically relevant landscape? *Landscape Ecology* 27(7):929–941.

Jenkins RKB, Keane A, Rakotoarivelo AR, *et al.* 2011. Analysis of patterns of bushmeat consumption reveals extensive exploitation of protected species in eastern Madagascar. *PLoS ONE* 6(12):e27570.

Jernvall J, Wright PC. 1998. Diversity components of impending primate extinctions. *Proceedings of the National Academy of Sciences, USA* 95:11279–11283.

Jiménez-Alfaro B, Draper D, Nogués-Bravo D. 2012. Modeling the potential area of occupancy at fine resolution may reduce uncertainty in species range estimates. *Biological Conservation* 147(1):190–196.

Joppa L, Loarie S, Pimm S. 2008. On the protection of "protected areas". *Proceedings of the National Academy of Sciences, USA* 105(18):6673.

Kamilar JM. 2009. Environmental and geographic correlates of the taxonomic structure of primate communities. *American Journal of Physical Anthropology* 139(3):382–393.

Kamilar JM, Muldoon KM, Lehman SM, Herrera JP. 2012. Testing Bergmann's rule and the resource seasonality hypothesis in Malagasy primates using GIS-based climate data. *American Journal of Physical Anthropology* 147(3):401–408.

Kremen C, Razafimahatratra V, Guillery RP, *et al.* 1999. Designing the Masoala National Park in Madagascar based on biological and socioeconomic data. *Conservation Biology* 13(5):1055–1068.

Kremen C, Cameron A, Moilanen A, *et al.* 2008. Aligning conservation priorities across taxa in Madagascar with high-resolution planning tools. *Science* 320(5873):222–226.

Laurance WF. 2000. Do edge effects occur over large spatial scales? *TREE* 15:134–135.

Lehman SM. 2006. Conservation biology of Malagasy strepsirrhines: a phylogenetic approach. *American Journal of Physical Anthropology* 130(2):238–253.

Lehman SM. 2013. Effects of altitude on the conservation biogeography of lemurs in south east Madagascar. In S Gursky, A Krzton, N Grows (eds.), *High Altitude Primates* (pp. 3–22). Springer Press, New York.

Lehman SM, Rajoanson A, Day S. 2006a. Edge effects and their influence on lemur distribution and density in southeast Madagascar. *American Journal of Physical Anthropology* 129(2):232–241.

Lehman SM, Rajoanson A, Day S. 2006b. Edge effects on the density of *Cheirogaleus major. International Journal of Primatology* 27(6):1569–1588.

Liow Lee H, Fortelius M, Lintulaakso K, Mannila H, Stenseth N. 2009. Lower extinction risk in sleep-or-hide mammals. *The American Naturalist* 173(2):264–272.

Louis Jr EE, Engberg SE, McGuire SM, *et al.* 2008. Revision of the mouse lemurs, *Microcebus* (Primates, Lemuriformes), of northern and northwestern Madagascar with descriptions of two new species at Montagne d'Ambre National Park and Antafondro Classified Forest. *Primate Conservation* 23:19–38.

MacArthur RH, Wilson EO. 1967. *The Theory of Island Biogeography.* Princeton University Press, Princeton.

Malcolm JR. 1994. Edge effects in central Amazonian forest fragments. *Ecology* 75(8):2438–2445.

McConnell W, Kull C. 2014. Deforestation in Madagascar: debates over the island's forest cover and challenges of measuring forest change. In I Scales (ed.), *Conservation and Environmental Management in Madagascar* (pp. 67–104). Earthscan, London.

McConnell WJ. 2002. Madagascar: emerald isle or paradise lost? *Environment: Science and Policy for Sustainable Development* 44(8):10–22.

Meyler SV, Salmona J, Ibouroi MT, *et al* 2012. Density estimates of two endangered nocturnal lemur species from northern Madagascar: new results and a comparison of commonly used methods. *American Journal of Primatology* 74(5):414–422.

Mittermeier RA, Konstant WR, Hawkins F, *et al.* 2006. *Lemurs of Madagascar.* Conservation International, Washington, DC.

Mittermeier R, Ganzhorn J, Konstant W, *et al.* 2008. Lemur diversity in Madagascar. *International Journal of Primatology* 29(6):1607–1656.

Mittermeier RA, Schwitzer C, Rylands AB, *et al.* 2012. *Primates in Peril: The World's 25 Most Endangered Primates, 2012–2014* (p. 40). Conservation and Science Foundation, Bristol.

Moat J, Smith P. 2007. Atlas of the Vegetation of Madagascar/Atlas de la Vegetation de Madagascar (text in English and French).

Murcia C. 1995. Edge effects in fragmented forests: implications for conservation. *Trends in Ecology & Evolution* 10(2):58–62.

Mutschler T, Feistner AT. 1995. Conservation status and distribution of the Alaotran gentle lemur *Hapalemur griseus alaotrensis*. *Oryx* 29(04):267–274.

Olivieri G, Zimmermann E, Randrianambinina B, *et al.* 2007. The ever-increasing diversity in mouse lemurs: three new species in north and northwestern Madagascar. *Molecular Phylogenetics and Evolution* 43(1):309–327.

Patel ER, Andrianandrasana LH. 2008. Low elevation silky sifakas (*Propithecus candidus*) in the Makira Conservation Site at Andaparaty-Rabeson: ranging, demography, and possible sympatry with red ruffed lemurs. *Lemur News* 13:18–22.

Quéméré E, Louis E, Ribéron A, Chikhi L, Crouau-Roy B. 2010. Non-invasive conservation genetics of the critically endangered golden-crowned sifaka (*Propithecus tattersalli*): high diversity and significant genetic differentiation over a small range. *Conservation Genetics* 11(3):675–687.

Rabinowitz D. 1981. Seven forms of rarity. In J Synge (ed.), *The Biological Aspects of Rare Plant Conservation* (pp. 205–217). Wiley, Chichester.

Radespiel U, Bruford MW. 2014. Fragmentation genetics of rainforest animals: insights from recent studies. *Conservation Genetics* 15(2):245–260.

Radespiel U, Olivieri G, Rasolofoson D, *et al.* 2008. Exceptional diversity of mouse lemurs (*Microcebus* spp.) in the Makira region with the description of one new species. *American Journal of Primatology* 70(11):1033–1046.

Radespiel U, Ratsimbazafy JH, Rasoloharijaona S, *et al.* 2012. First indications of a highland specialist among mouse lemurs (*Microcebus* spp.) and evidence for a new mouse lemur species from eastern Madagascar. *Primates* 53(2):157–170.

Rakotondravony R, Radespiel U. 2009. Varying patterns of coexistence of two mouse lemur species (*Microcebus ravelobensis* and *M. murinus*) in a heterogeneous landscape. *American Journal of Primatology* 71(11):928–938.

Rasoloarison RM, Weisrock DW, Yoder AD, Rakotondravony D, Kappeler PM. 2013. Two new species of mouse lemurs (Cheirogaleidae: *Microcebus*) from eastern Madagascar. *International Journal of Primatology*:1–15.

Ravaloharimanitra M, Ratolojanahary T, Rafalimandimby J, *et al.* 2011. Gathering local knowledge in Madagascar results in a major increase in the known range and number of sites for critically endangered greater bamboo lemurs (*Prolemur simus*). *International Journal of Primatology* 32:1–17.

Razafimanahaka JH, Jenkins RKB, Andriafidison D, *et al.* 2012. Novel approach for quantifying illegal bushmeat consumption reveals high consumption of protected species in Madagascar. *Oryx* 46(4):584–592.

Ries L, Sisk TD. 2004. A predictive model of edge effects. *Ecology* 85(11):2917–2926.

Rylands AB, Konstant WR. 2000. Primate conservation – global patterns, hotspots and wilderness areas. *Folia Primatologica* 71(5):365–366.

Schnell JK, Harris GM, Pimm SL, Russell GJ. 2013. Estimating extinction risk with metapopulation models of large-scale fragmentation. *Conservation Biology* 27(3):520–530.

Schwitzer C, Mittermeier RA, Davies AG, *et al.* (eds.) 2013. *Lemurs of Madagascar: A Strategy for Their Conservation 2013–2016*. IUCN SSC Primate Specialist Group, Bristol Conservation and Science Foundation, and Conservation International, Bristol.

Schwitzer C, Mittermeier RA, Johnson SE, *et al.* 2014. Averting lemur extinctions amid Madagascar's political crisis. *Science* 343(6173):842–843.

Seddon N, Butchart S, Tobias J, *et al.* 2000. Conservation issues and priorities in the Mikea Forest of south-west Madagascar. *Oryx* 34(4):287–304.

Steffens T, Lehman SM. Accepted. Landscape ecology of deforestation processes and lemur biogeography in Madagascar. In CA Shaffer, FL Dolins, J Hickey, N Nibbelink (eds.), *GPS and*

GIS for Primatologists: A Practical Guide to Spatial Analysis. Cambridge University Press, Cambridge.

Stevens NJ, O'Connor PM. 2006. Abiotic and biotic factors as predictors of species richness on Madagascar. In SM Lehman, JG Fleagle (eds.), *Primate Biogeography* (pp. 269–300). Plenum/ Kluwer Press, New York.

Tattersall I, 2007. Madagascar's lemurs: cryptic diversity or taxonomic inflation? *Evolutionary Anthropology* 16:12–23.

Trzcinski MK, Fahrig L, Merriam G. 1999. Independent effects of forest cover and fragmentation on the distribution of forest breeding birds. *Ecological Applications* 9(2):586–593.

Tscharntke T, Tylianakis JM, Rand TA, *et al.* 2012. Landscape moderation of biodiversity patterns and processes – eight hypotheses. *Biological Reviews* 87(3):661–685.

Valenta K, Burke RJ, Styler SA, *et al.* 2013. Colour and odour drive fruit selection and seed dispersal by mouse lemurs. *Scientific Reports* 3:2424.

Whitmore TC, Sayer JA. 1992. *Tropical Deforestation and Species Extinction.* Chapman & Hall, New York.

Willis K, Gillson L, Virah-Sawmy M. 2008. Nature or nurture: the ambiguity of C4 grasslands in Madagascar. *Journal of Biogeography* 35(10):1741–1742.

Woodroffe R, Ginsberg JR. 1998. Edge effects and the extinction of populations inside protected areas. *Science* 280(5372):2126–2128.

Zimmermann E, Radespiel U. 2014. Species concepts, diversity, and evolution in primates: lessons to be learned from mouse lemurs. *Evolutionary Anthropology: Issues, News, and Reviews* 23(1):11–1.

Index